Professor
Martin C. Michel
Academisch Medisch Centrum
Afd. Farmacologie & Farmacotherapie
Meibergdreef 15, 1105 AZ Amsterdam
The Netherlands
e-mail: m.c.michel@amc.uva.nl

With 81 Figures and 23 Tables

ISSN 0171-2004

ISBN 3-540-40581-X Springer-Verlag Berlin Heidelberg New York

Library of Congress Cataloging-in-Publication Data
Neuropeptide Y and related peptides / contributors M. Alfalah ... [et al.] ; editor Martin C. Michel. p. cm. – (Handbook of experimental pharmacology ; v. 162)
Includes bibliographical references and index
ISBN 3-540-40581-x (alk. paper)
1. Neuropeptide Y–Physiological effect. 2. Neuropeptide Y–Derivatives–Physiological effect. I. Alfalah, M. (Marwan) II. Michel, Martin C., 1959- III. Series. QP905.H3 vol. 162
[QP552.N38]   615'.5. s–dc22   [573.8'48]   20030556335'.5

This work is subject to copyright. All rights are reserved, whether the whole or part of the material is concerned, specifically the rights of translation, reprinting, re-use of illustrations, recitation, broadcasting, reproduction on microfilm or in any other way, and storage in data banks. Duplication of this publication or parts thereof is permitted only under the provisions of the German Copyright Law of September 9, 1965, in its current version, and permission for use must always be obtained from Springer-Verlag. Violations are liable for Prosecution under the German Copyright Law.

Springer-Verlag is a part of Springer Science+Business Media
springeronline.com

© Springer-Verlag Berlin Heidelberg 2004
Printed in Germany

The use of general descriptive names, registered names, etc. in this publication does not imply, even in the absence of a specific statement, that such names are exempt from the relevant protective laws and regulations and free for general use.

Product liability: The publishers cannot guarantee the accuracy of any information about dosage and application contained in this book. In every individual case the user must check such information by consulting the relevant literature.

Cover design: design & production GmbH, Heidelberg
Typesetting: Stürtz AG, 97080 Würzburg

Printed on acid-free paper   27/3150 hs – 5 4 3 2 1 0

Neuropeptide Y and Related Peptides

Contributors
M. Alfalah, A. Beck-Sickinger, S. Bedoui, A. Brennauer,
A. Buschauer, C. Carvajal, H. M. Cox, W. R. Crowley,
O. Della-Zuana, S. Dove, Y. Dumont, J.-L. Fauchere,
M. Feletou, J. P. Galizzi, M. Heilig, H. Herzog, S. von Hörsten,
T. Hoffmann, N. P. Hyland, S. P. Kalra, P. S. Kalra, T. Karl,
A. Kask, M. Lonchampt, N. R. Levens, M. C. Michel, K. Mörl,
R. Pabst, J. P. Redrobe, R. Quirion, K. Tatemoto, T. E. Thiele,
C. D. Wrann

Editor
Martin C. Michel

Springer

Handbook of Experimental Pharmacology

Volume 162

Editor-in-Chief
K. Starke, Freiburg i. Br.

Editorial Board
G.V.R. Born, London
M. Eichelbaum, Stuttgart
D. Ganten, Berlin
F. Hofmann, München
B. Kobilka, Stanford, CA
W. Rosenthal, Berlin
G. Rubanyi, Richmond, CA

Springer
*Berlin
Heidelberg
New York
Hong Kong
London
Milan
Paris
Tokyo*

Preface

A little more than 20 years have now passed since the original discovery of neuropeptide Y (NPY) by Tatemoto and colleagues. Meanwhile a Medline search for the keyword 'NPY' yields more than 7,000 hits, with a steady 400–500 per year in the last decade.

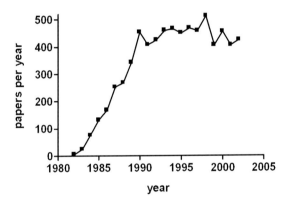

Fig. 1 Number of papers retrieved by year of publication from Medline under the keyword 'NPY'

Papers devoted exclusively to the closely related peptide YY and pancreatic polypeptide are not even included in this statistic. Thus, it is not surprising that the field of NPY research has made remarkable progress and is coming of age.

Phylogenetic research has shown that NPY is an evolutionary very old and highly conserved peptide. Members of the NPY family of peptides act on specific G-protein coupled receptors, of which five subtypes have been cloned in a variety of species and whose signal transduction is being unraveled. Cloning of the genes for NPY and its receptors has enabled the generation of transgenic and knockout animals. Using such tools we now understand major aspects of the physiological role of NPY not only in the brain but also, for example, in the peripheral tissues of the cardiovascular, gastrointestinal and immune system. While NPY largely exerts these effects as a neurotransmitter, the related peptide YY and pancreatic polypeptide predominantly act as hormones. Numerous peptide analogs and nonpeptide ligands acting on NPY receptors, often in a subtype-selective manner, have been described. They set the stage for the final act, the definition of the role of NPY in human health and disease and ultimately the manipulation of this system in treating patients.

The present volume addresses all of the above topics by established leaders in their respective areas. I would like to take this opportunity to thank all contributors for the time and effort they have spent in preparing their chapters and also Mrs. Susanne Dathe of Springer-Verlag who expertly supported the project. On behalf of all contributors I hope that experienced NPY aficionados will find new and useful additional information in this volume and that newcomers to the field will discover how much exciting research this still has to offer.

August 2003   Martin C. Michel, Amsterdam

List of Contributors

(Their addresses can be found at the beginning of their respective chapters.)

Alfalah, M. 23

Beck-Sickinger, A. G. 479
Bedoui, S. 23, 409
Brennauer, A. 505
Buschauer, A. 505

Carvajal, C. 101
Cox, H. M. 45, 389
Crowley, W. R. 185

Della-Zuana, O. 283
Dove, S. 505
Dumont, Y. 101

Fauchére, J.-L. 283
Félétou, M. 283
Fredriksson, R. 75

Galizzi, J.-P. 283

Heilig, M. 251
Herzog, H. 447
Hörsten, S. von 23, 409
Hoffmann, T. 23
Holliday, N. D. 45
Hyland, N. P. 389

Kalra, S. P. 221
Kalra, P. S. 221
Karl, T. 23
Kask, A. 101

Larhammar, D. 75
Larson, E. T. 75
Levens, N. R. 283
Lonchampt, M. 283

Michel, M. C. 45, 361
Mörl, K. 479
Morris, M. J. 327

Pabst, R. 23, 409

Quirion, R. 101

Redrobe, J. P. 101

Salaneck, E. 75

Tatemoto, K. 1
Thiele, T. E. 251

Westfall, T. C. 137
Wrann, C. D. 23

List of Contents

Neuropeptide Y: History and Overview 1
  K. Tatemoto

PP, PYY and NPY: Synthesis, Storage, Release and Degradation 23
  S. von Hörsten, T. Hoffmann, M. Alfalah, C. D. Wrann, T. Karl,
  R. Pabst, S. Bedoui

NPY Receptor Subtypes and Their Signal Transduction 45
  N. D. Holliday, M. C. Michel, H. M. Cox

Phylogeny of NPY-Family Peptides and Their Receptors 75
  D. Larhammar, R. Fredriksson, E. T. Larson, E. Salaneck

Neuropeptide Y and Its Receptor Subtypes in the Central Nervous System:
Emphasis on Their Role in Animal Models of Psychiatric Disorders 101
  J. P. Redrobe, C. Carvajal, A. Kask, Y. Dumont, R. Quirion

Prejunctional Effects of Neuropeptide Y and Its Role as a Cotransmitter ... 137
  T. C. Westfall

Neuroendocrine Actions Of Neuropeptide Y 185
  W. R. Crowley

NPY: A Novel On/Off Switch for Control of Appetite and Reproduction ... 221
  S. P. Kalra, P. S. Kalra

Behavioral Effects of Neuropeptide Y 251
  T. E. Thiele, M. Heilig

NPY Effects on Food Intake and Metabolism............................ 283
    N.R. Levens, M. Félétou, J.-P. Galizzi, J.-L. Fauchére, O. Della-Zuana,
    M. Lonchampt

Neuropeptide Y and Cardiovascular Function 327
    M.J. Morris

Neuropeptide Y and the Kidney....................................... 361
    M.C. Michel

NPY-Like Peptides, Y Receptors and Gastrointestinal Function 389
    N.P. Hyland, H.M. Cox

NPY and Immune Functions: Implications for Health and Disease 409
    S. Bedoui, R. Pabst, S. von Hörsten

Transgenic and Knockout Models in NPY Research 447
    H. Herzog

Structure–Activity Relationship
of Peptide-Derived Ligands at NPY Receptors.......................... 479
    K. Mörl, A.G. Beck-Sickinger

Structure–Activity Relationships
of Nonpeptide Neuropeptide Y Receptor Antagonists................... 505
    A. Brennauer, S. Dove, A. Buschauer

Subject Index... 547

Neuropeptide Y: History and Overview

K. Tatemoto

Department of Molecular Physiology, Institute for Molecular and Cellular Regulation, Gunma University, 371-8512 Maebashi, Japan
e-mail: tatekazu@showa.gunma-u.ac.jp

1	Introduction	2
2	Isolation and Primary Structures of NPY	2
2.1	Discovery of NPY	2
2.2	Primary Structure of NPY	3
2.3	NPY mRNA	4
3	Cellular Localization of NPY	4
3.1	NPY in the Peripheral Nervous System	4
3.2	NPY in the Central Nervous System	5
4	Studies on the Receptors and Physiological Functions of NPY (1982–1992)	5
4.1	Peripheral Actions of NPY	5
4.1.1	Cardiovascular Response	5
4.1.2	Hormone Secretion	6
4.2	Central Actions of NPY	6
4.2.1	Cardiovascular Response	6
4.2.2	Circadian Rhythms	6
4.2.3	Food Intake and Energy Expenditure	6
4.2.4	Hormone Secretion and Reproduction	7
4.2.5	Stress, Depression, Anxiety, and Pain	7
4.2.6	Seizures	8
4.3	NPY Receptor Binding and Intracellular Signaling	8
4.4	NPY Receptor Agonists and Antagonists	8
4.5	NPY Receptor Subtypes	8
5	Studies on the Receptors and Physiological Functions of NPY (1992–2002)	9
5.1	Cloning of NPY Receptor Subtypes	9
5.1.1	Y_1 Receptor	10
5.1.2	Y_2 Receptor	10
5.1.3	Putative Y_3 Receptor	10
5.1.4	Y_4 Receptor	10
5.1.5	Y_5 Receptor	11
5.1.6	y_6 Receptor	11
5.2	Selective NPY Receptor Agonists and Antagonists	11
5.3	NPY Receptor Subtypes and Their Physiological Functions	12
5.3.1	Cardiovascular Response	12
5.3.2	Circadian Rhythms	12
5.3.3	Food Intake and Energy Expenditure	12

5.3.4 Hormone Secretion and Reproduction. 13
5.3.5 Anxiety, Pain, Stress, and Depression 13
5.3.6 Seizures. 14
5.3.7 Ethanol Consumption 14

6   Conclusions and Future Studies 15

References 15

Abstract Neuropeptide Y (NPY) is a 36-amino acid peptide with structural similarities to peptide YY (PYY) and pancreatic polypeptide (PP). NPY, one of the most abundant neuropeptides known, is widely distributed throughout the central and peripheral nervous systems, while PYY and PP are predominantly distributed in the endocrine cells of the intestine and pancreas, respectively. Five NPY receptor subtypes denoted as Y_1, Y_2, Y_4, Y_5, and y_6 mediate the actions of NPY. NPY is involved in the regulation of diverse functions including food intake, blood pressure, circadian rhythms, stress, pain, hormone secretion, reproduction, and alcohol consumption. NPY has also been implicated in the pathophysiology of a number of diseases such as feeding disorders, seizures, hypertension, pain disorders, depression, and anxiety. This review will describe a brief history and an overview of the studies on NPY concerning the isolation, tissue distribution, receptor subtypes, receptor agonists and antagonists, physiological functions, and pharmacological activities.

Keywords Neuropeptide Y · Tissue distribution · Receptor subtype · Receptor antagonist · Review

1
Introduction

In this chapter, a brief history and an overview of the studies on neuropeptide Y (NPY) during the last 20 years are described. The main thrust of this chapter is to critically review the initial findings that have influenced later studies on NPY. Particular attention is focused on the studies concerning the receptors and physiological functions of NPY. The reader is referred to other excellent reviews in this book for more comprehensive discussion.

2
Isolation and Primary Structures of NPY

2.1
Discovery of NPY

In the last century, many neuropeptides and hormonal peptides were identified on the basis of specific biological responses mediated by them. Unlike most other known neuropeptides, however, NPY was first identified in brain extracts by its C-terminal tyrosine amide structure. In 1978, we developed a novel method for the detection of biologically active peptides based on the C-terminal amide structure that is a unique chemical feature of many peptide hormones and neuropeptides (Tatemoto and Mutt 1978). Since peptides with this structure are likely to be biologically active, the search for unknown peptide amides would result in the finding of novel peptides. We therefore carried out the isolation of previously unknown peptide amides from tissue extracts using a chemical method as the detection device.

In 1980, we isolated two novel peptide amides, which were designated peptide HI (PHI) and peptide YY (PYY) from porcine intestinal extracts (Tatemoto and Mutt 1980). Subsequently, we isolated a peptide with a C-terminal tyrosine amide from porcine brain extracts, which was named neuropeptide Y (Tatemoto et al. 1982). Using a similar approach, we isolated a series of other novel peptides such as galanin (Tatemoto et al. 1983), neuropeptide K (Tatemoto et al. 1985), and pancreastatin (Tatemoto et al. 1986).

2.2
Primary Structure of NPY

NPY is a linear polypeptide with 36 amino acid residues (Tatemoto 1982a). Since NPY contains many tyrosine (Y) residues in its structure, we named this peptide neuropeptide Y to distinguish it from PYY that possesses a very similar structure to NPY (Tatemoto 1982b). A comparison of the primary structures of NPY, PYY and pancreatic polypeptide (PP) reveals a high degree of sequence homology between NPY and PYY, with a lesser degree of homology between NPY and PP, as shown in Fig. 1. It was therefore proposed that NPY, PYY, and PP are members of a previously unrecognized peptide family (Tatemoto 1982a).

Later, human NPY was isolated from adrenal-medullary pheochromocytoma tissue. The primary structure of human NPY differs from that of the porcine peptide in only one position of the 36 residues (Corder et al. 1984). Subsequent studies identified the primary structures of NPY molecules from various animals, birds, frogs, and others (for a review see Larhammar et al. 1993). These

homology

NPY	Y P S K P D N P G E D A P A E D L A R Y Y S A L R H Y I N L I T R Q R Y*	100%
PYY	Y P A K P E A P G E D A S P E E L S R Y Y A S L R H Y L N L V T R Q R Y*	69%
PP	A P L E P V Y P G D D A T P E Q M A Q Y A A E L R R Y I N M L T R P R Y*	50%

Fig. 1 Comparison of the amino acid sequences of the porcine peptides, neuropeptide Y (*NPY*), peptide YY (*PYY*), and pancreatic polypeptide (*PP*). An *asterisk* indicates the amidated C-terminus. Identities are underlined

studies revealed that the structure of NPY has been strongly conserved throughout evolution.

2.3
NPY mRNA

Dixon and coworkers were the first to identify the sequence of a human cDNA encoding NPY from human pheochromocytoma cells. The 97-amino acid precursor has at least two processing sites, which would generate three peptides of 28 (signal peptide), 36 (NPY), and 30 (COOH-terminal peptide) amino acid residues (Minth et al. 1984). Subsequently, the structure of a human NPY gene identified from a human genomic DNA library was reported. The DNA sequences located within 530 bases of the start of transcription were found to be sufficient for transient expression in the two cell lines examined (Minth et al. 1986).

3
Cellular Localization of NPY

Since NPY was discovered by its chemical nature, no biological activity of the peptide was known when it was isolated. Therefore, we prepared a large quantity of natural NPY from more than 1,000 kg of porcine brains, and the natural NPY preparations thus obtained were sent to a number of laboratories in Europe and the USA to examine the biological activities of the peptide and to generate specific antisera against NPY for immunohistochemistry and radioimmunoassay studies. Between 1982 and 1985, the natural NPY preparations were used for many studies on the biological activity and localization of NPY, until synthetic NPY preparations became commercially available.

3.1
NPY in the Peripheral Nervous System

Lundberg et al. (1982) were the first to demonstrate NPY-like immunoreactivity in many peripheral neurons with a distribution mostly paralleling that of tyrosine hydroxylase (TH) and dopamine-beta-hydroxylase (DBH) containing neurons. Very high levels of NPY-like immunoreactivity were found in sympathetic ganglia and in tissues receiving a dense sympathetic innervation, such as the vas deferens, heart atrium, blood vessels, and spleen. NPY- and DBH-nerves had a roughly parallel occurrence in the heart, spleen, kidney, respiratory and urogenitary tracts, around blood vessels, and within visceral smooth muscle. NPY thus seems to be a major peptide in the sympathetic nervous system (Lundberg et al.1983). The presence of NPY-like immunoreactivity was also demonstrated in the neuronal elements of the gut and pancreas (Sundler. et al. 1983). NPY has generally been found in the sympathetic neurons, costored and coreleased with catecholamines, although NPY is also present in peripheral non-sympathetic neurons (for a review see MacDonald 1988).

3.2
NPY in the Central Nervous System

Bloom and his colleagues have shown that NPY-like immunoreactivity is widely and unevenly distributed in rat and human brains, and it is the most abundant neuropeptide known (Allen et al. 1983; Adrian et al. 1983). The highest concentrations of NPY were found in the paraventricular hypothalamic nucleus, hypothalamic arcuate nucleus, suprachiasmatic nucleus, median eminence, dorsomedial hypothalamic nucleus, and paraventricular thalamic nucleus (Chronwall et al. 1985). The extremely high concentrations and widespread distribution indicate important roles of NPY in many brain functions. Hokfelt and coworkers (1983) reported the coexistence of NPY-like immunoreactivity in catecholamine neurons in the medulla oblongata. During subsequent years, a number of in vitro studies have shown the coexistence of NPY not only with catecholamines but also with a variety of other neurotransmitters or neuropeptides (for a review see Everitt and Hokfelt 1989). It is suggested that the physiological roles of NPY in the central nervous system are very complex because of the interactions of NPY with other effector systems. More recently, localization of NPY mRNA by means of in situ hybridization was found to be comparable to that of NPY-immunoreactivity (Terenghi et al.1987; Gehlert et al. 1987).

4
Studies on the Receptors and Physiological Functions of NPY (1982–1992)

During the first 10 years of NPY research, a number of important biological activities of NPY in both the central and peripheral nervous systems were discovered, and the presence of NPY receptor subtypes, Y_1 and Y_2, was demonstrated using specific NPY receptor agonists.

4.1
Peripheral Actions of NPY

4.1.1
Cardiovascular Response

Lundberg and Tatemoto (1982) were the first to find a biological activity of NPY, demonstrating that NPY induced potent vasoconstriction, more potent than noradrenaline, which was resistant to alpha-adrenoceptor blockade. It was also shown that NPY caused strong contractions of cerebral arteries (Edvinsson et al. 1983), and that NPY produced an inhibition of colonic motility and a vasoconstriction of long duration (Hellstrom et al. 1985). NPY induced renal vasoconstriction and inhibited renin release by inhibiting adenylate cyclase in the vascular smooth muscle and renin-producing cells (Hackenthal et al. 1987). NPY produced potent pressor responses (Allen et al. 1984; Petty et al. 1984), while NPY (18–36), a C-terminal NPY fragment, exhibited substantial hypotensive action (Boublik et al. 1989).

Interestingly, NPY was found to prevent the blood pressure fall induced by endotoxin in conscious rats with adrenal medullectomy (Evequoz et al. 1988), and plasma levels of NPY were found to markedly increase in patients with septic shock (Watson et al. 1988). These data suggest that NPY plays a role in maintaining blood pressure during endotoxic shock.

4.1.2
Hormone Secretion

Allan et al. (1982) found that NPY inhibited the contraction of electrically stimulated mouse vas deferens, suggesting inhibitory actions of NPY on noradrenaline release at a pre-synaptic level. Subsequently, NPY was shown to depress the secretion of ^{3}H-noradrenaline and the contractile response evoked by field stimulation in the vas deferens (Lundberg and Stjarne 1984).

4.2
Central Actions of NPY

4.2.1
Cardiovascular Response

Fuxe et al. (1983) first demonstrated an effect of central administration of NPY. They found that NPY induced hypotension and bradypnea in the rat, suggesting the involvement of NPY in the cardiovascular and respiratory controls of the central nervous system.

4.2.2
Circadian Rhythms

Albers and Ferris (1984) found that microinjection of NPY into the suprachiasmatic region of the hypothalamus (SCN) phase-shifted the circadian rhythm of hamsters housed in constant light. It is suggested that NPY functions as a chemical messenger that is important for the light–dark cycle entrainment of circadian rhythms.

4.2.3
Food Intake and Energy Expenditure

In 1984, NPY was found, for the first time, to stimulate feeding behavior in rats (Clark et al. 1984; Levine and Morley 1984; Stanley and Leibowitz 1984). Since then, a number of studies have shown that NPY is the most potent orexigenic peptide identified to date. Later, cerebrospinal fluid NPY concentrations were found to be significantly elevated in anorectic and bulimic patients. These levels normalized in long-term weight-restored anorectic patients who had a return of normal menstrual cycles (Kaye et al. 1990). This suggests that NPY plays a role in eating disorders. It was also reported that NPY decreased rectal temperature after intracerebroventricular administration (Morioka et al. 1986). These observations suggest that NPY is involved in the regulation of food intake and energy expenditure.

4.2.4
Hormone Secretion and Reproduction

Kalra and Crowley (1984) found that central administration of NPY suppressed luteinizing hormone (LH) release in ovariectomized rats, while it stimulated LH release in ovariectomized rats pretreated with estrogen and progesterone. Kalra and coworkers found that NPY not only stimulated food intake but also inhibited sexual behavior in rats (Clark et al. 1985). NPY inhibited excitatory synaptic transmission in the hippocampus by acting directly at the terminal to reduce a calcium influx (Colmers et al. 1988). These observations suggest that NPY is involved in the central control of hormone and neurotransmitter release.

4.2.5
Stress, Depression, Anxiety, and Pain

Fuxe et al. (1983) demonstrated that central administration of NPY induced EEG synchronization. It is therefore suggested that NPY produces behavioral signs of sedation. Subsequently, Heilig and Murison (1987) found that intracerebroventricular administration of NPY protected against stress-induced gastric erosion in the rat. Stress-induced erosion was reduced by approximately 50% by NPY, suggesting the anti-stress action of NPY as a manifestation of its sedative properties. In addition, it was reported that the administration of NPY into the third ventricle of the brain enhanced memory retention. It is suggested that NPY modulates memory processes (Flood et al. 1987).

Interestingly, NPY-like immunoreactivity was found to be significantly lower in cerebrospinal fluid from patients with a major depressive disorder compared with healthy controls (Widerlov et al. 1988). It was also found that antidepressant drugs increased the concentrations of NPY-like immunoreactivity in the brain (Heilig et al. 1988). These observations support the hypothesis that NPY is involved in the pathophysiology of depressive illness.

Furthermore, Heilig et al. (1989) found that centrally administered NPY produced anxiolytic-like effects that were mediated through interactions with noradrenergic systems in animal anxiety models. In a hot plate test, spinally administered NPY produced a dose-dependent elevation in the nociceptive threshold in rats, suggesting the involvement of NPY in the mechanism of pain control (Hua et al. 1991).

4.2.6
Seizures

Marksteiner and Sperk (1988) observed significantly increased levels of NPY-like immunoreactivity in the frontal cortex of rats that had undergone strong limbic seizures induced by kainic acid. The increase could be prevented by early injection of an anticonvulsant (Marksteiner et al. 1990). These observations suggest that NPY is involved in the control of seizures.

4.3
NPY Receptor Binding and Intracellular Signaling

The specific binding of the iodinated NPY to membranes from the cerebral cortex was demonstrated. The binding of iodinated NPY was characterized by a Kd value of 0.38 nM (Unden et al. 1984). A study on autoradiographic localization of NPY receptors indicated that the receptors were discretely distributed in the rat brain with high densities found in areas such as the olfactory bulb, superficial layers of the cortex, ventral hippocampus, and area postrema (Martel et al. 1986).

NPY was found to be a potent inhibitor of cyclic AMP accumulation in feline cerebral blood vessels (Fredholm et al. 1985). NPY was shown to inhibit adenylate cyclase through a pertussis toxin-sensitive G protein (Kassis et al. 1987). In addition to inhibiting adenylate cyclase, NPY was found to elevate intracellular calcium (Motulsky and Michel 1988). It was also shown that guanine nucleotide-binding protein Go mediated the inhibitory effects of NPY on dorsal root ganglion calcium channels (Ewald et al. 1988).

4.4
NPY Receptor Agonists and Antagonists

Centrally truncated synthetic NPY agonists were synthesized and shown to be biologically active (Beck et al. 1989; Krstenansky et al. 1989). Fuhlendorff et al. (1990) reported that [Leu^{31}, Pro^{34}] NPY was a specific Y1 receptor agonist that could be useful in delineating the physiological importance of Y1 receptors. About the same time, the first NPY receptor antagonists, Ac-[3-(2,6-dichlorobenzyl)Tyr^{27}, D-Thr^{32}] NPY(27–36) and Ac-[3-(2,6-dichlorobenzyl) Tyr^{27,36}, D-Thr^{32}] NPY(27–36), designated PYX-1 and PYX-2, respectively, were synthesized based on the C-terminal structure of the NPY molecule (Tatemoto 1990). PYX-2 was found to block the stimulatory action of NPY on carbohydrate ingestion (Leibowitz et al. 1992).

4.5
NPY Receptor Subtypes

Wahlestedt et al. (1986) first suggested the presence of two receptor subtypes for NPY and its related peptides. They studied the effects of NPY, PYY, and the

C-terminal fragments of NPY or PYY on different smooth muscle preparations in vitro, and found that PYY(13–36) reproduced the NPY- and PYY-induced suppression of noradrenaline release. Thus, the C-terminal portion seems to be sufficient for exerting pre-junctional effects of NPY and PYY, while the whole sequence seems to be required for post-junctional effects. Later, Schwartz and coworkers showed that two subtypes of NPY/PYY-binding sites occurred in different cells, supporting the hypothesis of NPY receptor subtypes (Sheikh et al. 1989).

Since 1989, a number of the studies on NPY receptor subtypes have been published. Using selective receptor agonists, it was shown that Y_1 and Y_2 receptors were independently expressed in the brain and the majority of NPY receptors in the brain were of the Y_2 type (Aicher et al. 1991). It was found that the hypothalamic Y_1 receptors mediated the stimulatory effect of NPY on carbohydrate intake and meal size, while the Y_2 receptors had the opposite effect of suppressing carbohydrate intake (Leibowitz and Alexander 1991). Presynaptic inhibition by NPY observed in rat hippocampal slice was shown to be mediated by a Y_2 receptor (Colmers et al. 1991). Involvement of Y1 receptor subtype in the regulation of LH secretion was demonstrated by using NPY, NPY(2–36), [Leu^{31}, Pro^{34}] NPY, NPY(13–36) and other NPY fragments (Kalra et al. 1992).

5
Studies on the Receptors and Physiological Functions of NPY (1992–2002)

The main topics for the last 10 years have been the cloning of NPY receptor subtypes, Y_1, Y_2, Y_4, Y_5, and y_6, and subsequent studies on the biological functions of these receptor subtypes. Thus, in addition to studies on NPY-transgenic and deficient animals, a number of animals lacking specific NPY receptor subtypes were generated and the physiological functions of these animals were studied. Moreover, specific receptor agonists and antagonists for each NPY receptor subtype were developed and their physiological and pharmacological properties were evaluated. The synthesis of selective and potent receptor agonists and antagonists has provided useful tools to study the physiological functions of NPY receptor subtypes and to develop novel pharmacological treatments.

5.1
Cloning of NPY Receptor Subtypes

Recent advances in molecular biology have resulted in the identification of five NPY receptor subtypes, Y_1, Y_2, Y_4, Y_5, and y_6 receptors (for a review see Michel et al. 1998). These receptor subtypes were found to share only modest sequence homologies (30–50%). Moreover, each of the receptor subtypes seems to be characterized by a distinct tissue localization and unique pharmacological profile. The next section describes a brief history of the cloning of NPY receptor subtypes.

5.1.1
Y_1 Receptor

In 1992, the primary structures of rat and human Y1 receptors were identified (Krause et al. 1992; Herzog et al. 1992; Larhammar et al. 1992). The Y_1 receptor, the first NPY receptor to be cloned, was found to be a 384-amino acid protein belonging to a G protein-coupled receptor family. The functionality of the expressed NPY receptor was demonstrated by inhibition of adenylate cyclase and mobilization of intracellular calcium, both being characteristic of an NPY receptor. The distribution of Y_1 receptor expression correlated with that of NPY-immunoreactive nerves and the apparent actions of NPY in the intestine, kidney, and heart (Wharton et al. 1993).

5.1.2
Y_2 Receptor

The cloned Y_2 receptor consists of 381 amino acids, and has only 31% identity to the structure of the Y_1 receptor (Rose et al. 1995; Gerald et al. 1995, Gehlert et al. 1996; Rimland et al. 1996). The Y_2 receptor expressing cells have high affinity binding sites for NPY, PYY, and NPY(13–36), whereas [Leu^{31}, Pro^{34}] NPY binds with lower affinity. The Y_2 receptor is localized on a number of NPY-containing neurons in the brain, suggesting that this receptor has a characteristic of an autoreceptor (Caberlotto et al. 2000).

5.1.3
Putative Y_3 Receptor

The Y_3 receptor is distinguished from the other NPY receptors by its high affinity for NPY but relatively low affinity for PYY. However, evidence for the existence of such a subtype is not clear as the clone initially reported as a Y_3 receptor (Rimland et al. 1991) failed to confer NPY binding sites (Herzog et al. 1993; Jazin et al. 1993). Therefore, the evidence is not sufficient to grant the presence of a Y_3 receptor (Michel et al. 1998).

5.1.4
Y_4 Receptor

A unique feature of the Y_4 receptor is a high affinity for PP. Therefore, the Y_4 receptor is probably a PP receptor. The cloned human Y_4 receptor has 43% sequence homology with the human Y_1 receptor (Lundell et al. 1995; Bard et al. 1995; Yan et al. 1996). Both NPY and PYY have low affinities for this receptor. Y_4 receptor is present in the intestine, prostate, and pancreas (Lundell et al. 1995). The Y_4 receptor mRNA is sparsely expressed in the brain, except in the brainstem (Parker and Herzog 1999).

5.1.5
Y$_5$ Receptor

The cloning of a novel NPY receptor designated Y$_5$ receptor was reported (Gerald et al. 1996; Hu et al. 1996). The complementary DNA encoded a 456-amino-acid protein with less than 35% overall identity to the other known NPY receptors. [D-Trp^{32}] NPY had a high affinity for the Y$_5$ receptor, while it had low affinities for the other known NPY receptors. The Y$_5$ receptor, originally cloned as the 'feeding' receptor in the hypothalamus, was also found in the peripheral nervous system such as the testis, spleen, and pancreas (Statnick et al. 1998).

5.1.6
y$_6$ Receptor

The cloning of a novel NPY receptor proposed to be a Y$_5$ receptor was reported (Weinberg et al. 1996). However, other researchers reported the same clone as a PP receptor or Y$_{2b}$ receptor (Gregor et al. 1996; Matsumoto et al. 1996). To avoid confusion, it was renamed the y$_6$ receptor. The y$_6$ receptor gene is present in chicken, rabbit, cow, dog, mouse, and human, but it is completely absent in rat (Burkhoff et al. 1998). Sequence data revealed the y$_6$ gene to be the orthologue of the mouse Y$_5$ gene. Rabbits encode functional y$_6$ receptor, but the y$_6$ receptors in primates are functionally inactive due to a frameshift mutation occurring during early primate evolution (Matsumoto et al. 1996).

5.2
Selective NPY Receptor Agonists and Antagonists

Based on the C-terminal structure of the NPY molecule, the first nonpeptide Y$_1$ receptor antagonist BIBP 3226 was designed and synthesized (Rudolf et al. 1994), demonstrating that such a nonpeptide compound could be a useful tool for studying physiological functions and exploring therapeutic relevance.

Furthermore, synthesis of both peptide and nonpeptide Y$_1$ receptor antagonists such as [D-Tyr^{27,36}, D-Thr^{32}] NPY(27–36), SR120819A, 1229U91, BIBO3304, LY-357897, J-115814, and CP 617,906 have been reported. More recently, T4-[NPY(33–36)]4 and BIIE0246 have been described as selective Y$_2$ receptor antagonists. After the cloning of the Y$_5$ receptor, a number of Y$_5$ receptor antagonists including CGP71683A and L-152,804, and Y$_5$ receptor agonists such as [D-Trp^{34}] NPY and [Ala^{31}, Aib^{32}] NPY were synthesized (for reviews see Balasubramaniam 1997; Pheng and Regoli 2000; Parker et al. 2002). The Y$_1$ receptor antagonist 1229U91 has been shown to exhibit an agonist activity for the Y$_4$ receptor (Parker et al. 1998). However, no selective antagonist for the Y$_4$ receptor has yet been reported.

5.3
NPY Receptor Subtypes and Their Physiological Functions

The cloning of NPY receptor subtypes has made it possible to generate specific receptor subtype-deficient animals. The generation of such animals has provided unique models to examine the physiological functions of NPY. The next section focuses on the physiological functions of NPY and its receptor subtypes revealed by the use of receptor agonists and antagonists and genetically modified animals.

5.3.1
Cardiovascular Response

BIBP3226 antagonized vasoconstriction induced by NPY. This suggests that endogenous NPY acting on the Y_1 receptor is likely to account for the long-lasting component of sympathetic vasoconstriction in response to high-frequency stimulation (Malmstrom and Lundberg 1995). It was reported that the incubation of the subcutaneous arteries with Y_1 receptor antisense oligodeoxynucleotides attenuated NPY-induced vasoconstriction (Sun et al. 1996). Furthermore, Y_1 receptor-deficient mice showed a complete absence of blood pressure responses to NPY, suggesting the importance of Y_1 receptors in the NPY-mediated cardiovascular response (Pedrazzini et al. 1998).

However, it was also reported that the depressor effect of intrathecal NPY injection was primarily mediated by a Y_2 receptor (Chen and Westfall 1993). Furthermore, a Y_2 receptor agonist evoked vasoconstriction in the spleen, while a Y_2 receptor antagonist BIIE0246 antagonized the response. These suggest that the Y_2 receptor is also involved in NPY/PYY-evoked vasoconstriction (Malmstrom 2001).

5.3.2
Circadian Rhythms

NPY has been implicated in the phase shifting of circadian rhythms. Microinjection of a Y_2 receptor agonist produced phase advances that were significantly greater than those produced by the injection of a Y_1 receptor agonist. This suggests that NPY phase shifts circadian rhythms via the Y_2 receptor (Huhman et al. 1996; Golombek 1996). There is, however, some evidence that the Y_1/Y_5 receptors, in addition to the Y_2 receptor, may also be involved in the mechanism of NPY action by altering the levels of circadian clock-related genes (Fukuhara et al. 2001).

5.3.3
Food Intake and Energy Expenditure

NPY has been implicated to be a central stimulator of feeding behavior by interacting with a number of other hormones and neuroregulators that play roles in the regulation of body weight. A novel obese gene product, leptin, was found to regulate food intake by inhibiting the synthesis and release of NPY in the central nervous system (Stephens et al. 1995). It was reported that the mild obesity found in Y_1 receptor-deficient mice was caused by impaired insulin secretion and low energy expenditure (Kushi et al. 1998). Furthermore, NPY-induced food intake was remarkably reduced in Y_1-deficient mice (Kanatani et al. 2000). These results suggest the importance of Y_1 receptors in the regulation of food intake and body weight through the central control of energy expenditure.

It was found that the Y_5 receptor was also involved in NPY-induced food intake (Gerald et al. 1996). The Y_5 receptor-deficient mice responded significantly less to NPY-induced food intake than wild-type mice (Marsh et al. 1998). On the other hand, the results obtained using Y_2 receptor-deficient mice indicated an inhibitory role for the Y_2 receptor in the central regulation of body weight and food intake (Naveilhan et al. 1999). Hypothalamus-specific Y_2 receptor-deleted mice showed a significant decrease in body weight and a significant increase in food intake, suggesting an important role of hypothalamic Y_2 receptors in body weight regulation (Sainsbury et al. 2002). In addition, it was reported that peripheral injection of PYY(3–36) in rats inhibited food intake and reduced weight gain. PYY(3–36) also inhibited food intake in mice, but not in Y_2 receptor-deficient mice. This suggests that the anorectic effect requires the Y_2 receptor (Batterham et al. 2002).

5.3.4
Hormone Secretion and Reproduction

NPY has been known to be a putative neuroregulator of the reproductive axis in the central nervous system. A selective Y_5 agonist inhibited LH secretion, while the inhibitory action was fully prevented by Y_5 receptor antagonists (Raposinho et al. 1999). It was also shown that Y_5 receptor activation suppressed the reproductive axis in both virgin and lactating rats (Toufexis et al. 2002). These results suggest that the actions of NPY on the reproductive axis are predominantly mediated by the Y_5 receptor. On the other hand, using Y_1 receptor-deficient mice, crucial roles for the Y_1 receptor in controlling food intake, the onset of puberty, and the maintenance of reproductive functions were demonstrated (Pralong et al. 2002).

5.3.5
Anxiety, Pain, Stress, and Depression

It has been shown that NPY exhibits anxiolytic, antinociceptive, anti-stress, and anti-depressive actions. Involvement of the Y_1 receptor in the anxiolytic-like action of NPY was demonstrated (Wahlestedt et al. 1993; Heilig et al. 1993). NPY may produce not only an anxiolytic effect via the Y_1 receptor, but also an anxiogenic effect via the Y_2 receptor (Nakajima et al. 1998). It was reported that NPY transgenic mice displayed anxiolytic behaviors (Inui et al. 1998). Moreover, transgenic rats with hippocampal NPY overexpression were insensitive to restraint stress, had no fear suppression behavior, and displayed impaired spatial learning (Thorsell et al. 2000). It was also reported that Y_1 receptor-deficient mice developed hyperalgesia to acute pain, and showed a complete absence of the pharmacological analgesic effects of NPY (Naveilhan et al. 2001). These data suggest that NPY and its receptors are involved in the mechanisms of anxiety, stress, learning, and nociception.

Using an animal model of depression, alterations in the NPY levels and Y_1 receptor mRNA were observed after treatment with an anti-depressant drug (Caberlotto et al. 1998). When compared with healthy controls, the levels of NPY appeared to be low in patients who had recently attempted suicide. Patients who had repeatedly attempted suicide were found to have the lowest NPY levels (Westrin et al. 1999). These data suggest the possible involvement of NPY and Y_1 receptors in depression.

5.3.6
Seizures

NPY has been implicated to function as an endogenous anticonvulsant. It was reported that NPY-deficient mice were susceptible to seizures induced by a GABA antagonist (Erickson et al. 1996). Kainic acid-induced limbic seizures in NPY-deficient mice progressed uncontrollably and ultimately produced death in 93% of the mice, whereas intracerebroventricular NPY infusion could prevent such death (Baraban et al. 1997). Furthermore, the transgenic rats with NPY overexpression showed a significant reduction in the number and duration of kainic acid-induced seizures (Vezzani et al. 2002).

It was found that NPY, acting predominantly via Y_2 receptors, could dramatically inhibit epileptiform activity in vitro models of epilepsy (Klapstein and Colmers 1997). NPY was also found to potently inhibit seizures induced by kainic acid via Y_5 receptor (Woldbye et al. 1997). Moreover, mice lacking the Y_5 receptor were more sensitive to kainic acid-induced seizures (Marsh et al. 1999). In human epilepsy it is suggested that abundant sprouting of NPY fibers, concomitant upregulation of Y_2 receptors, and downregulation of Y_1 receptors in the hippocampus of patients with Ammon's horn sclerosis is involved in the anticonvulsant mechanism by the NPY system (Furtinger et al. 2001).

5.3.7
Ethanol Consumption

Thiele et al. (1998) first reported that NPY-deficient mice showed increased ethanol consumption, while transgenic mice with NPY overexpression had a lower preference for ethanol. These data suggest that alcohol consumption and resistance are inversely related to the NPY levels in the brain. Recently, it was reported that knockout mice lacking the Y_1 receptor showed increased ethanol consumption. It is suggested that the Y_1 receptor regulates voluntary ethanol consumption and some of the intoxicating effects caused by administration of ethanol (Thiele et al. 2002).

It was shown that blockade of central Y_2 receptors by a Y_2 receptor antagonist, BIIE0246, reduced ethanol self-administration in rats. It is therefore suggested that the Y_2 receptor is a candidate target for developing novel pharmacological treatments for alcoholism (Thorsell et al. 2002).

6
Conclusions and Future Studies

NPY has been shown to be involved in the regulation of diverse physiological functions and has been implicated in a variety of disorders such as anxiety, depression, obesity, epilepsy, and alcohol dependence. Thus, the NPY system has emerged as a potential drug target for a number of disorders.

During the last decade, the cloning of NPY receptor subtypes has made it possible to clarify the functional importance of the subtypes and to discover novel compounds with selective affinity to individual receptor subtypes. Indeed, a number of impressive advances have been made in the development of nonpeptide antagonists to NPY receptor subtypes. However, further studies are needed to clarify the potential of these compounds as useful drugs. In contrast, synthesis of nonpeptide NPY receptor agonists has not yet been successful, thereby hampering the development of drugs for the treatment of disorders such as anxiety, depression, pain disorders, and epilepsy. In addition, such an agonist may be of clinical importance for modulating the circadian-clock responses to light.

Advances in the development of orally-active nonpeptide NPY receptor agonists and antagonists that are capable of crossing the blood–brain barrier will facilitate our understanding of the physiological roles of NPY and will undoubtedly underscore the importance of NPY in the fields of pharmacology and clinical medicine.

References

Adrian TE, Allen JM, Bloom SR et al. (1983) Neuropeptide Y distribution in human brain. Nature 306:584–586

Aicher SA, Springston M, Berger SB et al. (1991) Receptor-selective analogs demonstrate NPY/PYY receptor heterogeneity in rat brain. Neurosci Lett 130:32–36

Albers HE, Ferris CF (1984) Neuropeptide Y: role in light-dark cycle entrainment of hamster circadian rhythms. Neurosci Lett 50:163–168

Allen JM, Adrian TE, Tatemoto K et al. (1982) Two novel related peptides, neuropeptide Y (NPY) and peptide YY (PYY) inhibit the contraction of the electrically stimulated mouse vas deferens. Neuropeptides 3:71–77

Allen JM, Rodrigo J, Yeats JC et al. (1984) Vascular distribution of neuropeptide Y (NPY) and effect on blood pressure. Clin Exp Hypertens A 6:1879–1882

Allen YS, Adrian TE, Allen JM et al. (1983) Neuropeptide Y distribution in the rat brain. Science 221:877–879

Balasubramaniam AA (1997) Neuropeptide Y family of hormones: receptor subtypes and antagonists. Peptides 18:445–457
Baraban SC, Hollopeter G, Erickson JC et al. (1997) Knock-out mice reveal a critical antiepileptic role for neuropeptide Y. J Neurosci 17:8927–8936
Bard JA, Walker MW, Branchek TA, Weinshank RL (1995) Cloning and functional expression of a human Y4 subtype receptor for pancreatic polypeptide, neuropeptide Y, and peptide YY. J Biol Chem 270:26762–26765
Batterham RL, Cowley MA, Small CJ et al. (2002) Gut hormone PYY(3–36) physiologically inhibits food intake. Nature 418:650–654
Beck A, Jung G, Gaida W et al. (1989) Highly potent and small neuropeptide Y agonist obtained by linking NPY 1–4 via spacer to alpha-helical NPY 25–36. FEBS Lett 244:119–122
Boublik J, Scott N, Taulane J et al. (1989) Neuropeptide Y and neuropeptide Y18–36. Structural and biological characterization. Int J Pept Protein Res 33:11–15
Burkhoff A, Linemeyer DL, Salon JA (1998) Distribution of a novel hypothalamic neuropeptide Y receptor gene and it's absence in rat. Brain Res Mol Brain Res 53:311–316
Caberlotto L, Fuxe K, Overstreet DH et al. (1998) Alterations in neuropeptide Y and Y1 receptor mRNA expression in brains from an animal model of depression: region specific adaptation after fluoxetine treatment. Brain Res Mol Brain Res 59:58–65
Caberlotto L, Fuxe K, Hurd YL (2000) Characterization of NPY mRNA-expressing cells in the human brain: co-localization with Y2 but not Y1 mRNA in the cerebral cortex, hippocampus, amygdala, and striatum. J Chem Neuroanat 20:327–337
Chen X, Westfall TC (1993) Depressor effect of intrathecal neuropeptide Y (NPY) is mediated by Y2 subtype of NPY receptors. J Cardiovasc Pharmacol 21:720–724
Chronwall BN, DiMaggio, DA, Massari VJ et al. (1985) The anatomy of neuropeptide Y-containing neurons in rat brain. Neuroscience 15:1159–1181
Clark JT, Kalra PS, Crowley WR, Kalra SP (1984) Neuropeptide Y and human pancreatic polypeptide stimulate feeding behavior in rats. Endocrinology 115:427–429
Clark JT, Kalra PS, Kalra SP (1985) Neuropeptide Y stimulates feeding but inhibits sexual behavior in rats. Endocrinology 117:2435–2442
Colmers WF, Lukowiak K, Pittman Q (1988) Neuropeptide Y action in the rat hippocampal slice: site and mechanism of presynaptic inhibition. J Neurosci 8:3827–337
Colmers WF, Klapstein GJ, Fournier A et al. (1991) Presynaptic inhibition by neuropeptide Y in rat hippocampal slice in vitro is mediated by a Y2 receptor. Br J Pharmacol 102:41–44
Corder R, Emson PC, Lowry PJ (1984) Purification and characterization of human neuropeptide Y from adrenal-medullary phaeochromocytoma tissue. Biochem J 219:699–706
Edvinsson L, Emson P, McCulloch J et al. (1983) Neuropeptide Y: cerebrovascular innervation and vasomotor effects in the cat. Neurosci Lett 43:79–84
Erickson JC, Clegg KE, Palmiter RD (1996) Sensitivity to leptin and susceptibility to seizures of mice lacking neuropeptide Y. Nature 381:415–421
Evequoz D, Waeber B, Aubert JF et al. (1988) Neuropeptide Y prevents the blood pressure fall induced by endotoxin in conscious rats with adrenal medullectomy. Circ Res 62:25–30
Everitt BJ, Hokfelt T (1989) The existence of neuropeptide Y with other peptides and amines in the central nervous system. In: Mutt V et al. (eds) Neuropeptide Y. Raven Press, New York, pp 61–71
Ewald DA, Sternweis PC, Miller RJ (1988) Guanine nucleotide-binding protein Go-induced coupling of neuropeptide Y receptors to Ca2+ channels in sensory neurons. Proc Natl Acad Sci USA 85:3633–3637
Flood JF, Hernandez EN, Morley JE (1987) Modulation of memory processing by neuropeptide Y. Brain Res 421:280–290

Fredholm BB, Jansen I, Edvinsson L (1985) Neuropeptide Y is a potent inhibitor of cyclic AMP accumulation in feline cerebral blood vessels. Acta Physiol Scand 124:467–469

Fuhlendorff J, Gether U, Aakerlund L et al. (1990) [Leu31, Pro34] neuropeptide Y: a specific Y1 receptor agonist. Proc Natl Acad Sci USA 87:182–186

Fukuhara C, Brewer JM, Dirden JC et al. (2001) Neuropeptide Y rapidly reduces Period 1 and Period 2 mRNA levels in the hamster suprachiasmatic nucleus. Neurosci Lett 314:119–122

Furtinger S, Pirker S, Czech T et al. (2001) Plasticity of Y1 and Y2 receptors and neuropeptide Y fibers in patients with temporal lobe epilepsy. J Neurosci 21:5804–5812

Fuxe K, Agnati LF, Harfstrand A et al. (1983) Central administration of neuropeptide Y induces hypotension bradypnea and EEG synchronization in the rat. Acta Physiol Scand 118:189–192

Gehlert DR, Chronwall BM, Schafer MP, O'Donohue TL (1987) Localization of neuropeptide Y messenger ribonucleic acid in rat and mouse brain by in situ hybridization. Synapse 1:25–31

Gehlert DR, Beavers LS, Johnson D et al. (1996) Expression cloning of a human brain neuropeptide Y Y2 receptor. Mol Pharmacol 49:224–228

Gerald C, Walker MW, Vaysse PJ et al. (1995) Expression cloning and pharmacological characterization of a human hippocampal neuropeptide Y/peptide YY Y2 receptor subtype. J Biol Chem 270:26758–26761

Gerald C, Walker MW, Criscione L et al. (1996) A receptor subtype involved in neuropeptide-Y-induced food intake. Nature 382:168–171

Golombek DA, Biello SM, Rendon RA, Harrington ME (1996) Neuropeptide Y phase shifts the circadian clock in vitro via a Y2 receptor. Neuroreport 7:1315–1319

Gregor P, Millham ML, Feng Y et al. (1996) Cloning and characterization of a novel receptor to pancreatic polypeptide, a member of the neuropeptide Y receptor family. FEBS Lett 381:58–62

Hackenthal E, Aktories K, Jakobs KH, Lang RE (1987) Neuropeptide Y inhibits renin release by a pertussis toxin-sensitive mechanism. Am J Physiol 252:F543–F550

Heilig M, Murison R (1987) Intracerebroventricular neuropeptide Y protects against stress-induced gastric erosion in the rat. Eur J Pharmacol 137:127–129

Heilig M, Wahlestedt C, Ekman R, Widerlov E (1988) Antidepressant drugs increase the concentration of neuropeptide Y (NPY)-like immunoreactivity in the rat brain. Eur J Pharmacol 147:465–467

Heilig M, Soderpalm B, Engel JA, Widerlov E (1989) Centrally administered neuropeptide Y (NPY) produces anxiolytic-like effects in animal anxiety models. Psychopharmacology (Berlin) 98:524–529

Heilig M, McLeod S, Brot M et al. (1993) Anxiolytic-like action of neuropeptide Y: mediation by Y1 receptors in amygdala, and dissociation from food intake effects. Neuropsychopharmacology 8:357–363

Hellstrom PM, Olerup O, Tatemoto K (1985) Neuropeptide Y may mediate effects of sympathetic nerve stimulations on colonic motility and blood flow in the cat. Acta Physiol Scand 124:613–624

Herzog H, Hort YJ, Ball H et al. (1992) Cloned human neuropeptide Y receptor couples to two different second messenger systems. Proc Natl Acad Sci USA 89:5794–5798

Herzog H, Hort YJ, Shine J, Selbie LA (1993) Molecular cloning, characterization, and localization of the human homolog to the reported bovine NPY Y3 receptor: lack of NPY binding and activation. DNA Cell Biol 12:465–471

Hokfelt T, Lundberg JM, Lagercrantz H et al. (1983) Occurrence of neuropeptide Y (NPY)-like immunoreactivity in catecholamine neurons in the human medulla oblongata. Neurosci Lett 36:217–222

Hu Y, Bloomquist BT, Cornfield LJ et al. (1996) Identification of a novel hypothalamic neuropeptide Y receptor associated with feeding behavior. J Biol Chem 271:26315–26319

Hua XY, Boublik JH, Spicer MA et al. (1991) The antinociceptive effects of spinally administered neuropeptide Y in the rat: systematic studies on structure-activity relationship. J Pharmacol Exp Ther 258:243–248

Huhman KL, Gillespie CF, Marvel CL, Albers HE (1996) Neuropeptide Y phase shifts circadian rhythms in vivo via a Y2 receptor. Neuroreport 7:1249–1252

Inui A, Okita M, Nakajima M et al. (1998) Anxiety-like behavior in transgenic mice with brain expression of neuropeptide Y. Proc Assoc Am Physicians 110:171–182

Jazin EE, Yoo H, Blomqvist AG et al. (1993) A proposed bovine neuropeptide Y (NPY) receptor cDNA clone, or its human homologue, confers neither NPY binding sites nor NPY responsiveness on transfected cells. Regul Pept 47:247–258

Kalra SP, Crowley WR (1984) Norepinephrine-like effects of neuropeptide Y on LH release in the rat. Life Sci 35:1173–1176

Kalra SP, Fuentes M, Fournier A et al. (1992) Involvement of the Y-1 receptor subtype in the regulation of luteinizing hormone secretion by neuropeptide Y in rats. Endocrinology 130:3323–3330

Kanatani A, Mashiko S, Murai N et al. (2000) Role of the Y1 receptor in the regulation of neuropeptide Y-mediated feeding: comparison of wild-type, Y1 receptor-deficient, and Y5 receptor-deficient mice. Endocrinology 141:1011–1016

Kassis S, Olasmaa M, Terenius L, Fishman PH (1987) Neuropeptide Y inhibits cardiac adenylate cyclase through a pertussis toxin-sensitive G protein. J Biol Chem 262:3429–3431

Kaye WH, Berrettini W, Gwirtsman H, George DT (1990) Altered cerebrospinal fluid neuropeptide Y and peptide YY immunoreactivity in anorexia and bulimia nervosa. Arch Gen Psychiatry 47:548–556

Klapstein GJ, Colmers WF (1997) Neuropeptide Y suppresses epileptiform activity in rat hippocampus in vitro. J Neurophysiol 78:1651–1661

Krause J, Eva C, Seeburg PH, Sprengel R (1992) Neuropeptide Y1 subtype pharmacology of a recombinantly expressed neuropeptide receptor. Mol Pharmacol 41:817–821

Krstenansky JL, Owen TJ, Buck SH et al. (1989) Centrally truncated and stabilized porcine neuropeptide Y analogs: design, synthesis, and mouse brain receptor binding. Proc Natl Acad Sci USA 86:4377–4381

Kushi A, Sasai H, Koizumi H et al. (1998) Obesity and mild hyperinsulinemia found in neuropeptide Y-Y1 receptor-deficient mice. Proc Natl Acad Sci USA 95:15659–15664

Larhammar D, Blomqvist AG, Yee F et al. (1992) Cloning and functional expression of a human neuropeptide Y/peptide YY receptor of the Y1 type. J Biol Chem 267:10935–10938

Larhammar D, Blomqvist AG, Soderberg C (1993) Evolution of neuropeptide Y and its related peptides. Comp Biochem Physiol 106 C:743–752

Leibowitz SF, Alexander JT (1991) Analysis of neuropeptide Y-induced feeding: dissociation of Y1 and Y2 receptor effects on natural meal patterns. Peptides 12:1251–1260

Leibowitz SF, Xuereb M, Kim T (1992) Blockade of natural and neuropeptide Y-induced carbohydrate feeding by a receptor antagonist PYX-2. Neuroreport 3:1023–1026

Levine AS, Morley JE (1984) Neuropeptide Y: a potent inducer of consummatory behavior in rats. Peptides 5:1025–1029

Lundberg JM, Tatemoto K (1982) Pancreatic polypeptide family (APP, BPP, NPY and PYY) in relation to sympathetic vasoconstriction resistant to alpha-adrenoceptor blockade. Acta Physiol Scand 116:393–402

Lundberg JM, Terenius L, Hokfelt T et al. (1982) Neuropeptide Y (NPY)-like immunoreactivity in peripheral noradrenergic neurons and effects of NPY on sympathetic function. Acta Physiol Scand 116:477–480

Lundberg JM, Terenius L, Hokfelt T, Goldstein M (1983) High levels of neuropeptide Y in peripheral noradrenergic neurons in various mammals including man. Neurosci Lett 42:167–172

Lundberg JM, Stjarne L (1984) Neuropeptide Y (NPY) depresses the secretion of ^{3}H-noradrenaline and the contractile response evoked by field stimulation, in rat vas deferens. Acta Physiol Scand 120:477–479

Lundell I, Blomqvist AG, Berglund MM et al. (1995) Cloning of a human receptor of the NPY receptor family with high affinity for pancreatic polypeptide and peptide YY. J Biol Chem 270:29123–29128

MacDonald JK (1988) NPY and related substances. Crit Rev Neurobiol 4:97–135

Malmstrom RE (2001) Vascular pharmacology of BIIE0246, the first selective non-peptide neuropeptide Y Y(2) receptor antagonist, in vivo. Br J Pharmacol 133:1073–1080

Malmstrom RE, Lundberg JM (1995) Neuropeptide Y accounts for sympathetic vasoconstriction in guinea-pig vena cava: evidence using BIBP 3226 and 3435. Eur J Pharmacol 294:661–668

Marksteiner J, Sperk G (1988) Concomitant increase of somatostatin, neuropeptide Y and glutamate decarboxylase in the frontal cortex of rats with decreased seizure threshold. Neuroscience 26:379–385

Marksteiner J, Prommegger R, Sperk G (1990) Effect of anticonvulsant treatment on kainic acid-induced increases in peptide levels. Eur J Pharmacol 81:241–246

Marsh DJ, Hollopeter G, Kafer KE, Palmiter RD (1998) Role of the Y5 neuropeptide Y receptor in feeding and obesity. Nature Med 4:718–721

Marsh DJ, Baraban SC, Hollopeter G, Palmiter RD (1999) Role of the Y5 neuropeptide Y receptor in limbic seizures. Proc Natl Acad Sci USA 96:13518–13523

Martel JC, St-Pierre S, Quirion R (1986) Neuropeptide Y receptors in rat brain: autoradiographic localization. Peptides 7:55–60

Matsumoto M, Nomura T, Momose K et al. (1996) Inactivation of a novel neuropeptide Y/peptide YY receptor gene in primate species. J Biol Chem 271:27217–27220

Michel MC, Beck-Sickinger A, Cox H et al. (1998) XVI International union of pharmacology recommendations for the nomenclature of neuropeptide Y, peptide YY, and pancreatic polypeptide receptors. Pharmacol Rev 50:143–150

Minth CD, Bloom SR, Polak JM, Dixon JE (1984) Cloning, characterization, and DNA sequence of a human cDNA encoding neuropeptide tyrosine. Proc Natl Acad Sci USA 81:4577–4581

Minth CD, Andrews PC, Dixon JE (1986) Characterization, sequence, and expression of the cloned human neuropeptide Y gene. J Biol Chem 261:11974–11979

Morioka H, Inui A, Inoue T et al. (1986) Neuropeptide Y decreases rectal temperature after intracerebroventricular administration in conscious dogs. Kobe J Med Sci 32:45–57

Motulsky HJ, Michel MC (1988) Neuropeptide Y mobilizes Ca2+ and inhibits adenylate cyclase in human erythroleukemia cells. Am J Physiol 255:E880–885

Nakajima M, Inui A, Asakawa A et al. (1998) Neuropeptide Y produces anxiety via Y2-type receptors. Peptides 19:359–363

Naveilhan P, Hassani H, Canals JM et al. (1999) Normal feeding behavior, body weight and leptin response require the neuropeptide Y Y2 receptor. Nature Med 5:1188–1193

Naveilhan P, Hassani H, Lucas G et al. (2001) Reduced antinociception and plasma extravasation in mice lacking a neuropeptide Y receptor. Nature 409:513–517

Parker RM, Herzog H. (1999) Regional distribution of Y-receptor subtype mRNAs in rat brain. Eur J Neurosci 11:1431–1448

Parker EM, Babij CK, Balasubramaniam A et al. (1998) GR231118 (1229U91) and other analogues of the C-terminus of neuropeptide Y are potent neuropeptide Y Y1 receptor antagonists and neuropeptide Y Y4 receptor agonists. Eur J Pharmacol 349:97–105

Parker E, Van Heek M, Stamford A (2002) Neuropeptide Y receptors as targets for anti-obesity drug development: perspective and current status. Eur J Pharmacol 440:173–187

Pedrazzini, Seydoux, Kunstner et al. (1998) Cardiovascular response, feeding behavior and locomotor activity in mice lacking the NPY Y1 receptor. Nature Med 4:722–726

Petty MA, Dietrich R, Lang RE (1984) The cardiovascular effects of neuropeptide Y (NPY). Clin Exp Hypertens A 6:1889–1892

Pheng LH, Regoli D (2000) Receptors for NPY in peripheral tissues bioassays. Life Sci 67:847–862

Pralong FP, Gonzales C, Voirol MJ et al. (2002) The neuropeptide Y Y1 receptor regulates leptin-mediated control of energy homeostasis and reproductive functions. FASEB J 16:712–714

Raposinho PD, Broqua P, Pierroz DD et al. (1999) Evidence that the inhibition of luteinizing hormone secretion exerted by central administration of neuropeptide Y (NPY) in the rat is predominantly mediated by the NPY-Y5 receptor subtype. Endocrinology 140:4046–4055

Rimland J, Xin W, Sweetnam P et al. (1991) Sequence and expression of a neuropeptide Y receptor cDNA. Mol Pharmacol 40:869–875

Rimland JM, Seward EP, Humbert Y et al. (1996) Coexpression with potassium channel subunits used to clone the Y2 receptor for neuropeptide Y. Mol Pharmacol 49:387–390

Rose PM, Fernandes P, Lynch JS et al. (1995) Cloning and functional expression of a cDNA encoding a human type 2 neuropeptide Y receptor. J Biol Chem 270:22661–22664

Rudolf K, Eberlein W, Engel W et al. (1994) The first highly potent and selective non-peptide neuropeptide Y Y1 receptor antagonist: BIBP3226. Eur J Pharmacol 271:R11–13

Sainsbury A, Schwarzer C, Couzens M et al. (2002) Important role of hypothalamic Y2 receptors in body weight regulation revealed in conditional knockout mice. Proc Natl Acad Sci USA 99:8938–8943

Sheikh SP, O'Hare MM, Tortora O, Schwartz TW (1989) Binding of monoiodinated neuropeptide Y to hippocampal membranes and human neuroblastoma cell lines. J Biol Chem 264:6648–6654

Stanley BG, Leibowitz SF (1984) Neuropeptide Y: stimulation of feeding and drinking by injection into the paraventricular nucleus. Life Sci 35:2635–2642

Statnick MA, Schober DA, Gackenheimer S et al. (1998) Characterization of the neuropeptide Y5 receptor in the human hypothalamus: a lack of correlation between Y5 mRNA levels and binding sites. Brain Res 810:16–26

Stephens TW, Basinski M, Bristow PK et al. (1995) The role of neuropeptide Y in the antiobesity action of the obese gene product. Nature 377:530–532

Sun XY, Zhao XH, Erlinge D et al. (1996) Effects of phosphorothioated neuropeptide Y Y1-receptor antisense oligodeoxynucleotide in conscious rats and in human vessels. Br J Pharmacol 118:131–136

Sundler F, Moghimzadeh E, Hakanson R et al. (1983) Nerve fibers in the gut and pancreas of the rat displaying neuropeptide-Y immunoreactivity. Intrinsic and extrinsic origin. Cell Tissue Res 230:487–493

Tatemoto K (1982a) Neuropeptide Y: complete amino acid sequence of the brain peptide. Proc Natl Acad Sci USA 79:5485–5489

Tatemoto K (1982b) Isolation and characterization of peptide YY (PYY), a candidate gut hormone that inhibits pancreatic exocrine secretion. Proc Natl Acad Sci USA 79:2514–2518

Tatemoto K (1990) Neuropeptide Y and its receptor antagonist. Ann New York Acad Sci. 611:1–6

Tatemoto K, Mutt V (1978) Chemical determination of polypeptide hormones. Proc Natl Acad Sci USA 75:4115–4119

Tatemoto K, Mutt V (1980) Isolation of two novel candidate hormones using a chemical method for finding naturally occurring polypeptides. Nature 285:417–418

Tatemoto K, Carlquist M, Mutt V (1982) Neuropeptide Y—a novel brain peptide with structural similarities to peptide YY and pancreatic polypeptide. Nature 296:659–660

Tatemoto K, Rokaeus A, Jornvall H et al. (1983) Galanin—a novel biologically active peptide from porcine intestine. FEBS Lett 164:124–128

Tatemoto K, Lundberg JM, Jornvall H, Mutt V (1985) Neuropeptide K: Isolation, structure and biological activities of a novel brain tachykinin. Biochem Biophys Res Commun 128:947–953

Tatemoto K, Efendic S, Mutt V et al. (1986) Pancreastatin, a novel pancreatic peptide that inhibits insulin secretion. Nature 324:476–478

Terenghi G, Polak JM, Hamid Q et al. (1987) Localization of neuropeptide Y mRNA in neurons of human cerebral cortex by means of in situ hybridization with a complementary RNA probe. Proc Natl Acad Sci USA 84:7315–7318

Thiele TE, Marsh DJ, Ste Marie L et al. (1998) Ethanol consumption and resistance are inversely related to NPY levels. Nature 396:366–369

Thiele TE, Koh MT, Pedrazzini T (2002) Voluntary alcohol consumption is controlled via the neuropeptide Y Y1 receptor. J Neurosci 22:RC208:1–6

Thorsell A, Michalkiewicz M, Dumont Y et al. (2000) Behavioral insensitivity to restraint stress, absent fear suppression of behavior and impaired spatial learning in transgenic rats with hippocampal neuropeptide Y overexpression. Proc Natl Acad Sci USA 97:12852–12857

Thorsell A, Rimondini R, Heilig M. (2002) Blockade of central neuropeptide Y (NPY) Y2 receptors reduces ethanol self-administration in rats. Neurosci Lett 332:1–4

Toufexis DJ, Kyriazis D, Woodside B (2002) Chronic neuropeptide Y Y5 receptor stimulation suppresses reproduction in virgin female and lactating rats. J Neuroendocrinol 14:492–497

Unden A, Tatemoto K, Mutt V, Bartfai T (1984) Neuropeptide Y receptor in the rat brain. Eur J Biochem 145:525–530

Vezzani A, Michalkiewicz M, Michalkiewicz T et al. (2002) Seizure susceptibility and epileptogenesis are decreased in transgenic rats overexpressing neuropeptide Y. Neuroscience 110:237–243

Wahlestedt C, Yanaihara N, Hakanson R (1986) Evidence for different pre-and post-junctional receptors for neuropeptide Y and related peptides. Regul Pept 13:307–318

Wahlestedt C, Pich EM, Koob GF, Yee F, Heilig M (1993) Modulation of anxiety and neuropeptide Y-Y1 receptors by antisense oligodeoxynucleotides. Science 259:528–531

Watson JD, Sury MR, Corder R et al. (1988) Plasma levels of neuropeptide tyrosine Y (NPY) are increased in human sepsis but are unchanged during canine endotoxin shock despite raised catecholamine concentrations. J Endocrinol 116:421–426

Weinberg DH, Sirinathsinghji DJ, Tan CP et al. (1996) Cloning and expression of a novel neuropeptide Y receptor. J Biol Chem 271:16435–16438

Westrin A, Ekman R, Traskman-Bendz L (1999) Alterations of corticotropin releasing hormone (CRH) and neuropeptide Y (NPY) plasma levels in mood disorder patients with a recent suicide attempt. Eur Neuropsychopharmacol 9:205–211

Wharton J, Gordon L, Byrne J et al. (1993) Expression of the human neuropeptide tyrosine Y1 receptor. Proc Natl Acad Sci USA 90:687–691

Widerlov E, Lindstrom LH, Wahlestedt C, Ekman R (1988) Neuropeptide Y and peptide YY as possible cerebrospinal fluid markers for major depression and schizophrenia, respectively. J Psychiatr Res 22:69–79

Woldbye DP, Larsen PJ, Mikkelsen JD et al. (1997) Powerful inhibition of kainic acid seizures by neuropeptide Y via Y5-like receptors. Nature Med 3:761–764

Yan H, Yang J, Marasco J et al. (1996) Cloning and functional expression of cDNAs encoding human and rat pancreatic polypeptide receptors. Proc Natl Acad Sci USA 93:4661–4665

PP, PYY and NPY:
Synthesis, Storage, Release and Degradation

S. von Hörsten[1] · T. Hoffmann[2] · M. Alfalah[3] · C. D. Wrann[3] · T. Karl[1] · R. Pabst[1]
S. Bedoui[1]

[1] Department of Functional and Applied Anatomy, OE 4120, Carl-Neuberg-Str.1,
 30625 Hannover, Germany
 e-mail: Hoersten.Stephan.von@MH-Hannover.de
[2] Probiodrug AG, 06120 Halle, Germany
[3] Department of Physiological Chemistry, School of Veterinary Medicine,
 30559 Hannover, Germany

1	Introduction	24
2	**Storage and Synthesis of PP, PYY and NPY**	26
2.1	Subcellular Storage in Vesicles	26
2.1.1	Large Vesicles	26
2.1.2	Small Vesicles	27
2.2	Posttranslational Modification	27
3	**Localization and Release of NPY, PYY and PP in the Periphery**	29
3.1	Cellular Sources for NPY and Release in the Periphery	29
3.1.1	Sympathetic Nerves	29
3.1.2	Adrenal Glands	30
3.1.3	Platelets	31
3.2	Cellular Sources of PP in the Periphery	32
3.2.1	Pancreatic PP Cells	32
3.2.2	PP in the Gastrointestinal Tract	32
3.3	Cellular Sources for PYY in the Periphery	32
3.3.1	PYY in the Gastrointestinal Tract	32
3.3.2	PYY in the Pancreas	33
3.3.3	PYY in Enteric Nerve Fibers	33
4	**The Release of Peptides by Exocytosis**	33
4.1	The Process of Exocytosis	33
4.2	Docking of the LDCV to the Cell Membrane	34
4.2.1	Fusion with the Cell Membrane	34
4.3	Molecular Regulation of the Release of PP, PYY and NPY	35
5	**Degradation of NPY, PP and PYY**	35
5.1	Limited Proteolysis of NPY and PYY	35
5.2	Cleavage of PP	37
5.3	In Vivo Metabolism of NPY Family Peptides	37
6	**Concluding Remarks**	38
	References	39

Abstract Peptides of the NPY family are synthesized as large precursor molecules in the endoplasmatic reticulum (ER), where posttranslational modification takes place, and from where they are translocated to the Golgi apparatus. After several structural and functional adjustments, two types of mature vesicle—large dense core vesicles and synaptic vesicles—serve as a storage depot. Notably, the expression of a gene for a peptide from the NPY family is not sufficient to ensure the production of mature peptides since several posttranslational steps are specifically involved and these steps themselves are subjected to specific regulatory processes. Similarly, after exocytotic release of NPY-like peptides, their local action depends on their concentration, their different receptor selectivity and the local expression of the different Y-receptors. Another major player in this complex network is found in the action of specific peptidases influencing half-life and receptor selectivity. At least aminopeptidase P, dipeptidyl-peptidase IV-like enzymes, specific endopeptidases, like meprin or neprilysin-like enzymes, and post-arginine hydrolyzing endoproteases are cleaving enzymes for NPY-like peptides. Due to a striking change of receptor specificity after N-terminal cleavage of NPY-like peptides, the development of inhibitors for NPY, PYY and PP cleaving peptidases is a complementary approach to the development of Y-receptor agonists or antagonists. In this chapter, we summarize key findings about synthesis, storage, release and localization of NPY family peptides, add recent findings on their degradation by specific enzymes and discuss implications for the interpretation of studies in future research.

Keywords Synthesis of NPY-like peptides · Storage of NPY-like peptides · Release of NPY-like peptides · Expression of NPY-like peptides · Degradation of NPY-like peptides · Peptidases of NPY-like peptides

1
Introduction

Current research on synthesis, storage, release and degradation of the neuropeptide Y (NPY)-like peptides pancreatic polypeptide (PP), peptide YY (PYY) and NPY provides a controversial picture. While there are not many recent studies available providing novel insights into synthesis, storage and release of NPY-like peptides, there are increasing numbers of data on specific peptidases, which mediate specific steps of limited proteolysis of these peptides. These are of great relevance because N-terminal degradation of NPY-like peptides results in changed receptor specificity. NPY itself, for example, represents one of the best, if not the best, substrates for the ectopeptidase dipeptidyl peptidase IV (DP IV) (Mentlein 1999). Cleavage of NPY by DP IV results in a specific loss of NPY Y_1 receptor affinity, while the remaining peptide NPY(3–36) is still active at the NPY Y_{2-5} receptors (Michel et al. 1998).

Peptides of the NPY family are synthesized as large precursor molecules in the ER. After posttranslational modification, precursor molecules are translocated to the Golgi apparatus, sorted in the trans Golgi network (TGN), and guided towards the secretory pathway. During these processes several posttranslatio-

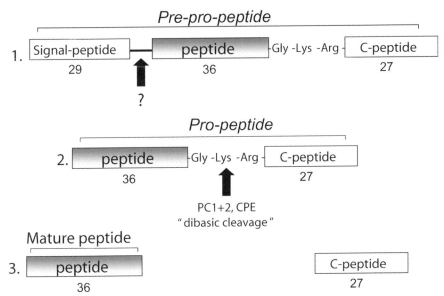

Fig. 1 Steps in the posttranslational modification of peptides from the NPY-family. *1*: The peptide is generated as a large precursor consisting of the so-called pre-pro-peptide. In this molecule the peptide is flanked by a c-peptide and a signal peptide. After enzymatic cleavage of the signal peptide the molecule is then referred to as pro-peptide. *2*: Enzymes such as prohormone convertase (*PC*) and carboxypetidase E (*CPE*) the pro-peptide cleave the pro-peptide at the dibasic site, thereby generating the mature peptide (*3*)

nal steps are specifically involved and these steps themselves are subjected to specific regulatory processes (Fig. 1). After exocytotic release of NPY-like peptides, the local action not only relies on their concentration, receptor selectivity, and the expression of Y-receptors, but also on the action of specific peptidases influencing half-life and receptor selectivity. The action of NPY-like peptides therefore is also influenced by the local distribution and concentration of, for example, aminopeptidase P and dipeptidyl-peptidase IV-like enzymes of which several are inducible by inflammatory processes. Moreover, these peptidases may also act on counter-regulatory peptides and endogenous inhibitors and competitive substrates influence their specific activity. Thus, the development of inhibitors for NPY, PYY and PP cleaving peptidases is a complementary approach to the development of Y-receptor agonists or antagonists.

The aim of this chapter is to summarize key findings about synthesis, storage, release and localization of NPY family peptides, and discuss recent findings on their degradation by specific enzymes.

2
Storage and Synthesis of PP, PYY and NPY

2.1
Subcellular Storage in Vesicles

Polarized cells such as neurons and epithelial cells maintain separate plasma membrane domains, each with a distinct protein and lipid composition and intracellular sorting mechanisms that recognize classes of proteins which ensure that specific vesicles are transported to the correct surface domain (Nillson 1992). Sorting proteins to the correct membrane is essential for their biological functions, since missorting often results in pathological conditions. Peptides of the NPY family are synthesized as large precursor molecules in the ER, from where they are translocated to the Golgi apparatus. Confined within a vesicle, the precursor molecules are sorted in the TGN and are guided towards the secretory pathway. After several structural and functional adjustments, two types of mature vesicles serve as a storage depot.

2.1.1
Large Vesicles

These vesicles have a diameter of 70–100 nm and are characterized by electron dense granules attached to the membrane of the vesicle. Therefore, they are referred to as large dense core vesicles (LDCV). The electron dense granules in LDCV have been recognized to contain protein and enzyme aggregates that are crucial for the biological activation of the precursor molecule. LDCV are found in neurons and endocrine cells.

After the formation of LDCV the majority is transported towards the plasma membrane or the axon (Pickel et al. 1995). This process is mediated by interactions with microtubuli, since treatment with vinblastine, a drug destroying microtubuli, dramatically decreases the number of neuropeptide containing LDCV found in the axon (Hemsén et al. 1991). The transport of LDCV from the TGN towards the plasmalemma is a rather slow process. D'Hooge et al. (1990) have demonstrated that NPY containing LDCV are transported with a velocity of about 5 mm/h.

Zhang et al. (1993) found that various combinations of peptides, presumably at varying concentrations, occur in the LDCV in a given nerve ending. Therefore, it is likely that individual LDCV produced in a neuron are heterogeneous with regard to peptide content and thus to the message that they transmit upon release. Indeed, peptide containing LDCV have been demonstrated to also comprise additional transmitters or hormones. Shortly, after NPY was discovered it became obvious that NPY is costored with catecholamines (Fried et al. 1985). Compelling evidence is now available to indicate that NPY is not only frequently costored with catecholamines, but also with a variety of other signaling molecules (see Table 1).

Table 1 Mediators and hormones that are costored with NPY

Mediator	Cell type/tissue	Species	References
Norepinephrine	Sympathetic neurons, vas deferens	Cattle, rat	de Potter et al. 1998; Fried et al. 1985
GABA	Spinal cord, brain	Fish, frog	Parker et al. 1998; de Rijk et al. 1992
VIP	Submucos neurons, perivascular nerves	Rat, cattle, pig	Cox et al. 1994; Majewski et al. 1995
Substance P	Submucos neurons, adrenal medulla	Guinea pig, cattle	Masuku et al. 1998; Bastiaensen et al. 1988
Opioid peptides (enkephalins, dynorphins)	perivascular nerves, retinal cells	Pig, human, guinea pig	Kong et al. 1990; Jotwani et al. 1994; Morris et al. 1985
Galanin	Dorsal horn	Rat	Zhang et al. 1993
ANP	Adrenal medulla	Fish, birds, reptiles, amphibian, rat	Wolfensberger et al. 1995
CGRP	Postganglionic sympathetic neurons	Pig	Majewski et al. 1992
Chromogranin A	Postganglionic sympathetic neurons	Human	Takiyyuddin et al. 1994

GABA, Gamma-aminobutyric acid; VIP, vasointestinal peptide; ANP, atrial natriuretic peptide; CGRP, calcitonin gene-related peptide.

2.1.2
Small Vesicles

In addition to the storage in LDCV, some members of the NPY family are also stored in smaller vesicles. With a diameter between 30 and 50 nm the so-called synaptic vesicles (SV) are the smallest known membrane-bound organelles (Dannies 1999).

2.2
Posttranslational Modification

As mentioned above, peptides of the NPY family are not synthesized in their final biologically active form. Instead, a larger and biologically inactive molecule is generated at first. Once the nascent peptide has passed all quality control mechanisms in the ER, it is transported onwards by vesicular transport (Rodriguez-Boulan 1992) or by tubular structure (Presley 1997) to the TGN. Most of secretory and membrane-bound proteins undergo several structural and posttranslational modifications including glycolysylation, fatty acid acylation, phosphorylation, amidation, or proteolytic cleavage (Matter 2000; Caplan 1997; Eipper 1992).

During intracellular travel along the secretory pathway, the precursor is submitted to successive enzymatic processing. The precursor contains additional amino acids at both the N terminus and the C terminus and only after specific

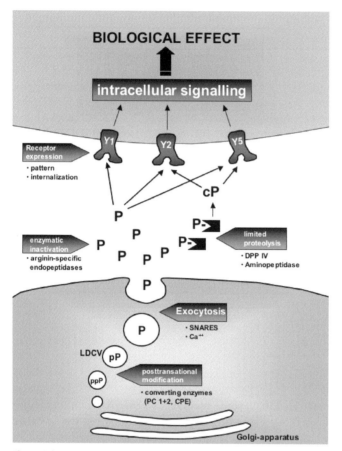

Fig. 2 Scheme to illustrate posttranslational modification, storage, release and enzymatic degradation of peptides of the NPY-family. Abbreviations: *ppP*, pre-pro-peptide; *pP*, pro-peptide; *P*, peptide; *cP*, cleaved peptide; *LDCV*, large dense core vesicle; *PC*, prohormone convertase; *CPE*, carboxypeptidase E; *SNARES*, soluble *N*-ethyl-maleimide-sensitive fusion protein; *DPP IV*, dipeptidyl peptidase IV

enzymes have cleaved both of these additional peptide sequences does the peptide become activated (see Fig. 2). This process is termed posttranslational modification and was first identified and characterized in the biosynthesis of insulin (for details see Lang 1999).

The precursors of PP, PYY and NPY contain about three times as many amino acids as the final biologically active peptide; for example pre-pro-PP and pre-pro-NPY consist of 96 amino acids, whereas the mature forms of PP and NPY consist of only 36 amino acids (Schwartz et al. 1981). Consistent with their proposed common evolutionary origin, the precursors of PP, PYY and NPY are organized into similar structural domains. The biologically active amino acid sequence is preceded by a peptide sequence referred to as signal peptide and followed by a Gly-Lys-Arg cleavage site and another peptide sequence at the C terminus (C-peptide) (Conlon 2002). The length of the signal peptide and the c-peptide varies slightly between different species (Sundler et al. 1993).

The activation of the pre-pro-peptides requires the proteolytic separation of the different structural domains on the one hand and an amidation at a central structural point on the other. Specific enzymes in different cellular compartments as well as in the LDCV mediate these reactions. So far little is known about the enzymes catalyzing the cleavage of the signal peptide, thereby generating pro-peptides. However, the reactions leading towards the generation of the mature NPY are much better characterized. Prohormone convertases (PC), in particular PC2 and PC3, and carboxypetidase E directly cleave the pro-peptides at a single dibasic site consisting of Gly-Lys-Arg (Paquet et al. 1996). The significance of the dibasic site for the proteolytic activation was recently documented by Brakch and colleagues (2002). By introducing a mutated version of pro-NPY into pituitary cells that are capable of generating mature NPY from normal pro-NPY, it was demonstrated that a single mutation at the dibasic site suppresses the formation of mature NPY dramatically. After cleaving, the molecule is converted to the mature peptide by enzymatic amidation. The so-called α-amidation is a two-step reaction catalyzed by two distinct catalytic domains within the responsible enzyme peptidylglycine-α-amidating monooxygenase (peptidyl-glycine-α-hyroxylating monooxygenase, peptidyl-α-hydroxylglycine amidating lyase) (Oyarce 1993; Eipper 1992). The enzymatic reaction depends on the presence of ascorbic acid (vitamin C), involves the reduction of the two copper atoms that are bound to its catalytic core and zinc is required for proper activity. When this process of posttranslational modification is completed, mature peptides are stored within the LDCV and can be released upon stimulation.

Thus, the capability of a certain cell to secrete biological active peptides depends on the activity and presence of specific enzymes. It is obvious, therefore, that the expression of a gene for a peptide from the NPY family is not sufficient to ensure the production of mature peptides.

3
Localization and Release of NPY, PYY and PP in the Periphery

3.1
Cellular Sources for NPY and Release in the Periphery

3.1.1
Sympathetic Nerves

Immuncytochemical studies have located NPY in postganglionic sympathetic nerve fibers throughout the body. Simultaneous immunostaining for NPY and tyrosine hydroxylase or dopamine hydroxylase, critical enzymes for the synthesis of catecholamines (CA), has provided evidence that NPY and norepinephrine (NE) coexist in many cell bodies in sympathetic ganglion and many perivascular fibers (reviewed by Lundberg, 1992). In fact, several studies have demonstrated that NPY and CA can be released simultaneously (Lundberg et al. 1989). Such a pattern of NPY and CA corelease was shown in stimulation of the sympathetic nerves in various preparations, both in vitro, for example in isolated perfused canine spleen and guinea pig heart, in situ in perfused pig spleen, and in vivo, in conscious calves, pithed guinea pigs, and rats (reviewed by Zukowska-Grojec et al. 1993). Interestingly, the process of corelease is regulated by several mechanisms. According to neurophysiological studies NPY is preferentially released under conditions of elevated neuronal activity (5–40 Hz), whereas the release of CA dominates under moderate stimulation (1–10 Hz) (Hökfelt et al. 1991). However, the duration of the nerve activation also modifies the composition of the sympathetic transmitter output. Prolonged sympathetic stimulation over 1 h decreases the content of NPY in sympathetic fibers innervating the spleen by 58% (Lundberg et al. 1989). Due to the rather slow velocity of the axonal transport of NPY of approximately 5 mm/h (D'Hooge et al. 1990), a complete re-supply of the nerve terminal with NPY can take up to 11 days (Lundberg et al. 1989).

3.1.2
Adrenal Glands

NPY can also be found in chromaffin cells of the adrenal medulla. There are studies indicating the presence of NPY in both epinephrine and norepinephrine-containing chromaffin cells. Additionally, NPY is contained in the nerve fibers derived from a plexus in the adrenal capsule and penetrating through adrenal cortex and medulla, many of which innervate blood vessels (Zukowska-Grojec et al. 1993). The adrenomedullary content varies markedly among species, with the highest levels found in the mouse and cat, and the lowest in pig adrenal glands. Furthermore, aging increases adrenomedullary content of NPY several fold. Several investigators have shown that the release of NPY can be evoked pharmacologically from the perfused bovine adrenal gland (Hexum et al. 1987), cultured chromaffin cells (Kataoka et al. 1985), and by direct splanchnic nerve stimulation (Briand et al. 1990).

In spite of these lines of evidence, the notion of NPY as a adrenomedullary hormone in physiological conditions is questionable. In vivo, the only known situation in which the adrenal medulla secretes large quantities of NPY into the bloodstream is in the pathological situation of pheochromocytoma (Corder et al. 1985; Pernow et al. 1986; Takahashi et al. 1987). In animals with normal adrenal medullas, increases in circulating plasma NPY levels evoked by stress (Zukowska-Grojec et al. 1988) correlate positively with norepinephrine, but not with epinephrine. This correlation indicates that NPY is derived from sympathetic nerves rather than from the adrenal glands. Furthermore adrenalectomy does not affect circulating levels of NPY at rest and upon immobilization stress in rats (Bernet et al. 1998). Therefore, under physiological conditions, the adrenomedullary NPY might play a role in the local microenvironment but does not have a systemic role.

3.1.3
Platelets

The presence and synthesis of NPY in extra neuronal tissues was first suggested by findings of NPY mRNA in megakaryocytes from rats and mice prone for certain autoimmune diseases (Ericsson et al. 1989). Subsequent studies revealed that rat platelets are a rich source of circulating NPY (Myers et al. 1988). Rat platelet-rich plasma, and plasma pellets prepared from it, contains approximately tenfold higher concentrations of NPY than platelet-poor plasma (Myers et al. 1988). Collagen, which evokes the secondary, irreversible stage of platelet aggregation and the associated release reaction, causes a dose-dependent release of NPY in parallel, as measured in platelet-poor plasmas prepared from the aggregating platelet suspension. Much less NPY is released during primary, reversible platelet aggregation, such as that induced by adenosine diphosphate in rat platelets prepared from citrated blood. Also, in vivo, circulating plasma NPY levels increase in a dose-dependent manner following intravenous injection of collagen in rats.

There are major variations in the platelet content of NPY among different species. In rats, high resting circulating NPY levels appear to result from platelet-derived NPY, and this may explain marked differences with other species, such as pigs, guinea pigs, rabbits, cats, dogs, and humans. The issue of whether the species with low circulating NPY platelet levels do not possess and/or release NPY has not been completely resolved. In contrast to rat megakaryocytes, no NPY mRNA was detected in normal human and pig bone marrow (Ericsson et al. 1991). It should be noted, however, that low quantities of NPY are detectable in human platelet-rich plasma and platelet pellet (Zukowska-Grojec et al. 1993), but the significance of these findings is unknown. High levels of NPY were also found in rabbit platelet-rich plasma and platelet pellets (Zukowska-Grojec et al. 1993). Conversely, Persson et al. (1989) were unable to detect NPY mRNA in rabbit spleen, although it is not certain whether the splenic expression of the NPY gene corresponds to that of megakaryocytes. On the one hand, the spleen contains large numbers of platelets, suggesting that data on NPY protein expression in this organ are always critically dependent on a proper perfusion aimed towards removing platelets. On the other hand, a high expression of the NPY Y_1 receptor is found in the spleen, being strongly suggestive for functional role of endogenously released NPY within the local microenvironment of this organ (Gehlert et al.1996).

Interestingly, it has been shown that although megakaryocytes from normal mice do not contain any NPY mRNA, autoimmune NZB mice express the NPY gene in megakaryocytes (Ericsson et al. 1989). Thus, it is possible that the NPY gene is expressed in platelets in all species, including humans, but is normally downregulated by unknown factors, and might be upregulated under some pathophysiological conditions. For example, in spontaneously hypertensive rats, platelet content of NPY has been shown to be several fold higher than in normotensive rats (Ogawa et al. 1989). Whether this "upregulation" of platelet-derived NPY is genetic and/or secondary to pathophysiological changes remains to be determined.

3.2
Cellular Sources of PP in the Periphery

3.2.1
Pancreatic PP Cells

PP is a hormone, which is released from the dominating endocrine cell type of the duodenal pancreas (Schwartz et al. 1983). Once PP had been localized to the pancreatic endocrine cells, it became of interest to identify the cell type. Using different histochemical staining techniques for the visualization of islet cells and a combination of light and electron microscopic immunocytochemistry, it became clear that the PP-containing cells were different from those that store insulin, glucagons, or somatostatin. In fact, PP cells represent the fourth cell type in the endocrine pancreas with the three others producing insulin (beta), glucagons (alpha), and somatostatin (delta). Morphologically these cells are found in the peripheral areas of the islet and are often scattered between glucagons and somatostatin cells (Ekblad and Sundler 2002). Several morphological features distinguish the PP cell from the other endocrine pancreatic cells. Clusters of PP cells can also be found within the exocrine portion of the pancreas (Böttcher et al. 1993), and PP cells are more frequent in the duodenal part of the pancreas (Orci et al. 1976). The different distribution within the pancreas is thought to reflect the fact that the pancreas derives from different primordia, one ventral and one dorsal, which fuse during embryonic development (Orci et al.1982).

3.2.2
PP in the Gastrointestinal Tract

PP cells occur also in the gastric mucosa as shown in the opossum, cat, and dog (Cox et al. 1998). Such cells are few and are considered to represent a subpopulation of gastrin producing cells (Ekblad and Sundler 2002). Cells immunoreactive for PP can also be found in the intestine, colon and rectum. Overall, there are large variations between different mammalian species, and the presence of the cells appears to be transient: in humans and rats, for example, PP cells appear in the gastric mucosa for a short postnatal period only (Tsutsumi et al. 1984).

3.3
Cellular Sources for PYY in the Periphery

3.3.1
PYY in the Gastrointestinal Tract

PYY is found in endocrine cells in the intestine of a wide range of mammalian and submammalian species (Ekblad and Sundler 2002). By light microscopy, PYY expressing cells are flask-shaped and are characterized by an accumulation of secretory granules at the basal site, which are the light microscopic equivalent of LDCV (Sundler et al. 1993). Even though PYY cells are found in both the upper and the lower gastrointestinal tract, only a few PYY cells are present in the gastric mucosa, whereas PYY cells become particularly numerous in the distal ileum and in the colon. Notably, the vast majority of intestinal PYY cells also contain other peptide hormones. Glicentin, also referred to as gut glucagon, and the glucagon-like peptides I and II (GLP I and II) have been reported to coexist with PYY (reviewed by Ekblad and Sundler 2002). There are also reports describing that PYY and serotonin occur together in endocrine cells of the human rectum (Horsch et al. 1994).

3.3.2
PYY in the Pancreas

PYY also occurs in the pancreas of mammalian as well as of submammalian species (Lundberg et al. 1983; El-Salhy et al. 1987; Böttcher et al. 1993). There is, however, a certain variation in the cellular localization in between different species. The predominant PYY containing cell types in the islet are glucagon cells (Sundler et al. 1993). Findings describing a very early expression of PYY in islet cell precursors during embryonic development have fostered speculation on a possible involvement of PYY in the process of islet cell differentiation and growth (Aponte et al. 1985).

3.3.3
PYY in Enteric Nerve Fibers

Notably, endocrine cells are not the only source for PYY in the gastrointestinal tract. PYY is also present in myenteric nerves. These fibers are quite numerous in the myenteric ganglia of the stomach and upper intestine, where they predominantly innervate smooth muscles (reviewed by Sundler et al. 1993). Immunohistochemical studies revealed that PYY-containing nerve fibers are distinct from autonomic fibers positive for NPY.

4
The Release of Peptides by Exocytosis

4.1
The Process of Exocytosis

As indicated above, peptide-containing LDCV are formed within the Golgi apparatus and are subsequently transported towards the cell membrane. Once arrived, they gather beneath the cell membrane in clusters and lie there waiting until a signal reaches the membrane and induces the vesicles to fuse with the cell membrane. Interestingly, LDCV are targeted to specific regions, called active zones, that are below the cell membrane . These active zones are in close proximity to ion channels within the cell membrane and reflect the location where the LDCV will dock and fuse with the membrane.

4.2
Docking of the LDCV to the Cell Membrane

The exocytotic release of storage vesicles is a complex phenomenon. Before exocytosis occurs, the vesicles undergo a "docking" reaction, which involves the interaction with specific proteins. The functional importance of these proteins, including soluble *N*-ethyl-maleimide-sensitive fusion protein (SNARE) and Munc18 (Rizo et al. 2002), was first shown by the observation that they constitute the specific targets for clostridial neurotoxins, which inhibits neurotransmitter release (Link et al. 1992; Blasi et al. 1993). Further studies have established that the docking event requires the assembly of SNARE complexes, which are composed of proteins that are anchored to each of the membranes that are destined to fuse (Chapman et al. 2002). The SNARE complex drives the LDCV into direct contact with the cell membrane. Subsequently, an energy wasting process is initiated to finally make the vesicles competent for the final step in exocytosis, the fusion with the cell membrane (Klenchin et al. 2000).

4.2.1
Fusion with the Cell Membrane

A key signal in initiating fusion with the cell membrane is the influx of Ca^{++}. With the depolarization of a neuron or the stimulation of endocrine cells by a secretagous, specific ion channels are activated and opened. Due to the gradient between the internal and external concentrations of free Ca^{++}, intracellular concentrations of Ca^{++} rapidly increase. This, in turn, directly affects proteins on the surface of LDCV present in active zones. Ca^{++} induces structural changes of the so-called C2 domain of synaptotagmin, the most abundant Ca^{++}-binding protein on LDCV surfaces (Chapman et al. 1994), which results in it forming a bond with specific proteins on the internal surface of the cell membrane. The LDCV is now tightly attached to the cell membrane via the SNARE complex on the one hand and the Ca^{++}-activated synaptotagmin on the other. The steps which follow are still not clear, but Chapman et al. (2002) have recently proposed a model, in which the C2 domains of synaptotagmin penetrate the target membrane and thereby initiate the fusion of the external layer of the vesicles with the internal layer of the cell membrane. Finally, both membranes fuse and the content of the vesicles is secreted to the outside of the cell.

After the secretion is completed, the inserted membrane part is taken back into the cytoplasm by endocytosis and remains ready to be recycled.

Exocytosis is one of the fastest things that animal cells can do (Almers et al. 1994). The release of neurotransmitters and hormones is known to proceed in fractions of milliseconds (0.5 ms) after the influx of Ca^{++} has occurred. This initial rapid burst results from exocytosis of vesicles that had been in a release-competent state at the time of the Ca^{++}-influx. Sustaining this response, however, is much slower, as additional vesicles need to undergo the preceding reaction before being competent to fuse with the cell membrane. Thus, distinct kinetics for the exocytotic release can be anticipated depending on the duration of the stimulus (Rettig and Neher 2002).

4.3
Molecular Regulation of the Release of PP, PYY and NPY

At the molecular level, ion currents regulate the release of NPY over the cell membrane. The rise of intracellular Ca^{++} via voltage-dependent N-type Ca^{++}-channels is a key event in the release of NPY and CA, indicating the exocytotic nature of the process. The Ca^{++}-dependent intracellular pathway can be activated by acetylcholine or increases in the extracellular concentrations of K^+. In addition to this classical pathway, Martire and colleagues (1997) demonstrated that decreases in extracellular Na+ concentrations also evoke a release of NPY. This effect, however, cannot be abolished by removal of extracellular Ca^{++}. Further testing revealed that the removal of extracellular Na^+ activates a Na^+/H^+ exchanger, which, in turn, decreases the intracellular pH. Since intracellular acidification has been demonstrated to stimulate the release of Ca^{++} from intracellular stores, it was concluded that the intracellular release of Ca^{++} accounts for the findings that removal of extracellular Na^+ evokes the release of NPY (Martire et al. 1997).

5
Degradation of NPY, PP and PYY

After release, the activity of the peptides depends on the one hand on the local distribution of the different Y-receptors, and on the other hand on the action of a couple of soluble and membrane bound peptidases. Hydrolysis by peptidases with broad specificity limits the half-life of the peptides by complete degradation. But beside this, in the case of the pancreatic polypeptide family, a limited proteolysis by highly specific peptidases was found, which results in a change of receptor selectivity of the peptides.

5.1
Limited Proteolysis of NPY and PYY

In vitro investigations identified a couple of arginine specific endoproteases, including urokinase-type plasminogen activator, plasmin, thrombin, and trypsin which are able to cleave NPY after all four arginines located in the C-terminal part of the peptide (*arrows* in Table 2) (Ludwig et al. 1996). This degradation leads to a complete loss of activity of NPY and PYY. Inactivation of NPY was also observed by endopeptidase-24.18 (E.C. 3.4.24.18) an NPY degrading enzyme (Price et al. 1991; Ludwig et al. 1995). In contrast to this, the limited pro-

Table 2 Main cleavage sites of peptides of the NPY family

	APP DPIV ⇓ ⇓		T NEP ⇓ ⇓	T T ⇓ ⇓	NEP NEP T T ⇓ ⇓ ⇓ ⇓
	1	10	20		30 36
NPY human, rat	Y/P/SKPDNPGEDAPAEDMAR/Y/YSALR/H			YIN	L/ITR/QR/Y
PYY rat	Y/P/AKPEAPGEDASPEELSR-Y-YASLR-H			YLN/L-VTR-QR-Y	
PYY human	Y/P/IKPEAPGEDASPEELNR-Y-YASLR-H			YLN/L-VTR-QR-Y	
PP rat	A-P/LEPMYPGDYATHEQRAQ Y-ETQLR-R-YIN			T LTR PR Y	
PP human	A-P/LEPVYPGDNATPEQMAQ Y-AADLR-R-YIN			M LTR PR Y	

teolysis catalyzed by aminopeptidase P (AP P, E.C. 3.4.11.9) or DP IV (E.C. 3.4.14.5) results in a release of only the N-terminal amino acid or dipeptide, respectively (Table 2, *arrows*; Mentelin et al. 1993; Grandt et al. 1993). In fact, the N-terminal tyrosine is a prerequisite for efficient Y_1-receptor binding (Michel et al. 1998). Hydrolysis by these exopeptidases abolishes Y_1-receptor activation whereas the remaining (2-36) or (3-36) peptides maintain their activity on the other Y-receptors. Nonspecific N-terminal degradation by other aminopeptidases, like aminopeptidase N or W, is prevented by proline in the penultimate position (Medeiros et al. 1993a, 1993b). In addition, prolines in position 5, 8 and 13 prevent further N-terminal degradation of the (3-36) peptide including further hydrolysis by DP IV. Furthermore it was demonstrated that NPY (3-36) could inhibit DP IV activity (Hoffmann et al. 1995).

Because of the amidated C-terminus the peptides seem to be also resistant to carboxypeptidase-catalysed hydrolysis, as exhibited by, for example, angiotensin converting enzyme. C-terminal degradation could be demonstrated for neutral endopeptidase (NEP, E.C. 3.4.24.11), which preferentially cleaves NPY between Tyr^{20} and Tyr^{21} and between Leu^{30} and Ile^{31} (Medeiros et al. 1996) and PYY at Asn^{29}-Leu^{30} (Table 2, *arrows*; Medeiros et al. 1994). Interestingly, an N-acetyleted 22-36 fragment was described as selective agonist for the rat intestinal PYY receptor or the neuronal Y_2 receptor (Balasubramaniam et al. 2000).

The experiments with purified enzymes were supported by investigations using cell lines and brush border membrane preparations. The product and inhibitor profiles determined during these experiments indicate that DP IV-like enzymes, AP P, NEP-like enzymes, endopeptidase 24.18 and post-arginine cleaving endoproteases are in general the main convertases of NPY and PYY. Depending on the enzymes expressed, different cell lines or membrane preparations produce a specific fragment pattern. Medeiros (1993) found that PYY is cleaved preferentially by NEP in a renal brush border membrane preparation whereas in the jejunal preparation, AP P- and DP IV-dependent cleavage was predominant (Medeiros et al. 1994). Similar results were observed for NPY (Medeiros et al. 1996). Using human smooth muscle cells only cleavage by AP P was found (Mentlein et al. 1996).

5.2
Cleavage of PP

Only very limited data are available concerning PP metabolism. Adamo and Hazelwood (1989) found that avian PP is degraded by a not further characterized soluble cytosolic endoprotease (Adamo and Hazelwood 1989), whereas Tasaka and colleagues (1989) described that thiol protease inhibitors inhibit the degradation of PP by an extract from rat submaxilliary glands (Tasaka et al. 1989). From the primary structure of PP it could be predicted that similar to NPY and PYY the conserved N-terminal part responsible for Y_1-receptor activation should be a substrate of AP P or DP IV. DP IV cleavage was shown by Nausch and coworkers (1990), whereas no experimental data about AP P are available so far. The same holds true for post-arginine cleavage and NEP-catalyzed hydrolysis. There are a number of amino acid exchanges around the cleavage sites in comparison to NPY and PYY necessitating an experimental approach to prove hydrolysis of PP by these enzymes. Concerning the post-arginine cleavage it could be predicted that there is a preferred cleavage after the dibasic Arg^{26}-Arg^{27}. On the other hand the monobasic cleavage site at position 20 should be lost because in PP there is a Gln instead of an Arg in NPY and PYY at this position. The degradation after Arg^{33} and Arg^{35} should be minimized by the proline in position 34.

5.3
In Vivo Metabolism of NPY Family Peptides

In general, the investigation of the in vivo release and metabolism of peptides is difficult. Separation techniques used for in vitro experiments are often limited by the sensitivity and specificity of assays, that is the detection of metabolites and by the high background of other compounds in biological samples. In addition, common antibody based detection methods often could not distinguish between the intact peptide and its hydrolysis products. Specific assays for active and truncated forms of peptides, as described for GLP-1 or glucose-dependent insulinotropic polypeptide, so far are not available for the pancreatic polypeptide family (Wolf et al. 2001). However, as long as 10 years ago, the first direct evidence for a proteolytic processing of NPY and PYY by DP IV was provided (Mentlein et al. 1993). In addition, other studies demonstrated that not only NPY and PYY but also NPY(3–36) (Grandt et al. 1996) as well as PYY(3–36) (Grandt et al. 1994a, 1994b) are abundantly present in mammals and are likely to be involved in energy metabolism via inhibition of exocrine pancreas function (Grandt et al. 1995) or other feeding associated processes (Gue et al. 1996; Lloyd et al. 1996). Recently, these studies became very important, since it was shown that the gut hormone PYY(3–36) physiologically inhibits food intake (Batterham et al. 2002). Thus, it is increasingly important to answer questions on the sources of these N-terminal truncated forms of NPY and PYY, on the intracellular and/or extracellular compartments where cleavage takes place, on the enzymes responsible, and on their regulation.

So far, there are some in vivo data available from experiments using peptidase inhibitors and/or NPY and truncated forms of NPY. Fujiwara demonstrated that the neutral endopeptidase inhibitor phosphoramidone enhances the inhibitory effect of NPY on acetylcholine output. In a model of acute inflammation (concanvalin A-induced paw edema), we recently demonstrated that the DP IV inhibitor Ile-Thiazolidide potentiates the pro-inflammatory effect of NPY, which in turn is mediated via the NPY Y_1 receptor (Dimitrijevic et al. 2002). These findings are among the first providing direct in vivo evidence that limited proteolysis by the highly specific peptidase DP IV results in prolonged Y_1 receptor selectivity of NPY. These findings agree with the very recent findings demonstrating that DP IV knock out mice (Marguet et al. 2000) and F344 rat substrains exhibiting a lack of DP IV enzyme activity (Karl et al. 2003a) both show an improved glucose tolerance. This phenomenon is most likely mediated via prolonged action of the incretin GLP-1, which is also a substrate of DP IV. Similarly, we were able to demonstrate that F344 rat substrains mutant for DP IV exhibit a phenotype of reduced anxiety, which is likely to be mediated by the DP IV substrates NPY or substance P (Karl et al. 2003b) and that central application of NPY results in more potent anxiolytic-like and sedative-like effects in those F344 substrains that lack DP IV enzyme activity (Karl et al. 2003c). The latter findings are most likely mediated by prolonged activation of central NPY Y_1 receptors, the predominantly anxiolytic-like acting receptor of NPY (Kask et al. 2002). Thus, there is now accumulating evidence that inhibition of DP IV is a novel and highly specific pharmacological tool for the potentiation of endogenous receptor specific effects mediated by substrates of DP IV including but not limited to NPY and GLP-1.

6
Concluding Remarks

Peptides of the NPY family are synthesized as large precursor molecules, which are translocated to the Golgi apparatus, and guided towards the secretory pathway. During these processes, several specifically regulated posttranslational steps are involved. After exocytotic release, NPY-like peptides are specifically subjected to several steps of limited proteolysis (summarized in Fig. 2). This aspect provides a novel mechanism for the regulation of NPY-like peptide mediated effects. So far, it can be concluded that the action of the peptides of the pancreatic polypeptide family depends on their local concentrations, their different receptor selectivity and the local expression of the different Y-receptors. A major player in this complex network is additionally found in the action of specific peptidases influencing half-life and receptor selectivity of these peptides. The action of the peptides therefore also depends on the local distribution and concentration of at least aminopeptidase P, DP IV-like enzymes, specific endopeptidases, like meprin or neprilysin-like enzymes, and post-arginine cleaving enzymes. Moreover, these peptidases may also act on counter-regulatory peptides and beyond it the specific activity of the peptidases could be influenced by endogenous inhibitors and competitive substrates. Nevertheless peptidases are interesting targets for drug development. The development of inhibitors for NPY, PYY and PP cleaving peptidases is a complementary approach to the development of Y-receptor agonists or antagonists.

References

Adamo ML, Hazelwood RL (1989) Tissue distribution of avian pancreatic polypeptide-degrading activity. Proc Soc Exp Biol Med 191:341–345

Almers W (1994) Synapses. How fast can you get? Nature 367:682–683

Aponte GW, Fink AS, Meyer JH, Tatemoto K, Taylor IL (1985) Regional distribution and release of peptide YY with fatty acids of different chain length. Am J Physiol 249:G745–G750

Balasubramaniam A, Tao Z, Zhai W, Stein M, Sheriff S, Chance WT, Fischer JE, Eden PE, Taylor JE, Liu CD, McFadden DW, Voisin T, Roze C, Laburthe M (2000) Structure-activity studies including a Psi(CH(2)-NH) scan of peptide YY (PYY) active site, PYY(22–36), for interaction with rat intestinal PYY receptors: development of analogues with potent in vivo activity in the intestine. J Med Chem 43:3420–3427

Bastiaensen E, De Block J, De Potter WP (1988) Neuropeptide Y is localized together with enkephalins in adrenergic granules of bovine adrenal medulla. Neuroscience 25:679–686

Batterham RL, Cowley MA, Small CJ, Herzog H, Cohen MA, Wren AM, Brynes AE, Low MJ, Ghatel MA, Cone RD, Bloom SR (2002) Gut hormone $PPY_{(3-36)}$ physiologically inhibits food intake. Nature 418:650–654

Bernet F, Dedieu J F, Laborie C, Montel V, Dupouy JP (1998) Circulating neuropeptide Y (NPY) and catecholamines in rat under resting and stress conditions. Arguments for extra-adrenal origin of NPY, adrenal and extra-adrenal sources of catecholamines. Neurosci Lett 250:45–48

Blasi J, Chapman ER, Link E, Binz T, Yamasaki S, De Camilli P, Sudhof TC, Niemann H, Jahn R (1993) Botulinum neurotoxin A selectively cleaves the synaptic protein SNAP-25. Nature 365:160–163

Böttcher G, Sjoberg J, Ekman R, Hakanson R, Sundler F (1993) Peptide YY in the mammalian pancreas: immunocytochemical localization and immunochemical characterization. Regul Pept 43:115–130

Brakch N, Allemandou F, Cavadas C, Grouzmann E, Brunner HR (2002) Dibasic cleavage site is required for sorting to the regulated secretory pathway for both pro- and neuropeptide Y. J Neurochem 81:1166–1175

Briand R, Yamaguchi N, Gagne J, Kimura T, Farley L, Foucart S, Nadeau R, de Champlain J (1990) Corelease of neuropeptide Y like immunoreactivity with catecholamines from the adrenal gland during splanchnic nerve stimulation in anesthetized dogs. Can J Physiol Pharmacol 68:363–369

Caplan M J (1997) Membrane polarity in epithelial cells: protein sorting and establishment of polarized domains. Am J Physiol 272:F425–429

Chapman ER, Jahn R (1994) Calcium-dependent interaction of the cytoplasmic region of synaptotagmin with membranes. Autonomous function of a single C2- homologous domain. J Biol Chem 269:5735–5741

Chapman E R (2002) Synaptotagmin: a Ca(2+) sensor that triggers exocytosis? Nature Rev Mol Cell Biol 3:498–508

Conlon J M (2002) The origin and evolution of peptide YY (PYY) and pancreatic polypeptide (PP). Peptides 23:269–278

Corder R, Lowry PJ, Emson PC, Gaillard RC (1985) Chromatographic characterisation of the circulating neuropeptide Y immunoreactivity from patients with phaeochromocytoma. Regul Pept 10:91–97

Cox HM, Rudolph A, Gschmeissner S (1994) Ultrastructural co-localization of neuropeptide Y and vasoactive intestinal polypeptide in neurosecretory vesicles of submucous neurons in the rat jejunum. Neuroscience 59:469–476

Cox HM (1998) Peptidergic regulation of intestinal ion transport. A major role for neuropeptide Y and the pancreatic polypeptides. Digestion 59:395–399

D'Hooge R, De Deyn PP, Verzwijvelen A, De Block J, De Potter WP (1990) Storage and fast transport of noradrenaline, dopamine beta-hydroxylase and neuropeptide Y in dog sciatic nerve axons. Life Sci 47:1851–1859

Dannies PS (1999) Protein hormone storage in secretory granules: mechanisms for concentration and sorting. Endocrinol Rev 20:3–21

De Potter WP, Partoens P, Schoups A, Llona I, Coen EP (1997) Noradrenergic neurons release both noradrenaline and neuropeptide Y from a single pool: the large dense cored vesicles. Synapse 25:44–55

de Rijk EP, van Strien FJ, Roubos EW (1992) Demonstration of coexisting catecholamine (dopamine), amino acid (GABA), and peptide (NPY) involved in inhibition of melanotrope cell activity in Xenopus laevis: a quantitative ultrastructural, freeze- substitution immunocytochemical study. J Neurosci 12:864–871

Eipper BA, Green CBR, Campbell TA, Stoffers DA, Keutmann HT, Mains RE, Ouafik L (1992) Alternative splicing and endoproteolytic processing generate tissue-spicific forms of pituitary petidylglycine α-amidating monooxygenase (PAM). J Biochem 267:4008–4015

Ekblad E, Sundler F (2002) Distribution of pancreatic polypeptide and peptide YY. Peptides 23:251–261

el Salhy M, Grimelius L, Emson PC, Falkmer S (1987) Polypeptide YY- and neuropeptide Y-immunoreactive cells and nerves in the endocrine and exocrine pancreas of some vertebrates: an. Histochem J 19:111–117

Ericsson A, Schalling M, McIntyre KR, Lundberg JM, Larhammar D, Seroogy K, Hökfelt T, Persson H (1987) Detection of neuropeptide Y and its mRNA in megakaryocytes: enhanced levels in certain autoimmune mice. Proc Natl Acad Sci USA 84:5585–5589

Ericsson A, Hemsen A, Lundberg J M, Persson H (1991) Detection of neuropeptide Y-like immunoreactivity and messenger RNA in rat platelets: the effects of vinblastine, reserpine, and dexamethasone on NPY expression in blood cells. Exp Cell Res 192:604–611

Fried G, Terenius L, Hökfelt T, Goldstein M (1985) Evidence for differential localization of noradrenaline and neuropeptide Y in neuronal storage vesicles isolated from rat vas deferens. J Neurosci 5:450–458

Gehlert DR, Gackenheimer SL (1996) Unexpected high density of neuropeptide Y Y_1 receptors in the guinea pig spleen. Peptides 17:1345–1348

Grandt D, Dahms P, Schimiczek M, Eysselein VE, Reeve JR Jr., Mentlein R (1993) [Proteolytic processing by dipeptidyl aminopeptidase IV generates receptor selectivity for peptide YY (PYY)]. Med Klin 88:143–145

Grandt D, Schimiczek M, Beglinger C, Layer P, Goebell H, Eysselein V E, Reeve JR Jr. (1994a) Two molecular forms of peptide YY (PYY) are abundant in human blood: characterization of a radioimmunoassay recognizing PYY 1–36 and PYY (3–36). Regul Pept 51:151–159

Grandt D, Schimiczek M, Rascher W, Feth F, Shively J, Lee TD, Davis MT, Reeve JR Jr., Michel MC (1996) Neuropeptide Y 3–36 is an endogenous ligand selective for Y2 receptors. Regul Pept 67:33–37

Grandt D, Schimiczek M, Struk K, Shively J, Eysselein VE, Goebell H, Reeve JR Jr. (1994b) Characterization of two forms of peptide YY, PYY(1–36) and PYY(3–36), in the rabbit. Peptides 15:815–820

Grandt D, Siewert J, Sieburg B, al Tai O, Schimiczek M, Goebell H, Layer P, Eysselein VE, Reeve JR Jr., Muller MK (1995) Peptide YY inhibits exocrine pancreatic secretion in isolated perfused rat pancreas by Y1 receptors. Pancreas 10:180–186

Gue M, Junien JL, Reeve JR Jr., Rivier J, Grandt D, Tache Y (1996) Reversal by NPY, PYY and 3–36 molecular forms of NPY and PYY of intracisternal CRF-induced inhibition of gastric acid secretion in rats. Br J Pharmacol 118:237–142

Hemsen A, Pernow J, Millberg BI, Lundberg JM (1991) Effects of vinblastine on neuropeptide Y levels in the sympathoadrenal system, bone marrow and thrombocytes of the rat. Agents Actions 34:429–438

Hexum TD, Majane EA, Russett LR, Yang HY (1987) Neuropeptide Y release from the adrenal medulla after cholinergic receptor stimulation. J Pharmacol Exp Ther 243:927–930

Hoffmann T, Reinhold D, Kahne T, Faust J, Neubert K, Frank R, Ansorge S (1995) Inhibition of dipeptidyl peptidase IV (DP IV) by anti-DP IV antibodies and non-substrate X-X-Pro- oligopeptides ascertained by capillary electrophoresis. J Chromatogr A 716:355–362

Horsch D, Fink T, Goke B, Arnold R, Buchler M, Weihe E (1994) Distribution and chemical phenotypes of neuroendocrine cells in the human anal canal. Regul Pept 54:527–542

Hökfelt T (1991) Neuropeptides in perspective: the last ten years. Neuron 7:867–879

Jotwani G, Itoh K, Wadhwa S (1994) Immunohistochemical localization of tyrosine hydroxylase, substance P, neuropeptide-Y and leucine-enkephalin in developing human retinal amacrine cells. Brain Res Dev Brain Res 77:285–289

Karl T, Chwalisz WT, Wedekind D, Hedrich H-J, Hoffmann T, Pabst R, von Hörsten S (2003a) Localization, transmission, spontaneous mutations, and variation of function of the *Dpp4* (Dipeptidyl-peptidase IV; CD26) gene in rats. Regul Pept 115:81–90

Karl T, Hoffmann T, Pabst R, von Hörsten S (2003b) Extreme reduction of dipeptidyl-peptidase IV activity in F344 rat substrains is associated with various behavioral differences. Physiol Behav 80:123–134

Karl T, Hoffmann T, Pabst R, von Hörsten S (2003c) Behavioral effects of centrally applied Neuropeptide Y in mutant F344 rat substrains with an extreme reduction in the dipeptidyl-peptidase IV activity. Pharmacol Biochem Behav 75:869–879

Kask A, Harro J, von Hörsten S, Redrobe JP, Dumont Y, Quirion R (2002) The neurocircuitry and receptor subtypes mediating anxiolytic-like effects of neuropeptide Y. Neurosci Biobehav Rev 26:259–283

Kataoka Y, Majane EA, Yang HY (1985) Release of NPY-like immunoreactive material from primary cultures of chromaffin cells prepared from bovine adrenal medulla. Neuropharmacology 24:693–695

Klenchin VA, Martin TF (2000) Priming in exocytosis: attaining fusion-competence after vesicle docking. Biochimie 82:399–407

Kong JY, Thureson-Klein A, Klein RL (1989) Differential distribution of neuropeptides and serotonin in pig adrenal glands. Neuroscience 28:765–775

Lang J (1999) Molecular mechanisms and regulation of insulin exocytosis as a paradigm of endocrine secretion. Eur J Biochem 259:3–17

Link E, Edelmann L, Chou JH, Binz T, Yamasaki S, Eisel U, Baumert M, Sudhof TC, Niemann H, Jahn R (1992) Tetanus toxin action: inhibition of neurotransmitter release linked to synaptobrevin proteolysis. Biochem Biophys Res Commun 189:1017–1023

Lloyd KC, Grandt D, Aurang K, Eysselein VE, Schimiczek M, Reeve JR Jr. (1996) Inhibitory effect of PYY on vagally stimulated acid secretion is mediated predominantly by Y1 receptors. Am J Physiol 270:G123–G127

Ludwig R, Lucius R, Mentlein R (1995) A radioactive assay for the degradation of neuropeptide Y. Biochimie 77:739–743
Ludwig R, Feindt J, Lucius R, Petersen A, Mentlein R (1996) Metabolism of neuropeptide Y and calcitonin gene-related peptide by cultivated neurons and glial cells. Brain Res Mol Brain Res 37:181–191
Lundberg J M, Terenius L, Hökfelt T, Goldstein M (1983) High levels of neuropeptide Y in peripheral noradrenergic neurons in various mammals including man. Neurosci Lett 42:167–172
Lundberg JM, Rudehill A, Sollevi A, Fried G, Wallin G (1989) Co-release of neuropeptide Y and noradrenaline from pig spleen in vivo: importance of subcellular storage, nerve impulse frequency and pattern, feedback regulation and resupply by axonal transport. Neuroscience 28:475–486
Lundberg JM (1996) Pharmacology of cotransmission in the autonomic nervous system: integrative aspects on amines, neuropeptides, adenosine triphosphate, amino acids and nitric oxide. Pharmacol Rev 48:113–178
Majewski M, Heym C (1992) Immunohistochemical localization of calcitonin gene-related peptide and cotransmitters in a subpopulation of post-ganglionic neurons in the porcine inferior mesenteric ganglion. Acta Histochem 92:138–146
Majewski M, Kaleczyc J, Sienkiewicz W, Lakomy M (1995) Existence and co-existence of vasoactive substances in nerve fibres supplying the abdomino-pelvic arterial tree of the female pig and cow. Acta Histochem 97:235–256
Marguet D, Baggio L, Kobayashi T, Bernard AM, Pierres M, Nielsen PF, Ribel U, Watanabe T, Drucker DJ, Wagtmann N (2000) Enhanced insulin secretion and improved glucose tolerance in mice lacking CD26. Proc Natl Acad Sci USA 97:6874–6879
Martire M, Preziosi P, Cannizzaro C, Mores N, Fuxe K (1997) Extracellular sodium removal increases release of neuropeptide Y-like immunoreactivity from rat brain hypothalamic synaptosomes: involvement of intracellular acidification. Synapse 27:191–198
Matter K (2000) Epithelial polarity: sorting out the sorters. Curr Biol 10: R39–R42
Medeiros MS, Turner AJ (1993) Processing and metabolism of peptide YY. Biochem Soc Trans 21 (Pt 3):248S
Medeiros MD, Turner AJ (1994) Processing and metabolism of peptide-YY: pivotal roles of dipeptidylpeptidase-IV, aminopeptidase-P, and endopeptidase-24.11. Endocrinology 134:2088–2094
Medeiros MS, Turner AJ (1996) Metabolism and functions of neuropeptide Y. Neurochem Res 21:1125–1132
Mentlein R, Dahms P, Grandt D, Kruger R (1993) Proteolytic processing of neuropeptide Y and peptide YY by dipeptidyl peptidase IV. Regul Pept 49:133–144
Mentlein R, Roos T (1996) Proteases involved in the metabolism of angiotensin II, bradykinin, calcitonin gene-related peptide (CGRP), and neuropeptide Y by vascular smooth muscle cells. Peptides 17:709–720
Mentlein R (1999) Dipeptidyl-peptidase IV (CD26)—role in the inactivation of regulatory peptides. Regul Pept 85:9–24
Michel MC, Beck-Sickinger A, Cox H, Doods HN, Herzog H, Larhammar D, Quirion R, Schwartz T, Westfall T (1998) XVI. International Union of Pharmacology recommendations for the nomenclature of neuropeptide Y, peptide YY, and pancreatic polypeptide receptors. Pharmacol Rev 50:143–150
Morris JL, Gibbins IL, Furness JB, Costa M, Murphy R (1985) Co-localization of neuropeptide Y, vasoactive intestinal polypeptide and dynorphin in non-noradrenergic axons of the guinea pig uterine artery. Neurosci Lett 62:31–37
Myers AK, Farhat MY, Vaz CA, Keiser HR, Zukowska-Grojec Z (1988) Release of immunoreactive-neuropeptide by rat platelets. Biochem Biophys Res Commun 155:118–122

Nausch I, Mentlein R, Heymann E (1990) The degradation of bioactive peptides and proteins by dipeptidyl peptidase IV from human placenta. Biol Chem Hoppe Seyler 371:1113–1118

Nilsson T, Warren G (1994) Retention and retrieval in the endoplasmic reticulum and the Golgi apparatus. Curr Opin Cell Biol. 6:517–521

Ogawa T, Kitamura K, Kawamoto M, Eto T, Tanaka K (1989) Increased immunoreactive neuropeptide Y in platelets of spontaneously hypertensive rats (SHR). Biochem Biophys Res Commun 165:1399–1405

Orci L (1982) Macro- and micro-domains in the endocrine pancreas. Diabetes 31:538–565

Oyarce AM, Eipper BA (1993) Neurosecretory vesicles contain soluble and membrane-associated monofunctional and bifunctional peptidylglycine alpha-amidating monooxygenase proteins. J Neurochem 60:1105–1114

Paquet L, Massie B, Mains RE (1996) Proneuropeptide Y processing in large dense-core vesicles: manipulation of prohormone convertase expression in sympathetic neurons using adenoviruses. J Neurosci 16:964–973

Parker D, Soderberg C, Zotova E, Shupliakov O, Langel U, Bartfai T, Larhammar D, Brodin L, Grillner S (1998) Co-localized neuropeptide Y and GABA have complementary presynaptic effects on sensory synaptic transmission. Eur J Neurosci 10:2856–2870

Pernow J, Lundberg JM, Kaijser L, Hjemdahl P, Theodorsson-Norheim E, Martinsson A, Pernow B (1986) Plasma neuropeptide Y-like immunoreactivity and catecholamines during various degrees of sympathetic activation in man. Clin Physiol 6:561–578

Persson H, Ericsson A, Hemsen A, Hökfelt T, Larhammar D, Lundberg JM, McIntyre KR, Schalling M (1989) Expression of NPY messenger RNA and peptide in non-neuronal cells. In: Mutt V, Fuxe K, Hökfelt T, Lundberg JM, (eds) Neuropeptide Y. Raven Press, New York, pp 43–50

Pickel VM, Chan J, Veznedaroglu E, Milner TA (1995) Neuropeptide Y and dynorphin-immunoreactive large dense-core vesicles are strategically localized for presynaptic modulation in the hippocampal formation and substantia nigra. Synapse 19:160–169

Presley JF, Cole NB, Schroer TA, Hirschberg K, Zaal K J, Lippincott-Schwartz J (1997) ER-to-Golgi transport visualized in living cells. Nature 389:81–85

Price JS, Kenny AJ, Huskisson NS, Brown MJ (1991) Neuropeptide Y (NPY) metabolism by endopeptidase-2 hinders characterization of NPY receptors in rat kidney. Br J Pharmacol 104:321–326

Rettig J, Neher E (2002) Emerging roles of presynaptic proteins in Ca++-triggered exocytosis. Science 298:781–785

Rizo J, Südhof TC (2002) Snares and Munc18 in synaptic vesicle fusion. Nature Rev Neurosci 3:641–653

Rodriguez-Boulan E, Powell SK (1992) Polarity of epithelial and neuronal cells. Annu Rev Cell Biol 8:395–427

Schwartz TW, Tager HS (1981) Isolation and biogenesis of a new peptide from pancreatic islets. Nature 294:589–591

Schwartz TW (1983) Pancreatic polypeptide: a hormone under vagal control. Gastroenterology 85:1411–1425

Sundler F, Böttcher G, Ekblad E, Hakanson R (1993) PP, PYY, and NPY: Occurence and Distribution in the Periphery. In: Colmers WF, Wahlestedt C (eds) Biology of Neuropeptide Y and Related Peptides. Humana Press Inc, Totowa, NJ, pp 157–196

Takahashi K, Mouri T, Itoi K, Sone M, Ohneda M, Murakami O, Nozuki M, Tachibana Y, Yoshinaga K (1987) Increased plasma immunoreactive neuropeptide Y concentrations in phaeochromocytoma and chronic renal failure. J Hypertens 5:749–753

Takiyyuddin MA, Brown MR, Dinh TQ, Cervenka JH, Braun SD, Parmer RJ, Kennedy B, O'Connor DT (1994) Sympatho-adrenal secretion in humans: factors governing catecholamine and storage vesicle peptide co-release. J Auton Pharmacol 14:187–200

Tasaka Y, Marumo K, Inoue Y, Hirata Y (1989) Degradation of 125I-glucagon, -pancreatic polypeptide and -insulin by acid saline extract of rat submaxillary gland and their protection by proteinase inhibitors. Endocrinol Jpn 36:47-53

Tsutsumi Y (1984) Immunohistochemical studies on glucagon, glicentin and pancreatic polypeptide in human stomach: normal and pathological conditions. Histochem J 16:869-883

Wolfensberger M, Forssmann WG, Reinecke M (1995) Localization and coexistence of atrial natriuretic peptide (ANP) and neuropeptide Y (NPY) in vertebrate adrenal chromaffin cells immunoreactive to TH, DBH and PNMT. Cell Tissue Res 280:267-276

Zhang X, Nicholas A P, Hokfelt T (1993) Ultrastructural studies on peptides in the dorsal horn of the spinal cord-I. Co-existence of galanin with other peptides in primary afferents in normal rats. Neuroscience 57:365-384

Zhang X, Nicholas AP, Hokfelt T (1995) Ultrastructural studies on peptides in the dorsal horn of the rat spinal cord-II. Co-existence of galanin with other peptides in local neurons. Neuroscience 64:875-891

Zukowska-Grojec Z, Konarska M, McCarty R (1988) Differential plasma catecholamine and neuropeptide Y responses to acute stress in rats. Life Sci 42:1615-1624

Zukowska-Grojec Z, Wahlestedt C (1993) Origin and actions of neuropeptide Y in the cardiovascular system. In: Colmers WF, Wahlestedt C (eds) Biology of Neuropeptide Y and Related Peptides. Humana Press Inc, Totowa, NJ, pp 157-196

NPY Receptor Subtypes and Their Signal Transduction

N. D. Holliday[1] · M. C. Michel[2] · H. M. Cox[1]

[1] Centre for Neuroscience Research, King's College London, Hodgkin Building, Guy's Campus, London, SE1 1UL, UK
e-mail: nicholas.2.holliday@kcl.ac.uk

[2] Department of Pharmacology and Pharmacotherapy, Academic Medical Centre, University of Amsterdam, Meibergdreef 15, 1105 AZ, Amsterdam, The Netherlands

1	Introduction	46
2	Agonist Activation of Y Receptors	46
2.1	Receptor Subtype Classification	46
2.2	Agonist Binding	48
2.3	Signalling Motifs	49
2.4	Dimerization	50
3	Intracellular Signalling	52
3.1	Coupling to Pertussis-Toxin Sensitive G Proteins and cAMP Inhibition	52
3.2	Elevations of Intracellular Ca^{2+} and Cross-Talk Between G_i and G_q	53
3.3	Modulation of Ion Channels	54
4	Y Receptor Regulation and Trafficking	55
4.1	Desensitization and β-Arrestin Interactions	55
4.2	Two Ways of Signal Termination—The Roles of Phosphorylation and Palmitoylation	58
4.3	Internalization	60
5	Long-Term Signalling by Y Receptors	64
5.1	Activation of Extracellular Signal Related Kinases	64
5.2	Control of Gene Expression	66
6	Conclusions	66
References		67

Abstract The neuropeptide Y family acts through five cloned G protein coupled receptors (Y_1, Y_2, Y_4, Y_5, y_6), all of which bind two or more of the endogenous peptides. At least three subtypes (Y_1, Y_2 and Y_5) may signal as homodimers, whose formation is independent of the presence of agonist or G protein heterotrimers. All the Y receptors display the full complement of pertussis-toxin sensitive $G_{i/o}$ signalling pathways (mediated by α or $\beta\gamma$ subunits) in a suitable cell context, including the inhibition of cAMP formation, intracellular Ca^{2+} mobilization and modulation of Ca^{2+} and K^+ channels. However recent studies have highlighted emerging differences in the desensitization and cellular trafficking of each subtype. Signalling of the Y_1 receptor is regulated by phosphorylation and palmitoylation at key C-terminal motifs, and on agonist stimulation this subtype rapidly associates with the inhibitory protein, β-arrestin 2. In contrast the interaction of the Y_2 receptor with β-arrestin 2 is much less pronounced, consistent with its reduced ability to undergo clathrin-mediated internalization compared to Y_1 and Y_4 subtypes. This observation may have additional consequences for the spatial and temporal organization of longer-term Y receptor signals, since β-arrestins have a wider role as scaffolds for protein kinase cascades. However current evidence suggests that Y_1, Y_2, Y_4 and Y_5 subtypes are all able to couple to mitogen activated protein kinase pathways, and as a consequence influence the control of gene expression and cell fate.

Keywords Y receptor · Dimerization · Desensitization · Internalization · Mitogen activated protein kinase

1
Introduction

The neuropeptide Y family of peptides including NPY, PYY, PP and active metabolites such as PYY(3–36) acts through a group of receptors belonging to the class A (rhodopsin-like) G protein coupled receptor (GPCR) family with seven transmembrane (TM) domains (Fig. 1). Five members of this family have been identified by molecular cloning and are designated Y_1, Y_2, Y_4, Y_5 and y_6. Unusually these receptors are distinguished both by their relative promiscuity for the endogenous ligands and the coupling of all subtypes to a single class of G protein ($G_{i/o}$). However rapid advances in the GPCR field have unmasked a new complexity in the way in which these receptors can interact with downstream signalling and regulatory components. The definition and basic signalling properties of the Y receptor subtypes have been reviewed comprehensively previously (Michel et al. 1998). Hence the present manuscript will summarize this information only briefly and focus on more recent data regarding the molecular understanding of Y receptors. In the context of these developments we will suggest how they may provide a means for different subtypes (which at first sight seem superficially similar) to tailor cell signalling responses to NPY-related ligands.

2
Agonist Activation of Y Receptors

2.1
Receptor Subtype Classification

Five subtypes of Y receptors are currently recognized and have been identified by molecular cloning (Michel et al. 1998; Michel 2002). These subtypes are products of the functional genes created by successive duplications of a single

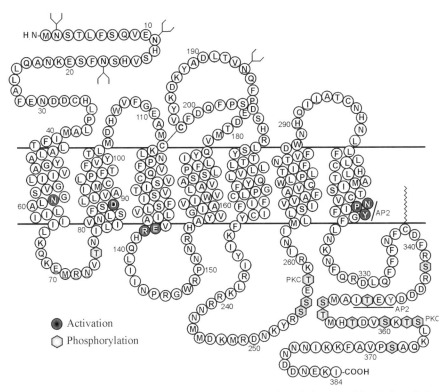

Fig. 1 The human Y_1 receptor sequence. *Highlighted* amino acids include extracellular N-glycosylation and disulphide bridge sites, residues which may form the activation switch, and potential recognition motifs for the AP-2 complex. The palmitoylated Cys, and Ser/Thr residues which may act as phosphorylation sites (including two PKC consensus motifs) are also indicated

ancestor (see the chapter by Larhammar, this volume) and differences in their affinities for endogenous and synthesized peptides provide an explanation for much of the pharmacology observed in endogenous cells and tissues.

The Y_1 receptor is a 384-amino acid protein that binds NPY and PYY with much higher affinity than PP, has high affinity for Pro^{34}-substituted fragments of NPY and PYY and low affinity for C-terminal fragments of NPY and PYY. Selective antagonists for the Y_1 receptor include BIBP 3226, BIBO 3304 and GR 231118. The Y_2 receptor is a 381-amino acid protein that also preferentially binds NPY and PYY over PP, but has high affinities for C-terminal fragments and low affinities for Pro^{34}-substituted analogues of NPY and PYY. BIIE 0246 is a selective antagonist for this subtype. The Y_4 receptor is a 375-amino acid protein that is specific for PP over PYY and NPY, but apart from this no specific agonists or antagonists have been reported for this subtype. The Y_5 receptor is a 445-amino acid protein that recognizes all three endogenous peptides, their long C-terminal fragments and Pro^{34} substituted ligands. CGP 71683A is a selective antagonist for this subtype. The y_6 receptor is functional in mouse and rabbit

(Gregor et al. 1996a; Rose et al. 1997; Weinberg et al. 1996), but is absent from the rat genome and is a pseudogene in primates. A frameshift mutation in the human coding sequence generates a six-TM domain protein unable to bind any peptides (Gregor et al. 1996a; Rose et al. 1997). The pharmacology of the mouse y_6 receptor is also controversial (Rose et al. 1997; Weinberg et al. 1996), but recent functional studies suggest a Y_1-like phenotype (Mullins et al. 2000).

Two further Y receptor subtypes were originally suggested to mediate responses in tissues where NPY and PYY were not equipotent (Michel et al. 1998). The peripheral 'PYY-preferring' receptor is in fact the Y_2 receptor (Cox and Tough 2000; Goumain et al. 2001). The second anomaly, where NPY responses were more potent than those of PYY, lead to the proposal of a Y_3 receptor. However, no such receptor has been cloned and the emergence of a specific molecular entity with such characteristics becomes increasingly unlikely following the sequencing of the human genome. Rather, this proposed site may involve one or a combination of known receptors, for example in rat colon (Pheng et al. 1999) and bovine adrenal chromaffin cells (Cavadas et al. 2001; Zhang et al. 2000). In addition some effects of NPY at high concentration, such as its action on mast cells, result from the direct stimulation of G proteins by its amphipathic α-helix (Mousli et al. 1995). Although a case may still be made for an unknown Y_3 subtype in the nucleus tractus solitaris (Glaum et al. 1997) the following discussion will be restricted to the cloned Y receptors. An overview table of these cloned receptors and their characteristics is presented in the chapter by Redrobe et al. in this volume.

2.2
Agonist Binding

Peptide affinities for the Y receptor subtypes have largely been determined in competition binding assays which use agonist radioligands (e.g. iodinated PYY, NPY and PP), although limited information is also available from the radiolabelled Y_1 antagonist BIBP 3226 (Entzeroth et al. 1995). GPCR agonists have low affinity for the inactive ground state of the receptor and bind selectively to a high affinity receptor conformation, most commonly the agonist-receptor-G protein complex (Kenakin 1996). Thus the number of binding sites attained in saturation curves will depend not only on total receptor number but also on the ability of the radiolabelled agonist to effect this conformational change (its efficacy). In addition, different radiolabelled agonists may preferentially target distinct receptor conformations, which can lead to large discrepancies in displacement curves amongst competing ligands (e.g. for the NK_1 receptor; Holst et al. 2001). [^{125}I]NPY and [^{125}I]PYY appear interchangeable in competition experiments using Y_1 and Y_2 receptors (although non-specific binding using [^{125}I]NPY is significantly greater), but the choice between [^{125}I]PP and [^{125}I]PYY may profoundly influence the purported pharmacology of other Y receptor subtypes. Thus PP binding to the y_6 receptor may only be detected if it is labelled with [^{125}I]PP {as in COS-7 (Gregor et al. 1996a) but not HEK293 cells where

[^{125}I]PYY was used; (Mullins et al. 2000)}. The relative affinities of PYY and NPY for the Y_4 receptor are very much dependent on the radioligand (e.g. for the mouse orthologue; Gregor et al. 1996b). In addition, [^{125}I]PYY recognized only 5%–60% of guinea-pig Y_4 receptor sites labelled by [^{125}I]PP, depending on levels of expression and host cell type (Berglund et al. 2001; Eriksson et al. 1998). Thus PYY might have a lower efficacy than PP, and a reduced ability to convert the Y_4 receptor pool to a high affinity conformation. Such differences could subsequently be masked by intervening amplification steps in cAMP assays, where PP and PYY are full agonists (Eriksson et al. 1998). Alternatively PP and PYY may select unique conformations of the guinea pig Y_4 receptor, raising the possibility that they could traffic receptor signals to different pathways (Kenakin 1996).

2.3
Signalling Motifs

In the predicted Y receptor sequences a number of critical amino acids are conserved, which may form the activation switch of class A GPCRs (Fig. 1). An Asn in TMI, an Asp in TMII and the NPXXY motif in TMVII form a polar pocket in the crystal structure of rhodopsin (Lu et al. 2002) that in the silent conformation holds the Arg (of the conserved E/DRY sequence) at the foot of TMIII. In models of receptor activation a shift of TMs VI and VII away from TMIII disrupts this contact site, aided by the protonation of the acidic amino acid adjoining the TMIII Arg. The role of the conserved Asp in the DRY motif was first investigated by Scheer et al. (1997), who observed that its mutation significantly increased the constitutive (i.e. agonist independent) activity of the α_{1B}-adrenoceptor, and has also been suggested in many other (though not all) GPCRs (Lu et al. 2002). All but one of the Y receptor subtypes retain the acidic TMIII residue (predominantly as Glu), but in Y_5 it has been replaced by Val, an interesting deviation which provides one of the distinctive characteristics of this receptor.

Contact between GPCR and G protein is established through the second and third intracellular loops, and perhaps also the recently identified juxtamembrane helix 8 following TMVII (Palczewski et al. 2000). All Y receptors share some features that may determine G protein coupling efficacy and fidelity (such as basic residues lining the N- and C-terminal portions of the third intracellular loop; Okamoto and Nishimoto 1992). However, the Y_5 receptor third intracellular loop (136 amino acids in the human sequence) is four times as long as the other subtypes. This portion of the Y_5 receptor may have expanded to include regulatory motifs normally carried in the C-terminal tail (as in the α_{2A}-adrenoceptor; Liggett et al. 1992), which is correspondingly much shorter than the Y_1, Y_2, Y_4 or y_6 sequences. However the underlying nucleotide code for the Y_5 third intracellular loop also provides (in the opposite direction) promoter B for the Y_1 receptor gene, a dual function dictated by the overlapping gene structure (Herzog et al. 1997). Thus this unusual feature of the Y_5 receptor may be conserved by a selection pressure unrelated to its function.

The C termini of all five Y receptors contains a Cys residue which may act as a palmitoylation site to create a fourth intracellular loop, by insertion of the fatty acid tails into the lipid bilayer (Qanbar and Bouvier 2003). Dynamic control of this post-translational modification could provide a mechanism for Y receptor regulation, because of its influence on proximal (e.g. helix 8) and distal C-terminal regions. The human Y_1, Y_2, Y_4 and y_6 subtypes contain similar numbers of intracellular Ser/Thr residues (8–12, predominantly in the C terminus) which could act as acceptors for phosphorylation by second messenger kinases [2–3 protein kinase C (PKC) consensus sites in each receptor] or GPCR kinases (GRKs) which target agonist-occupied receptors. In the Y_5 receptor clusters of Ser/Thr (26 in all) are concentrated in the third intracellular loop, and include multiple PKC (6) and protein kinase A (PKA, 2) motifs.

2.4
Dimerization

In recent years, the idea that GPCRs functioned as purely monomeric entities has been largely superceded in the face of mounting evidence that they form multimeric complexes (of which the simplest case is a dimer). Early biochemical experiments demonstrated that differentially tagged receptors could be co-immunoprecipitated as dimers, and that coexpression of mutant GPCRs deficient in either binding or signalling domains restored functional activity (Bouvier 2001). These studies have been complemented by the development of methods to study receptor oligomerization in living cells using fluorescent resonance energy transfer (FRET). Receptors are labelled with pairs of chromophores using fluorescent ligands, antibodies or fusion constructs containing variants of green fluoresecent protein (GFP). The excitation wavelength of the acceptor fluorophore is matched to the emission peak of the donor. When the donor is stimulated its excitation energy will be transferred to the acceptor molecule, provided that the two are close enough. The limit for FRET observance (between 50 and 100 Å) is of the same order of magnitude as the cross-sectional diameter of rhodopsin (Lu et al. 2002) and so its presence implies at least tight receptor clustering, if not dimeric units.

The ability of NPY and related peptides to form dimers in solution (Bader et al. 2001; see also chapter by Mörl and Beck-Sickinger, this volume) and the fact that dimeric ligands such as GR 231118 have much higher affinity for the Y_1 receptor than the monomers (Daniels et al. 1995) suggest that Y receptors may be suitable candidates for dimerization. To this end Dinger et al. (2003) have applied FRET to study homodimerization of Y_1, Y_2 and Y_5 receptors tagged at the C terminus with GFP pairs [e.g. cyan (CFP) and yellow (YFP)]. These fusion proteins retained the binding and functional properties of the wild-type receptors and when expressed alone in BHK cells no FRET was observed, even when a suitable acceptor GFP variant was cotransfected and present in the cytoplasm. However when the CFP and YFP fusion proteins of Y_1, Y_2 or Y_5 receptors were coexpressed, exciting at the CFP wavelength resulted in FRET and a weak yellow

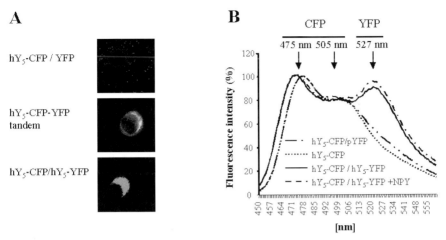

Fig. 2A, B Dimerization of the human Y_5 receptor. **A** Fluorescence images obtained with the FRET filter set from transiently transfected baby hamster kidney cells expressing the hY_5-CFP fusion construct and cytoplasmic YFP (negative control), the hY_5–CFP–YFP tandem (positive control) and hY_5-CFP/hY_5–YFP fusion proteins. **B** Emission spectra from the negative controls (hY_5-CFP ± cytoplasmic YFP) are overlaid with those from cells coexpressing the hY_5-CFP/hY_5–YFP FRET pair, in the absence or presence of 1 µM NPY. (Reproduced with permission from Dinger et al. 2003)

fluorescence from YFP, suggesting that each subtype can form homodimers (Fig. 2A). Figure 2B illustrates this FRET signal analysed in more detail by spectroscopy for the Y_5-CFP/YFP fusion protein pair. The spectra following excitation at 433 nm include not only the CFP emission peak (475 nm) and shoulder (505 nm) but a second peak at 525 nm corresponding to the YFP signal. By examining the 475/525 nm ratio, the effect of agonist exposure or disrupting the receptor G protein interaction (with GTPγS) was quantified. For each Y receptor subtype, the FRET signal was unaffected by NPY and GTPγS, illustrated for Y5-CFP/YFP in Fig. 2B. The constitutive signal appeared greater for Y_1 and Y_5 than for Y_2 receptors, but this may be a consequence of the Y–YFP acceptor fusion protein expression being proportionately more for the former subtypes, thus resulting in more Y-CFP/Y–YFP heterodimers able to display FRET compared to the Y-CFP/Y-CFP and Y-YFP/Y–YFP silent combinations. Thus Y_1, Y_2 and Y_5 receptors may form constitutive homodimers, and unusually it appears that dimerization is unaffected by agonist. At present the importance of homodimerization to Y receptor function and signalling can only be speculated, although it is necessary for the proper cell surface expression of some GPCRs (e.g. $GABA_B$) and the dimer may provide an increased contact interface between receptor and a single G protein heterotrimer (Bouvier 2001). In addition, the possibility of heterodimerization amongst Y receptor subtypes, or indeed Y receptors and GPCRs responding to different ligands (Bouvier 2001), may yet provide a further level of complexity in the study of Y receptor pharmacology.

3
Intracellular Signalling

3.1
Coupling to Pertussis-Toxin Sensitive G Proteins and cAMP Inhibition

Y receptor-mediated responses are almost always sensitive to inhibition by pertussis toxin indicating predominant if not exclusive coupling to G-proteins of the $G_{i/o}$ family (Michel et al. 1998). This G protein dependent pathway is highlighted by the ability of stable analogues of GTP (such as GTPγS) to disrupt the receptor–G protein complex and radiolabelled agonist binding (e.g. Feth et al. 1992; Freitag et al. 1995; Voisin et al. 2000), while stimulation of GTP$\gamma[^{35}S]$ binding may be used to monitor Y receptor activation in isolated membranes (Holliday and Cox 2003) and in autoradiography of brain slices (Primus et al. 1998). While the list of potential G protein independent pathways for GPCRs is expanding (Brzostowski and Kimmel 2001) there are as yet few clear instances where the $G_{i/o}$ inactivator pertussis toxin does not severely attenuate Y receptor responses.

At least five α subunits (plus the pertussis toxin insensitive $G_z\alpha$) and several possible $\beta\gamma$ complexes exist within the $G_{i/o}$ family, and a few studies have addressed the specificity of Y receptors for members of this group. The use of G protein antisera allowed an examination of Y_2 receptor coupling to specific $G\alpha$ subunits in SMS-KAN cells (Freitag et al. 1995), which suggested that $G_i\alpha$ but not $G_o\alpha$ contributed to cAMP responses. $G\alpha$ protein expression has also been targeted in cell lines expressing Y receptors. Dimethyl sulfoxide treatment down-regulated $G_{i3}\alpha$ over $G_{i2}\alpha$ in human erythroleukaemia (HEL) cells and inhibited endogenous Y_1 receptor Ca^{2+} but not cAMP signalling (Michel 1998). In renal proximal tubule cells the knockdown of $G_{i2}\alpha$ but not $G_{i3}\alpha$ (by antisense RNA expression) reduced the affinity of PYY for Y_2 receptors and eliminated cAMP and proliferative responses (Voisin et al. 1996). A complementary approach has been to increase the $G_{i/o}\alpha$ concentration, and in dorsal root ganglion neurones including purified $G_{o1}\alpha$ subunits in the patch pipette solution markedly increased Y_2 receptor inhibition of Ca^{2+} channel currents (Ewald et al. 1988). In addition *Xenopus* oocytes over-expressing $G_{i1}\alpha$ potentiated Y_1 receptor activation of G protein activated inwardly rectifying K^+ channels (GIRK1; Brown et al. 1995), an effect not replicated for another G_i coupled receptor (D_2 dopamine) or by introduction of cDNAs for three other $G\alpha$ (i2, i3 and o1). These early investigations have hinted at the ability of Y receptor subtypes to direct signalling through particular $G\alpha$ subunits, and it remains to be seen whether this choice may also be influenced by the agonist–receptor combination.

All cloned Y receptor subtypes activate $G_{i/o}$ proteins to inhibit adenylyl cyclase, which is frequently assessed as forskolin-stimulated cAMP accumulation (Bard et al. 1995; Gerald et al. 1995; Gerald et al. 1996; Herzog et al. 1992; Mullins et al. 2000). This response has also been observed ubiquitously in cells and tissues natively expressing NPY receptors (Michel et al. 1998). However, analogy to other $G_{i/o}$ coupled receptors suggests that lowering of intracellular cAMP concentrations may not be the most important mechanism in mediating the physiological effects of Y receptor stimulation (Limbird 1988).

3.2
Elevations of Intracellular Ca^{2+} and Cross-Talk Between G_i and G_q

Y receptor stimulation can cause elevations of intracellular Ca^{2+} concentrations as first shown for Y_1 receptors natively expressed in HEL cells (Motulsky and Michel 1988). Heterologously expressed Y_1, Y_2 and Y_4 receptors also elevate intracellular Ca^{2+} in a manner which is largely independent of the presence of extracellular Ca^{2+} (Bard et al. 1995; Gerald et al. 1995; Grouzmann et al. 2001; Selbie et al. 1995). The Ca^{2+} source is often an endoplasmic reticulum compartment sensitive to the Ca^{2+} ATPase pump inhibitor thapsigargin (Connor et al. 1997; Michel 1994; Selbie et al. 1995), although a thapsigargin and ryanodine insensitive store has also been described (Grouzmann et al. 2001). In contrast with the ubiquitous observance of cAMP signals, the Y receptor Ca^{2+} response is very much dependent on cell type. In epithelial cell lines for example, Y receptors only inhibit cAMP accumulation (Mannon et al. 1994) and as a consequence counteract coincident Ca^{2+} signals from other receptors (Bouritius et al. 1998). Moreover within an apparently homogeneous population both responding and non-responding cells may be observed (Grouzmann et al. 2001). In this context it is difficult to assess the importance of the only study which examined both cAMP and Ca^{2+} signalling pathways for the Y_5 receptor (Bischoff et al. 2001) and found that this subtype was not linked to Ca^{2+} mobilization. Whether this represents a true distinction between the Y_5 receptor and other subtypes, or is a reflection of the cell context (HEC-1B endometrial cells) remains to be determined.

In dissecting the potential Ca^{2+} signalling pathways, emphasis has been placed on the coupling of Y_1 receptors responsible for the vasopressor actions of sympathetically released NPY. In rabbit mesenteric small arteries, NPY promotes constriction through three mechanisms, all blocked by BIBP 3226 (Prieto et al. 2000). It reverses vasodilation caused by agents which elevate cAMP, but in addition increases intracellular Ca^{2+} and tension itself, and potentiates the actions of coreleased noradrenaline. Such synergy between α_1-adrenoceptor ($G_{q/11}$ coupled) and Y receptor ($G_{i/o}$) mediated events is characteristic of vascular smooth muscle cells (Racchi et al. 1999) and may also be true of other cell types. Indeed in SH-SY5Y cells activation of a $G_{q/11}$ coupled muscarinic receptor was necessary for Y_2 receptor Ca^{2+} responses (Connor et al. 1997). The interaction between $G_{q/11}$ and $G_{i/o}$ signalling has focussed attention on phospholipase Cβ (PLCβ), the enzyme which converts phosphotidylinositol 4,5-bisphosphate to the divergent messengers inositol (1,4,5) trisphosphate (IP$_3$) and diacylglycerol (DAG). IP$_3$ mobilizes Ca^{2+} from intracellular stores, while DAG activates PKC. In this pathway, PLCβ is a prime candidate for a coincidence detector because certain isoforms are activated not only by $G_{q/11}\alpha$, but also Ca^{2+} and $G_{i/o}\beta\gamma$ subunits (Exton 1996).

Selbie et al. (1995) studied these interactions using Chinese hamster ovary (CHO) cells cotransfected with cDNAs for α_1 and Y_1 receptors. Surprisingly they observed that the Ca^{2+} elevations in response to PYY alone preceded a slow rise in IP_3 production. Thus while Y_1 receptor Ca^{2+} mobilization from intracellular stores was pertussis-toxin sensitive, it appeared independent of the PLCβ pathway, and did not involve other possible intermediates (sphingosine-1-phosphate and cyclic ADP-ribose). Y receptors can apparently stimulate Ca^{2+} without a causative rise in IP_3 in vascular smooth muscle (Mihara et al. 1989) and controversially in HEL cells (Daniels et al. 1992; Motulsky and Michel 1988). More recent data suggest that the Y_2 receptor Ca^{2+} response in LN319 cells is not abolished by a PLCβ inhibitor (U73122), as well as being partly insensitive to pertussis toxin (Grouzmann et al. 2001). In contrast the synergy between phenylephrine (PE) and PYY stimulation in the CHO model was mediated by PLCβ, since co-addition of both agonists potentiated PE production of IP_3 (Selbie et al. 1995). The independent release of Ca^{2+} by PYY would be expected to contribute to this potentiation, but a role for $G_{i/o}\beta\gamma$ subunits has also been suggested because of the inhibitory actions of $G\beta\gamma$ scavengers such as transducin (Selbie et al. 1997). Although PE and PYY together caused at most additive peak Ca^{2+} responses, stimulation of PKC was markedly enhanced and led to pronounced synergy in the downstream activation of phospholipase A_2 and arachidonic acid production (Selbie et al. 1995, 1997). Clearly activated Y receptors can couple to PLCβ, and in many instances the basal $G_{q/11}$ tone provided by endogenous paracrine agents (e.g. ATP; Selbie et al. 1997) may make this a significant route to the generation of Ca^{2+} responses. However the evidence for an independent pathway leading to Ca^{2+} release from IP_3 sensitive stores is mounting, although the mechanisms at present remain elusive.

Interestingly an additional mechanism by which Y receptors promote synergistic signals has recently become apparent. NPY stimulates the recruitment of α_{1A}-adrenoceptors held in reserve in intracellular vesicles to the plasma membrane of renal cortical cells (Holtbäck et al. 1999), which consequently become sensitized to noradrenaline. Indeed this observation provides one explanation for the blockade of NPY responses in renal proximal tubular cells (stimulation of Na^+ K^+ ATPase activity) by α-adrenoceptor antagonists (Ohtomo et al. 1994). The unmasking of silent receptor populations may thus provide a more general mechanism for interactions between Y receptors and other GPCRs, which does not depend on the overlapping signals generated by downstream cAMP or Ca^{2+} pathways.

3.3
Modulation of Ion Channels

As for many other GPCRs, the activation of Y receptors rapidly alters the gating of ion channels in both excitable and non-excitable tissues. In several instances these responses can be attributed to the cAMP inhibition and Ca^{2+} pathways which influence PKA and PKC. For example PKC can activate L type Ca^{2+} channels, a mechanism of Ca^{2+} entry which makes a significant contribution to NPY constrictor responses in some types of vascular smooth muscle (Tanaka et al. 1995). Similarly Y receptor-mediated inhibition of PKA provides a sufficient explanation for its antisecretory actions in epithelial cells, through PKA-activated apical Cl⁻ channels (the cystic fibrosis conductance regulator) and basolateral K^+ conductances (Bouritius et al. 1998).

In neurones NPY exerts its effects through GIRK class K^+ channels or presynaptic inhibition of N- and P/Q-type voltage gated Ca^{2+} channels. Y_1, Y_2 and Y_4 receptors can each demonstrate this capacity to couple to K^+ and Ca^{2+} channels in transfected HEK293 cells (Sun et al. 1998). Ewald et al. (1988) described at least two phases of Y receptor signalling to presynaptic Ca^{2+} channels. A slow second messenger mediated phase was prevented by long-term treatments, which depleted PKC but there was also a more rapid modulation enhanced by the inclusion of purified $G_o\alpha$ in the patch pipette. There is now clear evidence supporting a role for G proteins in the fast component of $G_{i/o}$ GPCR modulation of both N- and P/Q-Ca^{2+} channels and GIRKs. The direct allosteric modulation of channel gating appears to be mediated by released $G\beta\gamma$ subunits (reviewed in Dascal 2001). The role of $G_{i/o}\alpha$ is much less well defined, but it certainly determines signalling specificity by providing a pool of $G\beta\gamma$, which is released only on stimulation of an appropriate GPCR. Interestingly $G_{i/o}\alpha$ can interact with GIRK subunits directly, and with Ca^{2+} channels through the participation of the accessory protein syntaxin, and it has been suggested that these are anchoring points which ensure close targeting of $G\beta\gamma$ subunits to their binding sites (Dascal 2001). Indeed $G_{i/o}$ coupled GPCRs can also form stable complexes with GIRKs, for which $G\beta\gamma$ are required to initiate but not to maintain the interaction (Lavine et al. 2002). Thus it is possible that Y receptors are also organized in multi-protein signalling complexes, allowing rapid transmission of G proteins between receptor and effector over tightly controlled spatial domains. It remains to be seen whether because of this close proximity Y receptors could contribute directly to $G_{i/o}$ independent modulation of downstream signals (perhaps accounting for the few instances of pertussis-toxin insensitive NPY effects; e.g. Lynch et al. 1994), and if so whether such directly interacting motifs are subtype-specific.

4
Y Receptor Regulation and Trafficking

4.1
Desensitization and β-Arrestin Interactions

The signalling of prototypic GPCRs such as the β_2-adrenoceptor is rapidly inhibited after prolonged agonist exposure (Ferguson 2001), and there is convincing evidence that at least some Y receptor subtypes also undergo homologous desensitization. NPY pretreatment inhibits subsequent Y_1 vasoconstrictor responses in isolated arterioles (Van Riper and Bevan 1991) and its long-term infusion also reduces NPY and sympathetic cardiovascular responses in vivo (Moriarty et al. 1993). Similarly in cells expressing endogenous or transfected Y_1 receptors (Gicquiaux et al. 2002; Michel 1994), prior exposure to NPY profoundly reduces Ca^{2+} or cAMP responses to a second agonist challenge. The extent to which other Y receptors desensitize is less clear cut, and may depend on cell type. Repeated NPY stimulation of Y_2 receptors in SH-SY5Y cells results in reproducible Ca^{2+} responses over an extended period, in contrast to profound homologous desensitization of endogenous somatostatin receptors observed in the same cells (Connor et al. 1997). However in LN319 cells even threshold concentrations of NPY substantially decrease subsequent Y_2 receptor-mediated Ca^{2+} increases (Grouzmann et al. 2001). Human Y_4 receptors recombinantly expressed in CHO cells have also been reported to be resistant to desensitization (Voisin et al. 2000), but a long-term conditioning PP treatment was generally used (24 h, as opposed to 10–30 min typically used). Other studies have found rapid desensitization of rabbit and human Y_4 receptors (Cox et al. 2001; Feletou et al. 1999). More generally the use of paired agonist additions to assess peptide GPCR desensitization can be problematic, since the high affinity (and thus low dissociation rate) of peptide ligands make adequate intervening wash steps essential. An alternative is to follow the time course of the receptor response, but this is difficult in standard second messenger assays measuring cAMP accumulation or a transient Ca^{2+} signal. However measurement of anion secretion in epithelial cell monolayers provides a continuous readout of G_i coupled receptor activity that broadly reflects the time profile of cAMP inhibition. In this way we have investigated the functional responses to transfected and endogenous Y_1 and Y_4 receptors (Cox et al. 2001; Cox and Tough 1995; Holliday and Cox 1996), all of which inhibit vasoactive intestinal polypeptide responses in a more transient manner as the agonist concentration is increased (Fig. 3). These time profiles differ substantially from those of other endogenous G_i coupled receptors, and do not result from general compensatory mechanisms following an antisecretory stimulus; thus they provide a signature demonstrating Y_1 and Y_4 receptor desensitization in these cells.

The mechanisms underlying Y receptor inactivation have yet to be rigorously examined. Desensitization of HEL cell NPY Ca^{2+} calcium responses was enhanced by activators of PKC (Daniels et al. 1992; Michel 1994), and the size of the response could be increased by tyrosine kinase inhibitors such as herbimycin (Michel 1994). Possible sites of action of these kinases could include many downstream stages of the signalling pathway, but a role for PKC is feasible in specific desensitization of Y receptors (all of which contain appropriate phosphoacceptor recognition sites), at least in cells where they are coupled to the PLCβ pathway. The phosphorylation of GPCRs by downstream second messenger kinases is well recognized as a signal terminator: for the G_s coupled β_2-adrenoceptor PKA phosphorylation may even switch signalling to a G protein with opposing effects (G_i, a somewhat controversial finding; Lefkowitz et al. 2002).

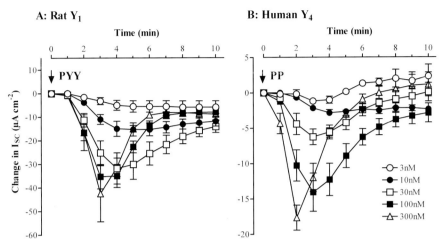

Fig. 3 Desensitization of Y_1 and Y_4 receptors in transfected epithelial cells. Epithelial layers stably expressing the rat Y_1 (HCA-7 Colony 1) or human Y_4 receptors (HT-29) were pretreated with the secretagogue vasoactive intestinal polypeptide (30 nM) to stimulate I_{SC}, followed by a single addition of PYY or PP. The time courses illustrated (n=3–8) demonstrate how the antisecretory responses to each receptor become more transient as the agonist concentration is increased

This type of desensitization derives certain key characteristics from the ability of PKA and PKC to phosphorylate both occupied and empty receptors. Because the amplification inherent in signalling pathways leads to kinase activation, GPCRs may become phosphorylated at low levels of agonist occupancy and thus threshold doses may be sufficient for full desensitization (as observed for Y_2 responses in LN319 cells; Grouzmann et al. 2001). In addition other unstimulated GPCRs may also be targeted, leading to heterologous desensitization (e.g. bradykinin inactivation of Y_2 receptors; Grouzmann et al. 2001), or Y_1 receptor inhibition of α_{2A}-adrenoceptor responses (Michel 1994).

In many cells and tissues where the predominant Y receptor response is an inhibition of cAMP (Gicquiaux et al. 2002; Holliday and Cox 1996) and second messenger kinases are not activated, desensitization may be initiated by the recruitment of GRKs to the plasma membrane through $G\beta\gamma$ subunits and their specific phosphorylation of activated receptors. In this case signal termination requires the subsequent binding of a member of the β-arrestin family member to restrict access to the receptor-G protein contact interface. Recently Berglund et al. (2003) confirmed the interaction of rhesus monkey Y receptor subtypes with β-arrestin 2, a promiscuous isoform with the ability to bind a broad spectrum of GPCRs. Using a FRET based approach they showed that Y_1 and Y_5 receptor–luciferase fusion proteins associate with GFP-tagged β-arrestin 2 rapidly ($t_{1/2}$ 3–5 min) on agonist stimulation in transfected HEK293 cells. However similar Y_2 and Y_4 receptor constructs exhibited a much lower FRET response [23% (Y_2) and 41% (Y_4) compared to Y_1] with slower kinetics ($t_{1/2}$ of 23 and 9 min respectively). Differential receptor association with β-arrestins thus provides one explanation for the emerging differences in desensitization and internalization (see below) of different Y receptor subtypes.

4.2
Two Ways of Signal Termination—The Roles of Phosphorylation and Palmitoylation

The rapid desensitization and β-arrestin binding exhibited by the Y_1 receptor suggests that its primary sequence contains unique intracellular motifs that are not shared by subtypes such as the Y_2 receptor. Candidates include the clustered organization of Ser/Thr amino acids in the C-terminal tail, since their phosphorylation could be the trigger for β-arrestin association. To investigate this possibility we have compared a haemagluttinin (HA) epitope-tagged rat Y_1 receptor with two truncated forms (T361STOP and S352STOP) that respectively remove four and eight of the 10 C-terminal Ser/Thr residues (N.D. Holliday and H.M. Cox, unpublished results). Both wild-type and the mutant HA-Y_1 receptors stimulated GTPγ^{35}S binding (a measurement of G protein activation; Holliday and Cox 2003) with similar potency and maximal response, using CHO clones with similar expression levels (Fig. 4). Thus truncation at or beyond Ser^{352} does not alter Y_1 receptor–G protein coupling. When CHO HA-Y_1 cells were pretreated with NPY (100 nM, 10 min) and extensively washed before being used in the GTPγ^{35}S assay, basal GTPγ^{35}S binding was not significantly increased compared to controls, nor was it inhibited to a greater extent by the Y_1 antagonist BIBO3304, confirming removal of the agonist. However, maximal GTPγ^{35}S responses to the second NPY exposure were reduced by 55%. A similar attenuation was observed in the HA-Y_1(T361*) clone, but not in pretreated HA-Y_1(S352*) membranes. In addition the S352* truncation substantially inhibited the NPY-induced phosphorylation of immunoprecipitated HA-Y_1 receptors (Fig. 4D). These studies suggest that the region between Ser^{352} and Thr^{361} may be a target for GRKs, and as such makes an important contribution to Y_1 receptor desensitization.

Because of its unique ability to radically alter the tertiary structure of the C-terminal tail, the dynamic palmitoylation of GPCRs has received considerable attention as a potential modulator of signalling (Kopin et al. 1988). The rat Y_1 receptor is acylated at Cys^{337} (Holliday and Cox 2003) and Cys^{337} mutation to Ser or Ala substantially inhibits receptor activation of G proteins in CHO membranes (Fig. 5A). Thus depalmitoylation (for example following agonist exposure) would be expected to inhibit Y_1 receptor signalling. However this 'desensitizing' effect is more subtle than traditional mechanisms involving phosphorylation. When expressed in epithelial cells, the differences in functional efficacy between Y_1 and Y_1(C337S) receptors are less pronounced because of the many intervening steps in the signalling pathway which amplify the stimulus. However in contrast with the transient Y_1 antisecretory responses, Y_1(C337S) receptors mediate sustained inhibition of anion secretion (Fig. 5B). This indicates that the agonist-occupied Y_1(C337S) receptor conformation is not only less able to cou-

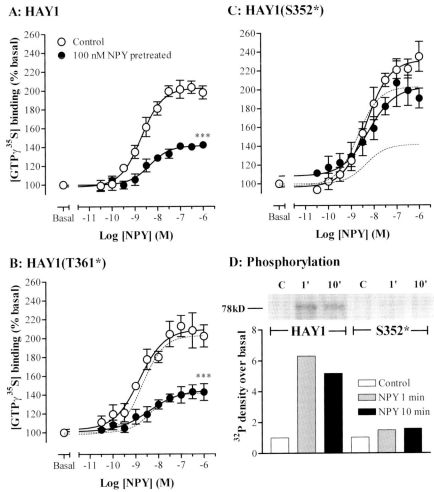

Fig. 4A–D Role of the Y_1 receptor C terminus in desensitization and phosphorylation. NPY GTPγ^{35}S responses were measured in CHO membranes stably expressing the HA-tagged Y_1 receptor ([^{125}I]PYY B_{max}: 5.4±1.5 pmol/mg, **A**), or the truncation mutants HA–Y_1(T361*) (B_{max}: 3.9±1.1 pmol/mg, **B**) and HA–Y_1(S352*) (B_{max}: 5.6±1.2 pmol/mg, **C**). Control concentration response curves were compared with those in which the cells had been pretreated with 100 nM NPY for 10 min (both n=4–5), with significant differences at 1 μM agonist indicated by ***P<0.001. In the example phosphorylation experiment in **D**, equal numbers of HA–Y_1 and HA–Y_1(S352*) receptors were immunoprecipitated from CHO clones labelled with 50 μCi ^{32}P$_i$, and resolved by SDS–PAGE. Increased radiolabelling of the 78-kDa receptor band, visualized by autoradiography of the dried gel (*upper panel*), was observed from CHO HA–Y_1 cell preparations pretreated with 1 μM NPY for 1 or 10 min, compared to unstimulated controls (**C**). Differences between the HA–Y_1 and HA–Y_1(S352*) responses are quantified in the histogram *below*, measuring relative density of the bands using Scion Image

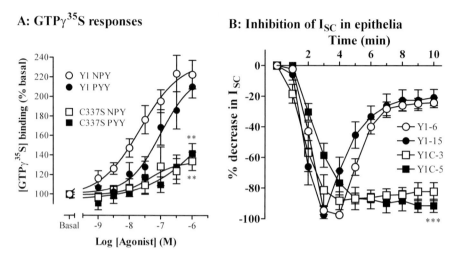

Fig. 5A, B Preventing palmitoylation reduces Y_1 receptor signalling efficacy. **A** NPY and PYY-stimulated GTPγ^{35}S binding in a CHO clone transfected with the Y_1 receptor was compared with responses to the non-palmitoylated Y_1(C337S) mutant expressed at similar levels (n=3–5), with significance at 1 µM agonist indicated by **$P<0.01$. **B** I_{SC} time courses to 100 nM PYY are shown for Colony 1 clones expressing the Y_1 (Y1-6, Y1-15) and Y_1(C337S) receptors (Y1C-3, Y1C-5), expressed as a percentage of the peak antisecretory response in VIP-stimulated epithelial layers (n=6–8). ***$P<0.001$. (Reproduced with permission from Holliday and Cox 2003)

ple to G proteins, but may also be resistant to phosphorylation by GRKs targeting the active state, in the same way that reduced desensitization is observed after partial agonist stimulation (Clark et al. 1999). By controlling receptor efficacy, palmitoylation may regulate the balance between G protein activation and events initiated by GRK phosphorylation and thereby profoundly influence longer term Y_1 receptor signalling.

4.3
Internalization

Many stimulated GPCRs are rapidly removed in vesicles from the plasma membrane, a process known as internalization (Ferguson 2001). Originally this was viewed as a mechanism of desensitization, as the sequestered receptors were no longer available to agonist. However the onset of desensitization is often more rapid than internalization and receptors may still be uncoupled from G proteins in the presence of internalization inhibitors (Pippig et al. 1995), suggesting that earlier events (e.g. phosphorylation and β-arrestin-binding) are more important in this respect. However internalization continues to be of interest in GPCR regulation as the first stage in a sorting pathway that may recycle receptors to the cell surface after the removal of the agonist, or target them for degradation ('down-regulation') in lyzosomes. Although several receptor internalization pathways have been described, the best characterized and most widespread for class A GPCRs involves the formation of a complex between receptor-bound β-arrestin and clathrin, a scaffolding protein whose cage-like multimers shape the endocytotic 'coated' vesicles. Clathrin-dependent endocytosis requires additional adaptor proteins (e.g. AP-2, which may interact with GPCR motifs (NPXXY, YXXΦ; Fig. 1) and the small G protein dynamin (Ferguson 2001).

Multiple complementary studies indicate that Y_1 receptors undergo internalization through the β-arrestin-clathrin pathway. Ultrastructural immunochemical localization of rat Y_1 receptors in dorsal root ganglion neurones clearly identifies receptor proteins associated with clathrin-coated vesicles and early endosomes (Zhang et al. 1999). Using confocal microscopy Fabry et al. (2000) found that Y_1 receptors in SK-N-MC cells rapidly clustered in response to stimulation with carboxy-fluorescein NPY, and temperature-dependent internalization of both radiolabelled and fluorescent ligands was observed in more conventional experiments which measured the increase in acid-resistant NPY binding to whole cells. In CHO cells stably expressing the cloned guinea-pig Y_1 receptor, internalization of $[^{125}I]$NPY is significantly greater than for $[^{125}I]$PYY (at peak, 10 min after agonist addition at 37°C; Parker et al. 2001, 2002b). $[^{125}I]$NPY sequestration was inhibited by agents that prevent clathrin-mediated endocytosis, phenylarsine oxide and hypertonic sucrose, and also by filipin III, a cholesterol chelating compound known for its disruptive influence on plasma membrane microdomains such as caveolae (Parker et al. 2002b).

The use of radiolabelled or fluorescent agonists has a disadvantage in that it is impossible to observe the unstimulated receptor. Gicquiaux et al. (2002) have addressed this problem by constructing a human Y_1 receptor tagged at the N-terminus with GFP (to avoid hindering potentially important C-terminal interactions). GFP–Y_1 receptors were observed at the cell surface of stably transfected HEK293 cells, and were partially redistributed to intracellular vesicles on stimulation with NPY. These vesicles could also be labelled with transferrin-Texas Red, a marker for clathrin dependent endocytosis, in accordance with a previous investigation in dorsal root ganglion neurones demonstrating partial colocalization of internalized transferrin and Y_1 receptors (Zhang et al. 1999). Interestingly no colocalization between GFP–Y_1 receptors and a lyzosomal marker (Lysotracker Red) was detected by confocal microscopy for up to 60 min after NPY stimulation, suggesting that the GFP–Y_1 protein was not sorted over this period to a degradative pathway. Gicquiaux et al. (2002) also made use of the pH sensing properties of GFP, whose fluorescence is inhibited in a more acidic environment. NPY, PYY and Pro^{34} substituted agonists (but not PP) decreased GFP fluorescence measured in cell suspensions (reaching a maximum 3–4 min after agonist addition at 37°C), an effect which was sensitive to Y_1 antagonists and pertussis toxin. This was interpreted as the transfer of the N-terminal GFP from an extracellular (pH 7.4) to an early endosomal (pH 6.0) compartment, as collapsing the pH gradient in endosomes (using monensin) or blocking internalization using 0.4 M sucrose or concanavalin A all inhibited the response. Moreover the subsequent recovery of the fluorescence intensity was suggested to result from fast trafficking of GFP–Y_1 receptors to recycling endo-

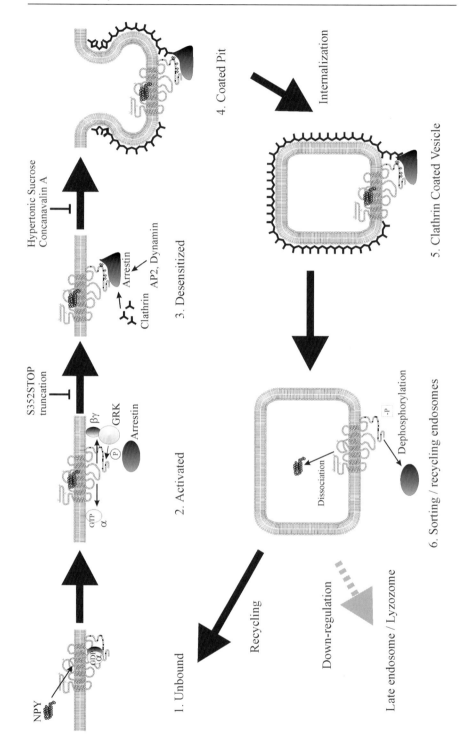

somes (pH 6.5) prior to their return to the cell surface. A mathematical model of the response over 30 min, used time constants for internalization and recycling very similar to those obtained for the transferrin receptor. Together the GFP–Y_1 experiments suggest that activated Y_1 receptors are internalized via a clathrin-dependent pathway, but may be rapidly recycled rather than degraded over the short term (Fig. 6). The only study to date that has investigated antagonist- (the homodimeric peptide GR 231118) and agonist-induced Y_1 receptor sequestration and internalization is that of Pheng et al. (2003). Using stably transfected HEK293 cells they observed internalization of both [^{125}I]GR 231118 and [^{125}I]Pro^{34}PYY and subsequently confirmed that the agonist induced receptor endocytosis and recycling via endosomes (a monensin, and pertussis toxin-sensitive mechanism) however, the antagonist also induced sequestration and in contrast this was long-lasting, toxin insensitive (G-protein independent) and only partially sensitive to hypertonic sucrose. Immunofluorescence labelling indicated subtly different patterns of Y_1 receptor intracellular localization following agonist (clustered and potentially pericentriolar endosomal labelling) or antagonist treatment (relatively diffuse cytoplasmic Y_1 staining). Thus Pheng et al. (2003) concluded that GR 231118 induced Y_1 receptor sequestration via a non-recycling (monensin-insensitive) pathway that was distinct from the clathrin dependent, agonist-induced Y_1 internalization.

Surprisingly no fluorescence decreases were observed using a GFP–Y_2 construct (Gicquiaux et al. 2002), consistent with the low internalization of [^{125}I]NPY in CHO cells expressing the guinea-pig Y_2 receptor (Parker et al. 2001) and the slow interaction of the rhesus monkey subtype with β-arrestin 2 (Berglund et al. 2003). Human Y_4 receptors tagged at the C terminus also fail to redistribute from the plasma membrane after 24 h' pretreatment with PP (Voisin et al. 2000), but this finding is more controversial since [^{125}I]PP is effectively and rapidly sequestered by rat Y_4 receptors in the same cell line (CHO; Parker et al. 2001, 2002a). Although more detailed investigations are required, the differences observed between Y_1 and Y_2 receptors are of particular interest. Despite similar selectivity for endogenous ligands and second messenger responses, these subtypes might be distinguished by the way in which they engage the trafficking machinery, with broader consequences for signalling mechanisms which use β-arrestins as a focal point.

Fig. 6 A model of Y_1 receptor internalization and recycling. Activation of the Y_1 receptor also promotes its phosphorylation and desensitization through β-arrestin binding, prevented in a truncation mutant (S352STOP) lacking the phosphoacceptor sites. The bound β-arrestin recruits the components required for internalization, resulting in clathrin-dependent endocytosis (blocked by hypertonic sucrose, concanavalin A). The results of Gicquiaux et al. (2002) suggest that the Y_1 receptor is rapidly recycled through sorting endosomes, in which dephosphorylation may occur, together with dissociation of the ligand. So far there is little evidence for a degradative pathway targeting the Y_1 receptor to lyzosomes

5
Long-Term Signalling by Y Receptors

5.1
Activation of Extracellular Signal Related Kinases

The rapid modulation of second messenger release and channel activity, followed by receptor inactivation processes, does not account for the full repertoire of GPCR signalling pathways. GPCR signals also integrate with mitogen-activated protein kinase (MAPK) cascades at numerous levels, including G proteins, second messenger-activated protein kinases and the β-arrestins (for a review, see Pierce et al. 2001). These cascades phosphorylate cytoplasmic and nuclear proteins to control gene expression and cell proliferation and differentiation, and thus provide a means by which GPCRs may influence long-term events traditionally ascribed to receptor tyrosine kinases such as the epidermal growth factor (EGF) receptor.

Y_1, Y_2, Y_4 and Y_5 receptors in transfected CHO cells all increase MAPK activity in immunoblotting assays for phosphorylated extracellular signal-related kinases 1/2 (ERK; Mannon and Raymond 1998; Mullins et al. 2002; Nakamura et al. 1995; Nie and Selbie 1998). In CHO Y_1 cells, the transient increase in ERK activation involves the Ras–Raf–MAPK kinase (MEK)–ERK cascade (Fig. 7), since PYY stimulates GTP binding to Ras and ERK activation can be inhibited by introduction of dominant negative mutant Ras and Raf proteins (Mannon and Raymond 1998). It is assumed that these downstream components also participate in responses to other Y receptors, although so far only a more general

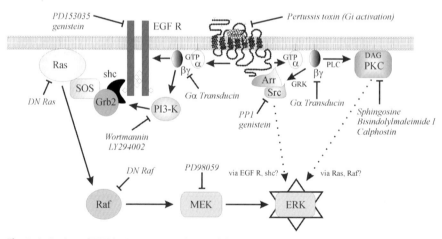

Fig. 7 Activation of ERK by Y receptors. This model (adapted from Pierce et al. 2001) illustrates the potential sites of action of various inhibitors of Y receptor MAPK signalling. The roles of the EGF receptor (EGF R, inhibited by PD153035) and Src (blocked by PP1) have so far been investigated only in IEC-6 cells transfected with the Y_1 receptor (Mannon and Mele 2000). *Arr*, β-arrestin; *DN*, dominant negative; *GRK*, G protein coupled receptor kinase

role of MEK has been demonstrated using the inhibitor PD98059 (Mannon and Raymond 1998; Mullins et al. 2002; Nie and Selbie 1998). MAPK signals have also been measured in endogenous Y_1 (Hansel et al. 2001; Keffel et al. 1999) and Y_5 receptor systems (Pellieux et al. 2000), but they are cell type dependent (for example, Y_5 receptors were unable to activate any MAPK in human endometrial carcinoma cells; Bischoff et al. 2001). In addition, Keffel et al. (1999) demonstrated that NPY activated ERK but not other MAPKs (p38, N-terminal Jun kinase) in HEL cells, suggesting that Y_1 receptor signalling can be tailored to a particular protein kinase cascade.

ERK activation by Y receptors depends on functional $G_{i/o}$ proteins and is prevented by pertussis toxin (Keffel et al. 1999; Mannon and Mele 2000; Mannon and Raymond 1998; Mullins et al. 2002). Free $G\beta\gamma$ subunits are crucial for Y_1 receptor signalling in transfected IEC-6 cells, since their sequestration with transducin $G\alpha$ prevented PYY- but not EGF-stimulated ERK activity (Mannon and Mele 2000). Downstream targets of $G\beta\gamma$ include phophoinositide-3-kinase (PI3-K; Hawes et al. 1996), and the PI3-K inhibitors wortmannin and LY294002 can partially suppress CHO Y_1 receptor ERK responses (Keffel et al. 1999; Mannon and Raymond 1998; Mullins et al. 2002; Nie and Selbie 1998). Mannon and Raymond (1998) also observed that activation of PKC and its effects on Raf or Ras were required in addition to the PI3-K Ras pathway. Their experiments prevented Y_1 receptor-stimulation of ERK in CHO cells by using PKC inhibitors or down-regulating the enzyme through long-term phorbol ester treatment. The specificity of these interventions has been questioned (Keffel et al. 1999) and in addition a permissive effect of basal PKC activity on the Ras–MAPK pathway cannot be excluded, particularly as NPY stimulation of PKC is difficult to detect in CHO hY_1 cells (Selbie et al. 1995). However in the IEC-6 cells Y_1 receptors stimulate transient translocation of PKC isoform ε from the cytosol to the membrane before peak ERK activation, and both processes are blocked by the PKC inhibitor sphingosine (Mannon and Mele 2000). Here the PKC mechanism may predominate over PI3-K-dependent events, since wortmannin and LY294002 are without effect. PKC has also been implicated in ERK signalling from transfected Y_2, Y_4 and Y_5 receptors (Mullins et al. 2002) and in endogenous cell systems. Y_5 receptors activated PKC in mouse cardiomyocytes, and PKC down-regulation abolished the resulting phosphorylation of ERK, p38 and N-terminal Jun kinase (Pellieux et al. 2000). Similarly, the chemically distinct PKC inhibitors calphostin and bisindolylmaleimide I, completely blocked Y_1 receptor-mediated MAPK activation in primary neuronal cultures (Hansel et al. 2001).

MAPK responses induced by native and transfected Y_1 receptors are also sensitive to tyrosine kinase inhibitors such as genistein (Hansel et al. 2001; Mannon and Mele 2000; Mannon and Raymond 1998). Interestingly more specific blockers suggest that PYY-stimulation of ERK in IEC6 cells requires autophosphorylation of the EGF receptor and downstream involvement of the cytoplasmic tyrosine kinase Src (Mannon and Mele 2000). One recent model of GPCR ERK responses (reviewed in Pierce et al. 2001 and depicted for the Y_1 receptor in Fig. 7) suggests that the transactivated EGF receptor (perhaps stimulated by release of a tethered EGF ligand) recruits the Shc-Grb2-SOS adapter complex linked to Ras, in addition to contributing to PKC activation through PLCγ. These elements would also serve as targets for convergent GPCR signals through PI3-K and Src, brought to the complex by its interaction with β-arrestins (Luttrell et al. 1999; Fig. 7). This first description of a role for receptor and cytoplasmic tyrosine kinases in Y_1 receptor signalling is therefore consistent with the emerging view of their participation in GPCR–MAPK responses.

5.2
Control of Gene Expression

Y receptor MAPK signalling provides one mechanism by which NPY and related peptides may exert long lasting effects on cells and tissues. In neuronal cells, they increase the expression of immediate early genes which encode transcription factors such as c-Fos (e.g. in hypothalamus; Batterham et al. 2002; Yokosuka et al. 1999). This may contribute to changes in plasticity by altering neuropeptide synthesis (Batterham et al. 2002; Sarkar and Lechan 2003) and influence the rhythmic control of circadian clock genes (Maywood et al. 2002). Similarly, epithelial PYY responses include long-term up-regulation of proteins involved in fat absorption (fatty acid binding protein; Halldén and Aponte 1997) and differentiation (such as the cytoskeletal associated protein CD63; Halldén et al. 1999). In addition, protein synthesis is increased following NPY stimulation of cardiac myocytes, resulting in hypertrophy (Goldberg et al. 1998). PYY and NPY are also proliferative for epithelial cells (Mannon 2002; Mannon and Mele 2000; Voisin et al. 1993), vascular smooth muscle (Zukowska-Grojec et al. 1998) and neuronal precursors (Hansel et al. 2001). Indeed NPY released by the olfactory epithelium is critical for the development of the full complement of olfactory neurones in adulthood (Hansel et al. 2001).

Determination of cell fate can be critically dependent on the time profile of ERK phosphorylation (Kao et al. 2001). Recent studies have also highlighted the increasing complexity in the way different GPCRs control the spatial localization of MAPK signalling, allowing specific targeting of cytoplasmic or nuclear ERK activity (Pierce et al. 2001). A key factor in this control is the length of time the GPCR remains associated with β-arrestin, which provides the scaffold for Src and other MAPK signalling proteins (e.g. c-Jun N-terminal kinase; Pierce et al. 2001). Thus future analysis of Y receptor MAPK activation may yet reveal significant differences between the subtypes (particularly the Y_1 and Y_2 receptors), with important consequences for the way in which they regulate long-term cellular responses.

6
Conclusions

There have been exciting developments in the field of GPCR signalling in recent years; from the concept that these receptors can act as dimers in tightly spatially organized complexes, to the mechanisms of their regulation and an increased awareness that proteins which terminate second messenger responses may in fact act as springboards for a new wave of longer term signalling. Armed with the molecular structure of the cloned subtypes, these advances represent an opportunity to probe Y receptor function in much greater detail using the latest techniques, as recent studies on dimerization (Dinger et al. 2003) and internalization (Gicquiaux et al. 2002; Berglund et al. 2003; Pheng et al. 2003) have illustrated. Future investigations could address the intriguing differences between the individual Y receptor intracellular domains which interact with G proteins and other signalling components, the potential for agonist trafficking (all Y receptors bind more than one endogenous peptide) and the complex pharmacology that may result from heterodimerization. In addition, the current evidence suggesting differential desensitization and internalization of Y receptors is an observation that has potentially far reaching consequences for the extent and type of signalling pathways they engage. By clearly defining the manner in which different Y receptors are regulated, our understanding of the individual physiological roles of each subtype will be significantly enhanced.

References

Bader R, Bettio A, Beck-Sickinger AG, Zerbe O (2001) Structure and dynamics of micelle-bound neuropeptide Y: comparison with unligated NPY and implications for receptor selection. J Mol Biol 305:307–329

Bard JA, Walker MW, Branchek TA, Weinshank RL (1995) Cloning and functional expression of a human Y_4 subtype receptor for pancreatic polypeptide, neuropeptide Y and peptide YY. J Biol Chem 270:26762–26765

Batterham RL, Cowley MA, Small CJ, Herzog H, Cohen MA, Dakin CL, Wren AM, Brynes AE, Low MJ, Ghatei MA, Cone RD, Bloom SR (2002) Gut hormone PYY(3–36) physiologically inhibits food intake. Nature 418:650–654

Berglund MM, Lundell I, Eriksson H, Söll R, Beck-Sickinger AG, Larhammar D (2001) Studies of the human, rat, and guinea pig Y_4 receptors using neuropeptide Y analogues and two distinct radioligands. Peptides 22:351–356

Berglund MM, Schober DA, Statnick MA, Mcdonald PH, Gehlert DR, Larhammar D (2003) The use of bioluminescence resonance energy transfer (BRET2) to study neuropeptide Y receptor agonist Induced β-arrestin 2 interaction. J Pharmacol Exp Therap: 306:147–156

Bischoff A, Püttmann K, Kötting A, Muscr C, Buschauer A, Michel MC (2001) Limited signal transduction repertoire of human Y_5 neuropeptide Y receptors expressed in HEC-1B cells. Peptides 22:387–394

Bouritius H, Oprins JC, Bindels RJ, Hartog A, Groot JA (1998) Neuropeptide Y inhibits ion secretion in intestinal epithelium by reducing chloride and potassium conductance. Pflugers Arch 435:219–226

Bouvier M (2001) Oligomerization of G-protein-coupled transmitter receptors. Nature Rev Neurosci 2:274–286

Brown NA, McAllister G, Weinberg D, Milligan G, Seabrook GR (1995) Involvement of G-protein alpha i1 subunits in activation of G-protein gated inward rectifying K^+ channels by human NPY1 receptors. Br J Pharmacol 116:2346–2348

Brzostowski JA, Kimmel AR (2001) Signaling at zero G: G-protein-independent functions for 7-TM receptors. Trends Biochem Sci 26:291–297

Cavadas C, Silva AP, Mosimann F, Cotrim MD, Ribeiro CA, Brunner HR, Grouzmann E (2001) NPY regulates catecholamine secretion from human adrenal chromaffin cells. J Clin Endocrinol Metab 86:5956–5963

Clark RB, Knoll BJ, Barber R (1999) Partial agonists and G protein-coupled receptor desensitization. Trends Pharmacol Sci 20:279–286

Connor M, Yeo A, Henderson G (1997) Neuropeptide Y Y_2 receptor and somatostatin sst_2 receptor coupling to mobilization of intracellular calcium in SH-SY5Y human neuroblastoma cells. Br J Pharmacol 120:455–463

Cox HM, Tough IR (1995) Functional characterization of receptors with affinity for PYY, NPY, [Leu^{31}, Pro^{34}] NPY and PP in a human colonic epithelial cell line. Br J Pharmacol 116:2673–2678

Cox HM, Tough IR, Zandvliet DW, Holliday ND (2001) Constitutive neuropeptide Y Y_4 receptor expression in human colonic adenocarcinoma cell lines. Br J Pharmacol 132:345–353

Daniels AJ, Matthews JE, Humberto VO, Lazarowski ER (1992) Characterization of the neuropeptide Y-induced intracellular calcium release in human erythroleukemic cells. Mol Pharmacol 41:767–771

Daniels AJ, Matthews JE, Slepetis RJ, Jansen M, Viveros OH, Tadepalli A, Harrington W, Heyer D, Landavazo A, Leban JJ, Spaltenstein A (1995) High-affinity neuropeptide Y receptor antagonists. Proc Natl Acad Sci USA 92:9067–9071

Dascal N (2001) Ion-channel regulation by G proteins. Trends Endocrinol Metab 12:391–398

Dinger MC, Bader JE, Kóbor AD, Kretzschmar AK, Beck-Sickinger AG (2003) Homodimerization of neuropeptide Y receptors investigated by fluorescence resonance energy transfer in living cells. J Biol Chem

Entzeroth M, Braunger H, Eberlein W, Engel W, Rudolf K, Wienen W, Wieland HA, Willim KD, Doods HN (1995) Labeling of neuropeptide Y receptors in SK-N-MC cells using the novel, nonpeptide Y_1 receptor-selective antagonist [^3H]BIBP3226. Eur J Pharmacol 278:239–242

Eriksson H, Berglund MM, Holmberg SK, Kahl U, Gehlert DR, Larhammar D (1998) The cloned guinea pig pancreatic polypeptide receptor Y_4 resembles more the human Y_4 than does the rat Y_4. Regul Pept 75–76:29–37

Ewald DA, Sternweis PC, Miller RJ (1988) Guanine nucleotide-binding protein G_o-induced coupling of neuropeptide Y receptors to Ca^{2+} channels in sensory neurons. Proc Natl Acad Sci USA 85:3633–3637

Exton JH (1996) Regulation of phosphoinositide phospholipases by hormones, neurotransmitters, and other agonists linked to G proteins. Annu Rev Pharmacol Toxicol 36:481–509

Fabry M, Langer M, Rothen-Rutishauser B, Wunderli-Allenspach H, Höcker H, Beck-Sickinger AG (2000) Monitoring of the internalization of neuropeptide Y on neuroblastoma cell line SK-N-MC. Eur J Biochem 267:5631–5637

Feletou M, Nicolas JP, Rodriguez M, Beauverger P, Galizzi JP, Boutin JA, Duhault J (1999) NPY receptor subtype in the rabbit isolated ileum. Br J Pharmacol 127:795–801

Ferguson SSG (2001) Evolving concepts in G protein-coupled receptor endocytosis: the role in receptor desensitization and signaling. Pharmacol Rev 53:1–24

Feth F, Rascher W, Michel MC (1992) Neuropeptide Y (NPY) receptors in HEL cells: comparison of binding and functional parameters for full and partial agonists and a non-peptide antagonist. Br J Pharmacol 105:71–76

Freitag C, Svendsen AB, Feldthus N, Lossl K, Sheikh SP (1995) Coupling of the human Y_2 receptor for neuropeptide Y and peptide YY to guanine nucleotide inhibitory proteins in permeabilized SMS-KAN cells. J Neurochem 64:643–650

Fuhlendorff J, Gether U, Aakerlund L, Langeland-Johansen N, ThØgersen H, Melberg SG, Olsen UB, Thastrup O, Schwartz TW (1990) [Leu^{31}, Pro^{34}] Neuropeptide Y: a specific Y_1 receptor agonist. Proc Natl Acad Sci USA 87:182–186

Gerald C, Walker MW, Criscione L, Gustafson EL, Batzl-Hartmann C, Smith KE, Vaysse PJJ, Durkin MM, Laz TM, Linemeyer DL, Schaffhauser AO, Whitebread S, Hofbauer KG, Taber RI, Branchek TA, Weinshank RL (1996) A receptor subtype involved in neuropeptide-Y-induced food intake. Nature 382:168–171

Gerald C, Walker MW, Vaysse PJJ, He C, Branchek TA, Weinshank RL (1995) Expression cloning and pharmacological characterization of a human hippocampal neuropeptide Y / peptide YY Y_2 receptor subtype. J Biol Chem 270:26758–26761

Gicquiaux H, Lecat S, Gaire M, Dieterlen A, Mely Y, Takeda K, Bucher B, Galzi JL (2002) Rapid internalization and recycling of the human neuropeptide Y Y_1 receptor. J Biol Chem 277:6645–6655

Glaum SR, Miller RJ, Rhim H, Maclean D, Georgic LM, MacKenzie RG, Grundemar L (1997) Characterization of Y_3 receptor-mediated synaptic inhibition by chimeric neuropeptide Y-peptide YY peptides in the rat brainstem. Br J Pharmacol 120:481–487

Goldberg Y, Taimor G, Piper HM, Schluter KD (1998) Intracellular signaling leads to the hypertrophic effect of neuropeptide Y. Am J Physiol 275: C1207–C1215

Goumain M, Voisin T, Lorinet AM, Ducroc R, Tsocas A, Roze C, Rouet-Benzineb P, Herzog H, Balasubramaniam A, Laburthe M (2001) The peptide YY-preferring receptor mediating inhibition of small intestinal secretion is a peripheral Y_2 receptor: pharmacological evidence and molecular cloning. Mol Pharmacol 60:124–134

Gregor P, Feng Y, DeCarr LB, Cornfield LJ, McCaleb ML (1996a) Molecular characterization of a second mouse pancreatic polypeptide receptor and its inactivated human homologue. J Biol Chem 271:27776–27781

Gregor P, Millham ML, Feng Y, DeCarr LB, McCaleb ML, Cornfield LJ (1996b) Cloning and characterization of a novel receptor to pancreatic polypeptide, a member of the neuropeptide Y receptor family. FEBS Lett 381:58–62

Grouzmann E, Meyer C, Burki E, Brunner H (2001) Neuropeptide Y Y_2 receptor signalling mechanisms in the human glioblastoma cell line LN319. Peptides 22:379–386

Halldén G, Aponte GW (1997) Evidence for a role of the gut hormone PYY in the regulation of intestinal fatty acid-binding protein transcripts in differentiated subpopulations of intestinal epithelial cell hybrids. J Biol Chem 272:12591–12600

Halldén G, Hadi M, Hong HT, Aponte GW (1999) Y receptor-mediated induction of CD63 transcripts, a tetraspanin determined to be necessary for differentiation of the intestinal epithelial cell line, hBRIE 380i cells. J Biol Chem 274:27914–27924

Hansel DE, Eipper BA, Ronnett GV (2001) Neuropeptide Y functions as a neuroproliferative factor. Nature 410:940–944

Herzog H, Darby K, Ball HJ, Hort YJ, Beck-Sickinger AG, Shine J (1997) Overlapping gene structure of the human neuropeptide Y receptor subtypes Y_1 and Y_5 suggests coordinate transcriptional regulation. Genomics 41:315–319

Herzog H, Hort YJ, Ball HJ, Hayes G, Shine J, Selbie LA (1992) Cloned human neuropeptide Y receptor couples to two different messenger systems. Proc Natl Acad Sci USA 89:5794–5798

Holliday ND, Cox HM (1996) The functional investigation of a human adenocarcinoma cell line, stably transfected with the neuropeptide Y Y_1 receptor. Br J Pharmacol 119:321–329

Holliday ND, Cox HM (2003) Control of signalling efficacy by palmitoylation of the rat Y_1 receptor. Br J Pharmacol 139:510–512

Holst B, Hastrup H, Raffetseder U, Martini L, Schwartz TW (2001) Two active molecular phenotypes of the tachykinin NK_1 receptor revealed by G-protein fusions and mutagenesis. J Biol Chem 276:19793–19799

Holtbäck U, Brismar H, DiBona GF, Fu M, Greengard P, Aperia A (1999) Receptor recruitment: a mechanism for interactions between G protein-coupled receptors. Proc Natl Acad Sci USA 96:7271–7275

Kao S, Jaiswal RK, Kolch W, Landreth GE (2001) Identification of the mechanisms regulating the differential activation of the MAPK cascade by epidermal growth factor and nerve growth factor in PC12 cells. J Biol Chem 276:18169–18177

Keffel S, Schmidt M, Bischoff A, Michel MC (1999) Neuropeptide-Y stimulation of extracellular signal-regulated kinases in human erythroleukemia cells. J Pharmacol Exp Ther 291:1172–1178

Kenakin T (1996) The classification of seven transmembrane receptors in recombinant expression systems. Pharmacol Rev 48:413–463

Kopin AS, Toder AE, Leiter AB (1988) Different splice site utilization generates diversity between the rat & human pancreatic polypeptide precursors. Arch Biochem Biophys 267:742–748

Lavine N, Ethier N, Oak JN, Pei L, Liu F, Trieu P, Rebois RV, Bouvier M, Hebert TE, Van Tol HH (2002) G protein-coupled receptors form stable complexes with inwardly rectifying potassium channels and adenylyl cyclase. J Biol Chem 277:46010–46019

Lefkowitz RJ, Pierce KL, Luttrell LM (2002) Dancing with different partners: protein kinase a phosphorylation of seven membrane-spanning receptors regulates their G protein-coupling specificity. Mol Pharmacol 62:971–974

Liggett SB, Ostrowski J, Chesnut LC, Kurose H, Raymond JR, Caron MG, Lefkowitz RJ (1992) Sites in the third intracellular loop of the α_{2A}-adrenergic receptor confer short term agonist-promoted desensitization. J Biol Chem. 267:4740–4746

Limbird LE (1988) Receptors linked to inhibition of adenylate cyclase: additional signaling mechanisms. FASEB J 2:2686–2695

Lu ZL, Saldanha JW, Hulme EC (2002) Seven-transmembrane receptors: crystals clarify. Trends Pharmacol Sci 23:140–146

Lynch JW, Lemos VS, Bucher B, Stoclet J-C, Takeda K (1994) A pertussis toxin-insensitive calcium influx mediated by neuropeptide Y_2 receptors in a human neuroblastoma cell line. J Biol Chem 269:8226–8233

Mannon PJ (2002) Peptide YY as a growth factor for intestinal epithelium. Peptides 23:383–388

Mannon PJ, Mele JM (2000) Peptide YY Y_1 receptor activates mitogen-activated protein kinase and proliferation in gut epithelial cells via the epidermal growth factor receptor. Biochem J 350:655–661

Mannon PJ, Mervin SJ, Sherrif-Carter KD (1994) Characterization of a Y_1-preferring NPY/PYY receptor in HT-29 cells. Am J Physiol 267: G901–G907

Mannon PJ, Raymond JR (1998) The neuropeptide Y/peptide YY Y_1 receptor is coupled to MAP kinase via PKC and Ras in CHO cells. Biochem Biophys Res Commun 246:91–94

Maywood ES, Okamura H, Hastings MH (2002) Opposing actions of neuropeptide Y and light on the expression of circadian clock genes in the mouse suprachiasmatic nuclei. Eur J Neurosci 15:216–220

Michel MC (1994) Rapid desensitization of adrenaline- and neuropeptide Y-stimulated Ca^{2+} mobilization in HEL-cells. Br J Pharmacol 112:499–504

Michel MC (2002) Neuropeptide Y receptor family. In: Ruffolo RR, Spedding M (eds) The IUPHAR Compendium of Receptor Characterization and Classification, 2nd edition, IUPHAR Media Ltd, London, pp 278–289

Michel MC (1998) Concomitant regulation of Ca^{2+} mobilization and G_{i3} expression in human erythroleukemia cells. Eur J Pharmacol 348:135–141

Michel MC, Beck-Sickinger AG, Cox HM, Doods HN, Herzog H, Larhammar D, Quirion R, Schwartz TW, Westfall T (1998) XVI. international union of pharmacology recommendations for the nomenclature of neuropeptide Y, peptide YY and pancreatic polypeptide receptors. Pharmacol Rev 50:143–150

Mihara S, Shigeri Y, Fujimoto M (1989) Neuropeptide Y-induced intracellular Ca^{2+} increases in vascular smooth muscle cells. FEBS Lett 259:79–82

Moriarty M, Potter EK, McCloskey DI (1993) Desensitization by neuropeptide Y of effects of sympathetic stimulation on cardiac vagal action in anaesthetized dogs. J Auton Nerv Syst 45:21–28

Motulsky HJ, Michel MC (1988) Neuropeptide Y mobilizes Ca^{2+} and inhibits adenylate cyclase in human erythroleukemia cells. Am J Physiol 255: E880–E885

Mousli M, Trifilieff A, Pelton JT, Gies J-P, Landry Y (1995) Structural requirements for neuropeptide Y in mast cell and G protein activation. Eur J Pharmacol 289:125–133

Mullins DE, Guzzi M, Xia L, Parker EM (2000) Pharmacological characterization of the cloned neuropeptide Y y_6 receptor. Eur J Pharmacol 395:87–93

Mullins DE, Zhang X, Hawes BE (2002) Activation of extracellular signal regulated protein kinase by neuropeptide Y and pancreatic polypeptide in CHO cells expressing the NPY Y_1, Y_2, Y_4 and Y_5 receptor subtypes. Regul Pept 105:65–73

Nakamura M, Sakanaka C, Aoki Y, Ogasawara H, Tsuji T, Kodama H, Matsumoto T, Shimizu T, Noma M (1995) Identification of two isoforms of mouse neuropeptide Y-Y_1 receptor generated by alternative splicing. J Biol Chem 270:30102–30110

Nie M, Selbie LA (1998) Neuropeptide Y Y_1 and Y_2 receptor-mediated stimulation of mitogen-activated protein kinase activity. Regul Pept 75/76:207–213

Ohtomo Y, Meister B, Hökfelt T, Aperia A (1994) Coexisting NPY and NE synergistically regulate renal tubular Na^+, K^+-ATPase activity. Kidney Int 45:1606–1613

Okamoto T, Nishimoto I (1992) Detection of G protein activator regions in M_4 subtype muscarinic, cholinergic and α_2-adrenergic receptors based upon characteristics in primary structure. J Biol Chem 267:8342–8346

Palczewski K, Kumasaka T, Hori T, Behnke CA, Motoshima H, Fox BA, Le Trong I, Teller DC, Okada T, Stenkamp RE, Yamamoto M, Miyano M (2000) Crystal structure of rhodopsin: A G protein-coupled receptor. Science 289:739–745

Parker MS, Lundell I, Parker SL (2002a) Internalization of pancreatic polypeptide Y_4 receptors: correlation of receptor intake and affinity. Eur J Pharmacol 452:279–287

Parker SL, Kane JK, Parker MS, Berglund MM, Lundell IA, Li MD (2001) Cloned neuropeptide Y (NPY) Y_1 and pancreatic polypeptide Y_4 receptors expressed in Chinese hamster ovary cells show considerable agonist-driven internalization, in contrast to the NPY Y2 receptor. Eur J Biochem 268:877–886

Parker SL, Parker MS, Lundell I, Balasubramaniam A, Buschauer A, Kane JK, Yalcin A, Berglund MM (2002b) Agonist internalization by cloned Y_1 neuropeptide Y (NPY) receptor in Chinese hamster ovary cells shows strong preference for NPY, endosome-linked entry and fast receptor recycling. Regul Pept 107:49–62

Pellieux C, Sauthier T, Domenighetti A, Marsh DJ, Palmiter RD, Brunner HR, Pedrazzini T (2000) Neuropeptide Y (NPY) potentiates phenylephrine-induced mitogen-activated protein kinase activation in primary cardiomyocytes via NPY Y_5 receptors. Proc Natl Acad Sci USA 97:1595–1600

Pheng LH, Perron A, Quirion R, Cadieux A, Fauchere JL, Dumont Y, Regoli D (1999) Neuropeptide Y-induced contraction is mediated by neuropeptide Y Y_2 and Y_4 receptors in the rat colon. Eur J Pharmacol 374:85–91

Pheng LH, Dumont Y, Fournier A, Chabot J-G, Beaudet A, Quirion R (2003) Agonist- and antagonist-induced sequestration/internalization of neuropeptide Y Y_1 receptors in HEK293 cells. Br J Pharmacol 139:695–704

Pierce KL, Luttrell LM, Lefkowitz RJ (2001) New mechanisms in heptahelical receptor signaling to mitogen activated protein kinase cascades. Oncogene 20:1532–1539

Pippig S, Andexinger S, Lohse MJ (1995) Sequestration and recycling of β_2-adrenergic receptors permit receptor resensitization. Mol Pharmacol 47:666–676

Prieto D, Buus CL, Mulvany MJ, Nilsson H (2000) Neuropeptide Y regulates intracellular calcium through different signalling pathways linked to a Y_1-receptor in rat mesenteric small arteries. Br J Pharmacol 129:1689–1699

Primus RJ, Yevich E, Gallager DW (1998) In vitro autoradiography of GTPγ[^{35}S] binding at activated NPY receptor subtypes in adult rat brain. Brain Res Mol Brain Res 58:74–82

Qanbar R, Bouvier M (2003) Role of palmitoylation/depalmitoylation reactions in G-protein-coupled receptor function. Pharmacol Ther 97:1–33

Racchi H, Irarrazabal MJ, Howard M, Moran S, Zalaquett R, Huidobro-Toro JP (1999) Adenosine 5'-triphosphate and neuropeptide Y are co-transmitters in conjunction with noradrenaline in the human saphenous vein. Br J Pharmacol 126:1175–1185

Rose PM, Lynch JS, Frazier ST, Fisher SM, Chung W, Battaglino P, Fathi Z, Leibel R, Fernandes P (1997) Molecular genetic analysis of a human neuropeptide Y receptor: the human homolog of the murine "Y$_5$" receptor may be a pseudogene. J Biol Chem 272:3622–3627

Sarkar S, Lechan RM (2003) Central administration of neuropeptide Y reduces alpha-melanocyte-stimulating hormone-induced cyclic adenosine 5'-monophosphate response element binding protein (CREB) phosphorylation in pro-thyrotropin-releasing hormone neurons and increases CREB phosphorylation in corticotropin-releasing hormone neurons in the hypothalamic paraventricular nucleus. Endocrinology 144:281–291

Scheer A, Fanelli F, Costa T, De Benedetti PG, Cotecchia S (1997) The activation process of the α_{1B}-adrenergic receptor: potential role of protonation and hydrophobicity of a highly conserved aspartate. Proc Natl Acad Sci USA 94:808–813

Selbie LA, Darby K, Schmitz-Peiffer C, Browne CL, Herzog H, Shine J, Biden TJ (1995) Synergistic interaction of Y$_1$-neuropeptide Y and α_{1b}-adrenergic receptors in the regulation of phospholipase C, protein kinase C and arachidonic acid production. J Biol Chem 270:11789–11796

Selbie LA, King NV, Dickenson JM, Hill SJ (1997) Role of G-protein βγ subunits in the augmentation of P2Y$_2$ (P$_{2U}$) receptor-stimulated responses by neuropeptide Y Y$_1$ G$_{i/o}$-coupled receptors. Biochem J 328:153–158

Sun L, Philipson LH, Miller RJ (1998) Regulation of K$^+$ and Ca^{++} channels by a family of neuropeptide Y receptors. J Pharmacol Exp Ther 284:625–632

Tanaka Y, Nakazawa T, Ishiro H, Saito M, Uneyama H, Iwata S, Ishii K, Nakayama K (1995) Ca^{2+} handling mechanisms underlying neuropeptide Y-induced contraction in canine basilar artery. Eur J Pharmacol 289:59–66

Van Riper DA, Bevan JA (1991) Evidence that neuropeptide Y and norepinephrine mediate electrical field-stimulated vasoconstriction of rabbit middle cerebral artery. Circ Res 68:568–577

Voisin T, Bens M, Cluzeaud F, Vandewalle A, Laburthe M (1993) Peptide YY receptors in the proximal tubule PKSV-PCT cell line derived from transgenic mice. J Biol Chem 268:20547–20554

Voisin T, Goumain M, Lorinet AM, Maoret JJ, Laburthe M (2000) Functional and molecular properties of the human recombinant Y$_4$ receptor: resistance to agonist-promoted desensitization. J Pharmacol Exp Ther 292:638–646

Voisin T, Lorinet AM, Maoret JJ, Couvineau A, Laburthe M (1996) G alpha i RNA antisense expression demonstrates the exclusive coupling of peptide YY receptors to G$_{i2}$ proteins in renal proximal tubule cells. J Biol Chem 271:574–580

Wahlestedt C, Yanaihara N, Håkanson R (1986) Evidence for different pre- and post-junctional receptors for neuropeptide Y and related peptides. Regul Pept 13:307–318

Weinberg DH, Sirinatsinghji DJS, Tan CP, Shiao L-L, Morin N, Rigby MR, Heavens RH, Rapoport DR, Bayne MI, Cascieri MA, Strader CD, Linemeyer DL, MacNeil DJ (1996) Cloning and expression of a novel neuropeptide Y receptor. J Biol Chem 271:16485–16488

Yokosuka M, Kalra PS, Kalra SP (1999) Inhibition of neuropeptide Y (NPY)-induced feeding and c-Fos response in magnocellular paraventricular nucleus by a NPY receptor antagonist: a site of NPY action. Endocrinology 140:4494–4500

Zhang P, Zheng J, Vorce RL, Hexum TD (2000) Identification of an NPY-Y_1 receptor subtype in bovine chromaffin cells. Regul Pept 87:9–13

Zhang X, Tong YG, Bao L, Hökfelt T (1999) The neuropeptide Y Y_1 receptor is a somatic receptor on dorsal root ganglion neurons and a postsynaptic receptor on somatostatin dorsal horn neurons. Eur J Neurosci 11:2211–2225

Zukowska-Grojec Z, Karwatowska-Prokopczuk E, Fisher TA, Ji H (1998) Mechanisms of vascular growth-promoting effects of neuropeptide Y: role of its inducible receptors. Regul Pept 75/76:231–238

Phylogeny of NPY-Family Peptides and Their Receptors

D. Larhammar[1] · R. Fredriksson[1] · E. T. Larson[1,2] · E. Salaneck[1]

[1] Department of Neuroscience, Unit of Pharmacology, Uppsala University, Box 593, 75124 Uppsala, Sweden
e-mail: Dan.Larhammar@neuro.uu.se
[2] Department of Biology, Northeastern University, 134 Mugar Hall, Boston, MA 02115, USA

1	Background	76
2	Evolution of NPY-Family Peptides	77
3	Origin of the NPY and PYY Genes by Block (or Chromosome) Duplication . .	81
4	Sequence Relationships of the Y Receptor Family	84
5	Origin of the Y Receptor Genes by Block (or Chromosome) Duplications . . .	86
6	Discussion	91
	References	95

Abstract All vertebrates investigated to date possess the related peptides neuropeptide Y and peptide YY. Additional duplicates of peptide YY have arisen in several lineages, for instance pancreatic polypeptide in tetrapods. The peptides bind to the Y family of G-protein coupled receptors which expanded dramatically before the radiation of jawed vertebrates (gnathostomes). First, an ancestral Y receptor gene generated a cluster consisting of the progenitors of Y_1, Y_2 and Y_5. Today these are only about 30% identical to each other. Subsequently, this chromosomal region (or the entire chromosome) was quadrupled whereupon differential gene losses in different classes of vertebrates ensued. Three Y_1-like genes sharing about 50% identity exist in mammals, chicken, and a shark, namely Y_1, Y_4 and Y_6. Teleost fishes seem to have lost Y_1 and Y_6 but have retained Y_4 (initially called Ya) and what probably is the fourth Y_1-like gene named Yb. The Y_5 gene is present in tetrapods as well as sharks, but not in zebrafish or pufferfish, and no duplicates of Y_5 seem to have survived. The Y_2 gene is present in all major vertebrate lineages. A duplicate of Y_2 called Y_7 has been found in amphibians, bony fishes and a shark, consistent with duplication before the origin of gnathostomes, but this gene has been lost in mammals. The evolutionary rates differ greatly between the various subtypes as well as for some specific subtypes across vertebrate classes. For instance, Y_4 has evolved more slowly in shark than in mammals. The degree of sequence conservation suggests that Y_1, Y_2 and Y_5 are all subjected to strong conservative selection pressures. Future comparative studies may be able to unravel how Y_2 and Y_1-Y_5 came to exert opposite effects on feeding in mammals.

Keywords Neuropeptide Y · G-protein coupled receptor · Gene duplication · Paralogon · Tetraploidy

1
Background

The neuropeptide Y system with its multiple peptide ligands and receptors in vertebrates offers a number of interesting features as well as challenges from an evolutionary point of view. One surprise that emerged several years ago was that the most divergent member of the peptide family, pancreatic polypeptide (PP), which was the first to be discovered (Kimmel et al. 1968, 1975), is actually the most recent member among the three major peptides in tetrapods. Its sequence divergence is due to a high rate of evolutionary change (Larhammar et al. 1993). The other two peptides, neuropeptide Y (NPY) and peptide YY (PYY), are present in all vertebrates investigated and evolve much more slowly (Cerdá-Reverter and Larhammar 2000; Larhammar 1996). Indeed, NPY is one of the most highly conserved neuropeptides known. Recent detailed studies of the chromosomal locations of the genes encoding NPY and PYY add further support for their origin from a common ancestral gene through a chromosome duplication event as will be described here.

Another remarkable feature of the NPY system is that the three major receptor subtypes, named Y_1, Y_2 and Y_5, are quite divergent from each other, sharing only approximately 30% identity. Yet, all three evolve slowly, suggesting great evolutionary age for all three subtypes (Larhammar et al. 2001). Furthermore, all three of these receptors have the ability to respond to both NPY and PYY which share 70% identity. Such receptor subtype diversity is unusual for peptide-binding receptors and may prove useful to identify the most important points of interaction between the peptide ligands and the various receptor subtypes.

The Y receptor family includes a number of Y_1-like receptors sharing approximately 50% identity with Y_1 and with one another, including Y_4 and Y_6 (the International Union of Pharmacology recommends designating the mammalian Y_6 receptor as y_6 receptor because it has not yet been documented as a biologically functional receptor in sufficient detail) in mammals, and the three fish receptors Ya, Yb and Yc (Larhammar et al. 2001; Lundell et al. 1997; Ringvall et al. 1997; Starbäck et al. 1999). The functional roles of these Y_1-like subtypes are still unknown. Their evolutionary inter-relationships have been somewhat confusing but now seem possible to disentangle thanks to detailed phylogenetic analyses of sequences from multiple vertebrate species in combination with data on chromosomal location. Also subtype Y_2 has a closer and recently discovered relative with approximately 50% identity called Y_7 (Fredriksson et al. 2003). Here we describe how the sequence and chromosome comparisons lead to the conclusion that a broad repertoire of Y receptors was already present in the ancestor of all jawed vertebrates (gnathostomes), and that the different evolutionary lineages radiating from this ancestor then lost different members of the Y receptor family. For example, Y_7 is present in shark, bony fishes and amphibians, but seems to have been lost in the lineage leading to mammals.

Recent studies have reported surprising differences in the tissue distribution of specific receptor subtypes across vertebrate classes. For instance, chicken Y_4 is more widespread than its mammalian ortholog (Lundell et al. 2002). Also, the Y_6 gene which is a pseudogene in primates and several other mammals, but not all (Starbäck et al. 2000), has a broad tissue distribution in a shark, whereas Y_1 has a more narrow distribution in shark than in mammals (Salaneck et al. 2003).

Thus, the families of NPY-like peptides and Y receptors contain examples of gene acquisition and loss, peptides and receptors evolving at variable rates across vertebrate classes and changes in tissue distribution. In short, the NPY system displays both extraordinary evolutionary conservation and surprisingly dynamic changes.

2
Evolution of NPY-Family Peptides

The NPY-family peptides in vertebrates are comprised of 36 amino acids, with a few exceptions (chicken PYY has 37 amino acids and Burmese python PP has 35 amino acids), and a carboxy-terminal amide group (Fig. 1). Thanks to sequence information from a large number of species it was possible to detect large differences in the evolutionary rates of the three peptides NPY, PYY and PP (Larhammar 1996; Larhammar et al. 1993). This led to the following scheme of gene duplications (Fig. 2) which has been corroborated by additional sequence information (for recent reviews, see Cerdá-Reverter and Larhammar 2000; Conlon 2002) as well as chromosomal locations of the genes (see below). The vertebrate ancestor most probably had a single NPY/PYY-like gene that was duplicated to generate NPY and PYY. A single peptide gene in a vertebrate ancestor is supported by the findings of single NPY-like genes in the fruitfly *Dro-*

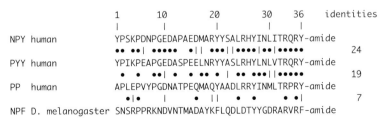

Fig. 1 Alignment of amino acid sequences for the three human NPY-family peptides and the *Drosophila melanogaster* NPY-like peptide called neuropeptide F (*NPF*; Brown et al. 1999). Residues identical to the top sequence (human NPY) are marked with *dots* and conservative replacements are marked with *vertical bars*

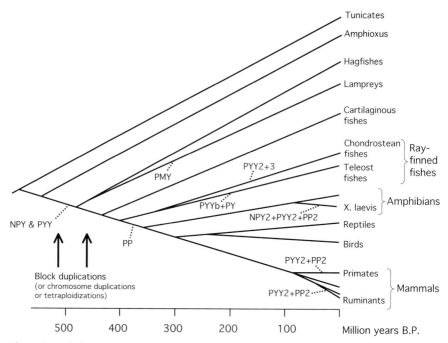

Fig. 2 Gene duplications in the NPY family during chordate evolution. The tree shows schematically the generally accepted phylogeny. Note that for some of the vertebrate classes, only the branches with NPY-family gene duplications are shown. The only Chondrostean fish that has been studied is a sturgeon. Duplicates of PYY and PP have arisen independently in several lineages although some of them were assigned the same names when they were first identified

sophila melanogaster (Brown et al. 1999) and the pond snail *Lymnaea stagnalis* (Tensen et al. 1998). However, the NPY-like gene seems to have been lost in the tunicate *Ciona intestinalis*, unless the gene is still missing from the database, but this lineage is known to have lost many genes (Dehal et al. 2002). In tetrapods, PYY was duplicated to generate PP. Many additional duplications of PYY have occurred independently in various vertebrate lineages (see below).

Vertebrate NPY is predominantly if not exclusively expressed in the nervous system, whereas PYY is expressed in endocrine cells as well as a few neuronal sites (Pieribone et al. 1992; Söderberg 2000; Söderberg et al. 1994). The findings that the *Drosophila melanogaster* NPY-related peptide NPF and the yellow fever mosquito *Aedes aegypti* NPY are present in both neurons and gut endocrine cells (Brown et al. 1999; Stanek et al. 2002) suggest that the ancestral vertebrate NPY/PYY-like peptide also had broad expression and that the gene duplication led to specialization or subfunctionalization as has been observed after other gene duplication events (Force et al. 1999). All vertebrates that have been investigated with regard to both the central nervous system and gut endocrine cells seem to express both NPY and PYY.

After the initial duplication that generated NPY and PYY, no additional duplications seem to have taken place for NPY except that the frog *Xenopus laevis* has two highly similar forms differing by a single conservative amino acid replacement (Griffin et al. 1994; van Riel et al. 1993). This is not surprising in light of its recent tetraploidization estimated to have taken place in the range 27–44 million years ago (Hughes and Hughes 1993). Future studies will tell whether any additional species or lineages that have undergone tetraploidization may have retained duplicates of NPY. In contrast, the evolutionary history of the PYY gene is replete with duplications (Fig. 2). The PP gene in tetrapods is a tandem copy of PYY within a distance of only a few kilobases and PP is present in all tetrapods investigated. Furthermore, the PYY-PP gene pair has been duplicated independently in primates (human and baboon) and artiodactyls (cattle and sheep) (Couzens et al. 2000; Herzog et al. 1995; Lewis et al. 1985). Whereas the original and functional human PYY and PP genes are located in 17q21.2 close to the HOXB cluster (Fig. 3), the duplicates PYY2 and PP2 (the gene is called PPY2) are nonfunctional and located close to the centromere of 17q (Couzens et al. 2000), abbreviated Hsa17q for *Homo sapiens* chromosome 17. The human PYY2 peptide is prematurely truncated and terminates after only 15 amino acids and the PP2 gene terminates even earlier in the signal peptide. The human and baboon PYY gene orthologs are 95% identical at the nucleotide level as are the PYY2 genes. The PYY-PYY2 gene identity within human (as well as the PP-PP2 gene identity) is lower, 91%–92%, suggesting that the gene duplication took place before the divergence of the lineages leading to human and baboon. Interestingly, the coding regions of the duplicates have diverged more rapidly than the introns, suggesting that there has been positive selection for peptide divergence, perhaps because an increased dose of PYY or PP due to the gene duplication was detrimental. Whereas it seems unlikely that the primate duplicates may have any functional roles, the cattle duplicate of PYY, also known by the name of seminalplasmin, is expressed in testis and has been reported to have antimicrobial activity and to inhibit lymphocyte proliferation (Reddy and Bhargava 1976) as well as calmodulin activity (Comte et al. 1986).

Duplication of PYY and PP has also occurred in *Xenopus laevis*, probably through its tetraploidization (see above). Interestingly, the two copies of PYY have diverged considerably more (six differences) than the two PP copies (only a single difference) (Conlon 2002; Kim et al. 2001). Analysis of the nucleotide sequences will be necessary to see whether there has been positive selection for PYY amino acid divergence. Duplications of PYY have also taken place in the pallid sturgeon *Scaphirhynchus albus*, which expresses three forms (Kim et al. 2000), probably as a result of tetraploidization(s) in this lineage of Chondrostean fishes. Teleost fishes too seem to have undergone tetraploidization before their extensive radiation (Taylor et al. 2003; Van de Peer et al. 2003), and consistent with this the pufferfish *Fugu rubripes* has two rather divergent PYY genes (D. Larhammar, unpublished results). The zebrafish *(Danio rerio)* genome database contains only a single PYY gene, but it is still unclear whether this is due to loss of the PYY duplicate or an incomplete database. A teleostean pancre-

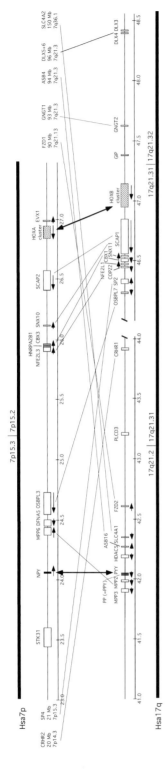

Fig. 3 Human chromosomal regions harboring the NPY and PYY-PP genes and the homeobox gene clusters HOXA and HOXB. Many additional genes and predicted genes are present in these regions, but have been omitted for clarity. Note that a portion of Hsa17q has been omitted between map positions 44 Mb and 46 Mb. The HOX clusters A and B consist of 11 and 10 genes, respectively. *Arrows* show the direction of transcription

atic peptide called PY has previously been found only in a subgroup of teleosts called Acanthomorpha (Cerdá-Reverter and Larhammar 2000; Cerdá-Reverter et al. 2000), but the genome databases show that this peptide is present also in zebrafish (D. Larhammar, unpublished results), suggesting that it arose in a teleost ancestor or even earlier. Finally, two PYY-like peptides, one of which is called PMY (=PYY1), have been sequenced in the river lamprey *Lampetra fluviatilis* (Conlon et al. 1991; Söderberg et al. 2000; Wang et al. 1999).

3
Origin of the NPY and PYY Genes by Block (or Chromosome) Duplication

The basal vertebrate gene duplication that generated NPY and PYY has been possible to trace because the two genes map close to the HOXA and HOXB clusters, respectively, in the human genome (Baker et al. 1995; Hort et al. 1995) as well as in zebrafish (Söderberg et al. 2000). This led to the hypothesis that the NPY/PYY ancestral gene was part of a large chromosomal block containing a HOX cluster as well as a few other genes, and that this whole block, or the entire chromosome, was duplicated before the radiation of vertebrates (Larhammar 1996; Söderberg et al. 2000). A detailed examination of these chromosomal regions in the human genome database provides further support for this hypothesis, as several additional genes have been discovered that belong to the same duplicated block (Fig. 3). However, it is also clear that many local rearrangements have subsequently taken place in each of the daughter chromosomes. For instance, either NPY or PYY (along with PP) has undergone inversion relative to its HOX cluster. As it has been reported that intrachromosomal rearrangements take place at a much higher rate than interchromosomal events (Pevzner and Tesler 2003; Postlethwait et al. 2000), one may analyze the entire chromosomal regions regardless of the exact gene order along the chromosomes. This has allowed identification of members on both chromosome 7 and 17 for a large number of gene families, and also on the two other HOX-bearing chromosomes 2 (HOXD) and 12 (HOXC) (Fig. 4). Such a set of paralogous chromosome regions has been named a paralogon (Coulier et al. 2000). In fact, the human genome contains several such paralogons (Lundin et al. 2003; Popovici et al. 2001) that seem to support the suggestion that the entire vertebrate genome underwent at least one and probably two tetraploidizations at an early stage and that the many new gene copies that were generated in these events facilitated the evolutionary novelties and the rapid radiation of the vertebrates (Garcia-Fernàndez and Holland 1994; Holland et al. 1994; Shimeld and Holland 2000).

No less than 24 gene families can be identified in the extended HOX regions shown in Fig. 4, most of which are in close proximity to NPY/PYY on at least one of the two chromosomes. Note that each of the HOX clusters consists of many genes that correspond to 13 paralogy classes. Additional gene families belonging to this paralogon have been identified through analyses of even wider regions of the chromosomes (Larhammar et al. 2002; Lundin et al. 2003; Panopoulou et al. 2003; Popovici et al. 2001). Very few of the gene families in

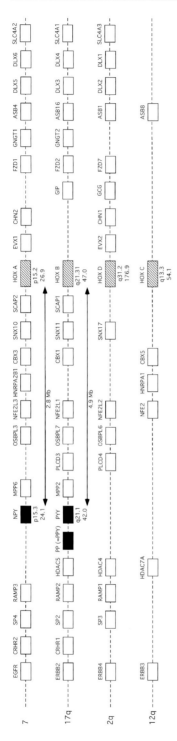

Fig. 4 Schematic representation of the two chromosomal regions containing the NPY and PYY-PP genes and the homeobox gene clusters HOXA and HOXB, as well as the two related chromosomal regions harboring the HOXC and HOXD clusters. The chromosomes are shown as *dashed lines* to indicate that they are not drawn to scale and that the physical gene order has been altered to highlight similarities between the four related chromosomes forming this paralogon. The NPY and PYY chromosomes Hsa7 and Hsa17 were used as starting point to identify gene families represented on two or more chromosomes in this paralogon. Many additional gene families exist that have representatives only on Hsa2 and Hsa12 or are located outside of the regions containing NPY-HOXA and PYY-HOXB. Abbreviations are those used in the Ensembl database

Fig. 4 are complete, i.e., they do not have the full quartet of members. Only the EGFR, NFE and ASB families still retain all four members as do only three of the 13 HOX subfamilies, showing that gene losses are extremely common after block duplications. It follows from the scheme in Fig. 4 that there once were additional copies of the ancestral NPY/PYY gene on chromosomes 2 and 12, but these seem to have been lost. It is still unclear which pairs of chromosomes are most closely related to each other, probably because the two proposed tetraploidizatons took place within a relatively short period of time. Thus, NPY and PYY may have arisen in the first tetraploidization while the second tetraploidization was followed by a loss of both the NPY duplicate and the PYY duplicate, i.e., two separate loss events (the two lamprey PYY-like peptides may possibly be products of this duplication, but it is also possible that they arose through a separate event in the lamprey lineage). Alternatively, the first tetraploidization may have been followed by a loss of one of the copies, whereupon the remaining copy was duplicated in the second tetraploidization to generate NPY and PYY. The latter alternative requires only one loss and is therefore more parsimonious, but both scenarios are possible as gene losses occur frequently as mentioned above.

Neuropeptide specialists will notice that other neuropeptide systems also expanded in the same duplication events, as the genes for glucagon (GCG) and GIP are located on chromosomes 2 and 17, respectively, and the genes for corticotropin-releasing hormone (CRH) receptors 1 and 2 are on chromosomes 17 and 7, respectively. The phylogenies and taxonomic distributions of these genes are consistent with origin in early vertebrate evolution.

It should be mentioned that a few criticisms of the block or chromosome duplication scenario have been published, arguing that the phylogenies of the gene families located on these chromosomes disagree with each other and are therefore inconsistent with block duplications (Hughes 1999; Hughes et al. 2001; Martin 1999, 2001). However, some of those analyses have included gene families that do not belong to the duplicated blocks. Also, those studies have extrapolated duplication events from phylogenetic trees that were based on a very limited number of species (mostly mammals), as recently pointed out (Larhammar et al. 2002). Furthermore, the critics have not taken into consideration the high rate of gene loss after such duplications (Gu and Huang 2002) as shown below for the Y receptor family.

As the NPY and PYY genes arose from a common ancestral gene by block duplication (rather than retrotransposition of the coding region only), they probably had identical promotor regions and regulatory elements immediately after the duplication event. The differences that we see today in tissue distribution most likely resulted from regulatory mutations. The block duplication therefore corroborates the notion that the gene duplication resulting in NPY and PYY was followed by subfunctionalization, i.e., specialization of the two daughter genes in neurons and (primarily) gut endocrine cells, respectively. Therefore, an amphioxus descendant of the NPY/PYY-like ancestral peptide, if still present, will be expected to have both neuronal and gut endocrine expression.

4
Sequence Relationships of the Y Receptor Family

Phylogenetic analyses of vertebrate Y receptors result in trees with three distinct clades displaying only approximately 30% identity to each other (Fig. 5), namely the Y_1 subfamily, the Y_2 subfamily, and Y_5. The Y_1 subfamily consists of the Y_1, Y_4 and Y_6 receptors and the teleost fish Ya, Yb and Yc receptors. These are approximately 50% identical among each other, except that teleost Yb and Yc are more closely related (75%). The second clade includes Y_2 and the newly discovered Y_7 receptor (Fredriksson et al. 2003). The third clade contains Y_5 sequences only, suggesting that no duplications of Y_5 have occurred, or rather that no Y_5 duplicates have been retained. The proposed Y_3 receptor (Michel et al.

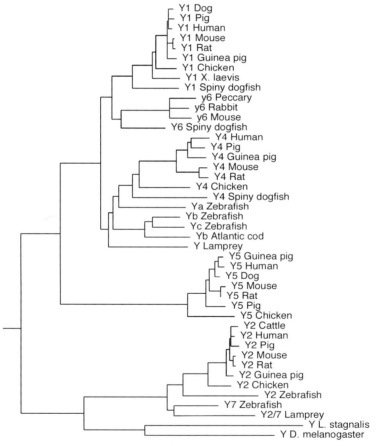

Fig. 5 Phylogenetic tree for full-length Y receptor sequences constructed with the Neighbor-Joining method. The human bradykinin B1 sequence was used as an outgroup to root the tree

1998) does not correspond to a separate gene product and the origin of this pharmacological binding profile remains to be accounted for.

The Y_1 receptor has been cloned in mammals, chicken (Holmberg et al. 2002), frogs and sturgeon (E.T. Larson and D. Larhammar, unpublished results), and a shark (Salaneck et al. 2003). However, Y_1 has not been found in any teleost fish despite extensive cloning efforts, suggesting that this subtype was lost in the ancestor of this lineage. Likewise, the Y_6 gene is present in mammals (except rat), chicken and frogs (R. Fredriksson and D. Larhammar, unpublished results), and a shark (Salaneck et al. 2003), but is missing in teleosts and may therefore share the fate of Y_1 of having been lost. This conclusion is based on the crucial discovery of Y_1 and Y_6 in the shark spiny dogfish, *Squalus acanthias* (Salaneck et al. 2003). The phylogenetic analyses consistently show that Y_1 and Y_6 are more closely related to each other than to Y_4 (Fig. 5). The Y_6 gene has a peculiar evolutionary history in mammals as it expressed in mouse and rabbit (Gregor et al. 1996; Matsumoto et al. 1996; Weinberg et al. 1996) but is a pseudogene in human and other primates, pig, and guinea pig (Gregor et al. 1996; Matsumoto et al. 1996; Rose et al. 1997; Starbäck et al. 2000; Wraith et al. 2000). The Y_4 gene has been found in mammals, chicken (Lundell et al. 2002), and shark (Salaneck et al. 2003), which shows that the Y_4 gene already existed by the time at which PP arose as a copy of PYY in tetrapods. Apparently, the Y_4 and PP sequences then started to drift and coevolve to the present state in mammals where PP is the favored ligand for Y_4, particularly in rat and mouse.

The zebrafish Ya, Yb and Yc receptors have, ever since their discovery, been difficult to reconcile with the mammalian receptor subtypes, both from sequence comparisons and the limited chromosomal mapping data available at the time (Starbäck et al. 1999). The chromosomal region around zebrafish Ya showed some resemblance to Hsa4 (Y_1) and the Yc region was reminiscent of Hsa5 (Y_6), but today we know that neither of these was correct (see below). More extensive phylogenetic analyses with additional sequences from a shark (Salaneck et al. 2003) and an agnathan, the river lamprey *Lampetra fluviatilis* (Salaneck et al. 2001), suggest that Ya may be the telost ortholog of Y_4 (Fig. 5). These studies also indicate that Y_4/Ya and Yb/Yc form a clade within the Y_1 subfamily and that the lamprey sequence may be an ortholog of the common ancestor of these two lineages, a so-called pro-ortholog. The duplication that led to Yb and Yc has been difficult to date, probably due to high evolutionary rates for these genes. The Yb-like sequences from Atlantic cod and rainbow trout (Arvidsson et al. 1998; Larson et al. 2003) gave conflicting results, but the presence of both Yb and Yc in the pufferfish *Fugu rubripes* (D. Larhammar, unpublished results) suggests that the duplication may coincide with the basal teleost tetraploidization (Taylor et al. 2003; Van de Peer et al. 2003).

The Y_2 clade includes the recently discovered Y_7 receptor which was initially found in zebrafish (Fredriksson et al. 2003). Like the Y_1-subfamily duplications, the one leading to Y_2 and Y_7 also seems to have taken place before the radiation of gnathostomes, as the two genes have also been cloned in spiny dogfish (E.T. Larson, R. Fredriksson, E. Salaneck and D. Larhammar, unpublished results).

Both Y_2 and Y_7 have also been cloned in frogs (Fredriksson et al. 2003), showing that Y_7 was still present in early tetrapods. Thus, the gene was lost in the lineage leading to amniotes or even later in a mammalian ancestor. A Y_2/Y_7-like gene has been identified in the river lamprey (E. Salaneck and D. Larhammar, unpublished results), and although it groups with zebrafish Y_7 in Fig. 5, this difference is not statistically significant. Thus the lamprey sequence may be a pro-ortholog of gnathostome Y_2 and Y_7.

Finally, the Y_5 clade is somewhat more closely related to the Y_1 subfamily than to the Y_2 clade. This is consistent with the closer physical distance between Y_5 and Y_1 on human chromosome 4, just over 20 kilobases for the coding regions (Herzog et al. 1997), whereas the Y_2 gene is located 8.1 megabases closer to centromere (Fig. 6). Thus, both the sequence divergence and the physical chromosomal distances suggest that the first gene duplication of an ancestral Y gene led to Y_2 and the Y_1/Y_5 ancestor and that these drifted apart on the chromosome. Later, the Y_1/Y_5 ancestor was duplicated to generate Y_1 and Y_5. In the pig, a pericentric chromosomal inversion has placed Y_2 on the opposite arm relative to the Y_1–Y_5 pair (Wraith et al. 2000), and in the mouse there has been a translocation to a different chromosome (Lutz et al. 1997a, 1997b). However, it is also possible that the first duplication led to Y_1 and Y_5 and that these genes were kept together physically by overlapping or shared regulatory regions, and that Y_2 arose later from one of these but was allowed to diverge more rapidly. All known Y_5-like sequences follow the expected species phylogeny as shown in Fig. 6, which strongly suggests that they are true orthologs and that no gene duplicates have been retained. The Y_5 sequences display somewhat lower sequence identity than Y_1 or Y_2, but this is largely due to divergence of the large cytoplasmic loop between transmembrane regions 5 and 6 (Larhammar et al. 2001; Wraith et al. 2000).

A few invertebrate receptors seem to be pro-orthologs of the vertebrate Y receptors, namely one receptor from *Lymnaea stagnalis*, one from *Drosophila melanogaster* and one from the malaria mosquito *Anopheles gambiae* (Garczynski et al. 2002; Hill et al. 2002; Tensen et al. 1998). The two former are included in the phylogenetic tree in Fig. 5. In addition to their closer relationship to the vertebrate Y receptors than to other peptide receptors, they also respond to the NPY-like peptides found in these species (Fig. 1). However, the receptor in *Caenorhabditis elegans* previously reported to be involved in social feeding (de Bono and Bargmann 1998), NPR-1, is not a true NPY receptor as the ligand for this receptor is the nonapeptide AF9 which is not a member of the NPY family (Kubiak et al. 2003).

5
Origin of the Y Receptor Genes by Block (or Chromosome) Duplications

The chromosomal locations of the human and porcine Y receptors suggest that they belong to a set of related chromosomes thought to have arisen from a common ancestral chromosome (Larhammar et al. 2001; Wraith et al. 2000). In hu-

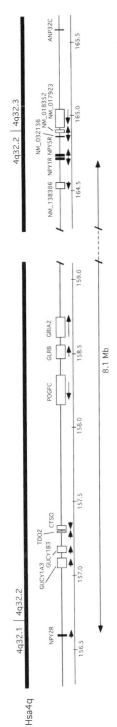

Fig. 6 Chromosomal region harboring the three human Y receptors Y_1, Y_2, and Y_5. The Y_1 receptor gene is called NPY1R, Y_2 is NPY2R, etc. Only a few of all additional genes are shown. *Arrows* indicate direction of transcription

mans this paralogon includes Hsa4 with the Y_1, Y_2 and Y_5 genes, Hsa5 with the Y_6 gene, and Hsa10 with the Y_4 gene (Fig. 7). A fourth related chromosome in this paralogon is Hsa8 (Lundin et al. 2003; Pébusque et al. 1998; Vienne et al. 2003; Wraith et al. 2000) although no Y receptor genes have been found on this chromosome (Wraith et al. 2000). The human genome database has allowed identification of several additional gene families represented on two to four of these four chromosomes, as shown in Fig. 7 which shows a total of 20 gene families that are likely to have duplicated in the same block duplication or chromosome doubling events. As for the HOX chromosomes described above for NPY and PYY–PP, all of these families seem to have lost one or more members, but it is also possible that the fourth member has been translocated to a different chromosome which shares only a short segment with this paralogon. The neuropeptide FF receptors and the tachykinin precursor genes also arose in these block duplications (Fig. 7).

Preliminary analyses of the still fragmentary pufferfish and zebrafish genome databases confirm linkage of many of these gene families with one another, although it is not yet possible to reconstruct the entire fish chromosomes. Importantly, the fish databases suggest that the Ya gene in pufferfish and zebrafish is located together with several genes that in human map to Hsa10, thereby adding further evidence that Ya is an ortholog of mammalian Y_4 in agreement with the sequence analyses described above. Such analyses also indicate that the fish Yb and Yc genes are located close to genes that in the human genome are located on Hsa8, thereby supporting the idea that the Yb/Yc ancestor is the fourth member of the original quartet of Y_1-subfamily genes and not an ortholog of Y_6 as initially proposed (Starbäck et al. 1999).

The location of Y_7 in this paralogon is supported by the mapping of Y_7 adjacent to a gene in the zebrafish genome whose ortholog is located on chromosome 5 in the human genome (Fredriksson et al. 2003). As expected, the zebrafish Y_2 gene is in a region which is syntenic with Hsa4 (Fredriksson et al. 2003). Thus, the ancestor of human chromosome 5 must once have had an Y_7 gene, but this was probably lost in the mammalian ancestor or even in the common ancestor of mammals and bird/reptiles, because no Y_7 gene has been found in the chicken genome.

As the chromosomal data from fishes provide additional support for the paralogon comprised of regions on Hsa4, 5, 8, and 10, they also strengthen the recently proposed duplication scheme for the complete family of Y receptors (Fredriksson et al. 2003; Salaneck et al. 2003). This scheme is shown in Fig. 8 and begins with two local duplications of an ancestral Y gene resulting in the ancestors of the three clades Y_1, Y_2 and Y_5. After the first chromosome duplication, the extra copies of Y_2 and Y_5 were probably lost. The four remaining genes were duplicated again in the next tetraploidization whereupon the new Y_5 duplicate was also lost. The resulting set of seven genes probably constituted the Y repertoire of the gnathostome ancestor. The mammalian lineage subsequently lost two of these, Y_7 and Yb, while the teleost fish lineage seems to have lost Y_1, Y_5, and Y_6. An additional duplication then took place in the lineage leading to

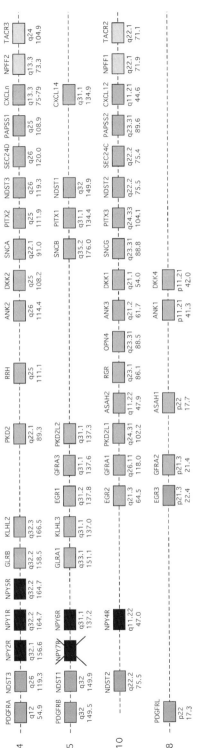

Fig. 7 Schematic representation of the three chromosomal regions containing the Y genes as well a fourth chromosome (Hsa8) that also belongs to this paralogon. The Y_1 receptor gene is called NPY1R, Y_2 is NPY2R, etc. The hypothetical position for the Y_7 gene thought to once have existed is shown and *crossed-through* on Hsa5. The chromosomes are shown as *dashed lines* to indicate that they are not drawn to scale and that the physical gene order has been altered to highlight similarities. Several additional gene families exist that have representatives on two or more of these chromosomes. Abbreviations are those used in the Ensembl database

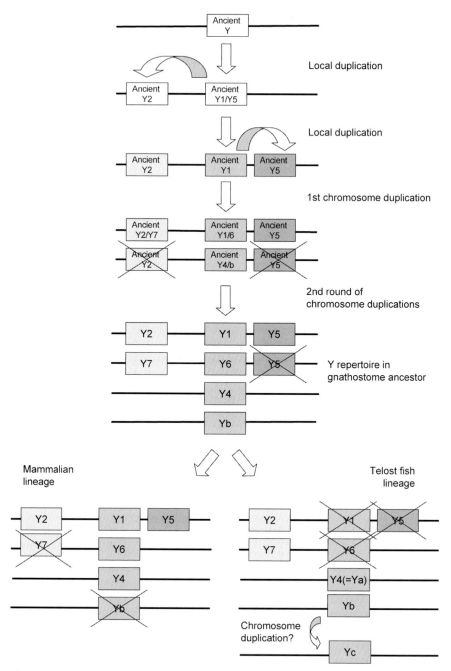

Fig. 8 Proposed gene duplication scheme for the Y receptor family. The *crossed-through* genes were probably lost after each duplication event

teleosts, generating Yc from Yb, presumably as a result of the early teleost tetraploidization (Taylor et al. 2003; Van de Peer et al. 2003).

6
Discussion

The combined results from sequence analyses and chromosomal mapping strongly support the notion that chromosome duplications in early vertebrate evolution, before the divergence of gnathostomes, gave rise to NPY and PYY and to an extensive repertoire of Y receptors. This initial receptor family consisted of three subfamilies with a total of seven members (Fig. 8): the Y_1 subfamily also included Y_4, Y_6 and Yb; the Y_2 subfamily included Y_7; and the Y_5 subfamily had just a single member. After these duplications, the receptor family underwent differential losses in the different vertebrate lineages, with loss of Yb and Y_7 in the mammalian lineage and subtypes Y_1, Y_6, and Y_5 in the lineage leading to teleost fishes. The latter have acquired an extra duplicate of Yb called Yc. In contrast to the many losses of receptor genes, the peptide family has undergone several additional and independent duplications of PYY in several lineages (Fig. 2), but the functional importance of these duplicates is still largely unknown.

To further narrow down the time periods for the losses of Y receptor genes in mammals and teleost fishes, we have tried to clone the different receptor subtypes in additional distantly related vertebrates. In chicken we have cloned the full-length coding regions for all five of the mammalian subtypes (Holmberg et al. 2002; Lundell et al. 2002; Salaneck et al. 2000). Y_6 appears to be functional (R. Fredriksson and D. Larhammar, unpublished results) although its properties and functional roles have yet to be defined. In frogs we have found by polymerase chain reaction (PCR) the subtypes Y_1, Y_2, Y_5, Y_6, and Y_7. The Y_4 receptor has not yet been found but is expected to be present. Work is in progress to define the Y gene repertoire in the sarcopterygian (flesh-finned) fish *Latimeria chalumnae*, the famous coelacanth. We are also studying basal actinopterygian (ray-finned) fishes such as bichir, sturgeon and gar to see when the Y_1, Y_5 and Y_6 genes were lost in the lineage leading to teleost fishes. As mentioned above, the Y_1 gene is present in sturgeon (E.T. Larson, E. Salaneck, and D. Larhammar, unpublished results) which diverged from the lineage leading to teleosts approximately 330 million years ago.

A particularly interesting evolutionary question is how Y_2, with its inhibitory effect on food intake (Batterham et al. 2002), acquired a role opposite to that of Y_1 and Y_5, both of which stimulate food intake (Pedrazzini et al. 2003). As Y_2 probably arose by a local duplication of an ancestral Y gene, it is possible that the duplicated region did not include all of the regulatory elements, or that Y_2 came under influence of other regulatory regions. Alternatively, it could be that Y_2 represents the function of the ancestral Y receptor while the Y_1/Y_5 ancestor became a stimulator of feeding. It should also be kept in mind that Y receptors influence many other physiological functions, for instance the circadian system, thus the roles in feeding may be a more recent addition; however, all vertebrates that have been investigated with regard to connections between the NPY system and food intake show increased feeding in response to NPY (mammals, chicken, sparrow, garter snake, and goldfish (Clark et al. 1984; Kuenzel et al. 1987; López-Patiño et al. 1999; Morris and Crews 1990; Richardson et al. 1995) and fasting leads to an upregulation of NPY mRNA (goldfish and salmon; Narnaware and Peter 2001; Silverstein et al. 1998) or to upregulation of NPY-binding sites (E.T. Larson and D. Larhammar, unpublished results).

The Y_1 and Y_5 receptors have also evolved some interesting differences. First, their anatomical distribution differs in that Y_1 is much more widespread than Y_5. Cells that express Y_5 always seem to express also Y_1 (Parker and Herzog 1999). Some species differences have been observed for Y_5 distribution, but its expression in feeding centers appears to be conserved across mouse, rat, guinea pig, and human (Holmberg et al., in press). Second, the recent development of subtype-selective agonists and antagonists have allowed functional studies in vivo that suggest subtype-specific effects on different aspects of feeding in guinea pigs (Lecklin et al. 2002, 2003). Stimulation of Y_1 increased the number of small meals, decreased the latency to the first meal, and reduced average meal size and duration (Lecklin et al. 2003). Thus, Y_1 stimulates the appetitive phase of feeding (the period before eating when the animal is looking for food) rather than the consummatory phase, i.e., biting and swallowing. Y_5 on the other hand increased the time spent on eating and the average meal size and duration, but did not affect the number of meals. Thus, Y_5 stimulates the consummatory phase of feeding (Lecklin et al. 2003). Future studies in other species will show whether this distinction between Y_1 and Y_5 is a general mammalian feature or tends to differ between species.

The appearance of PP by duplication of PYY seems to postdate the origin of its favored receptor in mammals, namely Y_4, thus the ligand–receptor pair of PP and Y_4 constitutes an interesting example of rapid ligand–receptor co-evolution. As PYY has been found to act as an inhibitor of feeding (Batterham et al. 2002, 2003a), it is interesting to note that also PP has been found to have such effects in rat and mouse (Asakawa et al. 2003) as well as in humans (Batterham et al. 2003b) and patients with the Prader–Willi syndrome (Bernston et al. 1993). Interestingly, the chicken Y_4 receptor binds PYY and PP with equal affinity (Lundell et al. 2002), suggesting that this species is still in a more ancestral state with regard to PYY and PP binding to Y_4.

The Y_6 receptor, too, seems to be in great evolutionary flux in mammals as it is clearly a pseudogene in several species. Furthermore, the inactivating mutations in the coding region appear to have occurred independently in primates, pig and guinea pig. This is further supported by the finding that a distant relative of the pig, the collared peccary, seems to have an intact open reading frame in the Y_6 gene (Wraith 1999). One possible explanation is that the Y_6 receptor had a rather minor role in the ancestor of mammals and then underwent further deterioration into a pseudogene in some lineages (Larhammar et al.

2001; Starbäck et al. 2000). In mouse and rabbit it may have regained some importance although its specific role is still unclear.

Our cloning of the three Y_1-subfamily receptors Y_1, Y_4 and Y_6 in spiny dogfish allowed studies of receptor mRNA distribution by reverse transcription (RT)-PCR. Surprisingly, the anatomical distribution differs greatly from that of mammals in that Y_4 has the widest distribution with expression in brain, retina, kidney and muscle, whereas Y_1 has narrow but strong expression in liver and kidney but was undetectable in brain (Salaneck et al. 2003). Y_6 mRNA was found in retina, kidney and the gastrointestinal tract. Likewise, the river lamprey Y_1-subfamily receptor, which is thought to be a pro-ortholog of Y_4 and Yb, was detected by RT-PCR in the central nervous system, the liver and the gonads (Salaneck et al. 2001). The Yb receptor in rainbow trout, on the other hand, could only be detected in hypothalamus and telencephalon and not in liver, kidney or intestine (Larson et al. 2003). Although these RT-PCR experiments were at best semi-quantitative and need to be complemented with in situ hybridizations, the results suggest that the anatomical distribution of a specific subtype may differ across classes of vertebrates, probably reflecting changes in the physiological roles of the receptor subtypes during evolution.

The pharmacological properties of the novel Y_7 subtype in zebrafish differ from the Y_2 receptor in mammals and chicken in that Y_7 does not respond to amino-terminally truncated fragments of NPY and PYY. Porcine NPY(13-36) has 50-fold lower affinity than intact NPY (Fredriksson et al. 2003). Thus, Y_2 remains unique among the Y receptors in its ability to respond to short fragments of NPY, in addition to the zebrafish Ya receptor which has a quite promiscuous pharmacological profile (Starbäck et al. 1999).

Y receptors from different species of mammals usually display indistinguishable or almost identical pharmacological profiles, even in comparisons across different orders of mammals. This great conservation was one reason why we decided to clone and express in vitro all five 'mammalian' subtypes in chicken whose ancestor began to diverge from the mammalian lineage approximately 300 million years ago. In addition to the interesting properties of the chicken Y_4 receptor described above, we also found that the chicken Y_2 receptor differed from its ortholog in mammals in the binding of the antagonist BIIE 0246, although the binding of the peptide radioligand was unaffected (Salaneck et al. 2000). After detailed sequence comparisons a few potentially important amino acid positions were selected for mutagenesis to identify the crucial positions. We introduced the residues found in chicken Y_2 into the human receptor, one at a time and in various combinations. Three chicken residues were thereby found to affect antagonist affinity, each of which reduced binding by approximately one order of magnitude (Berglund et al. 2002). To confirm the importance of these positions, the human residues were reciprocally introduced into the chicken Y_2 receptor whereby antagonist binding to this receptor could be achieved (Berglund et al. 2002). Further mutagenesis studies of the three positions suggested that one was important for direct interaction with the antagonist whereas the other two only needed to have sufficiently small side chains to allow space for the antagonist in the binding pocket. This illustrates the usefulness of evolutionary comparisons of species at appropriate evolutionary distances. The chicken Y_1 and Y_5 receptors, in contrast, were found to bind both agonists and antagonists with affinities very similar to those of the mammalian orthologs (Holmberg et al. 2002). Thus, these antagonists developed in mammals may be useful for in vivo studies of feeding in chicken.

The expansion of the NPY system by gene duplications in early vertebrate evolution by a mechanism of block or chromosome duplications is mirrored by similar expansions for other G-protein coupled receptor families. As briefly mentioned above, the CRH and NPFF receptors arose simultaneously with the duplications of the NPY and Y receptor chromosomes, respectively. The latter paralogon also contains the dopamine receptors D_{1A} and D_{1B} (=D_5) on Hsa5q and Hsa4p. These are the only D_1 receptor subtypes present in mammals whereas a third subtype, D_{1C}, is present in all other classes of gnathostomes. Thus, D_{1C} seems to have been lost in mammals (Le Crom et al. 2003). A fourth and seemingly equally old subtype, D_{1D}, has so far been found only in chicken. A scenario very similar to that of the Y receptors has recently been proposed for the adrenergic receptors, where an ancient local triplication led to the ancestors of α_1, α_2 and β, whereupon chromosome quadruplication and a few gene losses led to the three subtypes of each of these presently found in mammals (Ruuskanen et al. 2003). In addition, a fourth ancient α_2 subtype was discovered in zebrafish as well as several other teleost fishes (Ruuskanen et al. 2003). On top of that, many of these subtypes were found in duplicate due to the teleost tetraploidization. Another gene family with basal vertebrate duplications followed by several gene losses is the melanocortin receptor family (Klovins et al. 2003).

In conclusion, the complex families of NPY-like peptides and Y receptors have quite eventful evolutionary histories which are gradually beginning to reveal themselves thanks to combined comparative studies of sequences, chromosomal locations, anatomical distribution and pharmacological properties in a broad range of vertebrate species. The common ancestry of the different peptides and receptor subtypes helps explain their partially overlapping functions. The challenges ahead include studies of the NPY system in species that branched off before the origin of vertebrates, such as amphioxus, and the search for an evolutionary explanation to the opposing effects of NPY-Y_1-Y_5 and PYY(3–36)-Y_2 on food intake.

Acknowledgements. We are grateful to Ms. Christina Bergqvist for expert technical help. This work was supported by grants from the Swedish Research Council, Formas (The Swedish Research Council for Environment, Agricultural Sciences and Spatial Planning), the National Network for the Neurosciences (NNN) of the Swedish Strategic Funds (SSF), and Carl Trygger's Foundation.

References

Arvidsson A-K, Wraith A, Jönsson-Rylander A-C, Larhammar D (1998) Cloning of a neuropeptide Y/peptide YY receptor from the Atlantic cod: the Yb receptor. Regul Pept 75/76:39–43

Asakawa A, Inui A, Yuzuriha H, Ueno N, Katsuura G, Fujimiya M, Fujino MA, Niijima A, Meguid MM, Kasuga M (2003) Characterization of the effects of pancreatic polypeptide in the regulation of energy balance. Gastroenterology 124:1325–1336

Baker E, Hort YJ, Ball H, Sutherland GR, Shine J, Herzog H (1995) Assignment of the human neuropeptide Y gene to chromosome 7p15.1 by nonisotopic *in situ* hybridization. Genomics 26:163–164

Batterham RL, Cowley MA, Small CJ, Herzog H, Cohen MA, Dakin CL, Wren AM, Brynes AE, Low MJ, Ghatei MA, Cone RD, Bloom SR (2002) Gut hormone PYY3–36 physiologically inhibits food intake. Nature 418:650–654

Batterham RL, Cohen MA, Ellis SM, Le Roux CW, Withers DJ, Frost GS, Ghatei MA, Bloom SR (2003a) Inhibition of food intake in obese subjects by peptide YY3-36. N Engl J Med 349:941–948

Batterham RL, Le Roux CW, Cohen MA, Park AJ, Ellis SM, Patterson M, Frost GS, Ghatei MA, Bloom SR (2003b) Pancreatic polypeptide reduces appetite and food intake in humans. J Clin Endocrinol Metab 88:3989–3992

Berglund MM, Fredriksson R, Salaneck E, Larhammar D (2002) Reciprocal mutations of neuropeptide Y receptor Y2 in human and chicken identify amino acids important for antagonist binding. FEBS Lett 518:5-9

Bernston GG, Zipf WB, O'Dorsio TM, Hoffman JA, Chance RE (1993) Pancreatic polypeptide infusions reduce food intake in Prader-Willi syndrome. Peptides 14:497–503

Brown MR, Crim JW, Arata RC, Cai HN, Chun C, Shen P (1999) Identification of a Drosophila brain-gut peptide related to the neuropeptide Y family. Peptides 20:1035–1042

Cerdá-Reverter JM, Larhammar D (2000) Neuropeptide Y family of peptides: structure, anatomical expression, function, and molecular evolution. Biochem Cell Biol 78:371–392

Cerdá-Reverter JM, Martínez-Rodríguez G, Zanuy S, Carrillo M, Larhammar D (2000) Molecular evolution of the neuropeptide Y (NPY) family of peptides: cloning of NPY-related peptides from the sea bass *(Dicentrarchus labrax)*. Regul Pept 95:25–34

Clark JT, Kalra PS, Crowley WR, Kalra SP (1984) Neuropeptide Y and human pancreatic polypeptide stimulate feeding behavior in rats. Endocrinology 115:427–429

Comte M, Malnoë a, Cox JA (1986) Affinity purification of seminalplasmin and characterization of its interaction with calmodulin. Biochem J 240:567–573

Conlon JM (2002) The origin and evolution of peptide YY (PYY) and pancreatic polypeptide (PP). Peptides 23:269–278

Conlon JM, Bjørnholm B, Jørgensen FS, Youson JH, Schwartz TW (1991) Primary structure and conformational analysis of peptide methionine-tyrosine, a peptide related to neuropeptide Y and peptide YY isolated from lamprey intestine. Eur J Biochem 199:293–298

Coulier F, Popovici C, Villet R, Birnbaum D (2000) Meta*HOX* gene clusters. J Exp Zool 288:345–351

Couzens M, Liu M, Tüchler C, Kofler B, Nessler-Menardi C, Parker RMC, Klocker H, Herzog H (2000) Peptide YY-2 (PYY2) and pancreatic polypeptide-2 (PPY2): species-specific evolution of novel members of the neuropeptide Y gene family. Genomics 64:318–323

de Bono M, Bargmann CI (1998) Natural variation in a neuropeptide Y receptor homolog modifies social behavior and food response in *C. elegans*. Cell 94:679–689

Dehal P, Satou J, P D, Y S, RK C, J C, Degnan B DTA, Davidson B, Di Gregorio A, Gelpke M, Goodstein DM, Harafuji N, Hastings KE, Ho I, Hotta K, Huang W, Kawashima T, Lemaire P, Martinez D, Meinertzhagen IA, Necula S, Nonaka M, Putnam N, Rash S, Saiga H, Satake M, Terry A, Yamada L, Wang HG, Awazu S, Azumi K, Boore J, Branno M, Chin-Bow S, DeSantis R, Doyle S, Francino P, Keys DN, Haga S, Hayashi H, Hino K, Imai KS, Inaba K, Kano S, Kobayashi K, Kobayashi M, Lee BI, Makabe KW, Manohar C, Matassi G, Medina M, Mochizuki Y, Mount S, Morishita T, Miura S, Nakayama A, Nishizaka S, Nomoto H, Ohta F, Oishi K, Rigoutsos I, Sano M, Sasaki A, Sasakura Y, Shoguchi E, Shin-i T, Spagnuolo A, Stainier D, Suzuki MM, Tassy O, Takatori N, Tokuoka M, Yagi K, Yoshizaki F, Wada S, Zhang C, Hyatt PD, Larimer F, Detter C, Doggett N, Glavina T, Hawkins T, Richardson P, Lucas S, Kohara Y, Levine M, Satoh N, Rokhsar DS. (2002) The draft genome of *Ciona intestinalis*: Insights into chordate and vertbrate origins. Science 298:2157–2167

Force A, Lynch M, Pickett FB, Amores A, Yan Y, and Postlethwait J (1999) Preservation of duplicate genes by complementary, degenerative mutations. Genetics 151:1531–1545

Fredriksson R, Larson ET, Yan Y-L, Postlethwait JH, Larhammar D (2003) Novel neuropeptide Y Y2-like receptor subtype in zebrafish and frogs supports early vertebrate chromosome duplications. J Mol Evol (in press)

Garcia-Fernàndez J, Holland PWH (1994) Archetypal organization of the amphioxus *Hox* gene cluster. Nature 370:563–566

Garczynski SF, Brown MR, Shen P, Murray TF, Crim JW (2002) Characterization of a functional neuropeptide F receptor from *Drosophila melanogaster*. Peptides 23:773–780

Gregor P, Feng Y, DeCarr LB, Cornfield LJ, McCaleb ML (1996) Molecular characterization of a second mouse pancreatic polypeptide receptor and its inactivated human homologue. J Biol Chem 271:27776–27781

Griffin D, Minth CD, Taylor WL (1994) Isolation and characterization of the *Xenopus laevis* cDNA and genomic homologs of neuropeptide Y. Mol Cell Endocrinol 101:1–10

Gu X, Huang W (2002) Testing the parsimony test of genome duplications: a counterexample. Genome Res 12:1–2

Herzog H, Darby K, Ball H, Hort Y, Beck-Sickinger A, Shine J (1997) Overlapping gene structure of the human neuropeptide Y receptor subtypes Y1 and Y5 suggests coordinate transcriptional regulation. Genomics 41:315–319

Herzog H, Hort Y, Schneider R, Shine J (1995) Seminalplasmin: Recent evolution of another member of the neuropeptide Y gene family. Proc Natl Acad Sci USA 92:594–598

Hill CA, Fox AN, Pitts RJ, Kent LB, Tan PL, Chrystal MA, Cravchik A, Collins FH, Robertson HM, Zwiebel LJ (2002) G protein-coupled receptors in *Anopheles gambiae*. Science 298:176–178

Holland PWH, Garcia-Fernàndez J, Williams NA, Sidow A (1994) Gene duplications and the origins of vertebrate development. Development Supplement: 125–133

Holmberg SK, Mikko S, Boswell T, Zoorob R, Larhammar D (2002) Pharmacological characterization of cloned chicken neuropeptide Y receptors Y1 and Y5. J Neurochem 81:462–471

Holmberg SK, Johnson AE, Bergqvist C, Källström L, Larhammar D (in press) Distribution of neuropeptide Y receptor Y5 mRNA in the guinea pig brain. Regul Peptides

Hort Y, Baker E, Sutherland GR, Shine J, Herzog H (1995) Gene duplication of the human peptide YY gene (PYY) generated the pancreatic polypeptide gene (PPY) on chromosome 17q21.1. Genomics 26:77–83

Hughes AL (1999) Phylogenies of developmentally important proteins do not support the hypothesis of two rounds of genome duplication earlly in vertebrate history. J Mol Evol 48:565–576

Hughes AL, da Silva J, Friedman R (2001) Ancient genome duplications did not structure the human *Hox*-bearing cromosomes. Genome Res. 11:771–780

Hughes MK, Hughes AL (1993) Evolution of duplicate genes in a tetraploid animal, *Xenopus laevis*. Mol Biol Evol 10:1360–1369

Kim JB, Gadsbøll V, Whittaker J, Barton BA, Conlon JM (2000) Gastroenteropancreatic hormones (insulin, glucagon, somatostatin and multiple forms of PYY) from the pallid sturgeon, *Scaphirhynchus albus* (Acipenseriformes). Gen Comp Endocrinol 120:353–363

Kim JB, Johansson A, Conlon JM (2001) Anomalous rates of revolution of pancreatic polypeptide and peptide tyrosine-tyrosine (PYY) in a tetraploid frog, *Xenopus laevis* (Anura: Pipidae). Peptides 22:317–322

Kimmel JR, Hayden LJ, Pollock HG (1975) Isolation and characterization of a new pancreatic polypeptide hormone. J Biol Chem 250:9369–9374

Kimmel JR, Pollock HG, Hazelwood RL (1968) Isolation and characterization of chicken insulin. Endocrinology 83:1323–1330

Klovins J, Haitina T, Fridmanis D, Kilianova Z, Kapa I, Fredriksson R, Gallo-Payet N, Schiöth HB (2003) Determination of POMC/AGRP/MCR gene repertoire and synteny, as well as pharmacology and anatomical distribution of the MCRs in Fugu, clarifies the vertebrate origin of the melanocortin systm. (submitted for publication)

Kubiak TM, Larsen MJ, Nulf SC, Zantello MR, Burton KJ, Bowman JW, Modric T, Lowery DE (2003) Differential activation of 'social' and 'solitary' variants of the *C. elegans* GPCR NPR-1 by its cognate ligand AF9. J Biol Chem 278:33724–33729

Kuenzel WJ, Douglass LW, Davison BA (1987) Robust feeding following central administration of neuropeptide Y or peptide YY in chicks, *Gallus domesticus*. Peptides 8:823–828

Larhammar D (1996) Evolution of neuropeptide Y, peptide YY, and pancreatic polypeptide. Regul Pept 62:1–11

Larhammar D, Blomqvist AG, Söderberg C (1993) Evolution of neuropeptide Y and its related peptides. Comp Biochem Physiol 106C:743–752

Larhammar D, Lundin L-G, Hallböök F (2002) The human Hox-bearing chromosome regions did arise by block or chromosome (or even genome) duplications. 12:1910–1920

Larhammar D, Wraith A, Berglund MM, Holmberg SKS, Lundell I (2001) Origins of the multiple NPY-family receptors in mammals. Peptides 22:295–307

Larson ET, Fredriksson R, Johansson SRT, Larhammar D (2003) Cloning, pharmacology, and distribution of the neuropeptide Y-receptor Yb in rainbow trout. Peptides 24:385–395

Le Crom S, Kapsimali M, Barôme P-O, Vernier P (2003) Dopamine receptors for every species: gene duplications and functional diversification in craniates. J Struct Funct Genomics 3:161–176

Lecklin A, Lundell I, Paananen L, Wikberg JE, T. MP, D. L (2002) Receptor subtypes Y1 and Y5 mediate neuropeptide Y induced feeding in the guinea-pig. Br J Pharmacol 135:2029–2037

Lecklin A, Lundell I, Salmela S, Beck-Sickinger AG, D. L (2003) Agonists for neuropeptide Y receptors Y_1 and Y_5 stimulate different phases of feeding in guinea pigs. Br J Pharmacol 139:1433–1440

Lewis RV, Augustin JS, Kruggel W, Lardy HA (1985) The structure of caltrin, the calcium-transport inhibitor of bovine seminal plasma. Proc Natl Acad Sci USA 82:6490–6491

López-Patiño MA, Guijarro AI, Isorna E, Delgado MJ, Alonso-Bedate M, de Pedro N (1999) Neuropeptide Y has a stimulatory action on feeding behavior in goldfish (Carassius auratus). Eur J Pharmacol 377:147–53

Lundell I, Berglund MM, Starbäck P, Salaneck E, Gehlert DR, Larhammar D (1997) Cloning and characterization of a novel neuropeptide Y receptor subtype in the zebrafish. DNA Cell Biol 16:1357–1363

Lundell I, Boswell T, Larhammar D (2002) Chicken neuropeptide Y-family receptor Y4: a receptor with equal affinity for pancreatic polypeptide, neuropeptide Y and peptide YY. J Mol Endocrinol 28:225–235

Lundin LG, Hallböök F, Larhammar D (2003) Numerous groups of chromosomal regional paralogies strongly indicate two genome doublings at the root of the vertebrates. J Struct Funct Genomics 3:53–63

Lutz CM, Frankel WN, Richards JE, Thompson DA (1997a) Neuropeptide Y receptor genes on human chromosome 4q31-q-32 map to conserved linkage groups on mouse chromosomes 3 and 8. Genomics 41:498–500

Lutz CM, Richards JE, Scott KL, Sinha S, Yang-Feng TL, Frankel WN, Thompson DA (1997b) Neuropeptide Y receptor genes mapped in human and mouse: receptors with high affinity for pancreatic polypeptide are not clustered with receptors specific for neuropeptide Y and peptide YY. Genomics 46:287–290

Martin A (2001) Is tetralogy true? Lack of support for the 'one-to-four' rule. Mol Biol Evol 18:89–93

Martin AP (1999) Increasing genomic complexity by gene duplication and the origin of vertebrates. Am Naturalist 154:111–128

Matsumoto M, Nomura T, Momoses K, Ikeda Y, Kondou Y, Akiho H, Togami J, Kimura Y, Okada M, Yamaguchi T (1996) Inactivation of a novel neuropeptide Y/peptide YY receptor gene in primate species. J Biol Chem 271:27217–27220

Michel MC, Beck-Sickinger A, Cox H, Doods HN, Herzog H, Larhammar D, Quirion R, Schwartz T, Westfall T (1998) XVI. International Union of Pharmacology recommendations for the nomenclature of neuropeptide Y, peptide YY and pancreatic polypeptide receptors. Pharmacol Rev 50:143–150

Morris YA, Crews D (1990) The effects of exogenous neuropeptide Y on feeding and sexual behavior in the red-sided garter snake *(Thamnophis sirtalis parietalis)*. Brain Res 530:339–341

Narnaware YK, Peter RE (2001) Effects of food deprivation and refeeding on neuropeptide Y (NPY) mRNA levels in goldfish). Comp Biochem Physiol B 129:633–637

Panopoulou G, Hennig S, Groth D, Krause A, Poustka AJ, Herwig R, Vingron M, Lehrach H (2003) New evidence for genome-wide duplications at the origin of vertebrates using an amphioxus gene set and completed animal genomes. Genome Res 13:1056–1066

Parker RM, Herzog H (1999) Regional distribution of Y-receptor subtype mRNAs in rat brain. Eur J Neurosci 11:1431–1448

Pébusque M-J, Coulier F, Birnbaum D, Pontarotti P (1998) Ancient large-scale genome duplications: phylogenetic and linkage analyses shed light on chordate genome evolution. Mol Biol Evol 15:1145–1159

Pedrazzini T, Pralong F, Grouzmann E (2003) Neuropeptide Y: the universal soldier. Cell Mol Life Sci 60:350–377

Pevzner P, Tesler G (2003) Genome rearrangements in mammalian evolution: lessons from human and mouse genomes. Genome Res 13:37–45

Pieribone VA, Brodin L, Friberg K, Dahlstrand J, Söderberg C, Larhammar D, Hökfelt T (1992) Differential expression of mRNAs for neuropeptide Y-related peptides in rat nervous tissues: possible evolutionary conservation. J Neurosci 12:3361–3371

Popovici C, Leveugle M, Birnbaum D, Coulier F (2001) Homeobox gene clusters and the human paralogy map. FEBS Lett 491:237–242

Postlethwait JH, Woods IG, Ngo-Hazelett P, Yan Y-L, Kelly PD, Chu F, Hill-Force A, Walbot WS (2000) Zebrafish comparative genomics and the origins of vertebrate chromosomes. Genome Res 2000:1890–1902

Reddy ESP, Bhargava PM (1976) Seminalplasmin—an antimicrobial protein from bovine seminal plasma which acts in E. coli by specific inhibition of rRNA synthesis. Nature 279:725–728

Richardson RD, Boswell T, Raffety BD, Seeley RJ, Wingfield JC, Woods SC (1995) NPY increases food intake in white-crowned sparrows: effect in short and long photoperiods. Am J Physiol 268:R1418–R1422

Ringvall M, Berglund MM, Larhammar D (1997) Multiplicity of neuropeptide Y receptors: cloning of a third distinct subtype in the zebrafish. Biochem Biophys Res Commun 241:749–755

Rose PM, Lynch JS, Frazier ST, Fisher SM, Chung W, Battaglino P, Fathi Z, Leibel R, Prabhavathi F (1997) Molecular genetic analysis of a human neuropeptide Y receptor. The human homolog of the murine 'Y5' receptor may be a pseudogene. J Biol Chem 272:3622–3627

Ruuskanen J, Xhaard H, Marjamäki A, Salaneck E, Salminen T, Yan Y-L, Postlethwait JH, Johnson MS, Larhammar D, Scheinin M (2003) Early emergence of multiple a2-drenergic receptor subtypes by chromosomal duplications as revealed by cloning and mapping of five receptor subtype genes in the zebrafish (*Danio rerio*) - identification of a duplicated fourth α2-adrenergic receptor subtype. (in press)

Salaneck E, Ardell D, Larson ET, Larhammar D (2003) Three neuropeptide Y receptors in the spiny dogfish, *Squalus acanthias*, support chromosome doublings in early vertebrate evolution. Mol Biol Evol 20:1271–1280

Salaneck E, Fredriksson R, Larson ET, Conlon JM, Larhammar D (2001) A neuropeptide Y receptor Y1-subfamily gene from an agnathan, the European river lamprey. A potential ancestral gene. Eur J Biochem 268:6146–6154

Salaneck E, Holmberg SK, Berglund MM, Boswell T, Larhammar D (2000) Chicken neuropeptide Y receptor Y2: structural and pharmacological differences to mammalian Y2. FEBS Lett 484:229–234

Shimeld MS, Holland PWH (2000) Vertebrate innovations. Proc Natl Acad Sci USA 97:4449–4452

Silverstein JT, Breininger J, Baskin DG, Plisetskaya EM (1998) Neuropeptide Y-like gene expression in the salmon brain increases with fasting. Gen Comp Endocrinol 110:157–165

Söderberg C, Wraith A, Ringvall M, Yan YL, Postlethwait JH, Brodin L, Larhammar D (2000) Zebrafish genes for neuropeptide Y and peptide YY reveal origin by chromosome duplication from an ancestral gene linked to the homeobox cluster. J Neurochem 75:908–918

Söderberg C, Pieribone VA, Dahlstrand J, Brodin L, Larhammar D (1994) Neuropeptide role of both peptide YY and neuropeptide Y in vertebrates suggested by abundant expression of their mRNAs in a cyclostome brain. J Neurosci Res 37:633–640

Söderberg C, Wraith A, Ringvall M, Yan Y-L, Postlethwait J, Brodin L, Larhammar D (2000) Zebrafish genes for neuropeptide Y and peptide YY reveal origin by chromosome duplication from an ancestral gene linked to the homeobox cluster. J Neurochem 75:908–918

Stanek DM, Pohl J, Crim JW, Brown MR (2002) Neuropeptide F and its expression in the yellow fever mosquito, *Aedes aegypti*. Peptides 23:1367–1378

Starbäck P, Lundell I, Fredriksson R, Berglund MM, Yan Y-L, Wraith A, Söderberg C, Postlethwait JH, Larhammar D (1999) Neuropeptide Y receptor subtype with unique properties cloned in the zebrafish; the zYa receptor. Mol Brain Res 70:242–252

Starbäck P, Wraith A, Eriksson H, Larhammar D (2000) Neuropeptide Y receptor gene y6: multiple deaths or resurrection? Biochem Biophys Res Commun 277:264–269

Taylor JS, Braasch I, Frickey T, Meyer A, Van de Peer Y (2003) Genome duplication, a trait shared by 22,000 species of ray-finned fish. Genome Res 13:382–390

Tensen CP, Cox KJA, Burke JF, Leurs R, van der Schors RC, Geraerts WPM, Vreugdenhil e, van Heerikhuizen H (1998) Molecular cloning and characterization of an invertebrate homologue of a neuropeptide Y receptor. Eur J Neurosci 10:3409–3416

Van de Peer Y, Taylor JS, Meyer A (2003) Are all fishes ancient polyploids? J Struct Funct Genomics 2:65–73 van Riel MCHM, Tuinhof R, Roubos EW, Martens GJM (1993) Cloning and sequence analysis of hypothalamic cDNA encoding *Xenopus* preproneuropeptide Y. Biochem Biophys Res Commun 190:948-951

Vienne A, Rasmussen J, Abi-Rached L, Pontarotti P, Gilles A (2003) Systematic phylogenomic evidence of en bloc duplication of the ancestral 8p11.21-8p21.3-like region. Mol Biol Evol 20:1290-1298

Wang Y, Nielsen PF, Youson JH, Potter IC, Lance VA, Conlon JM (1999) Molecular evolution of peptide tyrosine-tyrosine: primary structure of PYY from the lampreys *Geotria australis* and *Lampetra fluviatilis*, bichir, python and desert tortoise. Regul Pept 79:103-108

Weinberg DH, Sirinathsinghji DJS, Tan CP, Shiao L-L, Morin N, Rigby MR, Heavens RH, Rapoport DR, Bayne ML, Cascieri MA, Strader CD, Linemeyer DL, MacNeil DJ (1996) Cloning and expression of a novel neuropeptide Y receptor. J Biol Chem 271:16435-16438

Wraith A (1999) Molecular Evolution of the Neuropeptide Y receptor family. Insights from mammals and fish. Uppsala University, Uppsala

Wraith A, Törnsten A, Chardon P, Harbitz I, Chowdhary BP, Andersson L, Lundin L-G, Larhammar D (2000) Evolution of the neuropeptide Y receptor family: gene and chromosome duplications deduced from the cloning of the five receptor subtype genes in pig. Genome Res 10:302-310

Neuropeptide Y and Its Receptor Subtypes in the Central Nervous System: Emphasis on Their Role in Animal Models of Psychiatric Disorders

J. P. Redrobe · C. Carvajal · A. Kask · Y. Dumont · R. Quirion

Douglas Hospital Research Centre, Institute of Neuroscience,
Mental Health and Addiction, 6875 LaSalle Blvd., Verdun, QC, H4H 1R3, Canada
e-mail: quirem@douglas.mcgill.ca

1	Introduction	102
2	Overview of NPY Receptor Subtypes	102
2.1	The Y_1 Subtype	102
2.2	The Y_2 Subtype	104
2.3	The Y_3 Subtype	104
2.4	The Y_4 Subtype	105
2.5	The Y_5 Subtype	105
2.6	The y_6 Subtype	106
3	Biological Effects of NPY and Related Peptides	106
3.1	General Overview	106
3.2	Role for NPY and Its Receptor Subtypes in Depression	108
3.3	Role for NPY and Its Receptor Subtypes in Anxiety	112
3.4	Role for NPY and Its Receptor Subtypes in Learning and Memory	117
3.5	Role for NPY and Its Receptor Subtypes in Seizure and Epilepsy	119
4	Conclusion and Future Directions	122
	References	123

Abstract In this chapter, we briefly review the pharmacology and distribution of the neuropeptide Y (NPY) Y_1, Y_2, Y_4 Y_5 and y_6 receptor proteins and mRNAs. NPY and its receptors are widely distributed in the central nervous system with varying concentrations being found throughout the limbic system and the hippocampal formation. Such brain structures have been implicated either in the modulation of emotional and cognition processing as well as in the pathogenesis of depressive and epileptic disorders. In fact, both preclinical and clinical studies lead us to suggest that NPY and its receptors may have a direct implication in psychiatric disorders such as depression and anxiety related illnesses. Furthermore, experimental studies also suggest that NPY and its receptors may play a role in the regulation of cognitive function associated with learning and memory and has anticonvulsant properties. Together, these data support the potential therapeutic usefulness of NPY in these disorders.

Keywords Neuropeptide Y · Depression · Anxiety · Learning and memory · Epilepsy

1
Introduction

Neuropeptide Y (NPY) was isolated from porcine brain more than two decades ago (Tatemoto and Mutt 1980). This 36-amino acid residue shares high sequence homology and structural identity with two other peptides, namely peptide YY (PYY) and the pancreatic polypeptide (PP) (Tatemoto et al. 1982). All these peptides have thus been included in the same peptide family called the NPY family (Table 1). NPY is one of the most abundant peptides found in the central nervous system (CNS) of all mammals, including humans (Chronwall et al. 1985; Chan-Palay et al. 1985a, 1986a), while PYY and PP are mostly found in endocrine cells of the intestine (Solomon 1985). Additionally, PYY has also been shown to be present in the brainstem and various hypothalamic nuclei (Ekman et al. 1986). These peptides, especially NPY and PYY, are among the most conserved peptides during evolution (Larhammar 1996a, 2001).

At present, five NPY receptor subtypes have been cloned and designated as Y_1, Y_2, Y_4, Y_5 and y_6 (Table 1). All belong to the seven-transmembrane G protein-coupled receptor of the rhodopsin family (Michel et al. 1998). Despite the fact that NPY and PYY have very high affinities for the cloned Y_1, Y_2, Y_5 and y_6 subtypes (Table 1) (Michel et al. 1998), these receptors display relatively low sequence identities between each other (about 30%–50%). In fact, NPY receptors appear to be the most divergent receptors among a given receptor family (Larhammar 1996b, 1997). Indeed, some NPY receptor subtypes have even higher homology for other families of G protein-coupled receptors (Elshourbagy et al. 2000; Bonini et al. 2000; Parker et al. 2000a, 2000b).

2
Overview of NPY Receptor Subtypes

The reader is invited to consult our recent reviews for more details on NPY receptors distribution and their pharmacology (Dumont et al. 2000b, 2002) and to refer to Table 1.

2.1
The Y_1 Subtype

The first NPY receptor clone was initially reported as an orphan receptor isolated by screening a rat forebrain cDNA library (Eva et al. 1990). Upon transfection into a cell line, this clone demonstrated a ligand selectivity profile that was typi-

Table 1 Neuropeptide Y Receptors: distribution, molecular, pharmacological and behavioral characteristics

	Y_1	Y_2	Y_4	Y_5	Y_6
Preferred endogenous ligand	neuropeptide Y (NPY) peptide YY (PYY)	NPY, PYY, NPY(3–36) PYY(3–36)	Pancreatic polypeptide (PPs)	NPY, PYY, NPY(3–36) PYY(3–36)	NPY, PYY, NPY(3–36) PYY(3–36), PP
Agonists	[Pro^{34}]NPY, [Pro^{34}]PYY, [Leu^{31},Pro^{34}]NPY, [Leu^{31},Pro^{34}]PYY	NPY(13–36), PYY(13–36) [Ahx$^{5-24}$, γ-Glu^2-ε-Lys^{20}]NPY, C2-NPY	[Leu^{31},Pro^{34}]NPY, [Leu^{31},Pro^{34}]PYY, GR231118	[Leu^{31},Pro^{34}]NPY, [Leu^{31},Pro^{34}]PYY, hPP, [Ala^{31},Aib^{32}]NPY, [hPP$_{1-17}$, Ala^{31},Aib^{32}]NPY	Not well defined
Antagonists	BIBP 3226, BIBO 3304 GR 231118, J104870 J115814, GI264879A	BIIE 0246		L152804, CGP 71683A	—
Radioligands	[^{125}I]-PYY, [^{125}I]-[Leu^{31}, Pro^{34}]PYY, [^{125}I]-GR231118	[^{125}I]-PYY, [^{125}I]-PYY(3–36)	[^{125}I]-PPs, [^{125}I]-PYY, [^{125}I]-[Leu^{31},Pro^{34}]PYY, [^{125}I]-GR231118	[^{125}I]-PYY, [^{125}I]-[Leu^{31},Pro^{34}]PYY, [^{125}I]hPP	[^{125}I]-PYY
Expression profile	Cerebral cortex, thalamus, brain stem, smooth muscle of blood vessels	Hippocampus, brain stem nuclei, hypothalamus, gastrointestinal tract, smooth muscle of blood vessels	Colon, small intestine, prostate, very low in brain, paraventricular hypothalamus, interpeduncular nucleus	Hippocampus, plexiform cortex of the olfactory bulb, suprachiasmatic and arcruate nuclei	Not fully characterizsed
Physiological function	Vasoconstriction, regulation of food intake, anxiety-related behaviors, regulation of transmitter release	Inhibition of glutamate release, inhibition of noradrenaline release, learning and memory	Possible in regulation of LH secretion	Possibly regulation of food intake	Unclear
Knockout phenotypes	Hyperalgesia, altered feeding in response to NPY, increased alcohol consumption	Increased body weight and food intake	Rescues fertility in ob/ob mice	Susceptible to seizures	No obvious phenotype

cal to that of the Y_1 receptor: $PYY \geq NPY \geq [Leu^{31}, Pro^{34}]NPY > NPY(2-36) >>$ human (h)PP>NPY(13-36) (Krause et al. 1992). Subsequently, the human (Herzog et al. 1992; Larhammar et al. 1992), mouse (m; Eva et al. 1992; Nakamura et al. 1995), guinea pig (gp; Berglund et al. 1999), porcine (p), dog (d; Malmstrom et al. 1998) and monkey (mo; Gehlert et al. 2001) Y_1 receptor cDNAs were isolated.

In rat brain, the localization of Y_1 receptor mRNA (Eva et al. 1990; Larsen et al. 1993; Tong et al. 1997; Parker and Herzog 1999) closely matches that of the Y_1 receptor protein (Dumont et al. 1990, 1993, 1996; Dumont and Quirion 2000; Gehlert et al. 1992; Schober et al. 1998) with predominant expression in the cerebral cortex, thalamus and brainstem nuclei. However, species differences have been noted concerning the distribution of the Y_1 receptor subtype (Dumont et al. 1997, 1998b, 2000b; Jacques et al. 1997; Statnick et al. 1997, 1998).

2.2
The Y_2 Subtype

Expression screening from cDNA libraries of neuroblastoma cells known as SMS-KAN (Rose et al. 1995a, 1995b), human hippocampus (Gerald et al. 1995) and human brain (Gehlert et al. 1996a) led to the isolation of a human cDNA receptor clone which possesses a pharmacological binding profile similar to that of the Y_2 receptor. This receptor has now been cloned from various species including the rat (r; St Pierre et al. 1998), bovine (b; Ammar et al. 1996), mouse (Nakamura et al. 1996), guinea pig, porcine, dog (Malmstrom et al. 1998) and monkey (Gehlert et al. 2001). High homology (90%–95%) is observed between species (Larhammar et al. 2001). Rather surprisingly, considering that NPY and PYY possess very high affinities for both the Y_1 and Y_2 receptor subtypes, the overall homology between the Y_1 and Y_2 receptors is only 31%.

In the rat CNS, Y_2 receptor mRNA and protein are abundantly expressed in the hippocampal formation and brainstem nuclei, while moderate levels of receptors are detected in various hypothalamic nuclei (Dumont et al. 1993, 1996, 1997, 2000a,b; Gustafson et al. 1997; Larsen and Kristensen 1998; Parker and Herzog 1999). In fact, the Y_2 receptor mRNA is discretely distributed in the rat brain and its localization is largely similar to that of the Y_2 receptor protein (Dumont et al. 2000b). However, as for the Y_1 subtype, species differences exist in the level of expression of the Y_2 receptor in various brain regions (Dumont et al. 1997, 1998b, 2000b).

2.3
The Y_3 Subtype

Various groups have proposed the existence of a receptor that possesses high affinity for NPY, but not PYY, in several assays including the rat brain (Grundemar et al. 1991), rat colon (Dumont et al. 1994), rat lung (Hirabayashi et al. 1996) and bovine adrenals (Wahlestedt et al. 1992). However, evidence for the existence of such a subtype is circumstantial as the clone initially reported as a Y_3 receptor

(Rimland et al. 1991) does not bind NPY (Herzog et al. 1993; Jazin et al. 1993) but actually belongs to the cytokine receptor family (Feng et al. 1996). The cloning of a genuine Y_3 receptor is still awaited. It may well be that the Y_3 receptor protein is a G protein-coupled receptor for which the expression at the cell surface is dependent on the presence of regulating activity modifying protein. Alternatively, the Y_3 receptor could exist as a dimer of any of the cloned NPY receptors. Further studies are required to verify these hypotheses.

2.4
The Y_4 Subtype

The use of sequence homology screening, with a Y_1 receptor probe, led to the isolation of a new human NPY receptor cDNA. This receptor was originally designated as either PP_1 (Lundell et al. 1995; Gehlert et al. 1996b) or Y_4 (Bard et al. 1995). Homologues of the Y_4 receptor have now been cloned from mouse (Gregor et al. 1996b), rat (Lundell et al. 1996; Yan et al. 1996) and guinea pig (Eriksson et al. 1998). Sequence homology between human and other species is one of the lowest (less than 75%) reported between different mammalian species (Larhammar et al. 2001). Moreover, the human Y_4 receptor protein has higher homology with the human Y_1 (43%) than the human Y_2 (34%) receptor (Larhammar 1996b).

In the rat brain, only low levels of expression of the Y_4 receptor mRNA have been detected thus far (Parker and Herzog 1999). Studies using [^{125}I]rPP and [^{125}I]hPP (Trinh et al. 1996) or [^{125}I]bPP (Gehlert et al. 1997; Whitcomb et al. 1997) have confirmed the restricted distribution of PP (Y_4-like) binding sites in the rat brain.

2.5
The Y_5 Subtype

In the 1990s, the existence of an atypical receptor subtype was proposed on the basis of the effects of NPY and long C-terminal fragments, such as NPY(2–36), on food intake. This receptor was referred to as the atypical Y_1 or 'feeding' receptor (Quirion et al. 1990; Stanley et al. 1992). More recently, the profile of another NPY receptor cloned from human and rat tissues was classified as the Y_5 subtype and was found to have a pharmacological profile similar to that of the atypical feeding receptor (Gerald et al. 1996; Hu et al. 1996). This Y_5 receptor has now been cloned from various species including mouse (Nakamura et al. 1997), dog (Borowsky et al. 1998), guinea pig (Lundell et al. 2001) and monkey (Gehlert et al. 2001).

In the rat brain, in situ hybridization signals of the Y_5 receptor mRNA are seen in the external plexiform layer of the olfactory bulb, anterior olfactory nuclei, hippocampus, suprachiasmatic and arcuate nuclei (Parker and Herzog 1999; Larsen and Kristensen 1998); a distribution pattern relatively similar to that observed for the Y_5 receptor protein, except for very low levels of Y_5 receptor protein in the arcuate nucleus (Dumont et al. 1998a, 2000b). In the human brain, strong in situ hybridization signals of Y_5 receptor mRNA have also been detected at the levels of the arcuate nucleus, a key structure implicated in feeding behaviors (Jacques et al. 1998; Statnick et al. 1998).

2.6
The y_6 Subtype

Three groups have reported the cloning of another NPY receptor in mice, rabbit, monkey and human tissues (Gregor et al. 1996a; Matsumoto et al. 1996; Weinberg et al. 1996) now known as the y_6 subtype (Michel et al. 1998). Upon transfection of the mouse and rabbit receptor clone into cell lines, distinct pharmacological profiles have been reported with similarities to the Y_2 (Matsumoto et al. 1996), Y_4 (Gregor et al. 1996a) or Y_5 (Weinberg et al. 1996) receptors. Surprisingly, transfection of the human y_6 receptor cDNA failed to be fully translated and hence to generate a functional receptor. It is now known that the y_6 receptor is not expressed in the rat (Burkhoff et al. 1998), while in human and primates, the cDNA contains a single base deletion resulting in the expression of a nonfunctional NPY receptor (Gregor et al. 1996a; Matsumoto et al. 1996). Pseudogenes have also been reported in the guinea pig (Starback et al. 2000) and pig (Wraith et al. 2000) while in the dog, it is expressed as a functional receptor (Borowsky et al. 1998).

3
Biological Effects of NPY and Related Peptides

3.1
General Overview

Numerous studies have addressed the physiological functions of NPY and its congeners in the central and peripheral nervous systems (Table 1). In fact, intracerebroventricular (icv) injections of NPY or PYY demonstrated that these peptides are among the most potent substances known, thus far, to stimulate feeding behaviors (Stanley and Leibowitz 1984; Inui 1999).

It has also been reported that NPY and PYY inhibit glutamatergic excitatory synaptic transmission (Colmers and Bleakman 1994; Vezzani et al. 1999), induce hypothermia (Esteban et al. 1989), decrease sexual behavior (Clark et al. 1985), shift circadian rhythms (Albers et al. 1984; Albers and Ferris 1984) and modulate various neuroendocrine secretions (Kalra and Crowley 1992).

Reducing the synthesis of NPY by icv administration of NPY antisense oligonucleotides has been shown to decrease food intake (Akabayashi et al. 1994) and attenuate progesterone-induced luteinizing hormone (LH) surge (Kalra et al. 1995). In addition, icv injection of an NPY Y_1 receptor antisense induced marked anxiogenic-like behaviors (Wahlestedt et al. 1993) and decreased body temperature (Lopez-Valpuesta et al. 1996).

It has been demonstrated that NPY-knockout mice exhibited mild spontaneous seizures (Erickson et al. 1996), increased alcohol consumption (Thiele et al.

1998) and anxiogenic-like behaviors (Bannon et al. 2000). Moreover, transgenic mice overexpressing NPY were shown to be low alcohol drinkers (Thiele et al. 1998), while transgenic rats displayed anxiolytic-like behaviors and memory impairments (Thorsell et al. 2000).

There is strong evidence for the involvement of the Y_1 receptor in feeding behaviors based on the use of recently developed nonpeptide Y_1 receptor antagonists. Such molecules include BIBO 3304 (Wieland et al. 1998), J104870 (Kanatani et al. 1999), J115814 (Kanatani et al. 2001), BIBP 3226 (Kask et al. 1998b), GR 231118 (Ishihara et al. 1998) and GI 264879A (Daniels et al. 2001). On the other hand, mice lacking Y_1 receptors do not display major abnormalities in feeding behavior (Pedrazzini et al. 1998). However, food intake induced by NPY is profoundly altered in these animals (Kanatani et al. 2000b). Additionally, Y_1-knockout mice develop hyperalgesia to acute thermal, cutaneous and visceral chemical pain, and exhibit mechanical hypersensitivity (Naveilhan et al. 2001) suggesting a role for the Y_1 receptor subtype in nociception. Furthermore, Y_1-knockout mice showed increased consumption of ethanol (Thiele et al. 2002), suggesting that the NPY Y1 receptor regulates voluntary ethanol consumption and some of the intoxicating effects caused by administration of ethanol.

The only consistent results obtained in the CNS, regarding the physiological role of the Y_2 receptor subtype, is its involvement in the inhibition of glutamate release (Colmers et al. 1987; Kombian and Colmers 1992; Colmers and Bleakman 1994). Early studies also suggested that NPY may facilitate learning and memory retention via action at the Y_2 receptor (Flood et al. 1987; Redrobe et al. 1999). It has also been reported that Y_2 receptor knockout mice displayed increased body weight and food intake (Naveilhan et al. 1999) suggesting that the Y_2 receptor subtype may negatively regulate feeding behaviors. In agreement with a role of Y_2 receptors in downregulating food intake, it has been reported that peripheral administration of PYY(3-36) in rats inhibited food intake and reduced weight gain (Batterham et al. 2002). Similar effects were observed in mice but not in the NPY Y_2 knockout. Furthermore, this group also reported a decrease in food intake in humans following peripheral administration of PYY(3-36) (Batterham et al. 2002). They suggested that peripheral PYY(3-36) may act through the arcuate nucleus Y_2 receptors to inhibit food intake. However, the recently developed hypothalamus specific Y_2 receptor knockout mice show a significant decrease in body weight and a significant increase in food intake in association with increased mRNA levels of NPY (Sainsbury et al. 2002a).

Limited information is currently available regarding the potential role of the Y_4 receptor subtype. Two reports have suggested its possible involvement in LH secretion (Jain et al. 1999; Raposinho et al. 2000). A recent study has shown that when genetically obese *ob/ob* mice are crossed with Y_4 receptor knockout mice, fertility is improved in the $Y_4^{-/-}$, *ob/ob* double knockout animals by 100% in males and 50% in females (Sainsbury et al. 2002b). These data suggest that under conditions of centrally elevated NPY, Y_4 receptor signaling may act to specifically inhibit reproductive function.

The precise physiological role of the Y_5 receptor subtype is still under debate. First described as the receptor involved in NPY-induced feeding (Gerald et al. 1996), most recent studies failed to provide clear evidence for such a role as Y_5 receptor knockout mice were shown to display normal feeding behavior (Marsh et al. 1998). Moreover, NPY-induced food intake is not altered in these animals (Kanatani et al. 2000b). Additionally, the effect of NPY on food intake was unaltered by L-152804, a Y5 antagonist, while that of bPP was reduced (Kanatani et al. 2000a). Interestingly, selective Y_5 agonists have been shown to stimulate food intake (McCrea et al. 2000; Cabrele et al. 2000, 2001).

In the following sections of this chapter, we will concentrate on evidence that suggests a putative role for NPY, and its receptors, in specific psychiatric disorders (not discussed elsewhere in this volume). We will review the possible role played by this peptide family in disorders such as depression, anxiety, learning and memory disruption and seizure/epilepsy.

3.2
Role for NPY and Its Receptor Subtypes in Depression

Preclinical data have consistently indicated a role for NPY in depression (Table 2). Such studies have used animal models widely considered to mimic, at

Table 2 Summary of preclinical evidence indicating a role for NPY in depression

Model	Effect	Reference
Electroconvulsive shock	↑ NPY gene expression ↑ NPY immunoreactivity ↑ Pre-pro-NPY mRNA ↑ Extracellular NPY	Heilig et al. 1988a Stenfors et al. 1995 Zachrisson et al. 1995a Husum et al. 2000
Lithium	↑ NPY immunoreactivity ↑ pre-pro-NPY mRNA	Zachrisson et al. 1995b
Antidepressants	↑ NPY immunoreactivity ↓ [^{3}H]-NPY binding sites ↑ NPY mRNA ↑ Y_1 receptor mRNA ↓ Y_2 receptor density ↑ ^{125}I-PYY binding sites	Heilig et al. 1988a Widdowson and Halaris 1991 Caberlotto et al. 1998 Widdowson and Halaris 1991 Husum et al. 2000
Flinders rats	↑↓ NPY mRNA ↑↓ NPY Y_1 mRNA ↑↓ NPY immunoreactivity ↑↓ NPY Y_1 binding sites = Y_2 binding sites	Caberlotto et al. 1998 Caberlotto et al. 1999
Olfactory bulbectomy	↓ Ambulation ↑ NA/5-HT ↓ Lymphocyte proliferation ↑ NPY gene expression	Song et al. 1996 Holmes et al. 1998
Forced swimming test	↓ Immobility time	Stogner and Holmes 2000 Redrobe et al. 2002

↑, Increase; ↓, decrease; ↑↓, increase/decrease depending on brain region; =, no change.

least in some respects, the behavioral, biochemical and neurochemistry of the clinical condition.

Experiments have shown that repeated, but not single, electroconvulsive shock stimulation in rats, considered as an animal model of electroconvulsive therapy in humans, increased NPY gene expression (Mikkelsen et al. 1994). Such treatment also markedly increased levels of NPY immunoreactivity in homogenates of hippocampal and cortical regions (Stenfors et al. 1995), pre-pro-NPY mRNA in the stratum oriens of the hippocampus (Zachrisson et al. 1995a) and hilus of the dentate gyrus (Mikkelsen et al. 1994). Moreover, it has recently been found that electroconvulsive shock stimulation significantly increased extracellular levels of NPY in the dorsal hippocampus of freely moving rats as determined by microdialysis, suggesting that such treatment led to an increased biosynthesis and release of NPY in this region (Husum et al. 2000). Similar results were obtained following treatment with lithium, a therapy often used for the pharmacotherapy of bipolar disorder (Husum et al. 2000). Previously, it has been shown that lithium increased levels of pre-pro-NPY mRNA and NPY immunoreactivity in hippocampal and cortical regions (Zachrisson et al. 1995b). Thus, it can be concluded that an animal model (i.e., electroconvulsive shock), incorporating characteristics similar to treatment employed in the clinic (i.e., electroconvulsive therapy), induces significant changes in NPY-ergic mediated systems.

In addition, chronic antidepressant treatment has been shown to increase NPY and NPY Y_1-type receptor mRNA levels (Caberlotto et al. 1998), and to reduce NPY Y_2-type receptor densities in certain brain regions (Widdowson and Halaris 1991). Similarly, chronic treatment with the tricyclic antidepressant imipramine, increased NPY immunoreactivity in frontal cortex (Heilig et al. 1988a) and decreased [^{3}H]NPY binding in frontal cortex and hippocampus of rats (Widdowson and Halaris 1991). In contrast, chronic treatment with the selective serotonin reuptake inhibitor (SSRI) citalopram did not induce any significant changes in NPY immunoreactivity in rat hippocampal homogenates after chronic treatment (Husum et al. 2000; Heilig and Ekman 1995). However, citalopram treatment did increase [^{125}I]PYY binding sites in the hippocampal formation, changes representative of a possible increase in expression, or decreased degradation, of NPY-sensitive receptors (Husum et al. 2000). Another possible mechanism may be related to the ability of citalopram to increase the affinity of NPY-sensitive receptors for the endogenous ligand (Husum et al. 2000) and hence increase NPY neurotransmission.

Interestingly, NPY-like immunoreactivity and NPY Y_1-type receptor binding sites were shown to be differentially altered, depending on the brain region studied, in the Flinders Sensitive Line (FSL) rats (Caberlotto et al. 1999), a purported genetic animal model of depression (Overstreet 1993). These animals display features similar to those observed in depressed patients: reduced basal motor activity (Overstreet and Russel 1982), elevated REM (rapid eye movement) sleep (Shiromani et al.1988) and increased immobility and anhedonia responses after stress exposure (Pucilowski et al.1993). NPY Y_2-type receptor mRNA expression was unchanged, suggesting that this subtype may not play such an important role as Y_1 receptors in this model. Moreover, treatment with the SSRI fluoxetine, attenuated changes in NPY-like receptor mRNA observed in the 'depressed' animals (Caberlotto et al. 1998).

Maternal deprivation is an animal model of depression/vulnerability to stress that posits that early life stress may cause changes in the CNS (e.g., hypothalamic–pituitary adrenal dysregulation) that are associated with an increased risk of adult life depressive psychopathology (Holsboer 2000). Using this model in rats for 3 h per day during postnatal days 2–14, NPY levels were shown to be reduced in the hippocampus and striatum and increased in the hypothalamus (Husum et al. 2002; Jimenez-Vasquez et al. 2001). However, if lithium treatment was used on days 50–83, the changes in NPY-like immunoreactivity induced by maternal deprivation were not observed in the hippocampus and striatum while NPY levels were further increased in the hypothalamus (Husum and Mathe 2002). Consequently, early life stress has long-term effects on NPY in the CNS and may be a factor in the development of depression; possibly through an increased vulnerability to stress.

Additional evidence of a role for NPY in depressive disorders is found in studies using the olfactory bulbectomized (OB) rat model of depression. Subchronic icv administration of NPY attenuated the increase in ambulation, rearing, grooming and defecation scores consistently found when OB animals are tested in the open field (Song et al. 1996). Treatment with NPY also increased noradrenaline and serotonin levels in the amygdala and hypothalamus (Song et al. 1996). In addition, NPY reversed the suppression of lymphocyte proliferation seen following OB (Song et al. 1996) as well as in depressed patients (Kronfol and House 1989). Another study demonstrated that OB caused long-term increases in the expression of the NPY gene in the olfactory/limbic system, suggesting that NPY plasticity may play some role in this model (Holmes et al. 1998).

More recently, it has been shown that NPY displayed antidepressant-like activity in the rat forced swimming test (Stogner and Holmes 2000). These results have since been confirmed in the mouse version of this test (Redrobe et al. 2002). The forced swimming test is an acute animal model widely used for the screening of potential antidepressant drugs (Porsolt et al.1977). Intracerebroventricular NPY administration significantly reduced immobility time in a dose-dependent manner, as did [Leu^{31}Pro^{34}]PYY (a preferential Y_1/Y_5 agonist) (Redrobe et al. 2002). In contrast, BIBP 3226 and BIBO 3304 (selective Y_1 antagonists; Doods et al. 1996; Wieland et al. 1998) and NPY(13–36) (a preferential Y_2 agonist) did not display any activity at the doses tested. However, pretreatment with BIBP 3226 and BIBO 3304 significantly blocked the anti-immobility effects of NPY (Redrobe et al. 2002). Hence, these results suggest that the development of synthetic NPY Y_1 agonists may serve as a new family of pharmacotherapeutic agents for the treatment of depressive disorders.

In contrast to the multitude of preclinical data implicating a role for NPY in depressive disorder, clinical studies have generated somewhat inconsistent find-

Table 3 Summary of clinical evidence indicating a role for NPY in depressive illnesses

Model	Effect	Reference
CSF	↓ NPY levels	Widerlov et al. 1988
		Heilig et al. 1990
	= NPY levels	Berrettini et al. 1987
Platelet poor plasma	↓ NPY levels	Hashimoto et al. 1996
		Nilsson et al. 1996
Platelets	↑ NPY immunoreactivity	Nilsson et al. 1996
Suicide (depression)	↓ NPY in frontal cortex	Widdowson et al. 1992
	↓ NPY caudate nucleus	
	= NPY in frontal cortex	Ordway et al. 1995
Antidepressant treatment	= CNS NPY levels	Widdowson et al. 1992
Electroconvulsive therapy	↑ NPY levels in CSF	Mathé et al. 1996

↑, Increase; ↓, decrease; =, no change.

ings (Table 3). Several studies have demonstrated decreased NPY levels in the cerebrospinal fluid (CSF) (Widerlov et al. 1988; Heilig and Widerlov 1990) and platelet-poor plasma (Hashimoto et al. 1996; Nilsson et al. 1996) of depressed patients, when compared to healthy control subjects. These authors also found that NPY immunoreactivity in the platelets of depressed patients was significantly increased (Nilsson et al. 1996). These results suggested that NPY release may be reduced, or that metabolism of the peptide is increased, in the CNS of depressed subjects. However, other studies involving patients suffering from major affective disorders failed to reveal any significant changes in CSF levels of NPY (Berrettini et al. 1987). Methodological differences have been suggested to account for the differences in the findings from these studies (Heilig and Widerlov 1990). In addition, these authors showed CSF NPY levels to be negatively correlated to scores of anxiety in clinically depressed patients (Heilig and Widerlov 1990), suggesting a possible link between low concentrations of NPY and predisposition to anxiety-related or stress-induced depression.

Further inconsistencies are found in studies where initial analysis revealed decreased NPY concentrations in the frontal cortex and caudate nucleus of suicide victims, which appeared to be particularly evident in subjects affected by major depression (Widdowson et al. 1992). However, studies published 3 years later (Ordway et al. 1995), demonstrated no significant differences in frontal cortex NPY levels between control subjects and subjects who were deemed affected by major depression. The authors used a rigorous methodology in the second study and did not only rely on a coroner's report which, they suggested, could have resulted in wrongful diagnosis in the initial study (Ordway et al. 1995).

The question as to whether increased concentrations of CNS NPY, as reported in several animal studies following antidepressant administration, may play some role in the clinical pharmacotherapeutic efficacy of these drugs has also been raised (Widdowson et al. 1992). Post-mortem NPY measurements from suicide victims who tested positive for antidepressant drugs at autopsy did not demonstrate any significant increase in CNS NPY concentrations when compared to subjects who tested negative (Widdowson et al. 1992). However, the sample size of suicide victims who tested positive for antidepressant drugs was rather small. On the other hand, various antidepressant treatments, including electroconvulsive treatment, have indeed been demonstrated to increase NPY immunoreactivity levels in the CSF of depressed patients (Mathé et al. 1996). Therefore, on the basis of these studies, a role for NPY in the efficacy of antidepressant drugs cannot be ruled out.

One of the major problems that may be pivotal in explaining the somewhat inconsistent results obtained from clinical studies to date is related to post-mortem recuperation of samples. It has been shown that several post-mortem parameters can influence the stability of mRNAs and proteins in the human brain (Caberlotto and Hurd 2001). It was shown that measurement of NPY Y_2 mRNA expression was significantly affected by post-mortem delay. Moreover, it was demonstrated that prefrontal cortical Y_2 mRNA expression was negatively correlated with post-mortem interval (Caberlotto and Hurd 2001). These findings are consistent with previous animal studies that demonstrated that post-mortem interval could influence NPY binding sites (Caberlotto et al. 1997). Therefore, it is suggested that the factors that differentially affect NPY receptor mRNA and protein stability should be investigated further.

Overall, the clinical data do suggest some involvement of NPY systems in the pathogenesis of affective disorders. However, the available clinical evidence has not yet revealed a correlation between depressive disorder and polymorphic alleles of the NPY gene, suggesting that any dysregulation of the NPY system could possibly result as a secondary event (Detera-Wadleigh et al. 1987). Only studies incorporating recently developed DNA-array techniques will serve to confirm, or not, a direct influence of depressive disorder on the NPY gene and NPY receptor genes or vice versa.

3.3
Role for NPY and Its Receptor Subtypes in Anxiety

NPY has also been implicated in the pathogenesis of anxiety disorders, based on the findings showing NPY-induced anxiolytic activity in animal models widely used for the screening of anxiolytic-like compounds/molecules (Table 4).

The open field test is an animal model often used to measure behavioral changes induced by anxiolytic or anxiogenic-like compounds (Harro 1993). Theoretically, anxiogenic drugs should increase the time that is spent close to the wall of the apparatus (thigmotaxis), yet in the case of NPY Y_1 receptor mediated anxiogenesis, there seems to be poor correlation between experimental anxiety and this measure (von Horsten et al. 1998). The time spent in the center of the arena, together with the number of lines crossed in the central area, may provide some information on innate fearfulness. Interestingly, NPY deficient mice were found to be less active in the central part of the open field and

Table 4 Summary of preclinical evidence indicating a role for NPY in anxiety

Model	Effect	Reference
Open field		
NPY knockout mice	↑ Anxiogenesis	Bannon et al. 2000
NPY icv (acute)	↓ Overall activity	Heilig and Murison 1987
		Heilig et al. 1988b
		Jolicoeur et al. 1991, 1995
		Song et al. 1996
NPY icv (chronic)	Reversal of OB rat deficits	Song et al. 1996
BIBO 3304 icv	↑ Defecation	Kask et al. 2000
Plus maze		
NPY icv	↑ Time on open arms	Heilig et al. 1989
NPY (l. coeruleus)	↑ Time on open arms	Kask et al. 1998
NPY(13–36) (l. coeruleus)	↑ Time on open arms	
PYY icv	↑ Time on open arms	Broqua et al. 1996
NPY(2–36) icv	↑ Time on open arms	
[Leu^{31},Pro^{34}]NPY icv	↑ Time on open arms	
NPY(13–36)	↓ Time on open arms	Nakajima et al. 1998
Light–dark box		
NPY icv	↑ Transitions	Pichet et al. 1993
Social interaction		
NPY (amygdala)	↑ Social behavior	Sajdyk et al. 1999
C2-NPY (amygdala)	↓ Social behavior	Sajdyk et al. 2002
NPY (septum)	↑ Social behavior	Kask et al. 2001
Geller–Seifter		
NPY icv	Anticonflict/anxiolytic	Britton et al. 1997, 2000
PP, NPY(13–36) icv	No effect	Britton et al. 1997
Vogel conflict		
NPY icv	Anticonflict/anxiolytic	Heilig et al. 1989
Fear pt. Startle		
NPY icv	↓ Startle response	Broqua et al. 1995
PYY icv	↓ Startle response	
NPY(2–36) icv	↓ Startle response	
[Leu^{31},Pro^{34}]NPY icv	↓ Startle response	
NPY(13–36) icv	No effect	

presented reduced rearing behavior, suggesting that these animals were more anxious (Bannon et al. 2000). The number of line crossings may not be an ideal measure to assess anxiety-related behavior in the open field test. Many anxiogenic-like compounds may also increase freezing responses and consequently decrease exploratory activity. In addition, it may be difficult to dissociate increased anxiety from nonspecific changes in locomotor activity. More recently, NPY Y_2 receptor knockout mice displayed increase preference for the central area of the open field when compared to $Y_2^{+/+}$ animals without changes in locomotor activity given that total entries did not differ between groups

(Redrobe et al. 2003a). This study suggests that NPY Y_2 receptors may play an inhibitory role and may be involved in the regulation of anxiety-like behavior by NPY.

Colonic motility is affected by stress, and when exposed to novel fear-provoking environment, rodents often urinate and defecate (Brady and Nauta 1955). The number of defecations is thought to correlate with emotionality (Hall 1934) and this measure is often used, in addition to horizontal and vertical activity (rearing) in the open field test (Abel 1991). NPY causes dose-dependent suppression in open field activity when the icv dose of NPY exceeds 5 µg (Heilig and Murison 1987; Heilig et al. 1988b; Jolicoeur et al. 1991, 1995), suggesting a nonspecific sedative effect. On the other hand, chronic daily administration with a nonsedative dose of NPY (7 days) has been shown to antagonize the increase in ambulation, rearing, grooming, and defecation that occurs when OB rats are exposed to a novel stressful environment (Song et al. 1996). Intracerebral administration of BIBO 3304, a selective NPY Y_1 receptor antagonist, has been shown to increase defecation in a novel open-field test (Kask and Harro 2000), results which are consistent with the emotionality-reducing role of NPY. The effects of intracerebral administration of NPY on locomotor activity depend on the site of injection. Microinjection of NPY into the frontal cortex has been shown to increase locomotor activity (Smialowski et al. 1992). Intrahippocampal injection (CA1) of NPY did not modulate locomotor activity, but inhibited amphetamine-induced increases in sniffing and rearing and, to a lesser extent, the number of line crossings (Smialowska et al. 1996).

The elevated plus-maze test is one of the most widely used animals model of anxiety (Hogg 1996) and its popularity owes much to the relative simplicity of the test. The test is based on a conflict between the natural aversion of rodents for open spaces and the drive to explore a novel environment (Harro 1993). Typically, the time spent on the open arms and the numbers of entries onto the open and closed arms of the maze are recorded. The percentage of open arm entries relative to the number of total arm entries is considered to be the superlative measure reflecting innate fearfulness. Few investigators have tried to incorporate additional measures such as activity on open arms or measures reflecting risk-assessment behavior such as rearing, approaches towards the open part (peeking) or stretch–attended postures and head dipping (Harro 1993; Rodgers et al. 1999). NPY, administered icv, decreased the preference of the rats for the closed arms and also increased the time spent on open arms (Heilig et al. 1989). Higher doses of NPY (exceeding 2 nmol) suppressed the entries to both closed and open arms, consistent with the sedative action of NPY observed at high doses (Heilig and Murison 1987). These findings were confirmed in a study that demonstrated the anxiolytic-like activity of PYY, NPY(2–36), and [Leu^{31}, Pro^{34}]NPY [but not NPY(13–36)] (Broqua et al. 1996). Anxiolytic-like effects in the elevated plus-maze have also been observed after direct administration of NPY and NPY(13–36) into the locus coeruleus in rats (Kask et al. 1998a). On the other hand, the Y_2-type receptor agonist, NPY(13–36), has also been shown to induce anxiogenic-like effects in this model, using mice, when administered icv

(Nakajima et al. 1998). Additionally, Y_2 receptor knockout mice displayed anxiolytic-like behavior in the elevated plus-maze as they made more entries into and spent significantly more time on the open arms of the maze when compared to their wild-type controls (Redrobe et al. 2003a). Thus, NPY Y_2 receptors may play an inhibitory role in the anxiolytic-like effects of NPY.

Using another anxiety model based on exploration, the light/dark compartment test, it was found that icv NPY increased the number of transitions between the two compartments (Pich et al. 1993), a validated measure of anxiolytic activity in this test. This effect of NPY was expressed in a lower dose in spontaneously hypertensive rats compared to the normotensive Wistar-Kyoto rats (Pich et al. 1993). Interestingly, in this study, NPY increased the activity of animals in the dark compartment in both rat strains, even though the effect of NPY in the open field was, as usually reported, to reduce activity (Pich et al. 1993). This suggests that the sedative effects of NPY, frequently reported after only marginally higher doses than those eliciting anxiolytic activity, are context-specific, but this possibility has received little attention.

Social interaction has been pharmacologically validated as an experimental model of anxiety (File 1980; File and Hyde 1978). Typically, rats are either tested in a novel, brightly or dimly lit environment (high–unfamiliar and low–unfamiliar, respectively) or under familiar testing conditions (high–familiar and low–familiar). The time spent in active social behavior, as well as locomotor activity, is recorded.

The effects of NPY in this model have also been studied and it has been shown that NPY increased social behavior when it was microinjected into the basolateral nucleus of the amygdala (Sajdyk et al. 1999) and into the caudal dorsolateral septum (Kask et al. 2001). Intracerebroventricular NPY can also reverse deficits of social behavior that are induced by the selective noradrenergic neurotoxin DSP-4 (Kask et al. 2000). Thus, NPY has anxiolytic-like effects in this paradigm of anxiety in rats, and several brain regions appear to mediate NPY-induced anxiolysis. This anxiolytic-like effect of NPY can even be observed without concomitant restoration of general locomotor activity in animals with lesioned projections of the locus coeruleus (Kask et al. 2000), suggesting that these effects are independent of vigilance and arousal.

Most recently, it has been suggested that the Y_2-type NPY receptor may mediate anxiety-like behaviors in the amygdala (Sajdyk et al. 2002). These authors injected a Y_2 receptor agonist, C2-NPY, directly into the basolateral nucleus of the amygdala and found decreased times spent in social interaction, and these effects were dose dependent (Sajdyk et al. 2002). However, this study is lacking in some respects as the authors did not attempt to reverse the anxiogenic-like effects of this molecule with a selective Y_2 receptor antagonist, such as BIIE 0246 (Doods et al. 1999).

Several versions of behavioral tests based on conflict of motivations have been developed. In all these tests, subjects are deprived of food or water. In Vogel punished drinking tests, subjects can be tested immediately, whereas in Geller–Seifter tests, rats are trained to associate defined cues with shock that comes during a punished phase. Thus, the Vogel test is based on unconditioned suppression of behavior, whereas the Geller–Seifter test requires previous training (Pollard and Howard 1979). In both tests, the treatment with anxiolytic drugs increases the number of accepted shocks during the punished phase. Intracerebroventricular injection of NPY consistently produces dose-dependent anticonflict/anxiolytic-like effects in the Geller–Seifter test of operant responding (Britton et al. 1997, 2000; Heilig et al. 1992), an established animal model of anxiety especially suitable for detecting the effects of benzodiazepine-like anxiolytics. Similarly, icv NPY markedly increased the number of electric shocks accepted in the Vogel's punished drinking test (Heilig et al. 1989). At the doses used, NPY was reported not to affect pain sensitivity in a shock threshold test, or thirst. Thus the anti-conflict effects can be considered related to a reduction of anxiety (Heilig et al. 1989). In a modified Geller–Seifter conflict paradigm, neither PP nor NPY(13–36) increased responding during the punished phase, suggesting that the effects of NPY are mediated via NPY Y_1 receptor activation (Britton et al. 1997). The robust effect of NPY in punished responding procedures suggests that nonpeptide NPY agonists could serve as potent anxiolytics alternative to the benzodiazepines. However, given the fact that very low doses of NPY (0.04 nmol) have antinociceptive effects in the periaqueductal gray matter in the rat (Wang et al. 2000), and NPY has potent NPY Y_1 receptor-mediated antinociceptive effects in mice (Broqua et al. 1996; Naveilhan et al. 2001) and spontaneously hypertensive rats (Pich et al. 1993), the anti-conflict effect of NPY in punished responding paradigms needs further characterization. It must also be taken into account that NPY treatment, in most of the studies, had a clear tendency to increase unpunished responding. Even though the proportion of this effect is lower than the increase in punished responding, it should be considered that the absolute levels of responding are very much lower in punished than in unpunished conditions.

Startle is an adaptive response to acoustic stimuli that enables the organism to avoid, or reduce, the risk of an injury by a predator (Koch 1999). The acoustic startle can be modified by a variety of stimuli and neuronal pathways underlying acoustic startle have been well characterized (Davis 1984). In fear-potentiated startle, an acoustic stimulus is paired with an aversive intervention such as foot-shock or air-puff, and after training, the conditioning stimulus alone is capable of elevating startle amplitude.

It has been shown that icv injections of NPY, PYY, NPY(2–36) and [Leu^{31}Pro^{34}]NPY inhibited fear-potentiated startle, whereas NPY(13–36) had no effect (Broqua et al. 1995). Treatment with monosodium glutamate (MSG) in the neonatal period produces damage to the arcuate nucleus and decreases NPY levels in this region (Abe et al. 1990; Kerkerian and Pelletier 1986; Meister et al. 1989), whereas in adult rats MSG treatment increases NPY content in hypothalamus (Tirassa et al. 1995). MSG-induced hypothalamic lesions do not affect NPY content in the median eminence (Meister et al. 1989). However, these lesions do lead to complete disappearance of agouti-related peptide-immunoreactivity (an indirect selective marker for arcuate nucleus-derived NPY terminals), in the ventral part of the periaqueductal gray matter and in the parabrachial nucleus of the brainstem (Broberger et al. 1998). MSG treatment has been shown to increase acoustic startle response (Yang et al. 2000). Collectively, these data suggest that NPY released from nerve terminals arising from the arcuate nucleus may dampen the expression of acoustic startle response. Whether such a reduction in startle amplitude reflects the decrease in anxiety levels or simply the reduced response to sensory stimuli is not known.

3.4
Role for NPY and Its Receptor Subtypes in Learning and Memory

The hippocampal Y_2 receptor has been implicated in facilitating learning and memory processes with increases in memory retention induced by NPY (Flood et al. 1987; Flood and Morley 1989). It has also been shown that NPY can reverse amnesia induced by protein synthesis inhibitors or the cholinergic antagonist, scopolamine (Flood et al. 1987).

Additional experiments involved direct injection of NPY in discrete brain regions (Flood et al. 1989). This work demonstrated that the effects of NPY on cognitive function may be region specific, i.e., inducing differential effects depending on the injection site (Table 5). Injection of NPY into the rostral hippocampus and the septal area enhanced memory retention, whereas NPY injection into the amygdaloid body and the caudal hippocampus induced amnesia. In further support of a physiological role for NPY in cognitive behaviors, it was shown that a passive immunization with NPY antibodies, injected into the responsive hippocampal regions, induced amnesia (Flood et al. 1989). Recently, NPY Y_2 receptor knockout mice displayed a deficit on the probe trial in the Morris water maze task when compared to NPY Y_2 wild-type animals. The $Y_2^{-/-}$ mice also exhibited a marked deterioration in object memory 6 h, but not 1 h after an initial exposure in the object recognition test (Redrobe et al. 2003b). These data suggest that NPY Y_2 receptors may play a facilitatory role in learning and memory. Further studies investigating the effects of the more recently developed NPY antagonists on learning and memory behaviors are warranted. For example, experiments using the nonpeptide Y_2 receptor antagonist, BIIE 0246 (Doods et al. 1999) should provide key evidence as to the role played by this receptor subtype in cognition.

Table 5 Comparative effect of NPY and NPY antibodies on learning and memory (adapted from Flood and Morley 1989; Flood et al. 1987, 1989)

Injection site	NPY	NPY antibodies
Dorsal hippocampus	Enhanced	Amnesic
Ventral hippocampus	Amnesic	Enhanced
Amygdala	Amnesic	Enhanced
Septum	Enhanced	Amnesic
Thalamus	No effect	No effect

On the other hand, it has recently been demonstrated that icv injection of NPY attenuated long-term potentiation (LTP) and inhibited KCl-induced glutamate release in synaptosomal preparations from dentate gyrus (Whittaker et al. 1999). These data are in further support of the work by St. Pierre et al. (2000) in which it was suggested that NPY Y_1 receptors, present on glutamatergic neurons, may act as heteroreceptors to regulate the release of this transmitter believed to play an important role in cognition. In addition, it was shown that, when incubated with NPY or following icv injection, synaptosomal activity of the stress-activated kinase, c-Jun NH_2-terminal kinase (JNK) was increased. It was suggested that activation of JNK may explain the inhibitory effects of NPY on LTP. However, the mechanisms by which NPY may activate JNK remain unknown (Whittaker et al. 1999).

The hippocampal formation is associated with learning and memory processes and is an area severely affected in Alzheimer's disease (AD) (Terry and Davies 1980). Several studies have reported significant decreases in NPY-like immunoreactivity in cortical, amygdaloid and hippocampal areas in AD brains (Beal et al. 1986; Chan-Palay et al. 1985b, 1986b). NPY-like immunoreactivity has also been detected within neuritic plaques present in the brains of patients with AD (Chan-Palay et al. 1986b). Moreover, the levels of [^{3}H]NPY binding sites are apparently reduced in the temporal cortex and hippocampus of patients suffering from AD (Martel et al. 1990). These data suggest that the degenerative processes occurring in AD may involve changes in NPY-related innervation. Interestingly, a major loss in NPY-like immunoreactive neurons has been reported in aged rats especially in cortical areas, the caudate putamen and the hippocampus (Cha et al. 1996 1997; Huh et al. 1997, 1998). Furthermore, injection of the immunotoxin 192TrkA-saporin, which induces degeneration of cholinergic basal forebrain neurons, resulted in 33%–60% decreases in NPY-like immunoreactivity in the frontoparietal and occipital cortices as compared to age matched control rats (Zhang et al. 1998). Thus, the loss of NPY-like immunoreactivity seen in AD may be a consequence of the well established cholinergic deficits. However, the direct impact of NPY losses on cognitive behaviors in AD remains to be established and further studies are warranted in this regard.

More recently, the generation of an NPY-transgenic rat has offered an attractive model for study of the effects of this peptide on learning and memory processing (Thorsell et al. 2000). This model is rather unconventional in several respects, which may render it advantageous. Firstly, the choice of species (rat) allows for a direct comparison with functional effects previously recorded following exogenous NPY administration. In addition, the construct used in generating these animals contains the normal intronic sequence thought to express the major regulatory elements normally controlling NPY expression (Larhammar 1997; Michalkiewicz and Michalkiewicz 2000). The expression of the transgene may therefore be regulated in a manner similar to that of endogenous NPY.

NPY transgenic rats were shown to display deficits in both the acquisition and retention of a spatial memory task, namely the Morris water maze (Thorsell et al. 2000). Concurrent anatomical mapping studies revealed a restricted, but highly significant, hippocampal NPY overexpression which was accompanied by a profound downregulation of NPY Y_1 receptors in NPY transgenic subjects (Thorsell et al. 2000). Attempts to understand the mechanisms behind these behavioral deficits lead us to the glutamatergic system, which is a crucial factor in learning and memory processing (Vizi and Kiss 1998). In vitro studies have shown that NPY, within the hippocampal formation, acts mainly on Y_2 sites to reduce presynaptic Ca^{2+} entry, inhibit glutamatergic transmission, and hence suppress the formation of LTP (Qian et al. 1997; Colmers et al. 1988; Whittaker et al. 1999). The studies performed in the Morris maze are in agreement with this in vitro work, and suggest an enhanced NPY Y_2-mediated inhibition of excitatory synaptic activity in the hippocampus.

3.5
Role for NPY and Its Receptor Subtypes in Seizure and Epilepsy

It has been shown that hippocampal NPY expression is increased following various treatments that induce seizure, e.g., electroconvulsive shocks (Kragh et al. 1994; Mikkelsen et al. 1994; Wahlestedt et al. 1990), electrical kindling (Rizzi et al. 1993; Schwarzer et al. 1995) and kainic acid administration (Sperk et al. 1992). This increase is also apparent in the entorhinal cortex of chronically epileptic rats (Vezzani et al. 1996a-c), spontaneously epileptic rats (Sadamatsu et al. 1995) and Ihara's epileptic rats (Takahashi et al. 1997). The increase in NPY-like immunoreactivity in the hippocampus could be a compensatory antiseizure mechanism. Thus, it has been suggested that NPY may act as an endogenous anticonvulsant (Greber et al. 1994; Sperk et al. 1992, 1996; Sperk 1994; Vezzani et al. 1999) (see Table 6).

Indeed, administration of NPY attenuated epileptiform-like activity in various models of epilepsy (Klapstein and Colmers 1997; Woldbye et al. 1996) and reduced seizure activity following kainic acid administration to NPY-deficient mice (Baraban et al. 1997). Furthermore, in vivo microdialysis in the dorsal hippocampus (Husum et al. 1998) and in entorhinal hippocampal (slices Vezzani et al. 1996a-c) revealed increases in the release of NPY in the kainic acid model of epilepsy. These changes in NPY release are associated with apparent increases of [^{125}I]PYY binding sites within the first 6-24 h following the kainic acid injection (Kofler et al. 1997; Roder et al. 1996). Moreover, under masking binding assay conditions, [^{125}I]PYY/[Leu^{31},Pro^{34}]NPY-sensitive (Y_1-like) sites are reduced (Gobbi et al. 1996, 1998, 1999; Kofler et al. 1997) while [^{125}I]PYY/[Leu^{31},Pro^{34}]NPY-insensitive (Y_2-like) binding is increased (Gobbi et al. 1998; Roder et al. 1996) (Table 6).

In accordance with these variations in [^{125}I]PYY binding, in situ hybridization studies have shown decreases in Y_1 receptor mRNA and increases in Y_2 receptor mRNA in the granule cell layer of the hippocampus of kindled rats (Gobbi et al. 1998). Isotherm saturation [^{125}I]PYY(3-36) binding revealed that stratum oriens (CA1) B_{max} values were not modified within the first 6 h following a kainic acid injection, only the apparent affinity (K_d) was increased

Table 6 Models of seizure and their effects on hippocampal NPY

Model	Effect on NPY	References
Electroconvulsive shock	↑ mRNA ↑ Immunoreactivity	Kragh et al. 1994; Mikkelsen et al. 1994; Wahlestedt et al. 1990
Electrical kindling	↑ mRNA ↑ immunoreactivity	Rizzi et al. 1993; Schwarzer et al. 1995, 1996
Kainic acid	↑ mRNA ↑ Immunoreactivity ↑ NPY release ↑ Y_2 receptor binding ↓ Y_1 receptor binding ↓ Y_5 receptor binding	Bregola et al. 2000; Sperk 1994
Chronic epileptic rat	↑ mRNA ↑ Immunoreactivity	Vezzani et al. 1996a, b, c
Spontaneous epileptic rat	↑ mRNA ↑ Immunoreactivity	Sadamatsu et al. 1995
Ihara epileptic rat	↑ mRNA ↑ Immunoreactivity	Takahashi et al. 1997
NPY knockout mice	↑ Seizure susceptibility	Bannon et al. 2000
NPY Y5 knockout mice	No spontaneous seizures ↑ Susceptibility to kainic acid NPY-induced effects absent	Marsh et al. 1999

(Schwarzer et al. 1998). In the dentate gyrus, no change in either K_d or B_{max} values of [^{125}I]PYY(3–36) binding were observed at 6 h, while marked increases in B_{max} (800%) were noted 24–48 h post kainate injection (Schwarzer et al. 1998). Taken together, these data suggest that NPY plays a major role in the modulation of neuronal activity in the hippocampus, especially during seizures.

Early electrophysiological and pharmacological studies suggested that the anticonvulsant-like activity of NPY may be mediated by an NPY Y_2-mediated mechanism (Colmers et al. 1988, 1991; Bleakman et al. 1992). These studies implicated an NPY Y_2-mediated suppression of excitatory transmission and, consequently, an inhibition of preseynaptic glutamate release (Colmers et al. 1988, 1991; Bleakman et al. 1992). More recently, studies with rat hippocampal slices suggest the involvement of both NPY Y_2 and Y_5 receptor subtypes (Bijak 1999).

Interestingly, NPY(3–36), [Leu^{31},Pro^{34}]NPY and hPP [but not NPY(13–36)] were shown to inhibit kainic acid induced seizures, suggesting that the anticonvulsive activity of NPY is mediated by a Y_5-like receptor subtype (Woldbye et al. 1997). A marked decrease in Y_5 receptor binding was recently observed in experimental rat models of epilepsy (kainate and/or kindling). Binding experiments were performed using [^{125}I][Leu^{31},Pro^{34}]PYY in the presence of 1 µM BIBP 3226 (Bregola et al. 2000). Y_5 receptor binding levels remained low (50%

reduction) in rats killed 7 days after the last stimulus-evoked seizures (Bregola et al. 2000). These data suggest that a long-lasting decrease in Y_5 receptor levels may contribute to the development of epileptic hyperexcitability.

The anticonvulsant properties of BIBP 3226 (Y_1 receptor antagonist) do suggest that the Y_1 receptor subtype may also have a permissive role in seizure modulation (Gariboldi et al. 1998). Studies using the recently developed Y_1 receptor antagonist, BIBO 3304 (less toxic than BIBP 3226) are warranted in order to confirm the possible implication of the Y_1 receptor in seizure modulation.

Most interestingly, the development of mutant mice lacking NPY (knockout mice) (Baraban et al. 1997) has confirmed that NPY plays a critical role in the control of seizure. It was shown that these mice were more susceptible to kainic acid induced seizures, compared to their wild-type littermates. Seizures in knockout mice progressed uncontrollably and ultimately resulted in the death of 93% of NPY-deficient mice (Baraban et al. 1997). The icv infusion of NPY, before kainic acid administration, prevented death (Baraban et al. 1997). It has been suggested that kainic acid-induced seizures are a result of excess excitation of neurons, leading to synchronized epileptiform discharge in the dentate gyrus, which then leads to excitatory epileptiform discharge throughout the hippocampal formation (Sperk 1994).

Other genetic approaches used to investigate the mechanisms by which NPY exerts anticonvulsant effects include the recent generation of mice lacking the NPY Y_5 receptor (Marsh et al. 1998). These studies showed that NPY Y_5 knockout did not exhibit spontaneous seizure-like activity; however these animals were more sensitive to kainic acid-induced seizures (Marsh et al. 1999). Electrophysiological analysis of hippocampal slice preparations from mutant mice revealed normal function, although the anti-epileptic effects of exogenously applied NPY were absent. Collectively, these results point to a role for the NPY Y_5 receptor in the mediation of the anti-epileptic-like activity of NPY. However, this receptor subtype does not seem important for normal hippocampal function or the control of normal excitatory signaling (Marsh et al. 1999).

More recently it has been suggested that the role played by NPY may vary depending to the type of seizure induced, and the NPY receptor subtype activated (Reibel et al. 2001). These authors showed that agonists of NPY Y_1, Y_2 and Y_5 receptors reduced seizure-like activity in hippocampal cultures. In addition, icv administration of NPY or NPY Y_5-like agonists reduced the expression of focal seizures produced by a single electrical stimulation of the hippocampus (Reibel et al. 2001). Conversely, NPY receptor agonists were found to increase the duration of generalized, hippocampal-independent, seizures induced by pentylenetrazol (Reibel et al. 2001). Taken together, these studies suggest a role not only for NPY Y_5 receptors in seizure modulation, but also that other NPY receptor subtypes in brain structures besides the hippocampus that may be involved in initiation, propagation and seizure-control. For example, a recent report revealed that exogenously applied NPY to rat layer V pyramidal neurons has a long-lasting increase in Ca^{2+}-dependent inhibitory synaptic transmission as well as prolonged decreases in the amplitude of evoked monosynaptic inhibitory postsynaptic currents in interneurons (Bacci et al.2002); suggesting powerful anticonvulsant effects of NPY in the neocortex through decreases in excitability in cortical circuits.

Thus, it is suggested that NPY acts as an endogenous anticonvulsant on the basis of the cumulative evidence discussed above. However, the precise receptor subtype(s) involved remains to be fully established. Accordingly, the eventual clinical use of NPYergic molecules in seizure control is still awaited.

4
Conclusion and Future Directions

It is rather evident that NPY may play a role in the pathophysiology of depression, and thus may represent a potential novel target for the treatment of this illness, probably via the development of selective nonpeptide NPY Y_1 receptor agonists.

Preclinical studies have demonstrated multi-level changes in NPY immunoreactivity, NPY receptor mRNA expression and NPY receptor subtype functioning both in animal models considered relevant for the study of depression (whether they be pharmacological, behavioral or genetic), as well as following treatment with pharmacologically distinct antidepressant drugs. The results, to date, suggest that the NPY Y_1 receptor subtype seems to be mostly involved, while the Y_2 receptor subtype may play a more minor role. Little is known about the role of the other NPY receptor subtypes in animal models of depression. However, as selective nonpeptide ligands for these subtypes, together with transgenic animals are becoming more accessible, further investigation into the possible roles of these subtypes is warranted.

In addition to the preclinical data, there is a plethora of clinical evidence suggesting a role for NPY in depression. However, differences in experimental design, wrongful patient diagnosis, small sample size, difficulties with sample availability and comorbidity of depression with other psychiatric disorders could overshadow the global conclusions drawn when the clinical data are reviewed.

Amongst the many neuropeptides implicated in anxiety, NPY (along with corticotrophin releasing factor) is most prominent as a peptide for which there is most extensive and consistent evidence emerging from different models of anxiety. As revealed by studies using the first subtype-selective nonpeptide antagonists of NPY, endogenous levels of this peptide play a role in reducing anxiety and could thus serve as a physiological stabilizer of neural activity in circuits involved in arousal and anxiety.

NPY appears to have a universal role in this regard in several brain regions, but the mediation of NPY anxiety-related effects do not seem to be restricted to a single NPY receptor subtype. Further characterization of NPY receptors and neural pathways mediating the effects of NPY on anxiety are still needed, as drugs related to NPY receptor subtypes remain an attractive target for the treatment of anxiety-related disorders.

In addition to the role played by NPY in mood-related disorders, it is also clear that hippocampal NPY plays a major role in the modulation of seizure, with the Y_1, Y_2 and Y_5 receptor subtypes being implicated. The role of NPY in seizure has also been underlined by the recent development of NPY and NPY Y_5-deficient mice, where these mice are more susceptible to seizures, and often die, following kainic acid administration; a phenomenon which is prevented by prior icv infusion of NPY. Further studies using more selective agonists and antagonists for the different NPY receptor subtypes will serve to further the knowledge of the role played by this peptide in seizure control. This work could then pave the way for the development of NPY-related medication for the treatment of epilepsy.

It is unfortunately evident that the investigation into the role played by NPY in learning and memory has been somewhat neglected since studies performed in the late 1980s. This early work demonstrated that NPY improved memory retention (possibly acting via Y_2 receptors) when administered following several amnesic molecules such as scopolamine and protein synthesis inhibitors. However, it was also shown that NPY may be beneficial or detrimental to cognitive processing depending on the injection site. Changes in NPY have also been documented in the brains of patients suffering from Alzheimer's disease. Now that more specific molecules and transgenic animals are available, the question of how NPY may modulate learning and memory needs further attention.

Recent advances in medicinal chemistry technology are beginning to yield more and more selective agonists and antagonists for each NPY receptor subtype. In addition, genetic manipulation has brought about the generation of NPY, Y_1, Y_2 and Y_5 receptor 'knockout' mice, together with NPY transgenic mice and rats. Exploitation of these new tools and models will hopefully result in therapeutic agents suitable for the treatment of the multitude of conditions in which NPY is thought to play a role.

References

Abe M, Saito M, Shimazu T (1990) Neuropeptide Y in the specific hypothalamic nuclei of rats treated neonatally with monosodium glutamate. Brain Res Bull 24:289–291

Abel EL (1991) Behavior and corticosteroid response of Maudsley reactive and nonreactive rats in the open field and forced swimming test. Physiol Behav 50:151–153

Akabayashi A, Wahlestedt C, Alexander JT, Leibowitz SF (1994) Specific inhibition of endogenous neuropeptide Y synthesis in arcuate nucleus by antisense oligonucleotides suppresses feeding behavior and insulin secretion. Mol Brain Res 21:55–61

Albers HE, Ferris CF (1984) Neuropeptide Y: role in light-dark cycle entrainment of hamster circadian rhythms. Neurosci Lett 50:163–168

Albers HE, Ferris CF, Leeman SE, Goldman BD (1984) Avian pancreatic polypeptide phase shifts hamster circadian rhythms when microinjected into the suprachiasmatic region. Science 223:833–835

Ammar DA, Eadie DM, Wong DJ, Ma YY, Kolakowski LF, Jr., Yang-Feng TL, Thompson DA (1996) Characterization of the human type 2 neuropeptide Y receptor gene

(NPY2R) and localization to the chromosome 4q region containing the type 1 neuropeptide Y receptor gene. Genomics 38:392–398

Bacci A, Huguenard JR, Prince DA (2002) Differential modulation of synaptic transmission by neuropeptide Y in rat neocortical neurons. Proc Natl Acad Sci USA 99:17125–17130

Bannon AW, Seda J, Carmouche M, Francis JM, Norman MH, Karbon B, McCaleb ML (2000) Behavioral characterization of neuropeptide Y knockout mice. Brain Res 868:79–87

Baraban SC, Hollopeter G, Erickson JC, Schwartzkroin PA, Palmiter RD (1997) Knockout mice reveal a critical antiepileptic role for neuropeptide Y. J Neurosci 17:8927–8936

Bard JA, Walker MW, Branchek TA, Weinshank RL (1995) Cloning and functional expression of a human Y4 subtype receptor for pancreatic polypeptide, neuropeptide Y, and peptide YY. J Biol Chem 270:26762–26765

Batterham RL, Cowley MA, Small CJ, Herzog H, Cohen MA, Dakin CL, Wren AM, Brynes AE, Low MJ, Ghatei MA, Cone RD, Bloom SR (2002) Gut hormone PYY(3–36) physiologically inhibits food intake. Nature 418:650–654

Beal MF, Mazurek MF, Chattha GK, Svendsen CN, Bird ED, Martin JB (1986) Neuropeptide Y immunoreactivity is reduced in cerebral cortex in Alzheimer's disease. Ann Neurol 20:282–288

Berglund MM, Holmberg SK, Eriksson H, Gedda K, Maffrand JP, Serradeil-Le Gal C, Chhajlani V, Grundemar L, Larhammar D (1999) The cloned guinea pig neuropeptide Y receptor Y1 conforms to other mammalian Y1 receptors. Peptides 20:1043–1053

Berrettini WH, Doran AR, Kelsoe J, Roy A, Pickar D (1987) Cerebrospinal fluid neuropeptide Y in depression and schizophrenia. Neuropsychopharmacology 1:81–83

Bijak M (1999) Neuropeptide Y suppresses epileptiform activity in rat frontal cortex and hippocampus in vitro via different NPY receptor subtypes. Neurosci Lett 268:115–118

Bleakman D, Harrison NL, Colmers WF, Miller RJ (1992) Investigations into neuropeptide Y-mediated presynaptic inhibition in cultured hippocampal neurones of the rat. Br J Pharmacol 107:334–340

Bonini JA, Jones KA, Adham N, Forray C, Artymyshyn R, Durkin MM, Smith KE, Tamm JA, Boteju LW, Lakhlani PP, Raddatz R, Yao WJ, Ogozalek KL, Boyle N, Kouranova EV, Quan Y, Vaysse PJ, Wetzel JM, Branchek TA, Gerald C, Borowsky B (2000) Identification and characterization of two G protein-coupled receptors for neuropeptide FF. J Biol Chem 275:39324–39331

Borowsky B, Walker MW, Bard J, Weinshank RL, Laz TM, Vaysse P, Branchek TA, Gerald C (1998) Molecular biology and pharmacology of multiple NPY Y5 receptor species homologs. Regul Pept 75/76:45–53

Brady JV, Nauta JH (1955) Subcortical mechanisms in emotional behavior: the duration of affective changes following septal and habenular lesions in the albino rat. J Comp Physiol Psychol 48:412–420

Bregola G, Dumont Y, Fournier A, Zucchini S, Quirion R, Simonato M (2000) Decreased levels of neuropeptide Y5 receptor binding sites in two experimental models of epilepsy. Neuroscience 98:697–703

Britton KT, Akwa Y, Spina MG, Koob GF (2000) Neuropeptide Y blocks anxiogenic-like behavioral action of corticotropin-releasing factor in an operant conflict test and elevated plus maze. Peptides 21:37–44

Britton KT, Southerland S, Van Uden E, Kirby D, Rivier J, Koob G (1997) Anxiolytic activity of NPY receptor agonists in the conflict test. Psychopharmacology (Berl) 132:6–13

Broberger C, Johansen J, Johansson C, Schalling M, Hokfelt T (1998) The neuropeptide Y/agouti gene-related protein (AGRP) brain circuitry in normal, anorectic, and monosodium glutamate-treated mice. Proc Natl Acad Sci USA 95:15043–15048

Broqua P, Wettstein JG, Rocher MN, Gauthier-Martin B, Riviere PJ, Junien JL, Dahl SG (1996) Antinociceptive effects of neuropeptide Y and related peptides in mice. Brain Res 724:25–32

Broqua PP, Wettstein JJ, Rocher MM, Gauthier-Martin BB, Junien JJ (1995) Behavioral effects of neuropeptide Y receptor agonists in the elevated plus-maze and fear-potentiated startle procedures. Behav Pharmacol 6:215–222

Burkhoff A, Linemeyer DL, Salon JA (1998) Distribution of a novel hypothalamic neuropeptide Y receptor gene and it's absence in rat. Mol Brain Res 53:311–316

Caberlotto L, Fuxe K, Overstreet DH, Gerrard P, Hurd YL (1998) Alterations in neuropeptide Y and Y1 receptor mRNA expression in brains from an animal model of depression: region specific adaptation after fluoxetine treatment. Mol Brain Res 59:58–65

Caberlotto L, Fuxe K, Sedvall G, Hurd YL (1997) Localization of neuropeptide Y Y1 mRNA in the human brain: abundant expression in cerebral cortex and striatum. Eur J Neurosci 9:1212–1225

Caberlotto L, Hurd YL (2001) Neuropeptide Y Y(1) and Y(2) receptor mRNA expression in the prefrontal cortex of psychiatric subjects. Relationship of Y(2) subtype to suicidal behavior. Neuropsychopharmacology 25:91–97

Caberlotto L, Jimenez P, Overstreet DH, Hurd YL, Mathe AA, Fuxe K (1999) Alterations in neuropeptide Y levels and Y1 binding sites in the Flinders Sensitive Line rats, a genetic animal model of depression. Neurosci Lett 265:191–194

Cabrele C, Langer M, Bader R, Wieland HA, Doods HN, Zerbe O, Beck-Sickinger AG (2000) The first selective agonist for the neuropeptide Y Y5 receptor increases food intake in rats. J Biol Chem 275:36043–36048

Cabrele C, Wieland HA, Langer M, Stidsen CE, Beck-Sickinger AG (2001) Y-receptor affinity modulation by the design of pancreatic polypeptide/neuropeptide Y chimera led to Y5-receptor ligands with picomolar affinity. Peptides 22:365–378

Cha CI, Lee YI, Lee EY, Park KH, Baik SH (1997) Age-related changes of VIP, NPY and somatostatin-immunoreactive neurons in the cerebral cortex of aged rats. Brain Res 753:235–244

Cha CI, Lee YI, Park KH, Baik SH (1996) Age-related change of neuropeptide Y-immunoreactive neurons in the cerebral cortex of aged rats. Neurosci Lett 214:37–40

Chan-Palay V, Allen YS, Lang W, Haesler U, Polak JM (1985a) Cytology and distribution in normal human cerebral cortex of neurons immunoreactive with antisera against neuropeptide Y. J Comp Neurol 238:382–389

Chan-Palay V, Kohler C, Haesler U, Lang W, Yasargil G (1986a) Distribution of neurons and axons immunoreactive with antisera against neuropeptide Y in the normal human hippocampus. J Comp Neurol 248:360–375

Chan-Palay V, Lang W, Allen YS, Haesler U, Polak JM (1985b) Cortical neurons immunoreactive with antisera against neuropeptide Y are altered in Alzheimer's-type dementia. J Comp Neurol 238:390–400

Chan-Palay V, Lang W, Haesler U, Kohler C, Yasargil G (1986b) Distribution of altered hippocampal neurons and axons immunoreactive with antisera against neuropeptide Y in Alzheimer's-type dementia. J Comp Neurol 248:376–394

Chronwall BM, DiMaggio DA, Massari VJ, Pickel VM, Ruggiero DA, O'Donohue TL (1985) The anatomy of neuropeptide-Y-containing neurons in rat brain. Neuroscience 15:1159–1181

Clark JT, Kalra PS, Kalra SP (1985) Neuropeptide Y stimulates feeding but inhibits sexual behavior in rats. Endocrinology 117:2435–2442

Colmers WF, Bleakman D (1994) Effects of neuropeptide Y on the electrical properties of neurons. Trends Neurosci 17:373–379

Colmers WF, Klapstein GJ, Fournier A, St Pierre S, Treherne KA (1991) Presynaptic inhibition by neuropeptide Y in rat hippocampal slice in vitro is mediated by a Y2 receptor. Br J Pharmacol 102:41–44

Colmers WF, Lukowiak K, Pittman QJ (1987) Presynaptic action of neuropeptide Y in area CA1 of the rat hippocampal slice. J.Physiol (Lond) 383:285-299

Colmers WF, Lukowiak K, Pittman QJ (1988) Neuropeptide Y action in the rat hippocampal slice: site and mechanism of presynaptic inhibition. J Neurosci 8:3827-3837

Daniels AJ, Chance WT, Grizzle MK, Heyer D, Matthews JE (2001) Food intake inhibition and reduction in body weight gain in rats treated with GI264879A, a non-selective NPY-Y1 receptor antagonist. Peptides 22:483-491

Davis M (1984) The mammalian startle response. In: Eaton RC (ed) Neural mechanisms of startle behavior. Plenum Press, New York. pp 287-351

Detera-Wadleigh SD, de Miguel C, Berrettini WH, DeLisi LE, Goldin LR, Gershon ES (1987) Neuropeptide gene polymorphisms in affective disorder and schizophrenia. J Psychiatr Res 21:581-587

Doods H, Gaida W, Wieland HA, Dollinger H, Schnorrenberg G, Esser F, Engel W, Eberlein W, Rudolf K (1999) BIIE0246: a selective and high affinity neuropeptide Y Y2 receptor antagonist. Eur J Pharmacol 384: R3-R5

Doods HN, Wieland HA, Engel W, Eberlein W, Willim KD, Entzeroth M, Wienen W, Rudolf K (1996) BIBP 3226, the first selective neuropeptide Y1 receptor antagonist: a review of its pharmacological properties. Regul Pept 65:71-77

Dumont Y, Cadieux A, Doods H, Pheng LH, Abounader R, Hamel E, Jacques D, Regoli D, Quirion R (2000a) BIIE0246, a potent and highly selective non-peptide neuropeptide Y Y2 receptor antagonist. Br J Pharmacol 129:1075-1088

Dumont Y, Cadieux A, Pheng LH, Fournier A, St Pierre S, Quirion R (1994) Peptide YY derivatives as selective neuropeptide Y/peptide YY Y1 and Y2 agonists devoided of activity for the Y3 receptor sub-type. Mol.Brain Res 26:320-324

Dumont Y, Fournier A, Quirion R (1998a) Expression and characterization of the neuropeptide Y Y5 receptor subtype in the rat brain. J Neurosci 18:5565-5574

Dumont Y, Fournier A, St Pierre S, Quirion R (1993) Comparative characterization and autoradiographic distribution of neuropeptide Y receptor subtypes in the rat brain. J Neurosci 13:73-86

Dumont Y, Fournier A, St Pierre S, Quirion R (1996) Autoradiographic distribution of[125I][Leu31,Pro34]PYY and [125I]PYY(3-36) binding sites in the rat brain evaluated with two newly developed Y1 and Y2 receptor radioligands. Synapse 22:139-158

Dumont Y, Fournier A, St Pierre S, Schwartz TW, Quirion R (1990) Differential distribution of neuropeptide Y1 and Y2 receptors in the rat brain. Eur J Pharmacol 191:501-503

Dumont Y, Jacques D, Bouchard P, Quirion R (1998b) Species differences in the expression and distribution of the neuropeptide Y Y_1, Y_2, Y_4 and Y_5 receptors in rodents, guinea pig and primates brains. J Comp Neurol 402:372-384

Dumont Y, Jacques D, St Pierre JA, Quirion R (1997) Neuropeptide Y receptor types in the mammalian brain: Species differences and status in the human central nervous system. In: Grundemar L, Bloom SR (eds) Neuropeptide Y and drug development. Academic Press, London, pp 57-86

Dumont Y, Jacques D, St Pierre JA, Tong Y, Parker R, Herzog H, Quirion R (2000b) Neuropeptide Y, peptide YY and pancreatic polypeptide receptor proteins and mRNAs in mammalian brains. Handbook of chemical neuroanatomy, Vol 16, Peptide receptor, Part 1. Elsevier, London, UK, pp 375-475

Dumont Y, Quirion R (2000) [^{125}I]-GR231118: a high affinity radioligand to investigate neuropeptide Y Y_1 and Y_4 receptors. Br J Pharmacol 129:37-46

Dumont Y, Redrobe JP, Quirion R (2002) Neuropeptide Y receptors. In: Pangalos MN, Davies CH (eds) Understanding G protein-coupled receptors and their role in the CNS. Oxford University Press, Oxford, pp 372-401

Ekman R, Wahlestedt C, Bottcher G, Sundler F, Hakanson R, Panula P (1986) Peptide YY-like immunoreactivity in the central nervous system of the rat. Regul Pept 16:157-168

Elshourbagy NA, Ames RS, Fitzgerald LR, Foley JJ, Chambers JK, Szekeres PG, Evans NA, Schmidt DB, Buckley PT, Dytko GM, Murdock PR, Milligan G, Groarke DA, Tan KB, Shabon U, Nuthulaganti P, Wang DY, Wilson S, Bergsma DJ, Sarau HM (2000) Receptor for the pain modulatory neuropeptides FF and AF is an orphan G protein-coupled receptor. J Biol Chem 275:25965–25971

Erickson JC, Clegg KE, Palmiter RD (1996) Sensitivity to leptin and susceptibility to seizures of mice lacking neuropeptide Y [see comments]. Nature 381:415–421

Eriksson H, Berglund MM, Holmberg SK, Kahl U, Gehlert DR, Larhammar D (1998) The cloned guinea pig pancreatic polypeptide receptor Y4 resembles more the human Y4 than does the rat Y4. Regul Pept 75/76:29–37

Esteban J, Chover AJ, Sanchez PA, Mico JA, Gibert-Rahola J (1989) Central administration of neuropeptide Y induces hypothermia in mice. Possible interaction with central noradrenergic systems. Life Sci. 45:2395–2400

Eva C, Keinanen K, Monyer H, Seeburg P, Sprengel R (1990) Molecular cloning of a novel G protein-coupled receptor that may belong to the neuropeptide receptor family. FEBS Lett 271:81–84

Eva C, Oberto A, Sprengel R, Genazzani E (1992) The murine NPY-1 receptor gene. Structure and delineation of tissue-specific expression. FEBS Lett 314:285–288

Feng Y, Broder CC, Kennedy PE, Berger EA (1996) HIV-1 entry factor: functional cDNA cloning of a seven-transmembrane, G protein-couple receptor. Science 272:872–877

File SE (1980) The use of social interaction as a method for detecting anxiolytic activity of chlordiazepoxide-like drugs. J Neurosci Methods 2:219–238

File SE, Hyde JR (1978) Can social interaction be used to measure anxiety? Br J Pharmacol 62:19–24

Flood JF, Baker ML, Hernandez EN, Morley JE (1989) Modulation of memory processing by neuropeptide Y varies with brain injection site. Brain Res 503:73–82

Flood JF, Hernandez EN, Morley JE (1987) Modulation of memory processing by neuropeptide Y. Brain Res 421:280–290

Flood JF, Morley JE (1989) Dissociation of the effects of neuropeptide Y on feeding and memory: evidence for pre- and postsynaptic mediation. Peptides 10:963–966

Gariboldi M, Conti M, Cavaleri D, Samanin R, Vezzani A (1998) Anticonvulsant properties of BIBP3226, a non-peptide selective antagonist at neuropeptide Y Y1 receptors. Eur J Neurosci 10:757–759

Gehlert DR, Beavers LS, Johnson D, Gackenheimer SL, Schober DA, Gadski RA (1996a) Expression cloning of a human brain neuropeptide Y Y_2 receptor. Mol Pharmacol 49:224–228

Gehlert DR, Gackenheimer SL, Schober DA (1992) [Leu31-Pro34] neuropeptide Y identifies a subtype of ^{125}I-labeled peptide YY binding sites in the rat brain. Neurochem Int 21:45–67

Gehlert DR, Schober DA, Beavers L, Gadski R, Hoffman JA, Smiley DL, Chance RE, Lundell I, Larhammar D (1996b) Characterization of the peptide binding requirements for the cloned human pancreatic polypeptide-preferring receptor. Mol Pharmacol 50:112–118

Gehlert DR, Schober DA, Gackenheimer SL, Beavers L, Gadski R, Lundell I, Larhammar D (1997) [^{125}I]Leu31, Pro34-PYY is a high affinity radioligand for rat PP1/Y4 and Y1 receptors: evidence for heterogeneity in pancreatic polypeptide receptors. Peptides 18:397–401

Gehlert DR, Yang P, George C, Wang Y, Schober D, Gackenheimer S, Johnson D, Beavers LS, Gadski RA, Baez M (2001) Cloning and characterization of Rhesus monkey neuropeptide Y receptor subtypes. Peptides 22:343–350

Gerald C, Walker MW, Criscione L, Gustafson EL, Batzl-Hartmann C, Smith KE, Vaysse P, Durkin MM, Laz TM, Linemeyer DL, Schaffhauser AO, Whitebread S, Hofbauer KG, Taber RI, Branchek TA, Weinshank RL (1996) A receptor subtype involved in neuropeptide-Y-induced food intake. Nature 382:168–171

Gerald C, Walker MW, Vaysse PJ, He C, Branchek TA, Weinshank RL (1995) Expression cloning and pharmacological characterization of a human hippocampal neuropeptide Y/peptide YY Y_2 receptor subtype. J Biol Chem 270:26758–26761

Gobbi M, Gariboldi M, Piwko C, Hoyer D, Sperk G, Vezzani A (1998) Distinct changes in peptide YY binding to, and mRNA levels of, Y_1 and Y_2 receptors in the rat hippocampus associated with kindling epileptogenesis. J Neurochem 70:1615–1622

Gobbi M, Mennini T, Vezzani A (1999) Autoradiographic reevaluation of the binding properties of ^{125}I-[Leu31,Pro34]peptide YY and ^{125}I-peptide YY3–36 to neuropeptide Y receptor subtypes in rat forebrain. J Neurochem 72:1663–1670

Gobbi M, Monhemius R, Samanin R, Mennini T, Vezzani A (1996) Cellular localization of neuropeptide-Y receptors in the rat hippocampus: long-term effects of limbic seizures. Neuroreport 7:1475–1480

Greber S, Schwarzer C, Sperk G (1994) Neuropeptide Y inhibits potassium-stimulated glutamate release through Y2 receptors in rat hippocampal slices in vitro. Br J Pharmacol 113:737–740

Gregor P, Feng Y, DeCarr LB, Cornfield LJ, McCaleb ML (1996a) Molecular characterization of a second mouse pancreatic polypeptide receptor and its inactivated human homologue. J Biol Chem 271:27776–27781

Gregor P, Millham ML, Feng Y, DeCarr LB, McCaleb ML, Cornfield LJ (1996b) Cloning and characterization of a novel receptor to pancreatic polypeptide, a member of the neuropeptide Y receptor family. FEBS Lett 381:58–62

Grundemar L, Wahlestedt C, Reis DJ (1991) Neuropeptide Y acts at an atypical receptor to evoke cardiovascular depression and to inhibit glutamate responsiveness in the brainstem. J Pharmacol Exp Ther 258:633–638

Gustafson EL, Smith KE, Durkin MM, Walker MW, Gerald C, Weinshank R, Branchek TA (1997) Distribution of the neuropeptide Y Y^2 receptor mRNA in rat central nervous system. Mol Brain Res 46:223–235

Hall CS (1934) Emotional behavior in the rat: defecation and urination as measures of individual differences in emotionality. J Comp Psychol 18:385–403

Harro J (1993) Measurement of exploratory behavior in rodents. In: Conn PM (ed) Paradigms for the Study of Behavior. Methods in Neurosciences. Academic Press, San Diego, vol 14, pp 359–377

Hashimoto H, Onishi H, Koide S, Kai T, Yamagami S (1996) Plasma neuropeptide Y in patients with major depressive disorder. Neurosci Lett 216:57–60

Heilig M, Ekman R (1995) Chronic parenteral antidepressant treatment in rats: unaltered levels and processing of neuropeptide Y (NPY) and corticotropin-releasing hormone (CRH). Neurochem Int 26:351–355

Heilig M, McLeod S, Koob GK, Britton KT (1992) Anxiolytic-like effect of neuropeptide Y (NPY), but not other peptides in an operant conflict test. Regul Pept 41:61–69

Heilig M, Murison R (1987) Intracerebroventricular neuropeptide Y suppresses open field and home cage activity in the rat. Regul Pept 19:221–231

Heilig M, Soderpalm B, Engel JA, Widerlov E (1989) Centrally administered neuropeptide Y (NPY) produces anxiolytic-like effects in animal anxiety models. Psychopharmacology (Berl) 98:524–529

Heilig M, Wahlestedt C, Ekman R, Widerlov E (1988a) Antidepressant drugs increase the concentration of neuropeptide Y (NPY)-like immunoreactivity in the rat brain. Eur J Pharmacol 147:465–467

Heilig M, Wahlestedt C, Widerlov E (1988b) Neuropeptide Y (NPY)-induced suppression of activity in the rat: evidence for NPY receptor heterogeneity and for interaction with alpha-adrenoceptors. Eur J Pharmacol 157:205–213

Heilig M, Widerlov E (1990) Neuropeptide Y: an overview of central distribution, functional aspects, and possible involvement in neuropsychiatric illnesses. Acta Psychiatr Scand 82:95–114

Herzog H, Hort YJ, Ball HJ, Hayes G, Shine J, Selbie LA (1992) Cloned human neuropeptide Y receptor couples to two different second messenger systems. Proc Natl Acad Sci USA 89:5794–5798

Herzog H, Hort YJ, Shine J, Selbie LA (1993) Molecular cloning, characterization, and localization of the human homolog to the reported bovine NPY Y_3 receptor: lack of NPY binding and activation. DNA Cell Biol 12:465–471

Hirabayashi A, Nishiwaki K, Shimada Y, Ishikawa N (1996) Role of neuropeptide Y and its receptor subtypes in neurogenic pulmonary edema. Eur J Pharmacol 296:297–305

Hogg S (1996) A review of the validity and variability of the elevated plus-maze as an animal model of anxiety. Pharmacol Biochem Behav 54:21–30

Holmes PV, Davis RC, Masini CV, Primeaux SD (1998) Effects of olfactory bulbectomy on neuropeptide gene expression in the rat olfactory/limbic system. Neuroscience 86:587–596

Holsboer F (2000) The corticosteroid receptor hypothesis of depression. Neuropsychopharmacology 23:477–501

Hu Y, Bloomquist BT, Cornfield LJ, DeCarr LB, Flores-Riveros JR, Friedman L, Jiang P, Lewis-Higgins L, Sadlowski Y, Schaefer J, Velazquez N, McCaleb ML (1996) Identification of a novel hypothalamic neuropeptide Y receptor associated with feeding behavior. J Biol Chem 271:26315–26319

Huh Y, Kim C, Lee W, Kim J, Ahn H (1997) Age-related change in the neuropeptide Y and NADPH-diaphorase-positive neurons in the cerebral cortex and striatum of aged rats. Neurosci Lett 223:157–160

Huh Y, Lee W, Cho J, Ahn H (1998) Regional changes of NADPH-diaphorase and neuropeptide Y neurons in the cerebral cortex of aged Fischer 344 rats. Neurosci Lett 247:79–82

Husum H, Mathe AA (2002) Early life stress changes concentrations of neuropeptide Y and corticotropin-releasing hormone in adult rat brain. Lithium treatment modifies these changes. Neuropsychopharmacology 27:756–764

Husum H, Mikkelsen JD, Hogg S, Mathe AA, Mork A (2000) Involvement of hippocampal neuropeptide Y in mediating the chronic actions of lithium, electroconvulsive stimulation and citalopram. Neuropharmacology 39:1463–1473

Husum H, Mikkelsen JD, Mork A (1998) Extracellular levels of neuropeptide Y are markedly increased in the dorsal hippocampus of freely moving rats during kainic acid-induced seizures. Brain Res 781:351–354

Husum H, Termeer E, Mathe AA, Bolwig TG, Ellenbroek BA (2002) Early maternal deprivation alters hippocampal levels of neuropeptide Y and calcitonin-gene related peptide in adult rats. Neuropharmacology 42:798–806

Inui A (1999) Neuropeptide Y feeding receptors: are multiple subtypes involved? Trends Pharmacol Sci 20:43–46

Ishihara A, Tanaka T, Kanatani A, Fukami T, Ihara M, Fukuroda T (1998) A potent neuropeptide Y antagonist, 1229U91, suppressed spontaneous food intake in Zucker fatty rats. Am J Physiol Regul Integr Comp Physiol 274: R1500–R1504

Jacques D, Dumont Y, Fournier A, Quirion R (1997) Characterization of neuropeptide Y receptor subtypes in the normal human brain, including the hypothalamus. Neuroscience 79:129–148

Jacques D, Tong Y, Shen SH, Quirion R (1998) Discrete distribution of the neuropeptide Y Y5 receptor gene in the human brain: an in situ hybridization study. Mol Brain Res 61:100–107

Jain MR, Pu S, Kalra PS, Kalra SP (1999) Evidence that stimulation of two modalities of pituitary luteinizing hormone release in ovarian steroid-primed ovariectomized rats may involve neuropeptide Y Y_1 and Y_4 receptors. Endocrinology 140:5171–5177

Jazin EE, Yoo H, Blomqvist AG, Yee F, Weng G, Walker MW, Salon J, Larhammar D, Wahlestedt C (1993) A proposed bovine neuropeptide Y (NPY) receptor cDNA clone, or its human homologue, confers neither NPY binding sites nor NPY responsiveness on transfected cells. Regul Pept 47:247–258

Jimenez-Vasquez PA, Mathe AA, Thomas JD, Riley EP, Ehlers CL (2001) Early maternal separation alters neuropeptide concentrations in selected brain regions in selected brain regions in adult rats. Dev. Brain Res 131:149–152

Jolicoeur FB, Bouali SM, Michaud JN, Menard D, Fournier A, St Pierre S (1995) Structure-activity analysis of the motor effects of neuropeptide Y. Brain Res Bull 37:1–4

Jolicoeur FB, Michaud JN, Rivest R, Menard D, Gaudin D, Fournier A, St Pierre S (1991) Neurobehavioral profile of neuropeptide Y. Brain Res Bull 26:265–268

Kalra PS, Bonavera JJ, Kalra SP (1995) Central administration of antisense oligodeoxynucleotides to neuropeptide Y (NPY) mRNA reveals the critical role of newly synthesized NPY in regulation of LHRH release. Regul Pept 59:215–220

Kalra SP, Crowley WR (1992) Neuropeptide Y: a novel neuroendocrine peptide in the control of pituitary hormone secretion, and its relation to luteinizing hormone. Front Neuroendocrinol 13:1–46

Kanatani A, Hata M, Mashiko S, Ishihara A, Okamoto O, Haga Y, Ohe T, Kanno T, Murai N, Ishii Y, Fukuroda T, Fukami T, Ihara M (2001) A typical Y1 receptor regulates feeding behaviors: effects of a potent and selective Y1 antagonist, J-115814. Mol Pharmacol 59:501–505

Kanatani A, Ishihara A, Iwaasa H, Nakamura K, Okamoto O, Hidaka M, Ito J, Fukuroda T, MacNeil DJ, Van der Ploeg LH, Ishii Y, Okabe T, Fukami T, Ihara M (2000a) L-152,804: orally active and selective neuropeptide Y Y5 receptor antagonist. Biochem Biophys Res Commun 272:169–173

Kanatani A, Kanno T, Ishihara A, Hata M, Sakuraba A, Tanaka T, Tsuchiya Y, Mase T, Fukuroda T, Fukami T, Ihara M (1999) The novel neuropeptide Y Y(1) receptor antagonist J-104870: a potent feeding suppressant with oral bioavailability. Biochem Biophys Res Commun 266:88–91

Kanatani A, Mashiko S, Murai N, Sugimoto N, Ito J, Fukuroda T, Fukami T, Morin N, MacNeil DJ, Van der Ploeg LH, Saga Y, Nishimura S, Ihara M (2000b) Role of the Y1 receptor in the regulation of neuropeptide Y-mediated feeding: comparison of wild-type, Y1 receptor-deficient, and Y5 receptor-deficient mice. Endocrinology 141:1011–1016

Kask A, Eller M, Oreland L, Harro J (2000) Neuropeptide Y attenuates the effect of locus coeruleus denervation by DSP-4 treatment on social behaviour in the rat. Neuropeptides 34:58–61

Kask A, Harro J (2000) Inhibition of amphetamine- and apomorphine-induced behavioural effects by neuropeptide Y Y(1) receptor antagonist BIBO 3304. Neuropharmacology 39:1292–1302

Kask A, Nguyen HP, Pabst R, von Horsten S (2001) Neuropeptide Y Y1 receptor-mediated anxiolysis in the dorsocaudal lateral septum: functional antagonism of corticotropin-releasing hormone-induced anxiety. Neuroscience 104:799–806

Kask A, Rago L, Harro J (1998a) Anxiolytic-like effect of neuropeptide Y (NPY) and NPY(13–36) microinjected into vicinity of locus coeruleus in rats. Brain Res 788:345–348

Kask A, Rago L, Harro J (1998b) Evidence for involvement of neuropeptide Y receptors in the regulation of food intake: studies with Y1-selective antagonist BIBP3226. Br J Pharmacol 124:1507–1515

Kerkerian L, Pelletier G (1986) Effects of monosodium L-glutamate administration on neuropeptide Y-containing neurons in the rat hypothalamus. Brain Res 369:388–390

Klapstein GJ, Colmers WF (1997) Neuropeptide Y suppresses epileptiform activity in rat hippocampus in vitro. J.Neurophysiol 78:1651–1661

Koch M (1999) The neurobiology of startle. Prog Neurobiol 59:107–128

Kofler N, Kirchmair E, Schwarzer C, Sperk G (1997) Altered expression of NPY-Y1 receptors in kainic acid induced epilepsy in rats. Neurosci Lett 230:129–132

Kombian SB, Colmers WF (1992) Neuropeptide Y selectively inhibits slow synaptic potentials in rat dorsal raphe nucleus in vitro by a presynaptic action. J Neurosci 12:1086–1093

Kragh J, Tonder N, Finsen BR, Zimmer J, Bolwig TG (1994) Repeated electroconvulsive shocks cause transient changes in rat hippocampal somatostatin and neuropeptide Y immunoreactivity and mRNA in situ hybridization signals. Exp Brain Res 98:305–313

Krause J, Eva C, Seeburg PH, Sprengel R (1992) Neuropeptide Y1 subtype pharmacology of a recombinantly expressed neuropeptide receptor. Mol Pharmacol 41:817–821

Kronfol Z, House JD (1989) Lymphocyte mitogenesis, immunoglobulin and complement levels in depressed patients and normal controls. Acta Psychiatr Scand 80:142–147

Larhammar D (1996a) Evolution of neuropeptide Y, peptide YY and pancreatic polypeptide. Regul Pept 62:1–11

Larhammar D (1996b) Structural diversity of receptors for neuropeptide Y, peptide YY and pancreatic polypeptide. Regul Pept 65:165–174

Larhammar D (1997) Extraordinary structural diversity of NPY family receptors. In: Grundemar L, Bloom SR (eds) Neuropeptide Y and drug development. Academic Press, London, pp 87–105

Larhammar D, Blomqvist AG, Yee F, Jazin E, Yoo H, Wahlestedt C (1992) Cloning and functional expression of a human neuropeptide Y/peptide YY receptor of the Y1 type. J Biol Chem 267:10935–10938

Larhammar D, Wraith A, Berglund MM, Holmberg SK, Lundell I (2001) Origins of the many NPY-family receptors in mammals. Peptides 22:295–307

Larsen PJ, Kristensen P (1998) Distribution of neuropeptide Y receptor expression in the rat suprachiasmatic nucleus. Mol Brain Res 60:69–76

Larsen PJ, Sheikh SP, Jakobsen CR, Schwartz TW, Mikkelsen JD (1993) Regional distribution of putative NPY Y1 receptors and neurons expressing Y1 mRNA in forebrain areas of the rat central nervous system. Eur J Neurosci 5:1622–1637

Lopez-Valpuesta FJ, Nyce JW, Myers RD (1996) NPY-Y1 receptor antisense injected centrally in rats causes hyperthermia and feeding. Neuroreport 7:2781–2784

Lundell I, Blomqvist AG, Berglund MM, Schober DA, Johnson D, Statnick MA, Gadski RA, Gehlert DR, Larhammar D (1995) Cloning of a human receptor of the NPY receptor family with high affinity for pancreatic polypeptide and peptide YY. J Biol Chem 270:29123–29128

Lundell I, Eriksson H, Marklund U, Larhammar D (2001) Cloning and characterization of the guinea pig neuropeptide Y receptor Y5. Peptides 22:357–363

Lundell I, Statnick MA, Johnson D, Schober DA, Starback P, Gehlert DR, Larhammar D (1996) The cloned rat pancreatic polypeptide receptor exhibits profound differences to the orthologous receptor. Proc Natl Acad Sci USA 93:5111–5115

Malmstrom RE, Hokfelt T, Bjorkman JA, Nihlen C, Bystrom M, Ekstrand AJ, Lundberg JM (1998) Characterization and molecular cloning of vascular neuropeptide Y receptor subtypes in pig and dog. Regul Pept 75/76:55–70

Marsh DJ, Baraban SC, Hollopeter G, Palmiter RD (1999) Role of the Y5 neuropeptide Y receptor in limbic seizures. Proc Natl Acad Sci USA 96:13518–13523

Marsh DJ, Hollopeter G, Kafer KE, Palmiter RD (1998) Role of the Y5 neuropeptide Y receptor in feeding and obesity. Nature Med 4:718–721

Martel JC, Alagar R, Robitaille Y, Quirion R (1990) Neuropeptide Y receptor binding sites in human brain. Possible alteration in Alzheimer's disease. Brain Res 519:228–235

Mathé AA, Rudorfer MV, Stenfors C, Manji HK, Potter WC, Theodorsson E (1996) Effect of electroconvulsive treatment on somatostatin, neuropeptide Y, endothelin and neurokinin A concentrations in cerebrospinal fluid of depressed patients. Depression 3:250–256

Matsumoto M, Nomura T, Momose K, Ikeda Y, Kondou Y, Akiho H, Togami J, Kimura Y, Okada M, Yamaguchi T (1996) Inactivation of a novel neuropeptide Y/peptide YY receptor gene in primate species. J Biol Chem 271:27217–27220

McCrea K, Wisialowski T, Cabrele C, Church B, Beck-Sickinger A, Kraegen E, Herzog H (2000) 2-36[K4,RYYSA(19-23)]PP a novel Y5-receptor preferring ligand with strong stimulatory effect on food intake. Regul Pept 87:47-58

Meister B, Ceccatelli S, Hokfelt T, Anden NE, Anden M, Theodorsson E (1989) Neurotransmitters, neuropeptides and binding sites in the rat mediobasal hypothalamus: effects of monosodium glutamate (MSG) lesions. Exp Brain Res 76:343-368

Michalkiewicz M, Michalkiewicz T (2000) Developing transgenic neuropeptide Y rats. Methods Mol Biol 153:73-89

Michel MC, Beck-Sickinger A, Cox H, Doods HN, Herzog H, Larhammar D, Quirion R, Schwartz T, Westfall T (1998) XVI International Union of Pharmacology recommendations for the nomenclature of neuropeptide Y, peptide YY, and pancreatic polypeptide receptors. Pharmacol Rev 50:143-150

Mikkelsen JD, Woldbye D, Kragh J, Larsen PJ, Bolwig TG (1994) Electroconvulsive shocks increase the expression of neuropeptide Y (NPY) mRNA in the piriform cortex and the dentate gyrus. Mol Brain Res 23:317-322

Nakajima M, Inui A, Asakawa A, Momose K, Ueno N, Teranishi A, Baba S, Kasuga M (1998) Neuropeptide Y produces anxiety via Y2-type receptors. Peptides 19:359-363

Nakamura M, Aoki Y, Hirano D (1996) Cloning and functional expression of a cDNA encoding a mouse type 2 neuropeptide Y receptor. Biochim Biophys Acta 1284:134-137

Nakamura M, Sakanaka C, Aoki Y, Ogasawara H, Tsuji T, Kodama H, Matsumoto T, Shimizu T, Noma M (1995) Identification of two isoforms of mouse neuropeptide Y-Y1 receptor generated by alternative splicing. Isolation, genomic structure, and functional expression of the receptors. J Biol Chem 270:30102-30110

Nakamura M, Yokoyama M, Watanabe H, Matsumoto T (1997) Molecular cloning, organization and localization of the gene for the mouse neuropeptide Y-Y5 receptor. Biochim Biophys Acta 1328:83-89

Naveilhan P, Hassani H, Canals JM, Ekstrand AJ, Larefalk A, Chhajlani V, Arenas E, Gedda K, Svensson L, Thoren P, Ernfors P (1999) Normal feeding behavior, body weight and leptin response require the neuropeptide Y Y2 receptor. Nature Med 5:1188-1193

Naveilhan P, Hassani H, Lucas G, Blakeman KH, Hao JX, Xu XJ, Wiesenfeld-Hallin Z, Thoren P, Ernfors P (2001) Reduced antinociception and plasma extravasation in mice lacking a neuropeptide Y receptor. Nature 409:513-517

Nilsson C, Karlsson G, Blennow K, Heilig M, Ekman R (1996) Differences in the neuropeptide Y-like immunoreactivity of the plasma and platelets of human volunteers and depressed patients. Peptides 17:359-362

Ordway GA, Stockmeier CA, Meltzer HY, Overholser JC, Jaconetta S, Widdowson PS (1995) Neuropeptide Y in frontal cortex is not altered in major depression. J. Neurochem. 65:1646-1650

Overstreet DH (1993) The Flinders sensitive line rats: a genetic animal model of depression. Neurosci Biobehav Rev 17:51-68

Overstreet DH, Russel RW (1982) Selective breeding for sensitivity to DFP. Effects of cholinergic agonists and antagonists. Psychopharmacology 78:150-154

Parker R, Liu M, Eyre HJ, Copeland NG, Gilbert DJ, Crawford J, Sutherland GR, Jenkins NA, Herzog H (2000a) Y-receptor-like genes GPR72 and GPR73: molecular cloning, genomic organisation and assignment to human chromosome 11q21.1 and 2p14 and mouse chromosome 9 and 6. Biochim Biophys Acta 1491:369-375

Parker RM, Copeland NG, Eyre HJ, Liu M, Gilbert DJ, Crawford J, Couzens M, Sutherland GR, Jenkins NA, Herzog H (2000b) Molecular cloning and characterisation of GPR74 a novel G-protein coupled receptor closest related to the Y-receptor family. Mol Brain Res 77:199-208

Parker RM, Herzog H (1999) Regional distribution of Y-receptor subtype mRNAs in rat brain. Eur J Neurosci 11:1431-1448

Pedrazzini T, Seydoux J, Kunstner P, Aubert JF, Grouzmann E, Beermann F, Brunner HR (1998) Cardiovascular response, feeding behavior and locomotor activity in mice lacking the NPY Y1 receptor. Nature Med 4:722–726

Pich EM, Agnati LF, Zini I, Marrama P, Carani C (1993) Neuropeptide Y produces anxiolytic effects in spontaneously hypertensive rats. Peptides 14:909–912

Pollard GT, Howard JL (1979) The Geller-Seifter conflict paradigm with incremental shock. Psychopharmacology (Berl) 62:117–121

Porsolt RD, Bertin A, Jalfre M (1977) Behavioral despair in mice: a primary screening test for antidepressants. Arch Int Pharmacodyn Ther 229:327–336

Pucilowski O, Overstreet DH, Rezvani AH, Janowski DS (1993) Chronic mild stress-induced anhedonia: greater effect in a genetic animal model of depression. Physiol. Behav. 54:1215–1220

Qian J, Colmers WF, Saggau P (1997) Inhibition of synaptic transmission by neuropeptide Y in rat hippocampal area CA1: modulation of presynaptic Ca2+ entry. J Neurosci 17:8169–8177

Quirion R, Martel JC, Dumont Y, Cadieux A, Jolicoeur F, St Pierre S, Fournier A (1990) Neuropeptide Y receptors: autoradiographic distribution in the brain and structure-activity relationships. Ann NY Acad Sci 611:58–72

Raposinho PD, Broqua P, Hayward A, Akinsanya K, Galyean R, Schteingart C, Junien J, Aubert ML (2000) Stimulation of the gonadotropic axis by the neuropeptide Y receptor Y1 antagonist/Y4 agonist 1229U91 in the male rat. Neuroendocrinology 71:2–7

Redrobe JP, Dumont Y, Fournier A, Quirion R (2002) The Neuropeptide Y (NPY) Y1 Receptor Subtype Mediates NPY-induced Antidepressant-like Activity in the Mouse Forced Swimming Test. Neuropsychopharmacology 26:615–624

Redrobe JP, Dumont Y, Herzog H, Quirion R (2003a) Neuropeptide Y Y2 receptors mediate behaviour in two animal models of anxiety: evidence from Y2 receptor knockout mice. Brain Behav Res 141:251–255

Redrobe JP, Dumont Y, Herzog H, Quirion R (2003b) Recognition of a target location during spatial navigation and functional object recognition memory require neuropeptide Y Y2 receptors. Eur J Neuropsychopharmocol (in press)

Redrobe JP, Dumont Y, St Pierre JA, Quirion R (1999) Multiple receptors for neuropeptide Y in the hippocampus: putative roles in seizures and cognition. Brain Res 848:153–166

Reibel S, Nadi S, Benmaamar R, Larmet Y, Carnahan J, Marescaux C, Depaulis A (2001) Neuropeptide Y and epilepsy: varying effects according to seizure type and receptor activation. Peptides 22:529–539

Rimland J, Xin W, Sweetnam P, Saijoh K, Nestler EJ, Duman RS (1991) Sequence and expression of a neuropeptide Y receptor cDNA. Mol Pharmacol. 40:869–875

Rizzi M, Monno A, Samanin R, Sperk G, Vezzani A (1993) Electrical kindling of the hippocampus is associated with functional activation of neuropeptide Y-containing neurons. Eur J Neurosci 5:1534–1538

Roder C, Schwarzer C, Vezzani A, Gobbi M, Mennini T, Sperk G (1996) Autoradiographic analysis of neuropeptide Y receptor binding sites in the rat hippocampus after kainic acid-induced limbic seizures. Neuroscience 70:47–55

Rodgers RJ, Haller J, Holmes A, Halasz J, Walton TJ, Brain PF (1999) Corticosterone response to the plus-maze: high correlation with risk assessment in rats and mice. Physiol Behav 68:47–53

Rose PM, Fernandes P, Lynch JS, Frazier ST, Fisher SM, Kodukula K, Kienzle B, Seethala R (1995a) Cloning and functional expression of a cDNA encoding a human type 2 neuropeptide Y receptor. J Biol Chem 270:22661–22664

Rose PM, Fernandes P, Lynch JS, Frazier ST, Fisher SM, Kodukula K, Kienzle B, Seethala R (1995b) Cloning and functional expression of a cDNA encoding a human type 2 neuropeptide Y receptor [erratum]. J Biol Chem 270:29038

Sadamatsu M, Kanai H, Masui A, Serikawa T, Yamada J, Sasa M, Kato N (1995) Altered brain contents of neuropeptides in spontaneously epileptic rats (SER) and tremor rats with absence seizures. Life Sci 57:523–531

Sainsbury A, Schwarzer C, Couzens M, Fetissov S, Furtinger S, Jenkins A, Cox HM, Sperk G, Hokfelt T, Herzog H (2002a) Important role of hypothalamic Y2 receptors in body weight regulation revealed in conditional knockout mice. Proc Natl Acad Sci USA 99:8938–8943

Sainsbury A, Schwarzer C, Couzens M, Jenkins A, Oakes SR, Ormandy CJ, Herzog H (2002b) Y4 receptor knockout rescues fertility in ob/ob mice. Genes Dev 16:1077–1088

Sajdyk TJ, Schober DA, Smiley DL, Gehlert DR (2002) Neuropeptide Y-Y(2) receptors mediate anxiety in the amygdala. Pharmacol Biochem Behav 71:419–423

Sajdyk TJ, Vandergriff MG, Gehlert DR (1999) Amygdalar neuropeptide Y Y1 receptors mediate the anxiolytic-like actions of neuropeptide Y in the social interaction test. Eur J Pharmacol 368:143–147

Schober DA, Van Abbema AM, Smiley DL, Bruns RF, Gehlert DR (1998) The neuropeptide Y Y1 antagonist, 1229U91, a potent agonist for the human pancreatic polypeptide-preferring (NPY Y4) receptor. Peptides 19:537–542

Schwarzer C, Kofler N, Sperk G (1998) Up-regulation of neuropeptide Y-Y2 receptors in an animal model of temporal lobe epilepsy. Mol Pharmacol 53:6–13

Schwarzer C, Williamson JM, Lothman EW, Vezzani A, Sperk G (1995) Somatostatin, neuropeptide Y, neurokinin B and cholecystokinin immunoreactivity in two chronic models of temporal lobe epilepsy. Neuroscience 69:831–845

Shiromani PJ, Overstreet DH, Levy D, Goodrich CA, Campbell SS, Gillin JC (1988) Increased REM sleep in rats selectively bred for cholinergic hyperactivity. Neuropsychopharmacol 1:127–133

Smialowska M, Sopala M, Tokarski K (1996) Inhibitory effect of intrahippocampal NPY injection on amphetamine-induced behavioural activity. Neuropeptides 30:67–71

Smialowski A, Lewinska-Gastol L, Smialowska M (1992) The behavioural effects of neuropeptide Y (NPY) injection into the rat brain frontal cortex. Neuropeptides 21:153–156

Solomon TE (1985) Pancreatic polypeptide, peptide YY, and neuropeptide Y family of regulatory peptides. Gastroenterology 88:838–841

Song C, Earley B, Leonard BE (1996) The effects of central administration of neuropeptide Y on behavior, neurotransmitter, and immune functions in the olfactory bulbectomized rat model of depression. Brain Behav Immunol 10:1–16

Sperk G (1994) Kainic acid seizures in the rat. Prog Neurobiol. 4:1–32

Sperk G, Bellmann R, Gruber B, Greber S, Marksteiner J, Roder C, Rupp E (1996) Neuropeptide Y expression in animal models of temporal lobe epilepsy. Epilepsy Res (Suppl 12):197–203

Sperk G, Marksteiner J, Gruber B, Bellmann R, Mahata M, Ortler M (1992) Functional changes in neuropeptide Y- and somatostatin-containing neurons induced by limbic seizures in the rat. Neuroscience 50:831–846

St Pierre JA, Dumont Y, Nouel D, Herzog H, Hamel E, Quirion R (1998) Preferential expression of the neuropeptide Y Y1 over the Y2 receptor subtype in cultured hippocampal neurons and cloning of the rat Y2 receptor. Br J Pharmacol 123:183–194

St Pierre JA, Nouel D, Dumont Y, Beaudet A, Quirion R (2000) Association of neuropeptide Y Y1 receptors with glutamate-positive and NPY-positive neurons in rat hippocampal cultures Eur.J Neurosci 12:1319–1330

Stanley BG, Leibowitz SF (1984) Neuropeptide Y: stimulation of feeding and drinking by injection into the paraventricular nucleus. Life Sci 35:2635–2642

Stanley BG, Magdalin W, Seirafi A, Nguyen MM, Leibowitz SF (1992) Evidence for neuropeptide Y mediation of eating produced by food deprivation and for a variant of the Y1 receptor mediating this peptide's effect. Peptides 13:581–587

Starback P, Wraith A, Eriksson H, Larhammar D (2000) Neuropeptide Y receptor gene y6: multiple deaths or resurrections? Biochem Biophys Res Commun 277:264–269

Statnick MA, Schober DA, Gackenheimer S, Johnson D, Beavers L, Mayne NG, Burnett JP, Gadski R, Gehlert DR (1998) Characterization of the neuropeptide Y5 receptor in the human hypothalamus: a lack of correlation between Y5 mRNA levels and binding sites. Brain Res 810:16–26

Statnick MA, Schober DA, Gehlert DR (1997) Identification of multiple neuropeptide Y receptor subtypes in the human frontal cortex. Eur J Pharmacol 332:299–305

Stenfors C, Mathe AA, Theodorsson E (1995) Chromatographic and immunochemical characterization of rat brain neuropeptide Y-like immunoreactivity (NPY-LI) following repeated electroconvulsive stimuli. J Neurosci Res 41:206–212

Stogner KA, Holmes PV (2000) Neuropeptide-Y exerts antidepressant-like effects in the forced swim test in rats. Eur J Pharmacol 387: R9–R10

Takahashi Y, Sadamatsu M, Kanai H, Masui A, Amano S, Ihara N, Kato N (1997) Changes of immunoreactive neuropeptide Y, somatostatin and corticotropin-releasing factor (CRF) in the brain of a novel epileptic mutant rat, Ihara's genetically epileptic rat (IGER). Brain Res 776:255–260

Tatemoto K, Carlquist M, Mutt V (1982) Neuropeptide Y-a novel brain peptide with structural similarities to peptide YY and pancreatic polypeptide. Nature 296:659–660

Tatemoto K, Mutt V (1980) Isolation of two novel candidate hormones using a chemical method for finding naturally occurring polypeptides. Nature 285:417–418

Terry RD, Davies P (1980) Dementia of the Alzheimer type. Ann Rev Neurosci 3:77–95

Thiele TE, Koh MT, Pedrazzini T (2002) Voluntary alcohol consumption is controlled via the neuropeptide Y Y1 receptor. J Neurosci 22: RC208

Thiele TE, Marsh DJ, Marie L, Bernstein IL, Palmiter RD (1998) Ethanol consumption and resistance are inversely related to neuropeptide Y levels. Nature 396:366–369

Thorsell A, Michalkiewicz M, Dumont Y, Quirion R, Caberlotto L, Rimondini R, Mathe AA, Heilig M (2000) Behavioral insensitivity to restraint stress, absent fear suppression of behavior and impaired spatial learning in transgenic rats with hippocampal neuropeptide Y overexpression. Proc Natl Acad Sci USA 97:12852–12857

Tirassa P, Lundeberg T, Stenfors C, Bracci-Laudiero L, Theodorsson E, Aloe L (1995) Monosodium glutamate increases NGF and NPY concentrations in rat hypothalamus and pituitary. Neuroreport 6:2450–2452

Tong Y, Dumont Y, Shen SH, Quirion R (1997) Comparative developmental profile of the neuropeptide Y Y1 receptor gene and protein in the rat brain. Mol.Brain Res 48:323–332

Trinh T, Dumont Y, Quirion R (1996) High levels of specific neuropeptide Y/pancreatic polypeptide receptors in the rat hypothalamus and brainstem. Eur J Pharmacol 318: R1–R3

Vezzani A, Bendotti C, Rizzi M, Monno A, Tarizzo G, Samanin R (1996a) Functional activation of somatostatin and neuropeptide Y containing neurons in experimental models of limbic seizures. Epilepsy Res.Suppl 12:187–95

Vezzani A, Monhemius R, Tutka P, Milani R, Samanin R (1996b) Functional activation of somatostatin- and neuropeptide Y-containing neurons in the entorhinal cortex of chronically epileptic rats. Neuroscience 75:551–557

Vezzani A, Schwarzer C, Lothman EW, Williamson J, Sperk G (1996c) Functional changes in somatostatin and neuropeptide Y containing neurons in the rat hippocampus in chronic models of limbic seizures. Epilepsy Res 26:267–279

Vezzani A, Sperk G, Colmers WF (1999) Neuropeptide Y: emerging evidence for a functional role in seizure modulation. Trends Neurosci 22:25–30

Vizzi ES, Kiss JP (1998) Neurochemistry and pharmacology of the major hippocampal transmitter systems: synaptic and nonsynaptic interactions. Hippocampus 8:566–607

von Horsten S, Nave H, Ballof J, Helfritz F, Meyer D, Schmidt RE, Stalp M, Exton NG, Exton MS, Straub RH, Radulovic J, Pabst R (1998) Centrally applied NPY mimics immunoactivation induced by non-analgesic doses of met-enkephalin. Neuroreport 9:3881–3885

Wahlestedt C, Blendy JA, Kellar KJ, Heilig M, Widerlov E, Ekman R (1990) Electroconvulsive shocks increase the concentration of neocortical and hippocampal neuropeptide Y (NPY)-like immunoreactivity in the rat. Brain Res 507:65–68

Wahlestedt C, Pich EM, Koob GF, Yee F, Heilig M (1993) Modulation of anxiety and neuropeptide Y-Y1 receptors by antisense oligodeoxynucleotides. Science 259:528–531

Wahlestedt C, Regunathan S, Reis DJ (1992) Identification of cultured cells selectively expressing Y1-, Y2-, or Y3-type receptors for neuropeptide Y/peptide YY. Life Sci 50: L7–12

Wang JZ, Lundeberg T, Yu L (2000) Antinociceptive effects induced by intra-periaqueductal grey administration of neuropeptide Y in rats. Brain Res 859:361–363

Weinberg DH, Sirinathsinghji DJ, Tan CP, Shiao LL, Morin N, Rigby MR, Heavens RH, Rapoport DR, Bayne ML, Cascieri MA, Strader CD, Linemeyer DL, MacNeil DJ (1996) Cloning and expression of a novel neuropeptide Y receptor. J Biol Chem 271:16435–16438

Whitcomb DC, Puccio AM, Vigna SR, Taylor IL, Hoffman GE (1997) Distribution of pancreatic polypeptide receptors in the rat brain. Brain Res 760:137–149

Whittaker E, Vereker E, Lynch MA (1999) Neuropeptide Y inhibits glutamate release and long-term potentiation in rat dentate gyrus. Brain Res 827:229–233

Widdowson PS, Halaris AE (1991) Chronic desipramine treatment reduces regional neuropeptide Y binding to Y2-type receptors in rat brain. Brain Res 539:196–202

Widdowson PS, Ordway GA, Halaris AE (1992) Reduced neuropeptide Y concentrations in suicide brain. J Neurochem 59:73–80

Widerlov E, Lindstrom LH, Wahlestedt C, Ekman R (1988) Neuropeptide Y and peptide YY as possible cerebrospinal fluid markers for major depression and schizophrenia, respectively. J Psychiatr Res 22:69–79

Wieland HA, Engel W, Eberlein W, Rudolf K, Doods HN (1998) Subtype selectivity of the novel nonpeptide neuropeptide Y Y1 receptor antagonist BIBO 3304 and its effect on feeding in rodents. Br J Pharmacol 125:549–555

Woldbye DP, Larsen PJ, Mikkelsen JD, Klemp K, Madsen TM, Bolwig TG (1997) Powerful inhibition of kainic acid seizures by neuropeptide Y via Y5-like receptors. Nature Med 3:761–764

Woldbye DP, Madsen TM, Larsen PJ, Mikkelsen JD, Bolwig TG (1996) Neuropeptide Y inhibits hippocampal seizures and wet dog shakes. Brain Res 737:162–168

Wraith A, Tornsten A, Chardon P, Harbitz I, Chowdhary BP, Andersson L, Lundin LG, Larhammar D (2000) Evolution of the neuropeptide Y receptor family: gene and chromosome duplications deduced from the cloning and mapping of the five receptor subtype genes in pig. Genome Res 10:302–310

Yan H, Yang J, Marasco J, Yamaguchi K, Brenner S, Collins F, Karbon W (1996) Cloning and functional expression of cDNAs encoding human and rat pancreatic polypeptide receptors. Proc Natl Acad Sci USA 93:4661–4665

Yang FC, Connor J, Patel A, Doat MM, Romero MT (2000) Neural transplants. effects On startle responses in neonatally MSG-treated rats. Physiol Behav. 69:333–344

Zachrisson O, Mathe AA, Stenfors C, Lindefors N (1995a) Limbic effects of repeated electroconvulsive stimulation on neuropeptide Y and somatostatin mRNA expression in the rat brain. Mol Brain Res 31:71–85

Zachrisson O, Mathe AA, Stenfors C, Lindefors N (1995b) Region-specific effects of chronic lithium administration on neuropeptide Y and somatostatin mRNA expression in the rat brain. Neurosci Lett 194:89–92

Zhang ZJ, Lappi DA, Wrenn CC, Milner TA, Wiley RG (1998) Selective lesion of the cholinergic basal forebrain causes a loss of cortical neuropeptide Y and somatostatin neurons. Brain Res 800:198–206

Prejunctional Effects of Neuropeptide Y and Its Role as a Cotransmitter

T. C. Westfall

Department of Pharmacological and Physiological Science,
Saint Louis University School of Medicine, 1402 S. Grand Blvd., St. Louis,
MO 63104-1083, USA
e-mail: westfatc@slu.edu

1	Introduction	138
2	Role of Neuropeptide Y as a Sympathetic Cotransmitter	139
2.1	Colocalization of NPY in Sympathetic Nerves	139
2.2	Nerve Stimulation-Induced Release of NPY from Sympathetic Neurons	140
2.3	Physiological and Pharmacological Evidence for a Cotransmitter Role for NPY at the Sympathetic Neuroeffector Junction	141
3	Prejunctional Effects of NPY at the Sympathetic Neuroeffector Junction	144
3.1	Concept of Prejunctional/Presynaptic Modulation of Transmitter Release: Autoreceptors and Heteroreceptors	144
3.2	NPY-Mediated Modulation of Sympathetic Cotransmitter Release	145
3.2.1	NA Release	145
3.2.2	ATP Release	150
3.2.3	NPY Release	151
3.2.4	Mechanism(s) for Presynaptic/Prejunctional Modulation of Transmitter Release by NPY	151
3.2.5	NPY-Induced Modulation of Catecholamine Synthesis	153
3.3	NA and ATP-Induced Modulation of NPY Release	155
3.4	Preferential and Differential Modulation of Sympathetic Cotransmitter Release	157
3.5	NPY-Mediated Modulation of the Release of Other Neurotransmitters in the Periphery	159
3.5.1	Acetylcholine	159
3.5.2	Calcitonin Gene Related Peptide/Substance P	160
3.6	Distribution/Location of NPY in the CNS	160
3.7	NPY-Induced Modulation of Noradrenaline Neurotransmission in the CNS	161
3.8	NPY-Induced Modulation of DA Release in the CNS	163
3.9	NPY-Induced Modulation of Other Neurotransmitters in the CNS	164
3.9.1	GABA	164
3.9.2	Glutamate	165
3.9.3	Serotonin	166
3.9.4	Neuroendocrine Mediators	167
3.10	Autoreceptor-Mediated Modulation of NPY-ir Release in the CNS	167
	References	168

Abstract Neuropeptide Y (NPY) has been shown to be colocalized and coreleased with noradrenaline (NA) and adenosine-5′-triphosphate (ATP) in most sympathetic nerves in the peripheral nervous system, especially those innervating blood vessels. There is also convincing evidence that NPY exerts prejunctional modulatory effects on transmitter release and synthesis. Moreover, there are numerous examples of postjunctional interactions that are consistent with a cotransmitter role for NPY at various sympathetic neuroeffector junctions. This chapter will provide convincing evidence that NPY does function physiologically as the third sympathetic cotransmitter together with NA and ATP. The functions of NPY include: (a) direct postjunctional contractile effects; (b) potentiation of the contractile effects of the other sympathetic cotransmitters, NA and ATP; and (c) inhibitory modulation of the nerve stimulation-induced release of all three sympathetic cotransmitters. Studies with selective NPY-Y_1 antagonists provide evidence that the principal postjunctional receptor is of the Y_1 subtype although information is incomplete and other receptors are also present at some sites and may exert physiological actions. Studies with selective NPY-Y_2 antagonist suggest that the principle prejunctional receptor is of the Y_2 subtype both in the periphery and central nervous system (CNS). Again, there is evidence for a role for other NPY receptors and clarification awaits the further development of selective antagonists. NPY can also act prejunctionally to inhibit the release of acetylcholine, calcitonin gene related peptide (CGRP) and substance P. In the CNS, NPY exists as a cotransmitter with catecholamines in some neurons and with peptides and mediators in other neurons. A prominent action of NPY is the presynaptic inhibition of the release of various neurotransmitters including NA, dopamine (DA), gamma-aminobutyric acid, glutamate, serotonin as well as inhibition or stimulation of various neurohormones such as luteinizing hormone (LH), luteinizing hormone releasing hormone, vasopressin, oxytocin as well as others. Evidence also exists for stimulation of NA and DA release. NPY also acts on autoreceptors to inhibit the release of itself. There is evidence that NPY may use several mechanisms to produce its prejunctional/presynaptic effects including: inhibition of calcium channels, activation of potassium channels and perhaps regulation of the vesicle release complex at some point post calcium entry. NPY may also play a role in several pathophysiological conditions. The therapeutic manipulation of NPY release and the development of further selective agonists and antagonist should to be an important goal of future research in order to further understand the physiological and pathophysiological role of NPY.

Keywords Neuropeptide Y · Cotransmission · Noradrenaline · Adenosine-5′-triphosphate · Sympathetic neurotransmission · Catecholamine

1
Introduction

Although proposed as early as 1904 by Elliot, the first experimental evidence for chemical neurotransmission is credited to Otto Loewi who in 1921 demonstrated that stimulation of the vagus nerve innervating a perfused frog heart released an inhibitory substance he called 'vagus-stoff' and stimulation of the sympathetic nerve released an excitatory substance called 'accelerens-stoff'. We now know these substances as acetylcholine and adrenaline (epinephrine). The mediator released from mammalian sympathetic neurons was subsequently identified as noradrenaline (NA) (Euler 1946). For years it was widely assumed that a neuron had only one neurotransmitter and this was due to an incorrect interpretation of the work of Sir Henry Dale and called 'Dales Principle' (Eccles 1986). Today we know this is not the case and it is known that in the central nervous system (CNS), peripheral nervous system and enteric nervous system, most neurons contain multiple mediators (Koelle 1955; Changeux 1986; Morris and Gibbins 1992; Sneddon 1995) that can act as cotransmitters/comodulators. The major focus of the present chapter is to discuss the evidence that NPY acts as a cotransmitter in peripheral sympathetic neurons and plays an important role as a prejunctional modulator via its action on autoreceptors and heteroreceptors in both the central and peripheral nervous system.

2
Role of Neuropeptide Y as a Sympathetic Cotransmitter

Adenosine triphosphate has long been suspected of being a cotransmitter with NA in sympathetic nerves (Holton and Holton 1954; Burnstock et al. 1970; Sneddon and Westfall 1984; Westfall et al. 1991). For instance not only is ATP present with NA in both small and large dense core vesicles in sympathetic neurons (Stjärne 1964; Lagercrantz 1971; Euler 1972; Smith 1979; Fried 1981; Klein 1982; Lagercrantz and Fried 1982) there is biochemical, electrophysiological and pharmacological evidence that nerve impulses release ATP along with NA (Stjärne 1989). In addition there is evidence that ATP contributes to the postjunctional responses together with NA following sympathetic nerve stimulation. Currently it seems well accepted that ATP plays a role in sympathetic neuroeffector mechanisms as a cotransmitter with NA at selective neuroeffector junctions (see extensive reviews by Stjärne 1989; Westfall et al. 1991; Silinsky et al. 1998; Burnstock 1999; Westfall et al. 2003). More recently there is evidence that neuropeptide Y (NPY) may also act as a sympathetic cotransmitter at some sympathetic neuroeffector junctions. The cardinal features for the acceptance of a chemical substance as a neurotransmitter or cotransmitter is that the substance is costored, coreleased and acts postjunctionally as well as prejunctionally in a coordinated fashion with the other cotransmitters.

2.1
Colocalization of NPY in Sympathetic Nerves

Neuropeptide Y has been shown to be present in a variety of organs and tissues in the periphery of numerous species (for complete references see reviews by Potter 1991; Sundler et al. 1993). For instance, NPY levels have been identified in the heart and stellate ganglia of cat, rat, mouse, guinea pig, cattle, pig, dog and man. The peptide has also been reported to be present in the respiratory system and in perivascular nerves of the lung, pancreas, uterus, bladder, nasal mucosa, submandibular gland, gastrointestinal tract, thyroid gland, muscle, skin, eye and vas deferens.

Immunohistochemical techniques reveal that NPY is mainly localized with NA in sympathetic ganglia and sympathetic nerve fibers of most of these organs and tissues (Potter 1991). In particular the colocalization of NPY and NA is very striking in postganglionic sympathetic nerves in blood vessels. NPY has also been found in the adrenal medulla although there is marked species variation in the amount found. In the adrenal medulla NPY has been found in both adrenaline containing chromaffin cells (Allen al. 1983a; Fisher-Colbrie et al. 1986; Kuramoto et al. 1986) as well as noradrenaline containing chromaffin cells (Varndell et al.1984; Majane et al. 1985).

In sympathetic neurons, where colocalization of NPY and NA has been demonstrated, unlike ATP, NPY appears to be present only in large dense core vesicles together with NA and ATP (De Deyn et al. 1989) while most of the NA has been shown to be stored separately in small dense core vesicles with ATP (Fried et al. 1985a,b; DeQuid et al. 1985; Lundberg et al. 1989b). Additional evidence for the location of NPY in sympathetic nerves are the observations that surgical sympathectomy and chemical destruction of sympathetic nerves with 6-hydroxydopamine results in a loss of perivascular NPY immunoreactive nerve fibers (Lundberg et al. 1982; Furness et al. 1983; Edvinsson et al. 1983; Ekblad et al. 1984).

2.2
Nerve Stimulation-Induced Release of NPY from Sympathetic Neurons

Numerous investigators have demonstrated that stimulation of sympathetic nerves results in the corelease of both NA and NPY-ir, in vitro, in situ and in vivo in a variety of species (see reviews by Lundberg 1996; Pernow 1988; Lacroix 1989; Dahlof 1989; Malmstrom 1997). In addition we and others have demonstrated (Kasakov et al. 1988) the simultaneous release of NA, NPY and ATP (Fig. 1). Corelease of NA and NPY has been observed in the isolated guinea pig heart (Haass et al. 1989a,b) the perfused pig spleen (Lundberg et al. 1986, 1989a–c) cat spleen (Lundberg et al. 1984a), pig heart (Rudehill et al. 1986), dog gracilis muscle (Pernow 1988), dog spleen (Schoups et al. 1988) and rat mesenteric artery (Westfall et al. 2002). In addition corelease of NA and NPY has also been observed in vivo in conscious calves (Allen et al. 1984) and in the pithed guinea pig (Dählof et al. 1986) and the pithed rat (Zukowaska-Grojec et al. 1992). The release of both NA and NPY-ir depends on influx of extracellular Ca^{2+} through omega-conotoxin sensitive N-type channels (Franco-Cercede et al. 1989; Haass et al. 1989a,b) and the release of both mediators is inhibited by the neuronal blocking agent, guanethidine (Lundberg et al. 1984a). Physiological activation of the sympathetic nervous system has been shown to enhance NPY-ir

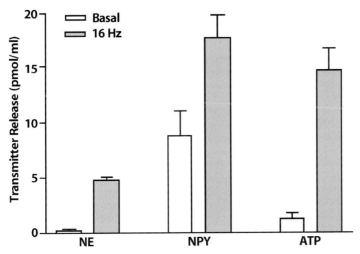

Fig. 1 Periarterial nerve stimulation induced overflow of NA, ATP and NPY-ir from the perfused mesenteric arterial bed. Periarterial nerves were stimulated for 90 s at a frequency of 16 Hz, supramaximal voltage. Superfusate effluents were continuously collected and analyzed for NA by high-pressure liquid chromatography coupled to electrochemical detection (HPLC-EC) (Chen and Westfall 1994), ATP by HPLC-flurometric detection (Levitt et al. 1984; Todorov et al. 1996) and NPY-ir by radioimmunoassay (DiMaggio et al. 1994; Chen and Westfall 1994)

in parallel with catecholamines both in animals exposed to stress (Zukowaska-Grojec et al. 1988; Zokowaska and Vaz 1988; Dahlöf et al. 1986; Han et al. 1998b) and by reflex-mediated sympathetic activation, as for example, by hemorrhagic hypotension or endotoxin shock. There is release of NPY-ir in man during graded and prolonged physical exercise. In both humans and animals, it is thought that the plasma NPY-ir is mainly derived from sympathetic nerves rather than the adrenal medulla since NPY-ir levels correlate better with NA levels than adrenaline levels (Lundberg et al. 1985a; Pernow et al. 1986a). In both animal and human studies, NPY release appears to require high frequency nerve stimulation or high frequency bursts compared to stimulation with a continuous low frequency given the same number of pulses.

2.3
Physiological and Pharmacological Evidence for a Cotransmitter Role for NPY at the Sympathetic Neuroeffector Junction

In addition to the observation that NPY is colocalized and coreleased with NA and ATP from sympathetic nerves and that it exerts prejunctional modulatory effects on transmitter release (discussed below), there are numerous examples of postjunctional interactions that are consistent with a cotransmitter role for NPY at various sympathetic neuroeffector junction.

Table 1 Preparations in which a nonadrenergic contractile response remains after reserpine or α-adrenoceptor blockade

Species	Preparation	References
Human	Saphenous vein	Racchi et al. 1999
Pig	Spleen	Lundberg et al. 1987; Malmström 1997
Pig	Nasal mucosa	Lundberg et al. 1987; Lacroix 1989
Pig	Kidney	Pernow and Lundberg 1989b
Dog	Gracilis muscle	Pernow 1988
Cat	Spleen	Lundberg et al. 1987
Cat	Nasal mucosa	Lundberg et al. 1987; Lacroix et al. 1988
Cat	Hindlimb	Pernow et al. 1988a
Cat	Submandibular gland	Lundberg and Tatemoto 1982
Guinea pig	Vas deferens	Stjärne 1989
Guinea pig	Vena cava	Malmström 1997
Guinea pig	Inferior vein	Smyth et al. 2000
Rat	Vas deferens	Stjärne 1989
Rat	Mesenteric artery	Westfall et al. 1987; Donoso et al. 1997
Rat	Anococcygens muscle	Hoyo et al. 2000
Rat	Tail artery	Bradley et al. 2003
Mouse	Vas deferens	Stjärne 1989

The presence of a nonadrenergic component of the contractile or vasoconstrictor response following sympathetic nerve stimulation has been observed in numerous vascular and non-vascular neuroeffector junctions of various species (Table 1). These include: human saphenous vein; pig spleen, nasal mucosa and kidney; dog gracilis muscle and hindlimb; cat spleen, nasal mucosa, hindlimb and submandibular salivary gland; guinea pig vas deferens, vena cava and inferior mesenteric vein; rat vas deferens, mesenteric artery, tail artery and anococcygens muscle; and mouse vas deferens among others. In these preparations it was repeatedly observed that the vasoconstrictor or contractile responses to sympathetic nerve stimulation could not be totally blocked by α-adrenoceptor antagonists implying that a nonadrenergic mediator or mediators contribute to the response. Two obvious candidates for the nonadrenergic component are ATP and NPY. At some neuroeffector junctions, such as the guinea pig and rat vas deferens it is clear that ATP is responsible for the fast contractile response following sympathetic nerve stimulation while NA is responsible for the more sustained phase (Westfall and Westfall 2001). This role for ATP has been discussed and summarized in many recent reviews (Stjärne 1989; Westfall et al. 1991; Burnstock 1999; Westfall et al. 2003).

At other neuroeffector junctions there is considerable evidence for NPY contributing to the nonadrenergic component of the contractile effect following nerve stimulation. In the majority of cases, NPY appears responsible for the slow and long lasting contractile or vasoconstrictor effect to sympathetic nerve stimulation in the presence of α-adrenoceptor and some cases β-adrenoceptor antagonists. In many preparations, such as dog gracilis muscle, sympathetic nerve stimulation-induced vasoconstriction is slow in onset, has a long duration and is most pronounced after high frequency stimulation (Pernow 1988). In addition it is sensitive to reserpine pretreatment. In this and other preparations the rapid and short lasting purinergic component of the response to nerve stimulation is not influenced by reserpine treatment (Muramatsu 1987; Warland and Burnstock 1987), and appears to involve low frequency nerve stimulation (Burnstock and Warland 1987; Westfall et al. 2003).

Further support for a role for NPY were the observations that there was inhibition of sympathetic nonadrenergic vasoconstriction after tachyphylaxis to NPY (Öhlen et al. 1990; Morris 1991) or in the presence of antiserum to NPY (Laher et al. 1994). In addition supersensitivity to NPY evoked vasoconstriction was produced in the pig nasal mucosa preparation after sympathetic denervation (Lacroix and Lundberg 1989). In many studies there were also a high level of correlation between sympathetically induced vasoconstriction and the overflow of NPY-ir detected in the local venous effluent (Lundberg et al. 1989c). There have also been many examples of NPY-mediated potentiation of the postjunctional contractile responses to NA and ATP, which is consistent with a cotransmitter role (Edvinsson et al. 1984; Westfall et al. 1987, 1990a, 1995).

The most convincing evidence for a cotransmitter role for NPY is a result of studies utilizing selective and potent NPY antagonists. There are now many observations demonstrating that the nonadrenergic, nonpurinergic contractile or vasoconstrictor responses to sympathetic nerve stimulation are attenuated by NPY-Y_1 antagonists. For instance this has been demonstrated in the guinea pig vena cava (Malmström and Lundberg 1995b) pig hindlimb, spleen kidney and nasal mucosa (Malmstrom 1997), rat mesenteric arterial bed (Han et al. 1998a; Donoso et al. 1997; Cortes et al. 1999), human saphenous vein (Racchi et al. 1999), and rat tail artery (Bradley et al. 2003). In some preparations, like the human saphenous vein (Racchi et al. 1999) the rat mesenteric arterial bed (Donoso et al. 1997) and rat tail artery (Bradley et al. 2003), it was necessary to block the response of all three cotransmitters (NA, ATP and NPY) with a 'cocktail' of appropriate antagonists (prazosin or phentolamine, suramin and BIBP 3226) in order to completely attenuate the vasoconstrictor response to sympathetic nerve stimulation. Figure 2 depicts a summary of experiments (carried out by Donoso et al. 1997) demonstrating the necessity to block the effect of all three mediators to totally attenuate sympathetic nerve stimulation-induced contraction. In summary, the presently available data suggests that NPY acts as a cotransmitter or comodulator with NA in the control of numerous peripheral vascular junctions as well as heart, spleen and vas deferens of several species. It appears to be most important under conditions of high sympathetic nerve activity. The contribution of NPY versus the other sympathetic cotransmitters NA and ATP varies between species, vascular bed and even vessels within each vascular bed as well as the impulse patterns of sympathetic activity. It also appears that all three cotransmitters have synergistic actions with each other for instance, despite the fact that NPY may have little direct effect on the contraction of vascular smooth muscle in some cases, neurogenically released NPY can act as a neuromodulator markedly potentiating the actions of the other two cotransmitters.

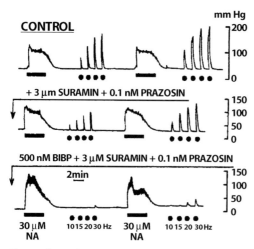

Fig. 2 Effects of simultaneous perfusion of the rat arterial mesenteric bed with either 3 μM suramin plus 0.1 nM prazosin, and 500 nM BIBP 3226, 3 μM suramin and 0.1 nM prazosin on the increase in perfusion pressure caused by electrical stimulation of the periarterial nerves. Representative polygraphic tracings show the increase in perfusion pressure caused by a 4-min pulse of 30 μM NA, followed by 10-s trains of 10, 15, 20 and 30 Hz (*left side*) or 30-s trains of electrical pulse (*right side*). The *upper* control trace shows a recording obtained in the absence of the antagonists. (Reproduced from Donoso et al. 1994 with permission)

3
Prejunctional Effects of NPY at the Sympathetic Neuroeffector Junction

3.1
Concept of Prejunctional/Presynaptic Modulation of Transmitter Release: Autoreceptors and Heteroreceptors

It is well established that receptors are located on soma, dendrites and axons of neurons, where they may respond to neurotransmitters or modulators released from the same neuron or from adjacent neurons or cells. Soma-dendritic receptors are those receptors located on, or near the cell body and dendrites, and when activated primarily modify functions of the soma-dendritic region such as protein synthesis and generation of action potentials. Presynaptic receptors, by convention, are those receptors presumed to be located on, in or near axon terminals or varicosities, and when activated modify functions of the terminal region such as synthesis and release of transmitters. Two main classes of presynaptic receptor have been identified: heteroreceptors are those presynaptic receptors that respond to neurotransmitters, neurohormones or neuromodulators released from adjacent neurons or cells or distant tissues, while autoreceptors are those receptors located on or close to those axon terminals of a neuron through which the neurons own transmitter can and under appropriate condi-

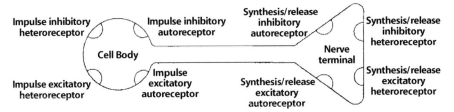

Fig. 3 A representation of the cell body and terminal region of a hypothetical neuron. Impulse modulating inhibitory and excitatory heteroreceptors or impulse modulating inhibitory or excitatory autoreceptors exist on the soma and dendrites of neurons. Likewise in the terminal region, there may be inhibitory or excitatory heteroreceptors and autoreceptors that regulate synthesis and/or release of the transmitter. (Reproduced from Westfall 1995 with permission)

tions may, modify transmitter synthesis or release. Figure 3 depicts a summary of various auto and heteroreceptors.

The basic evidence for presynaptic autoreceptors has been extensively discussed in numerous publications (for selected reviews see Boehm and Kubista 2002; Boehm and Huck 1997; Fuder and Muscholl 1995; Lundberg 1996; Chesselet 1984; Gillespie 1980; Langer 1981, 1988, 1997; Miller 1998; Starke 1977, 1981, 1987; Starke, Gothert and Kilbunger 1989; Stjärne 1989; Westfall, 1977, 1980, 1990b; Westfall and Martin, 1990; Westfall et al., 1986; von Kugelgen 1996). Briefly, agonists inhibit or enhance the release of the transmitter, while antagonists counteract the effect of the agonists and when given alone produce the opposite effects of the agonists. Although still not definitively proven, the locations of these receptors is thought to be presynaptic because the effects are obtained in the absence of cell bodies, can be seen in synaptosomal preparations and in neurons grown in tissue culture and by ligand binding and immunohistochemical studies where the receptors disappear upon denervation (Westfall 1990b). Evidence exists for the presence of presynaptic autoreceptors and heteroreceptors in all types of neurons in the CNS as well as in the periphery (Westfall 1990b, 1995). It is likely that both types of receptors are a basic feature of all neurons.

3.2
NPY-Mediated Modulation of Sympathetic Cotransmitter Release

3.2.1
NA Release

The evidence for presynaptic α_2 adrenoceptor-mediated autoreceptor modulation of NA release has been firmly established for many years (see reviews Westfall 1977, 1990b; Westfall et al. 1986; Starke 1977, 1987; Starke et al. 1989; Langer 1981, 1988, 1997). Since the discovery that ATP and NPY are colocalized and coreleased with NA in sympathetic neurons and behave as cotransmitters, as discussed above, several investigators have examined the effect of NPY on NA

Table 2 Examples of neuropeptide Y Induced inhibition of noradrenaline release from sympathetic neurons

Species	Tissue/preparation	References
Human	Submandibular artery	Lundberg et al. 1985
Human	Right atrium	Rump et al. 1997
Human	Renal cortex	Rump et al. 1997
Dog	Gracilis muscle	Pernow et al. 1988a; Kahan et al. 1988
Pig	Spleen	Lundberg et al. 1989c
Rabbit	Kidney	Rump et al. 1997
Rabbit	Oviduct	Samuelson and Dalsgaard 1985
Rabbit	Ear artery	Wong-Dusting and Rand 1988; Chernaeva and Charakchieva 1988
Guinea pig	Heart	Franco-Cerceda et al. 1985
Rat	Mesenteric artery	Westfall et al. 1987, 1988, 1990a
Rat	Femoral artery	Pernow et al. 1986b
Rat	Basilar artery	Pernow et al. 1986b; Lundberg et al. 1985c
Rat	Portal vein	Pernow et al. 1986b
Rat	Vas deferens	Dahlöf et al. 1985; Pernow et al. 1986b; Lundberg and Stjärne 1984; Stjärne et al. 1986
Mouse	Vas deferens	Stjärne et al. 1986; Lundberg and Stjärne 1984

release. The first evidence that NPY can indeed inhibit the evoked release of NA came from studies in which it was observed that NPY inhibited muscle contraction induced in various in vitro preparations by transmural nerve stimulation. For instance this inhibition was observed in the mouse vas deferens (Allen et al. 1982; Lundberg et al. 1982) rat uterus (Stjernquist 1983) and guinea pig urinary bladder (Lundberg et al. 1984b). Subsequently numerous overflow studies confirmed that NPY inhibited the evoked release of ^{3}H-NA or endogenous NA from numerous tissue preparations and species (see Table 2 and Fig. 4). In vitro preparations have included: human submandibular artery, right atrium and renal cortex; dog gracilis muscle; pig spleen; rabbit kidney, oviduct and ear artery; guinea pig heart and atrium; rat mesenteric artery, basilar artery, femoral artery, portal vein and vas deferens; mouse vas deferens. Similar types of studies have been carried out in situ in such preparations as the pithed rat (Dahlöf 1989) canine gracilis muscle (Pernow et al. 1988a; Kahan et al. 1988) and blood perfused pig spleen (Lundberg et al. 1989c).

The presence of α_1 or α_2 adrenoceptor antagonists did not alter the inhibitory effect of NPY on NA release (Westfall et al. 1987). The nature of the receptor mediating NPY-induced inhibition of NA release has been extensively investigated. Early studies by Wahlstedt et al. (1986, 1987) examined the effect of NPY and NPY analogs on the three well known actions of NPY depicted in Fig. 5: (1) direct postjunctional effects of NPY were studied on isolated blood vessels such as guinea pig iliac vein and cat middle cerebral artery; (2) the postjunctional potentiation effect of NPY on NA evoked vasoconstriction was studied in other blood vessels such as the rabbit femoral artery or vein or rabbit pulmonary artery; and (3) the prejunctional inhibitory effect on neurotransmitter release was studied in the rat vas deferens (Wahlestedt et al. 1986, 1987).

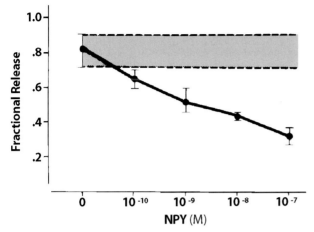

Fig. 4 The effect of NPY on the periarterial nerve stimulation induced release of NA from the perfused mesenteric arterial bed of the rat. Data are plotted as fractional release of endogenous NA (measured by HPLC-EC detection) versus the concentration of NPY. (Reproduced from Westfall et al. 1990a, with permission)

NPY and intestinal polypeptide (PYY) were active in all three types of assay while pancreatic polypeptide (PP) displayed much weaker activity. They observed that C-terminal fragments [e.g. NPY(13–36); PYY(13–36)] were essentially inactive in the assays for postjunctional activity, while they retained substantial activity prejunctionally. On the basis of these results Wahlstedt and colleagues proposed that NPY receptor subtypes exist (Wahlestedt et al. 1986). The nomenclature Y_1 and Y_2 was introduced and it was proposed that the whole NPY or PYY molecule was necessary to activate the Y_1 receptor while the Y_2 receptor could be selectively stimulated by the long C-terminal NPY or PYY fragments. This was a very useful and important hypothesis; however we now know that it was an oversimplification similar to the original α_1-α_2 adrenoceptor designation of postjunctional and prejunctional α-adrenoceptors (Westfall 1977). It is now known that Y_1 receptors are located prejunctionally as well as postjunctionally and Y_2 receptors are located postjunctionally as well as prejunctionally (McAuley and Westfall 1995; Malmström 2001; Lundberg and Modin 1995; Malmström et al. 1998, 2000; Malmström and Lundberg 1996a,b). Moreover, we now know there are at least five distinct NPY receptors that have been established by molecular approaches and have been cloned. As discussed in the chapter by Holliday et al. in this volume, these include Y_1 (Gerald et al. 1996; Krause et al., 1992; Larhammar et al. 1992; Pittito et al. 1994), Y_2 (Gehlert et al. 1996; Rose et al. 1995) Y_4 (Gerald et al. 1996; Lundell et al. 1995), Y_5 (Gerald et al. 1996) and Y_6 (Gregor et al. 1996; Weinberg et al. 1996). Another putative receptor subtype the Y_3 receptor has not been cloned but has been suggested based on functional and biochemical studies (McCullough and Westfall 1995, 1996; Norenberg et al. 1995).

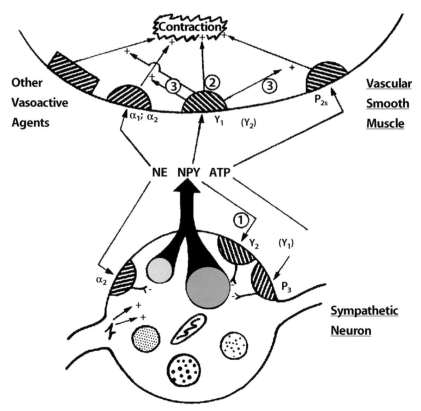

Fig. 5 A schema depicting the prejunctional and postjunctional effects of the three sympathetic co-transmitters NA, ATP and NPY following their release from a sympathetic nerve. In the case of NPY, it can exert direct postjunctional contractile effects or potentiate the contractile effect of NA, ATP or other vasoactive agents. Prejunctionally, NPY can inhibit the evoked release of NA, ATP or NPY itself

It is of interest that recently the location of Y_1 and Y_2 receptors in the periphery has been confirmed by using antibodies directed at the Y_1 and Y_2 receptor. It was observed that immunostaining for the Y_1 receptor was mainly seen in the smooth muscle layer of blood vessel and immunostaining for the Y_2 receptor was seen mostly in nerve cell bodies (Uddman et al. 2002).

Despite the existence of several NPY receptors there is a great deal of evidence that, at the sympathetic vascular neuroeffector junction, the Y_1 receptor is most important in mediating the direct contractile responses as well as the potentiation of contraction to NA as discussed above. On the other hand, the Y_2 receptor is thought to be the most prominent receptor mediating inhibition of NA release as well as the release of other transmitters such as acetylcholine from parasympathetic nerves (Potter 1991) and CGRP from sensory nerves (Kawasak et al. 1991). Data supporting this idea include the numerous studies demonstrating that agonists with some selectivity for Y_2 receptors, e.g. PYY(13–36);

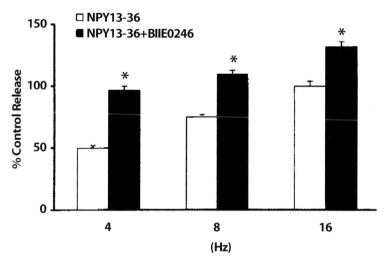

Fig. 6 The effect of NPY(13–36) alone or in the presence of BIIE 0246 on the periarterial nerve stimulation induced overflow of endogenous NA from the perfused mesenteric arterial bed of the rat. BIIE 0246 antagonized the inhibitory effect of NPY 13–36 at all three frequencies 4, 8 and 16 Hz

NPY(13–36), NPY(3–36), PYY(3–36), N-acetyl [Leu^{28}Leu^{31}]NPY(24–36), inhibit nerve stimulation-induced release of NA, ATP and NPY from a variety of preparations while nonselective agonists are less effective.

Recently, even more convincing evidence for the nature of the prejunctional Y_2 receptor was obtained following the use of the selective and potent Y_2 antagonist BIIE 0246. BIIE 0246 is the first potent and selective Y_2 antagonist developed (Dumont et al. 2000; Doods et al. 1999). The antagonistic and selectivity of BIIE 0246 for the Y_2 receptor has been demonstrated in several in vitro bioassays, and binding affinity studies. It possesses high affinity for the NPY Y_2 receptor and is devoid of affinity for the Y_1, Y_4 or Y_5 receptors or 60 other receptor types and enzymes in various binding assays (Dumont et al. 2000; Doods et al. 1999). In in vitro bioassays, no agonistic or antagonistic activities for BIIE 0246 were observed in isolated tissues in which NPY-induced effects are mediated by the activation of Y_1 or Y_4 subtypes. In contrast, it was observed to be a potent antagonist in the rat vas deferens and dog saphenous vein, two prototypical Y_2 assay systems (Pheng et al. 1997; Wahlstedt 1987).

Not only has BIIE 0246 been shown to attenuate the NPY-induced inhibition of contractions of the rat vas deferens and dog saphenous vein (Doods et al. 1999; Dumont et al. 2000), it has also been shown to attenuate the PYY-induced inhibition of the nerve stimulation evoked release of NA and NPY-ir from the pig kidney (Malmström 2002). Studies carried out in the author's laboratory demonstrate that BIIE 0246 antagonized the effect of NPY(13–36) to inhibit the transmural stimulation-induced release of NA from the perfused mesenteric arterial bed (Fig. 6) and caused a frequency-dependent enhancement of the nerve stimulation-induced increase in NA overflow in this preparation. This latter ob-

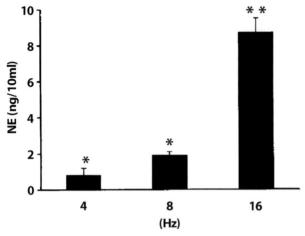

Fig. 7 The effect of BIIE 0246 (100 μM) on the periarterial nerve stimulation induced overflow of NA from the perfused mesenteric arterial bed of the rat. BIIE 0246 produced a significant increase in NA overflow over what was seen with nerve stimulation alone (at all three frequencies of nerve stimulation). The BIIE 0246 increase in NA overflow becomes greater as the frequency of nerve stimulation increases

servation is consistent with a physiological regulation by NPY sympathetic neurotransmission in this preparation (Fig. 7).

It was also shown that BIIE 0246 antagonized functional responses attributed to Y_2 receptors at cholinergic and purinergic peripheral autonomic neuroeffector junctions by the Y_2 selective agonist N-acetyl (leu^{28}Leu^{31})NPY(24–36) (Smith-White et al. 2001). Finally, BIIE 0246 attenuated the NPY-induced inhibition of the evoked release of NA in hippocampal and hypothalamic slices in the brain (King et al. 2000; Weiser et al. 2000).

3.2.2
ATP Release

It appears that activation of NPY receptors not only inhibits the release of NA but also the cotransmitter ATP. For instance in the mouse vas deferens NPY has been shown to inhibit the nerve stimulation evoked release of NA as well as the rapid stimulus evoked excitatory junction potentials (EJP) in smooth muscle cells that are caused by ATP (Stjärne et al. 1986, 1989). In addition in at least one overflow study it has been reported that NPY inhibits the evoked release of ATP and NA to a similar degree in the guinea pig vas deferens (Ellis and Burnstock 1990). There are also studies demonstrating that activation of $α_2$-adrenoceptor autoreceptors decreases the evoked release of ATP (Bohmann et al. 1995; Hammond et al. 1988; Sperlagh and Vizi 1992; Driessen et al. 1993; for a review see von Kügelgen 1996).

3.2.3
NPY Release

Activation of prejunctional NPY autoreceptors can also inhibit the evoked release of NPY itself in addition to NA and ATP. It was demonstrated by Pernow and Lundberg (1989a) that the evoked release both of NPY and NA from the pig kidney was attenuated by PYY. A similar NPY autoreceptor-mediated inhibition of NPY release was seen in nerve growth factor (NGF)-differentiated PC12 cells (Chen et al. 1997). In this preparation PYY(13–36), a selective NPY Y_2 agonist, produced a concentration-dependent attenuation of the K^+-evoked release of NPY-ir as well as catecholamines. Further evidence for the autoreceptor being that of the Y_2 subtype was the observation that the selective Y_2 receptor antagonist BIIE 0246 attenuated the PYY-induced inhibition of the nerve stimulation-induced release of NPY in the pig kidney (Malmström et al. 2002).

3.2.4
Mechanism(s) for Presynaptic/Prejunctional Modulation of Transmitter Release by NPY

The mechanism(s) of action for the presynaptic inhibitory modulation of transmitter release by NPY has been the subject of intense investigation but remains incompletely understood. This aspect of NPY as well as the action of other presynaptic receptors, has been recently reviewed (Langer 1997; Miller 1998; Boehm and Kubista 2002). There is evidence that NPY may use several mechanisms to produce its presynaptic modulatory effects including: inhibition of Ca^{2+} channels, activation of K^+ channels and regulation of the vesicle release complex.

Inhibition of Ca^{2+} currents by NPY would directly suppress Ca^{2+} influx into nerve terminals and inhibit transmitter release. There is considerable evidence for the idea that NPY-Y_2-induced inhibition of transmitter release is pertussis toxin sensitive, and acts by inhibiting of N-type voltage gated Ca^{2+} channels. NPY has been shown to attenuate the stimulation-induced increase in $[Ca^{2+}]_i$ in the NGF-differentiated PC12 cell (Chen et al. 1994), the KCl-induced rise in cytosolic Ca^{2+} in cultured rat superior cervical ganglion cells (Oellerich et al. 1994; Plummer et al. 1991; Foucart et al. 1993) and the K^+-induced increase in depolarization calcium transients in cultured sympathetic neurons (Hirning et al. 1990). The Y_2 receptor-mediated inhibition of N-type voltage gated channels has been further demonstrated in dorsal root ganglion (Bleakman et al. 1991), neuroblastoma cells (McDonald et al. 1995; Lynch et al. 1994) nodose ganglion (Bleakman et al. 1991), superior cervical ganglia cocultured with atrial myocytes (Toth et al. 1993) and NGF-differentiated PC12 cells (McCullough et al. 1998a,b). In addition to inhibiting N-type voltage channels, NPY was observed to inhibit P/Q channels as well as N channels in the hippocampus (Qian et al. 1997) and to inhibit both L and N channels in NGF-differentiated PC12 cells (McCullough and Westfall 1996; McCullough et al. 1998a,b).

It is clear, therefore, that activation of NPY-Y_2 receptors can inhibit the stimulation-induced release of transmitter via inhibition of N-type voltage gated channels. However, this may not be the only mechanism. There is also evidence that NPY inhibits transmitter release by activation of K^+ channels that would increase the conductance of the terminal making incoming action potentials less effective in activating Ca^{2+} channels. Evidence for NPY-induced activation of K^+-channels has been observed in desheathed mouse vas deferens, where excitatory junctional currents (EJC) were used to monitor ATP release from sympathetic nerves (Stjärne et al. 1989). In this preparation it was shown that tetraethylammonium and 4-aminopyridine increased EJCs and blocked the ability of NPY to decrease EJCs, suggesting that inhibition of ATP by NPY was due to activation of K^+ channels. In amphibian sympathetic ganglia, NPY activates an inwardly rectifying K^+-channel in C-cells (Zidichouski et al. 1990). The ability of NPY to decrease sensory neurotransmitter release in the guinea pig was sensitive to charybdotoxin, an agent that blocks Ca^{2+}-activated K^+ channels (Stretton et al. 1992).

In neurons acutely isolated from the rat arcuate nucleus, NPY was observed to regulate both Ca^{2+} and K^+ currents and produced a change in the pattern of electrical signaling in these cells (Sun and Miller 1999).

Although most of the available evidence points to the NPY-Y_2 receptor as the most important and prominent receptor modulating transmitter release, there is evidence for a role for additional NPY receptors. Sun and Miller (1999) observed that in the rat hypothalamic slice preparation, NPY analogs targeting different types of NPY receptors could inhibit both excitatory (glutamate) and inhibitory (GABA)-mediated synaptic transmission. In fact it was suggested that all of the major types of NPY receptors could inhibit the evoked release of glutamate. Bitran et al. (1999) provided evidence that the inhibitory effect of NPY to inhibit the evoked release of NA in rat hypothalamic slices may be mediated by the Y_1 and Y_5 receptor as well as the Y_2. Studies carried out in the NGF-differentiated PC12 cell also suggest that there exists a complex role for NPY in the modulation of sympathetic neurotransmission. It was observed that NPY and PP act through at least four receptor subtypes to modulate N and L-type channels (McCullough et al. 1998b). It has also been observed that NPY analogs with different affinity for NPY receptors inhibits NA synthesis in the rat mesenteric arterial bed (Westfall et al. 2003; see below). In addition to inhibition of Ca^{2+} currents and activation of K^+ currents there is some suggestion that NPY may also modulate sympathetic cotransmitter release at some point post calcium entry.

The molecular mechanisms involved in release and regulation of release of neurotransmitter has been the subject of intense study and there are several excellent reviews that have addressed the subject (Boehm and Kubista 2002; Miller 1998). Briefly, it is known that the majority of transmitters are stored in vesicles that are localized at special release sites, close to the presynaptic membrane, and ready for release to occur upon the appropriate stimulus. The vesicles initially dock and then are primed. It appears that there is the formation of a multiprotein complex that attaches and fastens the vesicle to the plasma membrane close to other signaling elements. The complex involves proteins from both the vesicular membrane and the presynaptic neuronal membrane, as well as other components that help link the two together. It is established that these various synaptic proteins, including the plasma membrane proteins syntaxin, and synaptosomal protein 25 kDa (SNAP-25), and the vesicle membrane protein, or synaptobrevin, form a complex. This complex interacts in an ATP-dependent manner with the soluble proteins N-ethylmaleimide sensitive fusion protein (NSF) and soluble NSF attachment proteins (SNAPs). The ability of synaptobrevin, syntaxin 1 and SNAP-25 to bind SNAPs has led to their designation as SNAP receptors (SNAREs). It has been hypothesized that most if not all intracellular fusion events are mediated by SNARE-interactions (Boehm and Kubista 2002; Miller 1998). Important evidence supporting the involvement of SNARE proteins (SNAP-25, syntaxin, synaptobrevin) in transmitter release comes from the fact that botulinum neurotoxins and tetanus toxin, which potently block neurotransmitter release, proteolyse these three proteins. This raises the question as to whether these proteins are associated with mechanisms of prejunctional receptor-induced modulation of transmitter release. Evidence exists that there is an interaction between the muscarinic cholinergic autoreceptor and the SNARE complex that is modulated by depolarization levels (Linial et al. 1997). It has also recently been observed that botulinum neurotoxin A can selectively inhibit the release of acetylcholine or NA without, or only minimally, affecting the release of the cotransmitter peptides vasoactive intestinal polypeptide or NPY from parasympathetic or sympathetic neurons, respectively (Morris et al. 2001; 2002). An important area of future research will be studies of the interactions between presynaptic NPY receptors and SNARE proteins as a mechanism for mediating inhibition of transmitter release.

3.2.5
NPY-Induced Modulation of Catecholamine Synthesis

In addition to producing an inhibition of the evoked release of NA, ATP and NPY from a variety of sympathetic neurons, there is evidence that NPY can also alter the synthesis of catecholamines (McCullough and Westfall 1995, 1996; McCullough et al. 1998a,b). Utilizing the NGF-differentiated PC12 cell, it was observed that NPY-inhibited depolarization stimulated catecholamine synthesis as determined by in situ measurement of dihydroxyphenylalanine (DOPA) production in the presence of the decarboxylase inhibitor m-hydroxybenzylhydrazine (NSD-1015). Following differentiation with NGF, the PC12 cell resembles sympathetic neurons and has been a useful model to study cellular aspects of NPY function concomitant with catecholamine and NPY neurotransmission (Chen et al. 1997). The inhibition by NPY was concentration dependent and was prevented by pretreatment with pertussis toxin in a similar way as for the inhibition of K^+-induced release of catecholamine from these cells. This suggests the involvement of a GTP-binding protein of the G_i or G_o subtype. The NPY analog [Leu^{31} Pro^{34}]NPY also caused inhibition of DOPA production but was less potent than

NPY itself, while PYY and NPY(13–36) had no significant effect. This pattern is most consistent with the involvement of the NPY-Y_3 receptor subtype.

Other studies utilize multiple selective Ca^{2+}-channel and protein kinase agonists and antagonists to elucidate the mechanisms by which NPY modulates catecholamine synthesis (McCullough and Westfall 1996). The L-type Ca^{2+} channel blocker, nifedipine, inhibited the depolarization-induced stimulation of DOPA production by approximately 90% and attenuated the inhibitory effect of NPY. In contrast, the N-type Ca^{2+} channel blocker, ω-conotoxin GVIA, did not inhibit either the stimulation of DOPA production or the effect of NPY. Antagonism of Ca^{2+} calmodulin-dependent protein kinase (CaM kinase) greatly inhibited the stimulation of DOPA production by depolarization and prevented the inhibitory effect of NPY, whereas alterations in the cyclic adenosine monophosphate-dependent protein kinase pathway modulated DOPA production but did not prevent the inhibitory effect of NPY. Stimulation of Ca^{2+} phospholipid-dependent protein kinase (PKC) with phorbol-12-myristate 13-acetate (PMA) did not affect the basal rate of DOPA production in NGF-differentiated PC12 cells but did produce a concentration-dependent inhibition of depolarization-stimulated DOPA production. In addition, NPY did not produce further inhibition of DOPA production in the presence of PMA, and the inhibition by both PMA and NPY was attenuated by the specific PKC inhibitor chelerythrine. These results indicate that NPY inhibits Ca^{2+} influx through L-type voltage-gated channels, possibly through a PKC-mediated pathway, resulting in attenuation of the activation of CaM kinase and inhibition of depolarization-stimulated catecholamine synthesis.

Using whole-cell patch-clamp recording, the effect of NPY on Ba^{2+} currents in the NGF-differentiated PC12 cells was examined in order to demonstrate that activation of the different receptor subtypes results in differential Ca^{2+} channel modulation. NPY was found to inhibit the voltage-activated Ba^{2+} current in a reversible fashion with an ED50 of 13 nM. Experiments using selective NPY analogs revealed that most of this inhibition was mediated by the Y_2 and Y_3 receptor subtypes. It was observed that the Y_2 effect was confined to inhibition of N-type Ca^{2+} channels. NPY also inhibited L-type Ca^{2+} channels in these cells, as demonstrated both by occlusion of the effects of nifedipine and by reduction of BAY K 8644-enhanced tail currents. The inhibition of L-channels was prevented by the PKC inhibitors chelerythrine and PKC(19–36) and was mimicked by the PKC activator PMA. The results of these studies indicate that in NGF-differentiated PC12 cells, NPY acts via Y_2 receptors to inhibit N-type Ca^{2+} channels concomitant with catecholamine and NPY-ir release and via Y_3 receptor and a PKC-dependent pathway to inhibit L-type Ca^{2+} channels and catecholamine synthesis.

In more recent studies multiple NPY, PYY and PP analogs were used to further define the receptor subtype involved in this Ca^{2+} channel modulation (McCullough et al. 1998b). It was observed that NPY and PP modulate Ca^{2+} channels through Y_1, Y_2, Y_3 and Y_4 receptors. In addition it was observed that these receptors are differentially coupled to N, L and Non-N, Non-L Ca^{2+} channel subtypes. Therefore it would appear that there is an intriguing and complex role for NPY and PP in the modulation of sympathetic neurotransmission.

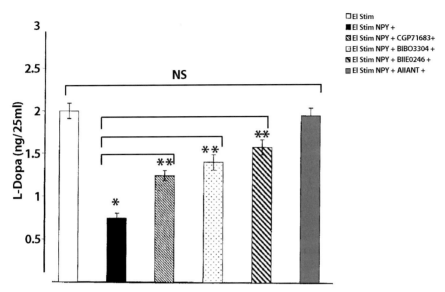

Fig. 8 The inhibitory effect of NPY(10^{-7} M) on the periarterial nerve stimulation induced increase in NA synthesis in the absence or presence of the Y antagonist BIBO 3304, the Y_2 antagonist BIIE 0246, the Y_5 antagonist CGPT 1683A or a combination of all three. NA synthesis was determined by measuring the increase in DOPA formation in the presence of the aromatic L-amino acid decarboxylase inhibitor NSD 1015 (see McCullough and Westfall 1995). All three of the antagonists attenuated the inhibitory effect of NPY but it required a combination of all three to completely block the effect of NPY

A similar inhibition of catecholamine synthesis by NPY and related analogs has also been observed in the perfused mesenteric arterial bed (TC Westfall, in press) and in the rat striatum (Adewale et al. 2001); although stimulation of synthesis has also been seen in the striatum depending on the receptor subtype activated (Adewale 2001). It appears that multiple NPY receptors contribute to the inhibitory effect of NPY in the mesenteric artery. It took a combination of the Y_1 antagonist BIBO 3304, the Y_2 antagonist BIIE 0246 and the Y_5 antagonist CGP 71683A to completely block the maximal inhibitory concentration of NPY. Figure 8 depicts a summary of the results using the three antagonists. Additional experiments are needed utilizing additional preparations, but the observation that catecholamine synthesis and release can be differentially modulated by NPY implies an additional level of control in the prejunctional regulation of sympathetic neurotransmission.

3.3
NA and ATP-Induced Modulation of NPY Release

In addition to NPY being able to inhibit its own release following stimulation of sympathetic nerves, activation of α-adrenoceptors also leads to inhibition of NPY-ir release. Activation of α_2-adenoceptors has been demonstrated to inhibit

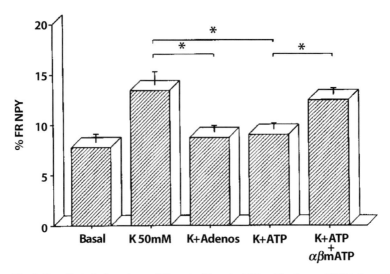

Fig. 9 The effect of adenosine or ATP on the K$^+$-evoked (50 mM) release of NPY-ir from NGF-differentiated PC12 cells. Adenosine (1 μM) or ATP (3 mM) significantly inhibited K$^+$-evoked release of NPY-ir. The effect of ATP was antagonized by α,β-methylene ATP (100 nM). (Reproduced from Chen et al. 1997 with permission)

nerve stimulation-induced release of NPY-ir as well as NA from the guinea pig heart (Haass et al. 1989a,b) and in the pithed guinea pig preparation (Dahlöf et al. 1986, 1991). Moreover blockade of α_2-adrenoceptors lead to an increase in the release of NPY-ir from the cat, dog and pig spleen (Lundberg et al. 1984a; Schoups et al. 1988; Lundberg et al. 1986a, 1989b), guinea pig heart (Archelos et al. 1987; Haass et al. 1989a,b) dog gracilis muscle (Pernow et al. 1988a; Kahan et al. 1988) guinea pig vas deferens and urinary bladder (Lundberg et al. 1984b). Nerve stimulation-induced release of NPY also appears to be modulated by prejunctional β_2-adrenoceptors known to enhance NA overflow (Westfall et al. 1979; Dahlof 1981). Activation of β-adrenoceptors with isoproterenol enhanced NPY-ir levels in the pithed guinea pig and this was blocked by propranolol (Dahlöf et al. 1991a; Lundberg et al. 1989a). Evidence is also available that endogenous angiotensin enhances the evoked release of NPY-ir (Pernow and Lundberg 1989a; Dahlöf et al. 1991b).

ATP and adenosine have been observed to inhibit the K$^+$-evoked release of NPY-ir from nerve growth factor differentiated PC12 cells (Fig. 9). The adenosine-induced inhibition of the evoked release of NPY-ir is consistent with the observation in the guinea pig heart that the adenosine analog cyclohexyladenosine significantly reduced the stimulated overflow of NPY (Haass et al. 1989). Thus it appears that all three sympathetic cotransmitters can produce autoreceptor-mediated inhibition of themselves and the other cotransmitters.

3.4
Preferential and Differential Modulation of Sympathetic Cotransmitter Release

There is emerging evidence that the three sympathetic cotransmitters NA, ATP and NPY can be differentially released and modulated under certain conditions. For instance, D. P. Westfall and colleagues argue that if ATP and NA originate from the same vesicles, then the ratio of released NA to purine should remain constant during stimulation, and the time course of release should be similar for both substances. However, they have observed that the release of ATP and NA from the sympathetic nerves of the guinea pig vas deferens differs markedly (Todorov et al. 1994). Release of NA reaches a peak between 30 s and 40 s and then remains constant for the duration of the stimulation, whereas the release of purines is transient reaching a peak at 20 s and then declines dramatically even with continued stimulation. There is also evidence for the differential modulation of sympathetic cotransmitters. For instance, clonidine inhibits the release of NA but not ATP from the myenteric plexus of the guinea pig ileum (Hammond et al. 1988). Activation of prejunctional α_2-adrenoceptors appears to depress the release of NA more than the release of ATP in several other preparations including the mouse vas deferens (Hammond et al. 1988; von Kugelgen et al. 1989), rabbit ileocolic artery (Bulloch and Starke 1990) and guinea pig vas deferens (Driessen et al. 1993). In contrast, the adenosine A_1 receptor-mediated inhibition of the release of ATP is more marked than the inhibition of NA release (Driessen et al. 1994). In the rat vas deferens angiotensin produces opposite effects on the release of ATP and NA, reducing the former and increasing the latter (Ellis and Burnstock 1989). Similarly, atrial natriuretic peptide inhibits the purinergic but not the adrenergic component of transmission in the guinea pig vas deferens (Todorov et al. 1996). In addition, pretreatment with the α_2-adrenoceptor antagonists idazoxan and yohimbine produces a greater increase in the overflow of NA than ATP suggesting that endogenously released NA has a greater influence on its own release than on the release of ATP (von Kugelgen 1996). Furthermore, it has been shown that endothelin 3 (ET-3, one of two endogenous endothelins) facilitates the release of ATP from the guinea pig vas deferens, which is converted to inhibition of release in the presence of BQ123, an antagonist for one of the two ET receptor subtypes (ET_A). On the other hand, NA release is not facilitated by ET-3 (Mutafova-Yambolieva and Westfall 1995). The same investigators have found that in the rat-tail artery both ATP and NA release are modulated in parallel by ET suggesting that the modulation of sympathetic neurotransmission can differ from tissue to tissue (Mutafova-Yambolieva and Westfall 1998). Finally, it is known that NPY release is resistant to low-frequency stimulation and requires either high frequency or a bursting pattern for release such as would take place during stress or intense sympathetic nerve stimulation (Lundberg et al. 1986, 1987; Haass et al. 1989a,b). Evidence for differential modulation of NA and NPY release has also been obtained (Hoang et al. 2002, 2003). It was observed that catestatin, sulprostone, iloprost, adrenomedullin and proadrenomedullin N-terminal 20 peptide (PAMP) inhibited the

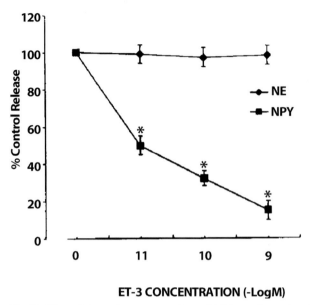

Fig. 10 Effect of increasing concentrations of ET-3 on the periarterial nerve stimulation induced (16 Hz, 90 s) overflow of NPY-ir or NA from the perfused mesenteric arterial bed of the rat. ET-3 produced an inhibition of the evoked release of NPY-ir but not NA. (Reproduced from Hoang et al. 2002 with permission)

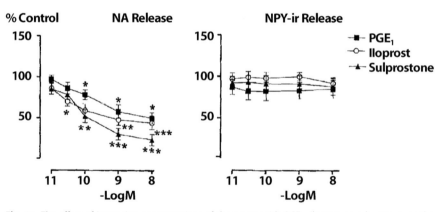

Fig. 11 The effect of increasing concentrations of the prostanoids PGE_1, iloprost or sulprostone on the periarterial nerve stimulation (16 Hz, 90 s) induced overflow of NPY-ir and NA from the perfused mesenteric arterial bed of the rat. All three prostanoids produced a concentration-dependent attenuation of the evoked release of NA but not NPY-ir. (Reproduced from Hoang et al. 2003 with permission)

evoked release of NA, but not NPY-ir, while endothelins (ET-1, ET-3 and STX6C) inhibited NPY-ir but not NA. Figure 10 depicts the ability of ET to inhibit the nerve stimulation-induced release of NPY-ir but not NA (Hoang et al. 2002) while Fig. 11 shows that the prostanoids PGE_1, iloprost and sulprostone can inhibit NA but not NPY-ir (Hoang et al. 2003).

3.5
NPY-Mediated Modulation of the Release of Other Neurotransmitters in the Periphery

3.5.1
Acetylcholine

In addition to prejunctional effects on sympathetic neurons, NPY appears to have prejunctional actions on other peripheral nerves. There is evidence that following its release from sympathetic neurons, NPY can inhibit the release of acetylcholine (ACh) and thus alter the postjunctional response seen following cholinergic neurotransmission. This has been seen at several neuroeffector junctions. The normal increase in contraction following stimulation of parasympathetic nerves in the uterine cervix (Stjernquist et al. 1983) trachea (Grundamer et al. 1988) and stomach (Potter 1991) were all attenuated by prior stimulation of sympathetic nerves or by the administration of NPY or NPY analogs. The vasodilation normally seen following parasympathetic nerve stimulation in the nasal mucosa of cat and dog was also decreased following sympathetic nerve stimulation or the administration of exogenous NPY (Lacroix et al. 1989, 1994).

Both the nasal vasodilator effect and increase in nasal secretions due to parasympathetic nerve stimulation were decreased in response to sympathetic nerve stimulation, an effect mimicked by NPY (Revington et al. 1997). Interactions between sympathetic and parasympathetic nerves have also been observed in the heart and have been particularly well characterized there. For instance, sympathetic nerve stimulation or exogenous NPY have been observed to attenuate the vagally mediated effects on heart rate, A-V conduction and atrial contraction (Potter 1985, 1987a,b; Warner and Levy 1989a,b; Warner et al. 1990; Yang and Levy 1993; Smith-White et al. 1999). A prejunctional site is suggested because the actions of NPY and sympathetic nerve stimulation inhibition of the vagally induced bradycardia was not seen following the exogenous administration of cholinomimetics (Potter 1985, 1987a). The effects were also not mimicked by the exogenous administration of NA or various peptides including: vasoactive intestinal polypeptide, somatostatin, neurotensin, vasopressin or substance P (Kilborn et al. 1986). Attenuation of the vagal effects could also be modified by the administration of agonists and antagonists (Hall and Potter 1988), reduced by guanethidine (Potter 1987) but persist following reserpine pretreatment (Moriarty et al. 1993). The overflow of NPY-ir into the coronary sinus blood or the plasma in response to stimulation of the cardiac sympathetic nerves correlated well with the subsequent attenuation of vagal neurotransmission elicited by sympathetic nerve stimulation (Potter 1987b; Warner et al. 1991). These prejunctional effects of NPY on parasympathetic neurons appear to be due to activation of Y_2 receptors since the effect is mimicked by the NPY-Y_2 agonist N-acetyl[Leu^{28,31}]NPY(24–36) and antagonized by the selective Y_2-antagonist BIIE 0246 (Smith-White et al. 2001). Further evidence for this important effect of NPY on Y_2 receptors is the observation that the ability of the Y_2

agonist N-acetyl (Leu$^{28-31}$) NPY(24–36) to increase the pulse interval evoked by cardiac vagal stimulation in control mice was absent when examined in Y_2 (–/–) receptor knockout mice (Smith-White et al. 2002).

3.5.2
Calcitonin Gene Related Peptide/Substance P

In addition to the evidence that NPY can modulate the release of the sympathetic cotransmitters NA, ATP and NPY as well as the release of ACh from cholinergic neurons, it appears that the peptide can also inhibit the release of transmitters from sensory vasodilator nerves. There is considerable evidence that CGRP is the major vasodilator transmitter released from sensory nerves in the mesenteric arterial bed (Kawasaki et al. 1991; Han et al. 1990a,b). Both Grundamar et al. (1990) and Kawasaki et al. (1991) have shown that NPY inhibited the neurogenic release of CGRP-ir induced by perivascular nerve stimulation in rat mesenteric arteries resulting in a reduction of neurogenically induced vasodilation. Kotecha (1998) observed that both intrinsic and extrinsic nerves supplying the intestinal blood vessels were subject to modulation by NPY and the Y_2 selective agonists N-acetyl [Leu^{28,31}]NPY(24–36). This investigator observed that the Y_2 receptor agonist had no effect on the relaxation produced by the exogenous application of ACh, but significantly reduced the relaxing action of vasodilator nerve stimulation to arterioles of the isolated submucosa of the guinea pig small intestine, which were precontracted with the thromboxane analog U46619. The Y_2 agonist significantly decreased the amplitude of EJPs evoked by perivascular nerve stimulation in normal arterioles and arterioles treated with the sensory neurotoxin, capsaicin. On the other hand, the Y_2 agonist failed to alter the amplitude of the constrictions obtained by perivascular nerve stimulation in normal arterioles, but significantly attenuated the amplitude of constriction in arterioles treated with capsaicin. This suggests that normally the Y_2 agonist acted to reduce the release of inhibitory transmitters (CGRP) from sensory nerves which masked the effects of reduced excitatory transmitter release from the sympathetic nerves. It will be of interest to know whether the NPY-induced inhibition of CGRP release and attenuation of neurogenic vasodilation occurs at other vascular neuroeffector junctions. In addition to inhibition of CGRP release NPY has also been observed to inhibit the release of substance P from rat sensory neurons (Ewald et al. 1988; Walker et al. 1988).

3.6
Distribution/Location of NPY in the CNS

NPY is widely distributed in the CNS of all species examined including guinea pig, mouse, rat, cat, dog, pigs, rabbits, primates and man (see Allen et al. 1983c; Hendry 1993; Potter 1991; Chromwall 1989; Everitt et al. 1989 for extensive reviews) and has been observed to be colocalized with several different neurotransmitters. NPY-ir has been found in cerebral cortex, hippocampus, olfactory bulb, striatum, septum, basal forebrain, amygdala, thalamus, hypothalamus, brain stem (e.g. pons and medulla) and spinal cord. The only area which lacks NPY-ir appears to be the cerebellum (Allen and Bloom 1986; Maccarrone and Jarrott 1986).

Of particular importance to this chapter are neurons in which NPY is colocalized with catecholamines. Everitt showed NPY-ir is colocalized with catecholamines in noradrenergic/adrenergic cell bodies in the A1/C1 cell groups in the rat ventrolateral medulla, as well as the noradrenergic cell bodies of the locus coeruleus (A6) and adrenergic cell bodies of the C2 cell group. NPY is also localized with catecholamines in A2 and C3 cell groups in the reticulum nuclei and in the nucleus tractus solitarius (Harfstrand et al. 1987) NPY/catecholamine neurons originating in the brain stem project to many hypothalmic nuclei as well as to the spinal cord (Blessing et al. 1986, 1987).

Colocalization of NPY with catecholamines has also been reported in the dorsal motor nucleus of the vagus, the nucleus of the tractus solitarius and the lateral reticular nucleus (Hökfelt et al. 1983a,b). Colocalization was not found in the A5 or A7 noradrenergic cells groups nor in the dopaminergic cells of the mesencephalon and forebrain (A8-A15), although there is a close juxtaposition of NPY and DA neurons in the striatum.

3.7
NPY-Induced Modulation of Noradrenaline Neurotransmission in the CNS

In a similar fashion as to what has been observed in peripheral sympathetic neurons, evidence exists for a NPY-induced modulation of NA neurotransmission in the CNS. It has been reported that NPY can inhibit the evoked release of NA both in vivo and in vitro as well as decrease NA turnover in various brain regions. Initial studies demonstrated that NPY potentiated the inhibitory effect of α_2-adrenoceptor agonists on the K^+-evoked release of NA from synaptosomes prepared from the medulla oblongata, hypothalamus and frontal cortex (Martire et al. 1986, 1989a,b). Others reported that NPY can directly inhibit K^+-evoked (Westfall et al. 1988, 1990a,c) and electrically induced overflow of ^{3}H-NA from slices of anterior and posterior hypothalamus (Westfall 1988, 1990a,c) hypothalamus (Yokoo et al. 1987) and the pineal gland (Simonneaux et al. 1994). There is regional specificity for this effect of NPY with an inhibitory effect being seen from slices obtained from hypothalamus, hippocampus, medulla oblongata and frontal cortex, but not the parieto-occipital cortex (Martire et al. 1993, 1995; Widdowson et al. 1991). Table 3 represents a summary of studies demonstrating NPY-induced inhibition of the evoked release of NA or ^{3}H-NA in the brain.

There is evidence to suggest that NPY inhibits the evoked release of NA following the activation of NPY-Y_2 receptors. This was demonstrated with Y_2 selective agonists such as NPY(13–36) (Martire et al. 1993, 1995; Grouzmann et al. 1997) and antagonized by Y_2 antagonists such as $T_4[NPY(33-36)]_4$ (Grouzman et al. 1997). More recent studies, however, suggest that there may be an involvement of Y_1 and Y_5 receptors in addition to Y_2 (Hastings et al. 2001; Bitran et al.

Table 3 Selected examples of inhibition of the evoked release of ^{3}H-NA or NA from brain slices or synaptosomes by NPY analogs

Preparation	Brain region	Mode of stimulation	Agonists utilized	References
Rat synaptosomes	Medulla oblongata	Potassium	NPY + clonidine	Martire et al. 1986; Martire et al. 1989a
Rat synaptosomes	Hypothalamus; frontoparietal cortex	Potassium	NPY + clonidine	Martire et al. 1989b
Rat slices	Anterior; posterior hypothalamus	Potassium	NPY	Westfall et al. 1988,1990a
Rat synaptosomes	Medulla oblongata; hypothalamus Hippocampus parieto-cortex	Potassium	NPY(13–36)	Martire et al. 1993 Martire et al. 1995
Rat slices	Hypothalamus; cerebral cortex	Electrical	NPY; NPY + UK 14,304	Yokoo et al. 1987
Rat slices	Hypothalamus	Electrical	NPY;NPY+ UK 14,304	Tsuda et al. 1989, 1990,1992
Rat slices	Hypothalamus	Potassium	NPY; GR 231118	Hastings et al. 2001
Rat slices	Hypothalamus	Electrical	NPY, Leu^{31}Pro^{34}NPY; NPY(13–36); PYY(3–36); GW 1229	Bitran et al. 1999
Rat slices	Hypothalamus	Potassium	NPY(13–36); T$_4$[NPY(33–36)]$_4$	Grouzmann et al. 1997
Rat slices	Hypothalamus; hippocampus; frontal cortex parieto-occipital cortex	Potassium	NPY; NPY + clonidine	Widdowson et al. 1991

1999). This is suggested by the ability of analogs with Y_1 and Y_5 receptor activity to mimic the effect of NPY as well as the ability of Y_1 antagonists to attenuate these effects (Hastings et al. 2001; Bitran et al. 1999). In vivo studies are consistent with the idea that NPY inhibits NA neurotransmission in the brain. For instance it was observed that the intracerebroventricular (ICV) injection of NPY resulted in a decrease in NA turnover in the brainstem, hypothalamus, midbrain and hippocampus (Vallejo et al. 1987). Using microdialysis techniques, the ICV administration of Leu^{31}Pro^{34} NPY produced a significant reduction in basal NA concentrations in the paraventricular hypothalmic nucleus (Hastings et al. 1997). NPY has also been demonstrated to potentiate the inhibitory effect of NA in the rat locus coeruleus using intracellular electrophysiological recordings in a pontine slice preparation containing the locus coeruleus (Illes et al. 1993). The inhibitory effect of NPY and the α_2-adrenoceptor agonist UK 14,304 on NA release has been reported to be significantly attenuated in slice preparations obtained from the Spontaneously Hypertensive Rat compared to WKY (Tsuda et al. 1989, 1990), although this has not been universally observed (Westfall et al. 1988, 1990a,b). This would suggest that NPY and α_2-adrenoceptors might be involved in the regulation of central catecholamine activity in this form of genetic hypertension.

Despite the many observations that NPY inhibits the evoked release of NA from selective brain regions, there is also some evidence for a facilitatory effect on NA transmission in the CNS (Pavia et al. 1995). These investigators reported that the ICV administration of NPY produced a doubling of the K^+-induced release of NA into the microdialysis cannula placed in the paraventricular region of the hypothalamus. Because of the regional specificity of the NPY-induced inhibitory effect in the brain, it is possible that both the inhibitory and the facilitatory effect may be produced depending on the brain region.

3.8
NPY-Induced Modulation of DA Release in the CNS

Although NPY and DA are not colocalized in the same neurons, both are present in the basal ganglia and limbic systems in very high abundance (Allen et al. 1983; Adrian et al. 1983). It is also known that NPY and DA neurons are in close juxtaposition in structures such as the striatum. Therefore it is possible that NPY and DA may exert modulatory influences on their respective neurotransmission. It has even been suggested that there is a reciprocal regulation between DA and NPY in these structures (Vuillet et al. 1989; Tatsuoka et al. 1987; Wahlestedt et al. 1991). The nature of the interaction between NPY and DA is far from clear. It has been observed that the administration of NPY results in a decrease in DA turnover in the brainstem and striatum in a manner similar to the effect of NPY on NA (Vallejo et al. 1987). Consistent with this observation are the studies of Tsuda et al. (1997) who observed that NPY significantly inhibited the electrically induced release of ^{3}H-DA in rat striatal slices. This is consistent with what has been seen in NGF-differentiated PC12 cells where NPY has been shown to inhibit the K^+-evoked release of DA (DiMaggio et al. 1995; Chen et al. 1994). In contrast, NPY has been observed to enhance N-methyl-D-aspartate stimulated DA release in rat striatal slices (Ault and Weiling 1997, Ault et al. 1998). These effects of NPY could be mimicked by the Y_2 agonist NPY(13–36) but not by PYY or Leu^{31}Pro^{34}NPY. Moreover there is evidence that NPY can inhibit or stimulate DA synthesis depending on the type of NPY receptor activated (Adewale et al. 2001). Consistent with these results, others have reported that NPY enhanced the turnover of DA in rat striatum (Beal 1986; Heilig et al. 1990). In addition Kerkrian-Le Goff (1992) observed the ICV injection of NPY increased the release of DA as measured by voltametry.

Currently, the effect of NPY on DA neurotransmission in the basal ganglia must still be considered an open question. Similarly unresolved are the effects of DA on NPY neurotransmission. It has been hypothesized that DA can exert an inhibitory effect on the functional activity of NPY containing neurons. This is supported by the finding that destruction of the nigrostriatal dopaminergic system by 6-hydroxydopamine leads to enhanced neuronal density and NPY levels in striatal regions (Kerkerian et al. 1986; Moukhles et al. 1992). An increase in NPY-ir was also observed in the rat cerebral cortex after treatment of rats with reserpine, haloperidol and SCH 23390. These agents can interrupt DA function by depletion of DA levels (e.g., reserpine) or blocking DA receptors (haloperidol and SCH 23390). These results were interpreted to indicate an inhibitory regulation of NPY-ir expression by DA neurons. It is suggested that when NPY neurons are deprived of this tonic inhibitory influence, the level of NPY-ir increases, most likely as a result of enhanced peptide synthesis. An increase in NPY-ir has also been observed in the locus coeruleus after reserpine or haloperidol treatment (Smialowska and Legutko 1990, 1992).

It has also been observed that the administration of the DA agonists apomorphine or *n*-propylnorapomorphine produced a concentration-dependent inhibition of the K^+-evoked release of NPY-ir from the NGF-differentiated PC12 cell as well as NPY-ir from slices of nucleus accumbens and striatum. This effect could be blocked by the D2 receptor antagonist, eticlopride D_2 which is consistent with a D_2 mechanism (Cao and Westfall 1993, 1995; Cao et al. 1995).

3.9
NPY-Induced Modulation of Other Neurotransmitters in the CNS

3.9.1
GABA

In addition to the colocalization of NPY with catecholamines, NPY is known to be colocalized with peptides (Chronwall et al. 1984) and other neurotransmitters such as GABA in the hypothalamus (Francois-Bellan et al. 1990) and other brain regions (Hendry et al. 1984; Morris 1989; Bendotti et al. 1990; Cox et al. 1995, 1997). Several investigators have provided evidence that NPY can mediate inhibition of presynaptic GABA release. Chen and van den Pol (1996) have obtained direct electrophysiological evidence that NPY does indeed mediate inhibitory GABAergic neurotransmission in the brain. Using single neuron microcultures of supra chiasmatic nucleus (SCN) neurons of the hypothalamus, they observed that both NPY-Y_1 and Y_2 receptors exist in the same SCN neurons and that both receptors mediate the inhibition of presynaptic GABA release and postsynaptic whole cell calcium currents.

The effect of NPY receptors on GABAergic transmission appears to differ however with the brain region and neuronal cell type. In addition, it seems reasonable to suspect that NPY receptors can be functionally segregated in different brain regions. This appears to be the case in the thalamus where NPY-Y_1 receptors were observed to be predominantly expressed at the somata and dendrites while the Y_2 receptors are located at recurrent and feed forward GABAergic terminals (Sun et al. 2001). Activation of Y_1 receptors appear to reduce the excitability of neurons in both the thalamic reticular nucleus and the adjacent ventral basal complex, while activation of Y_2 receptors decreases GABA release in the feed-forward GABAergic terminals in the ventral basal complex via inhibition of Ca^{2+} influx from voltage gated Ca^{2+} channels (Sun et al. 2001).

Further evidence of the differing effects of NPY in different brain nuclei are the observations in the hippocampus where NPY has been shown to affect functions at excitatory nerve terminals by selectively inhibiting glutamatergic release without a direct effect on GABA release (Klapstein and Colmers 1993).

3.9.2
Glutamate

Just as GABA is perhaps the most important inhibitory neurotransmitter in the brain, the amino acid glutamate is known to be one of the prominent excitatory neurotransmitters. While both neurotransmitters enjoy a ubiquitous distribution in the brain, a key region for NPY–glutamate interactions appears to be the hippocampal formation. Although NPY is not colocalized with glutamate in the hippocampus, NPY is contained in interneurons throughout the hippocampus (Kohler et al. 1986). Electron microscopic studies indicate the presence of NPY immunoreactive varicosities in immediate apposition to glutamatergic presynaptic terminals (Milner and Vezendaroglu 1992). There is abundant evidence from both electrophysiological and overflow studies that NPY inhibits the evoked release of glutamate from hippocampal neurons. For instance electrophysiological studies demonstrate that NPY exerts a presynaptic inhibitory effect on the glutamatergic transmission at Shaffer collateral/C1 synapses and at mossy fiber/C3 synapses of the hippocampus (Colmers et al. 1985, 1988; Haas et al. 1987; Klapstein and Colmers 1993).

Other investigators have observed that NPY produces a direct inhibitory effect on the potassium evoked overflow of glutamate from hippocampal slices or synaptosomes or preparations of subregions of the hippocampus such as dentate gyrus, C1 or C3 (Greber et al. 1994; Whittaker 1999; Silva et al. 2001). Studies from both the electrophysiological and overflow studies suggest that the principal NPY receptor mediating these effects is the Y_2 subtype. For instance, Gerber et al. (1994) observed that the inhibitory effects of NPY were mimicked by PYY and NPY(13–36) but not Leu^{31}Pro^{34}NPY. A similar conclusion was presented by Silva et al. (2001), however the possibility still exists for participation by Y_1 receptors or other NPY receptors. These latter investigators showed that both NPY Y_1 and Y_2 receptors can modulate (inhibit) intracellular calcium concentrations $[Ca^{2+}]_i$ and glutamate release from hippocampal synaptosomes. They claim that the Y_2 receptor played the predominant role both in the modulation of Ca^{2+}-dependent and -independent release, while admitting that the physiological relevance of the presynaptic Y_1 receptors in the dentate gyrus and C3 subregions remains to be clarified. Experiments carried out with the selective Y_2 antagonist BIIE 0246 provide very strong support for a role for the Y_2 receptor (Weiser et al. 2000). The effect of BIIE 0246 on the electrophysiological properties of NPY in rat hippocampus was determined as well as the affinity of the antagonist for rat hippocampal NPY Y_2 receptors. It was observed that BIIE 0246 was able to block NPY-mediated effects in the hippocampus and demonstrates that the presynaptic receptor in the hippocampus is the Y_2 subtype. The inhibitory effects of NPY on the evoked release of glutamate are not limited to the hippocampus as NPY has also been shown to inhibit glutamate release in the striatum (Ellis and Davies 1994).

Emerging evidence continues to indicate that NPY plays an important physiological role to limit excitability within the hippocampus (see reviews by Colmers and Bleakman 1994; Klapstein and Colmers 1993, 1997). Moreover, there is evidence that NPY plays an important role during pathological hyperactivity, such as that which occurs during seizures (Patrylo et al. 1999). It has been shown that NPY and its receptors are significantly affected by seizures, most notably in the hippocampus. NPY neurotransmission may be specifically enhanced by both the increased release of NPY and the increase in the number of presynaptic Y_2 receptors (Schwarzer et al. 1998; Furtinger et al. 2001). It has been hypothesized that this leads to a more efficient inhibition of glutamate release (Vezzani et al. 1999). It has been observed that exogenous NPY has anticonvulsant actions in several experimental models (Vezzani et al. 1999). In contrast to the inhibitory effects of NPY via Y_2 receptors, there is also evidence that NPY has proconvulsant actions mediated via Y_1 receptors (Vezzani et al. 1999). It would appear that NPY receptors represent an important target for antiepileptic drugs. As pointed out by Vezzani et al. (1999), a vast amount of work remains in what is clearly an exciting and important area of NPY research.

3.9.3
Serotonin

It has been reported that NPY, PYY and PP all attenuated the electrically and potassium evoked release of ^{3}H-serotonin from slices or synaptosomes prepared from rat cerebral cortex (Schlicker et al. 1991). Since the NPY-induced inhibition of the evoked release of ^{3}H-serotonin was seen in slices in which the propagation of action potentials along the axons was inhibited by tetrodotoxin, and inhibition of overflow was seen in synaptosomes a presynaptic site of action is likely and an involvement of interneurons is excluded. The inhibitory effect by NPY was not due to activation of α_2-adrenoceptor since the inhibitory effect of NPY on ^{3}H-serotonin overflow was not attenuated by the α-adrenoceptor antagonist, phentolamine and the inhibitory effect obtained at the maximal concentration of the α_2 agonist, clonidine was further increased by NPY.

The role of NPY in modulating serotonin neurotransmission remains an open question, however, because there are a lack of reports replicating the findings of Schlicker et al. (1991) and other investigators have failed to see an inhibitory effect by NPY on the K^+-evoked ^{3}H-serotonin overflow from synaptosomes of the rat hypothalamus and frontoparietal cortex (Martire et al. 1989a). The nature of the receptor subtype modulating the inhibitory effect of NPY is also unknown and awaits experiments utilizing selective agonists and antagonists.

3.9.4
Neuroendocrine Mediators

As has been repeatedly stated in this chapter and others, NPY is probably the most abundant peptide in the nervous systems and is present in high amounts in key endocrine and neuroendocrine regulatory centers in the hypothalamus and pituitary. A discussion of the effects of NPY on the secretion of various neurohormones is beyond the scope of this chapter and the reader is referred to the chapters by Crowley and by Kalra and Kalra in this volume as well as the numerous reviews in the literature. Suffice to say, there is support for a role of NPY acting in a paracrine, autocrine or neuroendocrine fashion in modulating the secretion of luteinizing hormone, luteinizing hormone releasing hormone, corticotropin releasing hormone, vasopressin, oxytocin, growth hormone, growth hormone releasing hormone, melanocyte stimulating hormone, follicle stimulating hormone, thyrotropin releasing hormone, adrenocorticotrophic hormone and thyroid stimulating hormone.

3.10
Autoreceptor-Mediated Modulation of NPY-ir Release in the CNS

In addition to the many effects of NPY on the evoked release of other neurotransmitters and neurohormones including colocalized NA, there is evidence for autoreceptors modulating the release of NPY. It has been observed that activation of α_2-adrenoceptors with clonidine results in a significant inhibition of NPY-ir release from slices of rat hypothalamus while the α_2-adrenoceptor antagonist enhanced stimulated overflow (Ciarleglio et al. 1993; Westfall et al. 1990a).

There is also evidence that NPY acting on autoreceptors can inhibit its own release (King et al. 1999, 2000). The Y_2 agonists NPY(13–36) and N-acetyl (Leu^{28}Pro^{31}) NPY(24–36) and the Y_4/Y_5 agonist human pancreatic polypeptide (hPP) all significantly reduced both basal and K$^+$-evoked NPY-ir release (King et al. 1999). The inhibitory effect of NPY(13–36) was partially prevented by the weak Y_2 antagonist T_4-[NPY(33–36)]$_4$ but strongly blocked by the selective Y_2 antagonist BIIE 0246 (King et al. 2000). The hPP- and rPP-induced inhibition of release was not affected by the Y_5 antagonist CGP71682A or the Y_1 antagonist BIBP 3226. The selective Y_1, Y_5 or weak Y_2 antagonist had no effect on either basal or K$^+$ stimulated release when administered alone (King et al. 1999). In contrast the Y_2 antagonist BIIE 0246, which exhibits more than 100-fold higher affinity at Y_2 receptors over T_4 [NPY(33–36)]$_4$, was observed to enhance NPY-ir release from the hypothalamus slices following depolarization (King et al. 2000). It would appear, therefore, that these data support the role of presynaptic Y_2 receptors in regulating the release of NPY from the hypothalamus. The precise role for Y_4 receptors in mediating autoreceptor-mediated inhibition of NPY-ir release must wait the availability of Y_4 selective antagonists.

Acknowledgements. The author's research has been supported in part by grants NIH-NHLBI HL-0602600 and NIGMS-GM08306.

References

Adewale S, Macarthur H, Westfall TC (2001) Modulation of catecholamine synthesis by neuropeptide Y in rat Striatal slices. Exp Biol 15:H453.7

Adrian TE, Allen JM, Bloom SR, Ghatei MA, Rossor M.N, Roberts GW, Crow TJ, Tatemoto K, Polak JM (1983) Neuropeptide Y distribution in human brain. Nature 306:584–586

Allen JM, Adrian TE, Polak JM, Bloom SR (1983a) Neuropeptide Y in the adrenal gland. J Auton Nerv Syst 9:599–563

Allen JM, Bloom SR (1986) Neuropeptide Y: a putative neurotransmitter. Neurochem Int 8:1–8

Allen JM, Adrian TE, Tatemoto K, Polak JM, Hughes J, Bloom SR (1982) Two novel related peptides, neuropeptide Y and peptide YY inhibit the contraction of the electrically stimulated mouse vas deferens. Neuropeptides 3:71–77

Allen JM, Bircham PMM, Edwards AV, Tatemoto K, Bloom SR (1983b) Neuropeptide Y (NPY) reduces myocardial perfusion and inhibits the force of contraction of the isolated perfused rabbit heart. Regul Peptides. 6:247–253

Allen JM, Birchan MM, Bloom SR, Edwards AV (1984) Release of neuropeptide Y in response to splann nerve stimulation in the conscious calfs. J Physiol (Lond) 357:405–408

Allen YS, Adrian TE, Allen JM, Tatemoto K, Crow TJ, Bloom SR, Polak JM (1983c) Neuropeptide Y distribution in the rat brain. Science 221–879

Archelos J, Xlang JZ, Reinecke M, Lang RC (1987) Regulation of release and function of neuropeptides in the heart. J Cardiovas Pharm 10 (Suppl 12):45–50

Ault DT, Werling LL (1997) Differential modulation of NMDA stimulated ^{3}H-dopamine release from rat striatum by neuropeptide Y and σ receptor ligands. Brain Res 760:210–217

Ault DT, Radeff JM, Werling LL (1998) Modulation of ^{3}H-dopamine release from rat nucleus accumbens by neuropeptide Y may involve a sigma 1-like receptor. J Pharmacol Exp Ther 284:553–560

Beal F, Frank RC, Ellison DW, Martin JB (1986) The effect of neuropeptide Y on Striatal catecholamines. Neurosci Lett 71:118–123

Bendotti C, Hohmann C, Furlough G, Reeves R, Coyl JT, Oster-Granite NL (1990) Developmental expression of somatostatin in mouse brain: In situ hybridization. Brain Res Develop 53:26–39

Bitran M, Tapia W, Eugenin E, Orio P, Boric MP (1999) Neuropeptide Y induced inhibition of noradrenaline release in rat hypothalamus: role of receptor subtype and nitric oxide. Brain Res 851:87–93

Bleakman D, Comers WF, Fournier A, Miller RJ (1991) Neuropeptide Y inhibits Ca^{2+} influx into cultured dorsal root ganglion neurons of the rat via Y_2 receptor. Br J Pharmacol 103:1781–1789

Blessing WW, Howe PRC, Johi TH, Oliver JR, Willoughby JO (1986) Distribution of tyrosine hydroxylase and neuropeptide Y-like immunoreactive neurons in rabbit medulla oblongata with attention to colocalizaiton studies, presumptive adrenaline containing perikarya ad vagal preganglionic cells. J Comp Neurol 248:285–300

Blessing WW, Oliver JR, Hodgson AJ, Joh TH, Willoughby JO (1987) Neuropeptide Y-like immunoreactive C1 neurons in the rostral ventrolateral medulla of the rabbit projects to sympathetic preganglionic neurons in the spinal cord. J Aut Nerv Syst 18:121–129

Boehm S, Huck S (1997) Receptors controlling transmitter release from sympathetic neurons in vitro. Program Neurobiol 51:225–242

Boehm S and Kubista H (2002) Fine tuning of sympathetic transmitter release via inotropic and metabutropic presynaptic receptors. Pharmacol Rev 54:43–99

Bohmann C, Rump LA Schaible U, von Kugelgen I (1995) α-Adrenoceptor modulation of norepinephrine and ATP release in isolated kidneys of spontaneously hypertensive rats. Hypertension 25:1224–1231

Bradley E, Law A, Bell D, Johnson CD (2003) Effects of varying impulse number on co-transmitter contributions to sympathetic vasoconstriction in rat tail artery. Am J Physiol Heart Circ Physiol 284:H2007–H2014

Bulloch JM, Starke K (1990) Presynaptic α_2-autoinhibition in a vascular neuroeffector junction where ATP and noradrenaline act as co-transmitters. Br J Pharmacol 99:279–284

Bultmann R, Von Kugelgen I, Starke K (1991) Contraction-mediating α_2-adrenoceptors in the mouse vas deferens. Naunyn-Schmiedeberg's Arch Pharmacol 343:623–632

Burnstock G (1999) Purinergic cotransmission. Brain Res Bull 50:355–357

Burnstock G, Warland JJT (1987) A pharmacological study of the rabbit saphenous artery in vitro: a vessel with a large purinergic contractile response to sympathetic nerve stimulation. Br J Pharmacol 90:111–120

Burnstock G, Campbell G, Satchell D, Smythe A (1970) Evidence that adenosine triphosphate or a related nucleotide is the transmitter substance released by non-adrenergic inhibitory nerves in the gut. Br J Pharmacol 40:668–688

Cao GH, Westfall TC (1993) Effects of dopamine agonists on the release of neuropeptide Y immunoreactivity from the nucleus accumbens and striatum of rat brain. Soc Neurosci Abs 19:1173

Cao GH, Houston D, Westfall TC (1995) Mechanism underlying dopamine receptor mediated inhibition of neuropeptide Y release from pheochromocytoma cells. Soc Neurosci Abs 21:1378

Cao GH, Westfall TC (1995) Dopamine D_2 receptor-mediated inhibitio of neuropeptide Y (NPY) release from pheochromocytoma (PC12) cells. Physiologist 38(5):A260

Changeux JP (1986) Coexistence of neuronal messengers and molecular selection. Prog Brain Res 63:373–403

Chen G, von den Pol AN (1996) Multiple NPY receptors coexist in pre- and postsynaptic sites: Inhibition of GABA release in isolated self-innervating SCN neurons. J Neurosci 16:7711–7724

Chen X, Westfall TC (1994) Modulation of intracellular calcium transients and dopamine release by neuropeptide Y in PC12 cells. Am J Physiol 266 (Cell Physiol 35):C784–C793

Chen X, DiMaggio D, Han S-P, Westfall TC (1997) Autoreceptor induced inhibition of neuropeptide Y release from PC12 cells is mediated by Y_2 receptors. Am J Physiol 273 (Heart Circ Physiol 42):H1737–1744

Chernaeva L and Charakchieva S (1988) Leucine-enkephalin- and neuropeptide Y-modulation of (^{3}H)-noradrenaline release in the oviduct of mature and juvenile rabbits. Gen Pharmacol 19:137–142

Chesselet MF (1984) Presynaptic regulation of neurotransmitter release in the brain: facts and hypothesis. Neuroscience 12:347–375

Chronwall BM (1989) Anatomical distribution of NPY and NPY messenger RNA in rat brain. In: Mutt V, HöKfelt T, Fuxe K, Lundberg J (eds) Neuropeptide Y. Raven Press, New York, pp 51–60

Chromwall BM, Chase TN, O'donohue T (1984) Coexistence of neuropeptide Y with somatostatin in rat and human cortical and rat hypothalamic neurons. Neurosci Lett 52:213–217

Ciarleglio A, Bienfeld MC, Westfall TC (1993) Pharmacological characterization of the release of neuropeptide Y-like immunoreactivity from the rat hypothalamus. Neuropharmacol 8:819–825

Colmers WB, Lukowiak K, Pittman QJ (1985) Neuropeptide Y reduces orthodromically evoked population spike in rat hippocampal CA1 by a possibly presynaptic mechanism. Brain Res 346:404–408

Colmers WF, Bleakman D (1994) Effects of neuropeptide Y on the electrical properties of neurons. Trends Neurosci 17:373–379

Colmers WF, Klapstein GJ, Fournier A, St-Pierre S, Treherne KA (1991) Presynaptic inhibition by neuropeptide Y in rat hippocampal slice in vitro is mediated by a Y_2 receptor. Br J Pharmacol 102:41–44

Colmers WF, Lukowiak K, Pittmann QJ (1988) Neuropeptide Y action in the hippocampal slice: site and mechanism of presynaptic inhibition. J Neurosci 8:3827–3837

Cortes V, Donoso MV, Brown N, Fanjul R, Lopez C, Fournier A, Huidobro-Toro JP (1999) Synergism between neuropeptide Y and norepinephrine highlights sympathetic cotransmission: studies in rat arterial mesenteric bed with neuropeptide Y, analogs and BIBP 3226. J Pharm Exp Ther 289:1313–1322

Cox CL, Huguenard JR, Prince DA (1995) Cholecystokinin depolarizes rat thalamic reticular neurons by suppressing a K^+ conductance. J Neurophysiol 74:990–1000

Cox CL, Huguenard JR, Prince DA (1997) Peptidergic modulation of intrathalamic circuit activity in vitro actions of cholecytokinin. J Neurosci 17:70–82

Dahlöf C (1981) Studies on β-adrenoceptor mediated facilitation of sympathetic neurotransmission. Acta Physiol Scand Suppl 500:1–147

Dahlöf C, Dahlöf P, Tatemoto K, Lundberg JM (1985) Neuropeptide Y (NPY) reduces field stimulation evoked release of noradrenaline and enhances force of contraction in the rat portal vein. Naunyn Schmiedeberg's Arch Pharmacol 328:327–330

Dahlöf C, Dahlöf P, Lundberg JM (1986) α_2-adrenerceptor mediated inhibition of nerve stimulation evoked release of neuropeptide Y immunoreactivity in the pithed guinea pig. Eur J Pharmacol 131:279–283

Dahlöf PL (1989) Modulatory interactions of neuropeptide Y on sympathetic neurotransmission. Acta Physiol Scand Suppl 586:1–85

Dahlöf P, Lundberg JM, Dahlöf C (1991a) α- and β-adrenoceptor mediated effects of nerve stimulation evoked release of neuropeptide Y (NPY)-like immunoreactivity in the pithed guinea-pig. J Auton Nervous System 35:199–210

Dahlöf P, Lundberg JM, Dahlöf C (1991b) Effect of angiotensin II and captopril on plasma levels of neuropeptide Y (NPY)-like immunoreactivity in the pithed guinea-pig. Neuropeptides 18:171–180

DeDeyn PP, Pickut BA, Vorzwijvelen A, D'Hodge A, Annart W, DePotter WP (1989) Subcellular distribution and axonal transport of noradrenaline, dopamine β hydroxylase and neuropeptide Y in dog splenic nerve. Neurochem Int 15:39–47

DeQuidt ME, Richardson PJ, Emson PC (1985) subcellular distribution of neuropeptide Y like immunoreactivity in guinea pig neocortex. Brain Res 335:354–359

DiMaggio DA, Farah JM, Westfall TC (1995) Effects of differentiation on neuropeptide receptors and response in pheochromocytoma cells. Endocrinol 134:719–721

Donoso MV, Steiner M, Huidobro-Toro JP (1997) BIBP 3226, suramin and prazosin identify neuropeptide Y, adenosine-5′-triphosphate and noradrenaline as sympathetic cotransmitter in the rat arterial mesenteric bed. J Pharmacol Exp Ther 282:691–698

Doods H, Balda W, Wieland HA, Dollinger H, Schnorrenberg G, Essen T, Engel W, Eberlein W, Rudolf K (1999) BIIE 0246: a selective and high affinity neuropeptide Y Y_2 receptor antagonist. Eur J Pharmacol 384:R3–R5

Driessen B, Von Kugelgen I, Starke K. (1994) P_1-purinoceptor mediated modulation of neural noradrenaline and ATP release in guinea pig vas deferens. Naunyn-Schmiedeberg's Arch Pharmacol 350:42–48

Driessan B, Von Kugelgen I, Starke K (1993) Neural ATP release and its α_2 adrenoceptor mediated modulation in guinea pig vas deferens. Naunyn-Schmideberg's Arch Pharmacol 348:358–366

Drumheller A, Bouoli SM, Fournier A, St-Pierre S, Jolicoeur FB (1994) Neurochemical effects of neuropeptide Y and NPY 2–36. Neuropeptides 27:271–296

Dumont Y, Cadienx A, Doods H, Pheng LH, Abounader R, Hamel E, Jaquqes D, Regole D, Quirion R (2000) BIIE 0246 – A potent and highly selective non-peptide neuropeptide Y Y_2 antagonist. Br J Pharmacol 129:1075–1088

Eccles JC (1986) Chemical transmission and Dale's principle. Prog Brain Res 68:3–13

Edvinsson L, Ekblad E, Hadanson R, Wahlestedt C (1984) Neuropeptide Y potentiates the effects of various vasoconstrictor agents in rabbit blood vessels. Br J Pharmacol 83:519–525

Edvinsson L, Emson P, McCulloch J, Tatemoto K, Uddman R (1983) Neuropeptide Y: cerebrovascular innervation and vasomotor effects in the cat. Neurosci Lett 43:79–84

Ewald DA, Sternweis PC, Miller RJ (1988) Guamine nucleotide-binding protein C_o induced coupling of neuropeptide Y receptors to Ca^{2+} channels in sensory neurons. Proc Natl Acad Sci USA 85:3633–3637

Ekblad E, Edvinsson L, Wahlestedt C, Uddman R, Hakanson R, Sundler F (1984) Neuropeptide Y co-exists and co-operates with noradrenaline in perivascular nerve fibers. Regul Pept 8:225–235

Elliott TR (1904) On the action of adrenalin. J Physiol (Lond) 31:20–31

Ellis Y, Davies JA (1994) The effects of neuropeptides on the release of neurotransmitter amino acids from rat striatum. Neuropeptides 26:65–69

Ellis JL, Burnstock G (1990) Neuropeptide Y neuromodulation of sympathetic co-transmission in the guinea pig vas deferens. Br J Pharmacol 100:457–462

Ellis JL, Burnstock G (1989) Angiotensin neuromodulation of adrenergic and purinergic co-transmission in the guinea pig vas deferens. Br J Pharmacol 97:1157–1164

Euler USv (1946) A specific sympathomimetic ergone in adrenergic nerve fibers (sympathetic) and its relationship to adrenaline and noradrenaline. Acta Physiol Scand 12:73–97

Euler USv (1972) Synthesis, uptake and storage of catecholamines in adrenergic nerves. The effect of drugs. In: Blaschko H, Muschull E (eds) Catecholamine. Handbook of Experimental Pharmacology vol 33. Springer, Berlin, pp 186–230

Everitt BJ, Hökfelt T (1989) The coexistence of neuropeptide Y with other peptides and amines in the central nervous system. In: Mutt V, Hökfelt T, Fuxe K, Lundberg JM (eds) Neuropeptide Y pp 61–71

Fisher-Colbrie R, Diez-Guerra J, Emson PC, Winkler H (1986) Bovine chromaffin granules: immunological studies with antisera agonist NPY, Met-enkephalin and bombesin. Neuroscience 18:167–174

Foucart S, Bleakman D, Bindokas VP, Miller RJ (1993) Neuropeptide Y and pancreatic polypeptide reduce calcium currents in acutely dissociated neurons from adult rat superior cervical ganglia. J Phrmacol Exp Ther 265:903–909

Franco-Cereceda A, Lundberg JM, Dahlof D (1985) Neuropeptide Y and sympathetic control of heart contractility and coronary vascular tone. Acta Physiol Scand 124:361–369

Franco-Cereceda A, Sarra A, Lundberg JM (1989) Differential release of calcitonin-gene related peptide and neuropeptide Y from the isolated heart by capsaicin, ischemia, nicotine, bradykinin and oubain. Acta Physiol Scand 135:173–187

Francois-Bellan A, Kachidian P, Dusticier G, Tonon MC, Vaudry H, Bosler O (1990) GABA neurons in the rat suprachismatic nucleus: involvement in chemospecific synaptic circuitry and evidence for GAD-peptide colocalization. J Neurocytol 19:937–947

Fried G (1981) Small noradrenergic vesicles isolated from rat vas deferens biochemical and morphological characterization. Acta Physiol Scand (Suppl)493:1–25

Fried G, Terenius L, Hokfelt T, Goldstein M (1985b) Evidence for differential localization of noradrenaline and neuropeptide Y in neuronal storage vesicles isolated from rat vas deferens. J Neurosci 5:440–458

Fried G, Lundberg JM, Theodorsson-Norheim E (1985a) Subcellular storage and axonal transport of neuropeptide Y in relation to catecholamines in the cat. Acta Physiol Scand 125:145–154

Fuder H, Muscholl E (1995) Heteroreceptor-mediated modulation of noradrenaline and acetylcholine release from peripheral nerves. Rev Physiol Biochem Pharmacol 126:265–412

Funess JB, Costa M, Emson PD, Hakanson R, Moghimzadeh E, Sunder F, Taylor IL, Chance RE (1983) Distribution pathways and reactions to drug treatment of nerves with neuropeptide Y and pancreatic polypeptide like immunoreactivity in the guinea pig digestive tract. Cell Tiss Res 234:71–92

Furtinger J, Pirker S, Czech T, Baumgartner C, Ransmayr G, Sperk G (2001) Plasticity of Y_1 and Y_2 receptors and neuropeptide Y fibers in patients with temporal lobe epilepsy. J Neurosci 21:5804–5812

Gehlert DR, Beavers LS, Johnson O, Gackenheimer S, Schober DA, Gadski A (1996) Expression cloning of a human brain neuropeptide Y2 receptor. Mol Pharmacol 49:224–228

Gerald C, Walker MW, Criscione L, Gustofson EL, Batzl-Hartman C, Hofbauer KG, Tober RI, Branchek TA, Weinshank RL (1996) A receptor subtype involved in neuropeptide Y induced food intake. Nature 382:168–171

Gillespie JS (1980) Presynaptic receptors in the autonomic nervous system. In: Szekers L (ed) Adrenergic activators and inhibitors. Handbook of experimental pharmacology vol 54. Springer-Verlag Berlin, pp 353–425

Greber S, Schwarzer C, Sperk G (1994) Neuropeptide Y inhibits potassium stimulated glutamate release through Y_2 receptors in rat hippocampal slices in vitro. Br J Pharmacol 113:737–740

Gregor P, Feng Y, DeCarr LB, Cornfield LJ, McCalebi ML (1996) Molecular characterization of a second mouse pancreatic polypeptide receptor and its inactivated human homologue. J Biol Chem 271:2776–27781

Grouzmann E, Bucling T, Martire M, Cannizzaro C, Dorner B, Razaname A, Mutter M (1997) Characterization of a selective antagonist of neuropeptide Y at the Y_2 receptor. J Biol Chem 272:7699–7706

Grundamar L, Widmark E, Waldesk B, Hakamson R (1988) Neuropeptide Y: prejunctional inhibition of vagally induced contraction in the guinea pig trachea. Regul Pept 23:309–313

Grundamer L, Grundstrom N, Johansson IGM, Andersson RCG, Hakasson R (1990) Suppression by NPY of capsasin-sensitive sensory nerve-mediated contractions in guinea pig air ways. Br J Pharm 99:473–476

Haas HL, Herman A, Greene RW, Chan-Palay V (1987) Action and location of neuropeptide tyrosine (Y) on hippocampal neurons of the rat in slice preparations. J Comp Neurol 257:208–215

Haass M (1989) Characterization and presynaptic modulation of stimulation-evoked exocytotic co-release of NA and NPY in guinea pig heart. Naunyn-Schmiedeberg's Arch Pharmacol 339:71–78

Haass M, Hock M, Richardt G, Schöming A (1989a) NPY differentiates between exocytoxic and nonexocytotic NA release in guinea pig heart. Naunyn-Schmiedeberg's Arch Pharm 340:509–515

Haass M, Cheng B, Richardt G, Lang RE, Schöming H (1989b) Characterization and presynaptic modulation of stimulation-evoked exocytotic co-release of noradrenaline and neuropeptide Y in guinea pig heart. Naunyn-Schmiedeberg's Arch Pharmacol 339:71–78

Hafstrand A, Fuxe K, Terenius L, Kalla M (1987) Neuropeptide Y-immunoreactive perikarya and rat nerve terminals in the rat medulla oblongata relationship to cytoarchitecture and catecholaminergic cell groups. J Comp Neurol 260:20-35

Hall GT, Potter EK (1988) Alpha-adrenoceptor mediated modulation of sympathetic effects on the action of the vagus nerve at the heart. Neurosci Lett 30 (Suppl): S70

Hammond JR, MacDonald WF, White TD (1988) Evoked secretion of (^{3}H)noradrenaline and ATP from nerve varicosities isolated from the myenteric plexus of the guinea pig ileum. Can J Physiol Pharmacol 66:560-575

Han S-P, Yang, C-L, Chen X, Naes L, Cox BF, Westfall TC (1998a) Direct evidence for the role of neuropeptide Y in sympathetic nerve stimulation induced vasoconstriction. Am J Physiol 274(Heart Circ Physiol 43):H290-294

Han S-P, Chen X, Cox B, Yang C-L, Wu YM, Naes L, Westfall TC (1998b) Role of neuropeptide Y in cold stress induced hypertension. Peptides 19:351-358

Hasting JA, McClure-Sharp JM, Morris MJ (2001) NPY Y1 receptors exert opposing effects in corticotropin releasing factor and noradrenaline overflow from rat hypothalamus in vitro. Brain Res 890:32-37

Hastings JA, Pavia JM, Morris MJ (1997) Neuropeptide Y and [Leu^{31},Pro^{34}] neuropeptide Y potentiate potassium induced noradrenaline release in the paraventricular nucleus of the aged rat. Brain Res 750:301-304

Heilig M, Vecsul L, Wahlestedt C, Alling C, Widerlov E (1990) Effects of centrally administered neuropeptide Y and NPY 13-36 on the brain monoaminergic system of the rat. J Neural Tran 79:193-208

Hendry SH, Jones EG, DeFelipe J, Schmechel P, Brandon C, Emson PC (1984) Neuropeptide containing neurons of the cerebral cortex are also GABAergic. Proc Natl Acad Sci USA 81:6526-6530

Hendry SHC (1993) Organization of neuropeptide Y neurons in the mammalian central nervous system. In: Colmers WF, Wahlestedt C (eds) The biology of Neuropeptide Y and related peptides. Humana Press, Totowa, pp 65-156

Hirning LA, Fox AP, Miller RJ (1990) Inhibition of calcium currents in cultured myenteric neurons by neuropeptide Y: evidence for direct receptor/channel coupling. Brain Res 532:120-130

Hoang D, Macarthur H, Gardner A, Westfall TC (2002) Endothelin induced modulation of neuropeptide Y and norepinephrine release from the rat mesenteric bed. Am J Physiol Heart Circ Physiol 283:H1523-H1530

Hoang D, Macarthur H, Gardner A, Yang C-L, Westfall TC (2003) Prostanoid induced modulation of neuropeptide Y and noradrenaline release from the rat mesenteric bed. Autonomic and Autacoid Pharmacol 23:141-147

Hokfelt T, Lundbert JM, Lagercrantz H, Tatemoto K, Mutt V, Lindberg J, Terenius L, Everitt B, Fuxe K, Agnati L, Goldstein M (1983a) Occurrence of neuropeptide Y-like immunoreactivity in catecholamine neurons in the human medulla oblongata. Neurosci Lett 36:217-222

Hokfelt T, Lundberg JM, Tatemoto K, Mutt V, Terenius L, Polak J, Blooms Sasek C, Elde R, Goldstein M (1983b) Neuropeptide Y and FMRF amide neuropeptide like immunoreactivities in catecholamine neurons of the rat medulla oblongata in catecholamine neurons of the rat medulla oblongata. Acta Physiol Scand 117:315-318

Holton FA, Holton P (1954) The capillary dilator substance in dry powders of spinal roots: a possible role of adenosine triphosphate in chemical transmission from nerve endings. J Physiol (Lond) 126:124-140

Hoyo Y, McGrath JC and Vila E (2000) Evidence for Y1-receptor mediated facilitatory, modulatory cotransmission by NPY in the rat anoccygeus muscle. J Pharm Exp Ther 294:38-44

Illes P, Finta EP, Nieber (1993) Neuropeptide Y potentiates via Y$_2$-receptors the inhibitory effect of noradrenaline in rat locus coeruleus neurons. Naunyn Schmiedeberg's Arch Pharmacol 348:546-548

Jonsson G (1983) Chemical lesioning techniques, monoamine neurotoxins In: Bjorkland A, Hökfelt J (eds) Methods in chemical neuroanatomy, vol. 1 Elsevier, Amsterdam, pp 463–507

Kahan T, Pernow J, Schwieler J, Lundberg JM, Hjemdahl P, Wallin BG (1988) Involvement of neuropeptide Y in sympathetic vascular control of skeletal muscle in vivo. J Hypertens 6 (Suppl)4:532–534

Kasakov L, Ellis J, Kirkpatrick K, Milner P, Burnstock (1988) Direct evidence for the concomitant release of noradrenaline, adenosine 5'-triphosphate and neuropeptide Y from sympathetic nerves supplying the guinea pig vas deferens. J Auton Nerv System 22:75–82

Kawasaki H, Nuki C, Sauto A, Takasaki K (1991) NPY modulates neurotransmission of CGRP-containing vasodilator nerves in rat mesenteric arteries. Am J Physiol 261:H683–690

Kerkerian L, Bosler O, Pelletier G, Nieouillon A (1986) Striatal neuropeptide Y neurons are under the influence of the nigrostriatal dopamine pathway: immunohistochemical evidence. Neurosci Lett 66:106–112

Kerkerian L, Le-Goff L, Forni C, Samuel D, Bloc A (1992) Intracerebroventricular administration of neuropeptide Y affects parameters of dopamine, glutamate and GABA activities in rat striatum. Brain Res Bull 28:187–193

Kilborn MJ, Potter KA, McCloskey DJ (1986) Effects of periods of conditioning stimulation and of neuropeptides on vagal action at the heart. J Auton Nervous System 17:131–142

King PJ, Williams G, Doods H, Widdowson PS (2000) Effect of a selective neuropeptide Y Y_2 receptor antagonist BIIE 0246 on neuropeptide Y release. Eur J Pharmacol 396:1–3

King PJ, Widdowson PS, Doods HN, Williams G (1999) Regulation of neuropeptide Y release by neuropeptide Y ligands and calcium channel antagonists in hypothalamic slices. J Neurochem 73:641–646

Klapstein G, Colmers WF (1997) Neuropeptide Y suppresses epileptic-form activity in rat hippocampus in vitro. J Neurophysiol 76:1651–1661

Klapstein GJ, Conners WF (1993) On the sites of presynaptic inhibition by neuropeptide Y in rat hippocampus in vitro. Hippocampus 3:103–111

Klein RL (1982) Chemical composition of the large noradrenergic vesicles. In: Klein RL, Lagercrantz H, Zimmermann H (eds) Neurotransmitter vesicles. Academic Press, London pp. 133–174

Koelle GB (1955) The histochemical identification of acetylcholinesterase in cholinergic, adrenergic and sensory neurons. J Pharmacol Exp Ther 114:167–184

Kohler C, Erickson LG, Davies S, Chan-Palay V (1986) Neuropeptide Y innervation of the hippocampal region in the rat and monkey brain. J Comp Neurol 244:384–400

Kotecha N (1998) Modulation of submucosal arterilar tone by NPY-Y_2 receptors in the guinea pig – small intestine. J Auton Ner Sys 70:157–163

Krause J, Eva C, Seeburg PH and Sprengel R (1992) Neuropeptide Y1 subtype pharmacology of a recombinantly expressed neuropeptide receptor. Mol Pharmacol 41:817–821

Kuramoto H, Kondo H, Fujita T (1986) Neuropeptide tyrosine like-immunoreactivity in adrenal chromaffin cells and intradrenal nerve fibers of rats. Anat Rec 214:321–328

Lacroix JS, Lundberg JM (1989) Adrenergic and neuropeptide Y supersensitivity in denervated nasal mucosa vasculature of the pig. Eur J Pharmacol 169:125–136

Lacroix JS, Stjärne L, Anggard A, Lundberg JM (1988) Sympathetic vascular control of the pig nasal mucosa (2): Reserpine-resistant, non-adrenergic nervous responses in relation to neuropeptide Y and ATP. Acta Physiol Scand 133:183–197

Lacroix JS, Ulman LG, Potter EK (1994) Modulation by neuropeptide Y of parasympathetic nerve-evoked nasal vasodilation by Y_2 prejunctional receptor. Br J Pharmacol 113:479–484

Lacroix J-S (1989) Adrenergic and non-adrenergic mechanisms in sympathetic vascular control of the nasal mucosa. Acta Physiol Scand (Suppl) 581:1–63

Lagercrantz H (1971) Isolation and characterization of sympathetic nerve trunk vesicles. Acta Physiol Scand (Suppl) 82366)1-40

Lagercrantz H, Fried G (1982) Chemical Composition of the Small Noradrenergic Vesicles. In: Klein RL, Lagercrantz H, Zimmermann H (eds) Neurotransmitter vesicles. Academic, London, pp 175-188

Laher I, Germann P, Bevan JA (1994) Neurogenecally evoked cerebral artery constriction is mediated by neuropeptide Y. Can J Physiol Pharm 72:1086-1088

Langer SZ (1981) Presynaptic regulation of the release of catecholamines. Pharmacol Rev 32:337-362

Langer SZ (1997) 25 years since the discovery of presynaptic receptors: present knowledge and future perspectives. Trends Pharmacol Sci 18:95-99

Langer SZ, Lehmann J (1988) Presynaptic receptors on catecholamine neurons. In: Catecholamines. Handbook of Experimental Pharmacology vol 85. Trendelenbury IU, Werner N (eds) Springer-Verlag, Berlin pp 419-507

Larhammar L, Bloomqvist AG, Yee F, Jasin E, Yoo H, Wahlestedt C (1992) Cloning and functional expression of human neuropeptide Y/peptide YY receptor of the Y_1 type. J Biol Chem 267:10935-10938

Levitt B, Head RJ, Westfall DP (1984) High pressure chromatographic flurometric detection of adenosine and adenosine nucleosides: application to endogenous content and electrically induced relese of adenylpurines in guinea pig vas deferens. Anal Biochem 137:93-100

Linial M, Ilouz N, Parnas H (1997) Voltage-dependent interaction between the muscarinic ACh receptor and proteins of the exocytic machinery. J Physiol 504:251-258

Loewi O (1921) Uer Lumorale Ubertregbarkeit dere Herznervenwirkung. Pflügers Arch 189:239-242

Loewi O (1936) Quantitative and qualitative Untersuchungen über der sympathicusstoff. Zugleich XIV Mittüber humorale Ubertragbarkait der Herznervenwirkung. Pflügers Arch 237:504-514

Lundberg, JM, Pernow J, Franco-Cereceda A, Rudehill A (1987) Effects of antihypertensive drugs on sympathetic vascular control iln relation to NPY. J Card Pharm 10(Suppl 12):551-568

Lundberg JM (1996) Pharmacology of cotransmission in the autonomic nervous system: interactive aspects on amines, neuropeptides, adenosine triphosphate, amino acids and nitric oxide. Pharmacol Rev 48:113-178

Lundberg JM, Modin A (1995) Inhibition of sympathetic vasoconstriction in pigs in vivo by the neuropeptide Y-Y1 receptor antagonist BIBP 3226. Br J Pharmacol 116:2971-2982

Lundberg JM, Tatemoto K (1982) Pancreatic polypeptide family (APP, BPP, NPY, PYY) in relation to sympathetic vasoconstriction resistant to α-adrenoceptor blockade. Acta Physiol Scand 116:393-402

Lundberg JM, Anggard A, Theodorsson Norheim E, Pernow J (1984a) Guanethidine-sensitive release of NPY-Li in the cat spleen by sympathetic nerve stimulation. Neurosci Lett 52:175-180

Lundberg JM, Hua X-Y, Franco-Cereceda A (1984b) Effects of neuropeptide Y mechanical activity and neurotransmission in the heart, vas deferens and urinary bladder of the guinea pig. Acta Physiol Scand 121:325-332

Lundberg M nad Stjärne L (1984) Neuropeptide Y depresses the secretion of ^{3}H-noradrenaline and the contractile response evoked by field stimulation of the rat vas deferens. Acta Physiol Scand 120:477-479

Lundberg JM, Martinson A, Hemsen A, Theodorsson-Norheim E, Svederberg J, Ekblom E, Hjemdahl P (1985a) Corelease of neuropeptide Y and catecholamines during physical exercise in man. Biochem Biophys Res Comm 133:30-36

Lundberg JM, Pernow J, Tatemoto K, Dahlof C (1985c) Pre- and postjunctional effects of NPY on sympathetic control of rat femoral artery. Acta Physiol Scand 123:511-513

Lundberg JM, Rudehill A, Sollevi A (1989a) Pharmacological characterization of neuropeptide Y and noradrenaline mechanisms in sympathetic control of pig spleen. Eur J Pharmacol 163:103–113

Lundberg JM, Rudehill A, Sollevi A, Hamberger B (1989c) Evidence for co-transmitter role of neuropeptide Y in the pig spleen. Br J Pharmacol 96:675–687

Lundberg JM, Rudehill A, Sollevi A, Fried G, Wallin G (1989b) Co-release of neuropeptide Y an noradrenaline from pig spleen in vivo: importance of subcellular storage, nerve impulse frequency and pattern, feedback regulation and resupply by axonal transport. Neuroscience 28:475–486

Lundberg JM, Rudehill A, Sollevi A, Theodorosson-Norheim E, Hamberger B (1986) Frequency- and reserpine-dependent chemical coding of sympathetic transmission: differential release of noradrenaline and neuropeptide Y from pig spleen. Neurosci Lett 63:96–100

Lundberg JM, Saria A, Franco-Cereceda A, Theodorsson-Norheim E (1985d) Mechanisms underlying a change in the contents of neuropeptide Y in cardiovascular nerves and adrenal gland induced by sympatholytic drugs. Acta Physiol Scand 124:603–611

Lundberg JM, Terenus L, Hokfelt J, Martling C-R, Tatemoto K, Mutt V, Polak J, Bloom S, Goldstein M (1982) Neuropeptide Y (NPY)-like immunoreactivity in peripheral noradrenergic neurons and the effects of NPY on sympathetic function. Acta Physiol Scand 116:477–480

Lundberg JM, Torssell L, Sollevi A, Pernow J, Theodorsson-Norheim E, Anggard A, Hamberger, B (1985b) Neuropeptide Y and sympathetic vascular control in man. Regul Pept 13:41–52

Lundblad L, Anggard A, Saria A, Lundberg JM (1987) Neuropeptide Y and non-adrenergic sympathetic vascular control of the cat nasal mucosa. J Auton Nerv Sys 20:188–197

Lundell LA, Bloomqvist G, Berglung MM, Schober DA, Johnson D, Statnick MA, Gadake RA, Gehlert DR, Larbammar D (1995) Cloning of a human receptor of the NPY receptor family with high affinity for pancreatic polypeptide and peptide YY. J Biol Chem 270:29129–29128

Lynch JW, Lemos VS, Buchner B, Stoclet JC, Takeda K (1994) A pertussis toxin-insensitive calcium influx mediated by neuropeptide Y receptors in a human neuroblstoma cell line. J Biol Chem 269:8226–8233

Maccarrone C, Jarrott B (1986) Neuropeptide Y: a putative neurotransmitter (critique). Neurochem Int 87:13–22

Malmström RC, Lundberg JON, Weitzberg E (2002) Effects of the neuropeptide Y-Y_2 receptor antagonists BIIE 0246 on sympathetic transmitter release in the pig in vivo. Naunyn-Schmiedeberg's Arch Pharmacol 365:106–111

Malmström RC (1997) Neuropeptide Y Y_1 receptor mechanism in sympathetic vascular control. Acta Physiol. Scand Suppl 636:1–55

Malmström RC, Modin A, Lundberg JM (1996a) SR 120107A antagonizes neuropeptide Y1 receptor mediated sympathetic vasoconstriction in pigs in vivo. Eur J Pharmacol 305:145–154

Malmström RC, Balmer KC, Lundberg JM (1997) The neuropeptide Y (NPY) Y1 receptor antagonist BIBP 3226: equal effects on vascular responses to exogenous and endogenous NPY in the pig in vivo. Br J Pharmacol 121:595–603

Malmström RC, Hokfelt T, Bjorkman JA, Nihlen C, Bystrom M, Ekstrand AJ, Lundberg JM (1998) Characterization and molecular cloning of vascular neuropeptide Y receptor subtypes in pig and dog. Regul Pept 75–76, 55–70

Malmström RE (2001) Vascular pharmacology of BIIE 0246, the first selective non-peptide neuropeptide Y Y_2 receptor antagonist in vivo. Br J Pharmacol 133:1073–1080

Malmström RE, Lundberg JM (1995a) Neuropeptide Y accounts for sympathetic vasoconstriction in guinea pig vena cava: evidence using BIBP 3226 and 3435. Eur J Pharmacol 294:661–668

Malmström RE, Lundberg JM (1995b) Endogenous NPY acting on Y_1 receptors accounts for the long-lasting part of the sympathetic contraction in guinea pig vena cava: evidence using SR 120207A. Acta Physiol Scand 155:329–330

Malmström RE, Lundberg JM (1996b) Effects of the neuropeptide Y Y_1 receptor antagonist SR 120107A on sympathetic vascular control in pigs in vivo. Naunyn Schmiedberg's Arch Pharmacol 354:633–642

Malmström RE, Alexandersson A, Balmer KC, Weilizt J (2000) In vivo characterization of the novel neuropeptide Y Y1 receptor antagonist H409/22. J Cardiovas Pharmacol 36:516–525

Martire M, Pistritto G (1992) Neuropeptide Y interaction with the adrenergic transmission line: A study of its effect on alpha 2 adrenergic receptors. Pharmacol Res 25:203–215

Martire M, Fuxe K, Pistriffo P, Agnati LF (1986) Neuropeptide Y enhances the inhibitory effects of clonidine on ^{3}H-noradrenaline release in synaptosomes isolated from the medulla oblongata of the male rat. J Neural Trans 67:113–124

Martire M, Fuxe K, Pistritto G, Preziosi P, Agnati LF (1989a) Neuropeptide Y increased the inhibitory effects of clonidine on potassium evoked ^{3}H-noradrenaline but not ^{3}H-5 hydroxytryptamine release from synaptosomes of the hypothalamus and the frontoparietal cortex of the male Sprague Dawley rat. J Neural Trans 78:61–72

Martire M, Fuxe K, Pistritto P, Agnati LF (1989b) Reduced inhibitory effects of clonidine and neuropeptide Y on ^{3}H-noradrenaline release of the medulla oblongata of the spontaneously hypertensive rat. J Neural Tran 76:181–189

Martire M, Pistritto G, Mores N, Agnati LF, Fuxe K (1993) Region specific inhibition of potassium evoked ^{3}H-noradrenaline release from rat brain synaptosomes by neuropeptide Y 13–36: involvement of NPY receptors of the Y_2 type. Eur J Pharmacol 230:231–234

Martire M, Pistritto G, Morris N, Agnati LF, Fuxe K (1995) Presynaptic A_2 adrenoceptors and neuropeptide Y Y2 receptors inhibit ^{3}H-noradrenaline release from rat hypothalamic synaptosomes via different mechanisms. Neurosci Lett 188:9–12

McAuley M, Westfall TC (1992) Possible location and function of neuropeptide Y receptor subtypes in the rat mesenteric arterial bed. J Pharmacol Exp Ther 261:863–868

McCullough A, Westfall TC (1996) Mechanisms of catecholamine synthesis inhibition by neuropeptide Y: role of Ca^{2+} channels and protein kinases. J Neurochem 67:1090–1099

McCullough LA, Westfall TC (1995) Neuropeptide Y inhibits depolarization induced catecholamine synthesis in rat pheochromocytoma cells. Eur J Pharmacol 287:271–277

McCullough LA, Egan TM, Westfall TC (1998a) Neuropeptide Y receptors involved in calciuim channel regulation in PC12 cells. Regul Pept 75/76:101–107

McCullough LA, Egan TM, Westfall TC (1998b) Neuropeptide Y inhibition of calcium channels in PC12 pheochromocytoma cells. (Cell Physiol)274 (Cell Physiol 43):C1290–C1297

McDonald RL, Vaughan PFT, Beck-Sickinger AG, Peers C (1995) Inhibition of Ca^{2+} channel currents in human neuroblastoma (SH-SY5Y) cells by neuropeptide Y and a novel cyclic neuropeptide Y analogue. Neuropharmacol 34:1507–1514

Miller RJ (1998) Presynaptic receptors. Ann Rev Pharmacol Toxicol 38:201–207

Milner TA, Veznedaroglu E (1992) Ultrastructural localization of neuropeptide Y-like immunoreactivity in the rat hippocampal formation. Hippocampus 2:107–126

Modin A, Pernow J, Lundberg JM (1993) Sympathetic regulation of skeletal muscle blood flow in the pig: a non-adrenergic component likely to be mediated by neuropeptide Y. Acta Physiol Scand 148:1–11

Moriarity M, Potter EK, McCloskey DI (1993) Pharmacological seperation of cardio-accelerator and vagal inhibitory capacities of sympathetic nerves. J Auton Nerv System 43:7–16

Morris BJ (1989) Neuronal localization of neuropeptide gene expression in rat brain. J Compet Neurol 290:358–368

Morris JL (1991) Roles of neuropeptide Y and noradrenaline in sympathetic neurotransmission to the thoracic vena cava and aorta of guinea pig. Regul Pept 32:297–310

Morris JL, Gibbins IL (1992) Co-transmission and neuromodulation. In: Ed. Burnstock G, Hoyle CHV (eds) Autonomic neuroeffector mechanisms. Harwood Academic Publishers, Chur UK, pp 33–119

Morris JL, Jobling P, Gibbins IL (2001) Differential inhibition by Botulinum neurotoxin A of cotransmitters released from autonomic vasodilator neurons. Am J Physiol (Heart Circ Physiol) 281:H2124–H2132

Morris JL, Jobling P, Gibbins IL (2002) Botulinum neurotoxin A attenuates release of norepinephrine but not neuropeptide Y from vasoconstrictor nerves. Am J Physiol (Heart Circ Physiol) 283:H2627–H2635

Moukhles H, Nieoullon A, Daszuta A (1992) Early and widespread normalization of dopamine-neuropeptide Y interactions in the rat striatum after transplantation of fetal mesencephalon Cells. Neuroscience 47:781–792

Muramatsu I (1987) The effect of reserpine on sympathetic purinergic neurotransmission in the isolated mesenteric artery of the dog: a pharmacological study. Br J Pharmacol 91:467–474

Mutofova-Yambolieva VN, Westfall DP (1998) Inhibitory and facilitory presynaptic effects of endothelin on sympathetic co-transmission in the rat isolated tail artery. Br J Pharmacol 123:136–142

Mutofova-Yambolieva VNM, Westfall DP (1995) Endothelin-3 can both facilitate and inhibit transmitter release in the guinea pig vas deferens. Eur J Pharmacol 285:213–216

NorenbergW, Bek M, Limberger N, Takeda K, Illes K (1995) Inhibition of nicotinic acetylcholine receptor channels in bovine adrenal chromaffin cells by Y_3-type neuropeptide Y receptors via the adenylate cyclase/protein kinase A septum. Naunyn Schmiedeberg's Arch Pharmacol 351:337–347

Öhlen A, Persson MG, Lindbom L, Gustafsson LE, Hedqvist P (1990) Nerve induced nonadrenergic vasoconstriction and vasodilation in skeletal muscle. Am J Physiol 258:H1334–H1338

Oellerich WF, Schwartz DD, Malik KU (1994) Neuropeptide Y inhibits adrenergic transmitter release in cultured rat superior cervical ganglion cells by restricting the availability of calcium through a pertussis toxin-sensitive mechanism. Neuroscience 60:495–502

Patrylo P, van der Pol AN, Spencer DD, Williamson A (1999) NPY inhibits glutamatergic excitation in the epileptic human dentate gyrus. J Neurophysiol 82:470–483

Pavia JM, Hastings JA, Morris MJ (1995) Neuropeptide Y potentiation of potassium induced noradrenaline release in the hypothalamic paraventricular nucleus of the rat in vivo. Brain Res 690:108–111

Pernow J, Lundberg JM, Kaijser L, Hjemdahl P, Theodorsson-Norheim E, Martinsson A, Pernow B. (1986a) Plasma neuropeptide Y-like immunoreactivity and catecholamines during various degrees of sympathetic activation in man. Clin Physiol 6:561–578

Pernow J (1988) Corelease and functional interactions of neuropeptide Y and noradrenaline in peripheral sympathetic vascular control. Acta Physiol Scand Suppl 569:1–56

Pernow J, Lundberg JM (1989a) Modulation of noradrenaline and neuropeptide Y release in the pig kidney in vivo: involvement of alpha 2, NPY and angiotensin II receptors. Naunyn Schmiedeberg's Arch Pharmacol 340:379–385

Pernow J, Lundberg JM (1989b) Release and vasoconstrictor effects of neuropeptide Y in relation to nonadrenergic sympathetic control of renal blood flow in the pig. Acta Physiol Scand 136:507–517

Pernow J, Kahan J, Hjemdahl P, Lundberg JM (1988a) Possible involvement of neuropeptide Y in sympathetic vascular control of canine skeletal muscle. Acta Physiol Scand 132:43–50

Pernow J, Kahan T, Lundberg JM (1988b) Neuropeptide Y and reserpine resistant vasoconstriction evoked by sympathetic nerve stimulation in dog skeletal muscle. Br J Pharmacol 94:952–960

Pernow J, Saria A, Lundberg JM (1986b) Mechanisms underlying pre- and postjunctional effects of neuropeptide Y in sympathetic vascular control. Acta Physiol Scand 126:239–249

Petitto JM, Huang Z, McCarthy DB (1994) Molecular cloning of a NPY-Y$_1$ receptor cDNA from rat splenic lymphocytes: evidence of low levels of mRNA expression and [^{125}I]NPY binding sites. N Neuroimmunol 54:81–86

Pheng LH, Fournier A, Dumont Y, Quirion R, Regoli D (1997) The dog saphenous vein: a sensitive and selective preparation for the Y$_2$ receptor of neuropeptide Y. Eur J Pharmacol 327:163–167

Plummer MR, Rittenhause A, Kanevsky M, Hess P (1991) Neurotransmitter modulation of calcium channels in rat sympathetic neurons. J Neurosci 11:2339–2348

Potter EK (1985) Prolonged non-adrenergic inhibition of cardiac vagal action following sympathetic stimulation: neuromodulation by neuropeptide Y? Neuorsci Lett 54:117–121

Potter EK (1987c) Guanethidine blocks neuropeptide Y-like inhibitory action of sympathetic nerves on cardiac vagus. J Auton Nerv Syst 21:87–90

Potter EK (1987a) Presynaptic inhibition of cardiac vagal postganglionic nerves by neuropeptide Y. Neurosci Lett 83:101–106

Potter EK (1987b) Cardiac vagal action and plasma levels of neuropeptide Y following intravenous injection in the dog. Neurosci Lett 77:243–247

Potter EK (1991) Neuropeptide Y as an autonomic neurotransmitter. In: Bell, C (ed) Novel peripheral neurotransmitters. Pergamon Press, New York, pp 81–112

Qian J, Colmers WF, Saggu P (1997) Inhibition of synaptic transmission by neuropeptide Y in rat hippocampal area CA1: modulation of presynaptic Ca^{2+} entry. J Neurosci 7:8169–8177

Racchi H, Irarrazabal MJ, Howard M, Moran S, Zalaquett R, Huidobro-Toro JP (1999) Adenosine 5'-triphosphatae and neuropeptide Y are co-transmitters in conjunction with noradrenaline in the human saphenous vein Br J Pharmacol 126:1175–1185

Revington M, Lacroix JS, Potter EK (1997) Sympathetic and parasympathetic interaction in vascular and secretory control of the nasal mucosa of the anesthetized dog. J Physiol 505:823–831

Rose PM, Fernandes P, Lynch JS, Frazier ST, Fisher SM, Koduknia K, Kiemzel B, Seethla B (1995) Cloning and functional expression of a cDNA enclosiong a human type 2 neuropeptide receptor. J Biol Chem 270:22661–22664

Rudehill A, Sollevi A, Franco-Cereceda A, Lundberg JM (1986) Neuropeptide Y (NPY) and the pig heart: release and coronary vasoconstrictor effects. Peptides 7:821–826

Rump LC, Riess M, Schwertfeger E, Michel ML, Bohmann C, Schollmeyer P (1997) Prejunctional neuropeptide Y receptors in human kidney and atrium. J Cardiovas Pharmacol 29:656–661

Samuelson VE, Dalsgaard CJ (2985) Action and localization of neuropeptide Y in the human fallopian tube. Neurosci Lett 58:49–54

Schlicker E, Grob G, Fink K, Claser T, Gothert M (1991) Serotonin release in the rat brain cortex is inhibited by neuropeptide-Y but not affected by ACTH-24, antiogensin II, bradykinin and delta sleep-inducing peptide. Nauyn Schmiedeberg's Arch Pharmacol 343:117–122

Schoups A, Saxena A, Tombeur K, DePotter WP (1988) Facilitation of the release of norepinephrine and neuropeptide Y by alpha 2 adrenoceptor blocking agents idazoxan an hydergene in the dog spleen. Life Sci 43:517–523

Schwarzer C, Kofler N, Sperk G (1998) Upregulation of neuropeptide Y Y$_2$ receptors in an animal model of temporal lobe epilepsy. Mol Pharmacol 53:6–13

Serfozo P, Bartfai T, Vizi ES (1986) Presynaptic effects of neuropeptide Y on ^{3}H-noradrenaline and ^{3}H-acetylcholine release. Regul Pept 16:117–123

Silinsky EM, vonKügelgen I, Smith A, Westfall DP (1998) Functions of extracellular nucleotides in peripheral and central neuronal tissues. In: Tumer JT, Weisman GA, Fedan JS (eds) The P2 Nucleotide Receptors. Humana Press Inc., Totowa. pp 259–290

Silva AP, Carvalho AP, Carvalho CM, Malva JO (2001) Modulation of intracellular calcium changes and glutamate release by neuropeptide Y_1 and Y_2 receptors in the rat hippocampus: differential effects in CA1, CA3 and dendate gyrus. J Neurochem 79:286–296

Simonneaux V, Ouichou A, Craft C, Pevet P (1994) Presynaptic and postsynaptic effects of neuropeptide Y in the rat pineal gland. J Neurochem 62:2464–2471

Smialowska M, Legutko B (1990) Influence of reserpine administration on neuropeptide Y immunoreactivity in the locus coeruleus and caudate putamen nucleus of the rat brain. Neuroscience 36:411–415

Smialowska M, Legutko B (1992) Haloperidol induced increase in neuropeptide Y immunoreactivity in locus coeruleus of rat brain. Neuroscience 47:351–355

Smith AD (1979) Biochemical Studies of the Mechanism of Release. In: Paton DM (ed) The release of catecholamines from adrenergic neurons. Pergamon Press, New York, pp 1–15

Smith-White MA, Hardy TA, Brock JA, Potter EK (2001) Effects of a selective neuropeptide Y-Y_2 receptor antagonist BIIE0246 on Y_2 receptors at peripheral neuroeffector junctions. Br J Pharmacol 132:861–868

Smith-White MA, Herzog H, Potter EK (2002) Role of neuropeptide Y Y_2 receptors in modulation of cardiac parasympathetic neurotransmission. Regul Peptides 103:105–111

Smith-White MA, Wallace D, Potter EK (1999) Sympathetic parasympathetic interactions in the heart in the anesthetized rat. J Auton Nervous System 75:171–175

Smyth L, Bobalova J, Ward SM, Mutafova-Yambolieva VN (2000) Neuropeptide Y is a cotransmitter with norepinephrine in guinea pig inferior mesenteric vein. Peptides 21:835–843

Sneddon P (1995) Co-transmission. In: Powis DA, BunnSJ (eds) Neurotransmitter release and its modulation. Cambridge University Press, Cambridge, pp 22–37

Sneddon P, Westfall TC (1984) Pharmacological evidence that adenosine triphosphate and noradrenaline are cotransmitters in the guinea pig vas deferens. J Physiol (Lond) 347:561–580

Sperlagh B, Vizi ES (1992) Is the neuronal ATP release from guinea-pig vas deferens subject to α_2-adrenoceptor mediated modulation? Neuroscience 51:203–209

Starke K (1977). Regulation of noradrenaline release by presynaptic receptor systems. Rev Physiol Biochem Pharmacol 77:1–124

Starke K (1981) Presynaptic receptors. Annu Rev Pharmacol Toxicol 2l:7–30

Starke K (1987). Presynaptic α-autoreceptors. Rev Physiol Biochem Pharmacol 107:74–146

Starke K, Gothert M, Kiblinger H (1989) Modulation of neurotransmitter release by presynaptic: receptors. Physiol Rev 69:865–989

Stjärne L, Stjärne E, Mshgina M (1989) Does clonidine or neuropeptide Y-mediated inhibition of ATP secretion from sympathetic nerves operate primarily by increasing a potassium conductance? Acta Physiol Scand 136:137–138

Stjärne L (1964) Studies of catecholamine uptake, storage and release mechanisms. Acta Physiol Scand (Suppl) 64:1–228

Stjärne L (1989) Basic mechanisms and local modulation of nerve impulse induced secretion of neurotransmitters from individual sympathetic nerve varicosities. Rev Physiol Biochem Pharmacol 112:1–137

Stjärne L, Lundberg JM, Astrand P (1986) Neuropeptide Y—a cotransmitter with noradrenaline and adenosine 5'-triphosphate in the sympathetic nerves of the mouse vas deferens? A biochemical, physiological and electropharmacological study. Neuroscience 18:151–166

Stjernquist M, Emson P, Owman C, Sjoberg N-O, Sander F, Tatemoto (1983) Neuropeptide Y in the female reproductive tract of the rat, distribution of nerve fibers and motor effects. Neurosci Lett 39:279–284

Stretton D, Miura M, Belvisi MG, Banes P (1992) Calcium-activated potassium channels mediate prejunctional inhibition of peripheral sensory nerves. Br J Pharmacol 93:672–678

Sun L, Miller RJ (1999) Multiple NPY receptors regulate K^+ and Ca^{2+} channels in acutely isolated neurons from the rat accuate nucleus. J Neurophysiol 81:1391–1403

Sun QQ, Akk G, Huguenard JR, Prince DA (2001) Differential regulation of GABA release and neuronal excitability mediated by neuropeptide Y_1, and Y_2 receptors in rat thalamic neurons. J Physiol 531:81–94

Sundler F, Böttcher G, Ekblad E, Håkanson R (1993) PP, PYY and NPY. Occurrence and distribution in the periphery. In: Colmers WF, Wahlestedt C (eds) The biology of neuropeptide Y and related peptides. Humana Press, Totowa, pp 157–196

Tasuoka Y, Riskind PN, Beal MF, Martin JB (1987) The effect of amphetamine on the in vivo release of dopamine, somatostatin and neuropeptide Y from rat caudate nucleus. Brain Res 411:200–203

Timmermans PBM, Thoolen WM (1987) Aulureceptors in the central nervous system. Med Res Rev 7:307–332

Todorov LD, Mihaylova-Todorova S, Grariso Gr, Bjur RA, Westfall TC (1996) Evidence for the differential release of the cotransmitters ATP and noradrenaline from sympathetic nerves of the guinea vas defrens. J Physiol 496:731–748

Todorov LK, Bjur RA, Westfall DP (1994) Temporal dissociation of the release of the sympathetic co-transmitters ATP and noradrenaline. Clin Exp Pharm Physiol 21:931–932

Toth PT, Bindokas VP, Bleakman A, Colmers WF, Miller RJ (1993) Mechanism of presynaptic inhibitio nby neuropeptide Y at sympathetic nerve terminals. Nature 364:635–639

Tsuda J, Tsuda S, Goldstein M, Nishioi I, Masuyama Y (1992) Modulation of noradrenergic transmission by neuropeptide Y and presynaptic α_2-adrenergic receptors in the hypothalamus of Spontaneously Hypertensive rats. Jap Heart J 33:229–238

Tsuda K, Tsuda S, Goldstein M, Masayama Y (1990) Effects of neuropeptide Y on norepinephrine levels in hypothalamic slices of Spontaneously Hypertensive rat. Eur J Pharm 182:175–179

Tsuda K, Tsuda S, Nishioi I, Goldstein M, Masuyama Y (1997) Modulation of ^{3}H-dopamine release by neuropeptide Y in rat Striatal slices. Eur J Pharmacol 321:5–11

Tsuda K, Yokou H, Goldstein M (1989) Neuropeptide Y and galanin in norepinephrine release in hypothalamic slices. Hypertension 14:81–86

Uddman R, Moller S, Nilson T, Nystrom S, Estrand J, Edvinsson L (2002) Neuropeptide Y Y_1 and neuropeptide Y Y_2 receptors in human cardiovascular tissue. Peptides 23:927–834

Vallejo M, Carter DA, Biswas S, Lightman SL (1987) Neuropeptide Y alters monoamine turnover in the rat brain. Neurosci Lett 73:155–160

Varndel IA, Polak TM, Allen JM, Terenghi G, Bloom SR (1984) Neuropeptide tyrosine immunoreactivity in norepinephrine containing cells and nerves of the mammalian adrenal gland. Endocrinology 114:1460–1462

Vezzani A, Sperk G, Colmers WF (1999) Neuropeptide Y: emerging evidence for a functional role in seizure modulation. Trends Neurosci 22:22–30

Vizi S (1979). Presynaptic modulation of' neurochemical transmission. Prog Neurobiol 12:181–292

vön Kügelgen I, Bultmann R, Starke K (1989) Effects of suramin and $\alpha\beta$-methylene ATP indicate noradrenaline-ATP co-transmission in the response of the mouse vas deferens to single and low frequency pulses. Naunyn-Schmiedebergs Arch Pharmacol 340:760–763

vön Kügelgen I (1996) Modulation of neural ATP release through presynaptic recepteors. Semin Neurosci 8:247–257

Vuillet J, Kerkerian K, Salin P, Nieoullon A (1989) Ultrastructural features of NPY-containing neurons in the rat striatum. Brain Res 477:241–251

Wahestedt C, Edvisson E, Ekblad and Hakerson R (1987) Effects of neuropeptide Y at sympathetic neuroeffector junctions: Existence of Y_1 and Y_2 receptors In: Nobin A, Owens C (eds) Neuronal messengers in vascular function. Fernstrom Foundation Series vol 10. Elsevier, Amsterdam, pp 231–244

Wahlestedt C, Karoum F, Jaskiw G, Wyatt RJ, Larhammar D, Ekman R, Reis DJ (1991) Cocaine induced reduction of brain neuropeptide Y synthesis dependent on medial prefrontal cortex. Proc Natl Acad Sci USA 88:2078–2082

Wahlestedt C, Yanihara N, Hakanson (1986) Evidence for different pre and post-junctional receptors for neuropeptide Y and related peptides. Regul Pept 13:307–318

Walker MW, Ewald DA, Perney TM, Miller RJ (1988) Neuropeptide Y modulates transmitter release and Ca^{2+} currents in rat sensory neurons. J Neurosci 8:2438–2446

Warland JJI, Burnstock G (1987) Effects of reserpine an 6-hydroxydopamine on the adrenergic and purinergic components of sympathetic nerve responses of the rabbit saphenous artery. Br J Pharmacol 92:871–880

Warner MR and Levy MN (1989a) Neuropeptide Y as a putative modulation of vagal effects on heart rate. Circ Res 64:882–889

Warner MR, Levy MN (1989b) Inhibition of cardiac vagal effects by neurally released and exogenous neuropeptide Y. Circ Res 65:1536–1546

Warner MR, Levy MN (1990) Role of neuropeptide Y in neural control of the heart. J Cardiovas Electrophysiol 1:80–91

Warner MR, Senanayake P, Ferrario CM, Levy MN (1991) Sympathetic stimulation-evoked overflow of norepinephrine and neuropeptide Y from the heart. CIRC Res 69:455–465

Weinberg DH, Sirinatainghji S, Tsu CP, Shiso LL, Morin N, Rigby MR, Heavens RH, Rapoport R, Bayne ML, Casclerl MA, Strader CD, Linemeyer DL, MacNeil BJ (1996) Cloning and expression of a novel neuropeptide Y receptor. J Biol Chem 271:16485–16488

Weiser T, Wieland HA, Doods HN (2000) Effects of neuropeptide Y Y_2 receptor antagonist BIIE 0246 on presynaptic inhibition by neuropeptide Y in rat hippocampal slices. Eur J Pharmacol 404:133–136

Westfall DP, Todorov LD, Mihaylova-Todorova ST (2003) ATP as a cotransmitter in sympathetic nerves and its inactivation by releasable enzymes. J Pharmacol Exp Ther 303:439–444

Westfall DP, Dalzeil H, Forsyth K (1991) ATP as a neurotransmitter, cotransmitter and neuromodulator. In: Phillis J (ed) Adenosine and adenine nucleotides as regulators of cellular function. CRC Press Inc., Boca Raton, pp 295–305

Westfall TD, Westfall DP (2001) Pharmacological techniques for the in vitro study of the vas deferens. J Pharmacol Toxicol Methods 45:109–122

Westfall TC (1980) Neuroeffector mechanisms. Annu Rev Physiol 42:383–397

Westfall TC (1990b) The physiological operation of presynaptic inhibitory autoreceptors. Ann NY Acad Sci 604:398–413

Westfall TC (1995) Beneficial therapeutic intervention via manipulation of presynaptic modulatory mechanisms In: Powis DA, Bunn SJ (eds) Neurotransmitter release and its modulation. Cambridge University Press, Cambridge, pp 328–346

Westfall TC, Martin JR (1990) Presynaptic autoreceptors in the peripheral and central nervous system. In: Feigenbaum JJ, Hanani MM (eds) Presynaptic regulation of neurotransmitter release. Freund Publishing Co., London, pp 311–370

Westfall TC, Carpentier S, Chen X, Beinfeld M, Naes L, Meldrum M (1987) Pre and postjunctional effects of neuropeptide Y at the noradrenergic neuroeffector junction of the perfused mesenteric arterial bed of the rat. J Cardiov Pharmacol 10:716–722

Westfall TC, Han SP, Del Valle K, Curfman M, Ciarleglio A, Naes L (1990a) Presynaptic peptides an hypertension. NY Acad Sci 604:372–388

Westfall TC, Martin J, Chen X, Ciarleglio A, Carpentier S, Henderson K, Knuepfer M, Beinfeld M, Naes L (1988) Cardiovascular effects and modulation of noradrenergic neurotransmission following central and peripheral administration of neuropeptide Y. Synapse 2:299–307

Westfall TC, Yang CL, Curfman-Falvey M (1995) Neuropeptide Y-ATP interactions at the vascular sympathetic neuroeffector junction. J Cardiol Pharmacol 26:682–687

Westfall TC, Zhang S-Q, Carpentier S, Naes L, Meldrum J (1986) Local modulation of noradrenergic neurotransmission in blood vessels of normotensive and hypertensive animals. In: Margo A, Osswald A, Reis D, Vanhoutte P (eds) Central and peripheral mechanisms of cardiovascular regulation. Plenum Publishing, New York, pp 111–113

Westfall TC (1977) Local regulation of adrenergic neurotransmission. Physiol Rev 57:659–728

Westfall TC, Chen X, Ciarleglio A, Henderson K, DelValle K, Curfman-Falvey M, Naes L (1990c) In vitro effects of neuropeptide Y at the vascular neuroeffector junction. Ann NY Acad Sci 611:145–155

Westfall TC, Peach MJ, Tittermary V (1979) Enhancement of the electricallyinduced release of norepinephrine from the rat portal vein: Modulation by beta2-adrenoceptors. Eur J Pharmacol 58:67–74

Whittaker E, Vereker E, Lynch MA (1999) Neuropeptide Y inhibits glutamate release and long-term potentiation in rat dentate gyrus. Brain Res 827:229–233

Widdowson PS, Masten T, Halaris AE (1991) Interactions between neuropeptide Y and alpha 2-adrenoceptors in selected brain regions. Peptides 12:71–75

Wiley JW, Gross RA, MacDonald RL (1993) Agonists for neuropeptide Y receptor subtypes NPY-1 and NPY-2 have opposite effects on rat no dose neuron calcium currents. J Neurophysiol 70:324–330

Wong-Dusting HK, Rand MJ (1988) Pre and postjunctional effects of neuropeptide Y in the rabbit isolated ear artery. Clin Exp Pharm Physiol 15:411–418

Yang Y-P, Chiba S (2002) Antagonistic interaction between BIIE 0246, a neuropeptide Y Y_2-receptor antagonist and ω-conotoxin GVIA, a Ca^{2+} channel antagonist in presynaptic transmitter release in dog splenic arteries. Jap J Pharmacol 89:108–191

Yang T, Levy MN (1993) Effects of intense antecedent sympathetic stimulation on sympathetic neurotransmission in the heart. Circ Res 72:137–144

Yokoo H, Schlesinger DH, Goldstein M (1987) The effect of neuropeptide Y on stimulation-evoked release of ^{3}H-norepinephrine from rat hypothalamic and cerebral cortical slices. Eur J Pharmacol 143:283–286

Zidichouski JA, Chen H, Smith PA (1990) Neuropeptide Y activates inwardly-rectifying K^+ channels in C-cells of amphibian sympathetic ganglia. Neurosci Lett 117:123–128

Zukowska-Grojec Z, Vaz CA (1988) Role of neuropeptide Y in the cardiovascular response to stress. Synapse 2:293–298

Zukowska-Grojec Z, Konarska M, McCarty R (1988) Differential plasma catecholamine and neuropeptide Y responses to acute stress in rats. Life Sci 42:1615–1624

Zukowska-Grozecz, Shen GH, Deka-Starosta A, Myers KA, Kvetnansky R, McCarty R (1992) Neuronal, adrenomodullary and platelet derived neuropeptide Y response to stress in rats. In: Kvetnasky R, McCarty R, Axelrod J (eds). Stress: neuroendocrine and molecular approches. Gordon and Breach Science Publishers, New York, pp 197–209

Neuroendocrine Actions Of Neuropeptide Y

W. R. Crowley

Department of Pharmacology and Toxicology, University of Utah, 30 South 2000 East, Salt Lake City, UT, 84112, USA
e-mail: william.crowley@deans.pharm.utah.edu

1	Introduction	186
2	Anatomy of the Neuroendocrine NPY Systems	187
3	NPY and the Magnocellular Neuroendocrine System: Regulation of OT and VP Secretion	190
3.1	Anatomy	190
3.2	Effects of NPY on VP and OT Release	191
4	NPY and Regulation of PRL Secretion: Integration of Hormone Secretion and Energy Balance During Lactation	192
4.1	Anatomy	192
4.2	Effects On PRL Secretion: Central Sites and Mechanisms of Action	193
4.3	Effects on PRL Secretion: Actions in the Anterior Pituitary	194
4.4	Central Effects of NPY on Food Intake During Lactation	195
4.5	What Are the Physiological Signals for NPY Upregulation in Lactation?	196
4.6	Summary: NPY Integrates Neuroendocrine and Behavioral Responses During Lactation and in Periods of Negative Energy Balance	198
5	Effects of NPY on the Hypothalamic–Pituitary–Adrenal Axis	199
5.1	Anatomy	199
5.2	Effects on ACTH/Corticosterone Secretion	200
5.3	Physiological Significance of NPY Actions on the HPA Axis	201
5.4	Summary	203
6	Effects of NPY on the Hypothalamic–Pituitary–Thyroid Axis	203
6.1	Anatomy	203
6.2	Effects on TSH/thyroid Hormone Secretion	204
6.3	Physiological Significance	204
7	Effects of NPY on Growth Hormone Secretion	205
7.1	Anatomy	205
7.2	Effects of NPY on GH Secretion: Central Mechanisms	206
7.3	Effects of NPY on GH Secretion: Pituitary Mechanisms	206
7.4	Physiological Significance of NPY Action on GH: A Role in Short-Loop Feedback	207
7.5	Physiological Significance of NPY Action on GH: Role in Metabolic Adaptations to Nutritional Inadequacy	207
7.5.1	Relevance for Reproductive States	207

7.6 Role of NPY in the Actions of Leptin on GH Secretion 208
7.7 Summary 209

8   **Summary and Perspectives** 209

References 211

Abstract In the two decades since its discovery neuropeptide Y (NPY) has emerged as a major peptidergic neuromessenger in neuroendocrine regulation, affecting the secretion of every anterior and posterior pituitary hormone. Innervation of these neuroendocrine networks derives from the major NPY cell group in the hypothalamic arcuate nucleus and from noradrenergic and adrenergic cells in the lower brainstem, in which NPY serves a cotransmitter role. Regarding the magnocellular neuroendocrine system, NPY directly innervates vasopressin (VP)-secreting, and probably, oxytocin (OT)-secreting, cells, and stimulates the release of both peptides. Whether such effects are important for the release of OT or VP in response to specific physiological stimuli remains to be determined, however. Much more information is available regarding NPY effects on anterior pituitary hormone secretion. NPY, mainly deriving from arcuate neurons, inhibits the secretion of prolactin (PRL), thyroid stimulating hormone (TSH) and growth hormone (GH), and stimulates the secretion of adrenocorticotropic hormone (ACTH). For the most part, these effects are exerted by altering the neurosecretion of the cognate hypothalamic releasing and/or inhibiting hormone systems that regulate anterior pituitary secretion. In addition, an action at the anterior pituitary gland is important in the NPY inhibition of PRL secretion. In the case of PRL, TSH, and ACTH, the agouti-related peptide, which is coexpressed in arcuate NPY neurons, exerts the same effect as NPY, suggesting a postjunctional co-action of these two messengers. When considered together with the critical orexigenic effects of NPY, it is evident that the arcuate NPY system is important in integrating a variety of physiological, neuroendocrine and behavioral responses to conditions of nutritional inadequacy. Moreover, changes in circulating leptin, signaling the status of adipose tissue energy stores, appears to be a major factor in orchestrating these behavioral and neuroendocrine changes via its effects on hypothalamic NPY synthesis and release.

Keywords Adrenocorticotropic hormone · Growth hormone · Leptin · Prolactin · Thyroid stimulating hormone

1
Introduction

In the two decades since the discovery of neuropeptide Y (NPY), this peptide has been clearly established as a major signaling molecule in neuroendocrine regulation. In a major review published in 1992 (Kalra and Crowley 1992), we focused primarily on the critical actions of NPY on the hypothalamic–pituitary–gonadal axis, and touched only briefly on the effects of NPY on secretion of other anterior and posterior pituitary hormones. The critical role of NPY in regulation of gonadotropin secretion remains perhaps the best characterized neuroendocrine action of this peptide and is reviewed elsewhere in this volume (see chapter by Kalra and Kalra). This chapter will focus on NPY's actions on the secretion of other anterior pituitary hormones, as well as on the two magnocellular neurosecretory systems, and thus will update and extend our previous review in this regard.

It is possible to identify a number of common themes that apply to the actions of NPY in control of anterior pituitary hormone secretion. These include: (1) direct innervation of hypothalamic releasing hormone systems by NPY neurons; (2) actions of NPY primarily exerted within the brain, but sometimes involving effects at the anterior pituitary gland; (3) anatomical colocalization with other neuromessengers; (4) postjunctional interactions with other neuroendocrine messengers related to signal amplification; and (5) plasticity in NPY synthesis and release in response to alterations in specific physiological and endocrine conditions.

In addition, from the explosion of research on the critical role of NPY in energy balance during the past 20 years (Kalra et al. 1999; Schwartz et al. 2000; chapter by Levens, this volume), it is also now possible to place the neuroendocrine regulatory effects of NPY within an overarching physiological framework, namely, the integration of neural controls over energy balance with the secretion of anterior pituitary hormones in response to alterations in the physiological milieu. Furthermore, it is now abundantly clear that the adipocyte-derived hormone leptin plays a critical role in signaling the status of peripheral energy stores to both brain and anterior pituitary gland; thus, a state of negative energy balance reduces the secretion of leptin (Schwartz et al. 2002), and produces characteristic adaptive changes in the secretion of every anterior pituitary hormone as an overall homeostatic response. This chapter will review the essential involvement of hypothalamic NPY in this regulatory mechanism.

2
Anatomy of the Neuroendocrine NPY Systems

The anatomical distribution of NPY-expressing perikarya and associated nerve fiber and terminal networks throughout the central and peripheral nervous systems have been well documented using immunohistochemistry to detect NPY peptide and in situ hybridization, indicating cells that express the preproNPY mRNA (Chronwall et al. 1985; O'Donohue et al. 1985; DeQuidt and Emson 1986; Morris 1989). Important for this review are those NPY systems that provide input to the neuroendocrine regulatory centers in hypothalamus and basal forebrain. Results from both experimental approaches are in good agreement that NPY innervations to specific neuroendocrine systems within the hypothalamus derive from both intra- as well as extra-hypothalamic sources.

Regarding intra-hypothalamic systems, one of the most prominent NPY-positive cell groups in brain is located throughout the rostral–caudal extent of the arcuate nucleus in the medial basal hypothalamus (Chronwall et al. 1985; DeQuidt and Emson 1986; Morris 1989; Fuxe et al. 1989; Pelletier 1990). It is well established that this cell group plays a major role in energy balance (Kalra et al. 1999; Schwartz et al. 2000; chapter by Levens, this volume) and in a wide variety of neuroendocrine regulations, as reviewed previously (Kalra and Crowley 1992), and in detail in this chapter, through its distribution to various hypothalamic nuclei (see below). A second cell group resides in the hypothalamic dorsomedial nucleus and surrounding areas, and, while less prominent than the arcuate cells, also appears to play a role in control of feeding behavior, and perhaps in specific neuroendocrine controls. Smaller numbers of scattered NPY-immunopositive or preproNPY mRNA-expressing neuronal cell bodies have been reported in medial preoptic, anterior, periventricular and paraventricular nuclei; the specific neuroendocrine roles of these cells are unclear.

NPY-containing fibers and nerve terminals are present in abundance throughout the hypothalamus, but are particularly dense in the paraventricular nucleus, medial preoptic and arcuate nuclei, and in the internal zone of the median eminence (Chronwall et al. 1985; DeQuidt and Emson 1986; Morris 1989; Ciofi et al. 1991). The NPY distribution to the latter site is significant since NPY has been detected in hypophyseal portal plasma in amounts higher than the systemic circulation (McDonald et al. 1985), and levels of NPY in this vasculature vary, e.g., in conjunction with gonadotropin releasing hormone (GnRH)/ luteinizing hormone (LH) release (Sutton et al. 1988). With respect to LH and prolactin (PRL) secretion at least (Kalra and Crowley 1992; chapter by Kalra and Kalra, this volume), both hypothalamic and pituitary sites of action of NPY have been demonstrated, and we have suggested that NPY may be considered as a 'hypothalamic cohormone', analogous to its cotransmitter action with norepinephrine in the sympathetic nervous system (Crowley 1999).

Also noteworthy is the NPY projection to the paraventricular nucleus, which has been described as 'providing perhaps the most prominent chemically specified input to [this structure] yet described.' (Engler et al. 1999, p 468). As discussed in detail in this chapter, it is within this nucleus that NPY nerve terminals are positioned to influence multiple neuroendocrine systems, including the secretion of oxytocin (OT) and vasopressin (VP) from the neurohypophysis, as well as the releasing hormones for secretion of prolactin, adrenocorticotropic hormone (ACTH), and thyroid stimulating hormone (TSH) from the anterior pituitary gland. Further, as reviewed elsewhere (Kalra et al. 1999; Schwartz et al. 2000; chapter by Levens, this volume), the paraventricular nucleus is also the major site for the critical orexigenic actions of this peptide.

Complicating the issue of NPY innervation to the neuroendocrine hypothalamus is the fact that as much as 50% of the NPY input to various hypothalamic structures is provided by fibers ascending from noradrenergic and adrenergic cell groups in pons and medulla, in which NPY presumably acts in a cotransmitter role with norepinephrine or epinephrine. Detailed analysis of this colocalization (Everitt et al. 1984; Sawchenko et al. 1985; Harfstrand et al. 1987b) indicates that approximately 90% of the A1 noradrenergic cells (ventrolateral medulla),

10% of A2 noradrenergic cells (caudal nucleus tractus solitarius) and 25% of A6 noradrenergic cells (locus coeruleus), each of which projects into hypothalamus, coexpress NPY. NPY appears to be present in virtually 100% of the cells in adrenergic C1, C2 and C3 groups of the medulla (Everitt et al. 1984; Sawchenko et al. 1985; Harfstrand et al. 1987b).

Transection of ascending catecholaminergic axons significantly diminishes, but does not eliminate, NPY inputs to hypothalamic nuclei such as the paraventricular, medial preoptic, dorsomedial nuclei and median eminence (Sahu et al. 1988). Input specifically to the paraventricular nucleus has also been analyzed by combined retrograde labeling and dual-label immunohistochemistry (Sawchenko et al. 1985). Approximately 60% of the A1 noradrenergic cells projecting to paraventricular nucleus contain NPY, while only 15%–20% of the A2- or A6-derived input to this nucleus coexpress NPY. Most, if not all, of the C1, C2, and C3 adrenergic cells that innervate paraventricular nucleus also contain NPY (Sawchenko et al. 1985).

Thus, as reviewed below, co-action of NPY with adrenergic transmitters occurs in several neuroendocrine systems via these ascending inputs. However, colocalization with other potential neuromessengers also applies to the intrinsic hypothalamic arcuate NPY system. For example, virtually all arcuate NPY neurons coexpress the agouti-related peptide (AgRP) (Hahn et al. 1998, Chen et al. 1999), which also exerts powerful orexigenic effects (Kalra et al. 1999; Schwartz et al. 2000), and effects on pituitary hormone secretion (see below). In addition, a substantial proportion of arcuate NPY cells contain the amino acid transmitter, γ-aminobutyric acid, also a stimulant of feeding behavior (Horvath et al. 1997; Pu et al. 1999), although the possible interaction with NPY in neuroendocrine regulation has received scant attention.

Of immediate interest to this chapter is the more specific anatomical question of whether components of the various magnocellular (i.e., oxytocin, vasopressin) and parvicellular (hypothalamic releasing/inhibiting hormones) neuroendocrine systems are directly contacted by NPY nerve terminals deriving from either the hypothalamic or lower brainstem systems. This issue is addressed in individual sections below; however, as an overview, there is good evidence that both types of magnocellular neuroendocrine neurons, i.e., OT and VP neurosecretory cells, receive synaptic contacts from NPY-positive terminals. Moreover, in the parvicellular neuroendocrine system, cells synthesizing corticotropin-releasing hormone, thyrotropin-releasing hormone, growth hormone-releasing hormone, somatostatin and dopamine are also known to receive NPY innervation.

An interesting hallmark of the NPY networks within the neuroendocrine hypothalamus is anatomical and physiological/molecular plasticity as influenced by physiological state. For example, as discussed in detail below, during, and only during, lactation, NPY is coexpressed in a subpopulation of the tuberoinfundibular dopamine neurons of the arcuate nucleus (Ciofi et al. 1991), which are known to be the major system regulating secretion of PRL, and which normally do not contain NPY; associated with this novel expression is the appearance of NPY fibers in the external, periportal region of the median eminence, where they are not otherwise present. As a second example, OT and VP cells in the magnocellular nuclei coexpress NPY during dehydration, but do not do so under normal physiological circumstances (Hooi et al. 1989; Larsen et al. 1992a,b, 1993). Further, as reviewed in individual sections below, synthesis and release of NPY in the hypothalamus is profoundly affected by physiological manipulations or changes in the internal milieu that alter anterior and posterior pituitary hormone secretion.

3
NPY and the Magnocellular Neuroendocrine System: Regulation of OT and VP Secretion

3.1
Anatomy

Both the supraoptic nucleus, which consists entirely of magnocellular OT and VP cells, and the paraventricular nucleus, with its magnocellular and parvicellular divisions, receive prominent NPY input from both intra-hypothalamic and ascending adrenergic sources (Chronwall et al. 1985; DeQuidt and Emson 1986). Analysis at the electron microscopic level has demonstrated direct synaptic contacts of identified VP cells in the paraventricular nucleus by NPY-positive terminal boutons (Iwai et al. 1989; Kagotani et al. 1989a,b); a significant portion of these NPY terminals also contain a monoamine, most likely norepinephrine (Kagotani et al. 1989b). Similar studies on NPY innervation of supraoptic nucleus VP cells or of OT cells in either of the magnocellular nuclei have not been reported, although light microscopic analysis reveals NPY terminal projections in subregions of these nuclei containing both VP and OT cells (Sawchenko et al. 1985), and pharmacological studies also strongly suggest direct, or at least closely indirect, relationships (see below).

An increase in plasma osmolality, resulting, for example, from dehydration or salt loading, is a well known activator of both VP and OT synthesis and release in the rat (Leng et al. 1999). Such manipulations also induce a novel expression of NPY in both cell types, as well as an increase in NPY-positive nerve fibers in the neurohypophysis (Hooi et al. 1989; Larsen et al. 1992a,b, 1993). It has also been reported that the size and density of NPY nerve terminals in the neural lobe is increased after salt loading, and during lactation (Sands et al. 2001), although there are no reports of NPY expression in OT or VP cells during this physiological state. At present, however, the physiological actions of NPY when coexpressed with VP and OT in response to dehydration have not been defined. Although NPY has well characterized excitatory effects on the systemic release of these hormones (see below), it is unknown whether it is released along with these peptides into the systemic circulation to contribute to the physiological response to dehydration.

3.2
Effects of NPY on VP and OT Release

A number of reports employing in vivo and in vitro experimental models have been consistent in demonstrating stimulation by NPY of both VP and OT secretion in rats. Central microinjection of NPY into the paraventicular or supraoptic nucleus of male rats stimulates the release of VP (Willoughby et al. 1987; Leibowitz et al. 1988), and increases electrical activity of identified vasopressinergic cells (Day et al. 1985; Khanna et al. 1993). With respect to OT, Parker and Crowley (1993) reported that microinjection of NPY into either the paraventricular or supraoptic nucleus dose-dependently stimulates OT secretion in lactating female rats.

Similar effects on hormone secretion and electrical activity have been seen in in vitro preparations as well. For example, Kapoor and Sladek (2001) have reported that NPY elicits release of both OT and VP from a perfused hypothalamo-neurohypophysial explant system used extensively in that laboratory, and Sibbald and coworkers (1989) have shown that bath application of NPY to hypothalamic slices excited the activity of unidentified supraoptic nucleus cells. Limited pharmacological characterization implicates the Y_1 NPY receptor subtype in these excitatory actions (Khanna et al. 1993; Kapoor and Sladek 2001).

Several of these studies have also addressed the issue of NPY–norepinephrine interactions in control of OT and VP release; in general, these studies have shown that low concentrations of NPY that are without effect alone markedly augment the secretion of VP or OT in response to the α-1 adrenergic agonist phenylephrine (Sibbald et al. 1989; Parker and Crowley 1993; Kapoor and Sladek 2001). This suggests a facilitatory postjunctional interaction between NPY and norepinephrine, perhaps reflecting their cotransmitter roles in the magnocellular nuclei. Sladek and Kapoor (2001) have proposed that this synergism may reflect 'crosstalk' between the second messenger systems coupled to the α_1 and Y_1 receptors. Referring to this interaction, they have also proposed that "...conversion of transient to sustained responses represents the physiological importance of corelease of multiple neuroactive substances from a single nerve terminal." (Sladek and Kapoor 2001, p 205).

The foregoing is consistent with NPY action either directly upon, or in the immediate vicinity of, VP and OT perikarya in the magnocellular nuclei. A second potential site of action for NPY is in the neural lobe, which contains NPY-positive nerve fibers and receptors (Larsen et al. 1992b, 1993, 1994b). NPY increases basal and potassium depolarization-induced VP release from perfused rat neurohypophysis in vitro; pharmacological studies have suggested Y_2 NPY receptor mediation of this effect (Sheikh et al. 1998; Larsen et al. 1994b).

While these studies are in substantial agreement that NPY exerts a central action to enhance secretion of both VP and OT, notably lacking are studies in which various means of antagonizing NPY action, e.g., with receptor antagonists, immunoneutralization, or antisense molecules, have been used to examine the potential contribution of NPY to the release of these hormones in response to specific physiologic stimuli, such as dehydration or nursing during lactation. An early study by Vallejo and coworkers (1987) did show a blunting of the normal VP secretory response to hemorrhage in rats that were given subcutaneous injections of an NPY antiserum during the neonatal period; however, the mechanisms underlying this long lasting effect are not clear. Thus, at present, our understanding of the significance of NPY in regulation of the magnocellular neuroendocrine systems remains incomplete in the absence of any evidence for or against an obligatory role for this peptide on neurohypophyseal hormone secretion in defined physiological states.

4
NPY and Regulation of PRL Secretion:
Integration of Hormone Secretion and Energy Balance During Lactation

To date, the most extensively characterized action of NPY in neuroendocrine regulation of anterior pituitary hormone secretion is its critical excitatory effect on the secretion of luteinizing hormone, particularly important for control of ovulation (Kalra and Crowley 1992; see chapter by Kalra and Kalra, this volume). Recently, however, there has been a surge of interest in the effects of NPY during lactation, focusing on regulation of the secretion of PRL, the hormone responsible for milk synthesis and secretion, as well as the control over energy balance, which is dramatically altered during this physiological state. Indeed, hypothalamic NPY systems may play an important role in integrating the numerous neuroendocrine and behavioral adaptations that occur within lactation (Smith and Grove 2002).

4.1
Anatomy

Similar to the control over gonadotropin secretion (Kalra and Crowley 1992), NPY also affects PRL via actions exerted both centrally and at the level of the anterior pituitary gland. NPY systems directly and indirectly interact with several hypothalamic neuroendocrine networks that control secretion of PRL. The major hypothalamic hormonal control over PRL secretion is exerted by the tuberoinfundibular dopamine (TIDA) neurons of the arcuate nucleus, which exert a tonic inhibitory control over the anterior pituitary lactotrophe (PRL-secreting cell) (Ben-Jonathan 1985; Ben-Jonathan and Hnasko 2001; Martinez de la Escalera and Weiner 1992). Two components to this system are located dorsomedially and ventrolaterally within the arcuate nucleus (Fuxe et al. 1989; Ben-Jonathan and Hnasko 2001). Under conditions calling for increased PRL secretion, e.g., in response to the nursing stimulation of the offspring, release of dopamine into the hypophyseal portal vasculature is inhibited; this by itself constitutes a powerful stimulus for PRL secretion; in addition, removal of dopamine inhibition allows for the expression of the stimulatory effects of one or more of the PRL-releasing hormones (Martinez de la Escalera 1992). Candidates for

PRL-releasing hormones under different conditions include thyrotropin-releasing hormone (TRH) (Gershengorn 1985), vasoactive intestinal peptide (Abe et al. 1985), and the more recently identified PRL-releasing peptide (Hinuma et al. 1998). An additional level of control is provided by arcuate neurons that express proopiomelanocortin (POMC), the precursor to β-endorphin. This peptide exerts a stimulatory influence on secretion of PRL via: (1) inhibition of TIDA neurons and (2) stimulation of one or more PRL-releasing hormones (Crowley 1988).

Several lines of investigation suggest direct NPY interactions with the TIDA system. Dual-label immunocytochemistry at the light level has revealed an abundance of NPY-positive nerve terminals colocalized with tyrosine hydroxylase-positive cell bodies in the arcuate nucleus (Fuxe et al. 1989). Electron microscopic studies reported by Pelletier (1990) indicate the presence of NPY immunopositive nerve terminals "very close to" tyrosine hydroxylase-positive cells of the arcuate nucleus; some direct synaptic contacts (axosomatic, axodendritic and axoaxonic) between NPY terminals and tyrosine hydroxylase positive elements were also seen. In this case, tyrosine hydroxylase immunoreactivity is clearly marking TIDA cell bodies.

As reviewed in more detail below (Sect. 6.1), NPY-positive fibers also innervate TRH-positive perikarya localized in the paraventricular nucleus (Legradi and Lechan 1998); this innervation has been linked to control over TSH, but whether it is also involved in the control over PRL secretion is uncertain. Of potentially greater relevance is the direct innervation of POMC cells in the arcuate nucleus by NPY fibers (Csiffary et al. 1990; Horvath et al. 1992); this innervation may play a role not only in regulation of PRL secretion (see below), but also in control of food intake through influences on melanocortin signaling (Kalra et al. 1999). An excellent diagram of NPY interactions with the adjacent TIDA and POMC systems of the arcuate nucleus can be found in a chapter by Fuxe et al. (1989 p 119).

4.2
Effects On PRL Secretion: Central Sites and Mechanisms of Action

Several studies have shown that intracerebroventricular administration of NPY to male rats decreases circulating PRL and concomitantly, increases activity of the TIDA system as determined by measurements of dopamine turnover or levels of dopamine metabolites (Fuxe et al. 1989; Harfstrand et al. 1987; Hsueh et al. 2002). These findings are consistent with the view that NPY reduces PRL secretion through a direct interaction with the TIDA system that results in increased secretion of dopamine into the portal blood. Whether this action also occurs in females to mediate the suckling-induced release of PRL during lactation has not been reported. NPY also exerts a central action to reduce secretion of TSH, apparently through inhibition of the TRH neurons in the paraventricular nucleus (see Sect. 6.2 for details). It is conceivable that this could be another means by which PRL secretion is inhibited by NPY, but as yet, this has not been directly demonstrated. To date, there have been no reports on effects of AgRP, the peptidergic comessenger with NPY, on PRL secretion during lactation. However, central administration of this peptide impairs steroid-induced PRL surges (Schioth et al. 2001), consistent with the inhibitory effect of central NPY on PRL secretion.

4.3
Effects on PRL Secretion: Actions in the Anterior Pituitary

NPY also interacts with dopamine at the level of the anterior pituitary gland in the suppression of PRL secretion. Studies conducted in the author's laboratory on this action were spurred by the observations of Ciofi and coworkers (Ciofi et al. 1991, 1993), which showed in lactating rats and mice that NPY was expressed in a subpopulation of TIDA neurons, which do not contain NPY in any other reproductive condition or in males. Moreover, the distribution of NPY-positive fibers in the median eminence was altered to reflect this colocalization; thus, in addition to the NPY terminals in the internal zone of the median eminence (largely colocalized with norepinephrine/epinephrine), NPY-positive (also tyrosine hydroxylase-positive) terminals in lactating females were also present in the external zone in proximity to portal blood vessels (Ciofi et al. 1991, 1993).

Wang et al. (1996) have provided the most extensive test of the hypothesis that NPY affects PRL secretion directly and/or modulates the effect of dopamine at the anterior pituitary gland. In those studies, the addition of either NPY or dopamine alone to cultured pituitary cells resulted in dose-dependent reductions in PRL secretion, and the combination of dopamine and NPY at submaximal concentrations produced an additive inhibition. They also tested NPY–dopamine interactions in a removal paradigm that models to some extent the mechanisms underlying suckling-induced PRL release (Martinez de la Escalera and Weiner 1992). Cultured anterior pituitary cells were incubated in the presence of NPY or dopamine alone or in combination, followed by removal of the respective agents. It is well established that removal of dopamine in such a paradigm is a powerful signal for PRL release, and the same effect was seen with NPY. Interestingly, when pituitary cells were treated with both NPY and dopamine, the removal of both signals produced greater PRL release than removal of either alone.

In further studies by Wang et al. (1996), NPY also inhibited PRL release evoked by the physiological secretagog, TRH, and again the effect was additive with dopamine. TRH is an activator of the inositol phosphate/Ca^{2+} messenger system in anterior pituitary cells, and produces the classic 'spike and plateau' profile of both PRL secretion and cytosolic Ca^{2+} concentrations, with the initial, transient spike phase due to mobilization of intracellular Ca^{2+}, followed by the prolonged plateau in PRL secretion, mediated by influx of extracellular Ca^{2+} (Gershengorn 1985). Both phases of TRH-induced PRL release and rise in cytosolic Ca^{2+} were significantly inhibited by either dopamine or NPY alone, and were virtually abolished by the two in combination.

More extensive pharmacological analysis of the effects of NPY on Ca^{2+} signaling in the lactotrophe by Wang et al. (1996) has led to the conclusion that NPY selectively activates mechanisms that inhibit influx of extracellular Ca^{2+} through voltage-regulated channels, but does not alter the mobilization of intracellular Ca^{2+}; in contrast, dopamine inhibits both phases of Ca^{2+} signaling. Further, the pharmacological characterization of this effect of NPY points to mediation by the Y_5 NPY receptor subtype (Wang and Crowley, unpublished observations), which is expressed in anterior pituitary (Parker et al. 2000).

We have proposed that during lactation, NPY may serve as a hypothalamic 'cohormone' with dopamine to reinforce an inhibitory tone over PRL secretion (Crowley 1999). This may seem paradoxical in view of the overall high level of PRL secretion during lactation. Not widely appreciated, however, is the fact that PRL release in response to the suckling stimulus is actually pulsatile (Higuchi et al. 1983; Nagy et al. 1986a,b), consisting of irregularly timed, large mass pulses followed by relatively low levels of secretion. It may be the case that additional excitatory as well as inhibitory signals from the brain are integrated by the lactotrophe into the already pleiotropic regulation of PRL to help 'shape' this episodic secretion pattern.

The inhibitory action of NPY on PRL secretion thus features all of the hallmark themes mentioned above for NPY as a neuroendocrine peptide, including sites of action in both hypothalamus and pituitary, molecular and morphological plasticity, as well as colocalization and postjunctional interactions with other neuromessengers. However, recent investigations suggest that NPY exerts other physiological actions during lactation related to energy balance, which may be integrated with its neuroendocrine effects on PRL secretion into an overall physiological adaptive framework.

4.4
Central Effects of NPY on Food Intake During Lactation

In addition to the novel expression of NPY in TIDA neurons of the arcuate nucleus specifically during lactation, the synthesis of NPY, as reflected by preproNPY mRNA levels, is also increased in the adjacent major NPY cell group in this nucleus, as well as in the smaller group within the dorsomedial hypothalamic nucleus in lactating animals (Pelletier and Tong 1992; Smith 1993; Pape and Tramu 1996; Chen et al. 1999; Li et al. 1999a,b). This is associated with increased NPY peptide content in the paraventricular and dorsomedial nuclei and median eminence (Smith 1993; Malabu et al. 1994; Pickavance et al. 1996). It is well established that, via their projections to the paraventricular nucleus, these hypothalamic systems comprise critical orexigenic pathways (Kalra et al. 1999; Schwartz et al. 2000; see chapter by Levens, this volume). Because lactation is characterized by a marked, physiological hyperphagia resulting from the heightened energy demands associated with milk production (Fleming 1976; Williamson 1980; Munday and Williamson 1983; Vernon 1989; Barber et al.

1997), it is therefore highly suggestive that these systems become upregulated during this condition.

To directly test the hypothesis that NPY is an important mediator of lactational hyperphagia, recent studies form our laboratory have examined the effects of an NPY antagonist analog on diurnal and nocturnal food intake in lactating and nonlactating female rats (Crowley et al. 2003). When infused into the third cerebral ventricle over a 4-day period in nonlactating rats, the receptor antagonist reduced nocturnal feeding during the first 2 days of treatment, with recovery to normal levels of food intake thereafter. The same regimen had very little effect in lactating rats, however, leading us to test further whether the persistence of high levels of feeding in lactators despite NPY blockade could reflect the continued action of AgRP, the orexigenic cotransmitter with NPY, whose expression is also increased in lactation (Chen et al. 1999).

To examine this hypothesis, we administered the anorexigenic peptide derived from POMC, α-MSH. This peptide inhibits food intake via actions at the MC-4 receptor, which is also the receptor mediating the effects of AgRP (Schwartz et al. 2000); recent evidence suggests that AgRP is a competitive antagonist and an inverse agonist at this receptor (Haskell-Luevano and Monck 2001; Nijenhuis et al. 2002), and therefore, that α-MSH and AgRP are functional antagonists of each other. A 4-day infusion of α-MSH had effects similar to the NPY blocker in nonlactators and lactators. However, a marked inhibition of nocturnal and diurnal feeding behavior was seen when both the NPY receptor antagonist and α-MSH were administered into the third ventricle of lactating females (Crowley et al. 2003).

These findings suggest that both of the colocalized orexigenic peptides, NPY and AgRP, are important in mediating the hyperphagia of lactation, but that removal of the inhibitory melanocortin tone may be a prerequisite for their effects to be expressed. It is intriguing that a similar mechanism underlies suckling-induced PRL secretion in lactation, i.e., a removal of strong dopaminergic inhibitory tone, which allows for the effects of various PRL-releasing peptides. Studies on expression of the NPY, AgRP and POMC genes during lactation support this hypothesis. Thus, NPY and AgRP mRNAs are increased, while POMC mRNA levels are decreased, in the arcuate nucleus during lactation in rats (Pelletier and Tong 1992; Smith 1993; Pape and Tramu 1996; Chen et al. 1999; Li et al. 1999a,b).

4.5
What Are the Physiological Signals for NPY Upregulation in Lactation?

The studies described above indicate that the marked upregulation of hypothalamic NPY systems during lactation is significant for the neuroendocrine regulation of PRL secretion, as well as to control of food intake and energy demands of lactation. An equally interesting question concerns identification of the physiological signals associated with lactation that evoke the increased NPY (and AgRP) gene expression. The novel appearance of NPY in TIDA neurons and the enhanced expression of NPY mRNA in the arcuate and dorsomedial cell groups in the hypothalamus depend upon the suckling stimulus, as several studies have shown a reduction in NPY mRNA levels in these structures, as well as diminished NPY immunoreactivity in median eminence nerve terminals, after removal of the litters (Ciofi et al. 1993; Wang et al. 1996; Pelletier and Tong 1992; Smith 1993; Pape and Tramu 1996; Li et al. 1999a,b); conversely, litter replacement restores preproNPY mRNA levels in arcuate cells (Smith 1993; Pape and Tramu 1996).

It is possible that this effect of suckling reflects a transynaptic action via neural pathways activated by the suckling stimulus that impinges upon various cell groups in the arcuate nucleus. The possibility also exists that the effect of suckling is secondarily mediated by an action of a hormone released by suckling. However, a possible central action of PRL in this regard has been ruled out. This was an attractive possibility in view of the observations that TIDA neuronal activity is activated by PRL, which targets these cells as a component of a short loop feedback regulation of its own secretion (Moore 1987). However, neither blockade of PRL secretion with the dopamine agonist bromocriptine, nor immunoneutralization of PRL, mimicked the effect of pup removal to decrease NPY mRNA expression in the arcuate nucleus (Pelletier and Tong 1992; Pape and Tramu 1996; Li et al. 1999b). Similarly, neither blockade of suckling-induced PRL secretion with bromocriptine, nor elevation of PRL release in nonlactating rats with the dopamine receptor antagonist haloperidol, affected the appearance NPY-like immunoreactivity in TIDA nerve endings (P. Ciofi and W.R. Crowley, unpublished results). On the other hand, Smith and coworkers have presented evidence that PRL is responsible for the suckling-induced upregulation of NPY mRNA expression in the cells of the dorsomedial nucleus (Li et al. 1999b).

An alternative possibility relates to the changes in two important hormonal signals in lactation that are known to influence NPY expression in the hypothalamus as a component of regulation of energy balance. Thus, reductions in circulating insulin and in leptin occur in response to food deprivation, and current concepts of energy balance suggest that this provides a critical neuroendocrine signal for upregulation of NPY expression in the arcuate nucleus and release within the hypothalamus (Kalra et al. 1999; Schwartz et al. 2000). Secretion of these two hormones is also reduced in lactation, probably as a consequence of the heightened energy demands of milk synthesis and secretion that create a negative energy balance (Williamson 1980; Vernon 1989; Madon et al. 1990; Malabu et al. 1994; Pickavance et al. 1996, 1998; Brogan et al. 1999; Smith and Grove 2002). Recent studies in this laboratory have tested the hypothesis that the hypo-insulinemia and/or hypo-leptinemia of lactation are responsible for the increased expression of NPY in the hypothalamus.

To examine this hypothesis, we administered insulin or leptin in lactating rats via Alzet Osmotic minipumps implanted subcutaneously in regimens designed to return insulin and leptin to their higher nonlactating levels (Crowley and Ramoz 2002). The observations of reduced plasma concentrations of insulin and leptin were confirmed. Administration of insulin via the pumps actually increased leptin as well as insulin to the nonlactating level, consistent with the action of insulin to stimulate leptin secretion, and decreased the concentrations of NPY in paraventricular nucleus and median eminence to the nonlactating level. Treatment with leptin via the pumps restored plasma leptin concentrations to normal without having any effect on insulin, and also normalized NPY concentrations in these two regions. Interestingly, leptin replacement, but not insulin replacement, was associated with a reduction in food intake as well.

4.6
Summary: NPY Integrates Neuroendocrine and Behavioral Responses During Lactation and in Periods of Negative Energy Balance

These recent findings indicate that the reduction in leptin, and perhaps insulin, secretion during lactation provides the important signal for increased NPY synthesis in the arcuate nucleus during lactation. The stimulatory influence of suckling on NPY expression and peptide content may involve a complex cascade of events as depicted in Fig. 1, in which the suckling stimulus of the offspring evokes PRL secretion, which in turn stimulates synthesis and release of milk. Milk synthesis and consumption by the offspring constitutes a heavy energy drain on the lactating female that results in decreased circulating concentrations of leptin. This provides an important signal to hypothalamic orexigenic systems (upregulation of NPY and AgRP expression) and anorexigenic pathways (down-

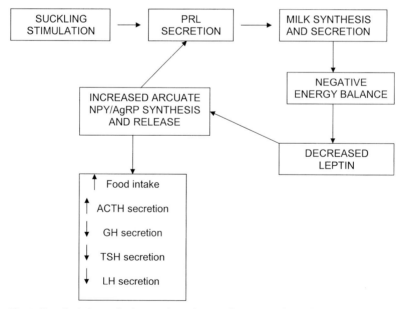

Fig. 1 Hypothesis for mechanisms and significance of NPY upregulation during lactation

regulation of POMC/α-MSH expression) that mediate the hyperphagia of lactation. In addition, NPY, perhaps interacting with these other messengers, contributes to the complex pleiotropic regulation of PRL secretion. Further, as discussed below, NPY may contribute to several other neuroendocrine changes during lactation, including reductions in secretion of thyroid hormone (see Sect. 6) and gonadotropins (Smith and Grove 2002), and activation of adrenal hormone release (see Sect. 7). Such actions may be integrated with the critical actions of this peptide on energy balance, thereby helping to integrate critical neuroendocrine, metabolic and behavioral adaptations during lactation.

Observations in nonlactating rats suggest that the mechanisms proposed above may have general applicability as a component of widespread neuroendocrine adaptations to negative energy balance. Thus, periods of food restriction are associated with decreased secretion of a number of pituitary hormones, including PRL (Campbell et al. 1977; Ahima et al. 1996; Hsueh et al. 2002), aimed most likely towards the conservation of energy. As noted above, this also is associated with decreased circulating leptin and upregulated hypothalamic NPY. While not yet tested directly for PRL secretion, the inhibitory effect of NPY on PRL could manifest itself under these circumstances. That leptin may be the primary peripheral hormonal signal for the NPY inhibition of PRL is suggested by the observations that chronic administration of leptin to food deprived rats prevents both the upregulation of hypothalamic NPY expression (Ahima et al. 1996; Korner et al. 2001) and the reduction in circulating PRL (Watanobe et al. 2001). As reviewed in the following sections on ACTH, GH and TSH secretion, the central role of leptin–NPY signaling in response to conditions of nutritional inadequacy is a recurring and overarching theme in the neuroendocrine actions of NPY.

5
Effects of NPY on the Hypothalamic–Pituitary–Adrenal Axis

5.1
Anatomy

In terms of neuroendocrine regulation, studies on the influence of NPY on the hypothalamic–pituitary–adrenal (HPA) axis are perhaps second only to the hypothalamic–pituitary–gonadal axis in number, and several of the common themes of NPY action identified above apply to ACTH/corticosterone secretion. As reviewed extensively (Engler et al. 1999), hypothalamic control over ACTH secretion is governed primarily by the peptide factor, corticotropin-releasing hormone (CRH), which is synthesized by neurosecretory cells whose perikarya are localized within a parvicellular subdivision of the paraventricular nucleus of the hypothalamus. A second means of stimulatory control over ACTH secretion is provided in some species by VP, which is coexpressed in many CRH neurosecretory cells in parvicellular paraventricular nucleus, distinct from the magnocellular system discussed above, and which augments CRH action on ACTH secretion (Engler et al. 1999). Superimposed upon these hypothalamic hormonal controls are circadian rhythmicity, responsiveness to stressful stimuli, steroid hormone feedback regulation, and a variety of neurotransmitter inputs (Bradbury et al. 1991; Dallman et al. 1994).

Perhaps not surprisingly, given the dense NPY projection to the paraventricular nucleus already discussed above, studies at the light microscopic level uniformly demonstrate an abundance of NPY-positive axons in the immediate vicinity of CRH cell bodies (Wahlestedt et al. 1987; Liposits et al. 1988; Li et al. 2000), and electron microscopy reveals synaptic contacts between NPY-positive boutons and CRH-positive dendrites and perikarya (Liposits et al. 1988). From anterograde (Li et al. 2000) and retrograde (Baker and Herkenham 1995) tracing, it is clear that many NPY-positive nerve endings near CRH cells derive from the arcuate nucleus; additionally, NPY undoubtedly influences the CRH/vasopressin neurosecretory system via ascending the noradrenergic and adrenergic projections discussed above (Engler et al. 1999).

5.2
Effects on ACTH/Corticosterone Secretion

Studies conducted in rats, sheep and dogs are consistent in showing a central stimulatory action of NPY on ACTH secretion, accomplished most likely through direct effects on the neurosecretion of CRH. For example, microinjection of NPY into either the ventricular system or directly into the paraventricular nucleus evokes a rapid release of ACTH and corticosterone in the rat (Wahlestedt et al. 1987; Leibowitz et al 1988; Albers et al. 1990; Inui et al. 1990; Suda et al. 1993; Tempel and Leibowitz 1993; Sainsbury et al. 1996, 1997b; Small et al. 1997). That this involves enhanced neurosecretion of CRH is suggested by the observations that central administration of NPY increases CRH immunoreactivity and mRNA levels in the paraventricular nucleus (Haas and George 1989; Suda et al. 1993) and by in vitro studies showing stimulatory effects of NPY on release of CRH (Tsagarakis et al. 1989; Hastings et al. 2001). Elegant studies conducted in sheep also provide definitive evidence that NPY exerts a central action to increase secretion of CRH and vasopressin into the hypophyseal portal system, leading to ACTH/corticoid secretion (Brooks et al. 1994; Liu et al. 1994). Unlike NPY effects on secretion of LH (Kalra and Crowley 1992; see chapter by Kalra and Kalra, this volume) or PRL (see Sect. 4), the stimulation of the HPA axis by NPY appears to be mediated exclusively by this central action; in vitro studies with anterior pituitary cells from rat or sheep have consistently shown no effect of NPY on either basal or CRH-induced release of ACTH (Brooks et al. 1994; Liu et al. 1994; Small et al. 1998).

Pharmacological studies in vivo have not been able to identify a specific NPY receptor subtype mediating these effects (Small et al. 1997), although in one study (Hastings et al. 2001), a Y_1 receptor antagonist prevented NPY-induced CRH release. On the other hand, Y_1 receptor immunoreactivity was not observed in CRH cell bodies (Li et al. 2000), while Y_5 receptor immunoreactivity is detectable in some CRH perikarya (Campbell et al. 2001).

The issue of co-action between NPY and its peptidergic comessenger, AgRP, in control of the HPA axis has received some attention. In one report (Dhillo et al. 2002), microinjection of AgRP into the paraventricular nucleus mimicked the effect of NPY in stimulating ACTH secretion; in addition, AgRP increased the release of CRH from hypothalamic explants in vitro, suggesting that, again like NPY, AgRP activates the neuroscretion of CRH. However, because in these studies α-MSH exerted the same effects as AgRP, the effects of AgRP may not be through competitive antagonist/inverse agonism at melanocortin receptors as described in Sect. 4 for feeding behavior.

It is well established that norepinephrine and epinephrine also act centrally to stimulate ACTH release (e.g., Leibowitz et al 1988; Engler et al. 1999). In view of the NPY colocalization in noradrenergic/adrenergic projections to the paraventricular nucleus discussed above, the question of whether NPY interacts with these adrenergic transmitters in control of the adrenal axis is an important one. This has not been addressed extensively, but the studies on NPY enhancement of CRH mRNA levels (Suda et al. 1993) and CRH release in vitro (Tsagarakis et al. 1989) showed that these responses to NPY were unaffected by blockade of α- or β-adrenergic receptors, suggesting that the effect of NPY might be exerted in parallel with the adrenergic transmitters, but does not depend upon their action.

5.3
Physiological Significance of NPY Actions on the HPA Axis

Is the stimulatory action of NPY on CRH/VP a component of the physiological neuroendocrine response to stress? At present, the literature does not contain reports in which this has been investigated using a pharmacological approach, investigating, for example, whether antagonism of NPY action might interfere with the ACTH/corticoid response to physiological stressors. An alternative strategy has been to examine whether changes in NPY synthesis or release are induced by a physiological stressor in conjunction with enhanced ACTH/corticosterone secretion. Although several studies have demonstrated that acute and chronic restraint stress increases the expression of NPY mRNA in the arcuate nucleus of rats (Conrad et al. 2000; Makino et al. 2000), another report demonstrates the opposite, i.e., a reduction in NPY mRNA levels in this nucleus (Krukoff et al. 1999) in response to restraint stress. Hence, it is difficult at present to form definite conclusions on this issue.

Another strategy to address the question of the physiological importance of NPY in relation to the HPA axis has been to examine the effects of endocrine manipulations such as adrenalectomy or treatment with glucocorticoids on various indices of NPY function. The reported effects of adrenalectomy have been inconsistent, with some (Dean and White 1990; White et al. 1990; Watanabe et al. 1995), but not other (Larsen et al. 1993; Akabayashi et al. 1994; Larsen et al.

1994; Baker and Herkenham 1995), studies demonstrating a decrease in preproNPY mRNA levels in the arcuate nucleus following removal of adrenal steroids. More consistently, however, a number of studies have demonstrated that administration of glucocorticoids significantly upregulates NPY mRNA levels in the arcuate nucleus (and reverses the decline observed after adrenalectomy) (Dean and White 1990; White et al. 1990; Wilding et al. 1993; White et al. 1994; Larsen et al. 1994a; Stack et al. 1995; Watanabe et al. 1995).

Several lines of evidence indicate that the stimulatory effect of glucocorticoids on NPY synthesis is likely to be exerted directly in the NPY neuron. Thus, at least one-half of the NPY cell bodies in the arcuate nucleus are immunopositive for the glucocorticoid receptor (Hisano et al. 1988; Harfstrand et al. 1989), and multiple glucocorticoid response elements are located in the 5' flanking region of the NPY gene; treatment of NPY gene-expressing cells with the glucocorticoid dexamethasone stimulates its transcription (Misaki et al. 1992). Further, the majority of lower brainstem noradrenergic/NPY and adrenergic/NPY cells also express glucocorticoid receptor immunoreactivity (Harfstrand et al. 1989). In this regard, it is highly interesting that glucocorticoid treatment increases levels of preproNPY mRNA in the noradrenergic cells of the locus coeruleus (Watanabe et al. 1995), which also are glucocorticoid receptor-positive (Harfstrand et al. 1989). Whether glucocortcoids also influence NPY expression in the other noradrenrgic/adrenergic cell groups has not been reported. Finally, selective type II receptor stimulation is sufficient to increase NPY mRNA levels (White et al. 1994; Watanabe et al. 1995).

Thus, it is apparent that neuroendocrine NPY neurons in hypothalamus and brainstem are major targets for glucocorticoid hormones, which directly upregulate NPY synthesis. However, given the stimulatory influence of NPY on CRH/ACTH/corticoid secretion, such effects cannot be components of adrenal steroid negative feedback control over the HPA axis (Bradbury et al. 1991; Dallman et al. 1994). Rather, several investigators have recently proposed that this relationship is more relevant to the neuroendocrine regulation of energy balance in an interaction with multiple peripheral hormonal signals, including insulin and leptin (Dallman et al. 1993, 1995). While discussion of such regulation is beyond the scope of this chapter (see chapter by Levens, this volume), several salient features of this interaction deserve emphasis.

First, circadian rhythms in NPY gene expression and peptide content in the arcuate–paraventricular circuit correlate positively with the rhythmicity in the HPA axis, with both systems activated in rats during the onset of dark hours, when feeding is initiated (Akabayashi et al. 1994). Second, it is well established that food deprivation is a major stimulus for NPY synthesis and release in the arcuate–paraventricular orexigenic circuit (reviewed in Dallman et al. 1993; Kalra et al. 1999; Schwartz et al. 2000), and concomitantly, to activation of the HPA axis (Dallman et al. 1993). It is controversial whether glucocorticoids are involved in the food deprivation-induced increased expression in hypothalamic NPY (Ponsalle et al. 1992; Hanson et al. 1997). However, administration of leptin to food-deprived animals does reverse both the increase in hypothalamic NPY expression and the activation of the HPA axis (Ahima et al. 1996).

Another intriguing connection is the dependence of NPY-induced feeding upon some as yet undefined action of glucocorticoids. For example, adrenalectomy or local application of a type II glucocorticoid antagonist abolish feeding induced by administration of NPY (or norepinephrine) into the paraventricular nucleus (Tempel and Leibowitz 1993). Moreover, the obesity syndrome seen in rats subjected to chronic intraventricular infusion of NPY is also prevented by adrenalectomy (Sainsbury et al. 1997a). The basis for this permissive action of glucocorticoids has not been identified, but could reflect some action on NPY receptors, signal transduction messenger systems and/or downstream effectors.

5.4
Summary

Thus, NPY exerts a seemingly straightforward, excitatory influence on ACTH and glucocorticoid release, mediated most likely entirely by a direct activation of CRH and perhaps VP secretion into the hypothalamo-pituitary portal vasculature. Several of the 'themes' of neuroendocrine NPY action are present in this system (direct effect on a releasing hormone; physiological plasticity), while others are not (pituitary site of action) or as yet uninvestigated (colocalization with other messengers, postjunctional signal amplification). Yet, the stimulatory effects of glucocorticoids on NPY expression suggest strongly that the relationship of NPY to the HPA axis and the physiological significance of this interaction are more complex. Dallman and coworkers (Dallman et al. 1993, 1995) have suggested that NPY might play a critical role in integrating the activity of the HPA axis with neural and neuroendocrine regulation of food intake and energy balance. The details of this integration remain to be worked out, but this concept is similar to that we have put forward above for NPY integration of the neuroendocrine and metabolic aspects of lactation (Sect. 4). Indeed, because the HPA axis is activated during lactation (Walker et al. 1992), this may be part of an overarching physiological role for NPY in the integration of anterior pituitary hormone secretion with controls over energy balance, with leptin an important peripheral hormonal signal to the HPA axis via NPY.

6
Effects of NPY on the Hypothalamic–Pituitary–Thyroid Axis

6.1
Anatomy

Compared with other anterior pituitary hormones, there have been fewer investigations on the action of NPY on the hypothalamic–pituitary–thyroid (HPT) axis, but the results obtained to date are consistent with several of the concepts developed above. The major hypothalamic regulator of TSH secretion is the tripeptide, TRH; cells synthesizing this peptide are found in several parvicellular subdivisions of paraventricular nucleus (Toni et al. 1990). Perhaps not surprisingly, a number of studies using both light and electron microscopy have demonstrated extensive innervation of TRH cell bodies by NPY immunoreactive boutons (Toni et al. 1990; Liao et al. 1991; Diano et al. 1998; Legradi and Lechan 1998); similar results have been observed in human brain as well (Mihaly et al. 2000). It is also evident that the vast majority of this input derives from the arcuate nucleus, rather than from ascending catecholaminergic systems, and co-expresses AgRP (Liao et al. 1991; Diano et al. 1998; Legradi and Lechan 1998; Mihaly et al. 2000).

6.2
Effects on TSH/thyroid Hormone Secretion

Several studies have consistently shown that central administration of NPY reduces circulating TSH and thyroid hormones (both T3 and T4) (Harfstrand et al. 1987a; Fekete et al. 2001, 2002). Concomitantly, NPY decreased the levels of preproTRH mRNA in paraventricular nucleus upon central administration (Fekete et al. 2001, 2002). Any significant action of NPY on TSH release at the level of the pituitary gland is unlikely, in that incubation of cultured anterior pituitary cells with even high concentrations of NPY had no effect on release of this hormone (Chabot et al. 1988).

Exactly the same results on circulating TSH and thyroid hormones have been reported to occur with AgRP, the peptidergic comessenger with NPY in arcuate–paraventricular projections, from intraventricular or intraparaventricular administration (Kim et al. 2000; Fekete et al. 2002). This is intriguing because of the findings cited above (Sect. 4) that AgRP is a functional antagonist/inverse agonist at receptors also stimulated by α-MSH, implying a parallel input to TRH neurons by this system. Indeed, anatomical and pharmacological studies by Fekete et al. (2000) reveal direct contacts of TRH cells by POMC/α-MSH-expressing nerve fibers, and stimulatory effects of α-MSH on the expression of the proTRH gene and the HPT axis. Consistent with these observations, α-MSH stimulates the release of TRH in vitro, and this effect is antagonized by AgRP (Kim et al. 2000, 2002).

6.3
Physiological Significance

Lacking thus far in the literature are studies evaluating the actions of manipulations such as thyroidectomy or administration of T3 or T4 on NPY synthesis, levels or release in the arcuate-paraventricular circuit; such studies could indicate whether the inhibitory action of NPY/AgRP on the HPT axis participates in the negative feedback regulation of TSH secretion by thyroid hormones. However, as with the interaction of NPY with PRL (Sect. 4) and the HPA axis reviewed in Sect. 5, the physiological significance of these effects of NPY may be more related to integration of the neuroendocrine regulation of energy balance.

Suppression of the HPT axis is a well known adaptation to food restriction and negative energy balance, most likely as part of an overall strategy for energy conservation (Ahima et al. 1996). As reviewed in Sect. 4, this is also a powerful stimulus for upregulation of the NPY/AgRP genes and for suppression of POMC expression within the arcuate nucleus. Thus, it is tempting to speculate that the inhibitory effects of NPY/AgRP on the HPT axis are components of these adaptive responses. Leptin has been suggested to be a prime coordinator of the compensatory changes in anterior pituitary hormone secretion that occur in negative energy balance (Ahima et al. 1996). Thus, in addition to reversing fasting-induced alterations in the HPT and HPA axes, leptin concomitantly reverses fasting-induced changes in NPY, AgRP and POMC gene expressions in arcuate neurons (Ahima et al. 1996), totally consistent with the effects of these peptides on the respective neuroendocrine states.

Further interrelating these various endocrine adaptations, it is intriguing to note that activity of the HPT axis is suppressed during lactation (Fukuda et al. 1980; Oberkotter and Rasmussen 1992), which, as reviewed in Sect. 4, is one of apparent negative energy balance despite the hyperphagia. Thus, it may also be the case that another important physiological action of NPY during lactation is to contribute to this neuroendocrine alteration as well.

7
Effects of NPY on Growth Hormone Secretion

7.1
Anatomy

Hypothalamic control over the secretion of growth hormone (GH) is accomplished via the coordinated neurosecretion of two hypothalamic factors, growth hormone-releasing hormone (GHRH) and the inhibitory regulator somatostatin, which together determine the characteristic episodic secretion of GH (Frohman et al. 1992). GHRH-synthesizing perikarya are localized predominantly within the hypothalamic arcuate nucleus (Everitt et al. 1986); while somatostatin neurons are more widespread, the hypophysiotropic population appears mainly within the periventricular nucleus at the level of the anterior hypothalamus (Kawano and Daikoku 1988).

Ciofi and coworkers (1987) reported that the majority of arcuate cells immunoreactive for GHRH also coexpressed NPY. However, they also made the important observation that these GHRH/NPY cells did not project to the external zone of the median eminence in proximity to the portal blood vessels, suggesting they might be involved in other functions. In contrast, a small group of GHRH-positive/NPY-negative cells located laterally in the arcuate nucleus did project to the periportal, external region of the median eminence. Perhaps not widely appreciated is the fact that many of these ventrolateral GHRH cells comprise part of the TIDA neuronal population (Fuxe et al. 1989), which, as reviewed in Sect. 4, is known to receive NPY innervation.

Light and electron microscopy has also revealed substantial direct innervation by NPY-positive fibers and boutons of somatostatin cells in the anterior periventricular nucleus; the NPY input evidently derived mainly from the arcuate nucleus (Hisano et al. 1990). Hence, the arcuate NPY neurons make contact with both limbs of the hypothalamic hormonal control over GH secretion.

7.2
Effects of NPY on GH Secretion: Central Mechanisms

A number of pharmacological studies have consistently shown that acute central administration of NPY inhibits GH secretion in both male and female rats (McDonald et al. 1985; Harfstrand et al. 1987a; Rettori et al. 1990; Suzuki et al. 1996; Carro et al. 1998); this effect was mimicked by either Y_1 or Y_2 NPY receptor agonists (Suzuki et al. 1996. Suppression of GH secretion has also been observed during chronic administration of NPY in male and female rats (Catzeflis et al. 1993; Pierroz et al. 1996). Conversely, central administration of anti-NPY serum elevates circulating GH, implying the existence of a tonic inhibitory tone exerted by NPY over GH secretion (Rettori et al. 1990). Consistent with the anatomical evidence cited above that NPY cells directly contact GHRH and somatostatin neurosecretory cells, in vitro studies have demonstrated that NPY inhibits the release of GHRH, but stimulates the release of somatostatin, from hypothalamic or median eminence fragments (Rettori et al. 1990; Korbonits et al. 1999). To date, there have been no reports on the effects of AgRP on GH secretion.

7.3
Effects of NPY on GH Secretion: Pituitary Mechanisms

While these studies are consistent with a relatively straightforward mechanism whereby NPY acts within the hypothalamus to both inhibit GHRH and activate somatostatin release, resulting in suppression of GH secretion, studies with cultured rat anterior pituitary cells show that NPY has the opposite effect on GH secretion at the level of the pituitary gland. Thus, in this experimental model, NPY produces a concentration-dependent increase in GH release (McDonald et al. 1985; Chabot et al. 1988). These converse central and pituitary actions have not been integrated into a comprehensive physiological explanation as yet. Complicating the picture further, NPY has been reported to inhibit GH secretion by human pituitary adenoma cells in culture (Adams et al. 1987). As yet there are no reports on receptor-messenger mechanisms underlying these disparate effects of NPY at the level of the pituitary gland.

7.4
Physiological Significance of NPY Action on GH:
A Role in Short-Loop Feedback

It is well established that GH regulates its own secretion via a short-loop negative feedback action exerted centrally, probably mainly via enhancement of somatostatin release (Frohman et al. 1992). Several studies suggest that hypothalamic NPY is important for this mechanism. For example, peripheral administration of GH to hypophysectomized rats induces the expression of c-Fos, a marker of neuronal activation, in the majority of NPY cells in arcuate nucleus, as well as in somatostatin cells in the periventricular nucleus (Kamagai et al. 1994). Further, nearly all of the arcuate NPY cells express GH receptor mRNA (Chan et al. 1996; Kamegai et al. 1996), implying a direct effect of GH on these cells. Moreover, hypophysectomy decreases preproNPY mRNA expression, while treatment with GH restores these levels to normal in rats (Chan et al. 1996). Given the stimulatory effect of NPY on release of somatostatin discussed above, it is tempting to speculate that GH short-loop negative feedback involves activation of NPY projections from the arcuate nucleus directly to the periventricular somatostatin population.

7.5
Physiological Significance of NPY Action on GH:
Role in Metabolic Adaptations to Nutritional Inadequacy

Similar to its actions on the HPA and HPT axes discussed above, the inhibitory effects of NPY on GH secretion may also be a component of the overall neuroendocrine adaptations to negative energy balance. Thus, we have emphasized that food deprivation is a major stimulus for the upregulation of NPY synthesis content and release in the hypothalamus; concomitantly, it is well established that the episodic secretion of GH is suppressed under these conditions (Tannenbaum et al. 1978). Although not tested directly, it would seem highly likely that the inhibitory effects of NPY are exerted on GH secretion under such conditions. This could serve to reduce the catabolic and protein synthetic actions of GH in the face of need for other energy stores.

7.5.1
Relevance for Reproductive States

A large literature attests to the fact that reproduction in both sexes, but particularly the processes of cyclicity, ovulation, pregnancy and lactation in females are profoundly disrupted during periods of food deprivation; similarly, sexual maturation as evidenced by onset of puberty is delayed under such conditions (see Wade and Schneider 1992 for review). Disruption of reproductive cyclicity, and ovulation as well as impairment of sexual development during periods of undernutrition may be mediated in large part via a suppression of the neurosecretion of gonadotropin-releasing hormone, resulting secondarily in reduced secretion of gonadotropins from the anterior pituitary (see Wade and Schneider 1992 for review). However, there is also considerable evidence that GH participates in the overall regulation of sexual development (Hull and Harvey 2001). A series of investigations by Aubert and coworkers has generated the intriguing hypothesis that the arcuate NPY system is important in suppressing activity of both the hypothalamic–pituitary–gonadal axis, as well as GH secretion, as adaptive responses to conditions of nutritional inadequacy.

That NPY might be involved in these adaptations is suggested by the following observations. First, chronic central administration of NPY reduces secretion of the gonadotropins as well as GH in rats (Catzeflis et al. 1993); this is consistent with a substantial literature showing that NPY can exert either inhibitory as well as stimulatory effects on GnRH/LH secretion depending upon the steroid hormone milieu as well as upon the mode of administration (see Kalra and Crowley 1992 for review). Secondly, chronic central administration of NPY to otherwise normal, immature rats significantly delays onset of puberty (Pierroz et al. 1995), and prevents sexual maturation in a model of food deprivation-induced impairment of puberty (Gruaz et al. 1993).

These considerations may also apply to NPY actions during lactation (see Sect. 4). GH contributes to the control of milk synthesis in a number of species (Hull and Harvey 2001), and to a certain extent is released by the suckling stimulus, although the dynamics are different than for PRL (Saunders et al. 1976; Nagy et al. 1986a,b). As with PRL, the suckling-induced release of GH is episodic (Nagy et al. 1986a,b), and though not yet tested directly, it is conceivable that NPY could help shape the pulsatile nature of GH secretion, as we have suggested for PRL (Wang et al. 1996). Further, although outside the immediate scope of this chapter, Smith and Grove (2002) in a recent review have summarized evidence that the chronic elevation of NPY during lactation could contribute to the suppression of gonadotropin secretion during lactation.

7.6
Role of NPY in the Actions of Leptin on GH Secretion

Related to this, arcuate NPY neurons appear to be critical targets for the peripheral hormone leptin in its actions on GH release, as well as on secretion of the other anterior pituitary hormones reviewed above. Leptin has little effect on GH secretion in normally fed rats but prevents the decline in GH secretion under conditions of food restriction (Casanueva and Dieguez 1999). Thus, the food deprivation-induced reduction in circulating leptin could remove an important stimulatory influence over GH secretion. That this might be accomplished via inhibition of the NPY system is suggested by findings that leptin replacement in fasting rats reverses the upregulation of NPY gene expression and concomitantly restores pulsatile GH secretion (Vuagnat et al. 1998). Further, central administration of anti-NPY serum reverses the suppression of GH secretion produced by immunoneutralization of leptin, and central administration of NPY prevents leptin-induced increases in circulating GH (Carro et al. 1998).

7.7
Summary

Thus, NPY effects on GH secretion are consistent with our general framework that NPY integrates neuroendocrine responses to nutritional inadequacy. NPY-induced inhibition of GH secretion is largely mediated by inhibition of GHRH and stimulation of somatostatin, and may be a component of GH-induced short-loop negative feedback. Further, food deprivation suppresses GH release, at least in part via the reduction in leptin secretion. This in turn, is highly likely to be mediated by removal of the leptin inhibition of arcuate NPY synthesis and release.

8
Summary and Perspectives

It is apparent that NPY neurons occupy a prominent position as a neuromessenger critical for a variety of neuroendocrine regulations. Anatomically, central NPY systems, intrinsic within the hypothalamus, as well as ascending from the brainstem in a comessenger role with norepinephrine and epinephrine, are in position to influence all of the major neuroendocrine systems. With regard to the magnocellular neuroendocrine hypothalamus, the initial observations that NPY stimulates secretion of VP and OT have not been pursued from a physiological perspective. Thus, it is still unknown whether NPY contributes to the release of these hormones in response to the well defined stimuli for their release, such as dehydration, hypovolemia or nursing during lactation. This is all the more disappointing in view of the remarkable physiological/molecular plasticity in this system, in which an increase in plasma osmolality, which is well known to increase OT and VP synthesis and secretion, also leads to a novel expression of NPY in both of these cell types. The mechanisms underlying, and the physiological significance of these observations have not been defined.

In contrast, considerable progress has been made in characterizing the actions of NPY on anterior pituitary hormone secretion. As summarized in Table 1, this chapter has attempted to identify several common themes in the neuroendocrine regulatory actions of NPY. First, NPY systems from the arcuate nucleus

Table 1 Summary of the effects of NPY on secretion of anterior pituitary hormones

AP hormone	Effect of NPY	Site of action		Co-action with AgRP	Target for leptin signaling
		Hyp	AP		
PRL	Inhibitory	Yes	Yes	Yes	?
GH	Inhibitory	Yes	Yes	?	Yes
TSH	Inhibitory	Yes	No	Yes	Yes
ACTH	Stimulatory	Yes	No	Yes	Yes

?, No published reports addressing this issue; Hyp, hypothalamic; AP, anterior pituitary gland.

and/or brainstem directly innervate all of the hypothalamic releasing/inhibiting hormone systems governing secretion of anterior pituitary hormones, and NPY effects on PRL, GH, TSH and ACTH secretion clearly are mediated via alterations in the neurosecretion of their cognate hypophysiotropic hormonal systems. In addition, NPY effects on PRL secretion (and LH; see chapter by Kalra and Kalra, this volume) also involve actions at the pituitary so prominent that NPY might be considered to be a 'coreleasing hormone' with dopamine and GnRH, respectively.

Second, it is common for NPY to interact postjunctionally with other comessengers released from the same nerve terminals. In the case of OT and VP, and to some extent LH (see chapter by Kalra and Kalra, this volume), the cotransmitter appears to be norepinephrine. However, for the other anterior pituitary hormones discussed in this chapter, recent work suggests a co-action with AgRP, present as a comesssenger in most of the arcuate NPY neurons, for PRL, TSH and ACTH secretion. Details of specific receptor subtype, second messenger coupling and potential intracellular 'crosstalk' between NPY- and AgRP-activated signal transduction pathways generally remain to be worked out in these systems. From work reviewed in this chapter on PRL secretion (Sect. 4) and previously on LH secretion (Kalra and Crowley 1992), the concept of signal amplification by NPY has emerged. By this is meant intracellular mechanisms responsible for augmenting target cell responses. For example, as reviewed in Sect. 4, our work suggests that NPY adds to the inhibitory effect of dopamine on PRL secretion by selectively augmenting dopamine inhibition of the influx of extracellular Ca^{2+}.

A third common theme for NPY regulation of anterior pituitary secretion is what we have termed 'physiological/molecular plasticity' in the synthesis and release of NPY especially in the arcuate-based circuits. This refers to the targeting of arcuate NPY neurons by specific physiological factors involved in anterior pituitary regulation. For example, with respect to feedback regulatory mechanisms, hypothalamic NPY neurons are targeted by adrenal glucocorticoids (Sect. 5) and GH (Sect. 6). Similar studies on the HPT axis have not yet been reported. During lactation, there is a general upregulation of NPY expression in the arcuate nucleus, including the novel expression in TIDA neurons that normally do not contain NPY; these depend in some way on the suckling stimulus.

Finally, as summarized in Fig. 2, arcuate NPY neurons appear to mediate the changes in anterior pituitary secretion that occur in conditions of food restriction and negative energy balance, namely suppression of gonadotropins (reviewed in Smith and Grove 2002), PRL, GH, and TSH, and an increase in ACTH. Further, it is possible to pinpoint the reduction in circulating leptin that occurs upon food deprivation as the key signal regarding the status of energy stores to each of the anterior pituitary hormones, via NPY systems based in the arcuate nucleus. Thus, when considered together with the major orexigenic action of NPY, one may place the neuroendocrine actions of NPY within the overall

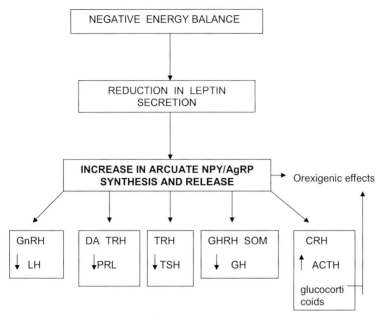

Fig. 2 Overall summary of NPY action to integrate behavioral and neuroendocrine responses to negative energy balance

framework that this system provides critical integration of the neuroendocrine and behavioral responses to a condition of negative energy balance.

References

Abe H, Engler D, Molitch ME, Bollinger-Gruber J, Reichlin S (1985) Vasoactive intestinal peptide is a physiological mediator of prolactin release in the rat. Endocrinology 116:1383–1390

Adams EF, Venetikou MS, Woods CA, Lacoumenta S, Burrin JM (1987) Neuropeptide Y directly inhibits growth hormone secretion by human pituitary somatotropic tumours. Acta Endocrinol 115:149–154

Ahima RS Prabakaran D, Mantzoros C, Qu D, Lowell B, Maratos-Flier E, Flier JS (1996) Role of leptin in the neuroendocrine response to fasting. Nature 382:250–252

Akabayashi A, Watanabe Y, Wahlestedt C, McEwen BS, Paez X, Leibowitz SF (1994) Hypothalamic neuropeptide Y, its gene expression and receptor activity: relation to circulating corticosterone in adrenalectomized rats. Brain Res 665:201–212

Albers RE, Ottenweller JE, Liou SY, Lumpkin MD, Anderson ER (1990) Neuropeptide Y in the hypothalamus: effect on corticosterone and single unit activity. Am J Physiol 258:R376–R382

Baker RA, Herkenham M (1995) Arcuate neurons that project to the hypothalamic paraventricular nucleus: neuropeptidergic identity and consequences of adrenalectomy on mRNA levels in the rat. J Comp Neurol 358:518–530

Barber MC, Clegg RH, Travers MT, Vernon RG (1997) Lipid metabolism in the lactating mammary gland. Biochim Biophys Acta 1347:101–126

Ben-Jonathan N (1983) Dopamine: a prolactin inhibiting hormone. Endocrinol Rev 6:564–589
Ben-Jonathan N, Hnasko R (2001) Dopamine as a prolactin (PRL) inhibitor. Endocrinol Rev 22:724–763
Bradbury MJ, Akana SF, Cascio CS, Levin N, Jacobson L, Dallman MF (1991) Regulation of basal ACTH secretion by corticosterone is mediated by both type I (MR) and type II (GR) receptors in rat brain. J Steroid Biochem Mol Biol 40:133–142
Brogan RS, Mitchell SE, Trayhurn P, Smith MS (1999) Suppression of leptin during lactation: Contribution of the suckling stimulus vs. milk production. Endocrinology 140:2621–2627
Brooks AN, Howe DC, Porter DW, Naylor AM (1994) Neuropeptide-Y stimulates pituitary-adrenal activity in fetal and adult sheep. J Neuroendocrinol 6:161–166
Campbell RE, ffrench-Mullen JMH, Cowley MA, Smith MS, Grove KL (2001) Hypothalamic circuitry of neuropeptide Y regulation of neuroendocrine function and food intake via the Y5 receptor subtype. Neuroendocrinology 74:106–119
Campbell CA, Kurcz M, Marshall S, Meites J (1977) Effects of starvation in rats on serum levels of follicle-stimulating hormone, luteinizing hormone, thyrotropin, growth hormone and prolactin; response to LH-releasing hormone and thyrotropin-releasing hormone. Endocrinology 100:580–587
Carro E, Seoane LM, Senaris R, Considine RV, Casanueva FF, Dieguez C (1998) Interaction between leptin and neuropeptide Y on in vivo growth hormone secretion. Neuroendocrinology 68:187–191
Casanueva FF, Dieguez C (1999) Neuroendocrine regulation and actions of leptin. Front Neuroendocrinol 20:317–363
Catzeflis C, Pierroz DD, Rohner-Jeanrenaud F, Rivier JE, Sizonenko PC, Aubert ML (1993) Neuropeptide Y administered chronically into the lateral ventricle profoundly inhibits both the gonadotropic and somatotropic axes in intact adult female rats. Endocrinology 132:224–234
Chabot J-C, Enjalbert A, Pelletier G, Dubois PM, Morel G (1988) Evidence for a direct action of neuropeptide Y in the rat pituitary gland. Neuroendocrinology 47:511–517
Chan YY, Steiner RA, Clifton DK (1996) Regulation of hypothalamic neuropeptide-Y neurons by growth hormone in the rat. Endocrinology 137:1319–1325
Chen P, Li C, Haskell-Luevano C, Cone RD, Smith MS (1999) Altered expression of agouti-related protein and its co-localization with neuropeptide Y in the arcuate nucleus of the hypothalamus during lactation. Endocrinology 140:2645–2650
Chronwall BM, DiMaggio DA, Massari VJ, Pickel VM, Ruggiero DA, O' Donohue TL. (1985) The anatomy of neuropeptide Y-containing neurons in rat brain. Neuroscience 15:1159–1181
Ciofi P, Crowley WR, Pillez A, Schmued LL, Tramu G, Mazzuca M (1993) Plasticity in expression of immunoreactivity for neuropeptide Y, enkephalins and neurotensin in the hypothalamic tubero-infundibular dopaminergic system during lactation in mice. J Neuroendocrinol 5:599–602
Ciofi P, Fallon JH, Croix, D, Polak JM, Tramu G (1991) Expression of neuropeptide Y precursor immunoreactivity in the hypothalamic dopaminergic tubero-infundibular system during lactation in rodents. Endocrinology 128:823–834
Conrad CD, McEwen BS (2000) Acute stress increases neuropeptide Y mRNA within the arcuate nucleus and hilus of the dentate gyrus. Mol Brain Res 79:102–109
Crowley WR (1988) Role of endogenous opioid neuropeptides in the physiological regulation of luteinizing hormone and prolactin secretion. In: Negro-Vilar A and Conn PM (eds) Peptide hormones: Effects and mechanisms of action III. CRC Press, Boca Raton, FL, pp. 79–116
Crowley WR (1999) Toward multifactorial regulation of anterior pituitary hormone secretion. News Physiol Sci 14:54–58

Crowley WR, Ramoz G (2002) Up-regulation of hypothalamic neuropeptide Y during lactation in rats: Mediation by hypo-insulinemia and hypo-leptinemia. Abstracts of the 32nd Meeting of the Society for Neuroscience, #134.3

Crowley WR, Ramoz G, Hurst B (2003) Evidence for involvement of neuropeptide Y and melanocortin systems in the hyperphagia of lactation in rats. Pharm Biochem Behav 74:417–424

Csiffary A, Gorcs TJ, Palkovits M (1990) Neuropeptide Y innervation of ACTH-immunoreactive neurons in the arcuate nucleus of rats: A correlated light and electron microscopic double immunolabeling study. Brain Res 506:215–222

Dallman MF, Akana SF, Levin N, Walker CD, Bradbury MJ, Suemaru S, Scribner KS. (1994) Corticosteroids and the control of function in the hypothalamo-pituitary-adrenal axis. Ann NY Acad Sci 746:22–31

Dallman MF, Akana SF, Srack AS, Hanson ES, Sebastian RJ (1995) The neural network that regulates energy balance is responsive to glucocorticoids and insulin and also regulates HPA axis responsivity at a site proximal to CRF neurons. Ann NY Acad Sci 771:730–742

Dallman MF, Strack AM, Akana SF, Bradbury MJ, Hanson ES, Scribner KA, Smith M (1993) Feast and famine: critical role of glucocorticoids with insulin in daily energy flow. Front Neuroendocrinol 4:303–347

Day TA, Jhamandas JH, Renaud LP (1985) Comparison between the actions of avian pancreatic polypeptide, neuropeptide Y and norepinephrine on the excitability of rat supraoptic vasopressin neurons. Neurosci Lett 62:181–185

De Quidt ME, Emson PC (1986) Distribution of neuropeptide Y-like immunoreactivity in the rat central nervous system. II. Immunohistochemical analysis. Neuroscience 15:1149–1157

Dean RG, White BD (1990) Neuropeptide Y expression in rat brain: effects of adrenalectomy. Neurosci Lett 114:339–344

Dhillo WS, Small CJ, Seal LJ, Kim MS, Stanley SA, Murphy KG, Ghatei MA, Bloom SR (2002) The hypothalamic melanocortin system stimulates the hypothalamo-pituitary-adrenal axis *in vitro* and *in vivo* in male rats. Neuroendocrinology 75:209–216

Diano S, Naftolin F, Goglia F, Horvath TL (1998) Segregation of the intra-and extrahypothalamic neuropeptide Y and catecholaminergic inputs on paraventricular neurons, including those producing thyrotropin-releasing hormone. Regul Pept 75/76:117–126

Engler D, Redei E, Kola I (1999) The corticotropin-release inhibitory factor hypothesis: A review of the evidence for the existence of inhibitory as well as stimulatory hypophysiotropic regulation of adrenocorticotropin secretion and biosynthesis. Endocrinol Rev 20:460–500

Everitt BJ, Hokfelt T, Terenius L, Tatemoto K, Mutt V, Goldstein M (1984) Differential coexistence of neuropeptide Y (NPY)-like immunoreactivity with catecholamines in the central nervous system of the rat. Neuroscience 11:443–462

Fekete C, Kelly J, Mihaly E, Sarkar S, Rand WM, Legradi G, Emerson CH, Lechan RM (2001) Neuropeptide Y has a central inhibitory action on the hypothalamic-pituitary-thyroid axis. Endocrinology 142:2606–2613

Fekete C, Legradi G, Mihaly E, Huang QH, Tatro JH, Rand WM, Emerson CH, Lechan RM (2000) α-Melanocyte-stimulating hormone is contained in nerve terminals innervating thyrotropin-releasing hormone-synthesizing neurons in the hypothalamic paraventricular nucleus and prevents fasting-induced suppression of prothyrotropin-releasing hormone gene expression. J Neurosci 20:1550–1558

Fekete C, Sarkar S, Rand WM, Harney JW, Emerson CH, Bianco AC, Lechan RM (2002) Agouti-related protein (AGRP) has a central inhibitory action on the hypothalamic-pituitary-thyroid (HPT) axis: comparison between the effect of AGRP and neuropeptide Y on energy homeostasis and the HPT axis. Endocrinology 143:3846–3853

Fleming AS (1976) Control of food intake in the lactating rat: role of suckling and hormones. Physiol Behav 17:841–848

Frohman LA, Downs TR, Chomczynski P (1992) Regulation of growth hormone secretion. Front Neuroendocrinol 13:344–405

Fukuda H, Ohshima K, Mori M, Kobayashi I, Greer MA (1980) Sequential changes in the pituitary-thyroid axis during pregnancy and lactation in the rat. Endocrinology 107:1711–1716

Fuxe K, Agnati LF, Harfstrand A, Eneroth P, Cintra A, Tinner B, Merlo Pich E, Aronsson M, Bunnemann B, Lang R, Ganten D (1989) Studies on the neurochemical mechanisms underlying the neuroendocrine actions of neuropeptide Y. In: Mutt V, Hokfelt T, Fuxe K, Lundberg JM (eds) Neuropeptide Y (Karolinska Institute Nobel Conference Series) Raven Press, New York, pp 115–136

Gershengorn MC (1985) Thyrotropin-releasing hormone action: Mechanism of calcium-mediated stimulation of prolactin secretion. Rec Progr Horm Res 41:607–646

Gruaz NM, Pierroz DD, Rohner-Jeanrenaud F, Sizonenko PC, Aubert ML (1993) Evidence that neuropeptide Y could represent a neuroendocrine inhibitor of sexual maturation in unfavorable metabolic conditions in the rat. Endocrinology 133:1891–1894

Haas DA, George SR (1989) Neuropeptide Y-induced effects on hypothalamic corticotropin-releasing factor content and release are dependent on noradrenergic/adrenergic neurotransmission. Brain Res 498:333–338

Hahn TM, Breninger JF, Baskin DG, Schwartz MW (1998) Coexpression of Agrp and NPY in fasting-activated hypothalamic neurons. Nature Neurosci 1:271–272

Hanson ES, Dallman MF (1995) Neuropeptide Y (NPY) may integrate responses of hypothalamic feeding systems and the hypothalamic-pituitary adrenal axis. J Neuroendocrinol 7:273–279

Hanson ES, Levin N, Dallman MF (1997) Elevated corticosterone is not required for the rapid induction of neuropeptide Y gene expression by an overnight fast. Endocrinology 138:1041–1047

Harfstrand A, Cintra A, Fuxe K, Aronsson M, Wikstrom A-C, Okret S, Gustafsson J-A, Agnati LF (1989) Regional differences in glucocorticoid receptor immunoreactivity among neuropeptide Y immunoreactive neurons of the rat brain. Acta Physiol Scand 135:3–9

Harfstrand A, Eneroth P, Agnati L, Fuxe K (1987a) Further studies on the effects of central administration of neuropeptide Y on neuroendocrine function in the male rat: Relationship to hypothalamic catecholamines. Regul Pept 17:167–179

Harfstrand A, Fuxe K, Terenius L, Kalia M (1987b) Neuropeptide Y-immunoreactive perikarya and nerve terminals in the rat medulla oblongata: relationship to cytoarchitecture and catecholaminergic cell groups. J Comp Neurol 260:20–35

Haskell-Luevano C, Monck EK (2001) Agouti-related protein functions as an inverse agonist at a constitutively active brain melanocortin-4 receptor. Regul Pept 99:1–7

Hastings JA, McClure-Sharp JM, Morris MJ (2001) NPY Y1 receptors exert opposite effects on corticotropin releasing factor and noradrenaline overflow from the rat hypothalamus in vitro. Brain Res 890:32–37

Higuchi T, Honda K, Fukuoka T, Negoro H, Hosono Y, Nishida E (1983) Pulsatile secretion of prolactin and oxytocin during nursing in the lactating rat. Endocrinol Japon 30:353–359

Hinuma S, Habata Y, Fujii R, Kawamata Y, Hosoya M, Fukusumi S, Kitada C, Masuo Y, Asano T, Matsumoto H, Sekiguchi M, Kurokawa T, Nishimura O, Onda H, Fujino M (1998) A prolactin-releasing peptide in the brain. Nature 393:272–276

Hisano S, Kagotani Y, Tsuruo Y, Daikoku S, Chihara K, Whitnall MH (1988) Localization of glucocorticoid receptor in neuropeptide Y-containing neurons in the arcuate nucleus of the rat hypothalamus. Neurosci Lett 95:13–18

Hooi SC, Richardson GS, McDonald JK, Allen JM, Martin JB, Koenig JI (1989) Neuropeptide Y (NPY) and vasopressin (AVP) in the hypothalamo-neurohypophysial axis of salt-loaded or Brattleboro rats. Brain Res 486:214–220

Horvath TL, Bechmann I, Naftolin F, Kalra SP, Leranth C (1997) Heterogeneity in the neuropeptide Y-containing neurons of the rat arcuate nucleus: GABAergic and non-GABAergic subpopulations. Brain Res 756:283–286

Horvath TL, Naftolin F, Kalra SP, Leranth C (1992) Neuropeptide-Y innervation of ß-endorphin-containing cells in the rat mediobasal hypothalamus: A light and electron microscopic double immunostaining analysis. Endocrinology 131:2461–2467

Hsueh Y-C, Cheng S-M, Pan J-T (2002) Fasting stimulates tuberoinfundibular dopaminergic neuronal activity and inhibits prolactin secretion in oestrogen-primed ovariectomized rats: Involvement of orexin A and neuropeptide Y. J Neuroendocrinol 14:745–752

Hull KL, Harvey S (2001) Growth hormone: roles in female reproduction. J Endocrinol 168:1–23

Inui A, Inoue T, Nakajima M, Okita M, Sakatani N, Okimura C, Chihara K, Baba S. (1990) Effect of neuropeptide Y in the control of adrenocorticotropic secretion in the dog. Brain Res 504:211–215

Iwai C, Ochai H, Nakai Y (1989) Electron-microscopic immunohistochemistry of neuropeptide Y immunoreactive innervation of vasopressin neurons in the paraventricular nucleus of rat hypothalamus. Acta Anat 136:279–284

Kagotani Y, Tsuruo Y, Hisano S, Daikoku S, Chihara K (1989a) Axons containing neuropeptide Y innervate arginine vasopressin-containing neurons in the rat paraventricular nucleus. Dual electron microscopic immunolabeling. Histochemistry 91:273–281

Kagotani Y, Tsuruo Y, Hisano S, Daikoku S, Chihara K (1989b) Synaptic regulation of paraventricular arginine vasopressin-containing neurons by neuropeptide Y-containing monoaminergic neurons in rats. Electron microscopic triple labeling. Cell Tiss Res 257:269–278

Kalra SP, Crowley WR (1992) Neuropeptide Y: A novel neuroendocrine peptide in the control of pituitary hormone secretion, and its relation to luteinizing hormone. Front Neuroendocrinol 13:1–46

Kalra SP, Dube MG, Pu S, Xu B, Horvath TL, Kalra PS (1999) Interacting appetite-regulating pathways in the hypothalamic regulation of body weight. Endocrinol Rev 20:68–100

Kamegai J, Minami S, Sugihara H, Hasegawa O, Higuchi H, Wakabayashi I (1996) Growth hormone receptor gene is expressed in neuropeptide Y neurons in hypothalamic arcuate nucleus of rats. Endocrinology 137:2109–2112

Kamegai J, Minami S, Sugihara H, Higuchi H, Wakabayashi I (1994) Growth hormone induces expression of the c-fos gene on hypothalamic neuropeptide-Y and somatostatin in hypophysectomized rats. Endocrinology 135:2765–2771

Kapoor JR, Sladek CD (2001) Substance P and NPY differentially potentiate ATP and adrenergic stimulated vasopressin and oxytocin release. Am J Physiol 280:R69–R78

Kawano H, Daikoku S (1988) Somatostatin-containing neuron systems in the rat hypothalamus: retrograde tracing and immunohistochemical studies. J Comp Neurol 271:293–299

Khanna S, Sibbald JR, Day TA.(1983) Neuropeptide Y modulation of A1 noradrenergic neuron input to supraoptic vasopressin cells. Neurosci Lett 161:60–64

Kim MS, Small CJ, Russell SH, Morgan DG, Abbott CR, al Ahmed SH, Hay DL, Ghatei MA, Smith DM, Bloom SR (2002) Effects of melanocortin receptor ligands on thyrotropin-releasing hormone release: Evidence for the differential role of melanocortin 3 and 4 receptors. J Neuroendocrinol 14:276–282

Kim MS, Small CJ, Stanley SA, Morgan DG, Seal LJ, Kong WM, Edwards CM, Abusana S, Sunter D, Ghatei MA, Bloom SR (2000) The central melanocortin system affects the hypothalamo-pituitary-thyroid axis and may mediate the effect of leptin. J Clin Invest 105:1005–1011

Korbonits M, Little JA, Forsling ML, Tringali G, Costa A, Navarra P, Trainer PJ, Grossman AB (1999) The effect of growth hormone secretagogues and neuropeptide Y on hypothalamic hormone release from acute rat hypothalamic explants. J Neuroendocrinol 11:521–528

Korner J, Savontaus E, Chua SC, Leibel RL, Wardlaw SL (2001) Leptin regulation of *Agrp* and *Npy* mRNA in the rat hypothalamus. J Neuroendocrinol 13:959–966

Krukoff TL, MacTavish D, Jhamandas JH (1999) Effects of restraint stress and spontaneous hypertension on neuropeptide Y neurons in the brainstem and arcuate nucleus. J Neuroendocrinol 11:715–719

Larsen PJ, Jessop DS, Chowdrey HS, Lightman SL (1992a) Neuropeptide Y messenger ribonucleic acid in the magnocellular hypothalamo-neurohypophysial system of the rat is increased during osmotic stimulation. Neurosci Lett 138:23–26

Larsen PJ, Jessop DS, Chowdrey HS, Lightman SL, Mikkelsen JD (1994a) Chronic administration of glucocorticoids directly up regulates prepro-neuropeptide Y and Y1-receptor mRNA levels in the arcuate nucleus of the rat. J Neuroendocrinol 6:153–159

Larsen PJ, Jukes KE, Chowdrey HS, Lightman SL, Jessop DS (1994b) Neuropeptide-Y potentiates the secretion of vasopressin from the neurointermediate lobe of the rat pituitary gland. Endocrinology 134:1635–1639

Larsen PJ, Mikkelsen JD, Jessop DS, Lightman SL, Chowdrey HS (1993) Neuropeptide Y mRNA and immunoreactivity in hypothalamic neuroendocrine neurons: Effects of adrenalectomy and chronic osmotic stimulation. J Neurosci 13:1138–1147

Larsen PJ, Sheikh SP, Mikkelsen JD (1992b) Osmotic regulation of neuropeptide Y and its binding sites in the magnocellular hypothalamo-neurohypophysial pathway. Brain Res 573:181–189

Legradi G, Lechan RM (1998) The arcuate nucleus is the major source for neuropeptide Y-innervation of thyrotropin-releasing hormone neurons in the hypothalamic paraventricular nucleus. Endocrinology 139:3262–3270

Leibowitz, SF, Sladek C, Spencer L, Tempel D (1988) Neuropeptide Y, epinephrine and norepinephrine in the paraventricular nucleus: Stimulation of feeding and the release of corticosterone, vasopressin and glucose. Brain Res Bull 21:905–912

Leng G, Brown CH, Russell JA. (1999) Physiological pathways regulating the activity of magnocellular neurosecretory cells. Progr Neurobiol 57:625–655

Li C, Chen P, Smith MS (1999a) The acute suckling stimulus induces expression of neuropeptide Y (NPY) in cells in the dorsomedial hypothalamus and increases NPY expression in the arcuate nucleus. Endocrinology 140:1645–1652

Li C, Chen P, Smith MS (1999b) Neuropeptide Y and tuberoinfundibular dopamine activities are altered during lactation: role of prolactin. Endocrinology 140:118–123

Li, C, Chen P, Smith MS (2000) Corticotropin releasing hormone neurons in the paraventricular nucleus are direct targets for neuropeptide Y neurons in the arcuate nucleus: an anterograde tracing study. Brain Res 854:122–129

Liao N, Bulant M, Nicolas P, Vaudry H, Pelletier G (1991) Anatomical interactions of pro-opiomelanocortin (POMC)-related peptides, neuropeptide Y (NPY) and dopamine beta-hydroxylase (D beta H) fibers and thyrotropin-releasing hormone (TRH) neurons in the paraventricular nucleus of rat hypothalamus. Neuropeptides 18:63–67

Liposits Z, Sievers L. Paull WK (1988) Neuropeptide-Y and ACTH-immunoreactive innervation of corticotropin releasing factor (CRF)-synthesizing neurons in the hypothalamus of the rat. An immunohistochemical analysis at the light and electron microscopic levels. Histochemistry 88:227–234

Liu JP, Clarke IJ, Funder JW, Engler D (1994) Studies on the secretion of corticotropin-releasing factor and arginine vasopressin into the hypophysial-portal circulation of conscious sheep. II. The central noradrenergic and neuropeptide Y pathways cause immediate and prolonged hypothalamic-pituitary-adrenal activation. Potential involvement in the pseudo-Cushing's syndrome of endogenous depression and anorexia nervosa. J Clin Invest 93:1439–1450

Madon RJ, Ensor DM, Flint DJ (1990) Hypoinsulinemia in the lactating rat is caused by a decreased glycaemic stimulus to the pancreas. J Endocrinol 125:81–88

Makino S, Baker RA, Smith MA, Gold PW (2000) Differential regulation of neuropeptide Y mRNA expression in arcuate nucleus and locus coeruleus by stress and antidepressants. J Neuroendocrinol 12:387–395

Malabu UH, Kilpatrick A, Ware M, Vernon RG, Williams G (1994) Increased neuropeptide Y concentrations in specific hypothalamic regions of lactating rats: Possible relationship to hyperphagia and adaptive changes in energy balance. Peptides 15:83–87

Martinez de la Escalera G, Weiner RI (1992) Dissociation of dopamine from its receptor as a signal in the pleiotropic hypothalamic regulation of prolactin secretion. Endocr Rev 13:241–255

McDonald JK, Koenig JI, Gibbs DM, Collins P, Noe BD (1989) High concentrations of neuropeptide Y in pituitary portal blood of rats. Neuroendocrinology 46:538–541

McDonald JK, Lumpkin MD, Samson WK, McCann SM (1985) Neuropeptide Y affects secretion of luteinizing hormone and growth hormone in ovariectomized rats. Proc Natl Acad Sci USA 82:561–564

Mihaly E, Fekete C, Tatro JB, Liposits Z, Stopa EG, Lechan RM (2000) Hypophysiotropic thyrotropin-releasing hormone-synthesizing neurons in human hypothalamus are innervated by neuropeptide Y, agouti-related protein, and alpha-melanocyte-stimulating hormone. J Clin Endocrinol Metab 85:2596–2603

Misaki N, Higuchi H, Yamagata K, Miki N (1992) Identification of glucocorticoid responsive elements (GREs) at far upstream of rat NPY gene. Neurochem Intl 21185–21189

Moore KE (1987) Interactions between prolactin and dopaminergic neurons. Biol Reprod 36:47–58

Morris BJ. (1989) Neuronal localization of neuropeptide Y gene expression in rat brain. J Comp Neurol 290:358–368

Munday MH, Williamson DH (1983) Diurnal variations in food intake and in lipogenesis in mammary gland and liver of lactating rats. Biochem J 214:183–187

Nagy G, Kacsoh B, Halasz B (1986a) Episodic prolactin and growth hormone secretion not related to the actual suckling activity in lactating rats. J Endocrinol 111:137–142

Nagy G, Kacsoh B, Kanyicska B, Toth BE, Korausz E (1986b) Separation and suckling-induced changes in serum growth hormone levels of lactating rats. Endocrinol Exp 20:217–222

Nijenhuis WAJ, Oosterom J, Adan RAH (2002) AgRP (83–132) acts as an inverse agonist on the human melanocortin-4 receptor. Mol Endocrinol 15:164–171

O'Donohue TL, Chronwall BM, Pruss RM, Mezey E, Kiss JZ, Eiden LE, Massari VJ, Tessel RE, Pickel VM, DiMaggio DA, Hotchkiss AJ, Crowley WR, Zukowska-Grojec Z. (1985) Neuropeptide Y and peptide YY neuronal and endocrine systems. Peptides 6:755–768

Oberkotter LV, Rasmussen KM (1992) Changes in plasma thyroid hormone concentrations in chronically food-restricted female rats and their offspring during suckling. J Nutr 122:435–441

Pape JR, Tramu G (1996) Suckling-induced changes in neuropeptide Y and proopiomelanocortin gene expression in the arcuate nucleus of the rat: Evaluation of a putative intervention of prolactin. Neuroendocrinology 63:540–549

Parker MS, Wang JJ, Fournier A, Parker SL (2000) Upregulation of pancreatic polypeptide-sensitive neuropeptide Y (NPY) receptors in estrogen-induced hypertrophy of the anterior pituitary gland in the Fischer-344 rat. Mol Cell Endocrinol 264:239–249

Parker SL, Crowley WR (1993) Central stimulation of oxytocin release in the lactating rat: Interaction of neuropeptide Y with α-1 adrenergic mechanisms. Endocrinology 132:658–666

Pelletier G (1990) Ultrastructural localization of neuropeptide Y in the hypothalamus. Ann NY Acad Sci 611:232–246

Pelletier G, Tong Y (1992) Lactation but not prolactin increases the levels of pre-proNPY messenger RNA in the rat arcuate nucleus. Mol Cell Neurosci 3:286–290

Pickavance L, Tadayyon M, Williams G, Vernon RG (1998) Lactation suppresses diurnal rhythm of serum leptin. Biochem Biophys Res Commun 248:196–199

Pickavance L., Dryden S., Hopkins D, Bing C, Frankish H, Wang Q, Vernon RG, Williams G (1996) Relationship between hypothalamic neuropeptide Y and food intake in the lactating rat. Peptides 17:577–582

Pierroz DD, Catzeflis C, Aebi AC, Rivier JE, Aubert ML (1996) Chronic administration of neuropeptide Y into the lateral ventricle inhibits both the pituitary-testicular axis and growth hormone and insulin-like growth factor I secretion in intact adult male rats. Endocrinology 137:3–12

Pierroz DD, Gruaz NM, d'Alieves V, Aubert ML (1995) Chronic administration of neuropeptide Y into the lateral ventricle starting at 30 days of life delays sexual maturation in the female rat. Neuroendocrinology 61:293–300

Ponsalle P, Srivastava LS, Uht RM, White JD (1997) Glucocorticoids are required for food-deprivation-induced increases in hypothalamic neuropeptide-Y expression. J Neuroendocrinol 4:585–591

Pu S, Jain MR, Horvath TL, Diano S, Kalra PS, Kalra SP (1999) Interactions between neuropeptide Y and γ-aminobutyric acid in stimulation of feeding; a morphological and pharmacological analysis. Endocrinology 140:933–940

Rees LH, Besser GM, Grossman A (1989) Neuropeptide-Y stimulates CRF-41 release from rat hypothalami in vitro. Brain Res 502:167–170

Rettori V, Milenkovic L, Aguila MC, McCann SM (1990) Physiologically significant effect of neuropeptide Y to suppress growth hormone release by stimulating somatostatin discharge. Endocrinology 126:2296–2301

Sahu A, Kalra SP, Crowley WR, Kalra PS (1988) Evidence that NPY-containing neurons in the brainstem project into selected hypothalamic nuclei: implication in feeding behavior. Brain Res 457:376–378

Sainsbury A, Cusin I, Rohner-Jeanrenaud F, Jeanrenaud B (1997a) Adrenalectomy prevents the obesity syndrome produced by chronic central neuropeptide Y infusion in normal rats. Diabetes 46:209–214

Sainsbury A, Rohner-Jeanrenaud F, Cusin I, Zakrzewska KE, Halban PA, Gaillard RC, Jeanrenaud B (1997b) Chronic central NPY infusion in normal rats: status of the hypothalamo-pituitary-adrenal axis, and vagal mediation of hyperinsulinemia. Diabetologia 40:1269–1277

Sainsbury A, Rohner-Jeanrenaud F, Grouzmann E, Jeanrenaud B (1996) Acute intracerebroventricular administration of neuropeptide Y stimulates corticosterone output and feeding but not insulin output in normal rats. Neuroendocrinology 63:318–326

Sands SA, Le Mon D, Chronwall BM (2001) Lactation and salt loading similarly alter neuropeptide Y, but differentially alter somatostatin, in separate sets of rat neural lobe axons. Peptides 18:1045–1050

Saunders A, Terry LC, Audet I, Brazeau P, Martin JB (1976) Dynamic studies of growth hormone and prolactin secretion in the female rat. Neuroendocrinology 21:193–203

Sawchenko PE, Swanson LW, Grzanna R, Howe PR, Bloom SR, Polak JM. (1985) Colocalization of neuropeptide Y immunoreactivity in brainstem catecholaminergic neurons that project to the paraventricular nucleus of the hypothalamus. J Comp Neurol 241:138–153

Schioth HB, Kakiyaki Y, Kohsaka A, Suda T, Watanobe H (2001) Agouti-related peptide prevents steroid-induced luteinizing hormone and prolactin surges in female rats. Neuroreport 12:687–690

Schwartz, MW, Woods SC, Porte D, Seely RJ, Baskin DG (2000) Central nervous system control of food intake. Nature 404:661–671

Sheikh SP, Feldthus N, Orkild H, Goke R, McGregor GP, Turner D, Moler M, Stuenkel EL (1998) Neuropeptide Y2 receptors on nerve endings from the rat neurohypophysis regulate vasopressin and oxytocin release. Neuroscience 82:107–115

Sibbald JR, Wilson BK, Day TA (1989) Neuropeptide Y potentiates excitation of supraoptic neurosecretory cells by noradrenaline. Brain Res 499:164–168

Sladek CD, Kapoor JR (2001) Neurotransmitter/neuropeptide interactions in the regulation of neurohypophysial hormone release. Exp Neurol 171:200–209

Small CJ, Morgan DG, Meeran K, Heath MM, Gunn I, Edwards CM, Gardiner J, Taylor GM, Hurley JD, Rossi M, Goldstone AP, O'Shea D, Smith DM, Ghatei MA, Bloom SR (1997) Peptide analogue studies of hypothalamic neuropeptide Y receptor mediating pituitary adrenocorticotrophic hormone release. Proc Natl Acad Sci USA 94:11686–11691

Small TJ, Todd JF, Ghatei M, Smith DM, Bloom SR (1998) Neuropeptide Y (NPY) actions on the corticotroph cell of the anterior pituitary gland are not mediated by a direct effect. Regul Pept 75/76:301–307

Smith MS (1993) Lactation alters neuropeptide-Y and proopiomelanocortin gene expression in the arcuate nucleus of lactating rats. Endocrinology 133:1258–1265

Smith MS, Grove KL (2002) Integration of reproductive function and energy balance: lactation as a model. Front Neuroendocrinol 23:225–226

Strack AS, Sebastian RJ, Schwartz MW, Dallman MF (1995) Glucocorticoids and insulin: reciprocal signals for energy balance. Am J Physiol 268:R142–R149

Suda T, Tozawa F, Iwai I, Sato Y, Sumitomo T, Nakano Y, Yamada M, Demura H. (1993) Neuropeptide Y increases the corticotropin-releasing factor messenger ribonucleic acid level in the rat hypothalamus. Mol Brain Res 18:311–315

Sutton S, Toyama T, Otto S, Plotsky P (1988) Evidence that neuropeptide Y (NPY) released into pituitary portal blood participates in priming gonadotrops to the effects of gonadotropin releasing hormone (GnRH). Endocrinology 123:1208–1210

Suzuki N, Okada K, Minami S, Wakabayashi I (1996) Inhibitory effect of neuropeptide Y on growth hormone secretion in rats is mediated by both Y1- and Y2-receptor subtypes and abolished after anterolateral deafferentation of the medial basal hypothalamus. Regul Pept 65:145–151

Tannenbaum GS, Epelbaum J, Colle E, Brazeau P, Martin JB (1978) Antiserum to somatostatin reverses starvation-induced inhibition of growth hormone but not insulin secretion. Endocrinology 102:1909–1914

Tempel DL, Leibowitz SF (1993) Glucocorticoid receptors in PVN: Interactions with NE, NPY and Gal in relation to feeding. Am J Physiol 265:E794–E800

Toni R, Jackson IM, Lechan RM (1990) Neuropeptide-Y-immunoreactive innervation of thyrotropin-releasing hormone-synthesizing neurons in the rat hypothalamic paraventricular nucleus. Endocrinology 126:2444–2453

Tsagarakis S, Rees LH, Besser GM, Grossman A (1989) Neuropeptide-Y stimulates CRF-41 release from rat hypothalami in vitro. Brain Res 502:167–170

Vallejo M, Carter DA, Dey-Guerra FJ, Emson PC, Lightman SL (1987) Neonatal administration of a specific neuropeptide Y antiserum alters the vasopressin response to haemorrhage and the hypothalamic content of noradrenaline in rats. Neuroendocrinology 45:507–509

Vernon RG (1989) Endocrine control of metabolic adaptation during lactation. Proc Nutr Soc 48:23–32

Vuagnat BA, Pierroz DD, Lalaoui M, Englaro P, Pralong FP, Blum WF, Aubert ML (1998) Evidence for a leptin-neuropeptide Y axis for the regulation of growth hormone secretion in the rat. Neuroendocrinology 67:291–300

Wade GN, Schneider JE (1992) Metabolic fuels and reproduction in female mammals. Neurosci Biobehav Rev 16:235–272

Wahlestedt C, Skagerberg G, Ekman R, Heilig M, Sundler F, Hakanson R (1987) Neuropeptide Y (NPY) in the area of the hypothalamic paraventricular nucleus activates the pituitary-adrenal axis in the rat. Brain Res 41:33–38

Walker CD, Lightman SL, Steele MK, Dallman MF (1992) Suckling is a persistent stimulus to the adrenocortical system of the rat. Endocrinology 130:115–125

Wang J,. Ciofi P, Crowley WR (1996) Neuropeptide Y suppresses prolactin secretion from rat anterior pituitary cells: Evidence for interactions with dopamine through inhibitory coupling to calcium entry. Endocrinology 137:587–594

Watanabe Y, Akabayashi A, McEwen BS (1995) Adrenal steroid regulation of neuropeptide Y mRNA: differences between dentate hilus and locus coeruleus and arcuate nucleus. Mol Brain Res 28:135–140

Watanabe H, Schioth HB, Suda T (2001) Stimulation of prolactin secretion by chronic, but not acute, administration of leptin in the rat. Brain Res 887:426–431

White BD, Dean RG, Edwards GL, Martin RJ. (1994) Type II corticosteroid receptor stimulation increases NPY gene expression in basomedial hypothalamus of rats. Am J Physiol 266:R1523–R1529

White BD, Dean RG, Matin RJ (1990) Adrenalectomy decreases neuropeptide Y mRNA levels in the arcuate nucleus. Brain Res Bull 25:711–715

Wilding JPH, Gilbey SG, Lambert PD, Ghatei MA, Bloom SR (1993) Increases in neuropeptide Y content and gene expression in the hypothalamus of rats treated with dexamethasone are prevented by insulin. Neuroendocrinology 57:581–587

Williamson DH (1980) Integration of metabolism in tissues of the lactating rat. FEBS Lett 117 (Suppl):K93–K105

Willoughby JO, Blessing WW (1987) Neuropeptide Y injected into the supraoptic nucleus causes secretion of vasopressin in the unanesthetized rat. Neurosci Lett 75:17–22

NPY: A Novel On/Off Switch for Control of Appetite and Reproduction

S. P. Kalra · P. S. Kalra

Departments of Neuroscience and Physiology and Functional Genomics,
College of Medicine, University of Florida McKnight Brain Institute, 100 S. Newell Drive,
Gainesville, FL 32610-0244, USA
e-mail: skalra@ufbi.ufl.edu

1	Introduction	222
2	NPY, a Physiological Appetite Transducer	223
2.1	NPY Synthesis, Release and Appetite	223
2.2	NPY Receptors and Appetite	224
2.3	NPY Network and Appetite	226
2.3.1	Distinct Roles of Two Subpopulations of NPY Producing Neurons	226
2.4	Interaction of NPY with Coproduced Neurotransmitters in Induction of Appetite	227
2.4.1	NPY Interaction with Other Components of Appetite Regulating Networks	227
2.5	High and Low Abundance of NPYergic Signaling Evokes Hyperphagia and Obesity: A New Insight	228
2.6	Regulation of NPY Secretion and Appetite by Afferent Neural and Hormonal Pathways	229
3	NPY, a Physiological Regulator of Reproduction	231
3.1	A Unique Bimodal Participation by NPY in Neuroendocrine Control of Reproduction	232
3.1.1	Excitatory Afferent Signal for Ultradian and Cyclic GnRH Discharge	232
3.1.2	Excitatory Role of the Two Subpopulations of NPY Expressing Neurons	233
3.1.3	Interaction of the NPY Network with Other Components of the Hypothalamic GnRH Regulating Circuitry	234
3.2	Regulation of NPY Secretion by Afferent Neural and Hormonal Signals	235
3.3	NPY, an Inhibitory Afferent Signal	236
4	NPY and Sex Behavior	237
5	Is NPY an On/Off Switch for Control of Appetite and Reproduction?	237
	References	239

Abstract The information collated here supports the concept that neuropeptide Y (NPY) is an essential messenger molecule in integration of the innate appetitive drive and the urge to reproduce. There is a spatial and temporal specificity of NPY action in the hypothalamus wherein two distinct NPY pathways receive and transduce external and internal environmental information to sustain nutritional homeostasis and reproductive functions. A distinct subset of NPY expressing neurons serves as the on/off switch to stimulate appetite but inhibit reproduction when unfavorable nutritional environments and hormonal, neural, metabolic and genetic disturbances persist. This capability of NPYergic signaling to reciprocally modify the two instinctual urges with the aid of distinct NPY receptor subtypes and intricate connectivities with varied peptidergic pathways, and with the cooperation of chemically diverse coexpressed neurotransmitters, has uncovered multiple loci vulnerable to therapeutic interventions to treat eating disorders and obesity without adversely affecting fertility.

Keywords Hypothalamus · Neural circuitry · Neuropeptides · Arcuate nucleus-paraventricular nucleus · Medial preoptic area-median eminence

1
Introduction

The search for energy source and conducive environments to reproduce are the two most significant intrinsic driving forces for the survival of the species. These innate drives are directed by the dynamics of ever-changing evolutionary pressures. Only in the 1980s was it appreciated that neural processes that receive and transduce external environmental signals and appropriately propagate neurogenic impulses to drive the organism toward the source of food and prepare it for the prolonged, heavy energy cost of reproduction, lactation and raising of progeny are neurochemically linked in the hypothalamus (Kalra et al. 1988a, 1989). The advent of this insight can be traced to the early clinical observations and experimental findings that established the hypothalamus as an integrator of these life sustaining physiological processes, and later, to the discovery of a spectrum of neurotransmitters and neuromodulators that relay messages within the hypothalamic integrative network (Bray 1984, 1998; Kalra 1993; Kalra and Kalra 1983). The realization that distinct transmitter(s) encode integration among various hypothalamic circuitries triggered research on two parallel fronts, the chemical characterization of messenger molecules and a deeper understanding of their pleiotropic actions at cellular and molecular levels. The neurobiology of neuropeptide Y (NPY) in integration of energy homeostasis and various neuroendocrine systems is one such example. The focus of this article is to collate information accumulated since 1984 when the appetite and pituitary gonadotropin release stimulating effects of NPY were first reported (Clark et al. 1984; Kalra and Crowley 1984a,b). The thesis advanced shortly thereafter that this molecule, elaborated by a cluster of neurons in the hypothalamus, can simultaneously evoke appetitive drive while suppressing the sexual behavior drive directly, and reproduction indirectly by regulating efflux of hypothalamic gonadotropin releasing hormone (GnRH) to stimulate gonadotropin release, ignited intense multidisciplinary research that uncovered the chemical nature of the crosstalk between these instinctual urges (Clark et al.1984, 1985; Crowley

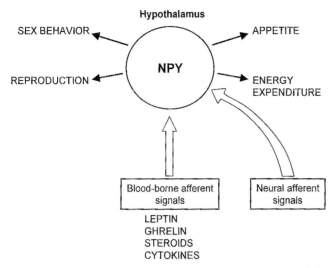

Fig. 1 Neuropeptide Y plays a pivotal role in the hypothalamus in regulating two neuroendocrine functions, reproduction and sexual behavior (*left*) and appetite and energy expenditure to maintain body energy homeostasis (*right*). In addition to an interconnected circuitry in the hypothalamus (for details, see text), distinct afferent pathways play an important role in regulating NPY release at target sites. Afferent messages from the periphery are relayed by multisynaptic neural pathways through the spinal cord, brainstem, and lateral hypothalamus and converge onto NPY neurons in the basal hypothalamus. NPY neurons also receive hormonal signals that cross the blood–brain barrier. (With permission from Kalra and Kalra 2003)

and Kalra 1988; Kalra et al. 1988a, 1989). The outcome of these investigations is presented here in four sections dealing with the pivotal roles of hypothalamic NPY in regulating appetite within the hypothalamic circuitry that regulates energy homeostasis, pituitary gonadotropin secretion within the context of neuroendocrine control, and sexual behavior within the context of neural and endocrine-based motivational behavior (Fig. 1). Finally, we present a heuristic model that integrates interactions between the circuitries regulating appetite and reproduction at the neuroanatomical, neurochemical and molecular biology levels under the influence of varied external forces.

2
NPY, a Physiological Appetite Transducer

2.1
NPY Synthesis, Release and Appetite

Although the role of the ventromedial hypothalamus–lateral hypothalamus (VMH–LtH) as the neuroanatomical basis of appetitive behavior was extensively evaluated for over 50 years (Bray 1984, 1998; Kalra 1997; Kalra et al. 1999), the possibility that a neurochemical signal originating outside the VMH–LtH complex could evoke behavior was not envisioned until our findings that intracerebroventricular (ICV) administration of NPY and human pancreatic peptide, members of the pancreatic polypeptide family, stimulated feeding in sated rats (Clark et al. 1984, 1987b). Subsequent investigations showed that NPY administration evoked episodic feeding and either repeated daily injections or continuous infusion evoked relentless hyperphagia, abnormal weight gain and obesity (Catzeflis et al. 1993; Kalra et al. 1988b; Kalra and Kalra 1990, 1996). That these animals did not develop tolerance to NPY was the earliest indication that NPY may be a physiologically relevant orexigenic signal. NPY was established as the first known physiological appetite transducer by subsequent reports from our laboratory showing increased NPY peptide levels and efflux from nerve terminals in the paraventricular nucleus (PVN) in response to experimentally induced hyperphagia, such as that elicited by fasting, restricted feeding or diabetes, and conversely, extinction of hyperphagia by suppression of NPY stores or blockade of postsynaptic receptor activation by passive immunoneutralization or receptor antagonist (Dube et al. 1992, 1994; Kalra 1996; Kalra et al. 1989, 1991a,b, 2000a; Sahu et al. 1988c, 1990b, 1992d; Xu et al. 1996a; Yokosuka et al. 1998, 1999, 2001). Further, enhanced NPY synthesis in the arcuate nucleus (ARC), as shown by increased NPY mRNA and peptide expression preceding its release selectively in the PVN, identified the ARC–PVN axis as the primary neural substrate in propagation of appetitive drive. This promulgation replaced the long-held assumption based on indirect evidence, obtained from indiscriminate destruction of neural tissue by electrolytic lesions or neurotoxins, that the VMH–LtH axis mediated the daily pattern of phagia (Bray 1998; Kalra et al. 1988a, 1989, 1999).

2.2
NPY Receptors and Appetite

Four of the five cloned receptors (R) for NPY and related peptides, and visualized in the hypothalamus, have been invoked in the hypothalamic integration of energy homeostasis. Whereas pharmacological evidence endorsed an important role for Y_1R and Y_5R (Balasubramaniam 1997; Campbell et al. 2001; Daniels et al. 1995, 2002; Dumont et al. 1997; Hokfelt et al. 1998, 2000; Inui 1999, Kalra et al. 1989; Larhammar et al. 2001; Magni 2003; Xu et al. 1998), physiological studies designed to identify the NPYR that participate in physiologically relevant paradigms, such as fasting (Kalra 1996; Kalra et al. 1998b, 1991a; Silva et al. 2002; Turnbull et al. 2002; Zammaretti et al. 2001), suppression of food intake by natural anorexigenic cytokines, such as ciliary neurotrophic factor (Kalra 2001; Kalra et al. 1998d; Xu et al. 1998), and interruption of neural signaling by neurotoxin in the ventromedial nucleus (VMN, Kalra et al. 1998b), have implicated the Y_1R as the primary mediator of appetite stimulation by endogenously released NPY in the PVN and neighboring sites.

Germline deletion of Y_1R in mice produced a modest but significant decrease in food intake and the fasting-induced feeding was markedly attenuated, a response reminiscent of that observed in NPY knockout mice (Bannon et al. 2000; Broberger and Hokfelt 2001; Erickson et al. 1996; Hollopeter et al. 1998; Inui 1999, 2000; Pedrazzini et al. 1998; Pralong et al. 2002). Interestingly, despite diminutions in energy intake, these mice developed late onset increase in weight gain and adiposity resulting from a marked reduction in activity level and energy expenditure. However, Kushi et al. (1998) observed late onset obesity in association with increased thermogenic energy expenditure and unaffected food intake in Y_1R-knockout mice. Germline deletion of Y_5R also elicited late onset obesity without changing feeding behavior (Marsh et al. 1998). Germline knockout approaches have identified many pitfalls, such as compensatory mechanisms arising during development and secondary reorganization of the appetite-regulating network (ARN), both at molecular and cellular levels. We propose that these reorganizations implement a new signaling modality that retains the instinctual appetitive drive but has multiple deficiencies in the ARN leading to imbalance in energy homeostasis. Because of the evolutionary advantage of appetitive instinctual drive for survival of the species, the outcome of these rearrangements is increased energy deposition in the form of fat rather than anorexia and inanition.

Whereas participation of the Y_5R in physiological events that evoke phagia remains uncertain, accumulated evidence invoked Y_1R in mediating the orexigenic effects of endogenous NPY in two ways (Fuxe et al. 1997; Kalra et al. 1999). Seemingly, Y_1R located on target neurons in the magnocellular PVN (mPVN) evoke appetite by transducing intracellular signaling in response to NPY released from projections of neuronal clusters in the ARC and brain stem (BS) (Sahu et al. 1988a,d; Smith and Grove 2002; Yokosuka et al. 1999). This drive is reinforced by a concurrent restraint on anorexigenic melanocortin signaling in the PVN, caused by activation of Y_1R on proopiomelanocortin (POMC) neurons in the ARC by NPY (Broberger et al. 1997; Cowley et al. 1999, 2001; Horvath et al. 1992a,b, 2001b). These revelations, coupled with reports implying Y_1R participation in appetite stimulation by melanin concentrating hormone (MCH) and orexins (ORX), neuropeptides produced in the LtH (Chaffer and Morris 2002; Dube et al. 1999a; Jain et al. 2000; Lopez et al. 2002; Sahu 2002), strengthen our original hypothesis that the Y_1R is the most important mediator of appetite expression within the hypothalamic interconnected orexigenic and anorexigenic pathways (Kalra et al. 1991a).

Perhaps the most recent advance is the knowledge that NPY not only drives but also terminates appetite expression by an autofeedback mechanism through the Y_2R abundantly expressed by the ARC NPY neurons (Broberger et al. 1997). The findings that: (i) NPY infusion inhibited NPY mRNA expression by activation of Y_2R (Pu et al. 2000a), NPY release in vitro and feeding in vivo were inhibited by Y_2R agonists (King et al. 2000), NPY and the coexpressed agouti related peptide (AgrP) gene expression were upregulated, while POMC and cocaine and amphetamine regulating transcript (CART) mRNA expression were downregulated in conditional Y_2R knockout mice (Sainsbury et al. 2002a) together imply autoregulation of NPY synthesis and release via the Y_2R on ARC NPY neurons in the ARC. However, the findings that whereas germline Y_2R knockout mice developed hyperphagia, increased bodyweight and adiposity (Naveilhan et al. 1999, 2002), the hypothalamus specific Y_2 knockout mice displayed a transient decrease in body weight concomitant with increased food intake (Sainsbury et al. 2002a), question the physiological relevance of Y_2R in the daily patterning of appetite. As discussed earlier, it is highly likely that these phenotype differences between the two experimental models are the outcome of disparate reorganizations of hypothalamic circuitries engaged in control of energy intake and expenditure.

Finally, Sainsbury et al. (2002b) reported that Y_4R knockout mice gained weight at a slower rate than wild-type mice, but that fat mass and food intake were significantly decreased only in female Y_4R knockouts. These new findings imply that significant amounts of Y_4R expressed in the PVN may, in some unknown manner, take part in signal transmission within the orexigenic NPYergic circuitry (Campbell et al. 2003; Kalra et al. 1999; Yokosuka et al. 1999, 2001).

2.3
NPY Network and Appetite

2.3.1
Distinct Roles of Two Subpopulations of NPY Producing Neurons

Two subpopulations of NPY producing neurons in the brain, one located in the BS and the other in the ARC, participate in different manners in the overall brain control of ingestive behavior (Chronwall 1988; Everitt and Hokfelt 1989). Nearly 50% of NPY peptide found in various hypothalamic sites, including the PVN and ARC, is derived from the BS subpopulation of neurons that coexpress adrenergic transmitters (Sahu et al. 1988d). Complete elimination of these BS projections was found to induce ligand deficiency at target sites, enhanced sensitivity to NPY and adrenergic transmitters that gradually imposed an increase in dark-phase food consumption, weight gain and adiposity (Sahu et al. 1988a; Pu et al. 2003). A similar progressive increase in weight gain along with a selective dark-phase hyperphagia occurred after interruption of axonal transport of adrenergic transmitters by the neurotoxin, 6-hydroxydopamine (Kalra et al. 1998a). Cumulatively, a deficiency of NPY and adrenergic transmitter levels by interruption of BS projections to the hypothalamus elicited dark-phase hyperphagia leading to increased rates of weight gain and fat deposition.

A large body of evidence also implicated NPY released from the ARC NPY projections in the PVN and neighboring sites in eliciting the daily dark-phase phagia (Bai et al. 1985; Chronwall 1988; Everitt and Hokfelt 1989). Fasting, food restriction and diabetic hyperphagia result from upregulation of NPY synthesis in the ARC and hypersecretion in the form of high amplitude and high frequency episodic release in the PVN (Dube et al. 1992; Kalra et al. 1991b; Sahu et al. 1988c, 1990b, 1992c, 1997). Participation of the ARC NPY subpopulation in triggering dark-phase phagia is apparently mandatory because a rise in gene expression in the ARC preceded the dark-phase NPY hypersecretion and phagia, a response blocked by passive immunoneutralization with NPY antibodies (Dube et al. 1994; Sahu et al. 1992d; Xu et al. 1996a, 1999).

2.4
Interaction of NPY with Coproduced Neurotransmitters in Induction of Appetite

A less appreciated but important feature of the two subpopulations of NPY expressing neurons in the brain is that each subpopulation coexpresses chemically diverse neurotransmitters that act synergistically with NPY to augment appetite by engaging distinct cellular and molecular mechanisms. Noradrenergic and adrenergic transmitters coexpressed in BS NPY neurons and coreleased in the PVN synergistically augment feeding via α-adrenergic receptors (Allen et al. 1985; Clark et al. 1987a, 1988; Everitt and Hokfelt 1989). Although the latency to feeding elicited by adrenergic transmitters is relatively shorter and the feeding response itself is short lived, co-action with NPY produced longer lasting ingestive behavior (Allen et al. 1985).

On the other hand, ARC NPY neurons coexpressing AgrP and γ-aminobutyric acid (GABA) are synaptically linked to POMC expressing neurons in the ARC and also project to NPY targets sites in the mPVN (Broberger et al. 1997; Fuxe et al. 1997; Hahn et al. 1992a, 1997, 1998; Ovesjo et al. 2001; Smith and Grove 2002; Yokosuka et al. 1999). Recent experimental and electrophysiological evidence indicated that NPY acting via Y_1R and GABA via $GABA_AR$ located on POMC neurons, synergistically inhibit α-MSH and CART efflux in the PVN (Cowley et al. 1999, 2001; Pronchuk et al. 2002). The possibility that the second coexpressed transmitter AgrP, either alone via MC3-R (Chen et al. 2000) or co-acting with NPY and GABA (Cowley et al. 1999, 2001; Horvath et al. 1997; Pu et al. 1997b, 1999b), may modulate melanocortin signaling in the ARC–PVN axis has not yet been experimentally validated. The synergistic effects of NPY and GABA through distinct receptor sites along with the antagonistic affects of AgrP at MC4R to curb melanocortin restraint at the mPVN targets, are believed to trigger and sustain nocturnal ingestive behavior in rodents and, possibly the discrete meal pattern in human and subhuman primates (Cowley et al. 1999; Pu et al. 1997b, 1999b).

2.4.1
NPY Interaction with Other Components of Appetite Regulating Networks

Morphological and pharmacological studies designed to decipher the crosstalk among various components of the ARN clearly showed that the hypothalamic NPY system interacts with both anorexigenic and orexigenic peptide pathways in the hypothalamus (Kalra 1997; Kalra et al. 1990, 1996a, 1999; Kalra and Horvath 1998; Kalra and Kalra 1984, 1996, 2003). Inhibition by corticotropin releasing hormone (CRH) and CART of NPY-induced food intake has been shown to be exerted at targets sites located in the PVN (Kalra et al. 1999). On the other hand, because ARC NPY neurons communicate synaptically with the CART–POMC coexpressing neurons in the ARC and with CRH expressing neurons in the PVN, a modulatory influence of NPY on CART and CRH release is also possible (Kalra et al. 1999). The functional significance of these morphological links between NPY and CRH and CART effector pathways is unclear at present but one can infer that these synapses relay signals to suppress CRH and CART efflux, an effect likely to suppress the melanocortin restraint on satiety.

Orexigenic MCH and ORX expressing pathways emanating from the LtH, communicate with the NPY network in the ARC–PVN axis (Broberger and Hokfelt 2001; Horvath et al. 1998, 1999; Kalra et al. 1999). Since Y_1 and Y_5R receptor antagonists were shown to inhibit MCH- and ORX-induced feeding, it is likely that the appetite stimulating effects of these neuropeptides are conferred through NPY release in the PVN (Chaffer and Morris 2002; Dube et al. 2000a; Jain et al. 2000). In addition, morphological (Horvath et al. 1992a; Kalra et al. 1999) and experimental evidence that opiate receptor antagonists inhibit NPY-induced feeding and that NPY stimulates β-endorphin release (Horvath et al. 1992a; Kalra et al. 1995b, 1999) also implicate a role for NPY in the orexigenic effects of β-endorphin produced by POMC neurons.

2.5
High and Low Abundance of NPYergic Signaling Evokes Hyperphagia and Obesity: A New Insight

A review of evidence from various experimental paradigms and genetic models of obesity shows that NPY is a major component of the hypothalamic ARN and that any excursion from the normal patterns of NPY synthesis, release and receptor dynamics deranges the tightly regulated homeostatic communication and causes operational reorganizations in the ARN (Kalra 1997; Kalra et al. 1993, 1996a, 1999). Quite unexpectedly, we discovered that a common outcome of these diverse reorganizations was relentless hyperphagia accompanied by obesity but never anorexia, inanition and morbidity. This was apparent when continuous NPY receptor activation with ICV NPY infusion failed to downregulate NPYR as would normally be expected, although it did downregulate gonadotropin secretion in these animals (see Sect. 3). Instead, continuous receptor activation elicited unabated phagia and abnormal weight gain (Catzeflis et al. 1993; Kalra et al. 1988b; Kalra and Kalra 1996). Similar sustained hyperphagia accompanied by positive energy balance, blockable by passive immunoneutralization or NPYR antagonists, prevails in several genetic models of obesity in conjunction with high abundance of NPY at PVN targets (Dube et al 1999b; Inui 1999; Inui 2000; Kalra et al. 1988b). Fasting also upregulates NPY output to sustain long-lasting phagia (Kalra et al. 1991b; Sahu et al. 1988c).

On the other hand, reorganization within the ARN imposed by low abundance of NPY at target sites and/or deficit of any one of the four NPYRs also results in positive energy balance due either to hyperphagia, decreased energy expenditure or both (Kalra et al. 1997a, 1998b; Kushi et al. 1998; Naveilhan et al.

1999, 2002; Ovesjo et al. 2001; Pedrazzini et al. 1998; Pralong et al. 2002; Sahu et al. 1992c, Sainsbury et al. 2002a,b). A notable example is that contrary to expectations, hyperphagia and obesity-induced by either ablation of VMH or interruption of neural signaling in the VMN with neurotoxins, was associated with decreased NPY synthesis in the ARC and efflux in the PVN (Dube et al. 1995, 1999b; Jain et al. 1998, Kalra et al. 1997a, 1998b, 1999). A deficit in hypothalamic NPY levels produced by interruption of BS NPY neuronal projections also progressively increased food intake and weight gain (Sahu et al. 1988a; Pu et al. 2003). Similarly, germline ablation of either Y_1, Y_2 or Y_4R caused obesity (Kushi et al. 1998; Naveihan et al. 1999, 2002; Pedrazzini et al. 1998; Pralong et al. 2002; Sainsbury et al. 2002a; Silva et al. 2002).

A comparative analysis of the etiology of hyperphagia and increased weight gain in these experimental paradigms has identified various modalities of compensatory reorganization in the ARN. In VMH-lesioned rats, hyperphagia and obesity were attributed to development of leptin resistance in POMC neurons and diminution in anorexigenic melanocortin signaling (Choi et al. 1999; Dube et al. 1999b, 2000b). When neural signaling in the VMN was disrupted, hyperphagia and abnormal rate of weight gain was, in part, attributed to increased sensitivity to NPY due to increased abundance of Y_1R, a response blockable by Y_1R antagonist (Kalra et al. 1997a, 1998b; Pu et al. 1998a). Rapid development of resistance to leptin and increased availability of orexigenic galanin also played a role in the development of phagia and obesity in these rats (Kalra et al. 1998c; Kalra and Horvath 1998; Pu et al. 1999a). A similar low abundance of NPY along with development of increased sensitivity to NPY produced by gold thioglucose or postnatal monosodium glutamate treatment, has been shown to render rodents obese (see review by Kalra et al. 1999).

2.6
Regulation of NPY Secretion and Appetite by Afferent Neural and Hormonal Pathways

It has been known for a long time that diminution in energy stores in the body and the clock-driven meal patterns are powerful signals to stimulate NPY secretion and appetite (Kalra et al. 1999). Only recently it has become apparent that disparate neural and hormonal signals to the hypothalamic NPY system play important roles in the NPY driven expression of appetite. ORX and MCH expressing pathways between the LtH and ARC constitute neural afferents to the NPY system in the ARC–PVN axis (Horvath et al. 1998, 1999). Evidence that stimulation of feeding by ORX and MCH is mediated by NPYergic signaling (Chaffer and Morris 2002; Dube et al. 1999a; Jain et al. 2000) and fasting upregulates expression of these peptides, are consistent with this implication. However, whether peptidergic signaling from the LtH to NPY network is important in daily meal patterning remains to be ascertained.

The recent discovery of opposing effects of the hormonal signals adipocyte leptin and gastric ghrelin on NPY secretion has provided strong evidence in support of afferent hormonal signals being critical in the hypothalamic integration of energy homeostasis (Ahima et al. 2000; Friedman and Halaas 1998; Horvath et al. 2001; Kalra et al. 2003; Pu et al. 2000b; Tschop et al. 2000). Leptin secretion is episodic and sex specific (Bagnasco et al. 2002c). Circulating leptin levels are generally higher in females, and this sexually dimorphic response may result from stimulation of leptin secretion by ovarian hormones (Bagnasco et al. 2002c). It is believed that leptin normally inhibits feeding by engaging two major anorexigenic and orexigenic effector pathways in the hypothalamus (Friedman and Halaas 1998; Kalra et al. 1999; Kalra and Kalra 1996, 2003). Leptin inhibits NPY mRNA expression in the ARC and efflux in the PVN (Ahima et al. 2000; Kalra et al. 1999). In addition, leptin stimulates energy expenditure by activating leptin target neurons in the medial preoptic area (MPOA) that, in turn, engage the autonomic nervous system and ARC NPY producing neurons (Bagnasco et al. 2002a; Billington et al. 1994; Chen et al. 1998; Kalra et al. 1999). In fact, increased leptin availability in hypothalamic target sites by enhanced leptin transgene expression with the aid of gene therapy, resulted in long-term suppression of weight gain with or without a decrease in food intake and increased nonthermogenic energy expenditure. This long-lasting suppression of weight gain by hypothalamic leptin expression was found to be mediated by NPYergic signaling (Bagnasco et al. 2002a; Beretta et al. 2002; Dhillon et al. 2001a,b, Dube et al. 2002; Kalra and Kalra 2002).

On the other hand, ghrelin is orexigenic and promotes adiposity (Horvath et al. 1997; Tschop et al. 2000). Secretion of ghrelin, like leptin, is episodic and subject to modulation by negative energy balance (Bagnasco et al. 2002b). Ghrelin is hypersecreted prior to mealtime and in genetic models of obesity in rodents and human (Cummings et al. 2001; Horvath et al. 2001a; Tschop et al. 2000, 2001). Whereas anorexigenic and orexigenic effector pathways are leptin targets, the ARC NPY network is the only target for ghrelin (Bagnasco et al. 2002b; Kalra et al. 2003). Ghrelin stimulates NPY secretion and ghrelin induced feeding is blockable by Y_1R antagonists (Tschop et al. 2001). Fasting was found to increase frequency and amplitude of ghrelin discharge (Bagnasco et al. 2002b). Our previous studies showed that NPY also promoted episodic meal patterning coincident with increased pulsatile NPY secretion in the PVN. Fasting promoted high amplitude and high frequency pulse discharge selectively in the PVN (Kalra et al. 1991b, 1999). Accordingly, we propose that episodic ghrelin signals are an important afferent hormonal modality to stimulate episodic NPY secretion prior to mealtime (Kalra et al. 2003).

With respect to the temporal interrelationships between anorexigenic leptin and orexigenic ghrelin, it seems that when energy stores are low, leptin secretion is decreased to curtail the restraint on the NPYergic network. Concomitantly, ghrelin hypersecretion augments NPYergic signaling. Thus, it is the synchronized curb on leptin restraint and increased ghrelin stimulus on NPYergic signaling that propagates and sustains the appetitive drive to replenish energy stores (Bagnasco et al. 2002b). In sum, rhythmic, reciprocal leptin and ghrelin signaling from the periphery, acting primarily at the hypothalamic NPY network, underlies the intermittent appetitive drive in the daily management of energy homeostasis (Kalra et al. 2003). Furthermore, even the anorectic effects of estrogens are mediated by NPYergic signaling (Bonavera et al. 1994a).

Consequently, a wealth of evidence accumulated in the past two decades is consistent with the earlier notion that the NPY network is the primary substrate in the hypothalamus that encodes appetite expression.

3
NPY, a Physiological Regulator of Reproduction

There are two basic operational elements in the hypothalamic control of reproduction in vertebrates (Kalra 1993; Kalra et al. 1997b; Kalra and Kalra 1983). GnRH produced by a network of neurons in the hypothalamus is the primary signal for release of the pituitary gonadotropins, luteinizing hormone (LH) and follicle stimulating hormone. GnRH is secreted into the hypophyseal portal veins in the median eminence (ME) in a periodic, pulsatile manner in both sexes.

Two oscillatory patterns of GnRH secretion have been identified. In general, GnRH is secreted in the form of low amplitude episodes that display a daily pattern and are subject to modulation by gonadal steroids, even during various stages of the reproductive cycle in females. Additionally in females, these regularly spaced, basal GnRH pulses are interrupted by an abrupt acceleration in the frequency of GnRH discharge culminating in the protracted preovulatory LH surge responsible for ovulation (Kalra 1993; Kalra et al. 1997b; Kalra and Kalra 1983). Research spanning three decades has shown that although a host of messenger molecules elaborated in the hypothalamus can affect the two patterns of GnRH secretion, intrahypothalamic afferent signals that impart a temporally related, modulatory influence on the two patterns of GnRH secretion are extremely limited (Kalra 1993; Kalra and Kalra 1983). Two reports published in 1984 demonstrated that members of the pancreatic polypeptide family, human pancreatic peptide and NPY exerted either an excitatory or inhibitory effect on pituitary LH secretion depending upon the gonadal steroid environment and mode of administration (Kalra and Crowely 1984a,b). This duality of effects, coupled with the demonstration that NPY can interact synergistically with GnRH at the level of pituitary gonadotrophs to amplify LH secretion (Crowley 1999; Crowley et al. 1987; Crowley and Kalra 1987, 1988), established NPY as a crucial afferent signal in the hypothalamus for regulating reproduction in mammals (Kalra et al. 1989). Subsequent investigations based on this premise formed the basis for designating NPY as the 'on/off' switch in governing the neuroendocrine control of reproduction (Kalra and Kalra 1996).

3.1
A Unique Bimodal Participation by NPY in Neuroendocrine Control of Reproduction

3.1.1
Excitatory Afferent Signal for Ultradian and Cyclic GnRH Discharge

Since the early report that an ICV injection of NPY stimulated pituitary LH release in ovarian steroid-primed ovariectomized rats (Kalra and Crowley 1984b), several laboratories undertook to delineate the precise nature of NPY participation in governing the two modalities of GnRH secretion (Kalra 1993; Kalra et al. 1997b; Kalra and Kalra 1983, 1996, 2003; Woller et al. 1992). NPY serves as an excitatory signal to stimulate LH secretion in two ways, by promoting the basal pattern of GnRH release into the hypophyseal portal system (Allen et al. 1987; Crowley and Kalra 1987; Kalra et al. 1995a; Xu et al. 1993, 1996b) and by potentiating the GnRH-induced LH release from the pituitary gonadotrophs. Experimental evidence showed that the excitatory effects of NPY on GnRH release were dependant upon the modality of NPY receptor activation (Kalra and Kalra 1996, 2003; Kalra et al. 1990, 2000b; 1990, Sahu et al. 1987a). Intermittent but not continuous infusion of NPY at a frequency similar to that which occurs normally stimulated GnRH release. There is concordance between basal pulsatile NPY release and GnRH discharge in the MPOA–ME axis. That each NPY pulse evokes a corresponding GnRH episode is affirmed by the finding that passive immunoneutralization of NPY in gonadectomized animals abolished GnRH pulsatility (Kalra and Kalra 2003; Kalra et al. 2000b; Woller et al. 1992; Xu et al. 1993, 1996c). Interestingly, NPY and galanin co-act to facilitate the basal pattern of episodic GnRH secretion (Xu et al. 1996c). The observation that NPY antisense deoxyoligonucleotides administration diminished the basal pattern of LH secretion suggested the importance of the newly formed NPY pool in basal GnRH release (Kalra et al. 1995a). NPY Y_1R and Y_5R and galanin R antagonists, likewise, inhibited basal LH secretion (Allen et al. 1985, 1987; Toufexis et al. 2002). Thus, NPY along with galanin constitutes an obligatory afferent excitatory signal for basal pulsatile GnRH secretion.

NPY signaling is also mandatory for induction of the preovulatory GnRH surge (Kalra 1993; Kalra et al. 1997b). NPY mRNA in the ARC and NPY peptide stores in the ME increased prior to onset of the preovulatory LH surge on proestrus and remained elevated during most of the LH surge period (Sahu et al. 1995, 1989). As seen in the case of basal GnRH secretion, preovulatory GnRH hypersecretion also critically depended upon augmented NPY hypersecretion because passive immunoneutralization with NPY antibodies decreased NPY stores and blocked the LH surge (Jain et al. 1999; Kalra et al. 1992; Kalra 1993).

An added feature of NPY participation in the preovulatory LH surge is that the excitatory effects of NPY are also exerted at two levels in the hypothalamic–pituitary–gonadotroph axis. Protracted hypersecretion from NPY nerve terminals synapsing on GnRH perikarya and dendrites in the MPOA and from nerve terminals in the ME elicit GnRH hypersecretion, and NPY released along with GnRH into the hypophyseal portal system acts at the level of pituitary gonadotrophs to amplify GnRH stimulation of LH release (Bauer-Dantoin et al. 1992; Crowley and Kalra 1988; Kalra 1993; Leupen et al. 1997). The cellular and molecular sequelae underlying this synergistic interplay between NPY and GnRH at the level of pituitary gonadotrophs were reviewed recently by Crowley (1999).

These antecedent neurosecretory events associated with the preovulatory GnRH–LH surge are facilitated by the timely interplay of ovarian steroids and the daily clock resident in the suprachiasmatic nuclei (SCN, Kalra 1993). We showed that NPY neurons in the ARC express steroid receptors (Sar et al. 1990), gonadal steroids stimulate NPY synthesis and storage in the ME nerve terminals (Kalra 1993; Sahu et al. 1987b, 1992b) and the neurosecretory response to gonadal steroids is sex-specific (Sahu et al. 1992a). Increased NPY synthesis and storage in the hypothalamus, in association with the preovulatory surge, is ovarian steroid-dependent because these events fail to occur during the estrogen-deficient diestrous stage of the estrous cycle, but they can be reproduced by appropriate sequential ovarian steroid-replacement in ovariectomized rats (Sahu et al. 1994a). Also, the facilatory effects of gonadal steroids are markedly attenuated in aged female and male rats (Kalra et al. 1993; Sahu et al. 1988b; Sahu and Kalra 1998). In NPY knockout mice, the cyclic LH surge is attenuated (Xu et al. 2000a).

The temporal NPY antecedent neurosecretory events facilitated by ovarian steroids are entrained to the timing device in the SCN because SCN ablation blocked the cyclic GnRH surge (Kalra 1993). Seemingly, two clock driven sequential chains of events are responsible for the cyclical ovulatory surge. The expression of the first culminates in increased NPY synthesis in the ARC and accumulation of NPY and GnRH in the ME nerve terminals in anticipation of the impending protracted discharge of NPY locally within the ME, and of NPY and GnRH into the hypophyseal portal system. The expression of the second clock-driven event apparently triggers NPY release to accelerate GnRH secretion for the preovulatory LH surge (Kalra 1993).

3.1.2
Excitatory Role of the Two Subpopulations of NPY Expressing Neurons

There is evidence to show that both BS and ARC subpopulations of NPY neurons participate in the neuroendocrine control of reproduction in different ways. Retrograde tracing studies and bilateral neural transection of BS projections to the hypothalamus identified innervations of the MPOA–ME axis by BS neurons (Everitt and Hokfelt 1989; Sahu et al. 1988a). As detailed in earlier sections, NPY and coexpressed adrenergic neurotransmitters released in the MPOA–ME axis from BS projections, provide an ancillary α-adrenergic driving input for the basal and cyclic GnRH discharge (Allen et al. 1985, 1987). Adrenergic transmitters synergize also with NPY to amplify LH release. Thus, it is highly plausible that extrahypothalamic NPY and adrenergic innervations to the MPOA–ARC–ME axis constitute the back up, reinforcing excitatory stimulus to sustain the two modalities of GnRH secretion under varied environmental challenges (Kalra 1993).

Our recent anatomical and pharmacologic investigations have uncovered a novel mechanism responsible for stimulation of GnRH release by the ARC NPY subpopulation. Seemingly, the coexpressed GABA acts in the MPOA to curb NPY-induced basal and cyclic GnRH release (Horvath et al. 2001b). Furthermore, these inhibitory effects of GABA are mediated by $GABA_A$ receptors, which together with NPY Y_4R attenuate the excitatory action of NPY on GnRH neurons (Horvath et al. 2001b; Jain et al. 1999; Sainsbury et al. 2002b). Thus, NPY and GABA coexpressing neurons in the ARC are a component of the pulse generator network that emits both excitatory and inhibitory signals. The operational complexity of the NPY network is also illustrated by the fact that NPY and GABA coexpressing neurons synapse on β-endorphin expressing POMC neurons in the ARC (Horvath et al. 1995, 1992b). It seems that increased NPY and GABA input to POMC neurons diminish the inhibitory opioid tone thus triggering the cyclic GnRH surge (Allen and Kalra 1986). Taken together, the revelations that activation of $GABA_AR$ on GnRH neurons can restrain NPY-induced LH release through Y_4R, and GABA concomitantly curtails the opioid restraint imposed by ARC POMC neurons, shed new light on the interactive crosstalk between an excitatory neuropeptide and an inhibitory amino acid within the hypothalamic network that regulates the two modalities of GnRH secretion.

3.1.3
Interaction of the NPY Network with Other Components of the Hypothalamic GnRH Regulating Circuitry

Of the various neuropeptidergic neurotransmitter pathways in the hypothalamus that may regulate NPY influence on GnRH secretion (Kalra 1993), endogenous opioid peptides, primarily β-endorphin, are the most prominent. Endogenous opioid peptides inhibit basal and cyclic gonadotropin release (Horvath et al. 1992a; Kalra 1993). Stimulation of gonadotropin release by pharmacological blockade of opiate receptors, suggested the existence of an inhibitory opioid tone on GnRH output (Allen and Kalra 1986). Although the precise underlying mechanisms of the tonic inhibitory influence have not been fully characterized, there is evidence to show that a restraint on this opioid tone may play a role in basal episodic GnRH secretion (Xu et al. 1993) and in triggering the preovulatory NPY and GnRH surges (Sahu et al. 1990a; Xu et al. 1996c).

The galanin producing network in the hypothalamus has also been shown to play a role in controlling of the two modalities of GnRH secretion (Kalra 1993). A galanin antagonist inhibited basal and preovulatory LH surge and NPY-induced LH release (Sahu et al. 1994b), thereby suggesting that galanin (GAL) may, in part, mediate the excitatory effects of NPY. Both NPY and GAL are secreted in a pulsatile manner and each is capable of driving GnRH pulsatility (Xu et al. 1993, 1996b). Further, passive immunoneutralization of GAL and NPY with antibodies in concentrations that individually exerted no detectable impact on LH secretion, markedly suppressed LH pulsatility when infused together. Since there is considerable concordance between GAL, NPY, GnRH and LH episodes, it is highly likely that GAL and NPY are, indeed, responsible for the propagation and sustenance of GnRH pulsatility as well as the preovulatory surge. A study of the morphological relationships between GAL and NPY pathways revealed that the NPY network is synaptically linked with GAL neurons in the MPOA-ARC-ME axis (Horvath et al. 1995, 1996). Thus, the results of various pharmacological and physiological investigations together with the demonstration of morphological connectivity clearly demonstrated that information transfer from the hypothalamic pulse and surge generator mobilizes a NPY→GAL line of communication in a rostro-caudal direction in the MPOA-ARC-ME axis.

3.2
Regulation of NPY Secretion by Afferent Neural and Hormonal Signals

Recently, two distinct neurotransmitter and hormonal pathways regulating NPY secretion have been identified. Nitric oxide is a ubiquitous gaseous intra- and intercellular messenger produced in the vicinity of GnRH neurons in the hypothalamus and has been shown to play a role in the basal and cyclic release of GnRH (Bonavera et al. 1993, 1994b; Pu et al. 1997a, 1998c). Nitric oxide stimulated NPY release in the hypothalamus and NPY stimulated GnRH release by direct action via Y_1/Y_5R on GnRH perikarya in the MPOA and nerve terminals in the ME. This finding implied an upstream neural system regulating NPY→GnRH information flow. Additionally, morphological evidence of a bidirectional line of communication between the LtH ORX and the ARC NPY neurons and stimulation of LH release by ORX implicated NPY also in ORX-induced LH stimulation (Pu et al. 1998b).

The cluster of NPY neurons in the ARC are also the neuroanatomical substrate in the gonadal steroid feedback effects on GnRH secretion. The existence of estrogen receptors in the ARC NPY neurons, decreases in hypothalamic NPY gene expression and release after gonadectomy, and restoration of NPY output by steroid replacement in gonadectomized rats are consistent with a direct regulatory control of NPY efflux by gonadal steroids (Kalra 1993; Sahu et al. 1992b; Sar et al. 1990). In addition, ovarian steroids upregulate NPY Y_1R in the hypothalamus (Leupen et al. 1997; Xu et al. 2000b; Zammaretti et al. 2001). Dependency on ovarian steroids of NPYergic signaling to evoke the cyclic GnRH surge is illustrated by the demonstration that replication of the cyclic sequential ovarian steroid milieu in ovariectomized rats reinstated NPY synthesis in ARC perikarya, NPY storage in the ME, and NPY Y_1R expression in the hypothalamus in synchrony with the clock-triggered NPY surge to evoke GnRH and LH surges (Sahu et al. 1994a).

Leptin has been shown to play a permissive role in the hypothalamic control of gonadotropin secretion (Beretta et al. 2002; Caprio et al. 2001; Chehab 1997;

Cheung et al. 2000). Leptin administration blocked the decrease in LH release in response to experimentally induced fuel shortage (Chehab 1997; Kalra et al. 1996b, 1998d). These facilitatory effects resulted from leptin-induced normalization of NPYergic signaling (Kalra et al. 1996b, 1998d). It is interesting to note that the ARC NPY neurons coexpress leptin and estrogen receptors (Diano et al. 1998), estrogen inhibits feeding and promotes leptin secretion (Bonavera et al. 1994a; Bagnasco et al. 2002c). These observations imply an intracellular leptin and gonadal steroid interaction linking energy homeostasis and neuroendocrine control of reproduction.

3.3
NPY, an Inhibitory Afferent Signal

In the original paper published in 1984, we reported that ICV NPY administration rapidly suppressed LH release in ovariectomized rats (Kalra and Crowley 1984b). This suppressive action of NPY was later found to be mediated by activation of the inhibitory opioidergic signaling to GnRH network (Kalra et al 1995b; Kalra and Kalra 1983; Xu et al. 1993). A critical analysis of the information accumulated since then indicates that an exquisite quantitative and temporal relationship between the NPY and GnRH networks is necessary for the intermittent GnRH discharge to generate the gonadotropin release patterns optimal for sustenance of reproductive function. Diminution in excitatory NPY signaling to the GnRH network, as seen in aging rats, resulted in cessation of reproductive cycles in females and suppressed gonadotropin secretion in males (Kalra et al. 1993; Sahu et al. 1988b; Sahu and Kalra 1998). On the other hand, a departure from this tight relationship between NPY and GnRH, either of a quantitative nature or in the intermittent pattern of NPY signaling resulted in suppression of GnRH secretion. For example, while a bolus injection readily stimulated LH release, continuous NPY infusion was ineffective in stimulating gonadotropin secretion in ovarian steroid-primed ovariectomized rats, and in gonad-intact rats it resulted in suppression of gonadal function (Catzeflis et al. 1993; Kalra et al. 1997b; Kalra and Kalra 1996; Raposinho et al. 2003; Toufexis et al. 2002). Physiological challenges of high-energy demand as those manifest during lactation (Crowley et al. 2003; Smith 1993), the prepubertal period (El Majdoubi et al. 2000) or fasting (Bergendahl and Veldhuis 1995; Kalra et al. 1996b), upregulate hypothalamic NPY signaling and are associated with suppressed gonadotropin secretion. That it is the inappropriate NPY hypersecretion that suppresses GnRH secretion is affirmed by our findings that blockade of NPY upregulation reinstated gonadotropin secretion in fasted rats (Kalra et al. 1996b, 1998d). Also, diminished reproductive function in genetically obese models is a consequence of NPY hypersecretion because suppression of NPYergic signaling either by leptin or cytokine therapy (Kalra 2001), or germline deletion of NPY Y_1R or Y_4R in infertile *ob/ob* mice, restored gonadotropin secretion and fertility (Pralong et al. 2002; Sainsbury et al. 2002b).

In sum, bimodal operation, either to promote or suppress reproduction, is endowed in the NPYergic signaling in the hypothalamus. Seemingly, this is an important neurochemical modality that provides the organism with the option to either sustain or cease reproduction in response to nutritional challenges.

4
NPY and Sex Behavior

Clark et al. (1985) reported that central administration of NPY stimulated feeding but inhibited sexual behavior in rats. This observation of an inverse relationship between two behaviors drew considerable attention because clinical and animal studies have recognized that disorders of feeding behavior are invariably associated with altered reproductive function and behavior. NPY was found to suppress various components of copulatory behavior in sexually active male rats (Clark et al. 1985; Kalra et al. 1989). Indeed, doses that were least effective in stimulation of feeding, dramatically suppressed copulatory behavior. The majority of sexually active males mounted, but intromission frequency was reduced along with an almost complete absence of ejaculation in tests performed within 10 min of injection. Further tests showed that sexual motivation was adversely impacted because penile reflex, including erection, in ex copula tests was normal. Another member of the pancreatic polypeptide family, pancreatic polypeptide, while effective in stimulation of feeding was ineffective in suppressing copulatory behavior (Clark et al. 1985). Further, microinjection of NPY into various hypothalamic sites identified the MPOA as the site where NPY acted to decrease the ejaculatory component of sexual behavior. Poggolioli et al. (1990) confirmed the inhibitory effects of NPY on sexual behavior in males.

A similar inhibitory effect of central NPY administration on lordosis behavior was observed in female rats. In fact, increased NPY receptor activation virtually eliminated proceptive behavior induced by sequential estrogen–progesterone treatment in ovariectomized rats. It was also reported that central administration of NPY or Y_2/Y_5R agonists suppressed estrous behavior in Syrian hamsters (Corp et al. 2001). That upregulation of NPYergic signaling may, along with cessation of reproductive cycles, inhibit sexual behavior was indicated by studies using Zucker obese rats which display increased NPY levels in the hypothalamus (Marin-Bivens et al. 1998). Blocking the effects of NPY with a neutralizing dose of NPY antibody in ovarian steroid-primed Zucker rats resulted in enhanced proceptive behavior (Marin-Bivens et al. 1998).

5
Is NPY an On/Off Switch for Control of Appetite and Reproduction?

A wealth of information summarized in the preceding sections documents a dynamic relationship between reproduction and nutrient abundance. Chronic undernourishment during drought, short-term nutritional imbalance resulting from multiple causes, such as strenuous exercise, adversely impact reproduction. Even the high-energy cost of lactation is compensated by hyperphagia and cessation of reproductive cycles. Recent findings clearly demonstrated that propagation of the appetitive drive to meet these adverse environmental and physiological challenges is neurochemically driven. Is there a neurochemical switch in the brain that turns on appetite and turns off reproduction in response to environmental challenges that enforce loss of energy fuels from the body? Is there one or more than one neurochemical modality subserving the on/off switch?

Our concerted research endeavors during the past two decades to delineate the neurochemical pathways regulating reproduction and appetite have laid the foundation for the basic tenet that these two instinctual urges originate in the hypothalamus and are neurochemically linked within the hypothalamic hardwiring (Fig. 1). We have unraveled the spatial and temporal organization of the two circuitries and the commonality in neurochemical networks for information transmission either to elicit the appetitive behavior or to evoke release of the reproduction-sustaining neurohormone GnRH. In this context, the reports first of the excitatory influence of NPY on appetite and reproduction and the later identification of NPY neurons in the ARC as the bridge connecting the two circuitries, the ARC–PVN appetitive axis and the MPOA–ARC–ME reproductive axis, have been instrumental in guiding research to unravel the operation of this link at the cellular and molecular levels.

One feature of the MPOA–ARC–ME reproductive axis that distinguishes it from the ARC–PVN appetitive axis, is that two distinct timing mechanisms, the circadian SCN clock for cyclic NPY discharge in the female and pulse generator clock for ultradian NPY discharge in both sexes, govern its operation on a daily basis. The NPY network communicates with the GnRH network both directly and indirectly through GAL and opioid channels. The exquisite time-based operation is modulated by gonadal steroidal feedback and afferent neural LtH ORX input, and aided by diverse messengers coexpressed in NPY neurons. Any excursion from the norm in NPY signaling, either due to diminished or enhanced NPY efflux, disrupts the finely tuned operation of the two circuitries and turns off GnRH secretion resulting in cessation of reproduction.

The distinguishing feature of the ARC–PVN appetitive neuroaxis, is the existence of a multifactorial control for optimal operation. To impart the daily feeding pattern, it receives excitatory ORX and MCH neural inputs from LtH, an inhibitory neural input from the VMN and exquisite reciprocal and dynamic minute-to-minute regulatory feedback instructions from the hormones, leptin and ghrelin, for the timely discharge of NPY at PVN target sites. Additionally, within the ARC–PVN axis, there is a complex local regulatory network composed of projections from the ARC and BS NPY neuronal subpopulations that coexpress diverse messengers to assist in eliciting sustained feeding either by synergizing with NPY or by curtailing melanocortin restraint. Thus, a two-pronged action of NPY, direct stimulation through Y_1R/Y_5R and reduced melanocortin restrain, underlies the robust appetite stimulation. Recent studies have shown that diminution in leptin restraint and enhanced orexigenic ghrelin input, are the most

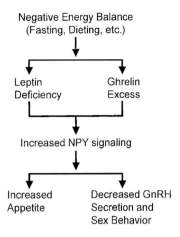

Fig. 2 Regulatory role of hypothalamic NPY in increasing appetite and concomitantly suppressing reproduction and sexual behavior in response to negative energy balance. For details, see text. (With permission from Kalra and Kalra 2003)

important afferent signaling modalities responsible for NPY hypersecretion and appetite expression in response to loss of energy.

Consequently, the most notable consequence of energy shortage is upregulation of NPYergic signaling to reliably sustain a robust appetitive drive for energy replenishment, and to simultaneously shut down the neuroendocrine mechanisms that require high energy, such as reproduction. This chain of the neuroendocrine and neural sequence linking energy homeostasis and reproduction is depicted in Fig. 2.

Apparently not all NPY producing neurons in the ARC engage in integration of energy balance and reproduction. We observed that whereas the entire subpopulation of ARC NPY neurons responded to energy deficit, only a subpopulation in the middle–caudal ARC participated in the control of cyclic gonadotropin surge (Kalra et al. 2000b; Pu et al. 1999c). Thus, a functional heterogeneity exists within the ARC neurons and a discrete subpopulation that evokes cyclic GnRH surge is susceptible to energy deficit and responds by turning off reproduction. The neuroendocrine regulatory processes for appetitive behavior and reproduction do not function in isolation but interact by employing common effector pathways. Information relayed by the hypothalamic NPYergic network is seemingly one such interactive communication channel.

Acknowledgements. This research was supported by grants from the National Institute of Health DK37273, HD 08634 and NS 32727. We also acknowledge the word processing assistance of Ms. Sandra Clark.

References

Ahima RS, Saper CB, Flier JS, Elmquist JK (2000) Leptin regulation of neuroendocrine systems. Front Neuroendocrinol 21:263–307

Allen LG, Crowley WR, Kalra SP (1987) Interactions between neuropeptide Y and adrenergic systems in the stimulation of luteinizing hormone release in steroid-primed ovariectomized rats. Endocrinology 121:1953–1959

Allen LG, Kalra PS, Crowley WR, Kalra SP (1985) Comparison of the effects of neuropeptide Y and adrenergic transmitters on LH release and food intake in male rats. Life Sci 37:617–623

Allen LG, Kalra SP (1986) Evidence that a decrease in opioid tone may evoke preovulatory luteinizing hormone release in the rat. Endocrinology 118:2375–2381

Bagnasco M, Dube MG, Kalra PS, Kalra SP (2002a) Evidence for the existence of distinct central appetite and energy expenditure pathways and stimulation of ghrelin as revealed by hypothalamic site-specific leptin gene therapy. Endocrinology 143:4409–4421

Bagnasco M, Kalra PS, Kalra SP (2002b) Ghrelin and leptin pulse discharge in fed and fasted rats. Endocrinology 143:726–729.

Bagnasco M, Kalra PS, Kalra SP (2002c) Plasma leptin levels are pulsatile in adult rats: effects of gonadectomy. Neuroendocrinology 75:257–263.

Bai FL, Yamano M, Shiotani Y, Emson PC, Smith AD, Powell JF, Tohyama M (1985) An arcuato-paraventricular and -dorsomedial hypothalamic neuropeptide Y- containing system which lacks noradrenaline in the rat. Brain Res 331:172–175

Balasubramaniam AA (1997) Neuropeptide Y family of hormones: receptor subtypes and antagonists. Peptides 18:445–457

Bannon AW, Seda J, Carmouche M, Francis JM, Norman MH, Karbon B, McCaleb ML (2000) Behavioral characterization of neuropeptide Y knockout mice. Brain Res 868:79–87

Bauer-Dantoin AC, McDonald JK, Levine JE (1992) Neuropeptide Y potentiates luteinizing hormone (LH)-releasing hormone- induced LH secretion only under conditions leading to preovulatory LH surges. Endocrinology 131:2946–2952

Beretta E, Dube MG, Kalra PS, Kalra SP (2002) Long-term suppression of weight gain, adiposity, and serum insulin by central leptin gene therapy in prepubertal rats: Effects on serum ghrelin and appetite-regulating genes. Pediatr Res 52:189–198

Bergendahl M, Veldhuis JD (1995) Altered pulsatile gonadotropin signaling in nutritional efficieny in the male. Trends Endocrinol Metab 6:145–159

Billington CJ, Briggs JE, Harker S, Grace M, Levine AS (1994) Neuropeptide Y in hypothalamic paraventricular nucleus: a center coordinating energy metabolism. Am J Physiol 266: R1765–R1770

Bonavera JJ, Dube MG, Kalra PS, Kalra SP (1994a) Anorectic effects of estrogen may be mediated by decreased neuropeptide- Y release in the hypothalamic paraventricular nucleus. Endocrinology 134:2367–2370

Bonavera JJ, Sahu A, Kalra PS, Kalra SP (1993) Evidence that nitric oxide may mediate the ovarian steroid-induced luteinizing hormone surge: involvement of excitatory amino acids. Endocrinology 133:2481–2487

Bonavera JJ, Sahu A, Kalra PS, Kalra SP (1994b) Evidence in support of nitric oxide (NO) involvement in the cyclic release of prolactin and LH surges. Brain Res 660:175–179

Bray GA (1984) Syndromes of hypothalamic obesity in man. Pediatr Ann 13:525–536

Bray GA (1998) Historical framework for the development of ideas about obesity. In: Bray GA, Bouchard C, James WPT (eds) Handbook of obesity, vol 1. New York, Marcel Dekker, pp 1–29

Broberger C, Hokfelt T (2001) Hypothalamic and vagal neuropeptide circuitries regulating food intake. Physiol Behav 74:669–682

Broberger C, Landry M, Wong H, Walsh JN, Hokfelt T (1997) Subtypes Y_1 and Y_2 of the neuropeptide Y receptor are respectively expressed in pro-opiomelanocortin- and neuropeptide-Y-containing neurons of the rat hypothalamic arcuate nucleus. Neuroendocrinology 66:393–408

Campbell RE, Ffrench-Mullen JM, Cowley MA, Smith MS, Grove KL (2001) Hypothalamic circuitry of neuropeptide Y regulation of neuroendocrine function and food intake via the Y_5 receptor subtype. Neuroendocrinology 74:106–119

Campbell RE, Smith MS, Allen SE, Grayson BE, French-Mullen JM, Grove KL (2003) Orexin neurons express a functional pancreatic polypeptide Y_4 receptor. J Neurosci 23:1487–1497

Caprio M, Fabbrini E, Isidori AM, Aversa A, Fabbri A (2001) Leptin in reproduction. Trends Endocrinol Metab 12:65–72

Catzeflis C, Pierroz DD, Rohner-Jeanrenaud F, Rivier JE, Sizonenko PC, Aubert ML (1993) Neuropeptide Y administered chronically into the lateral ventricle profoundly inhibits both the gonadotropic and the somatotropic axis in intact adult female rats. Endocrinology 132:224–234

Chaffer CL, Morris MJ (2002) The feeding response to melanin-concentrating hormone is attenuated by antagonism of the NPY Y(1)-receptor in the rat. Endocrinology 143:191–197

Chehab FF (1997) The reproductive side of leptin. Nature Med 3:952–953

Chen AS, Marsh DJ, Trumbauer ME, Frazier EG, Guan XM, Yu H, Rosenblum CI, Vongs A, Feng Y, Cao L, Metzger JM, Strack AM, Camacho RE, Mellin TN, Nunes CN, Min W, Fisher J, Gopal-Truter S, MacIntyre DE, Chen HY, Van Der Ploeg LH (2000) Inactivation of the mouse melanocortin-3 receptor results in increased fat mass and reduced lean body mass. Nature Genet 26:97–102

Chen XM, Hosono T, Yoda T, Fukuda Y, Kanosue K (1998) Efferent projection from the preoptic area for the control of non- shivering thermogenesis in rats. J Physiol 512:883–892

Cheung CC, Clifton DK, Steiner RA (2000) Perspectives on leptin's role as a metabolic signal for the onset of puberty. Front Horm Res. 26:87–105

Choi S, Sparks R, Clay M, Dallman MF (1999) Rats with hypothalamic obesity are insensitive to central leptin injections. Endocrinology 140:4426–4433

Chronwall BB (1988) Anatomical distribution of NPY and NPY messenger RNA in the brain. In: Mutt V, Fuxe K, Hokfelt T, Lundberg JD (eds) Neuropeptide Y. New York, Raven Press, pp 51–60

Clark JT, Gist RS, Kalra SP, Kalra PS (1988) Alpha 2-adrenoceptor blockade attenuates feeding behavior induced by neuropeptide Y and epinephrine. Physiol Behav 43:417–422

Clark JT, Kalra PS, Crowley WR, Kalra SP (1984) Neuropeptide Y and human pancreatic polypeptide stimulate feeding behavior in rats. Endocrinology 115:427–429

Clark JT, Kalra PS, Kalra SP (1985) Neuropeptide Y stimulates feeding but inhibits sexual behavior in rats. Endocrinology 117:2435–2442

Clark JT, Kalra SP, Kalra PS (1987a) Effects of a selective alpha 1-adrenoceptor agonist, methoxamine, on sexual behavior and penile reflexes. Physiol Behav 40:747–753

Clark JT, Sahu A, Kalra PS, Balasubramaniam A, Kalra SP (1987b) Neuropeptide Y (NPY)-induced feeding behavior in female rats: comparison with human NPY ([Met17]NPY), NPY analog ([norLeu4]NPY) and peptide YY. Regul Pept 17:31–39

Corp ES, Greco B, Powers JB, Bivens CL, Wade GN (2001) Neuropeptide Y inhibits estrous behavior and stimulates feeding via separate receptors in Syrian hamsters. Am J Physiol Regul Integr Comp Physiol 280: R1061–R1068.

Cowley MA, Pronchuk N, Fan W, Dinulescu DM, Colmers WF, Cone RD (1999) Integration of NPY, AGRP, and melanocortin signals in the hypothalamic paraventricular nucleus: evidence of a cellular basis for the adipostat. Neuron 24:155–163

Cowley MA, Smart JL, Rubinstein M, Cerdan MG, Diano S, Horvath TL, Cone RD, Low MJ (2001) Leptin activates anorexigenic POMC neurons through a neural network in the arcuate nucleus. Nature 411:480–484

Crowley WR (1999) Toward Multifactorial Hypothalamic Regulation of Anterior Pituitary Hormone Secretion. News Physiol Sci 14:54–58

Crowley WR, Hassid A, Kalra SP (1987) Neuropeptide Y enhances the release of luteinizing hormone (LH) induced by LH-releasing hormone. Endocrinology 120:941–945

Crowley WR, Kalra SP (1987) Neuropeptide Y stimulates the release of luteinizing hormone-releasing hormone from medial basal hypothalamus in vitro: modulation by ovarian hormones. Neuroendocrinology 46:97–103

Crowley WR, Kalra SP (1988) Regulation of luteinizing hormone secretion by neuropeptide Y in rats: hypothalamic and pituitary actions. Synapse 2:276–281

Crowley WR, Ramoz G, Hurst B (2003) Evidence for involvement of neuropeptide Y and melanocortin systems in the hyperphagia of lactation in rats. Pharmacol Biochem Behav 74:417–424

Cummings DE, Purnell JQ, Frayo RS, Schmidova K, Wisse BE, Weigle DS (2001) A preprandial rise in plasma ghrelin levels suggests a role in meal initiation in humans. Diabetes 50:1714–1719

Daniels AJ, Grizzle MK, Wiard RP, Matthews JE, Heyer D (2002) Food intake inhibition and reduction in body weight gain in lean and obese rodents treated with GW438014A, a potent and selective NPY-Y_5 receptor antagonist. Regul Pept 106:47–54

Daniels AJ, Matthews JE, Slepetis RJ, Jansen M, Viveros OH, Tadepalli A, Harrington W, Heyer D, Landavazo A, Leban JJ, Spaltenstein A (1995) High-affinity neuropeptide Y receptor antagonists. Proc Natl Acad Sci USA 92:9067–9071

Dhillon H, Kalra SP, Kalra PS (2001a) Dose-dependent effects of central leptin gene therapy on genes that regulate body weight and appetite in the hypothalamus. Mol Ther 4:139–145

Dhillon H, Kalra SP, Prima V, Zolotukhin S, Scarpace PJ, Moldawer LL, Muzyczka N, Kalra PS (2001b) Central leptin gene therapy suppresses body weight gain, adiposity and serum insulin without affecting food consumption in normal rats: a long-term study. Regul Pept 99:69–77

Diano S, Kalra SP, Sakamoto H, Horvath TL (1998) Leptin receptors in estrogen receptor-containing neurons of the female rat hypothalamus. Brain Res 812:256–259

Dube MG, Beretta E, Dhillon H, Ueno N, Kalra PS, Kalra SP (2002) Central leptin gene therapy blocks high fat diet-induced weight gain, hyperleptinemia and hyperinsulinemia: effects on serum ghrelin levels. Diabetes 51:1729–1736

Dube MG, Horvath TL, Kalra PS, Kalra SP (2000a) Evidence of NPY Y_5 receptor involvement in food intake elicited by orexin A in sated rats. Peptides 21:1557–1560

Dube MG, Kalra PS, Crowley WR, Kalra SP (1995) Evidence of a physiological role for neuropeptide Y in ventromedial hypothalamic lesion-induced hyperphagia. Brain Res 690:275–278

Dube MG, Kalra SP, Kalra PS (1999a) Food intake elicited by central administration of orexins/hypocretins: identification of hypothalamic sites of action. Brain Res 842:473–477

Dube MG, Pu S, Kalra SP, Kalra PS (2000b) Melanocortin signaling is decreased during neurotoxin-induced transient hyperphagia and increased body-weight gain. Peptides 21:793–801

Dube MG, Sahu A, Kalra PS, Kalra SP (1992) Neuropeptide Y release is elevated from the microdissected paraventricular nucleus of food-deprived rats: an in vitro study. Endocrinology 131:684–688

Dube MG, Xu B, Crowley WR, Kalra PS, Kalra SP (1994) Evidence that neuropeptide Y is a physiological signal for normal food intake. Brain Res 646:341–344

Dube MG, Xu B, Kalra PS, Sninsky CA, Kalra SP (1999b) Disruption in neuropeptide Y and leptin signaling in obese ventromedial hypothalamic-lesioned rats. Brain Res 816:38–46

Dumont T, Jacques D, St. Pierre JA, Quirion R (1997) Neuropeptide Y receptor types in the mammalian brain: species differences and status in the human central nervous system. In: Grundermar L, Bloom S (eds) Neuropeptide Y and drug development. Academic Press, New York, pp 57–86

El Majdoubi M, Sahu A, Ramaswamy S, Plant TM (2000) Neuropeptide Y: A hypothalamic brake restraining the onset of puberty in primates. Proc Natl Acad Sci USA 97:6179–6184

Erickson JC, Hollopeter G, Palmiter RD (1996) Attenuation of the obesity syndrome of ob/ob mice by the loss of neuropeptide Y. Science 274:1704–1707

Everitt BJ, Hokfelt T (1989) The coexistence of neuropeptide Y with other peptides and amines in the central nervous system. In: Mutt V, Fuxe K, Hokfelt T, Lundberg J (eds) Neuropeptide Y. New York, Raven Press, pp 61–72

Friedman JM, Halaas JL (1998) Leptin and the regulation of body weight in mammals. Nature 395:763–770

Fuxe K, Tinner B, Caberlotto L, Bunnemann B, Agnati LF (1997) NPY Y_1 receptor like immunoreactivity exists in a subpopulation of beta- endorphin immunoreactive nerve cells in the arcuate nucleus: a double immunolabelling analysis in the rat. Neurosci Lett 225:49–52

Hahn TM, Breininger JF, Baskin DG, Schwartz MW (1998) Coexpression of Agrp and NPY in fasting-activated hypothalamic neurons. Nature Neurosci 1:271–272

Hokfelt T, Broberger C, Xu ZQ, Sergeyev V, Ubink R, Diez M (2000) Neuropeptides—an overview. Neuropharmacology 39:1337–1356

Hokfelt T, Broberger C, Zhang X, Diez M, Kopp J, Xu Z, Landry M, Bao L, Schalling M, Koistinaho J, DeArmond SJ, Prusiner S, Gong J, Walsh JH (1998) Neuropeptide Y: some viewpoints on a multifaceted peptide in the normal and diseased nervous system. Brain Res Brain Res Rev 26:154–166

Hollopeter G, Erickson JC, Seeley RJ, Marsh DJ, Palmiter RD (1998) Response of neuropeptide Y-deficient mice to feeding effectors. Regul Pept 75/76:383–389

Horvath TL, Bechmann I, Naftolin F, Kalra SP, Leranth C (1997) Heterogeneity in the neuropeptide Y-containing neurons of the rat arcuate nucleus: GABAergic and non-GABAergic subpopulations. Brain Res 756:283–286

Horvath TL, Diano S, Sotonyi P, Heiman M, Tschop M (2001a) Minireview: ghrelin and the regulation of energy balance-a hypothalamic perspective. Endocrinology 142:4163–4169

Horvath TL, Diano S, van den Pol AN (1999) Synaptic interaction between hypocretin (orexin) and neuropeptide Y cells in the rodent and primate hypothalamus: a novel circuit implicated in metabolic and endocrine regulations. J Neurosci 19:1072–1087

Horvath TL, Diano S, van den Pol AN (1999) Synaptic interaction between hypocretin (orexin) and neuropeptide Y cells in the rodent and primate hypothalamus: a novel circuit implicated in metabolic and endocrine regulations. J Neurosci 19:1072–1087

Horvath TL, Kalra SP, Naftolin F, Leranth C (1995) Morphological evidence for a galanin-opiate interaction in the rat mediobasal hypothalamus. J Neuroendocrinol 7:579–588

Horvath TL, Naftolin F, Kalra SP, Leranth C (1992a) Neuropeptide-Y innervation of beta-endorphin-containing cells in the rat mediobasal hypothalamus: a light and electron microscopic double immunostaining analysis. Endocrinology 131:2461–2467

Horvath TL, Naftolin F, Leranth C (1992b) GABAergic and catecholaminergic innervation of mediobasal hypothalamic beta-endorphin cells projecting to the medial preoptic area. Neuroscience 51:391–399

Horvath TL, Naftolin F, Leranth C, Sahu A, Kalra SP (1996) Morphological and pharmacological evidence for neuropeptide Y-galanin interaction in the rat hypothalamus. Endocrinology 137:3069–3078

Horvath TL, Pu S, Dube MG, Diano S, Kalra SP (2001b) A GABA-neuropeptide Y (NPY) interplay in LH release. Peptides 22:473–481

Inui A (1999) Neuropeptide Y feeding receptors: Are multiple subtypes involved? Trends Pharmacol Sci 20:43–46

Inui A (2000) Transgenic approach to the study of body weight regulation. Pharmacol Rev 52:35–61
Jain MR, Dube MG, Kalra SP, Kalra PS (1998) Neuropeptide Y release in the paraventricular nucleus is decreased during transient hyperphagia induced by microinjection of colchicine into the ventromedial nucleus of rats. Neurosci Lett 256:21–24
Jain MR, Horvath TL, Kalra PS, Kalra SP (2000) Evidence that NPY Y_1 receptors are involved in stimulation of feeding by orexins (hypocretins) in sated rats. Regul Pept 87:19–24
Jain MR, Pu S, Kalra PS, Kalra SP (1999) Evidence that stimulation of two modalities of pituitary luteinizing hormone release in ovarian steroid-primed ovariectomized rats may involve neuropeptide Y Y_1 and Y_4 receptors. Endocrinology 140:5171–5177
Kalra PS, Bonavera JJ, Kalra SP (1995a) Central administration of antisense oligodeoxynucleotides to neuropeptide Y (NPY) mRNA reveals the critical role of newly synthesized NPY in regulation of LHRH release. Regul Pept 59:215–220
Kalra PS, Dube MG, Kalra SP (2000a) Effects of centrally administered antisense oligodeoxynucleotides on feeding behavior and hormone secretion. Methods Enzymol 314:184–200
Kalra PS, Dube MG, Xu B, Farmerie WG, Kalra SP (1998a) Evidence that dark-phase hyperphagia induced by neurotoxin 6- hydroxydopamine may be due to decreased leptin and increased neuropeptide Y signaling. Physiol Behav 63:829–835
Kalra PS, Dube MG, Xu B, Farmerie WG, Kalra SP (1998b) Neuropeptide Y (NPY) Y_1 receptor mRNA is upregulated in association with transient hyperphagia and body weight gain: evidence for a hypothalamic site for concurrent development of leptin resistance. J Neuroendocrinol 10:43–49
Kalra PS, Dube MG, Xu B, Kalra SP (1997a) Increased receptor sensitivity to neuropeptide Y in the hypothalamus may underlie transient hyperphagia and body weight gain. Regul Pept 72:121–130
Kalra PS, Kalra SP (2002) Obesity and metabolic syndrome: long-term benefits of central leptin gene therapy. In: Prous JR (ed) Drugs of Today. Barcelona, Prous Science 38:745–757
Kalra PS, Norlin M, Kalra SP (1995b) Neuropeptide Y stimulates beta-endorphin release in the basal hypothalamus: role of gonadal steroids. Brain Res 705:353–356
Kalra PS, Pu S, Edwards TG, Dube MG (1998c) Hypothalamic galanin is upregulated during hyperphagia. Ann N Y Acad Sci 863:432–434
Kalra S (1996) Is neuropeptide Y a naturally occuring appetite transducer? Curr Opin Endocrinol Diabetes 3:157–163
Kalra SP (1993) Mandatory neuropeptide-steroid signaling for the preovulatory luteinizing hormone-releasing hormone discharge. Endocrinol Rev 14:507–538
Kalra SP (1997) Appetite and body weight regulation: is it all in the brain? Neuron 19:227–230
Kalra SP (2001) Circumventing leptin resistance for weight control. Proc Natl Acad Sci USA 98:4279–4281
Kalra SP, Bagnasco M, Otukonyong EE, Dube MG, Kalra PS (2003) Rhythmic, reciprocal ghrelin and leptin signaling: new insight in the development of obesity. Regul Pept 111:1–11.
Kalra SP, Clark JT, Sahu A, Dube MG, Kalra PS (1988a) Control of feeding and sexual behaviors by neuropeptide Y: physiological implications. Synapse 2:254–257
Kalra SP, Clark JT, Sahu A, Kalra PS, Crowley WR (1989) Hypothalamic NPY: A local circuit is the control of reproduction and behavior. In: Hokfelt T, Lundberg JD, Mutt V, Fuxe K (eds) Neuropeptide Y. New York, Raven Press, pp 229–242
Kalra SP, Crowley WR (1984a) Differential effects of pancreatic polypeptide on luteinizing hormone release in female rats. Neuroendocrinology 38:511–513
Kalra SP, Crowley WR (1984b) Norepinephrine-like effects of neuropeptide Y on LH release in the rat. Life Sci 35:1173–1176

Kalra SP, Dube MG, Fournier A, Kalra PS (1991a) Structure-function analysis of stimulation of food intake by neuropeptide Y: effects of receptor agonists. Physiol Behav 50:5–9

Kalra SP, Dube MG, Kalra PS (1988b) Continuous intraventricular infusion of neuropeptide Y evokes episodic food intake in satiated female rats: effects of adrenalectomy and cholecystokinin. Peptides 9:723–728

Kalra SP, Dube MG, Pu S, Xu B, Horvath TL, Kalra PS (1999) Interacting appetite-regulating pathways in the hypothalamic regulation of body weight. Endocrinol Rev 20:68–100

Kalra SP, Dube MG, Sahu A, Phelps CP, Kalra PS (1991b) Neuropeptide Y secretion increases in the paraventricular nucleus in association with increased appetite for food. Proc Natl Acad Sci USA 88:10931–10935

Kalra SP, Fuentes M, Fournier A, Parker SL, Crowley WR (1992) Involvement of the Y-1 receptor subtype in the regulation of luteinizing hormone secretion by neuropeptide Y in rats. Endocrinology 130:3323–3330

Kalra SP, Horvath T, Naftolin F, Xu B, Pu S, Kalra PS (1997b) The interactive language of the hypothalamus for the gonadotropin releasing hormone (GNRH) system. J Neuroendocrinol 9:569–576

Kalra SP, Horvath TL (1998) Neuroendocrine interactions between galanin, opioids, and neuropeptide Y in the control of reproduction and appetite. Ann N Y Acad Sci 863:236–240

Kalra SP, Kalra PS (1983) Neural regulation of luteinizing hormone secretion in the rat. Endocrinol Rev 4:311–351

Kalra SP, Kalra PS (1984) Opioid-adrenergic-steroid connection in regulation of luteinizing hormone secretion in the rat. Neuroendocrinology 38:418–426

Kalra SP, Kalra PS (1990) Neuropeptide Y: A novel peptidergic signal for the control of feeding behavior. In: Pfaff DW, Ganten D (eds) Current topics in neuroendocrinology, vol 10. Berlin, Springer-Verlag, pp 192–197

Kalra SP, Kalra PS (1996) Nutritional infertility: the role of the interconnected hypothalamic neuropeptide Y-galanin-opioid network. Front Neuroendocrinol 17:371–401

Kalra SP, Kalra PS (2003) Neuropeptide Y. In: Henry HL, Norman AW (eds) Encyclopedia of hormones, vol 3. San Diego, Academic Press, pp 37–45

Kalra SP, Pu S, Horvath TL, Kalra PS (2000b) Leptin and NPY regulation of GnRH secretion and energy homeostasis. In: Bourguignon J, Pap TM (eds) The onset of puberty in perspective. Amsterdam, Elsevier Science, pp 317–327

Kalra SP, Sahu A, Dube MG, Bonavera JJ, Kalra PS (1996a) Neuropeptide Y and its neural conections in the etiology of obesity and associated neuroendocrine and behavioral disorders. In: Bray GA, Ryan DH (eds) Molecular and genetic aspects of obesity, vol 5. Baton Rouge, Louisiana State University Press, pp 219–232

Kalra SP, Sahu A, Kalra PS (1993) Ageing of the neuropeptidergic signals in rats. J Reprod Fertil Suppl 46:11–19

Kalra SP, Sahu A, Kalra PS, Crowley WR (1990) Hypothalamic neuropeptide Y: a circuit in the regulation of gonadotropin secretion and feeding behavior. Ann N Y Acad Sci 611:273–283

Kalra SP, Xu B, Dube M, Kaibara A, Martin D, Moldawer LL (1996b) Ciliary neurotropic factor is a potent anorectic cytokine acting through hypothalamic neuropeptide Y. 26th Annual Society for Neuroscience. Washington DC, November 1996 (abstract p 1352)

Kalra SP, Xu B, Dube MG, Moldawer LL, Martin D, Kalra PS (1998d) Leptin and ciliary neurotropic factor (CNTF) inhibit fasting-induced suppression of luteinizing hormone release in rats: role of neuropeptide Y. Neurosci Lett 240:45–49

King PJ, Williams G, Doods H, Widdowson PS (2000) Effect of a selective neuropeptide Y Y(2) receptor antagonist, BIIE0246 on neuropeptide Y release. Eur J Pharmacol 396: R1–R3

Kushi A, Sasai H, Koizumi H, Takeda N, Yokoyama M, Nakamura M (1998) Obesity and mild hyperinsulinemia found in neuropeptide Y-Y_1 receptor- deficient mice. Proc Natl Acad Sci USA 95:15659–15664

Larhammar D, Wraith A, Berglund MM, Holmberg SK, Lundell I (2001) Origins of the many NPY-family receptors in mammals. Peptides 22:295–307

Leupen SM, Besecke LM, Levine JE (1997) Neuropeptide Y Y_1-receptor stimulation is required for physiological amplification of preovulatory luteinizing hormone surges. Endocrinology 138:2735–2739

Lopez M, Seoane LM, Garcia Mdel C, Dieguez C, Senaris R (2002) Neuropeptide Y, but not agouti-related peptide or melanin-concentrating hormone, is a target peptide for orexin-A feeding actions in the rat hypothalamus. Neuroendocrinology 75:34–44

Magni P (2003) Hormonal control of the neuropeptide y system. Curr Protein Pept Sci 4:45–57

Marin-Bivens CL, Kalra SP, Olster DH (1998) Intraventricular injection of neuropeptide Y antisera curbs weight gain and feeding, and increases the display of sexual behaviors in obese Zucker female rats. Regul Pept 75/76:327–334

Marsh DJ, Hollopeter G, Kafer KE, Palmiter RD (1998) Role of the Y_5 neuropeptide Y receptor in feeding and obesity. Nature Med 4:718–721

Naveilhan P, Hassani H, Canals JM, Ekstrand AJ, Larefalk A, Chhajlani V, Arenas E, Gedda K, Svensson L, Thoren P, Ernfors P (1999) Normal feeding behavior, body weight and leptin response require the neuropeptide Y Y_2 receptor. Nature Med 5:1188–1193.

Naveilhan P, Svensson L, Nystrom S, Ekstrand AJ, Ernfors P (2002) Attenuation of hypercholesterolemia and hyperglycemia in ob/ob mice by NPY Y_2 receptor ablation. Peptides 23:1087–1091

Ovesjo ML, Gamstedt M, Collin M, Meister B (2001) GABAergic nature of hypothalamic leptin target neurones in the ventromedial arcuate nucleus. J Neuroendocrinol 13:505–516

Pedrazzini T, Seydoux J, Kunstner P, Aubert JF, Grouzmann E, Beermann F, Brunner HR (1998) Cardiovascular response, feeding behavior and locomotor activity in mice lacking the NPY Y_1 receptor. Nature Med 4:722–726

Poggioli R, Vergoni AV, Marrama D, Giuliani D, Bertolini A (1990) NPY-induced inhibition of male copulatory activity is a direct behavioural effect. Neuropeptides 16:169–172

Pralong FP, Gonzales C, Voirol MJ, Palmiter RD, Brunner HR, Gaillard RC, Seydoux J, Pedrazzini T (2002) The neuropeptide Y Y_1 receptor regulates leptin-mediated control of energy homeostasis and reproductive functions. FASEB J 16:712–714

Pronchuk N, Beck-Sickinger AG, Colmers WF (2002) Multiple NPY receptors Inhibit GABA(A) synaptic responses of rat medial parvocellular effector neurons in the hypothalamic paraventricular nucleus. Endocrinology 143:535–543

Pu S, Dhillon H, Moldawer LL, Kalra PS, Kalra SP (2000a) Neuropeptide Y counteracts the anorectic and weight reducing effects of ciliary neurotropic factor. J Neuroendocrinol 12:827–832

Pu S, Dube MG, Edwards TG, Kalra SP, Kalra PS (1999a) Disruption of neural signaling within the hypothalamic ventromedial nucleus upregulates galanin gene expression in association with hyperphagia: an in situ hybridization analysis. Brain Res Mol Brain Res 64:85–91

Pu S, Dube MG, Kalra PS, P. KS (2000b) Regulation of leptin secretion: Effects of aging on daily patterns of serum leptin and food consumption. Regul Pept 92:107–111

Pu S, Dube MG, Xu B, Kalra SP, Kalra PS (1998a) Induction of neuropeptide Y (NPY) gene expression in novel hypothalamic sites in association with transient hyperphagia and body weight gain. 80th Annual Meeting of the Endocrine Society. New Orleans, June 1998 (abstract p301)

Pu S, Horvath TL, Diano S, Naftolin F, Kalra PS, Kalra SP (1997a) Evidence showing that beta-endorphin regulates cyclic guanosine 3',5'- monophosphate (cGMP) efflux: anatomical and functional support for an interaction between opiates and nitric oxide. Endocrinology 138:1537-1543

Pu S, Jain MR, Horvath TL, Diano S, Kalra PS, Kalra SP (1999) Interactions between neuropeptide Y and gamma-aminobutyric acid in stimulation of feeding: a morphological and pharmacological analysis. Endocrinology 140:933-940

Pu S, Jain MR, Horvath TL, Diano S, Kalra PS, Kalra SP (1999b) Interactions between neuropeptide Y and gamma-aminobutyric acid in stimulation of feeding: a morphological and pharmacological analysis. Endocrinology 140:933-940

Pu S, Kalra PS, Kalra SP (1999c) Energy homeostasis and reproduction are regulated by different subpopulations of neuropeptide Y (NPY) neurons in the arcuate nucleus. 81st Annual Meeting of the Endocrine Society. San Diego, June 1999 (abstractp236)

Pu S, Jain MR, Kalra PS, Kalra SP (1998b) Orexins, a novel family of hypothalamic neuropeptides, modulate pituitary luteinizing hormone secretion in an ovarian steroid-dependent manner. Regul Pept 78:133-136

Pu S, Kalra PS, Kalra SP (1998c) Diurnal rhythm in cyclic GMP/nitric oxide efflux in the medial preoptic area of male rats. Brain Res 808:310-312

Pu S, Dube MG, Kalra, SP and Kalra PS (2003) Permanent interruption of information flow between hypothalamus and hindbrain produces obesity accompanied by a selective dark-phase hyperphagia and hyperinsulinemia. 33rd Annual Society for Neuroscience Meeting, New Orleans. November 2003 (abstract in press)

Raposinho PD, White RB, Aubert ML (2003) The melanocortin agonist Melanotan-II reduces the orexigenic and adipogenic effects of neuropeptide Y (NPY) but does not affect the NPY- driven suppressive effects on the gonadotropic and somatotropic axes in the male rat. J Neuroendocrinol 15:173-181

Sahu A (2002) Interactions of neuropeptide Y, hypocretin-I (orexin A) and melanin- concentrating hormone on feeding in rats. Brain Res 944:232-238

Sahu A, Crowley WR, Kalra PS, Kalra SP (1992a) A selective sexually dimorphic response in the median eminence neuropeptide Y. Brain Res 573:235-242

Sahu A, Crowley WR, Kalra SP (1990a) An opioid-neuropeptide-Y transmission line to luteinizing hormone (LH)- releasing hormone neurons: a role in the induction of LH surge. Endocrinology 126:876-883

Sahu A, Crowley WR, Kalra SP (1994a) Hypothalamic neuropeptide-Y gene expression increases before the onset of the ovarian steroid-induced luteinizing hormone surge. Endocrinology 134:1018-1022

Sahu A, Crowley WR, Kalra SP (1995) Evidence that hypothalamic neuropeptide Y gene expression increases before the onset of the preovulatory LH surge. J Neuroendocrinol 7:291-296

Sahu A, Crowley WR, Tatemoto K, Balasubramaniam A, Kalra SP (1987a) Effects of neuropeptide Y, NPY analog (norleucine4-NPY), galanin and neuropeptide K on LH release in ovariectomized (ovx) and ovx estrogen, progesterone-treated rats. Peptides 8:921-926

Sahu A, Dube MG, Kalra SP, Kalra PS (1988a) Bilateral neural transections at the level of mesencephalon increase food intake and reduce latency to onset of feeding in response to neuropeptide Y. Peptides 9:1269-1273

Sahu A, Jacobson W, Crowley WR, Kalra SP (1989) Dynamic changes in neuropeptide Y concentrations in the median eminence in neuropeptide Y concentrations in the median eminence in association with preovulatory luteinizing hormone (LH) release in the rat. J Neuroendo 1:83-87

Sahu A, Kalra PS, Crowley WR, Kalra SP (1988b) Evidence that hypothalamic neuropeptide Y secretion decreases in aged male rats: implications for reproductive aging. Endocrinology 122:2199-2203

Sahu A, Kalra PS, Kalra SP (1988c) Food deprivation and ingestion induce reciprocal changes in neuropeptide Y concentrations in the paraventricular nucleus. Peptides 9:83–86

Sahu A, Kalra SP (1998) Absence of increased neuropeptide Y neuronal activity before and during the luteinizing hormone (LH) surge may underlie the attenuated preovulatory LH surge in middle-aged rats. Endocrinology 139:696–702

Sahu A, Kalra SP, Crowley WR, Kalra PS (1988d) Evidence that NPY-containing neurons in the brainstem project into selected hypothalamic nuclei: implication in feeding behavior. Brain Res 457:376–378

Sahu A, Kalra SP, Crowley WR, O'Donohue TL, Kalra PS (1987b) Neuropeptide Y levels in microdissected regions of the hypothalamus and in vitro release in response to KCl and prostaglandin E2: effects of castration. Endocrinology 120:1831–1836

Sahu A, Phelps CP, White JD, Crowley WR, Kalra SP, Kalra PS (1992b) Steroidal regulation of hypothalamic neuropeptide Y release and gene expression. Endocrinology 130:3331–3336

Sahu A, Sninsky CA, Kalra PS, Kalra SP (1990b) Neuropeptide-Y concentration in microdissected hypothalamic regions and in vitro release from the medial basal hypothalamus-preoptic area of streptozotocin-diabetic rats with and without insulin substitution therapy. Endocrinology 126:192–198

Sahu A, Sninsky CA, Kalra SP (1997) Evidence that hypothalamic neuropeptide Y gene expression and NPY levels in the paraventricular nucleus increase before the onset of hyperphagia in experimental diabetes. Brain Res 755:339–342

Sahu A, Sninsky CA, Phelps CP, Dube MG, Kalra PS, Kalra SP (1992c) Neuropeptide Y release from the paraventricular nucleus increases in association with hyperphagia in streptozotocin-induced diabetic rats. Endocrinology 131:2979–2985

Sahu A, White JD, Kalra PS, Kalra SP (1992d) Hypothalamic neuropeptide Y gene expression in rats on scheduled feeding regimen. Brain Res Mol Brain Res 15:15–18

Sahu A, Xu B, Kalra SP (1994b) Role of galanin in stimulation of pituitary luteinizing hormone secretion as revealed by a specific receptor antagonist, galantide. Endocrinology 134:529–536

Sainsbury A, Schwarzer C, Couzens M, Fetissov S, Furtinger S, Jenkins A, Cox HM, Sperk G, Hokfelt T, Herzog H (2002a) Important role of hypothalamic Y_2 receptors in body weight regulation revealed in conditional knockout mice. Proc Natl Acad Sci USA 99:8938–8943

Sainsbury A, Schwarzer C, Couzens M, Jenkins A, Oakes SR, Ormandy CJ, Herzog H (2002b) Y_4 receptor knockout rescues fertility in ob/ob mice. Genes Dev 16:1077–1088

Sar M, Sahu A, Crowley WR, Kalra SP (1990) Localization of neuropeptide-Y immunoreactivity in estradiol- concentrating cells in the hypothalamus. Endocrinology 127:2752–2756

Silva AP, Cavadas C, Grouzmann E (2002) Neuropeptide Y and its receptors as potential therapeutic drug targets. Clin Chim Acta 326:3–25.

Smith MS (1993) Lactation alters neuropeptide-Y and proopiomelanocortin gene expression in the arcuate nucleus of the rat. Endocrinology 133:1258–1265

Smith MS, Grove KL (2002) Integration of the regulation of reproductive function and energy balance: lactation as a model. Front Neuroendocrinol 23:225–256

Toufexis DJ, Kyriazis D, Woodside B (2002) Chronic neuropeptide Y Y_5 receptor stimulation suppresses reproduction in virgin female and lactating rats. J Neuroendocrinol 14:492–497

Tschop M, Smiley DL, Heiman ML (2000) Ghrelin induces adiposity in rodents. Nature 407:908–913

Tschop M, Wawarta R, Riepl RL, Friedrich S, Bidlingmaier M, Landgraf R, Folwaczny C (2001) Post-prandial decrease of circulating human ghrelin levels. J Endocrinol Invest 24: RC19–RC21

Turnbull AV, Ellershaw L, Masters DJ, Birtles S, Boyer S, Carroll D, Clarkson P, Loxham SJ, McAulay P, Teague JL, Foote KM, Pease JE, Block MH (2002) Selective antagonism of the NPY Y_5 receptor does not have a major effect on feeding in rats. Diabetes 51:2441–2449

Woller MJ, McDonald JK, Reboussin DM, Terasawa E (1992) Neuropeptide Y is a neuromodulator of pulsatile luteinizing hormone- releasing hormone release in the gonadectomized rhesus monkey. Endocrinology 130:2333–2342

Xu B, Kalra PS, Farmerie WG, Kalra SP (1999) Daily changes in hypothalamic gene expression of neuropeptide Y, galanin, proopiomelanocortin, and adipocyte leptin gene expression and secretion: effects of food restriction. Endocrinology 140:2868–2875

Xu B, Kalra PS, Kalra SP (1996a) Food restriction upregulates hypothalamic NPY gene expression: loss of daily rhythmicity. 26th Annual Meeting of the Society for Neuroscience. Washington, DC, November 1996 (abstract p 1685

Xu B, Kalra PS, Moldawer LL, Kalra SP (1998) Increased appetite augments hypothalamic NPY Y_1 receptor gene expression: effects of anorexigenic ciliary neurotropic factor. Regul Pept 75/76:391–395

Xu B, Pu S, Kalra PS, Hyde JF, Crowley WR, Kalra SP (1996b) An interactive physiological role of neuropeptide Y and galanin in pulsatile pituitary luteinizing hormone secretion. Endocrinology 137:5297–5302

Xu B, Sahu A, Crowley WR, Leranth C, Horvath T, Kalra SP (1993) Role of neuropeptide-Y in episodic luteinizing hormone release in ovariectomized rats: an excitatory component and opioid involvement. Endocrinology 133:747–754

Xu B, Sahu A, Kalra PS, Crowley WR, Kalra SP (1996c) Disinhibition from opioid influence augments hypothalamic neuropeptide Y (NPY) gene expression and pituitary luteinizing hormone release: effects of NPY messenger ribonucleic acid antisense oligodeoxynucleotides. Endocrinology 137:78–84

Xu M, Hill JW, Levine JE (2000a) Attenuation of luteinizing hormone surges in neuropeptide Y knockout mice. Neuroendocrinology 72:263–271

Xu M, Urban JH, Hill JW, Levine JE (2000b) Regulation of hypothalamic neuropeptide Y Y_1 receptor gene expression during the estrous cycle: role of progesterone receptors. Endocrinology 141:3319–3327

Yokosuka M, Dube MG, Kalra PS, Kalra SP (2001) The mPVN mediates blockade of NPY-induced feeding by a Y_5 receptor antagonist: a c-FOS analysis. Peptides 22:507–514

Yokosuka M, Kalra PS, Kalra SP (1999) Inhibition of neuropeptide Y (NPY)-induced feeding and c-Fos response in magnocellular paraventricular nucleus by a NPY receptor antagonist: a site of NPY action. Endocrinology 140:4494–4500

Yokosuka M, Xu B, Pu S, Kalra PS, Kalra SP (1998) Neural substrates for leptin and neuropeptide Y (NPY) interaction: hypothalamic sites associated with inhibition of NPY-induced food intake. Physiol Behav 64:331–338

Zammaretti F, Panzica G, Eva C (2001) Fasting, leptin treatment, and glucose administration differentially regulate Y(1) receptor gene expression in the hypothalamus of transgenic mice. Endocrinology 142:3774–3782

Behavioral Effects of Neuropeptide Y

T. E. Thiele[1] · M. Heilig[2]

[1] Department of Psychology, University of North Carolina, Davie Hall, CB 3270, Chapel Hill, NC 27599-3270, USA
e-mail: thiele@unc.edu
[2] Department of NEUROTEC, M57 Huddinge University Hospital, Karolinska Institute, 14186 Stockholm, Sweden

1	Introduction	252
2	Actions of NPY Related to Stress, Anxiety, and Depression	253
2.1	Suppression of Locomotion and Sedation by Central NPY	253
2.2	Anxiolytic-Like Actions of NPY	254
2.2.1	NPY Is Active in a Wide Range of Anxiety Models	254
2.2.2	Receptor Pharmacology of the Anti-Stress Actions of NPY	255
2.2.3	Anatomical Substrates of the Anti-Stress Actions of NPY	257
2.2.4	'Phasic' vs. 'Tonic' Effects of Endogenous NPY in Stress and Fear	259
2.2.5	Expression of NPY and Behavioral Stress Responses	260
2.3	NPY and Depression	262
2.3.1	The Anxiety–Depression Spectrum	262
2.3.2	Effects of Antidepressant Treatments on the NPY System	262
2.3.3	Animal Models of Depression	264
2.3.4	Antidepressant-Like Effects of Central NPY	264
2.3.5	Human Findings: Central Nervous System	265
2.3.6	Human Findings: Peripheral Nervous System	266
2.4	Actions of NPY Related to Stress, Anxiety, and Depression: Summary	267
3	A Role for NPY in Ethanol-Seeking Behavior and Neurobiological Responses to Ethanol	267
3.1	Altered Ethanol Drinking Associated with Genetic Manipulation of NPY Signaling	267
3.2	Altered Ethanol Drinking Following Central Infusion of NPY and Related Compounds	268
3.3	Receptors Mediating the Effects of NPY on Ethanol Drinking	269
3.4	Brain Electrophysiological Responses to NPY and Ethanol Are Similar	270
3.5	Ethanol Administration Modulates Central NPY Signaling	271
3.6	NPY and Alcoholism: Human Genetic Studies	271
3.7	NPY and Alcohol: Summary	272
4	Behavioral Effects of NPY: Conclusions	273
References		274

Abstract Neuropeptide Y (NPY), a 36-amino acid neurotransmitter that is widely distributed throughout the nervous system, antagonizes behavioral consequences of stress/depression and attenuates ethanol-seeking behavior through actions within the brain. The anxiolytic actions of NPY were initially demonstrated by microinjection of exogenous ligands directly into the brain, and then by genetically modified animals. There is also evidence to suggest that altered NPY signaling may contribute to anxiety disorders and depression in humans. NPY reduces anxiety, at least in part, by acting on NPY Y_1 receptors in the lateral/basolateral nuclei of the amygdala. Importantly, NPY is anxiolytic across a wide range of animal models normally thought to reflect different aspects of emotionality and is more potent than other endogenous compounds. Thus, NPY likely acts on a common core mechanism of emotionality and behavioral stress. Recent research—genetic and pharmacological—suggests that NPY is also involved with voluntary ingestion of ethanol. Low levels (or the absence) of NPY promote high ethanol drinking, and enhanced NPY signaling prevents excessive ethanol consumption. Furthermore, brain tissue from alcoholics was found to have abnormally low levels of NPY when compared to control tissue. While the neuroanatomical substrate is unknown, NPY modulates ethanol intake by acting on the NPY Y_1 and Y_2 receptors. An interesting possibility is that high ethanol-seeking behavior is secondary to high levels of anxiety. Viewed this way, alterations of normal NPY activity may induce emotional disturbance, which in turn becomes a risk factor for alcoholism. Thus, drugs targeting the central NPY system may serve as useful therapeutic agents against stress-related disorders and alcoholism. Based on the current literature, compounds aimed at the Y_1 and/or Y_2 receptor(s) appear to be promising candidates.

Keywords Anxiety · Depression · Alcoholism · Neuropeptide Y · Behavior

1
Introduction

The present chapter focuses on two behavioral systems under the control of neuropeptide Y (NPY) signaling. NPY is expressed in several brain regions thought to be intimately involved with emotional balance, including the amygdala, hippocampus, and hypothalamus (see Chapter by Redrobe et al., this volume). In this chapter we present a review of literature showing how NPY integrates emotional behavior, including stress, anxiety, and depression (Heilig et al. 1993a; Heilig and Widerlov 1995). Systems involved in the regulation of emotionality are also likely to be central for alcoholism (Koob and Le Moal 1997). Recent research points to a role for NPY in ethanol-seeking behavior and neurobiological responses to ethanol (Heilig and Thorsell 2002; Thiele et al. 1998, 2002). These findings suggest that drugs targeting the central NPY system may serve as useful therapeutic agents against stress-related disorders and alcoholism. Food intake, one of the most well-studied behavioral systems modulated by NPY (Clark et al. 1984; Levine and Morley 1984), is described at length in the chapter by Levens in this volume.

2
Actions of NPY Related to Stress, Anxiety, and Depression

2.1
Suppression of Locomotion and Sedation by Central NPY

Shortly following its isolation, Fuxe and co-workers reported that central administration of NPY gave rise to long lasting synchronization of the EEG pattern (Fuxe et al. 1983). This type of effect is typically indicative of decreased arousal, and commonly seen upon administration of benzodiazepines or barbiturates at high, sedative doses. Following up on that lead, initial behavioral studies demonstrated that intracerebroventricular (icv) administration of high central NPY doses (1.0–5.0 nmol) produced behavioral sedation, manifested both as a decrease of home cage activity, and suppression of open-field activity. Sedative effects of NPY had some interesting and unusual characteristics. Of interest from a dependence point of view, no apparent tolerance was seen upon repeated administration. Furthermore, effects of a single icv dose lasted up to 3 days, but were still fully reversible (Heilig and Murison 1987b). Finally, in another similarity with sedative drugs, icv pretreatment with NPY largely prevented the formation of gastric erosions normally produced by cold water immersion stress (Heilig and Murison 1987a).

These findings indicated some shared properties between NPY and several classes of sedative compounds, including alcohol, benzodiazepines and barbiturates (see Sect. 10.3.4). This similarity was subsequently confirmed by EEG experiments in awake animals, where icv NPY decreased EEG power in general, but was found to particularly act on cortical areas and the amygdala, a pattern again resembling that produced by benzodiazepines (Ehlers et al. 1997). Observations suggesting similarities between these classes of compounds have more recently been extended by findings of anti-convulsant actions of NPY (Baraban et al. 1997; Woldbye et al. 1997), which possibly provide a mechanism for the anti-convulsant state seen following electroconvulsive treatment (Bolwig et al. 1999; Mikkelsen et al. 1994; Woldbye et al. 1996). Evidence for shared properties between NPY and sedative compounds was also provided by mutual substitution of NPY and ethanol with regard to electrophysiological effects (Ehlers et al. 1998b), and potentiation by NPY of barbiturate induced sleep (Naveilhan et al. 2001a, 2001b). The latter study mapped out NPY effects on sedation to the posterior hypothalamus, an area involved in the regulation of sleep–wake cycles.

Initial studies indicated that intact NPY was required for the behaviorally sedative effects of NPY, while the prototypical C-terminal fragment NPY(13–36) in fact produced the opposite effect (Heilig et al. 1988b). This was interpreted as evidence for Y_1-mediation of sedative NPY effects, although in retrospect, the data were clearly compatible with either a Y_1 or Y_5 mediated effect. More recently, supporting the original conclusion, Y_1 mediation of NPY-induced sedation was demonstrated using Y_1-receptor knockouts (Naveilhan et al. 2001a,). In contrast, anti-convulsant actions of NPY appear to be mediated through Y_5-receptors (Marsh et al. 1999).

2.2
Anxiolytic-Like Actions of NPY

2.2.1
NPY Is Active in a Wide Range of Anxiety Models

Based on similarities of NPY with benzodiazepines and barbiturates, it was hypothesized that lower doses of NPY might also mimic behavioral anti-stress and anti-anxiety actions of these compounds. This prediction has subsequently been confirmed in a highly consistent manner. Initial evidence along this line was obtained by analyzing the effects of icv NPY injections at nanomolar doses, in two different, pharmacologically validated animal models of anxiety: the elevated plus-maze and the Vogel punished drinking conflict test (Heilig et al. 1989). Subsequently, equivalent effects have been seen in numerous other models.

A wide range of animal anxiety models has an undisputable degree of predictive validity, but for most, this is based on their selective sensitivity to benzodiazepines. It is now being increasingly recognized that sensitivity to other classes of compounds, such as $5HT_{1A}$ ligands, may differ between models, and even between different protocols applied within one model, such as, for example, differences reported between single vs. repeated testing on the elevated plus-maze (File 1993). It has therefore been debated whether different animal models in the anxiety area reflect different aspects of clinical anxiety disorders, different disease entities, or perhaps are simply limited in their ability to model the relevant clinical disease (File 1993; Hogg 1996; McCreary et al. 1996; Rodgers 1997). Against this background, a remarkable feature of NPY is its ability to produce anti-stress/anti-anxiety effects in all models tested thus far, suggesting that it acts at a core process underlying fear- and stress-related behaviors and affect, common to different models and disorders.

To illustrate this, animal models in which NPY has demonstrated anxiolytic-like action can be categorized along some important dimensions. The Vogel (Vogel et al. 1971) and Geller–Seifter (Geller and Seifter 1960) tests, commonly referred to as 'conflict-tests', are the classical industry screening tests for benzodiazepine-like anti-anxiety action. They are both based on fear-induced suppression of behavior: in the former, drinking is suppressed by unconditioned fear of a mild electric shock, while the latter is based on cue-conditioned fear suppression of operant responding for a food pellet. In both tests, benzodiazepines reverse the fear-suppression of behavior, and so does icv administration of NPY, with high potency (Heilig et al. 1989, 1992). Results of the unconditioned Vogel test, where shock is actually delivered during the test session, could potentially be confounded by effects of NPY on nociception, in particular since antinociceptive effects of this peptide have been reported at different levels of the neuroaxis (Broqua et al. 1996; Hua et al. 1991; Mellado et al. 1996; Wang et al. 2000). However, control experiments suggest that icv administration of NPY at the doses and pretreatment intervals used to established its 'anti-conflict' effect, nociception is unaffected (Heilig et al. 1989). Another potential confound in the Vogel test is the motivation to drink, which, however, also seems unaffected at these conditions. In contrast, it is obvious that results obtained in the Geller–Seifter test represent a compound effect of anxiolytic-like and orexigenic action. However, as discussed below, an anatomical analysis based on site injections allows a clear separation of these two components.

A criticism of the classical conflict tests is that they do not probe the natural behavioral repertoire of the animal. The elevated plus-maze (Pellow et al. 1985) and the social interaction test (File 1980) are more attractive in that respect. The former is based on the conflict between the innate motivation to explore a novel environment, and its suppression by innate fear of open spaces, where presumably the probability of predator attack is increased. The latter model capitalizes on suppression of social interaction in a novel, exposed environment. In both models, classical anti-anxiety compounds counteract fear suppression of the respective spontaneous exploratory behavior, and so does NPY, with high potency and specificity (Broqua et al. 1995; Heilig et al. 1989; Sajdyk et al. 1999). A similar observation has been reported in the light–dark compartment test, which is conceptually related to the elevated plus-maze (Pich et al. 1993).

Finally, it can be noted that the tests described above are all based on fear-suppression of behavior. A potential weakness of this approach is that nonspecific, performance- or motivation-related drug effects might be interpreted as indicative of anxiety-modulating properties. The fear-potentiated startle model differs from most other anxiety paradigms in that it is based on fear potentiation rather than inhibition of behavior. It relies on anatomically well delineated circuitry to produce a well defined behavioral aspect of the integrated fear response, and is less sensitive to effects on motor performance or motivation (Davis et al. 1994). NPY effectively reverses the potentiation of the acoustic startle response which occurs upon presentation of an originally neutral stimulus, which during training has been paired with electric shock, making it a conditioned fear stimulus. This action of NPY occurs in the absence of effects on basal, unconditioned startle (Broqua et al. 1995). Consistent actions of NPY both in models based on response inhibition and response facilitation indicate that the observed behavioral actions are indeed specifically related to emotionality.

2.2.2
Receptor Pharmacology of the Anti-Stress Actions of NPY

From the outset, attempts to characterize the receptor subtype profile of the anti-stress effects of NPY have indicated a Y_1-like profile. Although this proposition was originally based on a very limited range of tools, a more refined pharmacological analysis seems to confirm it. Thus, initial experiments to address this issue were based on central administration of substituted peptide ligands, for the most part with agonist activity. The suggestion of Y_1-mediation was based on the observation that full length NPY peptide potently and dose dependently produced an anxiolytic-like effect in the elevated plus-maze, Vogel test (Heilig et al. 1989) and Geller–Seifter-test (Heilig et al. 1993a), while the C-terminal, presumably Y_2-selective fragment NPY(13–36) did not reproduce this action. When [Leu^{31}Pro^{34}]NPY was proposed as a selective Y_1-receptor agonist (Fuhlendorff et al. 1990), administration of this compound seemed to confirm the notion of Y_1 mediation, but it has since become apparent that [Leu^{31}Pro^{34}]NPY is best viewed as a 'non-Y_2' ligand, which is active at Y_1, Y_4 (PP) and Y_5 receptors. An involvement of Y_4-receptors appears unlikely for several reasons, and seems indeed to have been excluded using PP (Asakawa et al. 1999). Based on receptor distribution, it has been suggested that, similar to regulation of food intake, Y_5 receptors may be involved in emotionality in concert with Y_1 receptors (Parker and Herzog 1999). However, available functional data do not support this notion, as the selective Y_5 antagonist CGP 71683A does not influence anxiety-related behaviors in the social interaction test, the elevated plus-maze or the open field (Kask et al. 2001b).

Due to the weaknesses of peptide agonist studies, additional evidence for an involvement of Y_1 receptors was obtained using in vivo administration of antisense oligonucleotides targeting the Y_1 receptor. When administered icv, anti-Y_1 oligonucleotides selectively lowered the density of Y_1 binding sites, and this was accompanied by suppressed second messenger signaling in response to Y_1-stimulation in vitro, and behavioral effects on the elevated plus-maze opposite to those seen after NPY administration (Wahlestedt et al. 1993b). Furthermore, local anti-Y_1-oligo administration into the amygdala blocked the anxiolytic-like effects of NPY given icv (Heilig 1995).

The more recent arrival of subtype-selective, nonpeptide NPY receptor antagonists has allowed a more refined analysis of this issue. The first member in this class of compounds, BIBP 3226 has been extensively characterized, and acts as a Y_1-antagonist both in vitro and in vivo, while being devoid of affinity for Y_2, Y_4 and Y_5 receptors (Rudolf et al. 1994). In agreement with the agonist studies, BIBP 3226 has been reported to produce anxiogenic-like effects. Suppression of plus-maze exploration has been reported following icv administration of this antagonist in rats, an effect which was observed in the absence of locomotor actions, and was possible to reverse using the classical anxiolytic diazepam (Kask et al. 1996). Similar effects have also been seen following localized site injections (Kask et al. 2000; Kask and Harro 2000), which are discussed further below. However, while being selective among NPY receptor subtypes, BIBP 3226 has limited solubility, and is also clearly capable of producing nonreceptor mediated effects (Doods et al. 1996). For instance, reduction of food intake by BIBP 3226 is reproduced by its steric enantiomer BIBP 3435, which entirely lacks affinity for NPY receptors (Morgan et al. 1998). Similar data have been obtained in the Vogel conflict test (M. Heilig and C. Möller, unpublished results). Furthermore, it has recently been reported that BIBP 3226 has a moderate but non-negligible affinity for neuropeptide FF (NPFF) receptors (Mollereau et al. 2001). Taken together, these observations raise some concern regarding findings obtained using this tool, in particular since anxiety-related behaviors are highly sensitive to stressful side effects of drugs. However, it has been pointed out that nonspecific actions of BIBP 3226 probably require higher doses than those used in the published anxiety studies (Kask et al. 2002). The arrival of the structurally related Y_1 antagonist BIBO 3304 has improved our ability to resolve these issues, as this compound has increased solubility, higher affinity for Y_1 receptors, and appears to lack the nonspecific side effects (Wieland et al. 1998). Studies using this compound in relation to emotionality, stress and anxiety are limited, but to the extent they are available, they clearly confirm the results obtained using agonists, antisense oligonucleotides and BIBP 3226. For instance, anxiolytic-like effects of NPY in the social interaction test are blocked by intra-amygdala administration of BIBO 3304 (Sajdyk et al. 1999), and this compound also increased defecation during exploratory behavior in the open field, a variable normally correlated with the level of emotionality (Kask and Harro 2000).

Although activation of Y_1-receptors thus appears to be the major mechanism for mediating NPY's anxiolytic like actions, Y_2-receptors may also play an important role in the regulation of emotionality, and in fact offer the most attractive target for drug development efforts, as discussed below. NPY-Y_2 receptors are located presynaptically on NPY-ergic neurons, and control the release of endogenous NPY (King et al. 1999; King et al. 2000). Antagonism at these receptors would thus be expected to potentiate endogenous NPY release, and through this mechanism offer an 'NPY mimetic' without the requirement for developing a Y_1-agonist. Available data in an alcohol self-administration model provide indirect support for this notion (Thorsell et al. 2002), but data addressing the feasibility of this mechanism in relation to anxiety and depression are not yet available. Furthermore, a more direct involvement of Y_2-receptors has been suggested within the locus coeruleus, based on the observation that anxiolytic-like effects were produced here by low (10 pmol) doses of NPY microinjected into this structure, and that these were not reproduced by administration of [Leu^{31}, Pro^{34}]NPY, but were mimicked by microinjections of NPY(13–36). In addition, BIBP 3226 was ineffective when injected in this region (Kask et al. 2000). Some alternative interpretations of these findings are discussed below in the context of neuronal circuitry mediating NPY's anxiolytic-like actions.

2.2.3
Anatomical Substrates of the Anti-Stress Actions of NPY

The amygdala is crucial for emotional learning, and co-ordinates behavioral, autonomic and endocrine fear responses (Davis 1998; Fendt and Fanselow 1999; LeDoux 2000). For these reasons, this structure was an obvious initial candidate for mediating anxiolytic-like actions of NPY. In agreement with this hypothesis, site-specific injections of NPY and NPY analogs into the amygdala reproduced the anti-anxiety actions of NPY administered icv at approximately 10-fold lower doses, while other injection sites were ineffective in this respect (Heilig et al. 1993b; Sajdyk et al. 1999). Amygdala injections have also helped separate anxiolytic-like from appetitive effects of NPY, since they reproduce anxiolytic actions of NPY administered icv, without affecting feeding. The central amygdala constitutes an output relay for the functional consequences of amygdala activation by fearful stimuli (Davis 1998; Fendt and Fanselow 1999; LeDoux 2000; Swanson and Petrovich 1998), and was initially reported to be the amygdalar compartment within which anxiolytic-like actions are produced by amygdala injections of NPY (Heilig et al. 1993b). However, on re-analysis of these experiments, it is clear that the injection volumes used most likely did not allow a separation of effects mediated by different amygdala compartments. Subsequent work using microinjections of smaller volumes has prompted a re-evaluation of the data, suggesting that the lateral/basolateral complex in fact mediates anti-stress effects of NPY within the amygdala (Sajdyk et al. 1999).

The periaqueductal gray matter (PAG) is involved in the behavioral output of fear responses, with subcompartments differentially involved in defensive behaviors (Brandao et al. 1994; Fendt and Fanselow 1999). Its dorsolateral compartment (DPAG) has been suggested to tonically inhibit the amygdala. Microinjections of the Y_1-selective nonpeptide antagonist BIBP3226 within the DPAG have been reported to produce anxiogenic-like effects in the elevated plus-maze, with a degree of behavioral specificity demonstrated by unaffected open field behavior following the same treatment (Kask et al. 1998a). Similar effects were found in a separate report using the social interaction test (Kask et al. 1998b). A weakness of these studies was that saline treatment was used as control for BIBP 3226. Unlike the enantiomer BIBP 3435, which lacks affinity for NPY receptors but provides a control for nonreceptor mediated effects, this does not provide adequate control for nonspecific and toxic side effects of BIBP 3226 described above, which are most likely to present a problem when local intra-tissue injections are used. However, the latter of the two studies provided additional data using the di-peptide Y_1-antagonist 1229U91, which to the best of available knowledge is free of toxic side effects. The observation that 1229U91 mimicked the action of BIBP 3226 provides support for the conclusion proposed by the authors, that blockade of Y_1 receptors in the DPGA indeed is anxiogenic.

The septum has classically been described as a key component of a 'behavioral inhibition system', the activity of which was thought to be at the core of anxiety states, and also constitute the neural substrate of anxiolytic drug actions by benzodiazepines and related compounds. A cornerstone of this notion was the classical description of a 'septal syndrome', characterized by a dramatic activation of defensive behaviors (Gray 1983). A shift of attention away from this concept has occurred with the realization that the notion of a 'limbic system' essentially lacks anatomical as well as functional foundation (see e.g., Swanson and Petrovich 1998), that the amygdala plays a major role both in fear responses and in anxiolytic drug actions (Davis 1998; Fendt and Fanselow 1999; LeDoux 2000; Swanson and Petrovich 1998), and that septal lesion studies demonstrating effects on anxiety-related behaviors most likely reflect effects on fibers passing through this structure, most likely belonging to hippocampal output through Fornix Fimbriae (Lee and Davis 1997). However, septo-hippocampal circuits are likely to be important for fear-related behaviors. For instance, as discussed below, transgenic NPY overexpression within the hippocampus confers a behavioral phenotype which reproduces most if not all effects of icv administered NPY. This is consistent with observations that the dorsal hippocampus is an important component of neuronal circuitry controlling anxiety-related behaviors and stress responses (Andrews et al. 1997; Gonzalez et al. 1998). Furthermore, NPY microinjections into the lateral septum reproduced anxiolytic-like actions of icv administered NPY, and reversed the anxiogenic action of corticotrophin releasing factor (CRF). The anxiolytic-like action of NPY was clearly Y_1-receptor mediated, as it was blocked by the highly selective, nontoxic Y_1-receptor antagonist BIBO 3304 (Kask et al. 2001a). This study also demonstrated that NPY injections into the cholinergic medial septal nucleus were ineffective, contradicting the previously made suggestion that anxiolytic-like actions of NPY might be produced through interactions with cholinergic afferents to the hippocampus which originate from this structure (Zaborszky and Duque 2000).

Finally, the locus coeruleus has long been implicated in anxiety disorders and stress (Nestler et al. 1999; Sullivan et al. 1999), although its involvement may be restricted to symptom domains of pathological arousal and vigilance seen in these disorders, as a reflection of the normal physiology of this structure (Aston-Jones et al. 1991). As mentioned above, anxiolytic-like actions have been reported in the social interaction test after local injections of 10 pmol NPY into the locus coeruleus (Kask et al. 2000). Based on the ligand profile of this effect, and the inactivity of BIBP 3226 in this region, it was suggested that NPY-Y_2 receptors in this area mediate these actions. However, an alternative interpretation which needs to be considered is whether this reflects actions at presynaptic Y_2 receptors on A6 neurons in which NPY and norepinephrine are colocalized, ultimately inhibiting noradrenergic transmission in projection areas of these neurons.

2.2.4
'Phasic' vs. 'Tonic' Effects of Endogenous NPY in Stress and Fear

An important aspect of peptidergic transmission, introduced by Hökfelt and coworkers in the early eighties (Lundberg and Hökfelt 1983), is its activity dependence. According to this concept, at basal and moderate levels of activity, neurons preferentially release small synaptic vesicles, which primarily contain classical transmitters. In contrast, release of large, dense core vesicles in which classical and neuropeptide transmitters are colocalized requires high levels of activity and firing frequency. At the level of integrative physiology, this implies a role for neuropeptides as 'alarm-systems', which are not tonically active under normal circumstances, but are recruited under conditions of stress, noxious stimuli etc. The prediction then is that neuropeptide receptor antagonists under normal conditions should produce limited, if any, effects while important effects might still be expected under more extreme conditions. This prediction has been supported by the observations that CRF receptor antagonists as well as the opioid receptor antagonist naltrexone are largely inactive unless the organism is faced with a challenge (Gianoulakis et al. 1996; King et al. 1997; Koob et al. 1993; Spanagel and Zieglgansberger 1997). This type of property would obviously make neuropeptide receptors into attractive drug targets, promising highly restricted side effects.

However, the involvement of NPY in feeding regulation indicates that this principle does not uniformly apply to all peptidergic transmission. Kask and co-workers have recently discussed (Kask et al. 2002) whether NPY systems mediating anxiolytic-like effects are active in a 'tonic' manner, perhaps more appropriately labeled 'under normal circumstances', or 'phasically', that is, in response to specific challenges at the extremes or outside of the normal physiological range. On the basis of the microinjection studies reviewed above, these authors have proposed that NPY within the amygdala is only phasically active, while tonic activity is present in DPAG and possibly also the locus coeruleus.

Upon icv administration, a quantitatively dominant component of NPY-induced anxiolysis appears to be mediated by Y_1-receptors within the amygdala complex, since these effects of NPY are essentially blocked by intra-amygdala treatment with anti-Y_1-receptor oligonucleotides (Heilig 1995). This system does not appear to be active under basal conditions, as microinjections of the selective Y_1-antagonist BIBO 3304 alone were ineffective in this structure, while blocking effects of coadministered NPY (Sajdyk et al. 1999). Furthermore, as discussed below, transgenic overexpression studies have shown that hippocampal NPY signaling also requires activation by a stressor in order to become functionally relevant in relation to emotionality. Finally, studies in mouse mutants with an inactivation of the pre-pro-NPY gene support the overall notion that the role of endogenous NPY in regulation of emotionality falls in the activity dependent, 'alarm-system' category (Bannon et al. 2000). The one observation which is difficult to reconcile with this notion is the finding of anxiogenic effects following antisense-mediated suppression of Y_1-receptor expression (Wahlestedt et al. 1993a). It is, however, possible that the antisense-treatment per se was capable of sufficiently activating stress-related systems to demonstrate anti-stress actions of endogenous NPY (Heilig and Schlingensiepen 1996).

2.2.5
Expression of NPY and Behavioral Stress Responses

The pharmacological studies reviewed above have prompted the question whether regulation of endogenous NPY signaling, and in particular of NPY gene expression, could play a role in the modulation of behavioral stress responses. The reverse was first demonstrated: expression of NPY gene expression in amygdala and cortex is regulated by stress. Acute stress downregulates NPY expression within 1 h, with mRNA levels returning to normal within 10 h (Thorsell et al. 1998). Interestingly, with repeated stress exposure, leading to a behavioral habituation, this effect is reversed. Under these conditions, NPY expression is instead upregulated (Thorsell et al. 1999).

On the basis of the pharmacological and expression studies, we proposed that an upregulation of NPY expression may contribute to successful behavioral adaptation to stress. This extends a previously introduced hypothesis that NPY may act to 'buffer' behavioral effects of stress-promoting signals such as CRF (Heilig et al. 1994). Our hypothesis predicted that upregulated expression of NPY should render a subject less sensitive to anxiety-promoting effects of stress, a prediction potentially possible to test in a transgenic system. Although NPY transgenic mice have been described, only a limited phenotypic characterization for these was available (Thiele et al. 1998). Furthermore, the use of mice as a model limited the feasibility of phenotyping stress and anxiety responses in a manner allowing a direct comparison with the pharmacological data, all obtained in rats.

For these reasons, the generation of an NPY transgenic rat offered an attractive model for our studies. In addition to the species, which made it ideal for behavioral studies, this model had some potentially advantageous features in that it was based on the introduction of a 14.5-kb fragment of the rat NPY genomic sequence, containing normal intronic sequence elements, and flanked by an approximate 5-kb 5' sequence thought to contain the major regulatory elements normally controlling NPY expression (Larhammar et al. 1987). The expression of this transgene may thus be regulated in a manner similar to that of endogenous NPY.

In summary, results obtained in the NPY transgenic rat model have supported the initial hypothesis. No distinguishing anxiety-related phenotype was observed in the elevated plus-maze under basal, unstressed conditions. Likewise, locomotor activity and feeding were normal. However, a marked behavioral insensitivity to stress was found when stressful manipulations were introduced. In nontransgenic littermate controls, exposure to an established stressor—1 h of restraint—1 h prior to behavioral testing gave rise to an expected and marked anxiogenic effect on the plus-maze, manifested as a profound decrease of the percentage time spent exploring the open arms of the maze, and of the number of entries into the open arms. This behavioral consequence of stress was entirely absent in transgenic subjects. Furthermore, in the markedly stressful Vogel test, response inhibition is normally seen due to the delivery of electric shock upon drinking, and this was indeed observed in nontransgenic littermates. In contrast, and similar to benzodiazepines treated animals, response inhibition in this test was entirely absent in NPY transgenic subjects (Thorsell et al. 2000).

As mentioned above, pharmacological studies had demonstrated that activation of Y_1-receptors in the amygdala, DPAG and lateral septum reproduces the anti-stress effects found with icv administration of NPY ligands. However, in situ analysis of the transgenic rat model did not reveal significant overexpression within any of these locations. In fact, the only region where robust, approximately 100% overexpression was observed were hippocampal fields CA1–2, where also marked, presumably compensatory, downregulation of Y_1-binding was found.

The finding of restricted hippocampal overexpression has two major implications. Clearly, anti-stress effects of NPY can be produced through actions in more than one region, as also shown by the site-specific microinjections studies reviewed above. The involvement of the hippocampus should not come as a surprise, since this structure is intimately connected both with the septum and the amygdala, and also since direct functional data demonstrate an involvement of the hippocampus in aspects of fear learning (Fendt and Fanselow 1999), but also in unconditioned anxiety-related behaviors and stress responses (Andrews et al. 1997; Gonzalez et al. 1998). NPY has been shown to inhibit glutamate release within the hippocampus (McQuiston and Colmers 1996; Qian et al. 1997), and it is therefore of interest that the effects of hippocampal NPY overexpression resemble the actions of intrahippocampal administration of a metabotropic glutamate receptor antagonist (Chojnacka-Wojcik et al. 1997). Secondly, a restricted hippocampal overexpression in our system is likely to underlie the findings of normal feeding and alcohol intake in the transgenic subjects.

2.3
NPY and Depression

2.3.1
The Anxiety–Depression Spectrum

Symptoms of anxiety and depression commonly coexist, and both disorders are thought to reflect maladaptive changes in stress-responsive systems (Holsboer 2000). In fact, a latent class analysis of clinical symptoms has suggested that present classification of depressive and anxiety disorders may be artificial, and suggested the existence of a category labeled 'major depression–generalized anxiety disorder' (Sullivan and Kendler 1998). In agreement with this notion, genetic factors which confer increased vulnerability for both these disorders are largely the same (Kendler et al. 1992). Given the extensive evidence for an involvement of central NPY in stress and anxiety, it is therefore not unexpected that this system has also been implicated in depressive disorder, although less extensive data are available in this area.

2.3.2
Effects of Antidepressant Treatments on the NPY System

Treatment of experimental animals with clinically effective antidepressants was early reported to increase NPY peptide levels in several brain areas, with frontal cortex being the most consistent region (Heilig et al. 1988a). This was found both with desipramine, which preferentially acts at noradrenergic terminals, and citalopram, which is serotonin selective. Although a downregulation of NPY binding sites consistent with this type of effect was also reported following desipramine treatment (Widdowson and Halaris 1991), initial attempts to replicate the effects of chronic antidepressant treatment and extend them to the level of mRNA analysis were unsuccessful (Bellman and Sperk 1993; Heilig and Ekman 1995). However, a region-specific regulation of NPY and Y_1-receptor expression was then reported following chronic treatment with the serotonin-selective reuptake inhibitor (SSRI) fluoxetine, both in the Flinders Sensitive Line (FSL), a genetic model of depression (Overstreet et al. 1995; see below), and the corresponding control Flinders Resistant Line (FRL) (Caberlotto et al. 1998, 1999). In these studies, fluoxetine elevated NPY-like immunoreactivity in the hypothalamic arcuate nucleus and anterior cingulate cortex, and increased Y_1 binding sites in the medial amygdala and occipital cortex in both lines. In agreement with the peptide measures, an increase of the NPY mRNA hybridization signal was found in the arcuate nucleus of both strains. In other brain regions, fluoxetine administration caused a differential effect on the induction of NPY-related genes in the two rat strains: in hippocampus, NPY mRNA expression was increased in the FSL, but decreased in the FRL. In contrast, Y_1 mRNA levels tended to be decreased by fluoxetine in the nucleus accumbens of the FSL rats, but increased in the FRL. On the basis of these findings, an involvement of NPY was suggested in the antidepressant effect of fluoxetine.

As mentioned above, effects of antidepressant drugs on the NPY system have been inconsistent. Efficacy of fluoxetine in the absence of efficacy for desipramine could reflect selective interactions between serotonergic and NPY-ergic transmission. However, based on pharmacodynamic mechanisms, it is more difficult to interpret the findings that chronic treatment with a different SSRI, citalopram, failed to affect hippocampal NPY immunoreactivity (Heilig and Ekman 1995; Husum et al. 2000). The most striking difference between fluoxetine and other antidepressants, both within the SSRI and tricyclic categories, is fluoexetine's long half-life (Baumann 1996). The inconsistent effects of antidepressant treatments on NPY levels may well reflect pharmacokinetic factors, and in particular difficulties in maintaining sufficient plasma concentrations with the shorter-lived compounds.

Another established and effective antidepressant treatment, electroconvulsive shock (ECS), has been much more consistent in upregulating brain NPY levels, with the hippocampus as a seemingly central target. This was originally reported independently by two different groups, which both also demonstrated elevation of hippocampal NPY levels after repeated, but not single ECS, paralleling the requirements for clinical effect in depressed subjects (Stenfors et al. 1989; Wahlestedt et al. 1990). These findings have subsequently been replicated and extended (Husum et al. 2000; Mathe 1999; Mathe et al. 1997, 1998; Zachrisson et al. 1995), and indicate that this effect is robust both in 'normal' laboratory rats and in the genetically selected FSL and FRLs, represents an upregulation of pre-pro-NPY expression, and leads to increased extracellular availability of NPY peptide. Against the background of our behavioral findings in the transgenic rat model reviewed above, upregulated hippocampal NPY expression might be of importance both for therapeutic and amnesic effects of ECS (Husum et al. 2000). Finally, similar to what has been reported with ECS and some pharmacological antidepressant treatments, administration of the clinically established affective stabilizer lithium also leads to an upregulation of hippocampal NPY synthesis (Husum et al. 2000).

2.3.3
Animal Models of Depression

Further evidence for an involvement of NPY in depression comes from findings of differential NPY expression in two different genetic animal model of depression, the FSL described above (Caberlotto et al. 1998, 1999; Jimenez et al. 2000), and the Fawn Hooded rat (Mathe et al. 1998; Rezvani et al. 2002). In the latter line, ECS was more effective in increasing hippocampal NPY levels compared to 'nondepressed' subjects. The reported differences are only correlative, and do not establish a functional involvement of the altered NPY system in the behavioral phenotype. In addition, they are region-dependent in a manner which does not allow a simple interpretation. However, the hippocampus appears to be a consistent candidate structure for a possible functional involvement.

Olfactory bulbectomy in rats produces behavioral consequences which have been interpreted as indicative of a depression-like phenotype (Cairncross 1984). It is therefore of interest that some of these, normally interpreted as signs of increased irritability during open field exploration, are counteracted by subchronic administration of NPY (Song et al. 1996), while in the long run (2–4 weeks) NPY expression was upregulated in piriform cortex and the dentate hippocampal gyrus by olfactory bulbectomy, possibly as a compensatory mechanism (Holmes et al. 1998).

Early postnatal separation of rat pups induces a long-term phenotype characterized by increased responsiveness of the hypothalamic–pituitary–adrenal (HPA) axis (Plotsky and Meaney 1993), conferred through changes in maternal nursing behavior (Liu et al. 1997). Although not extensively validated pharmacologically as a model for detecting antidepressant drug action, this model is attractive in that it closely mimics dysregulations of the HPA axis thought to be at the core of pathophysiology in depressed humans (Holsboer 2000). It is therefore of interest that maternal separation has been shown to reduce NPY both in Sprague-Dawley (Jimenez-Vasquez et al. 2001) and in Wistar rats (Husum et al. 2002). Reductions of NPY-like immunoreactivity in this model were found within the hippocampus, consistent with the findings of increased hippocampal NPY following several types of antidepressant treatments. In addition, against the background of markedly higher prevalence of depression in females than in males, an interesting observation provided by the latter study is that of generally lower hippocampal NPY levels in female subjects.

2.3.4
Antidepressant-Like Effects of Central NPY

An important question is whether the NPY system may provide useful targets for novel antidepressant treatment. Although the concept of 'correcting' pathophysiological changes is an attractive one, this issue is in fact separate from the question of whether dysregulations of the NPY system contribute to the pathophysiology of depression or not. Initial evidence is presently available that NPY given icv produces antidepressant-like effects in the forced swimming test, both in rats (Stogner and Holmes 2000) and mice (Redrobe et al. 2002). These findings have been obtained in normal, 'nondepressed' animals. An important issue which awaits resolution is therefore whether central administration of NPY will be capable of rescuing a 'depressed' phenotype in this model. An indication that this may turn out to be the case is provided by the data on phenotypic rescue in the olfactory bulbectomy model (Song et al. 1996). Also, this issue may be less critical because it appears that, for reasons which are not understood, the forced swimming has predictive validity with regard to anti-depressant drug action also when carried out in normal, 'nondepressed' subjects.

2.3.5
Human Findings: Central Nervous System

An early study reported decreased levels of NPY in the cerebrospinal fluid (CSF) of patients with major depression, possibly reflecting decreased central availability of NPY (Widerlov et al. 1988). Within this patient population, a more detailed analysis revealed an inverse correlation between NPY levels in the CSF and ratings of anxiety symptoms. Independently, lower levels of NPY in brain tissue was also reported in suicide victims, and it was suggested that this decrease was most pronounced in subjects in whom evidence was available to suggest presence of major depression prior to death (Widdowson et al. 1992). These findings seemed to indicate that the NPY system might be primarily affected in depressive syndromes, and contribute to the clinical symptomatology.

However, neither the finding of reduced NPY levels in CSF of depressed subjects (Berrettini et al. 1987) nor in frontal cortex of depressed subjects (Ordway et al. 1995) was replicated in subsequent studies. The reason for this is unclear, but may have been related to the limited understanding of peptide processing and assay specificity issues in that early era, in particular since mass spectrometry studies have since shown that processing of NPY may differ between depressed subjects and normal controls, resulting in a different fragment pattern (Ekman et al. 1996). In a recent re-examination of this issue, we used a sample of therapy-refractory depressed patients of considerable size for this type of study ($n=50$), and analyzed CSF levels of monoamines, monoamine metabolites, and several neuropeptides. NPY was analyzed using an assay which has been extensively characterized with regard to specificity, and lacks cross-reactivity with C-terminal NPY fragments. The one robust difference between patients and controls using this approach was a highly significant, 30% reduction of CSF NPY content (Heilig and Ågren, in press). It remains to be established whether this reflects improved methodology, altered NPY signaling in subpopulations of depressed subjects only, or both.

Interestingly, the latter possibility is supported by the demonstration of suppressed NPY expression in human post mortem brain tissue in bipolar, but not unipolar affective disorder (Caberlotto and Hurd 1999). The genetics of these two disorders overlap (Karkowski and Kendler 1997). Is well known that a proportion of patients diagnosed with unipolar disorder in fact has a genetic vulnerability for the bipolar disease, but has not yet presented with their first manic episode, and may never do so. It is therefore possible that involvement of NPY is primarily related to bipolar traits, and that the discrepant CSF results are partly due to a varying proportion of this patient category in the different clinical samples.

2.3.6
Human Findings: Peripheral Nervous System

Central synthesis, release, and metabolism of NPY are difficult to access. Although CSF studies may provide a window on some of these processes, CSF levels obviously reflect a complex outcome of numerous processes, at different levels of the neuroaxis. In attempts to link NPY to psychiatric disorders, studies of the more easily accessible peripheral compartment have also been carried out. Presently available evidence suggests that peripheral NPY is simply a marker of sympathetic nervous system activity, which is unrelated to central NPY signaling of importance for emotionality and mood. However, it has been suggested that a functional polymorphism in the pre-pro-NPY gene leads to altered NPY expression and/or release (Kallio et al. 2001), and this kind of mechanism might affect central and peripheral NPY levels in a similar manner, making the latter a marker of the former.

In this context, it is of interest to note that peripheral NPY has been reported to be lowered in recent suicide attempters (Westrin et al. 1999), and correlate with personality variables in this patient population (Westrin et al. 1998). In a separate line of study, lowered peripheral NPY levels, both at baseline and upon stimulation of sympathetic neurons with the pre-synaptic autoreceptor blocker yohimbine, have been found in combat survivors with post-traumatic stress disorder (Rasmusson et al. 2000). In an interesting experimental follow-up of these findings, NPY was analyzed during exposure to interrogation, presumably a highly stressful experience, in special force as well as nonspecial force soldiers. The former had generally lower levels of plasma NPY. Furthermore, across both groups, a range of responses to this stressful situation was inversely correlated with plasma NPY. On the basis of these findings, it was suggested that peripheral NPY is a marker of stress-resilience (Morgan et al. 2000), possibly through a correlation with personality traits (Morgan et al. 2001). It remains to be established whether these findings reflect altered patterns of sympathetic activity, altered central NPY signaling, or both.

2.4
Actions of NPY Related to Stress, Anxiety, and Depression: Summary

Extensive evidence suggests that a distributed network of circuits utilizing NPY signaling, with the amygdala and hippocampus as core components, acts as an endogenous 'alarm system'. This system appears to be activated when the organism is faced with stressful challenges, and its functional activation counteracts the behavioral effects initially triggered in order to cope with the threatening stimulus. A possible function of this activation in the short term might be to appropriately terminate the necessary—but in the long run costly—behaviors activated by various threats to homeostasis, responses which to a large extent are mediated and coordinated by central release of CRF at multiple brain sites. In a longer perspective, regulation of NPY expression appears to serve as a mechanism for adapting the individual's behavioral responsiveness in response to chronic stress. From this conceptualization, it is apparent that a dysregulation of this system could easily contribute to the pathophysiology of anxiety and depressive disorders. Whether this is the case or not, effects of NPY related compounds in pharmacologically validated animal models of anxiolytic- and antidepressant-like drug action suggest that receptors of the NPY system offer attractive targets for drug development in search of novel treatments for anxiety, depression, and as discussed elsewhere, alcohol dependence (see Sect. 10.3). Although the desired action of potential therapeutic compounds would ideally seem to be an activation of the Y_1-subclass of NPY receptors, development of orally available, brain penetrant Y_1-agonists may represent a challenge which is not easily overcome. An attractive alternative strategy seems to be antagonism at central Y_2-receptors, which, through potentiation of endogenous NPY release, might achieve the same goal.

3
A Role for NPY in Ethanol-Seeking Behavior and Neurobiological Responses to Ethanol

3.1
Altered Ethanol Drinking Associated with Genetic Manipulation of NPY Signaling

The first genetic evidence linking NPY to alcoholism came from studies involving rats selectively bred for high alcohol drinking. Quantitative trait loci (QTL) analyses identified a region of chromosome 4 that significantly correlated with differences in alcohol drinking between the Indiana alcohol-preferring (P) and alcohol-nonpreferring (NP) rats. This chromosomal region includes the NPY precursor gene (Bice et al. 1998; Carr et al. 1998). Subsequent research found that P rats had low levels of NPY in the amygdala, frontal cortex, and hippocampus relative to NP rats, but higher levels of NPY in the hypothalamus and cingulate cortex (Ehlers et al. 1998a; Hwang et al. 1999). High alcohol-drinking (HAD) rats, bred by a similar strategy as that used to generate the P rats, also had low levels on NPY in the amygdala compared with low alcohol-drinking (LAD) rats, and had lower levels of NPY in hypothalamic nuclei (Hwang et al. 1999). Hwang et al. concluded that the high alcohol drinking by the P and HAD rats are best explained by low levels of NPY in the amygdala. It should be noted, however, that QTL analyses with HAD and LAD rats failed to confirm a role for the NPY precursor gene (Foroud et al. 2000). More recently, the low alcohol drinking ALKO Non-Alcohol line of rats was found to have high NPY mRNA in the hippocampal cornus ammons region and the dentate gyrus when compared with the high alcohol drinking (ALKO Alcohol; AA) line and nonselected Wistar rats. Additionally, NPY Y_2 receptor mRNA was reduced in the AA line, suggesting a role for the Y_2 receptor in modulating alcohol drinking (Caberlotto et al. 2001).

Genetically altered rodents have also been examined. Voluntary ethanol consumption and resistance to the intoxicating effects of ethanol were inversely related to NPY levels in knockout and transgenic mice (Thiele et al. 1998). However, ethanol-associated phenotypes were not consistently observed in NPY knockout mice and are dependent on the genetic background (Thiele et al. 2000b). Transgenic rats have NPY overexpression that is primarily limited to the CA1 and CA2 regions of the hippocampus. Relative to control animals, NPY transgenic rats were resistant to anxiety provoked by restraint-stress and showed impairment of spatial memory acquisition. However, the NPY transgenic rats showed normal voluntary ethanol drinking (Thorsell et al. 2000). Thorsell et al. suggested that NPY transgenic rats may ingest normal levels of alcohol because they fail to overexpress NPY in critical brain regions, such as the amygdala (see Sect. 10.2.2.5). Taken together, evidence from genetic animal models implies that low NPY signaling can promote high voluntary ethanol drinking while upregulation of NPY signaling can be protective against excessive consumption.

3.2
Altered Ethanol Drinking Following Central Infusion of NPY and Related Compounds

Several studies have used icv infusion of NPY to determine pharmacologically if NPY signaling regulates voluntary ethanol consumption. In the first attempt with Golden Hamsters, icv infusion of NPY did not reliably alter drinking of a 5% ethanol solution, as only one of six doses used caused significant increase of 60-min drinking (Kulkosky et al. 1988). More recently, Wistar rats were given icv infusion of various doses of NPY ranging from 2.5 to 15.0 μg in a within-subjects design. While 5.0 μg of NPY significantly increased consumption of a sucrose solution, none of the doses tested altered alcohol intake (Slawecki et al. 2000). Similarly, neither third ventricle infusion of NPY (Katner et al. 2002b) nor direct infusion of NPY into the amygdala (Katner et al. 2002a) altered ethanol drinking in Wistar rats. On the other hand, icv infusion of both 5.0 and 10.0 μg doses of NPY significantly reduced voluntary consumption of an 8% ethanol solution in P rats, but did not alter ethanol drinking of NP or outbred

Wistar rats (Badia-Elder et al. 2001). Furthermore, direct infusion of femtomolar doses of NPY into the paraventricular nucleus of the hypothalamus increased consumption of alcohol in Long-Evans rats, and this effect was blocked by pretreatment with the Y_1 receptor antagonist BIBP 3226 (Kelley et al. 2001). Thus, the ability of NPY to alter ethanol drinking depended on the genetic background of the rat and/or the site of NPY infusion. In some cases, enhancement of NPY signaling by administration of exogenous NPY reduced ethanol drinking.

3.3
Receptors Mediating the Effects of NPY on Ethanol Drinking

In the mouse, NPY acts through at least five receptor subtypes, namely the Y_1, Y_2, Y_4, Y_5 and y_6 receptors, all of which couple to heterotrimeric G proteins that inhibit production of cyclic AMP (see chapter by Redrobe et al., this volume). The Y_1 receptor is located postsynaptically and has been identified in several brain regions that are involved with neurobiological responses to ethanol, including the hippocampus, the hypothalamus, and the amygdala (Naveilhan et al. 1998; Ryabinin et al. 1997). Y_1 receptor knockout mice ($Y_1^{-/-}$) have been generated and grow and reproduce at normal rates despite slightly diminished daily food intake and reduced refeeding response to starvation. However, these animals develop late-onset obesity due to low energy expenditure (Pedrazzini et al. 1998) (see chapter by Herzog, this volume). Alcohol consumption by $Y_1^{-/-}$ mice and by normal $Y_1^{+/+}$ mice has recently been examined (Thiele et al. 2002). $Y_1^{-/-}$ mice showed increased consumption of solutions containing 3%, 6%, and 10% (v/v) ethanol but displayed normal consumption of sucrose and quinine solutions. Furthermore, $Y_1^{-/-}$ mice were less sensitive to the sedative effects of 3.5 and 4.0 g ethanol/kg as measured by more rapid recovery from ethanol-induced sleep, even though plasma ethanol levels did not differ significantly between the genotypes following a 3.5 g/kg dose. Finally, male $Y_1^{-/-}$ mice showed normal ethanol-induced ataxia on a rotarod test following administration of a 2.5 g/kg dose (Thiele et al. 2002).

Evidence suggests that the Y_2 receptor is a presynaptic autoreceptor and inhibits NPY release (Naveilhan et al. 1998). As described by Redrobe et al. (this volume), mutant mice lacking the Y_2 receptor ($Y_2^{-/-}$) have been shown to have increased food intake, body weight, and fat production but have a normal response to NPY-induced food intake (Naveilhan et al. 1999). It was hypothesized that if presynaptic Y_2 receptors are involved with modulating voluntary ethanol consumption and sensitivity, the $Y_2^{-/-}$ mice should exhibit ethanol-related phenotypes opposite to those found with the $Y_1^{-/-}$ mice. Thus, an absence of presynaptic inhibition of NPY release in $Y_2^{-/-}$ mice would augment NPY signaling, rendering mice with a similar phenotype as NPY overexpressing mice. Relative to wild-type ($Y_2^{+/+}$) mice, the $Y_2^{-/-}$ mice drank significantly less of solutions containing 3% and 6% ethanol, and had significantly lower ethanol preference at each concentration tested. On the other hand, $Y_2^{-/-}$ mice showed normal consumption of solutions containing either sucrose or quinine, normal time to recover from ethanol-induced sedation following 3.0 or 3.5 g/kg doses, and normal metabolism of ethanol following injection of a 3.0 g/kg dose (Thiele et al. 2000c).

Mutant mice lacking the NPY Y_5 receptor ($Y_5^{-/-}$) show late onset obesity and increased food intake, have reduced sensitivity to NPY, and are seizure prone (see chapter by Redrobe et al., this volume; Marsh et al. 1999). When given access to solutions containing ethanol, $Y_5^{-/-}$ mice drank normal amounts of 3, 6, 10, and 20% (v/v) ethanol, but had increased sleep time following administration of 2.5 or 3.0 g ethanol/kg. However, the $Y_5^{-/-}$ mice also showed high plasma ethanol levels relative to wild-type mice following injection of a 3.0 g/kg dose (Thiele et al. 2000b). Together, data from NPY receptor knockout mice suggest that voluntary consumption of ethanol is modulated by the Y_1 and Y_2 receptors, and that ethanol-induced sedation is modulated by Y_1, and perhaps Y_5, receptors. These results are consistent with several recent findings. First, like $Y_2^{-/-}$ mice which drink low amounts of ethanol, rats self-administer less ethanol following central infusion of a Y_2 receptor antagonist (Thorsell et al. 2002). Second, $Y_1^{-/-}$ mice are resistant to the sedative effects of ethanol, and recent studies found that $Y_1^{-/-}$ mice are resistant to sodium pentobarbital-induced sleep (Naveilhan et al. 2001a, 2001b).

3.4
Brain Electrophysiological Responses to NPY and Ethanol Are Similar

One of the first studies to suggest a role for NPY in modulating neurobiological responses to ethanol investigated the effects of icv infusion of NPY on brain electrophysiological responses in rats (see also Sect. 10.2.1). It was noted that icv infusion of NPY, as well as Y_1 receptor agonist, produced electrophysiological and behavioral profiles similar to those induced by anxiolytic drugs such as ethanol and benzodiazepines (Ehlers et al. 1997). Additionally, icv infusion of NPY and peripheral administration of ethanol to rats produced identical effects on event-related potential (ERP) profiles in response to auditory stimuli, both in cortex and amygdala. The effects of NPY and ethanol were additive (Ehlers et al. 1998b). Following 10–15 weeks of withdrawal from chronic exposure to ethanol vapor, icv infusion of NPY significantly decreased the amplitude of the N1 component of ERP in the amygdala of withdrawn Wistar rats when compared to controls, indicating that ethanol withdrawal augments brain sensitivity to NPY (Slawecki et al. 1999). Together, these data suggest that electrophysiological responses to ethanol and ethanol withdrawal may be mediated, in part, by NPY signaling. A related study comparing the P and NP rats showed opposite electrophysiological activity in the amygdala following icv infusion of NPY (Ehlers et al. 1999); this observation, together with those showing low NPY levels in P rats (Ehlers et al. 1998a) (see Sect. 10.3.1) strongly suggest that altered NPY signaling in the amygdala of P rats contributes to their high alcohol drinking.

3.5
Ethanol Administration Modulates Central NPY Signaling

There is also evidence that ethanol administration can influence NPY signaling. Relative to control animals that received an isocaloric diet as their sole source of calories, Long-Evans rats given access to a diet containing 6% ethanol for 12 weeks showed significant increases of NPY levels in the arcuate and ventromedial nuclei of the hypothalamus, the median eminence, and the suprachiasmatic nucleus (Clark et al. 1998). Additionally, peripheral injection of 1.5 and 3.5 g/kg ethanol caused activation of NPY-containing neurons in the ventrolateral medulla of Long-Evans rats (Thiele et al. 2000a). Wistar rats exposed to ethanol vapor for 14 h/day showed no differences in brain NPY expression after 7 weeks of exposure, but showed increased NPY expression in the hypothalamus 7 weeks after withdrawal from ethanol (Ehlers et al. 1998a). On the other hand, ethanol administration and withdrawal from ethanol have also been found to reduce NPY signaling. NPY mRNA levels in the arcuate nucleus of the hypothalamus were decreased when Sprague-Dawley rats were given a single peripheral injection of a 1.0 g/kg dose of ethanol (Kinoshita et al. 2000). More recently, Sprague-Dawley rats examined 24 h after withdrawal from a diet containing 9% ethanol (after 15 days of exposure) showed decreased NPY immunoreactivity in the cingulate gyrus, various regions of the cortex, the central and medial nuclei of the amygdala, and the paraventricular and arcuate nuclei of the hypothalamus (Roy and Pandey 2002). Thus, withdrawal from ethanol is associated with reduced central NPY signaling. Consistent with this observation, a recent report found that icv infusion of NPY significantly attenuated ethanol withdrawal responses in Wistar rats (Woldbye et al. 2002). The discrepancies between these studies (that is, either increases or decreases of NPY levels) may be related to rat strain differences, method of ethanol administration, technique for assessing NPY levels, or an interaction between these factors.

3.6
NPY and Alcoholism: Human Genetic Studies

A recent study used cDNA microarrays to examine the expression of approximately 10,000 genes in the frontal cortex and motor cortex of alcoholics and matched control samples. One of the most intriguing observations was that brain tissue from alcoholics had significantly lower NPY expression than brain tissue from controls (Mayfield et al. 2002). It is unclear if low NPY levels were present before the manifestation of alcoholism (thus potentially triggering the disease) or were the result of chronic alcohol use. Additionally, several groups have taken advantage of a known gene polymorphism to study the potential contribution of NPY to human alcoholism. In some individuals, a thymidine (T) to cytosine (C) polymorphism is present at the 1128 locus of the human *NPY* gene, resulting in a leucine-to-proline substitution (Leu7 to Pro7) in the signal part of pre-pro-NPY (Karvonen et al. 1998). Individuals with the *Leu7/Pro7* genotype have an average of 42% higher maximal increases of plasma NPY in response to physiological stress when compared with *Leu7/Leu7* individuals (Kallio et al. 2001). Interestingly, Finnish men with the Pro7 substitution reported 34% higher average alcohol consumption when compared to men not having this polymorphism (Kauhanen et al. 2000). Another report showed European-American men with diagnosed alcoholism had 5–5.5% Pro7 allele frequency, while non-alcoholics had a Pro7 allele frequency of only 2.0% (Lappalainen et al. 2002). This polymorphism has also been reported to have a lower frequency in type 2 alcoholics compared to controls (Ilveskoski et al. 2001). However, a more recent study found no difference of Pro7 allele frequency between diagnosed Caucasian alcoholics and ethnically matched controls from Finland and Sweden (Zhu et al. 2003). Furthermore, a meta-analysis performed by Zhu et al. found that while the Pro7 allele frequencies in alcoholics were similar in each report, the allele frequencies in nonalcoholic control groups were very different between studies (Ilveskoski et al. 2001; Lappalainen et al. 2002; Zhu et al. 2003). Zhu et al. suggested that recent work showing a complete lack of Pro7 polymorphism in Asians (Ding et al. 2002; Okubo and Harada 2001) demonstrates the fact that dramatic ethnic differences exist, and while careful controls were used, ethnic admixture may have biased control group frequencies in prior work (Lappalainen et al. 2002). Finally, the whole coding region and 5′-untranslated region of the *NPY* gene of Japanese male alcoholics and Japanese male control subjects were screened for polymorphic nucleotide substitutions. A significant C to T substitution at the 5,671 locus of the *NPY* gene was higher in alcoholic patients experiencing seizures, suggesting that this mutation may be involved with seizure during ethanol withdrawal (Okubo and Harada 2001).

3.7
NPY and Alcohol: Summary

A growing body of evidence suggests a role for NPY signaling in voluntary ethanol drinking and neurobiological responses to ethanol administration. While it may be premature to make conclusive statements about underlying mechanisms, a general pattern of results is emerging. First, normal neurobiological responses to ethanol involves central NPY signaling. This is evidenced by the observations that administration of ethanol and ethanol withdrawal altered central NPY levels and expression (Clark et al. 1998; Ehlers et al. 1998a; Kinoshita et al. 2000; Roy and Pandey 2002; Thiele et al. 2000a). Furthermore, ethanol and NPY administration produced similar and additive effects on brain electrophysiological activity (Ehlers et al. 1998b; Ehlers et al. 1997; Slawecki et al. 1999). Second, low NPY signaling in animal models predisposes high ethanol drinking (Ehlers et al. 1998a; Hwang et al. 1999; Thiele et al. 1998). Importantly, central administration of NPY reduced ethanol drinking in rats with a genetic predisposition towards ethanol preference (Badia-Elder et al. 2001), but did not affect ethanol consumption in 'normal' unselected animals (Katner et al. 2002a,b; Slawecki et al. 2000). Third, the NPY Y_1 and Y_2 receptors modulate the effects of NPY on ethanol drinking. Mutant mice lacking Y_1 receptor drank high amounts of ethanol (Thiele et al. 2002), and Y_2 receptor knockout mice showed suppressed ethanol drinking (Thiele et al. 2000c). Consistently, a Y_2 receptor antagonist attenuated ethanol self-administration in rats (Thorsell et al. 2002). Finally, human association research suggests that alterations of normal NPY function may be associated with alcoholism. Brain tissue from alcoholics was found to have abnormally low levels of NPY when compared to control tissue (Mayfield et al. 2002). While association studies linking polymorphism of the human *NPY* gene with alcoholism are intriguing (Ilveskoski et al. 2001; Lappalainen et al. 2002; Zhu et al. 2003), results have been mixed and further research is necessary before causation can be established.

In light of the current data, it is possible that central NPY activity is recruited in response to ethanol consumption, and that this NPY activation serves as a protective feedback mechanism to prevent high ethanol drinking. Animals with abnormally low NPY levels would not benefit from this feedback protection and drink excessive quantities of ethanol. Such a mechanism could also explain excessive drinking in alcoholics with low brain NPY expression. The current data suggest that drugs targeting the central NPY system may serve as useful therapeutic agents against alcoholism. Compounds aimed at the Y_1 and/or Y_2 receptor(s) appear to be promising candidates.

4
Behavioral Effects of NPY: Conclusions

The present chapter demonstrates that central NPY signaling is involved with the integration of emotional-associated behaviors and with drug-seeking behavior, specifically the ingestion of ethanol. While behavioral responses to stimuli associated with stressful events can be adaptive in the short term, long-term stress responses can be maladaptive and compromise the constitution of the organism. NPY serves as a homeostatic buffer by attenuating stress responses following prolonged activation. It is hypothesized that dysregulation of this system may contribute to the pathophysiology of anxiety and depression. Drug abuse disorders, including alcoholism, are thought to involve neurobiological systems that integrate emotionality (Koob and Le Moal 1997). A growing body of evidence suggests that alterations of NPY signaling can promote excessive ethanol drinking, both in animal models of alcoholism and in humans. An interesting possibility is that high ethanol-seeking behavior is secondary to high levels of anxiety. Viewed this way, alterations of normal NPY activity may induce emotional disturbance, which in turn becomes a risk for alcoholism. Further research is necessary to establish such a causal relationship.

References

Andrews N, File SE, Fernandes C, Gonzalez LE, Barnes NM (1997) Evidence that the median raphe nucleus-dorsal hippocampal pathway mediates diazepam withdrawal-induced anxiety. Psychopharmacology 130:228–234

Asakawa A, Inui A, Ueno N, Fujimiya M, Fujino MA, Kasuga M (1999) Mouse pancreatic polypeptide modulates food intake, while not influencing anxiety in mice. Peptides 20:1445–1448

Aston-Jones G, Chiang C, Alexinsky T (1991) Discharge of noradrenergic locus coeruleus neurons in behaving rats and monkeys suggests a role in vigilance. Prog Brain Res 88:501–520

Badia-Elder NE, Stewart RB, Powrozek TA, Roy KF, Murphy JM, Li TK (2001) Effect of neuropeptide Y (NPY) on oral ethanol intake in Wistar, alcohol-preferring (P), and -nonpreferring (NP) rats. Alcohol Clin Exp Res 25:386–390

Bannon AW, Seda J, Carmouche M, Francis JM, Norman MH, Karbon B, Mccaleb ML (2000) Behavioral characterization of neuropeptide Y knockout mice. Brain Res 868:79–87

Baraban SC, Hollopeter G, Erickson JC, Schwartzkroin PA, Palmiter RD (1997) Knockout mice reveal a critical antiepileptic role for neuropeptide Y. J Neurosci 17:8927–8936

Baumann P (1996) Pharmacology and pharmacokinetics of citalopram and other SSRIs. Int Clin Pharmacol 11 (Suppl)1:5–11

Bellman R, Sperk G (1993) Effects of antidepressant drug treatment on levels of NPY or prepro-NPY-mRNA in the rat brain. Neurochem Int 22:183–187

Berrettini WH, Doran AR, Kelsoe J, Roy A, Pickar D (1987) Cerebrospinal fluid neuropeptide Y in depression and schizophrenia. Neuropsychopharmacology 1:81–83

Bice P, Foroud T, Bo R, Castelluccio P, Lumeng L, Li TK, Carr LG (1998) Genomic screen for QTLs underlying alcohol consumption in the P and NP rat lines. Mamm Genome 9:949–955

Bolwig TG, Woldbye DP, Mikkelsen JD (1999) Electroconvulsive therapy as an anticonvulsant: a possible role of neuropeptide Y (NPY). J ECT 15:93–101

Brandao ML, Cardoso SH, Melo LL, Motta V, Coimbra NC (1994) Neural substrate of defensive behavior in the midbrain tectum. Neurosci Biobehav Rev 18:339–346

Broqua P, Wettstein JG, Rocher MN, Gauthier-Martin B, Junien JL (1995) Behavioral effects of neuropeptide receptor agonists in the elevated plus-maze and fear-potentiated startle procedure. Behav Pharmacol 6:215–222

Broqua P, Wettstein JG, Rocher MN, Gauthier-Martin B, Riviere PJ, Junien JL, Dahl SG (1996) Antinociceptive effects of neuropeptide Y and related peptides in mice. Brain Res 724:25–32

Caberlotto L, Fuxe K, Overstreet DH, Gerrard P, Hurd YL (1998) Alterations in neuropeptide Y and Y1 receptor mRNA expression in brains from an animal model of depression: region specific adaptation after fluoxetine treatment. Brain Res Mol Brain Res 59:58–65

Caberlotto L, Hurd YL (1999) Reduced neuropeptide Y mRNA expression in the prefrontal cortex of subjects with bipolar disorder. Neuroreport 10:1747–1750

Caberlotto L, Jimenez P, Overstreet DH, Hurd YL, Mathe AA, Fuxe K (1999) Alterations in neuropeptide Y levels and Y1 binding sites in the Flinders Sensitive Line rats, a genetic animal model of depression. Neurosci Lett 265:191–194

Caberlotto L, Thorsell A, Rimondini R, Sommer W, Hyytia P, Heilig M (2001) Differential expression of NPY and its receptors in alcohol-preferring AA and alcohol-avoiding ANA rats. Alcohol Clin Exp Res 25:1564–1569

Cairncross KD (1984) Olfactory bulbectomy as a model of depression. Animal models in psychopathology. Academic Press, Sydney, pp 99–128

Carr LG, Foroud T, Bice P, Gobbett T, Ivashina J, Edenberg H, Lumeng L, Li T-K (1998) A quantitative trait locus for alcohol consumption in selectively bred rat lines. Alcohol Clin Exper Res 22:884–887

Chojnacka-Wojcik E, Tatarczynska E, Pilc A (1997) The anxiolytic-like effect of metabotropic glutamate receptor antagonists after intrahippocampal injection in rats. Eur J Pharmacol 319:153–156

Clark JJ, Karla PS, Crowle WR, Karla SP (1984) Neuropeptide Y and human pancreatic polypeptide stimulate feeding behavior in rats. Endocrinology 115:427–429

Clark JT, Keaton AK, Sahu A, Kalra SP, Mahajan SC, Gudger JN (1998) Neuropeptide Y (NPY) levels in alcoholic and food restricted male rats: implications for site selective function. Regul Peptides 75/76:335–345

Davis M (1998) Are different parts of the extended amygdala involved in fear versus anxiety? Biol Psychiatry 44:1239–1247

Davis M, Rainnie D, Cassell M (1994) Neurotransmission in the rat amygdala related to fear and anxiety. Trends Neurosci 17:208–214

Ding B, Bertilsson L, Wahlestedt C (2002) The single nucleotide polymorphism T1128C in the signal peptide of neuropeptide Y (NPY) was not identified in a Korean population. J Clin Pharm Ther 27:211–212

Doods HN, Wieland HA, Engel W, Eberlein W, Willim KD, Entzeroth M, Wienen W, Rudolf K (1996) BIBP 3226, the first selective neuropeptide Y1 receptor antagonist: a review of its pharmacological properties. Regul Pept 65:71–77

Ehlers CL, Li TK, Lumeng L, Hwang BH, Somes C, Jimenez P, Mathe AA (1998a) Neuropeptide Y levels in ethanol-naive alcohol-preferring and nonpreferring rats and in Wistar rats after ethanol exposure. Alcohol Clin Exp Res 22:1778–1782

Ehlers CL, Somes C, Cloutier D (1998b) Are some of the effects of ethanol mediated through NPY? Psychopharmacology 139:136–44

Ehlers CL, Somes C, Lopez A, Kirby D, Rivier JE (1997) Electrophysiological actions of neuropeptide Y and its analogs: new measures for anxiolytic therapy? Neuropsychopharmacology 17:34–43

Ehlers CL, Somes C, Lumeng L, Li TK (1999) Electrophysiological response to neuropeptide Y (NPY): in alcohol-naive preferring and non-preferring rats. Pharmacol Biochem Behav 63:291–299

Ekman R, Juhasz P, Heilig M, Agren H, Costello CE (1996) Novel neuropeptide Y processing in human cerebrospinal fluid from depressed patients. Peptides 17:1107–1111

Fendt M, Fanselow MS (1999) The neuroanatomical and neurochemical basis of conditioned fear. Neurosci Biobehav Rev 23:743–760

File SE (1980) The use of social interaction as a method for detecting anxiolytic activity of chlordiazepoxide-like drugs. J Neurosci Methods 2:219–238

File SE (1993) The interplay of learning and anxiety in the elevated plus-maze. Behav Brain Res 58:199–202

Foroud T, Bice P, Castelluccio P, Bo R, Miller L, Ritchotte A, Lumeng L, Li TK, Carr LG (2000) Identification of quantitative trait loci influencing alcohol consumption in the high alcohol drinking and low alcohol drinking rat lines. Behav Genet 30:131–140

Fuhlendorff J, Gether U, Aakerlund L, Langeland-Johansen N, Thogersen H, Melberg SG, Olsen UB, Thastrup O, Schwartz TW (1990) [Leu31, Pro34]neuropeptide Y: a specific Y1 receptor agonist. Proc Natl Acad Sci USA 87:182–186

Fuxe K, Agnati LF, Harfstrand A, Zini I, Tatemoto K, Pich EM, Hokfelt T, Mutt V, Terenius L (1983) Central administration of neuropeptide Y induces hypotension bradypnea and EEG synchronization in the rat. Acta Physiologica Scandinavica 118:189–192

Geller I, Seifter J (1960) The effects of meprobamate, barbiturates,d-amphetamine and promazine on experimentally induced conflict in the rat. Psychopharmacologia 1:482–492

Gianoulakis C, De Waele JP, Thavundayil J (1996) Implication of the endogenous opioid system in excessive ethanol consumption. Alcohol 13:19–23

Gonzalez LE, File SE, Overstreet DH (1998) Selectively bred lines of rats differ in social interaction and hippocampal 5-HT1A receptor function: a link between anxiety and depression? Pharmacol Biochem Behav 59:787–792

Gray JA (1983) A theory of anxiety: the role of the limbic system. Encephale 9:161B–166B

Heilig M (1995) Antisense inhibition of neuropeptide Y (NPY)-Y1 receptor expression blocks the anxiolytic like action of NPY in amygdala and paradoxically increases feeding. Regul Peptides 59:201–205

Heilig M and Ågren H. J Psychiatr Res (in press)

Heilig M, Ekman R (1995) Chronic parenteral antidepressant treatment in rats: Unaltered levels and processing of neuropeptide Y (NPY) and Corticotropin-Releasing Hormone (CRH). Neurochem Int 26:351–355

Heilig M, Koob GF, Ekman R, Britton KT (1994) Corticotropin-releasing factor and neuropeptide Y: role in emotional integration. Trends Neurosci 17:80–85

Heilig M, McLeod S, Brot M, Heinrichs SC, Mensaghi F, Koob GF, Britton KT (1993a) Anxiolytic-like action of neuropeptide Y: mediation by Y1 receptors in amygdala, and dissociation from food intake effects. Neuropsychopharmacology 8:357–363

Heilig M, McLeod S, Koob GF, Britton KT (1992) Anxiolytic-like effect of neuropeptide Y (NPY), but not other peptides in an operant conflict test. Regul Pept 41:61–69

Heilig M, Murison R (1987a) Intracerebroventricular neuropeptide Y protects against stress-induced gastric erosion in the rat. Eur J Pharmacol 137:127–129

Heilig M, Murison R (1987b) Intracerebroventricular neuropeptide Y suppresses open field and home cage activity in the rat. Regul Pept 19:221–231

Heilig M, Schlingensiepen KH (1996) Antisense oligodeoxynucleotides as novel neuropharmacological tools for selective expression blockade in the brain. In: Latchman DS (ed) Genetic manipulation of the brain (Neuroscience Prespectives). Academic Press, London, pp 249–268

Heilig M, Soderpalm B, Engel JA, Widerlov E (1989) Centrally administered neuropeptide Y (NPY) produces anxiolytic- like effects in animal anxiety models. Psychopharmacology 98:524–529

Heilig M, Thorsell A (2002) Brain neuropeptide Y (NPY) in stress and alcohol dependence. Rev Neurosci 13:85–94

Heilig M, Wahlestedt C, Ekman R, Widerlov E (1988a) Antidepressant drugs increase the concentration of neuropeptide Y (NPY)-like immunoreactivity in the rat brain. Eur J Pharmacol 147:465–467

Heilig M, Wahlestedt C, Widerlov E (1988b) Neuropeptide Y (NPY)-induced suppression of activity in the rat: evidence for NPY receptor heterogeneity and for interaction with alpha-adrenoceptors. Eur J Pharmacol 157:205–213

Heilig M, Widerlov E (1995) Neurobiology and clinical aspects of neuropeptide Y. Crit Rev Neurobiol 9:115–136

Hogg S (1996) A review of the validity and variability of the elevated plus-maze as an animal model of anxiety. Pharmacol Biochem Behav 54:21–30

Holmes PV, Davis RC, Masini CV, Primeaux SD (1998) Effects of olfactory bulbectomy on neuropeptide gene expression in the rat olfactory/limbic system. Neuroscience 86:587–596

Holsboer F (2000) The corticosteroid receptor hypothesis of depression. Neuropsychopharmacology 23:477–501

Hua XY, Boublik JH, Spicer MA, Rivier JE, Brown MR, Yaksh TL (1991) The antinociceptive effects of spinally administered neuropeptide Y in the rat: systematic studies on structure-activity relationship. J Pharm Exp Ther 258:243–248

Husum H, Mikkelsen JD, Hogg S, Mathe AA, Mork A (2000) Involvement of hippocampal neuropeptide Y in mediating the chronic actions of lithium, electroconvulsive stimulation and citalopram. Neuropharmacology 39:1463–1473

Husum H, Termeer E, Mathe AA, Bolwig TG, Ellenbroek BA (2002) Early maternal deprivation alters hippocampal levels of neuropeptide Y and calcitonin-gene related peptide in adult rats. Neuropharmacology 42:798–806

Hwang BH, Zhang JK, Ehlers CL, Lumeng L, Li TK (1999) Innate differences of neuropeptide Y (NPY) in hypothalamic nuclei and central nucleus of the amygdala between selectively bred rats with high and low alcohol preference. Alcohol Clin Exp Res 23:1023–1030

Ilveskoski E, Kajander OA, Kehtimaki T, Kannus T, Karhunen PJ, Heinala P, Virkkunen M, Alho H (2001) Association of neuropeptide Y polymorphism with the occurrence of type 1 and type 2 alcoholism. Alcohol Clin Exp Res 25:1420–1422

Jimenez P, Salmi P, Ahlenius S, Mathe AA (2000) Neuropeptide Y in brains of the Flinders Sensitive Line rat, a model of depression. Effects of electroconvulsive stimuli and d-amphetamine on peptide concentrations and locomotion. Behav Brain Res 111:115–123

Jimenez-Vasquez PA, Mathe AA, Thomas JD, Riley EP, Ehlers CL (2001) Early maternal separation alters neuropeptide Y concentrations in selected brain regions in adult rats. Brain Res Devel Brain Res 131:149–152

Kallio J, Pesonen U, Kaipio K, Karvonen MK, Jaakkola U, Heinonen OJ, Uusitupa MI, Koulu M (2001) Altered intracellular processing and release of neuropeptide Y due to leucine 7 to proline 7 polymorphism in the signal peptide of preproneuropeptide Y in humans. FASEB J 15:1242–1244

Karkowski LM, Kendler KS (1997) An examination of the genetic relationship between bipolar and unipolar illness in an epidemiological sample. Psychiatr Genet 7:159–163

Karvonen MK, Pesonen U, Koulu M, Niskanen L, Laakso M, Rissanen A, Dekker JM, Hart LM, Valve R, Uusitupa MI (1998) Association of a leucine(7)-to-proline(7) polymorphism in the signal peptide of neuropeptide Y with high serum cholesterol and LDL cholesterol levels. Nature Med 4:1434–1437

Kask A, Eller M, Oreland L, Harro J (2000) Neuropeptide Y attenuates the effect of locus coeruleus denervation by DSP-4 treatment on social behaviour in the rat. Neuropeptides 34:58–61

Kask A, Harro J (2000) Inhibition of amphetamine- and apomorphine-induced behavioural effects by neuropeptide Y Y(1) receptor antagonist BIBO 3304. Neuropharmacology 39:1292–1302

Kask A, Harro J, von Horsten S, Redrobe JP, Dumont Y, Quirion R (2002) The neurocircuitry and receptor subtypes mediating anxiolytic-like effects of neuropeptide Y. Neurosci Biobehav Rev 26:259–283

Kask A, Nguyen HP, Pabst R, von Horsten S (2001a) Neuropeptide Y Y1 receptor-mediated anxiolysis in the dorsocaudal lateral septum: functional antagonism of corticotropin-releasing hormone-induced anxiety. Neuroscience 104:799–806

Kask A, Rago L, Harro J (1996) Anxiogenic-like effect of the neuropeptide Y Y-1 receptor antagonist BIBP3226 - antagonism with diazepam. Eur J Pharmacol 317: R3–R4

Kask A, Rago L, Harro J (1998a) Anxiogenic-like effect of the NPY Y1 receptor antagonist BIBP3226 administered into the dorsal periaqueductal gray matter in rats. Regul Pept 75/76:255–262

Kask A, Rago L, Harro J (1998b) NPY Y1 receptors in the dorsal periaqueductal gray matter regulate anxiety in the social interaction test. Neuroreport 9:2713–2716

Kask A, Vasar E, Heidmets LT, Allikmets L, Wikberg JE (2001b) Neuropeptide Y Y(5) receptor antagonist CGP71683A: the effects on food intake and anxiety-related behavior in the rat. Eur J Pharmacol 414:215–224

Katner SN, Slawecki CJ, Ehlers CL (2002a) Neuropeptide Y administration into the amygdala does not effect ethanol consumption. Alcohol 28:29–38

Katner SN, Slawecki CJ, Ehlers CL (2002b) Neuropeptide Y administration into the third ventricle does not increase sucrose or ethanol self-administration but does affect the cortical EEG and increases food intake. Psychopharmacology 160:146–154

Kauhanen J, Karvonen MK, Pesonen U, Koulu M, Tuomainen TP, Uusitupa MI, Salonen JT (2000) Neuropeptide Y polymorphism and alcohol consumption in middle-aged men. Am J Med Genet 93:117–121

Kelley SP, Nannini MA, Bratt AM, Hodge CW (2001) Neuropeptide-Y in the paraventricular nucleus increases ethanol self- administration. Peptides 22:515–522

Kendler KS, Neale MC, Kessler RC, Heath AC, Eaves LJ (1992) Major depression and generalized anxiety disorder. Same genes, (partly) different environments? Arch Gen Psychiatry 49:716–722

King AC, Volpicelli JR, Frazer A, O'Brien CP (1997) Effect of naltrexone on subjective alcohol response in subjects at high and low risk for future alcohol dependence. Psychopharmacology 129:15–22

King PJ, Widdowson PS, Doods HN, Williams G (1999) Regulation of neuropeptide Y release by neuropeptide Y receptor ligands and calcium channel antagonists in hypothalamic slices. J Neurochem 73:641–646

King PJ, Williams G, Doods H, Widdowson PS (2000) Effect of a selective neuropeptide Y Y(2) receptor antagonist, BIIE0246 on neuropeptide Y release. Eur J Pharmacol 396: R1–R3

Kinoshita H, Jessop DS, Finn DP, Coventry TL, Roberts DJ, Ameno K, Ijiri I, Harbuz MS (2000) Acute ethanol decreases NPY mRNA but not POMC mRNA in the arcuate nucleus. NeuroReport 11:3517–3519

Koob GF, Heinrichs SC, Pich EM, Menzaghi F, Baldwin H, Miczek K, Britton KT (1993) The role of corticotropin-releasing factor in behavioural responses to stress. Ciba Foundation Symposium 172:277–289

Koob GF, Le Moal M (1997) Drug abuse: hedonic homeostatic dysregulation. Science 278:52–58

Kulkosky PJ, Glazner GW, Moore HD, Low CA, Woods SC (1988) Neuropeptide Y: behavioral effects in the golden hamster. Peptides 9:1389–1393

Lappalainen J, Kranzler HR, Malison R, Price LH, Van Dyck C, Rosenheck RA, Cramer J, Southwick S, Charney D, Krystal J, Gelernter J (2002) A functional neuropeptide Y Leu7Pro polymorphism associated with alcohol dependence in a large population sample from the United States. Arch Gen Psychiatr 59:825–831

Larhammar D, Ericsson A, Persson H (1987) Structure and expression of the rat neuropeptide Y gene. Proc Natl Acad Sci USA 84:2068–2072

LeDoux JE (2000) Emotion circuits in the brain. Annu Rev Neurosci 23:155–184

Lee Y, Davis M (1997) Role of the septum in the excitatory effect of corticotropin-releasing hormone on the acoustic startle reflex. J Neurosci 17:6424–6433

Levine AS, Morley JE (1984) Neuropeptide Y: a potent inducer of consummatory behavior in rats. Peptides 5:1025–1029

Liu D, Diorio J, Tannenbaum B, Caldji C, Francis D, Freedman A, Sharma S, Pearson D, Plotsky PM, Meaney MJ (1997) Maternal care, hippocampal glucocorticoid receptors, and hypothalamic-pituitary-adrenal responses to stress. Science 277:1659–1662

Lundberg JM, Hokfelt T (1983) Coexistence of peptides and classical neurotransmitters. Trends Neurosci 6:325–333

Marsh DJ, Baraban SC, Hollopeter G, Palmiter RD (1999) Role of the Y5 neuropeptide Y receptor in limbic seizures. Proc Natl Acad Sci USA 96:13518–13523

Mathe AA (1999) Neuropeptides and electroconvulsive treatment. J ECT 15:60–75

Mathe AA, Gruber S, Jimenez PA, Theodorsson E, Stenfors C (1997) Effects of electroconvulsive stimuli and MK-801 on neuropeptide Y, neurokinin A, and calcitonin gene-related peptide in rat brain. Neurochem Res 22:629–636

Mathe AA, Jimenez PA, Theodorsson E, Stenfors C (1998) Neuropeptide Y, neurokinin A and neurotensin in brain regions of Fawn Hooded 'depressed', Wistar, and Sprague Dawley rats. Effects of electroconvulsive stimuli. Prog Neuropsychopharmacol Biol Psychiatr 22:529–546

Mayfield RD, Lewohl JM, Dodd PR, Herlihy A, Liu J, Harris RA (2002) Patterns of gene expression are altered in the frontal and motor cortices of human alcoholics. J Neurochem 81:802–813

McCreary AC, McBlane JW, Spooner HA, Handley SL (1996) 5-HT systems and anxiety: multiple mechanisms in the elevated X-maze. Polish J Pharmacol 48:1–12

McQuiston AR, Colmers WF (1996) Neuropeptide Y2 receptors inhibit the frequency of spontaneous but not miniature EPSCs in CA3 pyramidal cells of rat hippocampus. J Neurophysiol 76:3159–3168

Mellado ML, Gibert-Rahola J, Chover AJ, Mico JA (1996) Effect on nociception of intracerebroventricular administration of low doses of neuropeptide Y in mice. Life Sci 58:2409–2414

Mikkelsen JD, Woldbye D, Kragh J, Larsen PJ, Bolwig TG (1994) Electroconvulsive shocks increase the expression of neuropeptide Y (NPY) mRNA in the piriform cortex and the dentate gyrus. Brain Res Mol Brain Res 23:317–322

Mollereau C, Gouarderes C, Dumont Y, Kotani M, Detheux M, Doods H, Parmentier M, Quirion R, Zajac JM (2001) Agonist and antagonist activities on human NPFF(2) receptors of the NPY ligands GR231118 and BIBP3226. Br J Pharmacol 133:1–4

Morgan CA, III, Wang S, Rasmusson A, Hazlett G, Anderson G, Charney DS (2001) Relationship among plasma cortisol, catecholamines, neuropeptide Y, and human performance during exposure to uncontrollable stress. Psychosom Med 63:412–422

Morgan CA, III, Wang S, Southwick SM, Rasmusson A, Hazlett G, Hauger RL, Charney DS (2000) Plasma neuropeptide-Y concentrations in humans exposed to military survival training. Biol Psychiatry 47:902–909

Morgan DG, Small CJ, Abusnana S, Turton M, Gunn I, Heath M, Rossi M, Goldstone AP, O'Shea D, Meeran K, Ghatei M, Smith DM, Bloom S (1998) The NPY Y1 receptor antagonist BIBP 3226 blocks NPY induced feeding via a non-specific mechanism. Regul Peptides 75/76:377–382

Naveilhan P, Canals JM, Arenas E, Ernfors P (2001a) Distinct roles of the Y1 and Y2 receptors on neuropeptide Y-induced sensitization to sedation. J Neurochem 78:1201–1207

Naveilhan P, Canals JM, Valjakka A, Vartiainen J, Arenas E, Ernfors P (2001b) Neuropeptide Y alters sedation through a hypothalamic Y1-mediated mechanism. Eur J Neurosci 13:2241–2246

Naveilhan P, Hassani H, Canals JM, Ekstrand AJ, Larefalk A, Chhajlani V, Arenas E, Gedda K, Svensson L, Thoren P, Ernfors P (1999) Normal feeding behavior, body weight and leptin response require the neuropeptide Y Y2 receptor. Nature Med 5:1188–93

Naveilhan P, Neveu I, Arenas E, Ernfors P (1998) Complementary and overlapping expression of Y1, Y2 and Y5 receptors in the developing and adult mouse nervous system. Neuroscience 87:289–302

Nestler EJ, Alreja M, Aghajanian GK (1999) Molecular control of locus coeruleus neurotransmission. Biol Psychiatry 46:1131–1139

Okubo T, Harada S (2001) Polymorphism of the neuropeptide Y gene: an association study with alcohol withdrawal. Alcohol Clin Exp Res 25:59S–62S

Ordway GA, Stockmeier CA, Meltzer HY, Overholser JC, Jaconetta S, Widdowson PS (1995) Neuropeptide Y in frontal cortex is not altered in major depression. J Neurochem 65:1646–1650

Overstreet DH, Pucilowski O, Rezvani AH, Janowsky DS (1995) Administration of antidepressants, diazepam and psychomotor stimulants further confirms the utility of Flinders Sensitive Line rats as an animal model of depression. Psychopharmacologia 121:27–37

Parker RM, Herzog H (1999) Regional distribution of Y-receptor subtype mRNAs in rat brain. Eur J Pharmacol 11:1431–1448

Pedrazzini T, Seydoux J, Kunstner P, Aubert JF, Grouzmann E, Beermann F, Brunner HR (1998) Cardiovascular response, feeding behavior and locomotor activity in mice lacking the NPY Y1 receptor. Nature Med 4:722–726

Pellow S, Chopin P, File SE, Briley M (1985) Validation of open: closed arm entries in an elevated plus-maze as a measure of anxiety in the rat. J Neurosci Methods 14:149–167

Pich EM, Agnati LF, Zini I, Marrama P, Carani C (1993) Neuropeptide Y produces anxiolytic effects in spontaneously hypertensive rats. Peptides 14:909–912

Plotsky PM, Meaney MJ (1993) Early, postnatal experience alters hypothalamic corticotropin- releasing factor (CRF) mRNA, median eminence CRF content and stress-induced release in adult rats. Brain Res Mol Brain Res 18:195–200

Qian J, Colmers WF, Saggau P (1997) Inhibition of synaptic transmission by neuropeptide Y in rat hippocampal area CA1: modulation of presynaptic Ca2+ entry. J Neurosci 17:8169–8177

Rasmusson AM, Hauger RL, Morgan CA, Bremner JD, Charney DS, Southwick SM (2000) Low baseline and yohimbine-stimulated plasma neuropeptide Y (NPY) levels in combat-related PTSD. Biol Psychiatry 47:526–539

Redrobe JP, Dumont Y, Fournier A, Quirion R (2002) The neuropeptide Y (NPY) Y1 receptor subtype mediates NPY-induced antidepressant-like activity in the mouse forced swimming test. Neuropsychopharmacology 26:615–624

Rezvani AH, Parsian A, Overstreet DH (2002) The Fawn-Hooded (FH/Wjd) rat: a genetic animal model of comorbid depression and alcoholism. Psychiatr Genet 12:1–16

Rodgers RJ (1997) Animal models of 'anxiety': where next?. Behav Pharmacol 8:477–496

Roy A, Pandey SC (2002) The decreased cellular expression of neuropeptide Y protein in rat brain structures during ethanol withdrawal after chronic ethanol exposure. Alcohol Clin Exp Res 26:796–803

Rudolf K, Eberlein W, Engel W, Wieland HA, Willim KD, Entzeroth M, Wienen W, Beck-Sickinger AG, Doods HN (1994) The first highly potent and selective non-peptide neuropeptide Y Y1 receptor antagonist: BIBP3226. Eur J Pharmacol 271: R11–R13

Ryabinin AE, Criado JR, Henriksen SJ, Bloom FE, Wilson MC (1997) Differential sensitivity of c-Fos expression in hippocampus and other brain regions to moderate and low doses of ethanol. Mol Psychiat 2:32–43

Sajdyk TJ, Vandergriff MG, Gehlert DR (1999) Amygdalar neuropeptide Y Y-1 receptors mediate the anxiolytic-like actions of neuropeptide Y in the social interaction test. Eur J Pharmacol 368:143–147

Slawecki CJ, Betancourt M, Walpole T, Ehlers CL (2000) Increases in sucrose consumption, but not ethanol consumption, following ICV NPY administration. Pharmacol Biochem Behav 66:591–594

Slawecki CJ, Somes C, Ehlers CL (1999) Effects of chronic ethanol exposure on neurophysiological responses to corticotropin-releasing factor and neuropeptide Y. Alcohol Alcohol 34:289–299

Song C, Earley B, Leonard BE (1996) The effects of central administration of neuropeptide Y on behavior, neurotransmitter, and immune functions in the olfactory bulbectomized rat model of depression. Brain Behav Immunol 10:1–16

Spanagel R, Zieglgansberger W (1997) Anti-craving compounds for ethanol: new pharmacological tools to study addictive processes. Trends Pharmacol Sci 18:54–59

Stenfors C, Theodorsson E, Mathe AA (1989) Effect of repeated electroconvulsive treatment on regional concentrations of tachykinins, neurotensin, vasoactive intestinal polypeptide, neuropeptide Y, and galanin in rat brain. J Neurosci Res 24:445–450

Stogner KA, Holmes PV (2000) Neuropeptide-Y exerts antidepressant-like effects in the forced swim test in rats. Eur J Pharmacol 387: R9–R10

Sullivan GM, Coplan JD, Kent JM, Gorman JM (1999) The noradrenergic system in pathological anxiety: a focus on panic with relevance to generalized anxiety and phobias. Biol Psychiatr 46:1205–1218

Sullivan PF, Kendler KS (1998) Typology of common psychiatric syndromes. An empirical study. Br J Psychiatr 173:312–319

Swanson LW, Petrovich GD (1998) What is the amygdala? Trends Neurosci 21:323–331

Thiele TE, Cubero I, van Dijk G, Mediavilla C, Bernstein IL (2000a) Ethanol-induced c-Fos expression in catecholamine- and neuropeptide Y-producing neurons in rat brainstem. Alcohol Clin Exp Res 24:802–809

Thiele TE, Koh MT, Pedrazzini T (2002) Voluntary alcohol consumption is controlled via the neuropeptide Y Y1 receptor. J Neurosci 22: RC208

Thiele TE, Marsh DJ, Ste. Marie L, Bernstein IL, Palmiter RD (1998) Ethanol consumption and resistance are inversely related to neuropeptide Y levels. Nature 396:366–369

Thiele TE, Miura GI, Marsh DJ, Bernstein IL, Palmiter RD (2000b) Neurobiological responses to ethanol in mutant mice lacking neuropeptide Y or the Y5 receptor. Pharmacol Biochem Behav 67:683–691

Thiele TE, Naveilhan P, Ernfors P (2000c) Mutant mice lacking the Y2 neuropeptide Y (NPY) receptor consume less ethanol than wild-type mice. Alcohol Clin Exp Res 24:97A

Thorsell A, Carlsson K, Ekman R, Heilig M (1999) Behavioral and endocrine adaptation, and up-regulation of NPY expression in rat amygdala following repeated restraint stress. Neuroreport 10:3003–3007

Thorsell A, Michalkiewicz M, Dumont Y, Quirion R, Caberlotto L, Rimondini R, Mathe AA, Heilig M (2000) Behavioral insensitivity to restraint stress, absent fear suppression of behavior and impaired spatial learning in transgenic rats with hippocampal neuropeptide Y overexpression. Proc Natl Acad Sci USA 97:12852–12857

Thorsell A, Rimondini R, Heilig M (2002) Blockade of central neuropeptide Y (NPY) Y2 receptors reduces ethanol self-administration in rats. Neurosci Lett 332:1–4

Thorsell A, Svensson P, Wiklund L, Sommer W, Ekman R, Heilig M (1998) Suppressed neuropeptide Y (NPY) mRNA in rat amygdala following restraint stress. Regul Peptides 75/76:247–254

Vogel JR, Beer B, Clody DE (1971) A simple and reliable conflict procedure for testing anti- anxiety agents. Psychopharmacologia 21:1–7

Wahlestedt C, Blendy JA, Kellar KJ, Heilig M, Widerlov E, Ekman R (1990) Electroconvulsive shocks increase the concentration of neocortical and hippocampal neuropeptide Y (NPY)-like immunoreactivity in the rat. Brain Res 507:65–68

Wahlestedt C, Golanov E, Yamamoto S, Yee F, Ericson H, Yoo H, Inturrisi CE, Reis DJ (1993a) Antisense oligodeoxynucleotides to NMDA-R1 receptor channel protect cortical neurons from excitotoxicity and reduce focal ischaemic infarctions. Nature 363:260–263

Wahlestedt C, Pich EM, Koob GF, Yee F, Heilig M (1993b) Modulation of anxiety and neuropeptide Y-Y1 receptors by antisense oligodeoxynucleotides. Science 259:528–531

Wang JZ, Lundeberg T, Yu L (2000) Antinociceptive effects induced by intra-periaqueductal grey administration of neuropeptide Y in rats. Brain Res 859:361–363

Westrin A, Ekman R, Traskman-Bendz L (1999) Alterations of corticotropin releasing hormone (CRH) and neuropeptide Y (NPY) plasma levels in mood disorder patients with a recent suicide attempt. Eur Neuropsychopharmacol 9:205–211

Westrin A, Engstom G, Ekman R, Traskman-Bendz L (1998) Correlations between plasma-neuropeptides and temperament dimensions differ between suicidal patients and healthy controls. J Affect Disord 49:45–54

Widdowson PS, Halaris AE (1991) Chronic desipramine treatment reduces regional neuropeptide Y binding to Y2-type receptors in rat brain. Brain Res 539:196–202

Widdowson PS, Ordway GA, Halaris AE (1992) Reduced neuropeptide Y concentrations in suicide brain. J Neurochem 59:73–80

Widerlov E, Lindstrom LH, Wahlestedt C, Ekman R (1988) Neuropeptide Y and peptide YY as possible cerebrospinal fluid markers for major depression and schizophrenia, respectively. J Psychiatr Res 22:69–79

Wieland HA, Engel W, Eberlein W, Rudolf K, Doods HN (1998) Subtype selectivity of the novel nonpeptide neuropeptide Y Y1 receptor antagonist BIBO 3304 and its effect on feeding in rodents. Br J Pharmacol 125:549–555

Woldbye DP, Larsen PJ, Mikkelsen JD, Klemp K, Madsen TM, Bolwig TG (1997) Powerful inhibition of kainic acid seizures by neuropeptide Y via Y5-like receptors. Nature Med 3:761–764

Woldbye DP, Madsen TM, Larsen PJ, Mikkelsen JD, Bolwig TG (1996) Neuropeptide Y inhibits hippocampal seizures and wet dog shakes. Brain Res 737:162–168

Woldbye DPD, Ulrichsen J, Haugbol S, Bolwig TG (2002) Ethanol withdrawal in rats is attenuated by intracerebroventricular administration of neuropeptide Y. Alcohol 37:318–321

Zaborszky L, Duque A (2000) Local synaptic connections of basal forebrain neurons. Behav Brain Res 115:143–158

Zachrisson O, Mathe AA, Stenfors C, Lindefors N (1995) Limbic effects of repeated electroconvulsive stimulation on neuropeptide Y and somatostatin mRNA expression in the rat brain. Brain Res Mol Brain Res 31:71–85

Zhu G, Pollak L, Mottagui-Tabar S, Wahlestedt C, Taubman J, Vikkunen M, Goldman D, Heilig M (2003) NPY leu7pro and alcohol dependence in Finnish and Swedish populations. Alcohol Clin Exp Res 27:19–24

NPY Effects on Food Intake and Metabolism

N. R. Levens[1,4] · M. Félétou[1] · J.-P. Galizzi[2] · J.-L. Fauchére[3] · O. Della-Zuana[1]
M. Lonchampt[1]

[1] Division of Metabolic Diseases, Institut de Recherches Servier, 92150 Suresnes, France
e-mail: nigel.levens@biovitrum.com
[2] Division of Molecular and Cellular Pharmacology, Institut de Recherches Servier,
92150 Suresnes, France
[3] Division of Chemistry, Institut de Recherches Servier,
92150 Suresnes, France
[4] Head of Biology, Biovitrum AB, Stockholm, Sweden

1	Introduction	285
2	Organization of NPY-Containing Neural Pathways in the Brain That Control Food Intake and Peripheral Metabolic Processes	286
2.1	NPY-Containing Neural Pathways in the Brain	286
2.2	NPY-Containing Neurons in the ARC Are Influenced by Neurotransmitters and Hormones	287
2.3	NPY–Proopiomelanocortin Connections in the ARC	287
3	The Actions of Exogenous NPY on Food Intake	288
3.1	NPY Alters Food Intake and Peripheral Metabolism When Injected into Specific Regions of the Hypothalamus	288
3.2	NPY also Alters Food Intake and Metabolism when Injected into Extra-Hypothalamic Sites	289
3.3	NPY Increases the Motivation to Eat	290
3.4	NPY Enhances the Rewarding Properties of Food	291
3.5	Does NPY-Induced Food Intake Resemble a True Hunger State?	291
3.6	Can Injection of Exogenous NPY into the Brain Mimic the Actions of the Endogenous Peptide on Food Intake?	292
4	Relationship Between Brain NPY Levels and Energy Balance	292
4.1	Introduction	292
4.2	Changes in Hypothalamic NPY Levels Are Temporally Correlated with Food Intake	293
4.3	Changes in Energy Balance Alter Brain NPY Levels	293
4.4	Changes in Energy Balance Alter NPY Receptor Number in the Brain	294
4.5	Chronic NPY Infusion into the Brain Increases Food Intake and Leads to Obesity	295
4.6	NPY Overexpression in the Brain Leads to Overconsumption of a High-Calorie Diet and to Obesity	296
5	The Actions of Endogenous NPY on Food Intake	297
5.1	Introduction	297
5.2	Food Intake in the NPY Germline Receptor Knockout Mouse	297
5.3	Food Intake in Response to NPY Antisense ODNs	299

5.4	Food Intake in Response to Anti-NPY Antibodies.	299
6	**NPY Receptors**	300
6.1	Subtypes of NPY Receptors	300
6.2	Evidence for Obesity Linkage to NPY Receptor Subtypes	300
7	**Which NPY Receptor Subtype Mediates Food Intake in Response to Endogenous NPY?**	301
7.1	The NPY Y_1 Receptor	301
7.1.1	NPY Y_1 Subtype Selective Agonists.	301
7.1.2	NPY Y_1 Receptor Knockout.	304
7.1.3	NPY Y_1 Antisense ODNs	304
7.1.4	NPY Y_1 Subtype Selective Antagonists.	305
7.1.5	Are the Actions of Other Peptides on Food Intake Mediated by the NPY Y_1 Receptor?	307
7.2	The NPY Y_2 Receptor Subtype	308
7.2.1	NPY Y_2 Subtype Selective Peptide Agonists and Food Intake	308
7.2.2	The Gut Hormone PYY_{3-36}: A Physiological Agonist of the NPY Y_2 Receptor	309
7.2.3	NPY Y_2 Receptor Knockouts	309
7.2.4	NPY Y_2 Receptor Antisense.	310
7.2.5	NPY Y_2 Receptor Antagonists.	310
7.3	The NPY Y_4 Receptor Subtype	310
7.3.1	NPY Y_4 Subtype Selective Peptide Agonists and Food Intake	310
7.3.2	NPY Y_4 Receptor Knockout.	312
7.4	The NPY Y_5 Receptor Subtype	312
7.4.1	The NPY Y_5 Receptor	312
7.4.2	Knockout of the NPY Y_5 Receptor	312
7.4.3	Inhibition of the Production of the NPY Y_5 Receptor with Antisense ODNs	313
7.4.4	NPY Y_5 Subtype Selective Antagonists.	314
7.4.5	Examples of NPY Y_5 Antagonists That Reduce Food Intake	314
7.4.6	Examples of NPY Y_5 Antagonists That Do Not Affect Food Intake	315
8	**Conclusions**	315
	References	316

Abstract The aim of this review is to critically assess the evidence that neuropeptide Y (NPY) plays both an important role in the control of food intake and in the peripheral metabolic processes linked to the obese state. When given into the brain, NPY stimulates food intake and a variety of metabolic processes that promote fat deposition. The stimulation of food intake and body weight observed with chronic administration of the peptide is persistent and leads to obesity. Both the acute and chronic stimulation of food intake and body weight produced by NPY can be reproduced with selective NPY Y_1 and Y_5 agonists. Therefore, there is no doubt that exogenously administered NPY is a potent regulator of appetite and could be involved in the development and maintenance of obesity. However, questions remain as to whether the increase in food intake produced by exogenous NPY represents a true hunger state or is mediated by unrelated behavioral changes. Although brain NPY levels and food intake are temporally related, attempts to demonstrate a change in food intake after blockade of endogenous NPY have been mixed. Furthermore, although some studies have shown a change in food intake after blockade of NPY the conclusion that this peptide plays an important role in the control of food intake is difficult to fully accept because of the nonselective nature of the inhibitors used. Based on the available evidence our conclusion is that NPY probably plays a role in the day-to-day control of food intake. However, NPY is not a critical regulator of food intake. In its absence appetite can be controlled by a variety of other hormones and neurotransmitters. However, a definitive answer to the role played by NPY in the control of food intake and peripheral metabolism awaits the development of clean and selective inhibitors.

Keywords Neuropeptide Y · Food intake · Obesity · NPY receptor knockout · NPY receptor antagonists · NPY antisense oligodeoxynucleotides · NPY antibodies

Dr. J.-L. Fauchére is diseased.

1
Introduction

Neuropeptide Y (NPY) has been shown to strongly stimulate food intake when injected into either the hypothalamus or the ventricles of the brain. When NPY is infused continuously into the brain the increase in food intake is persistent and leads to increased fat mass and body weight. These observations have prompted the suggestion that NPY may play a contributing role in the development and maintenance of the obese state.

In support of this hypothesis, NPY levels are increased in the brain of the obese Zucker *fa/fa* rat and the obese *ob/ob* mouse. Blockade of the actions of NPY in the brain can decrease food intake and body weight in both of these monogenetic forms of obesity. These findings, among others, have led to the search for NPY receptor antagonists as a treatment for obesity. However, the actions of NPY are many and varied and mediated by at least five NPY receptor subtypes. Thus, the hope exists that the effects of NPY on food intake and body weight are mediated by a single NPY receptor subtype. In this case blockade of a single receptor would reduce appetite and minimize the risk of side effects.

The purpose of this review is to critically assess the evidence that NPY plays an important role in the control of food intake and the peripheral metabolic processes that underlie the obese state. The possibility that NPY may exert its actions on food intake and metabolism via a single receptor subtype and thus represent an important treatment for obesity will also be the subject of this review.

2
Organization of NPY-Containing Neural Pathways in the Brain That Control Food Intake and Peripheral Metabolic Processes

2.1
NPY-Containing Neural Pathways in the Brain

Although NPY-producing neurons are present at many sites in the brain, the most important for the control of food intake are those that arise in the brain stem and those located along the arcuate nucleus (ARC) and in the dorsomedial nucleus (DMN) of the hypothalamus (Fig. 1) (Everitt and Hökfelt 1989; Kalra and Kalra 1996; Chronwall 1989). The NPY-containing neurons coming from the brain stem innervate various hypothalamic sites including the ARC, ventromedial nucleus (VMN), the DMN and the paraventricular nucleus (PVN) (Everitt and Hökfelt 1989; Kalra and Kalra 1996; Sahu et al. 1988). The NPY-containing neurons present in the ARC project to the PVN as well as to the DMN of the hypothalamus (Baker and Herkenham 1995; Bai et al. 1985). The ARC, PVN and DMN play an important and central role in the control of food intake (Kalra et al. 1999). A more detailed account of the neuroanatomy of NPY-containing neurons is given in the chapter by Redrobe et al. in this volume.

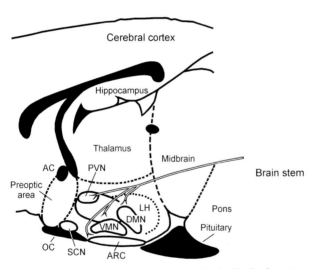

Fig. 1 Hypothalamic sites in the brain associated with the formation and actions of NPY. The figure represents a sagittal section near the midline of the rat brain. *AC*, anterior commissure; *OC*, optic chiasm; *PVN*, paraventricular nucleus; *SCN*, suprachiasmatic nucleus; *VMN*, ventromedial nucleus; *DMN*, dorsomedial nucleus; *LH*, lateral hypothalamus. NPY-containing neurons are present along the ARC and in the DMN. Other NPY-containing nerve fibers arise in the brain stem and innervate the PVN, the DMN, the VMN and the ARC. (Modified from Kalra et al. 1999)

2.2
NPY-Containing Neurons in the ARC Are Influenced by Neurotransmitters and Hormones

The NPY-containing neurons of the ARC have been shown to play an important role in the control of food intake and peripheral metabolic processes and are regulated by hormones and neurotransmitters. For example, ultrastructural studies have shown NPY-containing neurons in the ARC to be in synaptic contact with opiate- and catecholamine-containing neurons as well as with neurons containing other neuropeptides (Guy and Pelletier 1988; Guy et al. 1988; Li et al. 1993; Morton and Schwartz 2001). The neurotransmitters released in close proximity to NPY-containing neurons in the ARC–PVN projection influence their activity. For example, dopamine and serotonin, which inhibit and excite the activity of NPY-containing neurons respectively (Bina and Cincotta 2000; Dryden et al. 1996).

In the hypothalamus, hormonal factors influencing the activity of NPY-containing neurons include the gonadal steroids testosterone and estrogen as well as the adrenal glucocorticoids. Testosterone and glucocorticoids stimulate the activity of NPY-containing neurons, whereas estrogen in general inhibits their activity (Sahu et al. 1989, 1992; Bonavera et al. 1994; Hisano et al. 1988; Wilding et al. 1993). Most importantly, metabolic signals such as ghrelin, insulin and leptin have been shown to modulate the activity of NPY-containing neurons (Sahu et al. 1995; Wang et al. 2002; Kamegai et al. 2001; Schwartz et al. 1996; Lee and Morris 1998). For example ghrelin, a newly discovered peptide hormone released from the empty stomach stimulates the activity of NPY neurons (Wang et al. 2002; Kamegai et al. 2001). In contrast, both insulin and leptin produced in response to energy ingestion and energy excess inhibit NPY synthesis and release (Sahu et al. 1995; Schwartz et al. 1996; Lee and Morris 1998).

2.3
NPY–Proopiomelanocortin Connections in the ARC

Within the ARC, NPY-containing neurons also synthesize and secrete agouti gene related peptide (AgRP), (Fig. 2). AgRP, like NPY, stimulates food intake (Kim et al. 2002). The NPY/AgRP-containing neurons project to target sites in the PVN, DMN and VMN where they influence food intake and peripheral metabolic processes (target sites). In addition, these neurons innervate nerve cells that synthesize and secrete proopiomelanocortin (POMC), the precursor molecule of the potent anorexogenic peptide α-melomogcyte stimulating hormone (α-MSH) (Morton and Schwartz 2001; Cowley et al. 2001; Porte et al. 2002). Both the NPY/AgRP neurons and the POMC neurons express leptin receptors. In the model described in Fig. 2, leptin hyperpolarizes NPY/AgRP neurons inhibiting the release of these two peptides at their target sites. Hyperpolarization also inhibits GABA release, which disinhibits the activity of the POMC neurons

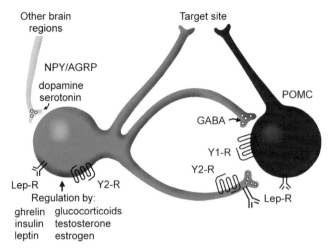

Fig. 2 Model representing the interaction between NPY/AgRP-containing neurons and POMC neurons in the arcuate nucleus of the rat. The NPY secreting neurons (*green*) are in synaptic contact with cells in the target site (i.e., the PVN) and POMC neurons that secrete α-MSH (*red*). The NPY Y_2 receptors have a presynaptic location and are autoreceptors inhibiting NPY release onto POMC neurons and at the target site. Both NPY and GABA inhibit the activity of the POMC neurons. For more details see text. (Modified from Cowley et al. 2001)

(Cowley et al. 2001). At the same time leptin acts directly on the POMC cells to stimulate α-MSH release.

This neural circuit has been postulated to play a central role in the control of food intake and peripheral metabolism. Clearly, the NPY neurons in this circuit can be activated and deactivated by the metabolic status of the animal via the actions of ghrelin, insulin and leptin. In turn, the activity of NPY-containing neurons can be modified by other hormones as well as by the activity of other brain regions that project to the ARC–PVN region. The role of the NPY receptors in this model will be discussed later.

The NPY-containing neurons arising in the brain stem appear to be activated by both glucose and peripheral hormones that are present in the blood perfusing this area. The control of the NPY-containing brain stem neurons will be also discussed, in context, later.

3
The Actions of Exogenous NPY on Food Intake

3.1
NPY Alters Food Intake and Peripheral Metabolism When Injected into Specific Regions of the Hypothalamus

Not surprisingly, the hypothalamic regions innervated by the ARC and the brain stem neural projections are influenced by the microinjection of NPY at these sites. The DMN, the VMN and particularly the PVN are considered to be the most important for mediating the effects of NPY on food intake (Kalra et al. 1999; Stanley et al. 1985; Stanley and Leibowitz 1985).

Injection of NPY into specific hypothalamic nuclei also alters peripheral metabolism. The neuroendocrine and metabolic effects of central NPY administration persist even when NPY-induced hyperphagia is matched by pair-feeding (Billington et al. 1994). The metabolic effects of NPY are most pronounced in the PVN. For example, after injection into this region, NPY activates the hypothalamic–pituitary–adrenal axis (Wahlestedt et al. 1987). NPY also modifies the activity of corticotrophin releasing factor containing cells in the PVN leading to hypothalamic adrenocorticotropic hormone release and ultimately to increased plasma corticosterone levels (Wahlestedt et al. 1987; Liposits et al. 1988). At the same time NPY suppresses the activity of the hypothalamic–pituitary–thyroid axis, firstly by altering the expression of pro-thyroid releasing hormone in the PVN and secondly through lowered thyroid stimulating hormone; this leads to decreased secretion of T3 and T4 into the plasma (Fekete et al. 2001). In addition, NPY increases central parasympathetic and decreases central sympathetic outflow from the brain (van Dijk et al. 1994). The fall in sympathetic activity is reflected in bradycardia, and a slight fall in blood pressure (van Dijk et 1994; Harland et al. 1988). UCP-1 expression in brown adipose tissue is also reduced as a consequence of the lowered sympathetic drive and contributes, at least in part, to the decreased body temperature characteristic of NPY administration into the PVN (Currie and Coscina 1995; Egawa et al. 1991; Kotz et al. 2000). Lipoprotein lipase activity in white adipose tissue is increased in response to NPY injection into the PVN (Billington et al. 1994). Finally, as a consequence of the decreased parasympathetic drive, both in-meal and post-meal insulin secretion are enhanced after injection of NPY into the PVN (van Dijk et al. 1994). These actions of NPY, although many and varied, act in concert to conserve and store energy in the form of fat.

3.2
NPY also Alters Food Intake and Metabolism when Injected into Extra-Hypothalamic Sites

Many brain regions outside of the hypothalamus are innervated with NPY-containing neurons (Everitt and Hökfelt 1989; Chronwall 1989). At several of these sites such as the cortex, the hippocampus and the hindbrain, local injection of NPY has been shown to stimulate food intake. The effects of NPY to influence food intake in these brain regions may be due to a local behavioral effect. Neural connections exist between the cortex, the hippocampus, the hindbrain and the hypothalamus. Therefore, local injection of NPY into the cortex or into the hippocampus could increase nerve traffic to the hypothalamus leading to altered food intake.

For example, injection of NPY into the cortex leads to changes in food intake as well as to increased carbohydrate utilization, fat synthesis and reduced energy expenditure (Cummings et al. 1998; McGregor et al. 1990). These effects are similar to those described above after injection of NPY into the PVN.

In the hippocampus peptide YY (PYY) (a close analog of NPY that also stimulates food intake) has been shown to increase the activity of acetylcholine-containing nerve terminals, which are believed to be important in modulating reward and arousal (Hagan et al. 1998; Hagan 2002). On the basis of this evidence, stimulation of NPY receptors in the hippocampus has been postulated to increase food intake both by modulating the activity of the hypothalamus and increasing the motivation to eat by increasing the perception of the rewarding properties of the food eaten (Hagan et al. 1998; Hagan 2002).

Finally, NPY has been shown to stimulate food intake when injected either into the 4th ventricle or directly into the hindbrain (Corpa et al. 2001; Steinman et al. 1994). The hindbrain receives both neural and chemical input from the periphery concerning the energy status of the animal. For example, after food consumption, chemical (i.e., plasma insulin, and glucose levels) and neural signals (i.e., vagal afferents) are received in the hindbrain from the periphery. This information is processed and passed to higher brain centers resulting in the cessation of eating.

Injection of NPY into the hindbrain has recently been shown to reduce the neuronal activity of this region in response to gastric distension (Schwartz and Moran 2002). Therefore, increased NPY levels in the hindbrain may contribute to enhanced food intake by suppressing the inhibitory signals from the periphery to the hypothalamus. Clearly there is much to be learnt about the factors controlling NPY release in different parts of the brain and how these changes are coordinated to ultimately change food intake behavior.

Over the past two decades a large number of studies have appeared in the literature demonstrating that injection of NPY into the brain leads to a robust and reliable increase in food intake. However, despite intensive study the underlying mechanisms through which NPY leads to increased appetite are not clearly understood. This is because food intake is a very complex behavior that is mediated by both physical and psychological factors. In consequence, NPY could induce hunger by activating the central neural pathways involved in the control of food intake. Alternatively, NPY could increase appetite by increasing food palatability, increasing the incentive/motivational properties of the food or producing behavioral activation. The following section will consider the mechanisms by which NPY may act to influence food intake.

3.3
NPY Increases the Motivation to Eat

Normal meal acquisition and consumption occurs in a tightly controlled series of behavioral events. At the beginning of each meal the animal must look for food. Once food is found it can be approached and placed into the mouth. This primary sequence of behaviors is known as the 'appetitive phase' of eating. After placement in the mouth the food is then chewed and swallowed. This secondary behavioral sequence is referred to as the 'consummatory phase' of eating.

It is clear that NPY could influence food intake either by increasing the appetitive and/or the consummatory phase of food intake. In fact, evidence has accumulated in recent years to show that injection of NPY into the ventricular system of the brain produces a motivation to eat which manifests, in part, as an enhancement of the appetitive phase of food intake (Flood and Morley 1991; Sederholm et al. 2002; Ammar et al. 2000; Woods et al. 1998). This behavioral activation is very strong, since after injection of NPY, animals shift their attention away from sexual activity to food seeking behavior (Ammar et al. 2000). In contrast to the appetitive phase, NPY has been shown to have no effect and even to inhibit the consummatory phase of eating (Sederholm et al. 2002; Ammar et al. 2000; Woods et al. 1998). Similar effects on the consummatory phase of food intake have been shown with cholecystokinin—a known satiety peptide (Mamoun et al. 1997). Thus it appears that the main actions of NPY are:

1. To direct the attention of the animal to food
2. To motivate the animals to find food and then to eat it

3.4
NPY Enhances the Rewarding Properties of Food

Injection of NPY into the brain also enhances reward. For example, injection of NPY into the hypothalamus forms a conditioned place preference and increases progressively the ratio responding for sucrose pellets—an assessment of motivational responding to rewarding stimuli (Brown et al. 1998, 2000). Furthermore, when presented with a choice between palatable sweet food and regular chow, animals centrally injected with NPY choose the palatable food and will even tolerate an intense foot shock to get it (Heilig et al. 1992).

3.5
Does NPY Induced Food Intake Resemble a True Hunger State?

The results of many of the studies above lead to the question of whether NPY-induced eating represents a true hunger state. For example, rats can readily discriminate between food deprivation and NPY or PYY injection into the brain (Jewett et al. 1991; Seeley et al. 1995). Furthermore, injection of NPY into the brain induces a conditioned taste aversion and produces motivation to eat that enhances the appetitive phase but does not change or may even decrease the consummatory phase of food intake (Flood and Morley 1991; Ammar et al. 2000; Woods et al. 1998; Levine et al. 1991). These are aspects of food intake that are different from deprivation-induced feeding and give us reason to believe that exogenous NPY acts primarily to increase the motivation of animals to eat by mechanisms that are not the same as either food deprivation or other types of metabolic challenge. Indeed it has been suggested that the actions of exogenously administered NPY are more in keeping with pathophysiological states such as binge eating than normal food intake (Hagan et al. 1998; Hagan 2002).

3.6
Can Injection of Exogenous NPY into the Brain Mimic the Actions of the Endogenous Peptide on Food Intake?

The above findings lead to the suggestion that endogenous NPY may control food intake through a coordinated series of activity in many parts of the brain. If this hypothesis is correct then injection of NPY into the ventricular system of the brain or directly into the hypothalamus may not exactly mirror the sites of action of the endogenous peptide to control food intake. For example, direct injection of NPY into the hypothalamus may activate only this site, while injection of NPY into the ventricular system may activate brain areas different from or additional to those activated by the endogenous peptide. For example, general activation of NPY receptors in the brain may be an explanation of the satiety and aversive consequences of cerebroventricular NPY injection. Therefore, many studies have attempted to investigate the effect on food intake of blocking the action of endogenous NPY.

4
Relationship Between Brain NPY Levels and Energy Balance

4.1
Introduction

The first step on the road to considering NPY as a physiologically important regulator of food intake would be to observe a direct correlation between increased brain NPY levels and food intake. However, it is important to note that in the majority of studies present in the literature only changes in brain NPY mRNA expression in response to alterations in energy balance have been reported. In contrast, only a relatively small number of investigations have actually measured the content or secretion of endogenous NPY. However, the final extracellular concentration of NPY is a finely regulated balance between the synthesis and the transport, release and degradation rates of the endogenous peptide. Therefore, altered NPY mRNA expression may not necessarily produce parallel changes in NPY peptide levels. Therefore, in the following section only those studies showing increased NPY concentrations in response to changes in energy balance will be considered. Furthermore, readers are strongly advised to determine whether increased mRNA levels are really representative of final peptide concentrations. In our view this is not always the case. For example, within 6 h of fasting a sustained elevation in NPY gene expression can be observed within the ARC (Dallman et al. 1999). However, in general at least 2–3 days are needed to see a significant increase in NPY protein levels in the PVN (Sahu et al. 1988). Clearly, correct model selection and timing is crucial to interpreting the outcome of interventions designed to observe changes in food intake after inhibition of the actions of NPY in the brain.

4.2
Changes in Hypothalamic NPY Levels Are Temporally Correlated with Food Intake

The release of NPY in the brain follows a daily rhythm. For example, NPY peptide levels are increased in the dorsomedial hypothalamus (contains the PVN) at the beginning of the dark phase when rats begin to eat the majority of their food (Jhanwar-Uniyal et al. 1990; Akabayashi et al. 1994). Additionally, in rats maintained on a scheduled feeding regimen, NPY release in the PVN is enhanced and temporally correlated with food intake (Kalra et al. 1991). The rate of NPY release from the PVN in these rats increases immediately before the onset of feeding and decreases progressively as the animals consume food. However, if food is withheld the rats continue to secrete high levels of NPY. Finally, NPY levels are increased in the PVH before the onset of hyperphagia in the diabetic rat (Sahu et al. 1997).

4.3
Changes in Energy Balance Alter Brain NPY Levels

Energy use by exercise and energy deprivation produced by long-term fasting, food restriction, lactation and uncontrolled diabetes (streptozotocin-treated and diabetic BB rats) are all associated with increased NPY protein levels in the hypothalamus (Sahu et al. 1988; Malabu et al. 1994; Abe et al. 1991; Lewis et al. 1993). For example, 2–3 days of fasting leads to significantly increased NPY levels in the hypothalamic medial preoptic area, and in the ARC and PVN (Sahu et al. 1988). NPY levels are also increased in several hypothalamic nuclei of the Zucker *fa/fa* rat (suprachiasmatic, PVN, ARC, VMN, DMV and anterior hypothalamus) (McKibbin et al. 1991). Figure 3 demonstrates that NPY levels are also elevated in the hypothalamic ARC and PVN of the adult Zucker diabetic fatty (*ZDF*) rat. In contrast to the effects of energy depletion and leptin deficiency, increased energy deposition induced by feeding a high fat diet is associated with decreased NPY levels in the hypothalamic ARC and PVN (Stricker-Krongrad et al. 1998).

Tantalizingly, long-term fasting has also been shown to increase NPY levels in the hippocampus and cortex and to decrease NPY levels in the striatum (Pages et al. 1993). These changes are selective, since NPY levels are not altered in either the caudate nucleus or the nucleus accumbens (Beck et al. 1990). It is unclear at the present time whether in the other models described above changes in NPY levels also occur in extrahypothalamic nuclei. However, it is clearly import to pursue this line of research, which will help to establish the brain areas in which NPY acts to produce the coordinated behavioral response that leads to a change in food intake.

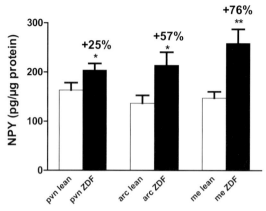

Fig. 3 Comparison of NPY levels in three hypothalamic nuclei of Zucker diabetic (ZDF/Gmi-*fa/fa*) and lean rats. In these experiments all rats (19 weeks old) were fed ad libitum and maintained on a 12-h light/dark schedule (6:00 am–6:00 pm light). The rats were sacrificed by decapitation between 9:00 am and 11:00 am and the brains rapidly removed and frozen. The hypothalamic area was cut into 300 μm slices. Sectioned hypothalamic areas were micro dissected into three regions (*PVN*, paraventricular nucleus; *ARC*, arcuate nucleus; *ME*, median eminence). The individual tissues were boiled for 10 min in 400 μl of 0.5 M acetic acid and sonicated to extract the NPY. Peptide concentrations were measured by enzyme immunoassay (Bachem) and the protein concentration measured using the BioRad protein assay. Results are expressed as mean±SE. Statistical analysis was performed by using Student's *t* test. *$P<0.05$, **$P<0.01$ (lean group $n=14$, ZDF group $n=15$)

4.4
Changes in Energy Balance Alter NPY Receptor Number in the Brain

The expression of G-protein coupled receptors on the cell surface is often diminished in response to an increase in the extracellular concentration of the interacting ligand. This is also the case for NPY where various models of energy depletion characterized by high levels of the endogenous peptide such as fasting, food restriction and uncontrolled diabetes are all associated with decreased NPY receptor density in various brain regions (Widdowson et al. 1997; Frankish et al. 1993). For example, after food restriction to 60% of normal over 10 days the concentration of NPY Y_2 and/or NPY Y_5 receptors in the hypothalamic lateral (perifornical) and dorsal areas, hypothalamic (VMN, ARC and DMV) nuclei, hippocampal CA3 region, centromedial amygdaloid nucleus and thalamic (paraventricular and reunions) nuclei are significantly diminished. In contrast, in these studies, food restriction did not lead to changes in NPY Y_1 receptor number (Widdowson et al. 1997). The downregulation of NPY receptors may explain the reduced ability of NPY to stimulate food intake in models of energy depletion (Parikh and Marks 1997).

In Zucker *fa/fa* rats, high brain NPY levels are associated with decreased NPY Y_2 and/or NPY Y_5 receptor number in the DMV and ARC nuclei as well as in the dorsal and lateral (perifornical) hypothalamic areas of obese rats as compared to lean rats (Widdowson 1997). In addition, there are significant reductions in binding to thalamic (reunions and centromedial nuclei) and in the hippocampal dentate gyrus of obese rats compared to lean rats. Again, as with food-restricted rats, no changes in NPY Y_1 receptor density were observed between lean and obese animals in these studies. In *ob/ob* mice that have high brain NPY levels similarly decreased NPY Y_5 receptor density has been shown in the hypothalamus, thalamus (midline group), cortex (cingulate, retrosplenial and granular) and hippocampus (Xin and Huang 1998).

In contrast, models of energy excess such as high fat/high sucrose feeding lead to decreased brain NPY levels and to increased NPY Y_5 and/or NPY Y_2 receptor density in the hypothalamus (lateral and dorsal areas), ARC and DMV nuclei, amygdaloid (medial and centromedial) nuclei and thalamic (centromedial and paraventricular) nuclei. Again, in these studies, no changes in NPY Y_1 receptor density were observed between control and dietary obese animals (Widdowson et al. 1997, 1999). High-fat-fed animals are hypersensitive to NPY-induced food intake in contrast to energy-depleted animals .

These studies add weight to the contention that energy depletion and decreased or absent leptin signaling lead to increased NPY release in different parts of the brain. The opposite conclusion can be drawn from the studies conducted in dietary obese animals.

4.5
Chronic NPY Infusion into the Brain Increases Food Intake and Leads to Obesity

The effects of NPY on food intake and peripheral metabolism are not just acute but are also observed with chronic administration. Thus, infusion of NPY into the brain ventricles of normal rats or mice leads to marked and sustained hyperphagia (Vettor et al. 1994; Beck et al. 1992; Raposinho et al. 2001). Body weight is also increased secondary to expansion of the adipose tissue mass (Fig. 4). Basal triglyceridemia, insulinemia, hypercorticoidemia and importantly a much greater insulin response to meal feeding are also observed after NPY infusion. Interestingly, NPY infusion results in a pronounced increase in insulin-stimulated glucose uptake by adipose tissue (increased fat mass) but also in a decrease in insulin-stimulated glucose uptake into skeletal muscle (Zarjevski et al. 1994). That is a state of insulin resistance is produced in skeletal muscle by central infusion of NPY. The increased skeletal muscle insulin resistance is most likely due to the concurrently high plasma triglyceride (Randle cycle) and insulin levels present in these animals. Thus, NPY-infused animals adopt a similar phenotype to the Zucker *fa/fa* rat and the *ob/ob* mouse both of which have high endogenous levels of the peptide within the central nervous system.

Removal of the adrenals has been shown to prevent and reverse all types of obesity. This action of adrenalectomy is thought to be due to the removal of glucocorticoids, which promote body weight gain and hyperinsulinemia. The obesity syndrome produced by chronic infusion of NPY into the brain can also be prevented by prior removal of the adrenals (Sainsbury et al. 1997). Conversely,

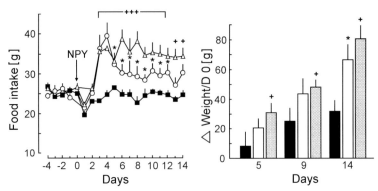

Fig. 4 Effect of chronic infusion of NPY into the brain on food intake and body weight. In these studies, male Sprague-Dawley rats were implanted with an indwelling catheter in the right lateral cerebroventricle. Each cannula was connected to an osmotic mini-pump (Alzet 2002) located beneath the skin of the back of the neck and which dispensed the artificial cerebrospinal fluid vehicle (■, 0.5 µl/h), NPY (○, 11 µg/24 h or ▲,△ 22 µg/24 h) for 14 days starting on day 0, the day on which the pumps were implanted. The results are expressed as mean ± SE from six rats in each group. The food intake and body weight results were analyzed by two-way ANOVA (treatment×time) with repeated measures on factor time; + and * significantly difference from the corresponding time period in the control group at $P<0.05$

following adrenalectomy, the NPY-induced obesity syndrome can be restored by the concurrent infusion of glucocorticoids into the brain (Zakrzewska et al. 1999).

The increase in plasma insulin produced by acute intracerebroventricular injection of NPY can also be prevented by adrenalectomy (Wisialowski et al. 2000). The ventromedial hypothalamus is believed to be involved in mediating NPY-induced hyperinsulinemia, insulin resistance and obesity. Under normal conditions glucocorticoids maintain NPY receptor number in this hypothalamic region. However, in the absence of glucocorticoids NPY Y_1 and NPY Y_5 receptor numbers fall. Decreased receptor number is believed to be the underlying mechanism through which NPY-induced insulin release is reduced following adrenalectomy.

4.6
NPY Overexpression in the Brain Leads to Overconsumption of a High-Calorie Diet and to Obesity

Transgenic mice have been produced that modestly overexpress NPY selectively within the central nervous system. In comparison to wild-type controls, adult transgenic mice maintained on a normal diet show reduced seizure susceptibility, behavioral signs of anxiety together with hypertrophy of the adrenal zona fasiculata, but have similar food intake and body weights (Vezzani et al. 2002; Inui et al. 1998). However, when fed a palatable high-calorie diet these mice display markedly increased food intake and body weight. Food intake returns to normal after 3 weeks but body weight remains elevated for as long as the mice are exposed to the high-calorie diet. After 13 weeks of exposure to the high-calorie diet, NPY overexpressing mice develop hyperinsulinemia and hyperglycemia along with increased body weight (Kaga et al. 2001).

The increase in food intake observed in NPY overexpressing mice consuming the high-calorie diet can be attenuated by intracerebroventricular administration of the NPY Y_1 antagonist BIBO 3304 but not by the NPY Y_5 antagonist L-152,804 (Kaga et al. 2001) (see Fig. 8). These results seem to imply that the hyperphagia induced by NPY over expression is mediated by NPY Y_1 receptors. However, this conclusion should be approached with caution since the selectivity of BIBO 3304 for the NPY Y_1 receptor has recently been questioned.

5
The Actions of Endogenous NPY on Food Intake

5.1
Introduction

In the brain, NPY levels and NPY receptor density are changed in response to alterations in energy balance, while overexpression of NPY is associated with an increase in the intake of high-calorie food. However, these are correlative results and a more direct means of linking altered levels of NPY to changes in food intake is needed.

Several approaches have been used to investigate the role of endogenously produced NPY in the control of food intake. These include germline knockout of NPY production, NPY antisense oligodeoxynucleotides (ODNs) and anti-NPY antibodies. However, there are limitations to the use of each of these approaches. For example, germline knockout removes NPY from all regions of the body while injection of NPY antibodies or NPY antisense ODNs into the ventricular system of the brain (the usual route of administration) will remove or interfere with the action of the peptide in all parts of the central nervous system. In the following section the approaches used to inhibit endogenous NPY in the brain together with a survey of the problems of interpretation inherent to each will be discussed.

5.2
Food Intake in the NPY Germline Receptor Knockout Mouse

Although NPY knockout animals express an anxiogenic-like phenotype and are hypoanalgesic they appear to be metabolically normal in every respect. For example, NPY-deficient mice have a normal feeding response to NPY and the melaninocortin-4 receptor agonist MTII (Marsh et al. 1999; Hollopeter et al. 1998). In addition, NPY-deficient mice have a normal hormonal profile, grow and eat normally and have a normal response to both diet and chemically induced obesity (high-fat diet, monosodium glutamate, gold thioglucose) (Palmiter et al.

1998; Erickson et al. 1997). Furthermore, in the majority of studies conducted with these animals they have also been shown to have a normal refeeding response to short-term fasting (Palmiter et al. 1998). However, one recent study has demonstrated a decrease in fasting-induced food intake in NPY knockout animals (Bannon et al. 2000). The difference between these studies appears to be due to whether wild-type or heterozygous animals were used as controls. Since these findings are not in agreement with an important physiological role for NPY in the control of food intake, how can they be interpreted? Two main explanations have been suggested. First, the clear lack of phenotype may be because NPY is not a critical feeding regulator. Alternatively, in the absence of NPY food intake may readily be maintained by the action of other neuropeptides. This is a distinct possibility, since neural pathways in the brain that control food intake are highly redundant and inhibition of one invariably leads over time to the return of normal feeding.

Although knockout of NPY production does not affect day-to-day food intake in normal animals this is not the case in either diabetic or *ob/ob* mice (Palmiter et al. 1998; Sindelar et al. 2002). For example, in both wild-type and NPY knockout mice streptozotocin administration results in pancreatic β-cell destruction and leads to uncontrolled diabetes. The diabetic state in wild-type animals is characterized by marked negative energy balance leading to compensatory hyperphagia. In contrast, food intake is not increased in NPY knockout mice, despite comparably decreased levels of body weight, fat content and plasma leptin levels between the two models (Sindelar et al. 2002). Within the ARC, NPY/AgRP-containing neurons innervate the PVN but also innervate and tonically inhibit the activity of POMC neurons (Fig. 2). Increased NPY/AgRP expression and decreased POMC expression has been suggested as the mechanism that underlies the hyperphagia of streptozotocin diabetic mice. In contrast, in these diabetic mice lacking NPY, AgRP levels remained elevated but POMC levels do not fall as they do in wild-type mice. Thus, the loss of NPY signaling and the failure to suppress POMC levels appear to be the reason for the absence of the hyperphagic response to uncontrolled diabetes in NPY knockout mice. In contrast, NPY-deficient mice respond to fasting with a much greater increase in hypothalamic AgRP and a greater decrease in POMC expression than diabetic mice. These changes in hypothalamic neuropeptides may explain why food intake is, in general, unaffected by fasting in the NPY knockout mouse.

Leptin signaling is absent in the *ob/ob* mouse and, as a consequence, hypothalamic NPY levels are high. Crossing the *ob/ob* mouse with the NPY knockout mouse partially reverses the obese phenotype (Palmiter et al. 1998). Together, these observations support the contention that NPY is also important for helping to maintain the obesity state of the *ob/ob* mouse.

Experiments performed in the knockout mouse suggest that there are situations (i.e., uncontrolled diabetes) in which NPY plays an important role in the control of food intake. However, the possibility that the true role of NPY in the control of food intake may be, at least in part, masked by changes occurring over the development period suggests that other methods are necessary to study the role of endogenous NPY in the control of food intake. In theory, a conceptually simple solution to the problem of food intake compensation in knockout animals has been found by inhibiting both the production and action of NPY. These approaches offer definite theoretical advantages over studies in knockout animals because formation of NPY is prevented only for a relatively brief period of time before study.

5.3
Food Intake in Response to NPY Antisense ODNs

Injection of oligodeoxynucleotides (ODNs) against NPY into the cerebroventricles or directly into the ARC has been shown to reduce food intake in several animal models (Akabayashi et al. 1994; Hulsey et al. 1995; Kalra and Kalra 2000). However, serious concerns have been raised about the use of ODNs to inhibit the endogenous production of NPY. These concerns stem partly from the lack of specificity and the toxicity of antisense molecules within the central nervous system (Dryden et al. 1998). This remains a particular problem when studying behaviors such as food intake, which can be strongly influenced by both toxic and nonselective interactions.

5.4
Food Intake in Response to Anti-NPY Antibodies

Injection of either monoclonal or polyclonal antibodies against NPY into the ventricular system of the brain has been shown to attenuate the hyperphagia observed following ventromedial hypothalamic (VMH)-lesions and after fasting-induced food intake in rats (Lambert et al. 1993; Dube et al. 1995). Interestingly, night phase food intake in lean rats and both day and night phase food intake in Zucker *fa/fa* rats are also reduced following administration of NPY antibodies into the brain (Dube et al. 1994; Marin-Bivens et al. 1998). Additional studies have shown that injection of NPY antibodies directly into the PVN can attenuate the hyperphagia following both 24 h of fasting and after 2-deoxy-D-glucose injections (He et al. 1998; Shibasaki et al. 1993).

Since hypothalamic NPY expression is increased in all of these models, the results suggest that NPY plays an important role in the control of food intake under a variety of conditions. However, closer examination of these studies reveals that in the majority only total food intake was measured, while in two cases only visual observation of the animals was performed. Therefore, it cannot be entirely certain that the reductions in food intake were behaviorally specific nor due to an aversive effect of the antibodies. In this regard injection of anti-NPY antibodies directly into the VMH of fasted mice has been shown to produce a large increase in amphetamine-like motor activity and decreased resting (Walter et al. 1994). These behavioral alterations are thought to explain the concurrent decrease in food intake observed in this study. Although injection of NPY antibodies is not always associated with increased activity, the results do suggest that care should be taken in the interpretation of studies where the possibly confounding influence of behavioral alterations, potential toxicity and antibody cross reactivity have not been rigorously eliminated.

The antibody approach has also been used to destroy NPY secreting neurons present in the ARC (Burlet et al. 1995). In these studies, monoclonal antibodies directed against an epitope of the NPY precursor molecule, the C flanking peptide, have been mixed with the cytotoxins ricin A and monensin. Injection of this complex into the ARC delivers the toxins to the NPY-containing neurons and leads to their destruction. Destruction of NPY-containing neurons in the ARC using this method reduces food intake after food deprivation and both food intake and body weight for 10 days after a single injection in free feeding rats.

6
NPY Receptors

6.1
Subtypes of NPY Receptors

Genes encoding five NPY receptor subtypes have been identified in mammals—NPY Y_1, Y_2, Y_4, Y_5 and y6 (Gehlert 1998). The pharmacology and signaling mechanisms of the NPY Y_1–Y_6 receptor subtypes are described in detail in chapters by Hyland and Cox, Michael, and Larhammar et al. in this volume.

Recently, an orphan receptor named GPR74 was identified that binds PYY with high affinity and has 25% identity with the NPY Y_1 receptor (Parker et al. 2000; Herzog, 2000). This receptor was originally designated as the NPY Y_7 receptor subtype but is now characterized as the NPFF2 receptor (Bonini et al. 2000; Elshourbagy et al. 2000; Liu et al. 2001). Nevertheless, the NPFF2 receptor is able to recognize specific NPY Y_1 receptor antagonists and NPY-related peptides such as frog PP (Mollereau et al. 2002). Whether association with accessory proteins, post-translational modifications or homo/heterodimerization events would allow NPFF2 receptor to recognize additional NPY peptides remains to be investigated (George 2002). The existence of additional NPY receptors awaits the achievement of the human genome sequencing and the characterization of the available orphan G-protein coupled receptors (Howard et al. 2001; Wise et al. 2002).

6.2
Evidence for Obesity Linkage to NPY Receptor Subtypes

Polymorphisms of NPY receptor subtypes have been evaluated for linkage with the obese state. In the majority of studies no obesity linkage between receptor mutations and obesity have been found (Herzog et al. 1993; Roche et al. 1997; Rosenkranz et al. 1998; Jenkinson et al. 2000). For example, neither the polymorphisms PstI (NPY Y_1/Y_5receptor), Glu-4-Ala (NPY Y_5 receptor) nor the silent polymorphism Gly-426-Gly (NPY Y_5 receptor) have associations with extremes of body weight. Conversely, three novel single nucleotide polymorphisms (P1, P2 and P3) located in the NPY Y_5 receptor gene noncoding regions had an association with extreme obesity in Pima Indians. Although these studies suggest that the NPY Y_5 receptor may be linked to some forms of obesity, it remains to be determined whether these mutations have functional consequences, are markers for another obesity locus or have relevance to human obesity in other populations.

Finally, a recent study identified four polymorphic variants within the common promotor and coding region of the NPY Y_1 and Y_5 receptors (Blumenthal et al. 2002). Among them, a cytosine to thymidine substitution in the untranslated region between the NPY Y_1 and Y_5 receptor genes was significantly associated with lower fasting triglyceride levels and higher high-density lipoprotein (HDL) concentrations in 306 obese subjects. Since NPY is known to increase both triglyceride and HDL levels by stimulating lipoprotein lipase activity in adipocytes, the authors suggest that this is a gain of function polymorphism and that variation in the NPY Y_1 and Y_5 cluster genes may contribute to the development of obesity and dyslipidemia.

7
Which NPY Receptor Subtype Mediates Food Intake in Response to Endogenous NPY?

Up to this point several lines of evidence have been presented that suggest that NPY Y_1, NPY Y_5 or a combination of these two receptors could mediate the actions of NPY on food intake and metabolism. In the final part of this chapter we will assess studies that have investigated the action of selective NPY receptor antagonists on food intake and peripheral metabolism in a variety of animal models. Considerable work has been performed in this area as it is clear that an NPY antagonist that is able to reduce food intake and ameliorate obesity would be a potential blockbuster drug for the pharmaceutical industry.

7.1
The NPY Y_1 Receptor

7.1.1
NPY Y_1 Subtype Selective Agonists

Pharmacologically, the NPY Y_1 receptor is characterized by a high affinity for NPY and PYY and a low affinity for pancreatic polypeptide (PP) and for N-terminally truncated analogs such as NPY(3–36) and NPY(13–36) (Wyss et al. 1998; Haynes et al. 1998). [Leu^{31},Pro^{34}]-NPY was the first modified NPY analog that showed apparent selectivity for the NPY Y_1 receptor (Table 1) and injection of this compound into the brain stimulated food intake in rodents (Wyss et al. 1998). However, [Leu^{31},Pro^{34}]-NPY was recently shown to also have significant affinity for the NPY Y_4 and NPY Y_5 receptor subtypes which clouds interpreta-

Table 1 Affinity of NPY peptide analogs for NPY receptor subtypes (IC$_{50}$ or K$_i$: nM)

	hNPY Y$_1$	hNPY Y$_2$	hNPY Y$_4$	hNPY Y$_5$
Non-selective				
HNPY	0.59±0.17	0.49±0.07	54.23±12.96	0.48±0.15
HPYY	2.36±0.50	0.22±0.05	47.58±11.62	0.45±0.11
NPY Y$_1$ receptor				
pLeu^{31}, Pro^{34} NPY	0.56±0.18	375.8±89.5	0.78±0.28	0.43±0.10
pPhe^7, Pro^{34} NPY	0.009	32.1	ND	34
[d-Arg^{25}]-NPY	0.9±0.2	11.6±2.9	74.6±21.3	43.4±4.2
[d-Hist^{26}]-NPY	2.0±0.3	29.0±4.7	20.1±1.9	34.6±3.8
Des-AA$^{11-18}$[Cys^{7,21},d-Lys^9 (Ac), d-His^{26},Pro^{34}]-NPY	1.2±0.1	801±41.4	31.4±1.3	2363±142
NPY Y$_2$ receptor				
C2-pNPY	1525	0.38	ND	6296
pNPY$_{13-36}$	76.35±14.18	2.0±6.51	192.28±27.89	42.08±7.89
hPYY$_{3-36}$	45.24±8.35	0.25±0.03	277.83±68.61	1.17±0.24
pNPY$_{18-36}$	214±3.9	9.3±0.09	ND	ND
NPY Y$_4$ receptor				
rPP	57.36±15.77	1247.2+345.3	0.06±0.01	24.93±3.64
hPP	29.83±8.72	1499.5±216.0	0.08±0.02	1.73±0.16
NPY Y$_5$ receptor				
[ala^{30}]NPY$_{3-36}$	>1,000	>1,000	314.8±100.9	14.5±6.10
[Ala^{31},Aib^{32}]pNPY	>700	>500	>1,000	6.0±2.4
[DTrp^{32}]hNPY*	>1,000	>1,000	>1,000	25.3±1.9
[DTrp^{34}]hNPY	6.49±0.044	5.43±0.041	7.08±0.059	7.53±0.068
CPP$^{1-7}$, NPY$^{19-23}$, Ala^{31}, Aib^{32}, Gen^{34} hPP	530±140	>500	51±5	0.24±0.096
[K^4,RYYSA$^{19-23}$]PP$_{2-36}$	0.87±0.26	1.95±0.32	0.0040±0.0029 (*25%)	0.029±0.008

tion of its mechanism of action (Wyss et al. 1998). More recently, highly selective NPY Y$_1$ receptor agonists have been reported such as [Phe^7,Pro^{34}]-pNPY, [D-Arg^{25}]-NPY, [D-His^{26}]-NPY, Des-AA$^{11-18}$[Cys^{7,21},D-Lys^9(Ac),D-His^{26},Pro^{34}]-NPY (Table 1) (Soll et al. 2001; Mullins et al. 2001). All three analogs markedly stimulate food intake when injected into the brain. Finally, when the affinity of modified NPY peptides for the NPY Y$_1$ receptor is compared with their ability to stimulate food intake after injection into the brain a significant positive correlation emerges (Fig. 5) (Wyss et al. 1998).

The available evidence strongly suggests that stimulation of the NPY Y$_1$ receptor with exogenous peptides increases food intake. This information leads immediately to the question of whether food intake changes when expression of the action of endogenous NPY acting through the NPY Y$_1$ receptor is prevented. The answer to this question is crucial since selective inactivation of the NPY Y$_1$ receptor still leaves all other NPY receptor subtypes open to the peptide (Fig. 6). The following section considers the evidence both for and against a physiologi-

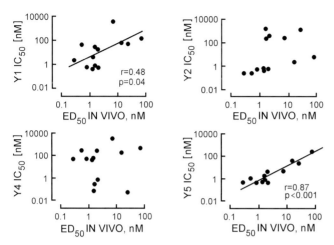

Fig. 5 Correlation between in vivo potency (ED$_{50}$ in nmol) and in vitro affinity (IC$_{50}$ in nmol) for the NPY Y$_1$, NPY Y$_2$, NPY Y$_4$ and NPY Y$_5$ receptor subtypes. (From Wyss et al. 1998, , with permission from Elsevier Science)

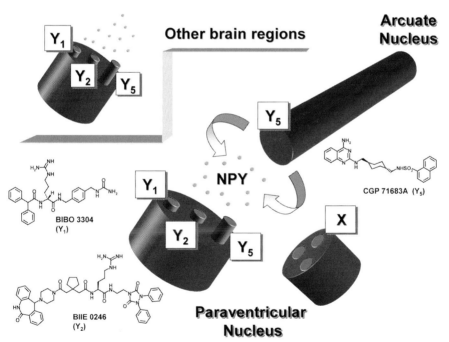

Fig. 6 Possible effects of NPY receptor stimulation on food intake. In the hypothalamus, release of NPY from the ARC leads to stimulation of NPY receptors in the PVN. The NPY Y$_1$, NPY Y$_2$ and NPY Y$_5$ receptors have both pre- and postsynaptic locations. Blockade of the NPY Y$_5$ receptor, for example, leaves the NPY Y$_1$ and the NPY Y$_2$ receptors open to stimulation. NPY may also activate NPY receptors in other areas of the brain involved in the control of food intake, for example in the cortex, hippocampus and hindbrain. NPY receptor antagonists may (but should not) have significant interaction with non-NPY receptors (x). (From Chamorro et al. 2002, Int J Obesity 26:281–298, with permission of the authors and the Nature Publishing Group)

cally important role for NPY acting through the NPY Y_1 receptor in the control of food intake.

7.1.2
NPY Y_1 Receptor Knockout

NPY Y_1 knockout mice present after 9 weeks of age the phenotype of increased body weight due to increased fat mass but with normal or slightly decreased food intake (Pedrazzini et al. 1998; Kushi et al. 1998; Burcelin et al. 2001). This seemingly paradoxical situation maybe explained by the high fasting plasma insulin levels present in these animals which would promote fatty acid uptake into white adipose tissue and lead to an increase in fat mass size. Fat deposition is also favored by the concurrent reduction in metabolic rate which is most likely due to the more frequent and longer inactivity periods characteristic of the NPY Y_1 knockout phenotype (Pedrazzini et al. 1998). The combined hyperinsulinemia and decreased metabolic rate would also explain why these animals loose weight less readily than wild-type animals when subjected to a period of calorie restriction (Kushi et al. 1998). The mechanistic basis of the decreased locomotion observed in the knockout mice has not been defined although increased anxiety levels in these animals maybe a plausible explanation (Kask et al. 2002).

Previous studies have shown that NPY released from pancreatic β-cells may act as a potent autocrine inhibitor of insulin secretion—perhaps acting through the NPY Y_1 receptor (Waeber et al. 1993). Thus, the fasting hyperinsulinemia observed in NPY Y_1 knockout animals may be due to removal of the inhibitory action of locally released NPY on insulin secretion (Kushi et al. 1998).

Although spontaneous food intake is unaltered or only slightly decreased in NPY Y_1 deficient mice, three lines of evidence suggest that this receptor subtype does in fact play an important role in mediating the effects of NPY on food intake. Thus, when compared to wild-type mice, the ability of NPY to stimulate food intake is reduced by up to 50% after knockout of the NPY Y_1 receptor (Kanatani et al. 2000). Furthermore, after fasting, when brain NPY expression is high, total food intake upon refeeding is markedly reduced in NPY Y_1 knockout animals (Pedrazzini et al. 1998). Finally, the hyperphagia observed in leptin deficient *ob/ob* mice is significantly diminished when these animals are crossed with NPY Y_1 knockout mice (Pralong et al. 2002). Thus, the unchanged spontaneous food intake observed in NPY Y_1 knockout animals may be due to the offsetting effect of the increased leptin and insulin levels present in these animals—both of which are potent appetite suppressing agents.

7.1.3
NPY Y_1 Antisense ODNs

The use of antisense ODNs directed against the NPY Y_1 receptor has not yielded consistent effects on food intake. For example, a treatment schedule with antisense ODNs that downregulated NPY Y_1 but not NPY Y_2 receptors in the brain induced anxiety but had no effect on spontaneous food intake (Wahlestedt et al. 1993). In contrast, all other studies using the NPY Y_1 antisense approach have produced a paradoxical increase in food intake, which appears to be model dependent. For example, in the majority of experiments NPY Y_1 antisense ODN increased spontaneous food intake (Lopez-Valpuesta et al. 1996; Lopez-Valpuesta et al. 1996a; Heilig 1995). However, in another study, antisense ODN treatment caused an increase in food intake in fasted but not in spontaneously feeding rats (Schaffhauser et al. 1998). The reason for the paradoxical increase in food intake after administration of NPY Y_1 antisense ODN is unclear although the nonselective nature of the antisense approach cannot be discounted. Perhaps, the observation in one study of higher body temperature coupled with diminished body weight suggests that increased metabolic rate is the cause of the hyperphagia (Lopez-Valpuesta et al. 1996).

7.1.4
NPY Y_1 Subtype Selective Antagonists

The often contradictory nature of the changes in food intake in the knockout and antisense treated animals does not allow a firm conclusion to be drawn regarding the role of the NPY Y_1 receptor in mediating the role of endogenous NPY in the control of food intake. Despite this, several pharmaceutical companies have attempted to develop NPY Y_1 receptor antagonists as appetite suppressing agents. The following section reviews key studies that have investigated the action of synthetic NPY Y_1 antagonists on food intake.

BIBP 3226 (Boehringer Ingelheim). BIBP 3226 was the first highly potent NPY Y_1 antagonist, which had low affinity for other NPY subtypes (Table 2, see also Fig. 8). When injected into the ventricles of the brain BIBP 3226 significantly inhibited the increase in food intake produced by NPY as well as the refeeding response to short-term fasting in normal animals (Kask et al. 1998). However, it was unclear in these experiments whether the changes in food intake observed in BIBP 3226 treated animals were due to direct effects on appetite or secondary to the anxiety, barrel rolling and catatonia observed with equivalent doses of the

Table 2 Specificity of selected NPY receptor antagonists for NPY receptor syubtypes (IC_{50}: nM)

	hNPY Y_1	hNPY Y_2	hNPY Y_4	hNPY Y_5
CGP 71683A	3,500	2,500	3,900	1.5
GW 438014A	>10,000	>10,000	>10,000	211
NPY 5RA-972	>10,000	>10,000	>10,000	3.1
L-152,804	>10,000	>10,000	>10,000	26
BIIE 0246	3,600	2	>10,000	300
GR 231118 (1229U91)	0.04	19	0.2	200
BIBP 3226	73	>10,000	5,000	NA
BIBO 3304	13	≥1,000	NA	≥1,000
J-115814	1.4	>10,000	620	6,000

compound. To complicate matters further it also appears that this molecule may also affect food intake by mechanisms that are unrelated to NPY Y_1 receptor inhibition (Morgan et al. 1998).

BIBO 3304 (Boehringer Ingelheim). In an attempt to overcome the problems encountered with BIBP 3226 a derivative compound BIBO 3304 was synthesized. BIBO 3304 has high affinity for the NPY Y_1 receptor but very low affinity for the NPY Y_2, NPY Y_4 and NPY Y_5 receptors (Table 2, see also Fig. 8). To avoid the production of secondary behaviors that could be produced by global exposure of the brain to BIBO 3304 the antagonist was injected directly into the PVN of the hypothalamus. Injection at this site inhibited both the refeeding response to fasting and the increase in food intake produced by NPY (Wieland et al. 1998). A possible indication of the improved specificity of BIBO 3304 over BIBP 3226 is the observation that the increase in food intake produced by either galanin or by norepinephrine is unaffected by BIBO 3304.

The above information obtained with NPY Y_1 antagonists has been taken as evidence for a role of this receptor subtype in the control of food intake. However, an exhaustive evaluation of the cross reactivity of these compounds with non-NPY receptors involved in the control of food intake has not been performed. In fact this is impossible because at this time not all of the receptors involved in the control of food intake have been identified. This problem is well illustrated by the recent finding that both BIBP 3226 and BIBO 3304, in addition to being NPY Y_1 antagonists, also have relatively high affinity for neuropeptide FF (NFF) receptors (Mollereau et al. 2001, 2002). Since NFF has been shown to influence food and water intake, the effects of BIBP 3226 and BIBO 3304 described above may be due, at least in part, to an effect on the NFF or on a yet undefined receptors (Sunter et al. 2001). The potential of individual compounds to produce changes in food intake by nonspecific mechanisms is clearly illustrated by the example of AMG 68, an NPY Y_1 receptor antagonist, that paradoxically lowers food intake in the NPY knockout mouse (Bannon et al. 2000).

J-115814 (Banyu). The most convincing evidence for the involvement of endogenous NPY acting through the NPY Y_1 receptor in the control of food intake has come from recently published experiments performed in NPY knockout mice using the low molecular weight antagonist J-115814. J-115814 has very high affinity for mouse NPY Y_1 receptors but very low affinity for the NPY Y_2, NPYY$_4$, NPY Y_5 receptors as well as for the mouse NPY Y_6 subtype (Table 2, see also Fig. 8) (Kanatani et al. 2001).

The most important experiments described in this series of studies are illustrated in Fig. 7. Injection of NPY (5 µg) into the brain produced a greater increase in food intake in wild-type mice than in NPY Y_1 knockout mice. This observation confirms the importance of the NPY Y_1 receptor subtype in mediating not all, but certainly the majority, of the orexigenic effects of exogenous NPY. Intraperitoneal injection of J-115814 at a dose which penetrated the brain in quantities sufficient to block the NPY Y_1 receptor significantly inhibited the in-

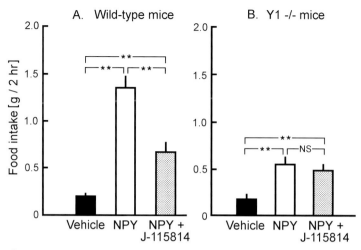

Fig. 7 Anorexogenic effect of J-115814 on NPY-induced feeding in wild-type and NPY Y_1 receptor knockout mice. The NPY Y_1 receptor antagonist J-115814 (30 mg/kg) was injected intraperitoneally 1 h before the intracerebroventricular injection of 5 μg NPY. (Modified from Kanatani et al. 2001, Mol Pharmacol 59:501–505)

crease in food intake produced by NPY in wild-type mice. Thus, the NPY Y_1 receptor is also responsible for mediating more than 50% of the orexigenic activity of endogenous NPY. Finally in these studies, J-115814 significantly lowered food intake in both *db/db* and normal mice during the dark phase, when hypothalamic NPY levels are reported to be high. These observations provide further evidence that the NPY Y_1 receptor is involved in the control of food intake.

Critical to the interpretation of these studies is the fact that J-115814 did not significantly alter spontaneous food intake, nor did it modify the increase in food intake produced by NPY in the NPY Y_1 receptor knockout mice (Kanatani et al. 2001a, 2001b). These observations strongly suggest that the actions of J-115814 can only be explained by an action of the compound on the NPY Y_1 receptor and not by nonspecific actions of the compound.

7.1.5
Are the Actions of Other Peptides on Food Intake Mediated by the NPY Y_1 Receptor?

Reports have appeared in the recent literature suggesting that the orexigenic effect of several peptides is mediated through an interaction of NPY with the NPY Y_1 receptor. For example, both BIBO 3304 and GR 231118, a structurally unrelated peptide NPY Y_1 antagonist, have been shown to strongly inhibit MCH-induced food intake (Chaffer and Morris 2002). Similarly, the effects of the gastric peptide ghrelin can be completely inhibited by the NPY Y_1 antagonist J-115814 (Shintani et al. 2001). However, based on the discussion presented above it is clear that unless the mode of action of the NPY Y_1 antagonists used is clearly understood there is a great risk of obtaining false conclusions in these studies.

In conclusion, some evidence exists to suggest that NPY acting through the NPY Y_1 receptor is important for modulating short-term food intake—for example after fasting. Even if this is so, concerns certainly remain about the ability of NPY Y_1 receptor blockers to reduce food intake without eliciting side effects such as increased anxiety or changed blood pressure, to name the more obvious examples.

7.2
The NPY Y_2 Receptor Subtype

7.2.1
NPY Y_2 Subtype Selective Peptide Agonists and Food Intake

Studies examining the consequences to food intake of stimulating the NPY Y_2 receptor have until now been complicated by the lack of truly selective ligands. However, modified NPY peptides with reasonable selectivity for the NPY Y_2 receptor, such as NPY(13–36), NPY(18–36) and C2-$_P$NPY, have either a weak effect or no effect on food intake in satiated rats (Table 1) (Wyss et al. 1998; Haynes et al. 1998). Furthermore, unlike the NPY Y_1 receptor subtype there is no correlation between the affinity of modified NPY analogs for the NPY Y_2 receptor and their ability to stimulate food intake during a 2-h period in satiated animals (Fig. 5) (Wyss et al. 1998).

While stimulation of the NPY Y_2 receptor does not affect food intake in satiated animals it is a different case in fasted animals where the highly selective NPY Y_2 receptor agonist N-acetyl [Leu^{28},Leu^{31}] NPY(24–36) has been shown to potently inhibit food intake after injection into the ARC (Potter et al. 1994; Batterham et al. 2002). The same result is also obtained when the marginally selective NPY Y_2 receptor agonist pPYY(3–36) is injected into the same brain region in fasted animals (Batterham et al. 2002).

These results are not surprising when the hypothalamic interactions that control food intake are taken into consideration (Fig. 2). The NPY Y_2 receptor has a predominantly presynaptic location in the hypothalamus where it has been postulated to act as an autoreceptor to inhibit both NPY and γ-aminobutyric acid (GABA) release from NPY-containing nerve terminals (Broberger et al. 1997). In satiated animals, stimulation of the NPY Y_2 receptor would have little effect on food intake because both NPY and GABA release are at low levels and the prominent drive in this case would be α-MSH release from POMC neurons. In contrast, in fasted animals NPY Y_2 receptor activation would both directly inhibit NPY release in the PVN and remove the tonic inhibition of GABA on the POMC neurons which would result in the release of α-MSH and decreased food intake.

7.2.2
The Gut Hormone PYY$_{3-36}$: A Physiological Agonist of the NPY Y$_2$ Receptor

The NPY Y$_2$ receptor agonist PYY(3-36) is released from the intestine in proportion to the number of calories ingested (Pedersen-Bjergaard et al. 1996). In satiated rats, in which NPY release in the hypothalamus is low, PYY(3-36) stimulates feeding—an effect most likely due to an interaction of the peptide with both NPY Y$_1$ and NPY Y$_5$ receptors (Table 1) (Wyss et al. 1998). However, when NPY release is high, for example after fasting and during the dark phase, PYY(3-36) acts to inhibit food intake (Batterham et al. 2002). The actions of PYY(3-36) to inhibit fasting-induced food intake are abolished in NPY Y$_2$ knockout mice. This observation would suggest that PYY(3-36) acts by suppressing the release of endogenous NPY via an interaction with the NPY Y$_2$ receptor. In human studies, infusion of PYY(3-36) at concentrations mimicking postprandial levels decreased hunger scores and total calorie intake (Batterham et al. 2002). Taken together these results suggest that PYY(3-36) is a physiologically important peptide that signals the brain to reduce food intake after meal ingestion—an effect apparently mediated by the NPY Y$_2$ receptor.

7.2.3
NPY Y$_2$ Receptor Knockouts

Conditional Knockout. The hypothesis that NPY acts through the NPY Y$_2$ receptor to decrease food intake by inhibition of neuronal NPY release would predict increased food intake as the phenotype of knockout of this receptor subtype. Indeed, after NPY Y$_2$ receptor deletion in the hypothalamus of male mice an increase in NPY/AgRP levels is observed in the ARC. This observation that may underlie the increased food intake observed in these animals (Sainsbury et al. 2002). Importantly, the increase in food intake observed in these studies was only transient and had returned to control levels 28 days after knockout of the NPY Y$_2$ receptor. The reason for this rebound is not clear, but may be driven by the associated increase in corticosterone and, the rise in the plasma levels of pancreatic polypeptide (PP) or the elevated POMC and cocaine and amphetamine regulating transcript (CART) (Sainsbury et al. 2002; Guimaraes et al. 2002; Katsuura et al. 2002).

Germline Knockout. The first germline knockout of the NPY Y$_2$ receptor produced the expected phenotype of increased food intake, fat mass and body weight (Naveilhan et al. 1999). Surprisingly, germline knockout of the NPY Y$_2$ receptor in another laboratory produced an animal with decreased food intake despite the expected increase in NPY/AgRP and decreased POMC/CART levels in the ARC (Sainsbury et al. 2002). Chronically elevated plasma PP levels were an associated characteristic of the second knockout phenotype. Since PP has been shown to be a peripherally acting anorexogenic substance the reduction in food intake in these later studies may be due to the chronically elevated increase in this peptide. The underlying reason for the differences observed between the two knockout strains is not known but could be due to differences in background strain, completeness of receptor knockout as well as the presence or absence of the *neo* gene used in the transfection process.

Crossing the NPY Y_2 receptor knockout mouse onto the *ob/ob* background has been shown to attenuate the increased adiposity, hyperinsulinemia, hyperglycemia as well as the increased activity of the hypothalamus–pituitary–adrenal axis present in these animals (Sainsbury et al. 2002). These observations would suggest that the phenotype of the *ob/ob* mouse is partially mediated by NPY acting through the NPY Y_2 receptor. Interestingly, the hyperphagia of the *ob/ob* mouse was unaffected by knockout of the NPY Y_2 receptor suggesting that other receptor subtypes, for instance the NPY Y_1 and NPY Y_5 receptor subtypes, are involved in mediating the hyperphagic actions of NPY in this model.

7.2.4
NPY Y_2 Receptor Antisense

To date there are no reported studies of the effect of NPY Y_2 receptor antisense administration on food intake.

7.2.5
NPY Y_2 Receptor Antagonists

Recently, a nonpeptide NPY Y_2 receptor antagonist, BIIE 0246 was described as a selective high affinity antagonist of the NPY Y_2 receptor (King et al. 2000) (Table 2, Fig. 8). In our hands this compound also appeared to have significant affinity for the NPY Y_5 receptor and for muscarinic receptors. If confirmed in other laboratories these findings suggest that considerable care should be taken when using this compound to investigate the role of the NPY Y_2 receptor in the control of food intake.

7.3
The NPY Y_4 Receptor Subtype

7.3.1
NPY Y_4 Subtype Selective Peptide Agonists and Food Intake

Although the NPY Y_4 receptor is expressed predominantly in the periphery, significant quantities of NPY Y_4 mRNA and specific binding sites have been found in many hypothalamic regions including the PVN. Injection of rat PP, the only known high affinity agonist of the NPY Y_4 receptor with reasonably good selectivity against other NPY receptor subtypes has little or no effect on food intake in either rats or mice (Wyss et al. 1998; Haynes et al. 1998). Furthermore, no correlation exists between the affinity of NPY analogs for the NPY Y_4 receptor and their ability to stimulate food intake (Fig. 5) (Wyss et al. 1998). Taken to-

NPY Effects on Food Intake and Metabolism   311

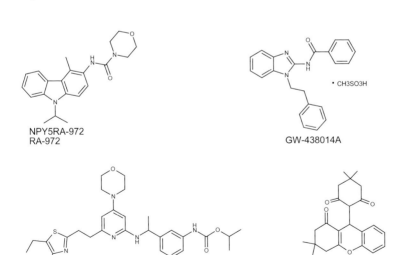

Fig. 8 Molecular structure of the nonpeptide Y receptor antagonists described in this review

gether these observations suggest that the NPY Y_4 receptor plays little role in the control of food intake.

7.3.2
NPY Y_4 Receptor Knockout

The NPY Y_4 receptor knockout in male mice presents a phenotype of decreased food intake, body weight and adiposity (Sainsbury et al. 2002a). The reason for the decrease in food intake appears not to be related to alterations in hypothalamic neuropeptide levels since NPY/AgRP and POMC/cocaine and CART levels in the ARC are unchanged in the knockout animals. Similar to the NPY Y_2 receptor knockout described above, a large increase in plasma PP concentrations was observed in NPY Y_4 knockout animals and this may be the underlying mechanism for the fall in food intake observed in these studies. Crossing the NPY Y_4 receptor knockout mouse with the *ob/ob* mouse has no effect on body weight and adiposity. Therefore, the NPY Y_4 receptor is not involved in mediating the actions of NPY to produce the phenotype of the *ob/ob* mouse.

7.4
The NPY Y_5 Receptor Subtype

7.4.1
The NPY Y_5 Receptor

Analogs of NPY have recently been synthesized that have both high affinity and selectivity for the rat NPY Y_5 receptor. Recent examples include [K^4, RYYSA$^{19-23}$]PP(2–36), [Ala^{31},Aib^{32}]pNPY, [D-Trp^{32}]hNPY, CPP(1–17), NPY(19–23), and [Ala^{31},Aib^{32},Gln^{34}]hPP. All of these peptides potently stimulate food intake and in the case of [K^4,RYYSA$^{19-23}$]PP(2–36) the effect is greater than that of the parent peptide (Parker et al. 2000; Cabrele et al. 2000; McCrea et al. 2000). Correlation studies have also been performed between the affinity of a variety of modified NPY peptides for the rat NPY Y_5 receptor with their ability to stimulate food intake (Wyss et al. 1998). In this case a significant positive correlation emerges which appears 'tighter' than for the NPY Y_1 receptor (Fig. 5).

7.4.2
Knockout of the NPY Y_5 Receptor

Knockout of the NPY Y_5 receptor subtype produces an animal of essentially normal phenotype that when young grows and feeds normally and has both a normal food intake response to fasting and to centrally administered leptin (Marsh et al. 1998). In this regard it is much like the NPY knockout mouse described above. Interestingly, with age the NPY Y_5 receptor knockout mice become hyperphagic and mildly obese. Since neither of these findings is in agreement with a role for the NPY Y_5 receptor in the control of food intake, how can they be interpreted? Two main explanations have been suggested. The early lack of phenotype may be because the NPY Y_5 receptor subtype is not a critical feeding receptor. In its absence food intake is maintained by the action of NPY on other receptor subtypes (the NPY Y_1 receptor for instance; see Fig. 6). Alternatively, during development the actions of NPY mediated by the NPY Y_5 receptor may be compensated for completely by other neural pathways as described above. The late developing obesity has been explained by postulating an exclusively presynaptic location for the NPY Y_5 receptor. In this position the NPY Y_5 subtype could act as an autoreceptor negatively inhibiting the release of NPY. Its absence in the NPY Y_5 knockout mouse would then leave the release of NPY unopposed—leading to overproduction of the peptide and to increased food intake (Fig. 6). Indeed, there is some evidence for the presence of a presynaptically located NPY Y_5 receptor, at least in the subiculum of the rat brain (Ho et al. 2000).

7.4.3
Inhibition of the Production of the NPY Y_5 Receptor with Antisense ODNs

The physiological role of endogenous NPY acting through the NPY Y_5 receptor has also been investigated using the antisense approach (Schaffhauser et al. 2000; Tang-Christensen et al. 1998). Repeated injections of antisense ODNs has been shown to reduce NPY-induced food intake, spontaneous food intake and to inhibit the refeeding response to an overnight fast in rats (Schaffhauser et al. 2000; Tang-Christensen et al. 1998). These data, obtained using antisense ODNs, contradict the conclusion of the knockout mouse that the NPY Y_5 receptor has no involvement in the control of fasting-induced and spontaneous food intake. Interestingly, in one of the above studies antisense ODNs directed against the NPY Y_5 receptor reduced total food intake but did not affect either the microstructure of food intake nor the food intake response to galanin, providing some evidence for the specificity of the approach (Schaffhauser et al. 1997). However, in general, concerns for the selectivity and toxicity of NPY Y_5 ODNs have not been rigorously pursued. Another criticism that has been leveled at the above studies is the failure to determine whether the antisense molecules actually entered their target cells to reduce NPY Y_5 receptor density. In addition, the effect of the antisense molecules on food intake was in some cases faster than the 60–96-h turnover expected for G-protein coupled receptors (3–4 days). While there may be explanations for this apparent discrepancy this finding does add to the problems in data interpretation.

In conclusion, while studies with antisense ODNs may point to a role for the NPY Y_5 receptor in the physiological control of food intake, in view of the above criticisms this conclusion needs to be verified using another approach.

7.4.4
NPY Y_5 Subtype Selective Antagonists

Considerable effort has been made by a variety of pharmaceutical companies to produce NPY Y_5 receptor antagonists as anti-obesity agents. Although thousands of compounds from many different chemical series have been synthesized it is still not entirely clear whether inhibition of the NPY Y_5 receptor can lead to reduced food intake. This lack of clarity is largely because the activity of many of these compounds have appeared in the patent literature with only a brief description of their binding affinity for the NPY Y_5 receptor and selectivity against other receptors. Although in the majority of cases decreased food intake has also been reported, the mechanism or mechanisms through which these compounds suppress appetite has not been systematically studied. The drugs could simply affect food intake by inducing illness or pain or could have significant cross reactivity with other receptors and channels, which could explain their effect on food intake. The following four compounds have recently been published in the literature and nicely illustrate the problems that have delayed a final conclusion as to whether blockade of the NPY Y_5 receptor has an effect on food intake.

7.4.5
Examples of NPY Y_5 Antagonists That Reduce Food Intake

CGP 71683A (Novartis) is perhaps the most widely studied NPY Y_5 antagonist. This compound has high affinity for the NPY Y_5 receptor with very low affinity for the NPY Y_1, Y_2 and Y_4 receptor subtypes. The compound is not orally active and therefore has been administered peripherally. Initial studies demonstrated that intraperitoneal injection of CGP 71683A strongly inhibited NPY-induced food intake. In both fasted and streptozotocin diabetic rats, animal models associated with elevated levels of hypothalamic NPY, CGP 71683A significantly suppressed the attendant hyperphagia. However, in subsequent studies, CGP 71683A was found to have equally high affinity for the serotonin reuptake recognition site and for cholinergic muscarinic receptors in the rat brain as for the NPY Y_5 receptor. A lower but still reasonably high affinity for α_2-adrenergic receptors was also observed—a finding in agreement with the original published studies on this compound. Since increased serotonin as well as changes in muscarinic and α-adrenergic receptor activity can affect food intake, these observations call into doubt the conclusion that the changes in food intake observed with CGP 71683A are due only to inhibition of NPY Y_5 receptors. In agreement with the above studies, CGP 71683A has recently been shown to reduce food intake equally in both wild-type and in NPY knockout mice. This latter result confirms that the activity of CGP 71683A resides in a mechanism or mechanisms of action other than NPY Y_5 receptor blockade.

GW438014A (GlaxoSmithKline) is a moderately potent but highly selective antagonist of the NPY Y_5 receptor subtype, which like CGP 71683A is not orally active (Table 2, Fig. 8). After intraperitoneal administration GW438014A penetrates into the brain and blocks both NPY and fasting-induced food intake. In animal models in which brain NPY levels have been shown to be high, such as during the dark phase in normal rats, in fasted *ob/ob* mice and in Zucker obese rats, GW438014A inhibits food intake. The effect of GW438014A on food intake does not appear to be due to the induction of pain or illness since the product doses not induce a conditioned taste aversion nor interact with a variety of other receptors.

7.4.6
Examples of NPY Y_5 Antagonists That Do Not Affect Food Intake

Both NPY5RA-972 (AstraZeneca) and L-152,804 (Takeda) are high affinity antagonists of the rat NPY Y_5 receptor with virtually no affinity for the NPY Y_1, NPY Y_2 and NPY Y_4 receptor subtypes or over 120 different receptors, channels and enzymes (Table 2, Fig. 8). Both compounds are orally active, penetrate into the brain at a concentration far in excess of their 50% inhibitory concentration (IC_{50}) values and readily inhibit the increase in food intake produced by selective NPY Y_5 agonists (Turnbull et al. 2002; Block et al. 2002; Kanatani et al. 2000). However, despite this favorable profile neither compound is able to prevent the increase in food intake produced by injection of NPY on spontaneous feeding or in free feeding, fasted or obese Zucker rats or dietary obese rats. These results have led to the conclusion that the blockade of the NPY Y_5 receptor does not have a major influence on feeding in rats.

8
Conclusions

When given into the brain, NPY stimulates food intake and a variety of metabolic processes that favor fat deposition. The stimulation of food intake and body weight observed with chronic administration of the peptide is persistent and leads to obesity. Both the acute and chronic stimulation of food intake and body weight produced by NPY can be reproduced with selective NPY Y_1 and Y_5 agonists. Therefore, there is no doubt that exogenously administered NPY is a potent regulator of appetite and could be involved in the development and maintenance of obesity. However, questions remain as to whether the increase in food intake produced by exogenous NPY represents a true hunger state or is mediated by unrelated behavioral changes.

Although brain NPY levels and food intake are temporally related, attempts to demonstrate a change in food intake after blockade of endogenous NPY have been mixed. Some studies have shown a change in food intake after blockade of the actions of NPY. However, the conclusion that NPY plays an important role in the control of food intake is difficult to fully accept because of the nonselective nature of the inhibitors used.

Based on the available evidence our conclusion is that NPY probably plays a role in the day-to-day control of food intake. However, NPY is not a critical regulator of food intake. Therefore in the absence of NPY food intake can be controlled by a variety of other hormones and neurotransmitters. A definitive answer to the role played by NPY in the control of food intake and metabolism awaits the development of clean and selective inhibitors.

References

Abe M, Saito M, Ikeda H, Shimazu T (1991) Increased neuropeptide Y content in the arcuato-paraventricular hypothalamic neuronal system in both insulin-dependent and non-insulin-dependent diabetic rats. Brain Res 539:223–227

Akabayashi A, Levin N, Paez X, Alexander JT, Leibowitz SF (1994) Hypothalamic neuropeptide Y and its gene expression: relation to light/dark cycle and circulating corticosterone. Mol Cell Neurosci 5:210–218

Akabayashi A, Wahlestedt C, Alexander JT, Leibowitz SF (1994) Specific inhibition of endogenous neuropeptide Y synthesis in arcuate nucleus by antisense oligonucleotides suppresses feeding behavior and insulin secretion. Brain Res Mol Brain Res 21:55–61

Ammar AA, Sederholm F, Saito TR, Scheurink AJ, Johnson AE, Sodersten P (2000) NPY-leptin: opposing effects on appetitive and consummatory ingestive behavior and sexual behavior. Am J Physiol Regul Integr Comp Physiol 278:R1627–R1633

Bai FL, Yamano M, Shiotani Y, Emson PC, Smith AD, Powell JF, Tohyama M (1985) An arcuato-paraventricular and -dorsomedial hypothalamic neuropeptide Y-containing system which lacks noradrenaline in the rat. Brain Res 331:172–175

Baker RA, Herkenham M (1995) Arcuate nucleus neurons that project to the hypothalamic paraventricular nucleus: neuropeptidergic identity and consequences of adrenalectomy on mRNA levels in the rat. J Comp Neurol 358:518–30

Bannon AW, Seda J, Carmouche M, Francis JM, Norman MH, Karbon B, McCaleb ML (2000) Behavioral characterization of neuropeptide Y knockout mice. Brain Res 868:79–87

Batterham RL, Cowley MA, Small CJ, Herzog H, Cohen MA, Dakin CL, Wren AM, Brynes AE, Low MJ, Ghatei MA, Cone RD, Bloom SR (2002) Gut hormone PYY(3–36) physiologically inhibits food intake. Nature 418:650–654

Beck B, Jhanwar-Uniyal M, Burlet A, Chapleur-Chateau M, Leibowitz SF, Burlet C (1990) Rapid and localized alterations of neuropeptide Y in discrete hypothalamic nuclei with feeding status. Brain Res 1990 528:245–249

Beck B, Stricker-Krongrad A, Nicolas JP, Burlet A (1992) Chronic and continuous intracerebroventricular infusion of neuropeptide Y in Long-Evans rats mimics the feeding behaviour of obese Zucker rats. Int J Obes Relat Metab Disord 16:295–302

Billington CJ, Briggs JE, Harker S, Grace M, Levine AS (1994) Neuropeptide Y in hypothalamic paraventricular nucleus: a center coordinating energy metabolism. Am J Physiol 266:R1765–R1770

Bina KG, Cincotta AH (2000) Dopaminergic agonists normalize elevated hypothalamic neuropeptide Y and corticotropin-releasing hormone, body weight gain, and hyperglycemia in ob/ob mice. Neuroendocrinology 71:68–78

Blasquez C, Jegou S, Friard O, Tonon MC, Fournier A, Vaudry H (1995) Effect of centrally administered neuropeptide Y on hypothalamic and hypophyseal proopiomelanocortin-derived peptides in the rat. Neuroscience 68:221–227

Block MH, Boyer S, Brailsford W, Brittain DR, Carroll D, Chapman S, Clarke DS, Donald CS, Foote KM, Godfrey L, Ladner A, Marsham PR, Masters DJ, Mee CD, O'Donovan MR, Pease JE, Pickup AG, Rayner JW, Roberts A, Schofield P, Suleman A, Turnbull

AV (2002) Discovery and optimization of a series of carbazole ureas as NPY5 antagonists for the treatment of obesity. J Med Chem 45:3509–3523

Blumenthal JB, Andersen RE, Mitchell BD, Seibert MJ, Yang H, Herzog H, Beamer BA, Franckowiak SC, Walston JD (2002) Novel neuropeptide Y1 and Y5 receptor gene variants: associations with serum triglyceride and high-density lipoprotein cholesterol levels. Clin Genet 62:196–202

Bonavera JJ, Dube MG, Kalra PS, Kalra SP (1994) Anorectic effects of estrogen may be mediated by decreased neuropeptide-Y release in the hypothalamic paraventricular nucleus. Endocrinology 134:2367–2370

Bonini JA, Jones KA, Adham N, Forray C, Artymyshyn R, Durkin MM, Smith KE, Tamm JA, Boteju LW, Lakhlani PP, Raddatz R, Yao WJ, Ogozalek KL, Boyle N, Kouranova EV, Quan Y, Vaysse PJ, Wetzel JM, Branchek TA, Gerald C, Borowsky B. (2000) Identification and characterization of two G protein-coupled receptors for neuropeptide FF.The Journal of Biological Chemistry 275 (50):39324–39331

Broberger C, Landry M, Wong H, Walsh JN, Hokfelt T (1997) Subtypes Y1 and Y2 of the neuropeptide Y receptor are respectively expressed in pro-opiomelanocortin- and neuropeptide-Y-containing neurons of the rat hypothalamic arcuate nucleus. Neuroendocrinology 66:393–408

Brown CM, Coscina DV, Fletcher PJ (2000) The rewarding properties of neuropeptide Y in perifornical hypothalamus vs. nucleus accumbens. Peptides 21:1279–1287

Brown CM, Fletcher PJ, Coscina DV (1998) Neuropeptide Y-induced operant responding for sucrose is not mediated by dopamine. Peptides 19:1667–1673

Burcelin R, Brunner H, Seydoux J, Thorensa B, Pedrazzini T (2001) Increased insulin concentrations and glucose storage in neuropeptide Y Y1 receptor-deficient mice. Peptides 2001 22:421–427

Burlet A, Grouzmann E, Musse N, Fernette B, Nicolas JP, Burlet C (1995) The immunological impairment of arcuate neuropeptide Y neurons by ricin A chain produces persistent decrease of food intake and body weight. Neuroscience 66:151–159

Cabrele C, Langer M, Bader R, Wieland HA, Doods HN, Zerbe O, Beck-Sickinger AG (2000) The first selective agonist for the neuropeptide YY5 receptor increases food intake in rats. J Biol Chem 275:36043–36048

Chaffer CL, Morris MJ (2002) The feeding response to melanin-concentrating hormone is attenuated by antagonism of the NPY Y(1)-receptor in the rat. Endocrinology 143:191

Chamorro S, Della-Zuana O, Fauchere JL, Feletou M, Galizzi JP, Levens N (2002) Appetite suppression based on selective inhibition of NPY receptors (review). Int J Obes Relat Metab Disord 26:281–298

Chronwall BB (1989) Anatomical distribution of NPY and NPY messenger RNA in the brain. In: Mutt V, Füxe K, Hökfelt T, Lundberg J (eds) Neuropeptide Y. Raven Press, New York, pp 50–60

Corpa ES, McQuade J, Krasnicki S, Conze DB (2001) Feeding after fourth ventricular administration of neuropeptide Y receptor agonists in rats. Peptides 22:493–499

Cowley MA, Smart JL, Rubinstein M, Cerdain MG, Diano S, Horvath TL, Cone RD, Low MJ (2001) Leptin activates anorexigenic POMC neurons through a neural network in the arcuate nucleus. Nature 411:480–484

Criscione L, Rigollier P, Batzl-Hartmann C, Rueger H, Stricker-Krongrad A, Wyss P, Brunner L, Whitebread S, Yamaguchi Y, Gerald C, Heurich RO, Walker MW, Chiesi M, Schilling W, Hofbauer KG, Levens N (1998) Food intake in free-feeding and energy-deprived lean rats is mediated by the neuropeptide Y5 receptor. J Clin Invest. 102:2136–2145

Criscione L, Schaffhauser AO, Guerini D, Rigolier P, Schmid H, Buhlmayer P (2001) Effects of the Y5 receptor antagonist CGO71683A on food intake in male Y5 receptor knock-out mice. Presented at the 6th International NPY Conference. Sydney, April 2001

Cummings SL, Truong BG, Gietzen DW (1998) Neuropeptide Y and somatostatin in the anterior piriform cortex alter intake of amino acid-deficient diets. Peptides 19:527–535

Currie PJ, Coscina DV (1995) Dissociated feeding and hypothermic effects of neuropeptide Y in the paraventricular and perifornical hypothalamus. Peptides 16:599–604

Dallman MF, Akana SF, Bhatnagar S, Bell ME, Choi S, Chu, A, Horsley C, Levin N; Meijer O, Soriano LR, Strack AM, Viau V (1999) Starvation: Early signals, sensors, and sequelae. Endocrinology 140:4015–4023

Daniels AJ, Grizzle MK, Wiard RP, Matthews JE, Heyer D (2002) Food intake inhibition and reduction in body weight gain in lean and obese rodents treated with GW438014A, a potent and selective NPY-Y5 receptor antagonist. Regul Pept 106:47–54

Della Zuana O, Sadlo M, Germain M, Feletou M, Chamorro S, Tisserand F, de Montrion C, Boivin JF, Duhault J, Boutin JA, Levens N (2001) Reduced food intake in response to CGP 71683A may be due to mechanisms other than NPY Y5 receptor blockade. Int J Obes Relat Metab Disord 25:84–94

Dryden S, Frankish HM, Wang Q, Williams G (1996) Increased feeding and neuropeptide Y (NPY) but not NPY mRNA levels in the hypothalamus of the rat following central administration of the serotonin synthesis inhibitor p-chlorophenylalanine. Brain Res 724:232–237

Dryden S, Pickavance L, Tidd D, Williams G (1998) The lack of specificity of neuropeptide Y (NPY) antisense oligodeoxynucleotides administered intracerebroventricularly in inhibiting food intake and NPY gene expression in the rat hypothalamus. J Endocrinol 157:169–175

Dube MG, Kalra PS, Crowley WR, Kalra SP (1995) Evidence of a physiological role for neuropeptide Y in ventromedial hypothalamic lesion-induced hyperphagia. Brain Res 690:275–278

Dube MG, Xu B, Crowley WR, Kalra PS, Kalra SP 1994) Evidence that neuropeptide Y is a physiological signal for normal food intake. Brain Res 646:341–344

Duhault J, Boulanger M, Chamorro S, Boutin JA, Della Zuana O, Douillet E, Fauchere JL, Feletou M, Germain M, Husson B, Vega AM, Renard P, Tisserand F (2000) Food intake regulation in rodents: Y5 or Y1 NPY receptors or both? Can J Physiol Pharmacol 78:173–185

Egawa M, Yoshimatsu H, Bray GA (1991) Neuropeptide Y suppresses sympathetic activity to interscapular brown adipose tissue in rats. Am J Physiol 260:R328–R334

Elshourbagy NA, Ames RS, Fitzgerald LR, Foley JJ, Chambers JK, Szekeres PG, Evans NA, Schmidt DB, Buckley PT, Dytko GM, Murdock PR, Milligan G, Groarke DA, Tan KB, Shabon U, Nuthulaganti P, Wang DY, Wilson S, Bergsma DJ, Sarau HM. (2000). Receptor for the pain modulatory neuropeptides FF and AF is an orphan G protein-coupled receptor. J Biolog Chem 275(34):25965–25971

Erickson JC, Ahima RS, Hollopeter G, Flier JS, Palmiter RD (1997) Endocrine function of neuropeptide Y knockout mice. Regul Pept 70:199–202

Everitt BJ, Hökfelt T (1989) The coexistance of neuropeptide Y with other peptides and amines in the central nervous system. In: Mutt V, Füxe K, Hökfelt T, Lundberg J (eds) Neuropeptide Y. Raven Press, New York, pp 61–72

Fekete C, Kelly J, Mihaly E, Sarkar S, Rand WM, Legradi G, Emerson CH, Lechan RM (2001) Neuropeptide Y has a central inhibitory action on the hypothalamic-pituitary-thyroid axis. Endocrinology 142:2606–2613

Flood JF, Morley JE (1991) Increased food intake by neuropeptide Y is due to an increased motivation to eat. Peptides 12:1329–1332

Frankish HM, McCarthy HD, Dryden S, Kilpatrick A, Williams G Frankish HM, McCarthy HD, Dryden S, Kilpatrick A, Williams G (1993) Neuropeptide Y receptor numbers are reduced in the hypothalamus of streptozotocin-diabetic and food-deprived rats:

further evidence of increased activity of hypothalamic NPY-containing pathways. Peptides 14:941–948

Gehlert DR (1998) Multiple receptors for the pancreatic polypeptide (PP-Fold) family: Physiological implications. Proc Soc Exp Biol Med 218:7–22

George SR, O'Dowd BF, Lee SP. (2002) G-protein-coupled receptor oligomerization and its potential for drug discovery. Nature Rev Drug Discov 1(10):808–20

Guimaraes R, Telles M, Coelho V, Mori R, Nascimento C, Ribeiro E (2002) Adrenalectomy abolishes the food-induced hypothalamic serotonin release in both normal and monosodium glutamate-obese rats. Brain Res Bull 58:363

Guy J, Pelletier G (1988) Neuronal interactions between neuropeptide Y (NPY) and catecholaminergic systems in the rat arcuate nucleus as shown by dual immunocytochemistry. Peptides 9:567–570

Guy J, Pelletier G, Bosler O (1988) Serotonin innervation of neuropeptide Y-containing neurons in the rat arcuate nucleus. Neurosci Lett 85:9–13

Hagan MM (2002) Peptide YY: a key mediator of orexigenic behavior. Peptides 23:377–382

Hagan MM, Castaneda E, Sumaya IC, Fleming SM, Galloway J, Moss DE (1998) The effect of hypothalamic peptide YY on hippocampal acetylcholine release in vivo: implications for limbic function in binge-eating behavior. Brain Res 805:20–28

Harland D, Bennett T, Gardiner SM (1988) Cardiovascular actions of neuropeptide Y in the hypothalamic paraventricular nucleus of conscious Long Evans and Brattleboro rats. Neurosci Lett 85(2):239–243

Haynes AC, Arch JRS, Wilson S, McCLue S, Buckingham RE (1998) Charactorization of the neuropeptide Y receptor that mediates feeding in the rat: a role for the Y5 receptor. Regul Pept 75/76:355–361

He B, White BD, Edwards GL, Martin RJ (1998) Neuropeptide Y antibody attenuates 2-deoxy-D-glucose induced feeding in rats. Brain Res 781:348–350

Heilig M (1995) Antisense inhibition of neuropeptide Y (NPY)-Y1 receptor expression blocks the anxiolytic-like action of NPY in amygdala and paradoxically increases feeding. Regul Pept 59:201–205

Heilig M, McLeod S, Koob GK, Britton KT (1992) Anxiolytic-like effect of neuropeptide Y (NPY), but not other peptides in an operant conflict test. Regul Pept 41:61–69

Herzog H (2000) NPY-Y7 receptor gene. International application published under the patent cooperation treaty (PCT). International publication number WO 00/00606

Herzog H, Selbie LA, Zee RYL, Morris BJ and Shine J (1993) Neuropeptide-Y Y1 receptor gene polymorphism: cross-sectional analyses in essential hypertension and obesity. Biochem Biophys Research Commun 196: 902–906

Hisano S, Kagotani Y, Tsuruo Y, Daikoku S, Chihara K, Whitnall MH (1988) Localization of glucocorticoid receptor in neuropeptide Y-containing neurons in the arcuate nucleus of the rat hypothalamus. Neurosci Lett 95:13–18

Ho MW, Beck-Sickinger AG, Colmers WF (2000) Neuropeptide Y(5) receptors reduce synaptic excitation in proximal subiculum, but not epileptiform activity in rat hippocampal slices. J Neurophysiol 83:723–734

Hollopeter G, Erickson JC, Seeley RJ, Marsh DJ, Palmiter RD (1998) Response of neuropeptide Y-deficient mice to feeding effectors. Regul Pept 75/76:383–389

Howard AD, McAllister G, Feighner SD, Liu Q, Nargund RP, Van der Ploeg LH, Patchett AA. (2001) Orphan G-protein-coupled receptors and natural ligand discovery. Trends Pharmacol Sci 22(3):132–140

Hulsey MG, Pless CM, White BD, Martin RJ (1995) ICV administration of anti-NPY antisense oligonucleotide: effects on feeding behavior, body weight, peptide content and peptide release. Regul Pept 59:207–214

Inui A, Okita M, Nakajima M, Momose K, Ueno N, Teranishi A, Miura M, Hirosue Y, Sano K, Sato M, Watanabe M, Sakai T, Watanabe T, Ishida K, Silver J, Baba S, Kasuga

M (1998) Anxiety-like behavior in transgenic mice with brain expression of neuropeptide Y. Proc Assoc Am Physicians 110:171–182

Jenkinson CP, Cray K, Walder K, Herzog H, Hanson R and Ravussin E (2000) Novel polymorphisms in the neuropeptide-Y Y5 receptor associated with obesity in Pima Indians. International J Obes 24:580–584

Jewett DC, Schaal DW, Cleary J, Thompson T, Levine AS (1991) The discriminative stimulus effects of neuropeptide Y. Brain Res 561:165–168

Jhanwar-Uniyal M, Beck B, Burlet C, Leibowitz SF (1990) Diurnal rhythm of neuropeptide Y-like immunoreactivity in the suprachiasmatic, arcuate and paraventricular nuclei and other hypothalamic sites. Brain Res 536:331–334

Kaga T, Inui A, Okita M, Asakawa A, Ueno N, Kasuga M, Fujimiya M, Nishimura N, Dobashi R, Morimoto Y, Liu IM, Cheng JT (2001) Modest overexpression of neuropeptide Y in the brain leads to obesity after high-sucrose feeding. Diabetes 50:1206–1210

Kalra PS, Kalra SP (2000) Use of antisense oligodeoxynucleotides to study the physiological functions of neuropeptide Y. Methods 22:249–254

Kalra SP, Dube MG, Pu S, Xu B, Horvath TL, Kalra PS (1999) Interacting appetite-regulating pathways in the hypothalamic regulation of body weight. Endocrinol Rev 20:68–100

Kalra SP, Dube MG, Sahu A, Phelps CP, Kalra PS (1991) Neuropeptide Y secretion increases in the paraventricular nucleus in association with increased appetite for food. Proc Natl Acad Sci USA 88:10931–10935

Kalra SP, Kalra PS (1996) Is neuropeptide Y a naturally occuring appetite transducer? Curr Opin Endocrinol Diabetes 3:157–163

Kamegai J, Tamura H, Shimizu T, Ishii S, Sugihara H, Wakabayashi I (2001) Chronic central infusion of ghrelin increases hypothalamic neuropeptide Y and Agouti-related protein mRNA levels and body weight in rats. Diabetes 50:2438–2443

Kanatani A, Fukami T, Ishihara A, Ishii Y, MacNeil DJ, Van der Ploeg LHT, Ihara M (2001) Anorexigenic effects of orally-active NPY antagonists: Participation of Y1 and Y5 receptors in feeding behavior. Presented at the 6th International NPY Conference. Sydney, April 2001

Kanatani A, Hata M, Mashiko S, Ishihara A, Okamoto O, Haga Y, Ohe T, Kanno T, Murai N, Ishii Y, Fukuroda T, Fukami T, Ihara M (2001) A typical Y1 receptor regulates feeding behaviors: effects of a potent and selective Y1 antagonist, J-115814. Mol Pharmacol 59:501–505

Kanatani A, Ishihara A, Iwaasa H, Nakamura K, Okamoto O, Hidaka M, Ito J, Fukuroda T, MacNeil DJ, Van der Ploeg LH, Ishii Y, Okabe T, Fukami T, Ihara M (2000) L-152,804: orally active and selective neuropeptide Y Y5 receptor antagonist. Biochem Biophys Res Commun 272:169–173

Kanatani A, Mashiko S, Murai N, Sugimoto N, Ito J, Fukuroda T, Fukami T, Morin N, MacNeil DJ, Vand der Ploeg LH, Saga Y, Nishimura S, Ihara M (2000) Role of the Y1 receptor in the regulation of neuropeptide Y-mediated feeding: comparison of wild-type, Y1 receptor-deficient, and Y5 receptor-deficient mice. Endocrinology 141:1011–1016

Kask A, Haro J, von HorstenS, Redrobe JP, Dumont Y, Quirion R (2002) The neurocircuitry and receptor subtypes mediating anxiolytic-like effects of neuropeptide Y. Neurosci Biobehav Rev 26:259–283

Kask A, Rago L, Harro J (1998) Evidence for involvement of neuropeptide Y receptors in the regulation of food intake: studies with Y1-selective antagonist BIBP3226. Br J Pharmacol 1998 124:1507–1515

Kask A, Vasar E, Heidmets LT, Allikmets L, Wikberg JE (2001) Neuropeptide Y Y(5) receptor antagonist CGP71683A: the effects on food intake and anxiety-related behavior in the rat. Eur J Pharmacol 2001 414:215–224

Katsuura G, Asakawa A, Inui A (2002) Roles of pancreatic polypeptide in regulation of food intake. Peptides 23:323–329

Kim MS, Rossi M, Abbott CR, AlAhmed SH, Smith DM, Bloom SR (2002) Sustained orexigenic effect of Agouti related protein may be not mediated by the melanocortin 4 receptor. Peptides 23:1069–1076

King PJ, Williams G, Doods H, Widdowson PS (2000) Effect of a selective neuropeptide Y Y(2) receptor antagonist, BIIE0246 on neuropeptide Y release. Eur J Pharmacol 396:R1–R3

Kotz CM, Wang CF, Briggs JE, Levine AS, Billington CJ. (2000) Effect of NPY in the hypothalamic paraventricular nucleus on uncoupling protein 1, 2 and 3 in the rat. Am J Physiol Regul Integr Comp Physiol 278:R494–R498

Kushi A, Sasai H, Koizumi H, Takeda N, Yokoyama M, Nakamura M (1998) Obesity and mild hyperinsulinemia found in neuropeptide Y-Y1 receptor-deficient mice. Proc Natl Acad Sci USA 95:15659–15664

Lambert PD, Wilding JP, al-Dokhayel AA, Bohuon C, Comoy E, Gilbey SG, Bloom SR (1993) A role for neuropeptide-Y, dynorphin, and noradrenaline in the central control of food intake after food deprivation. Endocrinology 133:29–32

Lee J, Morris MJ (1998) Modulation of neuropeptide Y overflow by leptin in the rat hypothalamus, cerebral cortex and medulla. Neuroreport 9:1575–1580

Levine AS, Kuskowski MA, Grace M, Billington CJ (1991) Food deprivation-induced vs. drug-induced feeding: A behavioral evaluation. Am J Physiol 260:R546–R552

Lewis DE, Shellard L, Koeslag DG, Boer DE, McCarthy HD, McKibbin PE, Russell JC, Williams G (1993) Intense exercise and food restriction cause similar hypothalamic neuropeptide Y increases in rats. Am J Physiol 264:E279–E284

Li S, Hisano S, Daikoku S (1993) Mutual synaptic associations between neurons containing neuropeptide Y and neurons containing enkephalin in the arcuate nucleus of the rat hypothalamus. Neuroendocrinology 57:306–313

Liposits Z, Sievers L, Paull WK (1988) Neuropeptide-Y and ACTH-immunoreactive innervation of corticotropin releasing factor (CRF)-synthesizing neurons in the hypothalamus of the rat. An immunocytochemical analysis at the light and electron microscopic levels. Histochemistry 88:227–234

Liu Q, Guan XM, Martin WJ, McDonald TP, Clements MK, Jiang Q, Zeng Z, Jacobson M, Williams DL Jr, Yu H, Bomford D, Figueroa D, Mallee J, Wang R, Evans J, Gould R, Austin CP (2001). Identification and characterization of novel mammalian neuropeptide FF-like peptides that attenuate morphine-induced antinociception. J Biolog Chem 276(40):36961–36969

Lopez-Valpuesta FJ, Nyce JW, Griffin-Biggs TA, Ice JC, Myers RD (1996) Antisense to NPY-Y1 demonstrates that Y1 receptors in the hypothalamus underlie NPY hypothermia and feeding in rats. Proc R Soc Lond B Biol Sci 263:881–886

Lopez-Valpuesta FJ, Nyce JW, Myers RD (1996) NPY-Y1 receptor antisense injected centrally in rats causes hyperthermia and feeding. Neuroreport 7:2781–2784

Malabu UH, Kilpatrick A, Ware M, Vernon RG, Williams G (1994) Increased neuropeptide Y concentrations in specific hypothalamic regions of lactating rats: Possible relationship to hyperphagia and adaptive changes in energy balance. Peptides 15:83–87

Mamoun AH, Bergstrom J, Sodersten P (1997) Cholecystokinin octapeptide inhibits carbohydrate but not protein intake. Am J Physiol 273:R972–R980

Marin-Bivens CL, Kalra SP, Olster DH (1998) Intraventricular injection of neuropeptide Y antisera curbs weight gain and feeding, and increases the display of sexual behaviors in obese Zucker female rats. Regul Pept 75/76:327–334

Marsh DJ, Hollopeter G, Kafer KE, Palmiter RD (1998) Role of the Y5 neuropeptide Y receptor in feeding and obesity. Nature Med 4:718–721

Marsh DJ, Miura GI, Yagaloff KA, Schwartz MW, Barsh GS, Palmiter RD (1999) Effects of neuropeptide Y deficiency on hypothalamic agouti-related protein expression and responsiveness to melanocortin analogues. Brain Res 848:66–77

McCrea K, Wisialowski T, Cabrele C, Church B, Beck-Sickinger A, Kraegen E, Herzog H (2000) 2-36[K4,RYYSA(19-23)]PP a novel Y5-receptor preferring ligand with strong stimulatory effect on food intake. Regul Pept 87:47-58

McGregor IS, Menendez JA, Atrens DM (1990) Metabolic effects of neuropeptide Y injected into the sulcal prefrontal cortex. Brain Res Bull 24:363-367

McKibbin PE, Cotton SJ, McMillan S, Holloway B, Mayers R, McCarthy HD, Williams G (1991) Altered neuropeptide Y concentrations in specific hypothalamic regions of obese (fa/fa) Zucker rats. Possible relationship to obesity and neuroendocrine disturbances. Diabetes 40:1423-1429

Mollereau C, Gouarderes C, Dumont Y, Kotani M, Detheux M, Doods H, Parmentier M, Quirion R, Zajac JM (2001) Agonist and antagonist activities on human NPFF(2) receptors of the NPY ligands GR231118 and BIBP3226. Br J Pharmacol 133:1-4

Mollereau C, Mazarguil H, Marcus D, Quelven I, Kotani M, Lannoy V, Dumont Y, Quirion R, Detheux M, Parmentier M, Zajac JM (2002) Pharmacological characterization of human NPFF(1) and NPFF(2) receptors expressed in CHO cells by using NPY Y(1) receptor antagonists. Eur J Pharmacol 451:245-256

Morgan DG, Small CJ, Abusnana S, Turton M, Gunn I, Heath M, Rossi M, Goldstone AP, O'Shea D, Meeran K, Ghatei M, Smith DM, Bloom S (1998) The NPY Y1 receptor antagonist BIBP 3226 blocks NPY induced feeding via a non-specific mechanism. Regul Pept 75/76:377-382

Morton GJ, Schwartz MW (2001) The NPY/AgRP neuron and energy homeostasis. In J Obes realt Metabo Disord Suppl 5:S56-S62

Mullins D, Kirby D, Hwa J, Guzzi M, Rivier J, Parker E (2001) Identification of potent and selective neuropeptide Y Y(1) receptor agonists with orexigenic activity in vivo. Mol Pharmacol 60:534-540

Naveilhan P, Hassani H, Canals JM, Ekstrand AJ, Larefalk A, Chhajlani V, Arenas E, Gedda K, Svensson L, Thoren P, Ernfors P (1999) Normal feeding behavior, body weight and leptin response require the neuropeptide Y Y2 receptor. Nature Med 5:1188-1193

Pages N, Orosco M, Rouch C, Yao O, Jacquot C, Bohuon C (1993) Refeeding after 72 hour fasting alters neuropeptide Y and monoamines in various cerebral areas in the rat.- Comp Biochem Physiol Comp Physiol 106:845-849

Palmiter RD, Erickson JC, Hollopeter G, Baraban SC, Schwartz MW (1998) Life without neuropeptide Y. Recent Prog Horm Res 53:163-199

Parikh R, Marks JL (1997) Metabolic and orexigenic effects of intracerebroventricular neuropeptide Y are attenuated by food deprivation. J Neuroendocrinol 9:789-795

Parker EM, Balasubramaniam A, Guzzi M, Mullins DE, Salisbury BG, Sheriff S, Witten MB, Hwa JJ (2000) D-Trp(34)] neuropeptide Y is a potent and selective neuropeptide Y Y(5) receptor agonist with dramatic effects on food intake. Peptides 21(3):393-399.

Parker RMC, Copeland NG, Eyre HJ, Liu M, Gilbert DJ, Crawford J, Couzens M, Sutherland GR, Jenkins NA, Herzog H (2000) Molecular cloning and characterisation of GPR74 a novel G-protein coupled receptor closest related to the Y-receptor family. Mol Brain Res 77:199-208

Pedersen-Bjergaard U, Host U, Kelbaek H, Schifter S, Rehfeld JF, Faber J, Christensen NJ (1996) Influence of meal composition on postprandial peripheral plasma concentrations of vasoactive peptides in man. Scand J Clin Lab Invest 56:497-503

Pedrazzini T, Seydoux J, Kunstner P, Aubert J-F, Grouzmann E, Beermann F, Brunner H-R (1998) Cardiovascular response, feeding behavior and locomotor activity in mice lacking the NPY Y1 receptor. Nature Med 4:722-726

Porte D Jr, Baskin DG, Schwartz MW (2002) Leptin and insulin action in the central nervous system. Nutr Rev 60:S20-S29

Potter EK, Barden JA, McCloskey MJ, Selbie LA, Tseng A, Herzog H, Shine J (1994) A novel neuropeptide Y analog, N-acetyl [Leu28,Leu31]neuropeptide Y-(24-36), with functional specificity for the presynaptic (Y2) receptor. Eur J Pharmacol 267:253-262

Pralong FP, Gonzales C, Voirol MJ, Palmiter RD, Brunner HR, Gaillard RC, Seydoux J, Pedrazzini T (2002) The neuropeptide Y Y1 receptor regulates leptin-mediated control of energy homeostasis and reproductive functions. FASEB J 16:712–714

Raposinho PD, Pierroz DD, Broqua P, White RB, Pedrazzini T, Aubert ML (2001) Chronic administration of neuropeptide Y into the lateral ventricle of C57BL/6J male mice produces an obesity syndrome including hyperphagia, hyperleptinemia, insulin resistance and hypogonadism. Mol Cell Endocrinol 185:195–204

Roche C, Boutin P, Dina C, Gyapay G, Basdevant A, Hager J, Guy-Grand B, Clement K, Froguel P (1997) Genetic studies of neuropeptide Y and neuropeptide Y receptors Y1 and Y5 regions in morbid obesity. Diabetologia 40:671–675

Rosenkranz K, Hinney A, Ziegler A, von Prittwitz S, Barth N, Roth H, Mayer H, Siegfried W, Lehmkuhl G, Poustka F, Schmidt M, Schäfer H, Remschmidt H, Hebebrand J (1998) Screening for mutations in the neuropeptide Y Y5 receptor gene in cohorts belonging to different weight extremes. Int J Obes 22:157–163

Sahu A, Dube MG, Phelps CP, Sninsky CA, Kalra PS, Kalra SP (1995) Insulin and insulin-like growth factor II suppress neuropeptide Y release from the nerve terminals in the paraventricular nucleus: a putative hypothalamic site for energy homeostasis. Endocrinology 136:5718–5724.

Sahu A, Kalra PS, Kalra S (1988) Food deprivation and ingestion induce reciprocal changes in neuropeptide Y concentrations in the paraventricular nucleus. Peptides 9:83–86

Sahu A, Kalra SP, Crowley WR, Kalra PS (1988) Evidence that NPY-containing neurons in the brainstem project into selected hypothalamic nuclei: implication in feeding behavior. Brain Res 457:376–378

Sahu A, Kalra SP, Crowley WR, Kalra PS (1989) Testosterone raises neuropeptide-Y concentration in selected hypothalamic sites and in vitro release from the medial basal hypothalamus of castrated male rats. Endocrinology 124:410–414

Sahu A, Phelps CP, White JD, Crowley WR, Kalra SP, Kalra PS (1992) Steroidal regulation of hypothalamic neuropeptide Y release and gene expression. Endocrinology 130:3331–3336

Sahu A, Sninsky CA, Kalra SP (1997) Evidence that hypothalamic neuropeptide Y gene expression and NPY levels in the paraventricular nucleus increase before the onset of hyperphagia in experimental diabetes. Brain Res 755:339–342

Sainsbury A, Cusin I, Rohner-Jeanrenaud F, Jeanrenaud B (1997) Adrenalectomy prevents the obesity syndrome produced by chronic central neuropeptide Y infusion in normal rats. Diabetes 46:209–214

Sainsbury A, Schwarzer C, Couzens M, Fetissov S, Furtinger S, Jenkins A, Cox HM, Sperk G, Hokfelt T, Herzog H (2002) Important role of hypothalamic Y2 receptors in body weight regulation revealed in conditional knockout mice. Proc Natl Acad Sci USA 99:8938–8943

Sainsbury A, Schwarzer C, Couzens M, Herzog H (2002) Y2 Receptor Deletion Attenuates the Type 2 Diabetic Syndrome of ob/ob Mice. Diabetes 51:3420–3427

Sainsbury A, Schwarzer C, Couzens M, Jenkins A, Oakes SR, Ormandy CJ, Herzog H (2002) Y4 receptor knockout rescues fertility in ob/ob mice. Genes Dev 16:1077–1088

Schaffhauser AO, Stricker-Krongrad A, Brunner L, Cumin F, Gerald C, Whitebread S, Criscione L, Hofbauer KG (1997). Inhibition of food intake by neuropeptide Y5 receptor antisense oligodeoxynucelotides. Diabetes 46:1792–1798.

Schaffhauser AO, Stricker-Krongrad A, Hofbauer KG (2000) NPY Y5 receptor subtype. Pharmacological characterization with antisense oligodeoxynucleotide screening strategy. Methods Mol Biol 153:129–150

Schaffhauser AO, Whitebread S, Haener R, Hofbauer KG, Stricker-Krongrad A (1998) Neuropeptide Y Y1 receptor antisense oligodeoxynucleotides enhance food intake in energy-deprived rats. Regul Pept 75/76:417–423

Schwartz GJ, Moran TH (2002) Leptin and neuropeptide y have opposing modulatory effects on nucleus of the solitary tract neurophysiological responses to gastric loads: implications for the control of food intake. Endocrinology 143:3779–3784

Schwartz MW, Baskin DG, Bukowski TR, Kuijper JL, Foster D, Lasser G, Prunkard DE, Porte Jr D, Woods SC, Seeley RJ, Weigle DS (1996) Specificity of leptin action on elevated blood glucose levels and hypothalamic neuropeptide Y gene expression in ob/ob mice. Diabetes 45:531–535

Sederholm F, Ammar AA, Sodersten P (2002) Intake inhibition by NPY: role of appetitive ingestive behavior and aversion. Physiol Behav 75:567–575

Seeley RJ, Benoit SC, Davidson TL (1995) Discriminative cues produced by NPY do not generalize to the interoceptive cues produced by food deprivation. Physiol Behav 58:1237–1241

Seeley RJ, Benoit SC, Davidson TL (1995) Discriminative cues produced by NPY do not generalize to the interoceptive cues produced by food deprivation. Physiol Behav 58:1237–1241

Seeley RJ, Payne CJ, Woods SC (1995) Neuropeptide Y fails to increase intraoral intake in rats. Am J Physiol 268:R423–R427

Shibasaki T, Oda T, Imaki T, Ling N, Demura H (1993) Injection of anti-neuropeptide Y y-globulin into the hypothalamic paraventricular nucleus decreases food intake in rats. Brain Res 601:313–316

Shintani M, Ogawa Y, Ebihara K, Aizawa-Abe M, Miyanaga F, Takaya K, Hayashi T, Inoue G, Hosoda K, Kojima M, Kangawa K, Nakao K (2001) Ghrelin, an endogenous growth hormone secretagogue, is a novel orexigenic peptide that antagonizes leptin action through the activation of hypothalamic neuropeptide Y/Y1 receptor pathway. Diabetes 50:227–232

Sindelar DK, Mystkowski P, Marsh DJ, Palmiter RD, Schwartz MW (2002) Attenuation of diabetic hyperphagia in neuropeptide Y–deficient mice. Diabetes 51:778–783

Soll RM, Dinger MC, Lundell I, Larhammar D, Beck-Sickinger AG (2001) Novel analogues of neuropeptide Y with a preference for the Y1-receptor. Eur J Biochem 268:2828–2837

Stanley BG, Chin AS, Leibowitz SF (1985) Feeding and drinking elicited by central injection of neuropeptide Y: evidence for a hypothalamic site(s) of action. Brain Res Bull 14:521–524

Stanley BG, Leibowitz SF (1985) Neuropeptide Y injected in the paraventricular hypothalamus: a powerful stimulant of feeding behavior. Proc Natl Acad Sci USA 82:3940–3943

Steinman JL, Gunion MW, Morley JE (1994) Forebrain and hindbrain involvement of neuropeptide Y in ingestive behaviors of rats. Pharmacol Biochem Behav 47:207–214

Stricker-Krongrad A, Cumin F, Burlet C, Beck B (1998) Hypothalamic neuropeptide Y and plasma leptin after long-term high-fat feeding in the rat. Neurosci Lett 254:157–160

Sunter D, Hewson AK, Lynam S, Dickson SL (2001) Intracerebroventricular injection of neuropeptide FF, an opioid modulating neuropeptide, acutely reduces food intake and stimulates water intake in the rat. Neurosci Lett 313:145–148

Tang-Christensen M, Kristensen P, Stidsen CE, Brand CL, Larsen PJ (1998) Central administration of Y5 receptor antisense decreases spontaneous food intake and attenuates feeding in response to exogenous neuropeptide Y. J Endocrinol 159:307–312

Turnbull AV, Ellershaw L, Masters DJ, Birtles S, Boyer S, Carroll D, Clarkson P, Loxham SJ, McAulay P, Teague JL, Foote KM, Pease JE (2002) Block MH Selective antagonism of the NPY Y5 receptor does not have a major effect on feeding in rats. Diabetes 51:2441–2449

van Dijk G, Bottone AE, Strubbe JH, Steffens AB (1994) Hormonal and metabolic effects of paraventricular hypothalamic administration of neuropeptide Y during rest and feeding. Brain Res 660:96–103

Vettor R, Zarjevski N, Cusin I, Rohner-Jeanrenaud F, Jeanrenaud B (1994) Induction and reversibility of an obesity syndrome by intracerebroventricular neuropeptide Y administration to normal rats. Diabetologia 37:1202–1208

Vezzani A, Michalkiewicz M, Michalkiewicz T, Moneta D, Ravizza T, Richichi C, Aliprandi M, Mule F, Pirona L, Gobbi M, Schwarzer C, Sperk G (2002) Seizure susceptibility and epileptogenesis are decreased in transgenic rats overexpressing neuropeptide Y. Neuroscience 110:237–243

Waeber G, Thompson N, Waeber B, Brunner HR, Nicod P, Grouzmann E (1993) Neuropeptide Y expression and regulation in a differentiated rat insulin-secreting cell line. Endocinology 133:1061–1067

Wahlestedt C, Pich EM, Koob GF, Yee F, Heilig M (1993) Modulation of anxiety and neuropeptide Y-Y1 receptors by antisense oligodeoxynucleotides. Science 259:528–531

Wahlestedt C, Skagerberg G, Ekman R, Heilig M, Sundler F, Hakanson R (1987) Neuropeptide Y (NPY) in the area of the hypothalamic paraventricular nucleus activates the pituitary-adrenocortical axis in the rat. Brain Res 417:33–38

Walter MJ, Scherrer JF, Flood JF, Morley JE (1994) Effects of localized injections of neuropeptide Y antibody on motor activity and other behaviors. Peptides 15:607–613

Wang L, Saint-Pierre DH, Tache Y. (2002) Peripheral ghrelin selectively increases FOS expression in neuropeptide Y synethesizing neurons in mouse hypothalamic arcuate nucleus. Neuro Sci Lett 325:47–51

Widdowson PS (1997) Regionally-selective down-regulation of NPY receptor subtypes in the obese Zucker rat. Relationship to the Y5 'feeding' receptor. Brain Res 758:17–25

Widdowson PS, Henderson L, Pickavance L, Buckingham R, Tadayyon M, Arch JR, Williams G (1999) Hypothalamic NPY status during positive energy balance and the effects of the NPY antagonist, BW 1229U91, on the consumption of highly palatable energy-rich diet. Peptides 20:367–372

Widdowson PS, Upton R, Henderson L, Buckingham R, Wilson S, Williams G (1997) Reciprocal regional changes in brain NPY receptor density during dietary restriction and dietary-induced obesity in the rat. Brain Res 774:1–10

Wieland HA, Engel W, Eberlein W, Rudolf K, Doods HN (1998) Subtype selectivity of the novel nonpeptide neuropeptide Y Y1 receptor antagonist BIBO 3304 and its effect on feeding in rodents. Br J Pharmacol 125:549–555

Wilding JP, Gilbey SG, Lambert PD, Ghatei MA, Bloom SR (1993) Increases in neuropeptide Y content and gene expression in the hypothalamus of rats treated with dexamethasone are prevented by insulin. Neuroendocrinology 57:581–587

Wise A, Gearing K, Rees S. (2002) Target validation of G-protein coupled receptors. Drug Disc Today 7:235–246

Wisialowski T, Parker R, Preston E, Sainsbury A, Kraegen E, Herzog H, Cooney G (2000) Adrenalectomy reduces neuropeptide Y-induced insulin release and NPY receptor expression in the rat ventromedial hypothalamus. J Clin Invest 105:1253–1259

Woods SC, Figlewicz DP, Madden L, Porte D Jr, Sipols AJ, Seeley RJ (1998) NPY and food intake: discrepancies in the model. Regul Pept 75/76:403–408

Wyss P, Stricker-Krongrad A, Brunner L, Miller J, Crossthwaite A, Whitebread S, Criscione L (1998) The pharmacology of neuropeptide Y (NPY) receptor-mediated feeding in rats characterizes better Y5 than Y1, but not Y2 or Y4 subtypes. Regul Pept 75/76:363–371

Xin XG, Huang XF (1998) Down-regulated NPY receptor subtype-5 mRNA expression in genetically obese mouse brain. Neuroreport 19:737–741

Zakrzewska KE, Sainsbury A, Cusin I, Rouru J, Jeanrenaud B, Rohner-Jeanrenaud F (1999) Selective dependence of intracerebroventricular neuropeptide Y-elicited effects on central glucocorticoids. Endocrinology 140:3183–3187

Zarjevski N, Cusin I, Vettor R, Rohner-Jeanrenaud F, Jeanrenaud B (1994) Intracerebroventricular administration of neuropeptide Y to normal rats has divergent effects on glucose utilization by adipose tissue and skeletal muscle. Diabetes 43:764–769

Neuropeptide Y and Cardiovascular Function

M. J. Morris

Department of Pharmacology, The University of Melbourne, 3010 Victoria, Australia
e-mail: mjmorris@unimelb.edu.au

1	Introduction	328
1.1	NPY Is Released on Sympathetic Activation	328
2	Effects of NPY in the Blood Vessel	329
2.1	Direct Constrictor Effects of NPY	330
2.2	Vasodilator Effects of NPY	330
2.3	Potentiation of Constriction by NPY	330
2.4	NPY Effects on Veins Versus Arteries	332
3	Receptors Involved in the Vascular Responses to NPY	332
3.1	NPY Receptor Subtypes	332
3.2	NPY Receptor Distribution	333
4	Cardiovascular Effects of NPY In Vivo	334
4.1	Animal Studies	334
4.2	Effect of NPY Administration in Man	335
4.3	Animal Models of Hypertension	335
5	NPY in the Mammalian Heart	336
5.1	Distribution and Cardiac Effects of NPY	336
5.2	Distribution of Cardiac NPY Receptors	337
5.3	Mitogenic Effects of NPY in Vessels and Cardiomyocytes	338
5.4	NPY and Atherosclerosis	338
6	NPY in the CNS	339
6.1	Localization of NPY in the Brain	339
6.2	NPY Receptors in the Brain	340
6.3	Central Cardiovascular Effects of NPY	341
6.4	Central Interactions Between NPY and Noradrenaline	341
6.5	Central NPY Content and Responsiveness in Hypertensive Animals	342
7	NPY in Cardiovascular Disease	343
8	Conclusions and Future Directions	345
References		346

Abstract Neuropeptide Y (NPY) is coreleased with the classical transmitter noradrenaline in both the peripheral and central nervous systems, and has been shown to potentiate the pressor effects of noradrenaline. In mammals NPY is released into the circulation in response to intense sympathetic stimulation, and it maintains blood pressure in the face of hemorrhage. Despite the abundance of NPY in nerves supplying blood vessels and the heart, it has been difficult to discern the extent of involvement of this peptide transmitter in minute-to-minute blood pressure control. Vasoconstrictor actions of NPY across a number of vascular beds are likely linked to Y_1 receptor activation; the magnitude of these effects is specific to the particular bed under examination. There is some evidence that raised plasma NPY levels are associated with hypertension, and increased cardiac NPY release is a hallmark of heart failure. NPY is a constrictor of coronary vessels and has been implicated in the development of atherosclerosis and in angiogenesis; these effects of NPY warrant further investigation. Neurons containing NPY also participate in central cardiovascular control. While it is difficult to judge the precise role of central NPY in cardiovascular regulation, in the brain NPY exerts depressor effects and modulates baroreceptor function. NPY released in states of sympathoexitation such as heart failure and panic disorder may contribute to the cardiovascular consequences of these conditions.

Keywords Atherosclerosis · Nucleus tractus solitarii · Heart failure · Hypertension · Vasoconstriction

1
Introduction

Neuropeptide Y (NPY) is a 36-amino acid peptide that is colocalized and coreleased with noradrenaline from sympathetic nerves. NPY is a member of the pancreatic polypeptide family, which also comprises peptide YY (PYY) and pancreatic polypeptide (PP); the three peptides share structural similarities and activate multiple receptors. To date five G-protein coupled NPY receptors have been cloned (Michel et al. 1998); their characteristics are discussed in detail elsewhere in this volume. Cardiovascular effects of NPY have been widely investigated since the discovery of the neuropeptide 20 years ago (Waeber et al. 1988). This reflects the wide distribution of NPY not only in blood vessels, heart and kidney, but in brain regions involved in the central regulation of cardiovascular function (Everitt et al. 1984; Lundberg et al. 1983; Tatemoto et al. 1982; Walker et al. 1991). This chapter will focus on the cardiovascular actions of NPY in mammals, in both the periphery and the central nervous system , and examine the evidence for alterations of NPY function in cardiovascular disease states.

1.1
NPY Is Released on Sympathetic Activation

NPY is one of the most abundant peptides in peripheral sympathetic nerves innervating cardiovascular structures. Soon after its discovery in extracts of porcine brain (Tatemoto et al. 1982) evidence accumulated to its distribution in peripheral sympathetic nerves surrounding blood vessels (Ekblad et al. 1984; Lundberg et al. 1983). Several findings implicated NPY in noradrenergic sympathetic vasoconstriction. NPY mimicked the component of sympathetic vasoconstriction that was resistant to α-adrenoceptor blockade (Lundberg and Tatemoto 1982) and subsequently experimenters applied a diverse range of investigative tools to better understand the cardiovascular effects of this transmitter. The discovery of NPY followed an increasing awareness and understanding of the important physiological role of nonclassical transmitters such as the neuropeptides and ATP in both the peripheral and central nervous systems.

NPY is synthesized in the nerve cell, and transported to the nerve terminals. Overflow of NPY was described from a variety of vascular beds following sympathetic nerve stimulation. Corelease of NPY and noradrenaline was demonstrated during electrical stimulation of the guinea pig heart (Haass et al. 1989) and dog skeletal muscle (Kahan et al. 1988) with noradrenaline release favored on a molar basis by around 1,000-fold. This is in line with the storage of NPY in large dense cored vesicles. There is some evidence that NPY is preferentially released at high stimulation frequencies (Kahan et al. 1988; Kennedy et al. 1997; Pernow et al. 1989), although other work suggests that NPY can be released along with noradrenaline at lower frequencies (de Potter et al. 1995).

Support for the concept of NPY as a sympathetic cotransmitter in man developed quickly with the observation of increased plasma concentrations of NPY when sympathetic nerves were activated by exercise (Morris et al. 1986b; Pernow et al. 1986) or other sympathetic stimuli such as cigarette smoking (Rudehill et al. 1989) and cold pressor test (Morris et al. 1986b). Studies across a variety of species have shown that marked increases in plasma NPY concentrations occur after intense sympathetic activation, and these may contribute to the cardiovascular sequelae. Thus in the conscious rat, activation by hemorrhage (Morris et al. 1987a), handling or electrical shock (Castagne et al. 1987) and cold exposure (Zukowska-Grojec et al. 1991) led to increased plasma NPY. In dogs, yohimbine (0.5 mg/kg) led to a 30% increase in blood pressure, accompanied by large increases in noradrenaline and NPY (Poncet et al. 1992). On the other hand sino-aortic denervation caused smaller increases in noradrenaline, and no significant changes in NPY (Poncet et al. 1992). NPY is also a major adrenal peptide and release of NPY may contribute to the cardiovascular sequelae of pheochromocytoma (Allen et al. 1983a).

2
Effects of NPY in the Blood Vessel

It is now clear the sympathetic nerves release NPY in addition to noradrenaline and ATP. Within the vascular system, the relative contribution of NPY to sympathetically mediated responses appears to vary considerably with the pattern of nerve activation, as well as the particular vascular bed under examination (Gibbins and Morris 2000). NPY is widely distributed throughout sympathetic nerves innervating mammalian blood vessels, particularly arteries (Ekblad et al. 1984; Franco-Cereceda and Liska 1998). The functional significance of vascular NPY-ergic innervation is supported by the correlation between immunohistochemical localization of NPY in sympathetic axons and constrictor responses to exogenously applied NPY (Morris 1994).

2.1
Direct Constrictor Effects of NPY

Many early studies investigated the vasoconstrictor effects of NPY in isolated blood vessels. NPY was shown to invoke vasoconstriction resistant to adrenergic blockade and dependent on calcium entry (Edvinsson et al. 1984; Lundberg and Tatemoto 1982; Mabe et al. 1985; Pernow et al. 1987). The constriction was slow in onset and long lasting. Contractile responses to NPY were shown to be independent of the endothelium (Pernow and Lundberg 1988).

Studies in vitro demonstrated that NPY is able to constrict a range of human vessels including those in skeletal muscle (Pernow and Lundberg 1988), uterus (Fried and Samuelson 1991) and meningeal and cerebral arteries (Jansen et al. 1992). Direct, long-lasting vasoconstrictor effects of NPY were also demonstrated in vivo in the human forearm (Clarke et al. 1991).

Several authors have highlighted the fact that direct contractile effects of NPY are less apparent in vitro than in vivo (Ekblad et al. 1984; Grundemar and Hogestatt 1992).

2.2
Vasodilator Effects of NPY

Some vasodilator neurons also express NPY (Gibbins and Morris 2000). Application of NPY, [Leu^{31},Pro^{34}]NPY or NPY(13–36) into the lumen of rat cerebral arteries led to dilation that was not affected by Y_1 receptor blockade, but abolished by inhibition of nitric oxide synthase (You et al. 2001). In this vessel NPY had constrictor effects mediated by Y_1 receptors on vascular smooth muscle, and endothelium-derived dilator effects (You et al. 2001). This suggests that NPY can dilate arteries in an endothelium-dependent process that may be highly regionally selective. The functional significance of this dilatory effect of NPY remains to be established, but increased cerebral blood flow was observed following NPY injection into the carotid artery of the cat (Kobari et al. 1993). Vasodilatory actions of NPY were recently described in small human cutaneous vessels (Nilsson et al. 2000).

2.3
Potentiation of Constriction by NPY

In addition to direct vasoconstrictor responses, NPY can also potentiate the contractile response to other pressor agents and to nerve stimulation. The potentiating effect appears particularly relevant in vessels where NPY does not exert direct contractile effects, including the rat femoral artery (Grundemar and Hogestatt 1992) and rabbit femoral and ear artery (Edvinsson et al. 1984; Saville et al. 1990). The enhancement of noradrenaline-evoked constriction of rat mesenteric artery by NPY was shown to be more prominent at low concentrations of noradrenaline (Fallgren et al.1993). The potentiation response to NPY persisted when calcium entry was attenuated (Chen et al. 1996; Fallgren et al. 1993), but chelation of intracellular calcium abolished NPY potentiation of noradrenaline (Fallgren et al. 1993) leading these authors to suggest that NPY initiates an intracellular calcium-sensitive mechanism that increases α-adrenoceptor sensitivity. Others working on the same vessels reported that NPY enhancement of noradrenaline-induced contraction involved a rise in intracellular calcium through an influx of calcium from an extracellular source (Andriantsitohaina et al. 1993).

Conflicting data have emerged regarding the dependence of an intact endothelium for the potentiating effects of NPY on the vasoconstriction elicited by nerve stimulation. For instance, in the rabbit ear artery endothelium-dependent (Daly and Hieble 1987) and -independent (Budai et al. 1989) effects were reported. Subsequent work in the rat mesenteric (Andriantsitohaina et al. 1991; Gustafsson and Nilsson 1990) and tail (Small et al. 1992) arteries reported no requirement for an intact endothelium, but supported the involvement of voltage-operated calcium channels in the potentiation, at least in the tail artery (Small et al. 1992). There is evidence that NPY can potentiate purinergic as well as adrenergic responses in the rabbit ear artery (Saville et al. 1990) and rat mesenteric artery (Westfall et al. 1996), thus this may contribute to the potentiation of nerve stimulation seen in vivo. In some preparations, such as the guinea pig saphenous artery, NPY seems to preferentially potentiate the purinergic component of the neural response (Cheung 1991).

In addition to the demonstrated potentiating effects of NPY in vitro, various groups have examined this question in vivo (Wahlestedt et al. 1990b). Early work demonstrated that NPY enhanced the blood pressure responses induced by α-adrenoceptor activation in pithed rats (Dahlof et al. 1985); this effect was extended to angiotensin, but not vasopressin, in work carried out in conscious rats (Aubert et al. 1988; Lopez et al. 1989). The NPY potentiation of the pressor actions of catecholamines, unlike its direct pressor actions, was nifedipine-independent (Lopez et al. 1989), leading to the suggestion that different mechanisms are involved. Subconstrictor amounts of NPY were shown to potentiate phenylephrine-induced constriction of human hand veins (Linder et al. 1996) and led to significantly greater pressor responses to the α_1-adrenoceptor agonist (Schuerch et al. 1998).

In addition to enhancement of vasoconstrictor substances, there is evidence that NPY can modulate the actions of vasodilator substances. Thus NPY was shown to inhibit relaxation to a range of vasodilators, including acetylcholine, vasoactive intestinal polypeptide and substance P, in rabbit coronary arteries (Han and Abel 1987), guinea pig uterine (Fallgren et al. 1989) and coronary arteries (Gulbenkian et al. 1992), and in the rat femoral artery (Grundemar and Hogestatt 1992). In rat mesenteric arteries NPY inhibited the release of calcitonin gene related peptide evoked by nerve stimulation, leading to decreased neurogenic vasodilation in the presence of NPY (Kawasaki et al. 1991).

2.4
NPY Effects on Veins Versus Arteries

There is strong immunohistochemical evidence that the pattern of NPY expression in nerves supplying blood vessels varies both with the segment of vessel and the species under examination (see Gibbins and Morris 2000 for review). Several experiments point to a difference between arteries and veins in the ability of NPY to exert direct constrictor effects or to potentiate vasoconstrictor stimuli.

In the guinea pig, NPY failed to contract the thoracic aorta, or to potentiate noradrenaline-induced contraction, but in the vena cava it was a more potent contractile agent than noradrenaline (Morris 1991). Thus different neurotransmitters appear responsible for sympathetic responses of the aorta and vena cava, in line with the different functions of these major vessels. Thus release of NPY leading to contraction of the vena cava would possibly only occur under conditions of strong sympathetic activation. Differences in responses to NPY were also described in the guinea pig mesenteric artery and vein (Smyth et al. 2000), with NPY exerting greater contractile effects in the vein than in the artery. Other work in canine vessels showed greater stimulation-induced release of NPY in saphenous and portal veins compared to mesenteric and popliteal arteries, but no contractile effect was observed in any vessel up to 100 μmol/l NPY and potentiating effects were confined to arteries (Hunter et al. 1996). In the rat we have demonstrated threefold higher NPY content in equivalent lengths of abdominal vena cava versus abdominal aorta (data not shown). There appear to be major variations between species and vessel type, with regard to the role of NPY in veins. The size of the vessel examined may be important as NPY preferentially constricted small arterioles (Ekelund and Erlinge 1997). It is likely that species differences in the pattern of NPY innervation of veins versus arteries also contribute to the different physiological effects of NPY in the two types of vessels.

3
Receptors Involved in the Vascular Responses to NPY

3.1
NPY Receptor Subtypes

The existence of multiple NPY receptor subtypes was proposed by Wahlestedt and colleagues in 1986 (Wahlestedt et al. 1986). At this time three NPY actions at sympathetic neuroeffector junctions were proposed: a direct postjunctional response (e.g., vasoconstriction); postjunctional potentiation of noradrenaline-induced constriction; and a postjunctional suppression of noradrenaline release (Wahlestedt et al. 1986). Long C-terminal amidated fragments of NPY were able to inhibit transmitter release, a prejunctional effect, while postjunctional effects required the whole NPY or PYY molecule. Thus prejunctional Y_2 and postjunctional Y_1 receptor subtypes were born (Wahlestedt et al. 1990a). While other receptor subtypes were subsequently characterized (Michel et al. 1998), Y_1 and Y_2 receptors remain the most relevant to the cardiovascular actions of NPY.

Much of the early experimental work investigating the receptor subtype specificity of cardiovascular effects of NPY relied on testing responses to selective agonists. In many beds Y_1 receptor activation was shown to exert vasoconstrictor effects (see Franco-Cereceda and Liska 1998; Malmstrom 2002; see Wahlestedt and Reis 1993 for a review). For instance, Y_1 selective agonists such as [Leu^{31}Pro^{34}]NPY preferentially contracted rat mesenteric (Chen et al. 1996; Evequoz et al. 1994), guinea pig uterine artery and thoracic vena cava (Morris and Sabesan 1994). In the skeletal muscle vascular bed of the pig, NPY(13–36) was tenfold less potent than [Leu^{31}Pro^{34}]NPY as a constrictor agent (Modin et al. 1993). Nonetheless it is clear that in some vessels Y_2 receptors can mediate vasoconstriction (Kotecha 1998) in addition to regulating transmitter release through a presynaptic action.

More recently, the development of subtype selective NPY receptor antagonists allowed confirmation of the relative contributions of different receptor subtypes to the cardiovascular actions of NPY. Thus several groups have examined the effect of Y_1 antagonists such as BIBP 3226 on vasoconstrictor effects of NPY. These experiments have revealed that in human cerebral arteries (Abounader et al. 1995; Nilsson et al. 1996), cat skeletal muscle arteries (Ekelund and Erlinge 1997) Y_1 receptors rather than Y_2 receptors mediate the vasoconstrictor effects of NPY. Studies using BIBP 3226 and SR 120107A revealed Y_1 mediated vasoconstrictor actions of NPY in pig kidney, spleen and hindlimb, as well as dog spleen and guinea pig vena cava (see Malmstrom 1997) and mesenteric artery and vein (Donoso et al. 1997; Doods et al. 1995; Smyth et al. 2000). Experiments using BIBP 3226 and 1229U91 conducted in whole animals also supported a major involvement of the Y_1 receptor in the pressor response seen after electrical stimulation of sympathetic nerves (Kennedy et al. 1997). BIBP 3226 also attenuated vasoconstriction induced by cold exposure in the rat, suggesting that NPY activation of Y_1 receptors may be important in the regulation of vascular tone during stress (Zukowska-Grojec et al. 1996).

3.2
NPY Receptor Distribution

For much of the past two decades the distribution of NPY receptors has been inferred from pharmacological studies investigating pressor responses using NPY and NPY receptor agonists. Another important experimental approach has been the mapping of NPY receptors on blood vessels of animals and man. Limited binding studies have been conducted, possibly due to high nonspecific binding. [^{3}H]propionyl-NPY bound to rabbit aortic membranes (Chang et al. 1988). ^{125}I-labeled NPY binding to pig aorta membranes was more readily displaced by NPY and PYY than NPY(13–36) or PP, demonstrating Y_1-like behavior (Shigeri et al. 1991).

More recently, autoradiography, immunohistochemistry and reverse transcriptase polymerase chain reaction (RT–PCR) techniques have been applied to improve our understanding of the distribution of NPY receptor subtypes within the cardiovascular system. Studies in humans detected Y_1 and Y_2 receptor mRNA in cerebral, meningeal and coronary arteries (Uddman et al. 2002). Immunohistochemical data suggested a higher density of NPY Y_1 receptors in smaller arteries (Uddman et al. 2002) in keeping with the reported constrictor effects in smaller vessels (Owen 1993).

The Y_4 receptor displays a pharmacological profile distinct from the other NPY receptors, with high affinity for PP; however, NPY can bind with low affinity. Not surprisingly, the role of Y_4 receptors in cardiovascular actions of NPY has received limited attention although Y_4 receptors were reported on rat arterial smooth muscle (Barrios et al. 1999). Anesthetized Y_4 receptor knockout mice were shown to have markedly lower blood pressure and reduced pressor response to NPY (Smith-White et al. 2002a); however, it is difficult to comment on the mechanism(s) underlying these changes, which may relate to central or peripheral effects.

4
Cardiovascular Effects of NPY In Vivo

4.1
Animal Studies

The cardiovascular effects of NPY have been investigated in a number of mammalian species. NPY exerts pressor effects in vivo, and potentiates the vasoconstriction to both nerve stimulation and adrenergic agents (Dahlof et al. 1985). Increases in blood pressure have been described in a range of species (Doods et al. 1996; Grundemar et al. 1992; Lopez et al. 1989; Minson et al. 1989; Warner and Levy 1989). As discussed in detail elsewhere in this volume, NPY has also been shown to exert constrictor effects in the kidney of several species, including rats and rabbits (Bischoff and Michel 1998; Gardiner et al. 1988; Minson et al, 1990b), and this may impact on cardiovascular control.

Increases in blood pressure induced by NPY were shown to be Y_1 receptor dependent due to the potency of Y_1 selective agonists (Grundemar et al. 1992; Potter and McCloskey 1992) and pressor responses to NPY and [Leu^{31}Pro^{34}]NPY were effectively inhibited by the Y_1 receptor antagonists BIBP 3226 (Doods et al. 1996) and SR 120819A (Serradeil-Le Gal et al. 1995) respectively. Further support for the importance of Y_1 receptors in mediating the pressor effects of NPY came with the observation of a lack of blood pressure response in mice lacking the Y_1 receptor (Pedrazzini et al. 1998). Administration of Y_1 receptor antagonists BIBP 3226 and BIBO 3304 was shown to attenuate the increase in blood pressure in anesthetized cats following carotid occlusion, suggesting an involvement of NPY in the baroreflex (Capurro and Huidobro-Toro 1999).

4.2
Effect of NPY Administration in Man

There has been considerable interest in determining whether NPY exerts cardiovascular effects in humans. Intravenous infusion of NPY led to reductions in renal and splanchnic blood flows in humans (Ahlborg et al. 1992). A dose-related increase in blood pressure (from 94 to 104 mmHg) accompanied intravenous injection of NPY (90–900 mol/kg) in human volunteers, whereas no significant increase was seen on continuous infusion of NPY; however, this led to smaller increments in plasma NPY (Ullman et al. 2002). Atrioventricular node conduction was also prolonged after an NPY bolus (Ullman et al. 2002).

4.3
Animal Models of Hypertension

Considerable attention has been directed at determining whether the distribution and density of NPY-containing nerves surrounding blood vessels may differ in animal models of hypertension. Immunohistochemical studies have reported increased density of NPY and noradrenergic fibers in the cerebral (Dhital et al. 1988) and mesenteric (Kawamura et al. 1989; Lee et al. 1988) artery of the spontaneously hypertensive rat. We have observed increased NPY content in the mesenteric and femoral artery, jugular vein and vena cava of spontaneously hypertensive rats compared to Wistar Kyoto control rats (M.J. Morris and J.M. Pavia, unpublished results). Exaggerated contractile responses to NPY have been demonstrated in the spontaneously hypertensive rat (Daly et al. 1988; Zukowska et al. 1993a) and infusion of NPY led to greater increases in blood pressure in this model relative to several control strains (Miller and Tessel 1991).

Changes in blood vessel NPY content may contribute to increased peripheral resistance in spontaneous hypertension, particularly during the development phase (Zukowska et al. 1993a). One study exploring the ability of NPY to affect noradrenergic transmission in the mesenteric artery of spontaneously hypertensive rats showed a greater potentiating effect of NPY on responses to nerve stimulation or exogenous noradrenaline (Westfall et al. 1990). Moreover the ability of NPY to decrease evoked release of noradrenaline was attenuated in the spontaneously hypertensive rat, suggesting that both prejunctional and postjunctional actions of NPY may be affected in this model (Westfall et al. 1990).

In addition to genetic hypertension, the role of NPY has been explored in other models of hypertension. No change in vascular NPY content was observed in one-kidney or two-kidney one-clip hypertension in the rat (Allen et al. 1986). Later studies demonstrated depletion of NPY in mesenteric and renal arteries after hypertension was induced by aortic coarctation (Ballesta et al. 1987). Increased plasma NPY was observed in rabbits with renal hypertension (Minson et al. 1990b).

5
NPY in the Mammalian Heart

5.1
Distribution and Cardiac Effects of NPY

Soon after its discovery NPY was identified as an important cardiac peptide (Gu et al. 1983). Radioimmunoassay studies demonstrated that NPY is present in high concentrations in the heart of several species, including humans (Gu et al. 1984; Onuoha et al. 1999), with high levels in the atria. NPY-containing sympathetic nerve fibers innervate coronary vessels and cardiomyocytes of rodents (Allen et al. 1986; Gu et al. 1984; Nyquist-Battie et al. 1994). NPY is found in the sinoatrial and atrioventricular nodes and atrioventricular bundles (Gu et al. 1984).

Sympathectomy led to a depletion of cardiac NPY in the rat (Corr et al. 1990; Lundberg et al. 1985; Maccarrone and Jarrott 1987; Morris et al. 1986a), confirming the presence of NPY in sympathetic nerves. However, NPY was not completely depleted by sympathetic denervation (Maccarrone and Jarrott 1987) in keeping with the presence of NPY in intrinsic cardiac nerves (Hassall and Burnstock 1987).

Electrical stimulation of sympathetic nerves induces release of NPY along with noradrenaline from the heart in the guinea pig (Haass et al. 1989) and dog (Warner et al. 1991). NPY was initially reported to exert negative inotropic and chronotropic effects, possibly linked to constriction (Allen et al. 1983b). In whole animal studies it is difficult to separate direct cardiac actions of the peptide from reflex responses to its pressor actions (Minson et al. 1989). In conscious dogs with atrioventricular block, a bradycardic response to NPY infusion was linked to reflex withdrawal of β-adrenergic tone (Boucher et al. 1994). A ventricular tachycardic effect was attributed to an action of NPY on receptors in the His bundle (Boucher et al. 1994). One study in anesthetized dogs reported that intravenous NPY administration induced hypertension due to peripheral vasoconstriction, with weak positive inotropic effects, and no effect on heart rate (Hashim et al. 1997). Overall NPY appears to exert little direct influence on heart rate.

There has been considerable interest in characterizing the effects on NPY in coronary vessels. In vitro studies in the coronary artery showed direct contractile effects of NPY, potentiation of noradrenaline-induced constriction in the rat (Prieto et al. 1991), and attenuation of noradrenaline-induced relaxation under some conditions in the rabbit (Corr et al. 1993; Han and Abel 1987). NPY-containing fibers also innervate coronary veins, and weak constrictor effects of NPY were reported (Gulbenkian et al. 1994).

Direct contractile effects of NPY on coronary arteries, resistant to adrenoceptor blockade, have been reported in isolated heart preparations as well as in vivo preparations from a number of species, including guinea pig (Franco-Cereceda et al. 1985; Rioux et al. 1986), pig (Rudehill et al. 1986) and dog (Aizawa et al. 1985; Ertl et al. 1993; Komaru et al. 1990; Martin et al. 1992; Maturi et al. 1989; Stack and Patterson 1991). Intracoronary injection of NPY in anesthetized dogs led to reductions in coronary blood flow, heart rate, and increased coronary vascular resistance (Ertl et al. 1993; Martin et al. 1992; Tanaka et al. 1997) that were resistant to α-adrenergic blockade (Stack and Patterson 1991) and appeared to involve thromboxane (Martin et al. 1992). During nerve stimulation in the presence of α-adrenoceptor blockade, NPY levels correlated with coronary vascular resistance; this relationship was abolished by treatment with BIBP 3226 (Tanaka et al. 1997). It is difficult to extrapolate from these animal studies into man; however, NPY appears likely to act as a constrictor of coronary vessels. The ability of NPY to induce coronary ischemia may depend on the degree of sympathetic activation and the activity of other constrictor agents, as well as the condition of the blood vessels.

Work in humans where NPY was injected into the coronary artery showed evidence of coronary vasospasm. Intracoronary injection of NPY into patients with angina pectoris induced myocardial ischemia (Clarke et al. 1987). Thus changes in local NPY release may have important cardiovascular consequences.

Another important cardiac action of NPY relates to the vagal inhibition following sympathetic stimulation first described in the dog by Potter (Potter 1985) and subsequently demonstrated in the rat (Potter et al. 1989). Neurally released and exogenous NPY attenuated the parasympathetic effects on atrial myocardium and the nodes (Warner and Levy 1989). Subsequent experiments in Y_2 receptor knockout mice showed that NPY modulation of cardiac parasympathetic neurotransmission was dependent on Y_2 receptors (Smith-White et al. 2002b). Work in humans showed that a prolonged period of exercise, known to lead to NPY release from the heart (Morris et al. 1997a), attenuated the baroreflex slowing of the heart (Ulman et al. 1997).

5.2
Distribution of Cardiac NPY Receptors

Binding studies in rat ventricular membranes using ^{125}I-labeled-NPY showed that NPY(13–36) could compete for binding, suggesting the presence of Y_2 receptors in addition to Y_1 (Balasubramaniam et al. 1990). [Leu^{31}Pro^{34}]NPY and NPY(13–36) were shown to exert opposing effects on contraction of rat cardiomyocytes, suggesting the presence of postjunctional Y_1 and Y_2 receptors (McDermott et al. 1997). Products corresponding to Y_1 and Y_2 mRNA were identified by RT–PCR in human myocardium and coronary arteries (Uddman et al. 2002). A recent immunohistochemical study has described the distribution of Y_1 and Y_2 receptors in the human heart (Jonsson-Rylander et al. 2003). Immunoreactivity against both Y_1 and Y_2 receptors was evident in atrial and ventricular tissue and nerve fibers. Overall, vessels were enriched with Y_2 receptors relative to Y_1 receptors (Jonsson-Rylander et al. 2003). Interestingly, a greater proportion of subendocardial vessels contained Y_1 receptors than those located in the subepicardium.

Changes in cardiac NPY release, content and receptors have been observed in patients with heart disease; these are discussed in Sect. 7.

5.3
Mitogenic Effects of NPY in Vessels and Cardiomyocytes

In addition to its effects on vascular tone, NPY exerts angiogenic effects at concentrations below those required for vasoconstriction (Zukowska-Grojec et al. 1998). These effects were first shown in rat vascular smooth muscle cells derived from aorta and vena cava, stimulating ^{3}H-thymidine incorporation (Zukowska-Grojec et al. 1993b) and DNA synthesis (Shigeri and Fujimoto 1993). Although originally ascribed to a Y_1 receptor, subsequent work in human umbilical vein endothelial cells showed effects on capillary tube formation that were not inhibited by the Y_1 antagonist BIBP 3226 (Zukowska-Grojec et al. 1998). The activity of the Y_1 agonist [Leu^{31}Pro^{34}]NPY in this assay was ascribed to its conversion to NPY(3–36) by dipeptidyl dipeptidase (Zukowska-Grojec et al. 1998). NPY and [Pro^{34}]NPY stimulated proliferation of human vascular smooth muscle cells in culture and NPY potentiated the proliferative action of noradrenaline (Erlinge et al. 1994), while effects of ATP were additive. Support for the involvement of Y_2 receptors in this response has come from two separate, recent studies in Y_2 receptor knockout mice where NPY-induced angiogenesis was impaired (Ekstrand et al. 2003; Lee et al. 2003). Based on experiments using antagonists these authors suggest that Y_5, and to a lesser extent Y_1 receptors may contribute to NPY-mediated angiogenesis (Lee et al. 2003). In support of this, Y_2 receptors were found to be widely expressed on endothelial cells of newly formed blood vessels (Ekstrand et al. 2003).

NPY stimulates protein synthesis in rat cardiomyocytes (Millar et al. 1994), and the initiation of hypertrophy of cardiomyocyte cultures was reported to involve Y_1 and Y_2 receptor mediated mechanisms (Nicholl et al. 2002). In cultured mice cardiomyocytes, NPY potentiated phenylephrine-induced activation of mitogen-activated protein kinase (MAPK). A Y_5 receptor was invoked in this response as the cells expressed primarily this receptor subtype, and NPY failed to potentiate MAPK phosphorylation in cardiomyocytes from Y_5 receptor knockout mice (Pellieux et al. 2000).

5.4
NPY and Atherosclerosis

In addition to mitogenic effects, NPY has recently been implicated in atherosclerosis. A leucine–proline substitution at position 7 in the signal part of the NPY gene correlated with elevated total and low density lipoprotein cholesterol and medial thickening of the carotid artery, particularly in obese subjects (Karvonen et al. 2001; Niskanen et al. 2000), although others reported a more limited association (Erkkila et al. 2002). This proline 7 substitution may accelerate atherosclerotic progression, and this finding suggests that the role of NPY in human atherogenesis warrants further attention.

In this regard, the potentiating effect of NPY on neurally mediated contractions of the mesenteric artery was greater in 12-month-old Watanabe heritable hyperlipemic rabbits, suggesting that atherosclerosis may be associated with modifications in the response to NPY (Stewart-Lee et al. 1992). Another study from the same group reported increased direct vasoconstrictor effects of NPY in coronary vessels from hyperlipidemic rabbits compared with normal rabbits (Corr et al. 1993).

6
NPY in the CNS

6.1
Localization of NPY in the Brain

NPY is abundant and widely distributed in the mammalian brain (Allen et al. 1983c), particularly in the paraventricular nucleus and arcuate nucleus of the hypothalamus (Everitt et al. 1984; Chronwall et al. 1985). The distribution of NPY appears similar in rodent and human.

Within the brain NPY is present in cell groups involved in blood pressure regulation including brainstem catecholamine cells (Chronwall et al. 1985; Hokfelt et al. 1983; Potter et al. 1988). The A2 (noradrenaline) and C2 (adrenaline) cell groups lie within the nucleus of the tractus solitarii (NTS), the site of termination of baroreceptor afferent neurons. Baroreceptor information passes from the NTS to the (depressor) caudal ventrolateral medulla A1 area. From here, inhibitory signals project to the rostral ventrolateral medulla, a region that mediates pressor actions via direct bulbospinal projections (Chalmers and Pilowsky 1991). NPY is extensively co-localized within brainstem adrenergic neurons that project to the paraventricular nucleus of the hypothalamus (Sawchenko et al. 1985), with partial overlap with noradrenaline containing cell groups such as A1 and A2.

As well as being a cotransmitter in ascending catecholaminergic projections relaying information to the hypothalamus, NPY-positive neurons in the rostral ventrolateral medulla project to the intermediolateral cell column of the spinal cord, and thus are implicated in the bulbospinal control of cardiovascular function (Blessing et al. 1987; Minson et al. 1990a; Tseng et al. 1993). The observation that increased NPY overflow into spinal cord perfusate accompanies the pressor response to both C1 stimulation (Morris et al. 1987b) and inhibition of vasodepressor neurons in the caudal medulla (Pilowsky et al. 1987) in the anesthetized rabbit, underlines the physiological significance of spinally projecting

NPY neurons. NPY positive cells have also been described in the rostral ventrolateral medulla of the human brain (Halliday et al. 1988).

Results of studies using a number of other approaches highlight the possible involvement of medullary NPY in tonic blood pressure control. Hemorrhage led to increased NPY mRNA expression in medullary catecholamine cell groups, particularly C1 and C2 (Chan and Sawchenko 1998). These changes correlated with the extent of hemorrhage, supporting the involvement of NPY in the autonomic responses to perturbations in blood pressure (or volume). In support of this, another study showed that hypotension led to increased activation (assessed by c-fos) of NPY-containing neurons in the NTS (McLean et al. 1999). Co-arctation induced hypertension was accompanied by increased NPY immunoreactivity in the NTS and ventral medulla, and altered affinity of ^{125}I-labeled PYY binding (Ferrari et al. 2002).

6.2
NPY Receptors in the Brain

Early studies characterizing the brain distribution of NPY receptors relied on nonspecific ligands such as ^{125}I-labeled PYY and -NPY (Martel et al. 1990; Ohkubo et al. 1990; Walker and Miller 1998). Differential distributions of Y_1 and Y_2 receptors were reported (Dumont et al. 1990; Wahlestedt and Reis 1993; Widdowson et al. 1993). Membrane binding experiments revealed predominantly Y_1 receptor binding in the cerebral cortex and Y_2 binding in the hippocampus (Larsen et al. 1993). Autoradiographic distribution studies demonstrated that Y_1 receptors are located in the cerebral cortex, thalamus and brainstem nuclei, with lower levels in hypothalamic areas (Dumont et al. 1993; Larsen et al. 1993). Putative Y_2 binding sites were widespread (Dumont et al. 1993); these receptors have been located in the dorsal vagal complex, where a proportion appears to be on vagal afferent terminals innervating the NTS (McLean et al. 1996). The advent of receptor subtype-selective ligands allowed a re-evaluation of the regional distribution of receptors for NPY (Dumont et al. 1996; Gobbi et al. 1999).

As further receptors became cloned, mRNA expression studies examined the regional subtype localization (Parker and Herzog 1999). These techniques revealed some species differences in the expression and distribution of Y_1, Y_2, Y_4 and Y_5 receptors across rodents, guinea pig and primate brain (Dumont et al. 1998).

In keeping with receptor autoradiographic studies, Y_1 receptor mRNA was described in cortex, hippocampus, thalamus, hypothalamus and numerous brainstem nuclei (Parker and Herzog 1999). The brain distribution of the Y_1 receptor was recently mapped using a terminally directed Y_1 antibody (Migita et al. 2001). Y_2 receptor mRNA is highly expressed in the amygdala, hypothalamus, thalamus, hippocampus and brainstem (Gustafson et al. 1997). Y_4 binding sites and Y_4 receptor mRNA have been reported in neurons of the dorsal vagal nucleus and the area postrema (Larsen and Kristensen 1997; Parker and Herzog 1999; Trinh et al. 1996). In the human and rat brain, Y_5 receptor mRNA is found in hippocampal regions and throughout the hypothalamic nuclei, with little expression in brainstem areas (Jacques et al. 1998; Parker and Herzog 1999).

6.3
Central Cardiovascular Effects of NPY

The early discovery of NPY in brainstem cell groups implicated in cardiovascular regulation prompted interest in the possible involvement of central NPY in blood pressure regulation. In contrast to the pressor effect of peripherally administered NPY, administration of NPY into the lateral ventricle, cisterna magna or by the intrathecal route, was generally shown to lead to hypotension (Fuxe et al. 1983; Harfstrand 1986; Scott et al. 1989; Westfall et al. 1988). In conscious rabbits the reduction in blood pressure was associated with reduced renal sympathetic nerve activity (Mastumura et al. 2000). Others have reported an increase in blood pressure after intracerebroventricular administration to urethane anesthetized rats (Hu and Dunbar 1997; Vallejo and Carter 1986) that was associated with reduced conductance in the iliac and mesenteric vessels (Hu and Dunbar 1997).

Site-specific cardiovascular changes were observed when NPY was administered into brain parenchyma. Injection of NPY into the posterior hypothalamus evoked a pressor response in the rat (Martin et al. 1988). Several groups have investigated the effects of NPY in the NTS, which is the site of termination of baroreceptor afferent fibers and an area rich in NPY binding sites (Harfstrand et al. 1986). Microinjection of NPY into the NTS elicits a dose-dependent fall in blood pressure and a long lasting reduction in heart rate (Barraco et al. 1990; Carter et al. 1985; Grundemar et al. 1991a,b; Morris and Pavia 1997; Tseng et al 1989). There is evidence that NPY administered within the NTS may also influence baroreceptor reflex function (Grundemar et al. 1991b; Shih et al. 1992).

Few studies have investigated the subtype of brain NPY receptors responsible for the cardiovascular actions of exogenously applied NPY. One study suggested that Y_1 receptors mediate a depressor effect in the NTS that can be counteracted by a Y_2 receptor selective ligand, NPY(13–36) (Yang et al. 1993). Within the NTS, Y_1, Y_2 and Y_4 receptors have been reported (see above), and we have autoradiographic evidence of Y_5-like binding (A. Stafford, P. Sexton and M.J. Morris, unpublished results).

6.4
Central Interactions Between NPY and Noradrenaline

NPY and noradrenaline are colocalized in ascending medullary neurons (Chronwall et al. 1985) and these transmitters appear to be coreleased in the central nervous system (Hastings et al. 1998). Various groups have examined the possibility of NPY–noradrenaline interactions at the level of the central nervous system. Postnatal treatment with NPY antiserum led to lasting effects on hypothalamic noradrenaline content (Vallejo et al. 1987). More acute effects have been reported: for instance α_2-adrenoceptors were shown to regulate binding characteristics of NPY receptors in the NTS (Harfstrand et al. 1989). There is evidence of modulatory effects of NPY on noradrenaline that are mediated by Y_1 (Hastings et al. 1997) and Y_2 (Illes et al. 1992) receptors. Moreover we have evidence of a functional interaction between NPY and α_2-adrenoceptors, as the depressor actions of NPY were significantly attenuated by the α_2-adrenoceptor antagonist yohimbine (Morris et al. 1997b). Anatomical data support an involvement of the α_2-adrenoceptor in the depressor effects of NPY in the NTS as both Y_1 and Y_2 receptors are expressed in the brainstem (see above) and there is overlap between Y_1 receptor immunoreactivity in α_2-adrenoceptor-containing regions, including A2 and C2 cell groups (Yang et al. 1996). Recent ultrastructural evidence suggests that in the NTS Y_1 receptors are located presynaptically on NPY or catecholamine-containing terminals, as well as postsynaptically on NPY or catecholamine-containing neurons that most likely receive terminals of nerves containing amino acid transmitters (Glass et al. 2002). Thus Y_1 agonists may modulate a variety of transmitters in the NTS, thereby affecting blood pressure.

The central cardiovascular effects of NPY may be mediated by the adipocyte-derived hormone leptin. Leptin can inhibit release of NPY not only in the hypothalamus, but also in the medulla oblongata (Lee and Morris 1998), and pretreatment with leptin attenuated the depressor response to NPY (Matsumura et al. 2000).

6.5
Central NPY Content and Responsiveness in Hypertensive Animals

Several experimenters have turned to hypertensive models to explore the role of central NPY in cardiovascular regulation. A potential role of endogenous NPY in hypertension was suggested following observations of altered concentrations of brain NPY in the spontaneously hypertensive rat. Conflicting findings have been reported, with reductions (Maccarrone and Jarrott 1985) and no change (Pavia and Morris 1994) in NPY immunoreactivity in the brain of the spontaneously hypertensive rat. Other workers described a higher density of NPY binding sites in the area postrema of the spontaneously hypertensive rat (Nakajima et al. 1987), and increased Y_2 receptor binding in the NTS (Aguirre et al. 1995) although this was not borne out in autoradiography studies of the dorsal vagal complex (McLean et al. 1996). Altered responses to NPY injected into the NTS of the spontaneously hypertensive rat (Takesako et al. 1994), and an attenuation of the depressor response to intrathecal NPY were described in this model (Westfall et al. 1988). The physiological significance of these findings is difficult to assess.

An age-related decrease in NPY activity has been reported in medullary C1 and C2 cell groups of the rat (Fuxe et al. 1987), which may contribute to altered cardiovascular responsiveness with ageing. However, no difference was observed in the blood pressure lowering effect of NPY injected into the NTS of

3-month-old versus 17-month-old Sprague-Dawley rats (Morris and Pavia 1997). Interestingly, age-related decreases in hypothalamic NPY peptide content (Pavia and Morris 1994) and NPY immunohistochemical staining of the paraventricular nucleus (Fuxe et al. 1987) was also observed, possibly reflecting reduced input from medullary NPY-containing regions.

Another approach used to investigate the role of NPY in cardiovascular regulation is the development of the NPY transgenic rat that overexpresses NPY. Resting mean arterial pressure was not affected in this animal despite elevated calculated vascular resistance. On the other hand, noradrenaline-induced pressor responses were exaggerated, and the cardiovascular response to hemorrhage was reduced in transgenic animals (Michalkiewicz et al. 2001). Overall this finding suggests that endogenous NPY tone may be important in cardiovascular regulation under certain conditions.

7
NPY in Cardiovascular Disease

NPY exerts direct pressor effects, and potentiates effects of other constrictors, actions that are now known to be largely Y_1 receptor mediated. Moreover this peptide is released into the circulation on sympathetic activation, thus many laboratories have examined the involvement of NPY in hypertension (for a review see Michel and Rascher 1995). Conflicting data presented regarding resting plasma NPY concentrations in human hypertension probably reflect methodological and patient sampling differences between laboratories. Some workers have suggested that elevated plasma NPY (and noradrenaline) may reflect increased sympathetic nervous activity in hypertensive patients (Solt et al. 1990). Others reported increases in NPY in both male and female hypertensive patients, with no elevation in noradrenaline (Wocial et al. 1995). We observed no elevation in plasma NPY in patients with untreated moderate hypertension (Chalmers et al. 1989), and the NPY response to sympathetic activation was similar to that observed in normotensive subjects (Morris et al. 1986b). Takahashi and colleagues (1989) reported no difference in NPY levels in hypertensive versus normotensive patients undergoing renal hemodialysis. Higher plasma NPY concentrations were observed in patients with untreated hypertension (Erlinge et al. 1992) and in those admitted to an emergency ward due to severe hypertension (Edvinsson et al. 1991); no correlation was observed between systolic or diastolic blood pressure and NPY (or noradrenaline). Another group reported elevated plasma NPY in adolescents with primary hypertension; these were maintained when blood pressure was lowered by β-adrenoceptor blockade (Lettgen et al. 1994). When patients with hypertension related to endocrine causes were examined, increases in NPY were observed in those with pheochromocytoma, but not primary hyperaldosteronism or Cushing's syndrome (Tabarin et al. 1992). The hypertension induced by dexamethasone in healthy volunteers was not associated with increases in circulating NPY (Whitworth et al. 1994). On balance, small changes in plasma NPY appear to occur in hypertension, and these appear to vary with the degree of the elevation of blood pressure.

It is difficult to speculate on the possible functional significance of alterations in circulating NPY, and few studies have investigated vascular responsiveness to NPY in hypertension. It has been reported that there is diminished contractile response to NPY in veins but not in arteries from patients with essential hypertension (Lind et al. 1997). Given the vascular bed-specific effects of this peptide, altered release of NPY may be important in various pathophysiological conditions. An age-related enhancement of the vasoconstrictive effects of NPY has been demonstrated in the rat and rabbit in vitro (Corr et al. 1993; Glenn and Duckles 1994), and enhanced pressor responses to infusion of NPY occurred in aged rats (Miller and Tessel 1991). It is not known whether age-related changes in NPY responsiveness occur in humans.

Plasma NPY was shown to be increased in patients undergoing exercise tests for the investigation of chest pain (Morris et al. 1986c) and in patients with coronary artery disease. The increase in NPY correlated with the degree and duration of ST-segment depression after exercise, suggesting NPY may contribute to myocardial ischemia (Gullestad et al. 2000).

It is difficult to interpret the significance of systemic NPY levels, given that these reflect overflow from numerous vascular beds, of a neuropeptide which has a long half-life in humans (Pernow et al. 1987). One useful approach to examine the pathophysiological significance of NPY in human disease is to examine regional NPY overflow across individual organs. For instance in patients with cardiac failure, overflow of NPY from the coronary sinus at rest was markedly elevated (Kaye et al. 1994), achieving levels seen during strenuous bicycle exercise in healthy subjects (Morris et al. 1986b). This is in line with the degree of cardiac sympathetic stimulation in this disease. Interestingly, patients with cardiac denervation following transplant showed a relative extraction of NPY compared to control subjects (Morris et al. 1997a).

Several groups of experimenters have examined plasma NPY in patients with heart failure. Increased plasma levels of NPY at rest were reported (Derchi et al. 1993; Hulting et al. 1990; Kaye et al. 1994; Maisel et al. 1989) while in other studies plasma NPY was not elevated relative to healthy control subjects (Dubois-Rande et al. 1992; Masden et al. 1993). NPY was correlated with the degree of decompensation of heart failure (Hulting et al. 1990) and with a reduction in stroke volume index (Ullman et al. 1993). In patients with dilated cardiomyopathy, depletion of NPY and noradrenaline content was observed in left and right ventricles, probably linked to increased adrenergic drive (Anderson et al. 1992). Sustained cardiac release of NPY may lead to receptor changes and the expression of ventricular NPY Y_1 receptor mRNA, assessed by RT–PCR, was also dramatically reduced in human heart patients relative to donors (Gullestad et al. 1998).

The observation of enhanced overflow of NPY during exercise under hypoxic conditions (Kaijser et al. 1990) suggests that release of NPY may be influenced by ischemia. This was directly investigated in patients undergoing aortic occlusion during coronary artery surgery, where the outflow of NPY and noradrenaline were enhanced, in keeping with increased sympathetic nerve activity (Franco-Cereceda et al. 1990). Increased release of NPY into the coronary sinus was demonstrated during the sympathetic activation that accompanied panic attacks (Esler et al. 2003) and this may contribute to the increased risk of sudden death and myocardial infarction in sufferers of this disorder.

Under certain conditions, increased concentrations of NPY may be an important determinant of vascular tone. NPY is increased following hemorrhage and endotoxic shock in animals (Corder et al. 1990; Morris et al. 1987a) and was shown to prevent the blood pressure decrease induced by endotoxin injection into adrenalectomized rats (Evequoz et al. 1988). NPY treatment improved survival in endotoxic shock, in contrast to other pressor agents (Hauser et al. 1993).

8
Conclusions and Future Directions

NPY exerts cardiovascular effects in both the peripheral and central nervous systems. The advent of receptor specific antagonists of NPY has allowed the delineation of site-specific effects of NPY. The lack of pressor response to NPY in animals lacking the Y_1 receptor is strongly supportive of a major role of this receptor in the constrictor effects of NPY. Despite the clear evidence of the importance of NPY in mediating vascular effects in its own right, and in modulating the responses to other vasoconstrictor or vasodilatory agents, the precise involvement of NPY in basal vascular tone remains difficult to judge. Work in animals indicates there is little doubt that NPY is involved in the cardiovascular effects of a number of pathophysiological states such as hypertension and heart failure, as well as the responses to stress of various kinds. NPY is a major cardiac peptide and has been shown to induce hypertrophy of cardiomyocytes. Evidence gathered using a number of approaches supports an involvement of medullary NPY in blood pressure control. Cardiovascular effects are observed following NTS microinjection, and activation of NPY cells in this region follows hemorrhage.

Work in recent years has confirmed the presence of NPY receptors in human vascular tissue and in central neurons involved in blood pressure control. Plasma levels of NPY appear increased in some forms of hypertension, and in heart failure marked increases are observed. Further work is required to examine whether bed-specific effects of NPY, e.g., in the coronary artery, may contribute to coronary events in humans. This may be particularly relevant given the recently described link between NPY and atherosclerosis.

Acknowledgements. Work in the author's laboratory was supported by the National Health and Medical Research Council, National Heart Foundation of Australia and The University of Melbourne. The author acknowledges the contributions of longstanding collaborators to the work outlined in this chapter.

References

Abounader R, Villemure JG, Hamel E (1995) Characterization of neuropeptide Y (NPY) receptors in human cerebral arteries with selective agonists and the new Y1 antagonist BIBP 3226. Br J Pharmacol 116: 2245–2250

Aguirre JA, Hedlund PB, Narvaez JA, Bunnemann B, Ganten D, Fuxe K (1995) Increased vasopressor actions of intraventricular neuropeptide Y-(13–36) in spontaneously hypertensive versus normotensive Wistar-Kyoto rats. Possible relationship to increases in Y2 receptor binding in the nucleus tractus solitarius. Brain Res 684:159–164

Ahlborg G, Weitzberg E, Lundberg JM (1992) Splanchnic and renal vasoconstriction during neuropeptide Y infusion in healthy humans. Clin Physiol 12:145–153

Aizawa Y, Murata M, Hayashi M, Funazaki T, Ito S, Shibata A (1985) Vasoconstrictor effect of neuropeptide Y (NPY) on canine coronary artery. Jpn Circ J 49:584–588

Allen JM, Adrian TE, Polak JM, Bloom SR (1983a) Neuropeptide Y (NPY) in the adrenal gland. J Auton Nerv Syst 9(2–3):559–563

Allen JM, Bircham PM, Edwards AV, Tatemoto K, Bloom SR (1983b) Neuropeptide Y (NPY) reduces myocardial perfusion and inhibits the force of contraction of the isolated perfused rabbit heart. Regul Pept 6:247–253

Allen JM, Godfrey NP, Yeats JC, Bing RF, Bloom SR (1986) Neuropeptide Y in renovascular models of hypertension in the rat. Clin Sci 70:485–488

Allen YS, Adrian TE, Allen JM, Tatemoto K, Crow TJ, Bloom SR, Polak JM (1983c) Neuropeptide Y distribution in the rat brain. Science 221:877–879

Anderson FL, Port JD, Reid BB, Larrabee P, Hanson G, Bristow MR (1992) Myocardial catecholamine and neuropeptide Y depletion in failing ventricles of patients with idiopathic dilated cardiomyopathy. Correlation with beta-adrenergic receptor downregulation. Circulation 85:46–53

Andriantsitohaina R, Stoclet JC, Bukoski RD (1991) Role of endothelium on the effects of neuropeptide Y in mesenteric resistance arteries of spontaneously hypertensive and Wistar-Kyoto normotensive rats. J Pharmacol Exp Ther 257:276–281

Andriantsitohaina R, Bian K, Stoclet JC, Bukoski RD (1993) Neuropeptide Y increases force development through a mechanism that involves calcium entry in resistance arteries. J Vasc Res 30: 309–314

Aubert JF, Waeber B, Rossier B, Geering K, Nussberger J, Brunner HR (1988) Effects of neuropeptide Y on the blood pressure response to various vasoconstrictor agents. J Pharmacol Exp Ther 246:1088–1092

Balasubramaniam A, Sheriff S, Rigel DF, Fischer JE (1990) Characterization of neuropeptide Y binding sites in rat cardiac ventricular membranes. Peptides 11:545–550

Ballesta J, Lawson JA, Pals DT, Ludens JH, Lee YC, Bloom SR, Polak JM (1987) Significant depletion of NPY in the innervation of the rat mesenteric, renal arteries and kidneys in experimentally (aorta coarctation) induced hypertension. Histochemistry 87:273–278

Barraco RA, Ergene E, Dunbar JC, el-Ridi MR (1990) Cardiorespiratory response patterns elicited by microinjections of neuropeptide Y in the nucleus tractus solitarius. Brain Res Bull 24: 465–485

Barrios VE, Sun J, Douglass J, Toombs CF (1999) Evidence of a specific pancreatic polypeptide receptor in rat arterial smooth muscle. Peptides 20:1107–1113

Bell D, Allen AR, Kelso EJ, Balasubramaniam A, McDermott BJ (2002) Induction of hypertrophic responsiveness of cardiomyocytes to neuropeptide Y in response to pressure overload. J Pharmacol Exp Ther 303:581–591

Bischoff A, Michel MC (1998) Renal effects of neuropeptide Y. Pflugers Arch. 435:443–453

Blessing WW, Oliver JR, Hodgson AH, Joh TH, Willoughby JO (1987) Neuropeptide Y-like immunoreactive C1 neurons in the rostral ventrolateral medulla of the rabbit

project to sympathetic preganglionic neurons in the spinal cord. J Auton Nerv Syst 18: 121-129
Boucher M, Chassaing C, Chapuy E, Lorente P (1994) Chronotropic cardiac effects of NPY in conscious dogs: interactions with the autonomic nervous system and putative NPY receptors. Regul Pept 54:409-415
Budai D, Vu HQ, Duckles SP (1989) Endothelium removal does not affect potentiation by neuropeptide Y in rabbit ear artery. Eur J Pharmacol 168:97-100
Capurro D, Huidobro-Toro JP (1999) The involvement of neuropeptide Y Y_1 receptors in the blood pressure baroreflex: studies with BIBP 3226 and BIBO 3304. Eur J Pharmacol 376:251-255
Carter DA, Vallejo M, Lightman SL (1985) Cardiovascular effects of neuropeptide Y in the nucleus tractus solitarius of rats: relationship with noradrenaline and vasopressin. Peptides 6:421-425
Castagne V, Corder R, Gaillard R, Mormede P (1987) Stress-induced changes of circulating neuropeptide Y in the rat: comparison with catecholamines. Regul Pept 19:55-63
Chalmers J, Pilowsky P (1991) Brainstem and bulbospinal neurotransmitter systems in the control of blood pressure. J Hypertens 9:675-694
Chalmers J, Morris M, Kapoor V, Cain M, Elliott J, Russell A, Pilowsky P, Minson J, West M, Wing L (1989) Neuropeptide Y in the sympathetic control of blood pressure in hypertensive subjects. Clin Exp Hypertens A 11(SuppL 1):59-66
Chan RK, Sawchenko PE (1998) Differential time- and dose-related effects of haemorrhage on tyrosine hydroxylase and neuropeptide Y mRNA expression in medullary catecholamine neurons. Eur J Neurosci 10:3747-3758
Chang RS, Lotti VJ, Chen TB (1988) Specific [^{3}H]propionyl-neuropeptide Y (NPY) binding in rabbit aortic membranes: comparisons with binding in rat brain and biological responses in rat vas deferens. Biochem Biophys Res Commun 151:1213-1219
Chen H, Fetscher C, Schafers RF, Wambach G, Philipp T, Michel MC (1996) Effects of noradrenaline and neuropeptide Y on rat mesenteric microvessel contraction. Naunyn Schmiedebergs Arch Pharmacol 353: 314-323
Cheung DW (1991) Neuropeptide Y potentiates specifically the purinergic component of the neural responses in the guinea pig saphenous artery. Circ Res 68:1401-1407
Chronwall BM, DiMaggio DA, Massari VJ, Pickel VM, Ruggiero DA, O'Donohue TL (1985) The anatomy of neuropeptide-Y-containing neurons in rat brain. Neuroscience 15:1159-1181
Clarke JG, Davies GJ, Kerwin R, Hackett D, Larkin S, Dawbarn D, Lee Y, Bloom SR, Yacoub M, Maseri A (1987) Coronary artery infusion of neuropeptide Y in patients with angina pectoris. Lancet 1:1057-1059
Clarke J, Benjamin N, Larkin S, Webb D, Maseri A, Davies G (1991) Interaction of neuropeptide Y and the sympathetic nervous system in vascular control in man. Circulation 83: 774-777
Corder R, Pralong FP, Gaillard R (1990) Comparison of hypotension-induced neuropeptide Y release in rats subjected to hemorrhage, endotoxemia, and infusions of vasodepressor agents. Ann NY Acad Sci 611:474-476
Corr LA, Aberdeen JA, Milner P, Lincoln J, Burnstock G (1990) Sympathetic and nonsympathetic neuropeptide Y-containing nerves in the rat myocardium and coronary arteries. Circ Res 66:1602-1609
Corr L, Burnstock G, Poole-Wilson P (1993) Effects of age and hyperlipidemia on rabbit coronary responses to neuropeptide Y and the interaction with norepinephrine. Peptides 14:359-364
Dahlof C, Dahlof P, Lundberg JM (1985) Neuropeptide Y (NPY): enhancement of blood pressure increase upon alpha-adrenoceptor activation and direct pressor effects in pithed rats. Eur J Pharmacol 109:289-292
Daly RN, Hieble JP (1987) Neuropeptide Y modulates adrenergic neurotransmission by an endothelium dependent mechanism. Eur J Pharmacol 138:445-446

Daly RN, Roberts MI, Ruffolo RR Jr, Hieble JP (1988) The role of neuropeptide Y in vascular sympathetic neurotransmission may be enhanced in hypertension. J Hypertens Suppl 6:S535–S538

De Potter WP, Kurzawa R, Miserez B, Coen EP (1995) Evidence against differential release of noradrenaline, neuropeptide Y, and dopamine-beta-hydroxylase from adrenergic nerves in the isolated perfused sheep spleen. Synapse 19:67–76

Derchi G, Dupuis J, de Champlain J, Rouleau JL (1993) Paradoxical decrease in circulating neuropeptide Y-like immunoreactivity during mild orthostatic stress in subjects with and without congestive heart failure. Eur Heart J 14:34–39

Dhital KK, Gerli R, Lincoln J, Milner P, Tanganelli P, Weber G, Fruschelli C, Burnstock G (1988) Increased density of perivascular nerves to the major cerebral vessels of the spontaneously hypertensive rat: differential changes in noradrenaline and neuropeptide Y during development. Brain Res 444:33–45

Donoso MV, Brown N, Carrasco C, Cortes V, Fournier A and Huidobro-Toro JP (1997) Stimulation of the sympathetic perimesenteric arterial nerves releases neuropeptide Y potentiating the vasomotor activity of noradrenaline: involvement of neuropeptide Y-Y_1 receptors. J Neurochem 69:1048–1059

Doods HN, Wienen W, Entzeroth M, Rudolf K, Eberlein W, Engel W, Wieland HA (1995) Pharmacological characterization of the selective nonpeptide neuropeptide Y Y_1 receptor antagonist BIBP 3226. J Pharmacol Exp Ther 275:136–142

Doods HN, Wieland HA, Engel W, Eberlein W, Willim KD, Entzeroth M, Wienen W, Rudolf K (1996) BIBP 3226, the first selective neuropeptide Y_1 receptor antagonist: a review of its pharmacological properties. Regul Pept 65:71–77

Dubois-Rande JL, Comoy E, Merlet P, Benvenuti C, Carville C, Hittinger L, Macquin-Mavier I, Bohuon C, Castaigne A (1992) Relationship among neuropeptide Y, catecholamines and haemodynamics in congestive heart failure. Eur Heart J 13:1233–1238

Dumont Y, Fournier A, St-Pierre S, Schwartz TW, Quirion R (1990) Differential distribution of neuropeptide Y_1 and Y_2 receptors in the rat brain. Eur J Pharmacol 191:501–503

Dumont Y, Fournier A, St-Pierre S, Quirion R (1993) Comparative characterization and autoradiographic distribution of neuropeptide Y receptor subtypes in the rat brain. J Neurosci 13:73–86

Dumont Y, Fournier A, St-Pierre S, Quirion R (1996) Autoradiographic distribution of [^{125}I]Leu^{31},Pro^{34}]PYY and [^{125}I]PYY3-36 binding sites in the rat brain evaluated with two newly developed Y_1 and Y_2 receptor radioligands. Synapse 22:139–158

Dumont Y, Jacques D, Bouchard P, Quirion R (1998) Species differences in the expression and distribution of the neuropeptide Y Y_1, Y_2, Y_4, and Y_5 receptors in rodents, guinea pig, and primates brains. J Comp Neurol 402:372–384

Edvinsson L, Ekblad E, Hakanson R, Wahlestedt C (1984) Neuropeptide Y potentiates the effect of various vasoconstrictor agents on rabbit blood vessels. Br J Pharmacol 83:519–525

Edvinsson L, Ekman R, Thulin T (1991) Increased plasma levels of neuropeptide Y-like immunoreactivity and catecholamines in severe hypertension remain after treatment to normotension in man. Regul Pept 32: 279–287

Ekblad E, Edvinsson L, Wahlestedt C, Uddman R, Hakanson R, Sundler F (1984) Neuropeptide Y co-exists and co-operates with noradrenaline in perivascular nerve fibres. Regul Pept 8:225–235

Ekelund U, Erlinge D (1997) In vivo receptor characterization of neuropeptide Y-induced effects in consecutive vascular sections of cat skeletal muscle. Br J Pharmacol 120:387–392

Ekstrand AJ, Cao R, Bjorndahl M, Nystrom S, Jonsson-Rylander AC, Hassani H, Hallberg B, Nordlander M, Cao Y (2003) Deletion of neuropeptide Y (NPY) 2 receptor in mice results in blockage of NPY-induced angiogenesis and delayed wound healing. Proc Natl Acad Sci USA 100:6033–6038

Erkkila AT, Lindi V, Lehto S, Laakso M, Uusitupa MI (2002) Association of leucine 7 to proline 7 polymorphism in the preproneuropeptide Y with serum lipids in patients with coronary heart disease. Mol Genet Metab. 75:260–264

Erlinge D, Ekman R, Thulin T, Edvinsson L (1992) Neuropeptide Y-like immunoreactivity and hypertension. J Hypertens 10:1221–1225

Erlinge D, Brunkwall J, Edvinsson L (1994) Neuropeptide Y stimulates proliferation of human vascular smooth muscle cells: cooperation with noradrenaline and ATP. Regul Pept 50:259–265

Ertl G, Bauer B, Becker HH, Rose G (1993) Effects of neurotensin and neuropeptide Y on coronary circulation and myocardial function in dogs. Am J Physiol 264:H1062–H1068

Esler M, Alvarenga M, Kaye D, Lambert G, Thompson J, Hastings J, Schwartz R, Morris M, Richards J (2003) Panic disorder. In: Primer on the autonomic nervous system (in press)

Evequoz D, Waeber B, Aubert JF, Fluckiger JP, Nussberger J, Brunner HR (1988) Neuropeptide Y prevents the blood pressure fall induced by endotoxin in conscious rats with adrenal medullectomy. Circ Res 62:25–30

Evequoz D, Grouzmann E, Beck-Sickinger AG, Brunner HR, Waeber B (1994) Differential vascular effects of neuropeptide Y(NPY) selective receptor agonists. Experientia 50:936–938

Everitt BJ, Hokfelt T, Terenius L, Tatemoto K, Mutt V, Goldstein M (1984) Differential coexistence of neuropeptide Y (NPY)-like immunoreactivity with catecholamines in the central nervous system of the rat. Neuroscience 11:443–462

Fallgren B, Ekblad E, Edvinsson L (1989) Co-existence of neuropeptides and differential inhibition of vasodilator responses by neuropeptide Y in guinea pig uterine arteries. Neurosci Lett 100:71–76

Fallgren B, Arlock P, Edvinsson L (1993) Neuropeptide Y potentiates noradrenaline-evoked vasoconstriction by an intracellular calcium-dependent mechanism. J Auton Nerv Syst 44:151–159

Ferrari MF, Almeida RS, Chadi G, Fior-Chadi DR (2002) Acute changes in ^{3}H-PAC and ^{125}I-PYY binding in the nucleus tractus solitarii and hypothalamus after a hypertensive stimulus. Clin Exp Hypertens 24:169–186

Franco-Cereceda A, Liska J (1998) Neuropeptide Y Y_1 receptors in vascular pharmacology. Eur J Pharmacol 349:1–14

Franco-Cereceda A, Lundberg JM, Dahlof C (1985) Neuropeptide Y and sympathetic control of heart contractility and coronary vascular tone. Acta Physiol Scand 124:361–369

Franco-Cereceda A, Owall A, Settergren G, Sollevi A and Lundberg JM, (1990) Release of neuropeptide Y and noradrenaline from the human heart after aortic occlusion during coronary artery surgery. Cardiovasc Res 24:242–246

Fried G, Samuelson U (1991) Endothelin and neuropeptide Y are vasoconstrictors in human uterine blood vessels. Am J Obstet Gynecol 164:1330–1336

Fuxe K, Agnati LF, Harfstrand A, Zini I, Tatemoto K, Pich EM, Hokfelt T, Mutt V, Terenius L (1983) Central administration of neuropeptide Y induces hypotension bradypnea and EEG synchronization in the rat. Acta Physiol Scand 118:189–192

Fuxe K, Agnati LF, Kitayama I, Zoli M, Janson AM, Harfstrand A, Vincent M, Kalia M, Goldstein M, Sassard J (1987) Evidence for discrete alterations in central cardiovascular catecholamine and neuropeptide Y immunoreactive neurons in aged male rats and in genetically hypertensive male rats of the Lyon strain. Eur Heart J 8(Suppl B):139–145

Gardiner SM, Bennett T, Compton AM (1988) Regional haemodynamic effects of neuropeptide Y, vasopressin and angiotensin II in conscious, unrestrained, Long Evans and Brattleboro rats. J Auton Nerv Syst 24:15–27

Gibbins IL, Morris JL (2000) Pathway specific expression of neuropeptides and autonomic control of the vasculature. Regul Pept 93:93–107

Glass MJ, Chan J, Pickel VM (2002) Ultrastructural localization of neuropeptide Y Y_1 receptors in the rat medial nucleus tractus solitarius: relationships with neuropeptide Y or catecholamine neurons. J Neurosci Res 67:753–765

Glenn TC, Duckles SP (1994) Vascular responses to neuropeptide Y in the rat: Effect of age. Aging Clin Exp Res 6:277–286

Gobbi M, Mennini T, Vezzani A (1999) Autoradiographic reevaluation of the binding properties of ^{125}I-[Leu^{31},Pro^{34}]peptide YY and ^{125}I-peptide YY$_{3-36}$ to neuropeptide Y receptor subtypes in rat forebrain. J Neurochem 72:1663–1670

Grundemar L, Hogestatt ED (1992) Unmasking the vasoconstrictor response to neuropeptide Y and its interaction with vasodilating agents in vitro. Eur J Pharmacol 221:71–76

Grundemar L, Wahlestedt C, Reis DJ (1991a) Neuropeptide Y acts at an atypical receptor to evoke cardiovascular depression and to inhibit glutamate responsiveness in the brainstem. J Pharmacol Exp Ther 258:633–638

Grundemar L, Wahlestedt C, Reis DJ. (1991b) Long-lasting inhibition of the cardiovascular responses to glutamate and the baroreceptor reflex elicited by neuropeptide Y injected into the nucleus tractus solitarius of the rat. Neurosci Lett 122:135–139

Grundemar L, Jonas SE, Morner N, Hogestatt ED, Wahlestedt C, Hakanson R (1992) Characterization of vascular neuropeptide Y receptors. Br J Pharmacol 105:45–50

Gu J, Polak JM, Adrian TE, Allen JM, Tatemoto K, Bloom SR (1983) Neuropeptide tyrosine (NPY)–a major cardiac neuropeptide. Lancet 1:1008–1010

Gu J, Polak JM, Allen JM, Huang WM, Sheppard MN, Tatemoto K, Bloom SR (1984) High concentrations of a novel peptide, neuropeptide Y, in the innervation of mouse and rat heart. J Histochem Cytochem 32:467–472

Gulbenkian S, Edvinsson L, Saetrum Opgaard O, Valenca A, Wharton J, Polak JM (1992) Neuropeptide Y modulates the action of vasodilator agents in guinea-pig epicardial coronary arteries. Regul Pept 40:351–362

Gulbenkian S, Saetrum Opgaard O, Barroso CP, Wharton J, Polak JM, Edvinsson L (1994) The innervation of guinea pig epicardial coronary veins: immunohistochemistry, ultrastructure and vasomotility. J Auton Nerv Syst 47:201–212

Gullestad L, Aass H, Ross H, Ueland T, Geiran O, Kjekshus J, Simonsen S, Fowler M, Kobilka B (1998) Neuropeptide Y receptor 1 (NPY-Y_1) expression in human heart failure and heart transplantation. J Auton Nerv Syst 70:84–91

Gullestad L, Jorgensen B, Bjuro T, Pernow J, Lundberg JM, Dota CD, Hall C, Simonsen S, Ablad B (2000) Postexercise ischemia is associated with increased neuropeptide Y in patients with coronary artery disease. Circulation 102:987–993

Gustafson EL, Smith KE, Durkin MM, Walker MW, Gerald C, Weinshank R, Branchek TA (1997) Distribution of the neuropeptide Y Y_2 receptor mRNA in rat central nervous system. Mol Brain Res 46:223–235

Gustafsson H, Nilsson H (1990) Endothelium-independent potentiation by neuropeptide Y of vasoconstrictor responses in isolated arteries from rat and rabbit. Acta Physiol Scand 138:503–507

Haass M, Cheng B, Richardt G, Lang RE, Schomig A (1989) Characterization and presynaptic modulation of stimulation-evoked exocytotic co-release of noradrenaline and neuropeptide Y in guinea pig heart. Naunyn Schmiedebergs Arch Pharmacol 339:71–78

Halliday GM, Li YW, Oliver JR, Joh TH, Cotton RG, Howe PR, Geffen LB, Blessing WW (1988) The distribution of neuropeptide Y-like immunoreactive neurons in the human medulla oblongata. Neuroscience 26:179–191

Han C, Abel PW (1987) Neuropeptide Y potentiates contraction and inhibits relaxation of rabbit coronary arteries. J Cardiovasc Pharmacol. 9:675–681

Harfstrand A (1986) Intraventricular administration of neuropeptide Y (NPY) induces hypotension, bradycardia and bradypnoea in the awake unrestrained male rat. Counteraction by NPY-induced feeding behaviour. Acta Physiol Scand 128:121–123

Harfstrand A, Fuxe K, Agnati LF, Benfenati F, Goldstein M (1986) Receptor autoradiographical evidence for high densities of ^{125}I-neuropeptide Y binding sites in the nucleus tractus solitarius of the normal male rat. Acta Physiol Scand 128:195–200

Harfstrand A, Fuxe K, Agnati L, Fredholm B (1989) Reciprocal interactions between alpha$_2$-adrenoceptor agonist and neuropeptide Y binding sites in the nucleus tractus solitarius of the rat. A biochemic and autoradiographic analysis. J Neural Transm 75:83–99

Hashim MA, Harrington WW, Daniels AJ, Tadepalli AS (1997) Hemodynamic profile of neuropeptide Y in dogs: effect of ganglionic blockade. Peptides 18:235–239

Hassall CJ, Burnstock G (1987) Immunocytochemical localisation of neuropeptide Y and 5-hydroxytryptamine in a subpopulation of amine-handling intracardiac neurones that do not contain dopamine beta-hydroxylase in tissue culture. Brain Res 422:74–82

Hastings JA, McClure-Sharp JM, Morris MJ (1998) In vitro studies of endogenous noradrenaline and NPY overflow from the rat hypothalamus during maturation and ageing. Naunyn Schmiedebergs Arch Pharmacol 357:218–224

Hastings JA, McClure-Sharp JM, Morris MJ (2001) NPY Y$_1$ receptors exert opposite effects on corticotropin releasing factor and noradrenaline overflow from the rat hypothalamus in vitro. Brain Res 890:32–37

Hastings JA, Pavia JM, Morris MJ (1997) Neuropeptide Y and [Leu^{31},Pro^{34}]neuropeptide Y potentiate potassium-induced noradrenaline release in the paraventricular nucleus of the aged rat. Brain Res 750:301–304

Hauser GJ, Myers AK, Dayao EK, Zukowska-Grojec Z (1993) Neuropeptide Y infusion improves hemodynamics and survival in rat endotoxic shock. Am J Physiol 265: H1416–H1423

Hieble JP, Duesler JG Jr, Daly RN (1989) Effects of neuropeptide Y on the response of isolated blood vessels to norepinephrine and sympathetic field stimulation. J Pharmacol Exp Ther 250:523–528

Hokfelt T, Lundberg JM, Lagercrantz H, Tatemoto K, Mutt V, Lindberg J, Terenius L, Everitt BJ, Fuxe K, Agnati L, Goldstein M (1983) Occurrence of neuropeptide Y (NPY)-like immunoreactivity in catecholamine neurons in the human medulla oblongata. Neurosci Lett 36:217–222

Hu Y, Dunbar JC (1997) Intracerebroventricular administration of NPY increases sympathetic tone selectively in vascular beds. Brain Res Bull 44:97–103

Hulting J, Sollevi A, Ullman B, Franco-Cereceda A, Lundberg JM (1990) Plasma neuropeptide Y on admission to a coronary care unit: raised levels in patients with left heart failure. Cardiovasc Res 24:102–108

Hunter LW, Tyce GM, Rorie DK (1996) Neuropeptide Y release and contractile properties: differences between canine veins and arteries. Eur J Pharmacol 313:79–87

Illes P, Finta EP, Nieber K (1993) Neuropeptide Y potentiates via Y$_2$-receptors the inhibitory effect of noradrenaline in rat locus coeruleus neurones. Naunyn Schmiedebergs Arch Pharmacol 348:546–548

Jacques D, Tong Y, Shen SH, Quirion R (1998) Discrete distribution of the neuropeptide Y Y$_5$ receptor gene in the human brain: an in situ hybridization study. Brain Res Mol Brain Res 61:100–107

Jansen I, Uddman R, Ekman R, Olesen J, Ottosson A, Edvinsson L (1992) Distribution and effects of neuropeptide Y, vasoactive intestinal peptide, substance P, and calcitonin gene-related peptide in human middle meningeal arteries: comparison with cerebral and temporal arteries. Peptides 13:527–536

Jonsson-Rylander AC, Nordlander M, Svindland A and Ilebekk A (2003) Distribution of neuropeptide Y Y_1 and Y_2 receptors in the postmortem human heart. Peptides 24:255–262

Kahan T, Pernow J, Schwieler J, Lundberg JM, Hjemdahl P, Wallin BG (1988) Involvement of neuropeptide Y in sympathetic vascular control of skeletal muscle in vivo. J Hypertens Suppl 6: S532–S534

Kaijser L, Pernow J, Berglund B, Lundberg JM (1990) Neuropeptide Y is released together with noradrenaline from the human heart during exercise and hypoxia. Clin Physiol 10:179–188

Karvonen MK, Valkonen VP, Lakka TA, Salonen R, Koulu M, Pesonen U, Tuomainen TP, Kauhanen J, Nyyssonen K, Lakka HM, Uusitupa MI, Salonen JT (2001) Leucine7 to proline7 polymorphism in the preproneuropeptide Y is associated with the progression of carotid atherosclerosis, blood pressure and serum lipids in Finnish men. Atherosclerosis 159:145–151

Kawamura K, Ando K, Takebayashi S (1989) Perivascular innervation of the mesenteric artery in spontaneously hypertensive rats. Hypertension 14:660–665

Kawasaki H, Nuki C, Saito A and Takasaki K (1991) NPY modulates neurotransmission of CGRP-containing vasodilator nerves in rat mesenteric arteries. Am J Physiol 261:H683–H690

Kaye DM, Lambert GW, Lefkovits J, Morris M, Jennings G, Esler MD (1994) Neurochemical evidence of cardiac sympathetic activation and increased central nervous system norepinephrine turnover in severe congestive heart failure. J Am Coll Cardiol 23:570–578

Kennedy B, Shen GH and Ziegler MG (1997) Neuropeptide Y-mediated pressor responses following high-frequency stimulation of the rat sympathetic nervous system. J Pharmacol Exp Ther 281:291–296

Kobari M, Fukuuchi Y, Tomita M, Tanahashi N, Yamawaki T, Takeda H, Matsuoka S (1993) Transient cerebral vasodilatory effect of neuropeptide Y mediated by nitric oxide. Brain Res Bull 31:443–448

Komaru T, Ashikawa K, Kanatsuka H, Sekiguchi N, Suzuki T, Takishima T (1990) Neuropeptide Y modulates vasoconstriction in coronary microvessels in the beating canine heart. Circ Res 67:1142–1151

Kotecha N (1998) Modulation of submucosal arteriolar tone by neuropeptide Y Y_2 receptors in the guinea-pig small intestine. J Auton Nerv Syst 70:157–163

Larsen PJ, Kristensen P (1997) The neuropeptide Y (Y_4) receptor is highly expressed in neurones of the rat dorsal vagal complex. Brain Res Mol Brain Res 48:1–6

Larsen PJ, Sheikh SP, Jakobsen CR, Schwartz TW, Mikkelsen JD (1993) Regional distribution of putative NPY Y_1 receptors and neurons expressing Y_1 mRNA in forebrain areas of the rat central nervous system. Eur J Neurosci 5:1622–1637

Lee EW, Grant DS, Movafagh S, Zukowska Z (2003) Impaired angiogenesis in neuropeptide Y (NPY)-Y_2 receptor knockout mice. Peptides 24:99–106

Lee J, Morris MJ (1998) Modulation of neuropeptide Y overflow by leptin in the rat hypothalamus, cerebral cortex and medulla. Neuroreport 9:1575–1580

Lee RM, Nagahama M, McKenzie R, Daniel EE (1988) Peptide-containing nerves around blood vessels of stroke-prone spontaneously hypertensive rats. Hypertension 11:I117–I1120

Lettgen B, Wagner S, Hanze J, Lang RE, Rascher W (1994) Elevated plasma concentration of neuropeptide Y in adolescents with primary hypertension. J Hum Hypertens 8:345–349

Lind H, Erilnge D, Brunkwall J, Edvinsson L (1997) Attenuation of contractile responses to sympathetic co-transmitters in veins from subjects with essential hypertension. Clin Auton Res 7:69–76

Linder L, Lautenschlager BM, Haefeli WE (1996) Subconstrictor doses of neuropeptide Y potentiate alpha$_1$-adrenergic venoconstriction in vivo. Hypertension 28:483–487

Lopez LF, Perez A, St-Pierre S, Huidobro-Toro JP (1989) Neuropeptide tyrosine (NPY)-induced potentiation of the pressor activity of catecholamines in conscious rats. Peptides 10:551–558

Lundberg JM, Tatemoto K (1982) Pancreatic polypeptide family (APP, BPP, NPY and PYY) in relation to sympathetic vasoconstriction resistant to alpha-adrenoceptor blockade. Acta Physiol Scand 116:393–402

Lundberg JM, Terenius L, Hokfelt T, Goldstein M (1983) High levels of neuropeptide Y in peripheral noradrenergic neurons in various mammals including man. Neurosci Lett 42:167–172

Lundberg JM, Saria A, Franco-Cereceda A, Hokfelt T, Terenius L, Goldstein M (1985) Differential effects of reserpine and 6-hydroxydopamine on neuropeptide Y (NPY) and noradrenaline in peripheral neurons. Naunyn Schmiedebergs Arch Pharmacol 328:331–340

Mabe Y, Perez R, Tatemoto K, Huidobro-Toro JP (1987) Chemical sympathectomy reveals pre- and postsynaptic effects of neuropeptide Y (NPY) in the cardiovascular system. Experientia 43:1018–1020

Maccarrone C, Jarrott B (1985) Differences in regional brain concentrations of neuropeptide Y in spontaneously hypertensive (SH) and Wistar-Kyoto (WKY) rats. Brain Res 345:165–169

Maccarrone C, Jarrott B (1987) Differential effects of surgical sympathectomy on rat heart concentrations of neuropeptide Y-immunoreactivity and noradrenaline. J Auton Nerv Syst 21:101–107

Madsen BK, Husum D, Videbaek R, Stokholm KH, Saelsen L, Christensen NJ (1993) Plasma immunoreactive neuropeptide Y in congestive heart failure at rest and during exercise. Scand J Clin Lab Invest 53:569–576

Maisel AS, Scott NA, Motulsky HJ, Michel MC, Boublik JH, Rivier JE, Ziegler M, Allen RS, Brown MR (1989) Elevation of plasma neuropeptide Y levels in congestive heart failure. Am J Med 86:43–48

Malmstrom RE (1997) Neuropeptide Y Y_1 receptor mechanisms in sympathetic vascular control. Acta Physiol Scand Suppl. 636:1–55

Malmstrom RE (2002) Pharmacology of neuropeptide Y receptor antagonists. Focus on cardiovascular functions. Eur J Pharmacol 447:11–30

Martel JC, Fournier A, St Pierre S, Quirion R (1990) Quantitative autoradiographic distribution of [^{125}I]Bolton-Hunter neuropeptide Y receptor binding sites in rat brain. Comparison with [^{125}I]peptide YY receptor sites. Neuroscience 36:255–283

Martin JR, Beinfeld MC, Westfall TC (1988) Blood pressure increases after injection of neuropeptide Y into posterior hypothalamic nucleus. Am J Physiol 254:H879–H888

Martin SE, Kuvin JT, Offenbacher S, Odle BM, Patterson RE (1992) Neuropeptide Y and coronary vasoconstriction: role of thromboxane A_2. Am J Physiol 263:H1045–H1053

Martire M, Pistritto G, Mores N, Agnati LF, Fuxe K (1993) Region-specific inhibition of potassium-evoked [^{3}H]noradrenaline release from rat brain synaptosomes by neuropeptide Y-(13-36). Involvement of NPY receptors of the Y_2 type. Eur J Pharmacol 230:231–234

Matsumura K, Tsuchihashi T, Abe I (2000) Central cardiovascular action of neuropeptide Y in conscious rabbits. Hypertension 36:1040–1044

Maturi MF, Greene R, Speir E, Burrus C, Dorsey LM, Markle DR, Maxwell M, Schmidt W, Goldstein SR, Patterson RE (1989) Neuropeptide-Y. A peptide found in human coronary arteries constricts primarily small coronary arteries to produce myocardial ischemia in dogs. J Clin Invest 83:1217–1224

McDermott BJ, Millar BC, Dolan FM, Bell D, Balasubramaniam A (1997) Evidence for Y_1 and Y_2 subtypes of neuropeptide Y receptors linked to opposing postjunctional effects observed in rat cardiac myocytes. Eur J Pharmacol 336:257–265

McLean KJ, Jarrott B, Lawrence AJ (1996) Neuropeptide Y gene expression and receptor autoradiography in hypertensive and normotensive rat brain. Brain Res Mol Brain Res 35:249–259

McLean KJ, Jarrott B, Lawrence AJ (1999) Hypotension activates neuropeptide Y-containing neurons in the rat medulla oblongata. Neuroscience 92:1377–1387

Michalkiewicz M, Michalkiewicz T, Kreulen DL, McDougall SJ (2001) Increased blood pressure responses in neuropeptide Y transgenic rats. Am J Physiol Regul Integr Comp Physiol 281:R417–R426

Michel MC, Rascher W (1995) Neuropeptide Y: a possible role in hypertension? J Hypertens 13:385–395

Michel MC, Beck-Sickinger A, Cox H, Doods HN, Herzog H, Larhammar D, Quirion R, Schwartz T, Westfall T (1998) XVI. International Union of Pharmacology recommendations for the nomenclature of neuropeptide Y, peptide YY, and pancreatic polypeptide receptors. Pharmacol Rev 50:143–150

Migita K, Loewy AD, Ramabhadran TV, Krause JE, Waters SM (2001) Immunohistochemical localization of the neuropeptide Y Y1 receptor in rat central nervous system. Brain Res 889:23–37

Millar BC, Schluter KD, Zhou XJ, McDermott BJ, Piper HM (1994) Neuropeptide Y stimulates hypertrophy of adult ventricular cardiomyocytes. Am J Physiol 266:C1271–C277

Miller DW and Tessel RE (1991) Age-dependent hyperresponsiveness of spontaneously hypertensive rats to the pressor effects of intravenous neuropeptide Y (NPY): role of mode of peptide administration and plasma NPY-like immunoreactivity. J Cardiovasc Pharmacol 18:647–656

Minson J, Llewellyn-Smith I, Neville A, Somogyi P, Chalmers J (1990a) Quantitative analysis of spinally projecting adrenaline-synthesising neurons of C1, C2 and C3 groups in rat medulla oblongata. J Auton Nerv Syst 30:209–220

Minson RB, McRitchie RJ, Morris MJ, Chalmers JP (1990b) Effects of neuropeptide Y on cardiac performance and renal blood flow in conscious normotensive and renal hypertensive rabbits. Clin Exp Hypertens A. 12:267–284

Minson RB, McRitchie RJ, Chalmers JP (1989) Effects of neuropeptide Y on the heart and circulation of the conscious rabbit. J Cardiovasc Pharmacol. 14:699–706

Modin A, Pernow J, Lundberg JM, (1993) Sympathetic regulation of skeletal muscle blood flow in the pig: a non-adrenergic component likely to be mediated by neuropeptide Y. Acta Physiol Scand 148:1–11

Morris JL (1991) Roles of neuropeptide Y and noradrenaline in sympathetic neurotransmission to the thoracic vena cava and aorta of guinea-pigs. Regul Pept 32:297–310

Morris JL (1994) Selective constriction of small cutaneous arteries by NPY matches distribution of NPY in sympathetic axons. Regul Pept 49:225–236

Morris JL, Sabesan S (1994) Comparison of the NPY receptors mediating vasoconstriction of the guinea-pig uterine artery and thoracic vena cava using a range of NPY analogues. Neuropeptides 26:21–28

Morris JL, Murphy R, Furness JB, Costa M (1986a) Partial depletion of neuropeptide Y from noradrenergic perivascular and cardiac axons by 6-hydroxydopamine and reserpine. Regul Pept 13:147–162

Morris MJ, Pavia JM (1997) Lack of effect of age on the cardiovascular response to neuropeptide Y injection in the rat nucleus tractus solitarius. Clin Exp Pharmacol Physiol 24:162–165

Morris MJ, Russell AE, Kapoor V, Cain MD, Elliott JM, West MJ, Wing LM, Chalmers JP (1986b) Increases in plasma neuropeptide Y concentrations during sympathetic activation in man. J Auton Nerv Syst 17:143–149

Morris MJ, Elliott JM, Cain MD, Kapoor V, West MJ, Chalmers JP (1986c) Plasma neuropeptide Y levels rise in patients undergoing exercise tests for the investigation of chest pain. Clin Exp Pharmacol Physiol 13:437–440

Morris M, Kapoor V, Chalmers J (1987a) Plasma neuropeptide Y concentration is increased after hemorrhage in conscious rats: relative contributions of sympathetic nerves and the adrenal medulla. J Cardiovasc Pharmacol 9:541–545

Morris MJ, Pilowsky PM, Minson JB, West MJ, Chalmers JP (1987b) Microinjection of kainic acid into the rostral ventrolateral medulla causes hypertension and release of neuropeptide Y-like immunoreactivity from rabbit spinal cord. Clin Exp Pharmacol Physiol 14:127–132

Morris MJ, Cox HS, Lambert GW, Kaye DM, Jennings GL, Meredith IT, Esler MD (1997a) Region-specific neuropeptide Y overflows at rest and during sympathetic activation in humans. Hypertension 29:137–143

Morris MJ, Hastings JA, Pavia JM (1997b) Central interactions between noradrenaline and neuropeptide Y in the rat: implications for blood pressure control. Clin Exp Hypertens 19:619–630

Nakajima T, Yashima Y, Nakamura K (1987) Higher density of ^{125}I-neuropeptide Y receptors in the area postrema of SHR. Brain Res 417:360–362

Nicholl SM, Bell D, Spiers J, McDermott BJ (2002) Neuropeptide Y Y_1 receptor regulates protein turnover and constitutive gene expression in hypertrophying cardiomyocytes. Eur J Pharmacol 441:23–34

Nilsson T, Cantera L, Edvinsson L (1996) Presence of neuropeptide Y Y_1 receptor mediating vasoconstriction in human cerebral arteries. Neurosci Lett 204:145–148

Nilsson T, Lind H, Brunkvall J, Edvinsson L (2000) Vasodilation in human subcutaneous arteries induced by neuropeptide Y is mediated by neuropeptide Y Y1 receptors and is nitric oxide dependent. Can J Physiol Pharmacol 78:251–255

Niskanen L, Karvonen MK, Valve R, Koulu M, Pesonen U, Mercuri M, Rauramaa R, Toyry J, Laakso M, Uusitupa MI (2000) Leucine 7 to proline 7 polymorphism in the neuropeptide Y gene is associated with enhanced carotid atherosclerosis in elderly patients with type 2 diabetes and control subjects. J Clin Endocrinol Metab 85:2266–2269

Nyquist-Battie C, Cochran PK, Sands SA and Chronwall BM (1994) Development of neuropeptide Y and tyrosine hydroxylase immunoreactive innervation in postnatal rat heart. Peptides 15:1461–1469

Ohkubo T, Niwa M, Yamashita K, Kataoka Y, Shigematsu K (1990) Neuropeptide Y (NPY) and peptide YY (PYY) receptors in rat brain. Cell Mol Neurobiol 10:539–552

Onuoha GN, Nicholls DP, Alpar EK, Ritchie A, Shaw C, Buchanan K (1999) Regulatory peptides in the heart and major vessels of man and mammals. Neuropeptides 33:165–172

Owen MP (1993) Similarities and differences in the postjunctional role for neuropeptide Y in sympathetic vasomotor control of large vs. small arteries of rabbit renal and ear vasculature. J Pharmacol Exp Ther 265:887–895

Parker RM, Herzog H (1999) Regional distribution of Y-receptor subtype mRNAs in rat brain. Eur J Neurosci 11:1431–1448

Pavia JM, Morris MJ (1994) Age-related changes in neuropeptide Y content in brain and peripheral tissues of spontaneously hypertensive rats. Clin Exp Pharmacol Physiol 21:335–338

Pedrazzini T, Seydoux J, Kunstner P, Aubert JF, Grouzmann E, Beermann F, Brunner HR (1998) Cardiovascular response, feeding behavior and locomotor activity in mice lacking the NPY Y1 receptor. Nature Med 4:722–726

Pellieux C, Sauthier T, Domenighetti A, Marsh DJ, Palmiter RD, Brunner HR, Pedrazzini T (2000) Neuropeptide Y (NPY) potentiates phenylephrine-induced mitogen-activated protein kinase activation in primary cardiomyocytes via NPY Y5 receptors. Proc Natl Acad Sci USA 97:1595–1600

Pernow J, Lundberg JM (1988) Neuropeptide Y induces potent contraction of arterial vascular smooth muscle via an endothelium-independent mechanism. Acta Physiol Scand 134:157–158

Pernow J, Lundberg JM, Kaijser L, Hjemdahl P, Theodorsson-Norheim E, Martinsson A, Pernow B (1986) Plasma neuropeptide Y-like immunoreactivity and catecholamines during various degrees of sympathetic activation in man. Clin Physiol 6:561–578

Pernow J, Lundberg JM, Kaijser L (1987) Vasoconstrictor effects in vivo and plasma disappearance rate of neuropeptide Y in man. Life Sci 40:47–54

Pernow J, Schwieler J, Kahan T, Hjemdahl P, Oberle J, Wallin BG, Lundberg JM (1989) Influence of sympathetic discharge pattern on norepinephrine and neuropeptide Y release. Am J Physiol 257:H866–H872

Pilowsky PM, Morris MJ, Minson JB, West MJ, Chalmers JP, Willoughby JO, Blessing WW (1987) Inhibition of vasodepressor neurons in the caudal ventrolateral medulla of the rabbit increases both arterial pressure and the release of neuropeptide Y-like immunoreactivity from the spinal cord. Brain Res 420:380–384

Poncet MF, Damase-Michel C, Tavernier G, Tran MA, Berlan M, Montastruc JL, Montastruc P (1992) Changes in plasma catecholamine and neuropeptide Y levels after sympathetic activation in dogs. Br J Pharmacol 105:181–183

Potter EK (1985) Prolonged non-adrenergic inhibition of cardiac vagal action following sympathetic stimulation: neuromodulation by neuropeptide Y. Neurosci Lett 54:117–121

Potter EK (1988) Neuropeptide Y as an autonomic neurotransmitter. Pharmacol Ther 37:251–273

Potter EK, McCloskey MJ (1992) [Leu^{31}, Pro^{34}] NPY, a selective functional postjunctional agonist at neuropeptide-Y receptors in anaesthetised rats. Neurosci Lett 134:183–186

Potter EK, Mitchell L, McCloskey MJ, Tseng A, Goodman AE, Shine J, McCloskey DI (1989) Pre- and postjunctional actions of neuropeptide Y and related peptides. Regul Pept 25:167–177

Prieto D, Benedito S, Simonsen U, Nyborg NC (1991) Regional heterogeneity in the contractile and potentiating effects of neuropeptide Y in rat isolated coronary arteries: modulatory action of the endothelium. Br J Pharmacol 102:754–758

Rioux F, Bachelard H, Martel JC and St-Pierre S (1986) The vasoconstrictor effect of neuropeptide Y and related peptides in the guinea pig isolated heart. Peptides 7:27–31

Rudehill A, Sollevi A, Franco-Cereceda A, Lundberg JM, (1986) Neuropeptide Y (NPY) and the pig heart: release and coronary vasoconstrictor effects. Peptides 7:821–826

Rudehill A, Franco-Cereceda A, Hemsen A, Stensdotter M, Pernow J, Lundberg JM (1989) Cigarette smoke-induced elevation of plasma neuropeptide Y levels in man. Clin Physiol 9:243–248

Saville VL, Maynard KI, Burnstock G (1990) Neuropeptide Y potentiates purinergic as well as adrenergic responses of the rabbit ear artery. Eur J Pharmacol 76:117–125

Sawchenko PE, Swanson LW, Grzanna R, Howe PR, Bloom SR, Polak JM (1985) Colocalization of neuropeptide Y immunoreactivity in brainstem catecholaminergic neurons that project to the paraventricular nucleus of the hypothalamus. J Comp Neurol 241:138–153

Schuerch LV, Linder LM, Grouzmann E, Haefeli WE (1998) Human neuropeptide Y potentiates alpha$_1$-adrenergic blood pressure responses in vivo. Am J Physiol 275:H760–H766

Scott NA, Webb V, Boublik JH, Rivier J, Brown MR (1989) The cardiovascular actions of centrally administered neuropeptide Y. Regul Pept 25:247–258

Serradeil-Le Gal C, Valette G, Rouby PE, Pellet A, Oury-Donat F, Brossard G, Lespy L, Marty E, Neliat G, de Cointet P, Maffrand, JP, Le Fur G (1995) SR 120819A, an orally-active and selective neuropeptide Y Y1 receptor antagonist. FEBS Lett 362:192–196

Shigeri Y and Fujimoto M (1993) Neuropeptide Y stimulates DNA synthesis in vascular smooth muscle cells. Neurosci Lett 149:19–22

Shigeri Y, Mihara S, Fujimoto M (1991) Neuropeptide Y receptor in vascular smooth muscle. J Neurochem 56:852–859

Shih CD, Chan JY, Chan SH (1992) Tonic suppression of baroreceptor reflex response by endogenous neuropeptide Y at the nucleus tractus solitarius of the rat. Neurosci Lett 148:169–172

Small DL, Bolzon BJ, Cheung DW (1992) Endothelium-independent potentiating effects of neuropeptide Y in the rat tail artery. Eur J Pharmacol 210:131–136

Smith-White MA, Herzog H, Potter EK (2002a) Cardiac function in neuropeptide Y Y_4 receptor-knockout mice. Regul Pept 110:47–54

Smith-White MA, Herzog H, Potter EK (2002b) Role of neuropeptide Y Y_2 receptors in modulation of cardiac parasympathetic neurotransmission. Regul Pept. 103:105–111

Smyth L, Bobalova J, Ward SM, Keef KD, Mutafova-Yambolieva VN (2000) Cotransmission from sympathetic vasoconstrictor neurons: differences in guinea-pig mesenteric artery and vein. Auton Neurosci 86:18–29

Solt VB, Brown MR, Kennedy B, Kolterman OG, Ziegler MG (1990) Elevated insulin, norepinephrine, and neuropeptide Y in hypertension. Am J Hypertens 3:823–828

Stack RK, Patterson RE (1991) Haemodynamic effects of intracoronary neuropeptide Y in dogs: resistance to alpha adrenergic blockade. Cardiovasc Res 25:757–763

Stewart-Lee AL, Aberdeen J, Burnstock G (1992) The effect of atherosclerosis on neuromodulation of sympathetic neurotransmission by neuropeptide Y and calcitonin gene-related peptide in the rabbit mesenteric artery. Eur J Pharmacol 216:167–174

Tabarin A, Minot AP, Dallochio M, Roger P, Ducassou D (1992) Plasma concentration of neuropeptide Y in patients with adrenal hypertension. Regul Pept 42:51–61

Takahashi K, Mouri T, Tachibana Y, Itoi K, Sone M, Nozuki M, Murakami O, Ohneda M, Yoshinaga K (1989) Neuropeptide Y and blood pressure in haemodialysis patients. Endocrinol Jpn 36:553–558

Takesako T, Takeda K, Kuwahara T, Takenaka K, Tanaka M, Itoh H, Nakata T, Sasaki S, Nakagawa M (1994) Alteration of response to neuropeptide Y in the nucleus tractus solitarius of spontaneously hypertensive rats. Hypertension 23(Suppl 1):I93–I96

Tanaka E, Mori H, Chujo M, Yamakawa A, Mohammed MU, Shinozaki Y, Tobita K, Sekka T, Ito K, Nakazawa H (1997) Coronary vasoconstrictive effects of neuropeptide Y and their modulation by the ATP-sensitive potassium channel in anesthetized dogs. J Am Coll Cardiol. 29:1380–1389

Tatemoto K, Carlquist M, Mutt V (1982) Neuropeptide Y–a novel brain peptide with structural similarities to peptide YY and pancreatic polypeptide. Nature 296:659–660

Trinh T, van Dumont Y, Quirion R (1996) High levels of specific neuropeptide Y/pancreatic polypeptide receptors in the rat hypothalamus and brainstem. Eur J Pharmacol 318:R1–R3

Tseng CJ, Mosqueda-Garcia R, Appalsamy M, Robertson D (1989) Cardiovascular effects of neuropeptide Y in rat brainstem nuclei. Circ Res 64:55–61

Tseng CJ, Lin HC, Wang SD, Tung CS (1993) Immunohistochemical study of catecholamine enzymes and neuropeptide Y (NPY) in the rostral ventrolateral medulla and bulbospinal projection. J Comp Neurol 334:294–303

Uddman R, Moller S, Nilsson T, Nystrom S, Ekstrand J, Edvinsson L (2002) Neuropeptide Y Y_1 and neuropeptide Y Y_2 receptors in human cardiovascular tissues. Peptides 23:927–934

Ullman B, Jensen-Urstad M, Hulting J, Lundberg JM (1993) Neuropeptide Y, noradrenaline and invasive haemodynamic data in mild to moderate chronic congestive heart failure. Clin Physiol 13:409–418

Ullman B, Pernow J, Lundberg JM, Astrom H, Bergfeldt L (2002) Cardiovascular effects and cardiopulmonary plasma gradients following intravenous infusion of neuropeptide Y in humans: negative dromotropic effect on atrioventricular node conduction. Clin Sci (Lond) 103:535–542

Ulman LG, Potter EK, McCloskey DI, Morris MJ (1997) Post-exercise depression of baroreflex slowing of the heart in humans. Clin Physiol, 17:299–309

Vallejo M, Carter DA, Diez-Guerra FJ, Emson PC, Lightman SL (1987) Neonatal administration of a specific neuropeptide Y antiserum alters the vasopressin response to haemorrhage and the hypothalamic content of noradrenaline in rats. Neuroendocrinology. 45:507–509

Vallejo M, Lightman SL (1986) Pressor effect of centrally administered neuropeptide Y in rats: role of sympathetic nervous system and vasopressin. Life Sci 38:1859–1866

Waeber B, Aubert JF, Corder R, Evequoz D, Nussberger J, Gaillard R, Brunner HR (1988) Cardiovascular effects of neuropeptide Y. Am J Hypertens 1:193–199

Wahlestedt C, Reis DJ (1993) Neuropeptide Y-related peptides and their receptors—are the receptors potential therapeutic drug targets? Annu Rev Pharmacol Toxicol 33:309–352

Wahlestedt C, Yanaihara N, Hakanson R (1986) Evidence for different pre-and post-junctional receptors for neuropeptide Y and related peptides. Regul Pept 13:307–318

Wahlestedt C, Grundemar L, Hakanson R, Heilig M, Shen GH, Zukowska-Grojec Z, Reis DJ (1990a) Neuropeptide Y receptor subtypes, Y_1 and Y_2. Ann N Y Acad Sci 611:7–26

Wahlestedt C, Hakanson R, Vaz CA, Zukowska-Grojec Z (1990b) Norepinephrine and neuropeptide Y: vasoconstrictor cooperation in vivo and in vitro. Am J Physiol 258:R736–R742

Walker MW, Miller RJ (1988) ^{125}I-neuropeptide Y and ^{125}I-peptide YY bind to multiple receptor sites in rat brain. Mol Pharmacol 34:779–792

Walker P, Grouzmann E, Burnier M, Waeber B (1991) The role of neuropeptide Y in cardiovascular regulation. Trends Pharmacol Sci 12:111–115

Warner MR, Senanayake PD, Ferrario CM, Levy MN (1991) Sympathetic stimulation-evoked overflow of norepinephrine and neuropeptide Y from the heart. Circ Res 69:455–465

Warner MR, Levy MN (1989) Inhibition of cardiac vagal effects by neurally released and exogenous neuropeptide Y. Circ Res 65:1536–1546

Westfall TC, Martin J, Chen XL, Ciarleglio A, Carpentier S, Henderson K, Knuepfer M, Beinfeld M, Naes L (1988) Cardiovascular effects and modulation of noradrenergic neurotransmission following central and peripheral administration of neuropeptide Y. Synapse 2:299–307

Westfall TC, Han SP, Knuepfer M, Martin J, Chen XL, del Valle K, Ciarleglio A, Naes L (1990) Neuropeptides in hypertension: role of neuropeptide Y and calcitonin gene related peptide. Br J Clin Pharmacol 30 (Suppl 1):75S–82S

Westfall TC, Yang CL, Rotto-Perceley D, Macarthur H (1996) Neuropeptide Y-ATP interactions and release at the vascular neuroeffector junction. J Auton Pharmacol 16:345–348

Whitworth JA, Williamson PM, Brown MA, Morris MJ (1994) Neuropeptide Y in cortisol-induced hypertension in male volunteers. Clin Exp Pharmacol Physiol 21:435–438

Widdowson PS (1993) Quantitative receptor autoradiography demonstrates a differential distribution of neuropeptide-Y Y1 and Y2 receptor subtypes in human and rat brain. Brain Res 631:27–38

Wocial B, Ignatowska-Switalska H, Pruszczyk P, Jedrusik P, Januszewicz A, Lapinski M, Januszewicz W, Zukowska-Grojec Z (1995) Plasma neuropeptide Y and catecholamines in women and men with essential hypertension Blood Press 4:143–147

Yang SN, Narvaez JA, Bjelke B, Agnati LF, Fuxe K (1993) Microinjections of subpicomolar amounts of NPY(13–36) into the nucleus tractus solitarius of the rat counteract the vasodepressor responses of NPY(1–36) and of a NPY Y_1 receptor agonist. Brain Res 621:126–132

Yang SN, Bunnemann B, Cintra A, Fuxe K (1996) Localization of neuropeptide Y Y_1 receptor-like immunoreactivity in catecholaminergic neurons of the rat medulla oblongata. Neuroscience 73:519–530

You J, Edvinsson L, Bryan RM, Jr. (2001) Neuropeptide Y-mediated constriction and dilation in rat middle cerebral arteries. J Cereb Blood Flow Metab 21:77–84

Zukowska-Grojec Z, Shen GH, Capraro PA, Vaz CA (1991) Cardiovascular, neuropeptide Y, and adrenergic responses in stress are sexually differentiated. Physiol Behav 49:771–777

Zukowska-Grojec Z, Golczynska M, Shen GH, Torres-Duarte A, Haass M, Wahlestedt C, Myers AK (1993a) Modulation of vascular function by neuropeptide Y during development of hypertension in spontaneously hypertensive rats. Pediatr Nephrol 7:845–852

Zukowska-Grojec Z, Pruszczyk P, Colton C, Yao J, Shen GH, Myers AK, Wahlestedt C (1993b) Mitogenic effect of neuropeptide Y in rat vascular smooth muscle cells. Peptides 14:263–268

Zukowska-Grojec Z, Dayao EK, Karwatowska-Prokopczuk E, Hauser GJ, Doods HN (1996) Stress-induced mesenteric vasoconstriction in rats is mediated by neuropeptide Y Y_1 receptors. Am J Physiol 270:H796–H800

Zukowska-Grojec Z, Karwatowska-Prokopczuk E, Rose W, Rone J, Movafagh S, Ji H, Yeh Y, Chen WT, Kleinman HK, Grouzmann E, Grant DS (1998) Neuropeptide Y: a novel angiogenic factor from the sympathetic nerves and endothelium. Circ Res 83:187–195

Neuropeptide Y and the Kidney

M. C. Michel

Department of Pharmacology and Pharmacotherapy, University of Amsterdam,
Postbus 22700, 1100 DE, Amsterdam, The Netherlands
e-mail: m.c.michel@amc.uva.nl

1	Introduction	362
2	Presence and Release of NPY in the Kidney	362
3	Renal NPY Receptors and Their Signal Transduction	364
4	Regulation of Renovascular Function by NPY	366
5	Regulation of Water and Electrolyte Excretion by NPY	370
6	A Functional Role for Endogenous NPY-Related Peptides in the Kidney? . . .	375
7	Regulation of Renal NPY Content and Responsiveness	376
8	NPY System in Renal Disease	378
9	Conclusions and Directions for Future Research	379
	References	380

Abstract Neuropeptide Y (NPY) is a co-transmitter of the sympathetic nervous system including the renal nerves. The kidney expresses NPY receptors, which can also be activated by peptide YY (PYY), a circulating hormone released from gastrointestinal cells. Five subtypes of NPY receptors have been cloned, among which Y_1, Y_2 and Y_5 appear to be involved in the regulation of renal function. NPY produces potent renal vasoconstriction in vitro in isolated interlobar arteries and in the isolated perfused kidney and in vivo upon intrarenal or systemic administration via a Y_1 receptor. Nevertheless glomerular filtration rate is altered only little if at all by NPY, indicating a greater effect on the vas efferens than the vas afferens. NPY can inhibit renin release via Y_1-like receptors. NPY can stimulate Na^+/K^+-ATPase in proximal tubules via Y_2 receptors and can antagonize the effects of vasopressin on isolated collecting ducts. It can also act prejunctionally to inhibit noradrenaline release via Y_2 receptors. Despite the profound reductions of renal blood flow, systemic NPY infusion can cause diuresis and natriuresis; this occurs largely independent of the pressure natriuresis mechanisms and is possibly mediated by an extrarenal Y_5 receptor. Studies with the conversion enzyme inhibitor, ramiprilat, and the bradykinin receptor antagonist, icatibant, indicate that bradykinin at least partly mediates diuretic NPY effects. NPY antagonists enhance basal renal blood flow but do not alter basal diuresis or natriuresis indicating that renovascular but not tubular NPY receptors may be tonically activated by endogenous NPY.

Keywords Kidney · Renal blood flow · Diuresis · Natriuresis · Hypertension · Uremia

1
Introduction

The kidney controls not only fluid and osmolar balance and metabolic end product excretion but is also crucial for long-term blood pressure regulation and has endocrine effects. Therefore, it can be considered to play a central role in the maintenance of homeostasis in the organism, and hence renal function is tightly controlled by a variety of neural and hormonal systems. Neuropeptide Y (NPY), the related peptide YY (PYY) and, perhaps, also pancreatic polypeptide (PP) can potentially regulate renal function in several direct and indirect ways. The direct NPY and PYY effects occur via NPY receptors in the renal vasculature and the tubules. The indirect effects can involve a number of different mechanisms. Thus, alterations of intrarenal hemodynamics may have secondary effects on tubular function whereas alterations of systemic hemodynamics, that is, blood pressure elevation (McDermott et al. 1993; Michel and Rascher 1995), could cause pressure natriuresis and/or baroreflex-mediated lowering of sympathetic tone in the kidneys. A lowering of sympathetic drive to the kidneys can also occur locally via prejunctional NPY receptors (Edvinsson et al. 1987). Stimulation of NPY receptor within or outside of the kidney may also affect the activity of various humoral systems including the renin–angiotensin and the atrial natriuretic peptide system. Finally, activation of NPY receptors in certain parts of the brain may affect renal function (Chen et al. 1990; Martin et al. 1989; Matsumura et al. 2000; Roberts et al. 2000). This chapter will analyze the complex interactions between the above mechanisms, with the exception of the central effects.

2
Presence and Release of NPY in the Kidney

NPY is found, for example, by immunohistochemistry, in the kidney of various species including rat (Ballesta et al. 1984; Chevendra and Weaver 1992; Knight et al. 1989; Reinecke and Forssmann 1988), mouse (Ballesta et al. 1984), hamster (Ballesta et al. 1984), guinea pig (Ballesta et al. 1984; Reinecke and Forssmann 1988), pig (Reinecke and Forssmann 1988), dog (Reinecke and Forssmann 1988), monkey (Ballesta et al. 1984; Norvell and MacBride 1989) and man (Ballesta et al. 1984; Grouzmann et al. 1994; Norvell and MacBride 1989). It can be detected in almost all renal sympathetic neurons (Chevendra and Weaver 1992; Knight et al. 1989). Whether neurons are the only source of renal NPY content is not fully clear. On the one hand, destruction of sympathetic nerve terminals by 6-hydroxy-dopamine dramatically decreases renal NPY content (Ballesta et al. 1984). Similarly, two studies failed to detect mRNA for NPY in human renal cells (Turman et al. 1997; Turman and Apple 1998). On the other hand, NPY mRNA was not only detected in rat kidney but also reported to colocalize with NPY immunoreactivity (Haefliger et al. 1999). Whether these conflicting results can be explained by species differences remains unclear, since human kidney was also reported to contain the C-terminal flanking peptide of NPY, which is formed as part of the processing of pre-pro-NPY (Grouzmann et al. 1994), although that may have been generated by local processing of pre-pro-NPY of neuronal origin.

The intrarenal localization of NPY-containing nerve endings has been studied in guinea pigs, monkeys and humans. Thus, studies in the guinea pig have detected NPY-containing nerves in all segments of the renal arterial plexus (with lowest concentration in medullary arteries) and in the juxtaglomerular region (Reinecke and Forssmann 1988). One study in humans and monkeys (*Macaca fascicularis*) has also detected NPY-containing nerves surrounding arteries (and to a lesser extent veins) in the juxtamedullary region and at the afferent and efferent arterioles of glomeruli (Norvell and MacBride 1989). While the localization in the juxtaglomerular region of the human kidney was confirmed in another study (Ballesta et al. 1984), a more recent report has suggested that within the human kidney NPY is found mainly around tubules with little abundance in the perivascular space and no presence in glomeruli (Grouzmann et al. 1994).

Release of NPY together with noradrenaline has been detected following renal nerve stimulation in pigs in vivo (Malmström et al. 2002a,b,; Modin et al. 1994) and with the isolated perfused rat kidney (Oellerich and Malik 1993). In reserpine-treated pigs the initial NPY release upon renal nerve stimulation is enhanced but with repeated stimulation trains NPY release gradually decreases indicating a possible feedback role of noradrenaline in the filling of neuronal NPY stores (Modin et al. 1994). NPY release in the porcine kidney can be inhibited by infusion of PYY (Malmström et al. 2002a; Modin et al. 1996; Pernow and Lundberg 1989a). This is blocked by the Y_2-selective antagonist BIIE 0246 (Malmström et al. 2002a). This antagonist also enhances NPY release upon nerve stimulation in the same preparation (Malmström et al. 2002b), indicating that the prejunctional Y_2 receptor is also activated by endogenous NPY. NPY release from the renal nerves can be enhanced by the α_2-adrenoceptor antagonist, yohimbine, or by angiotensin II whereas infusion of the converting enzyme inhibitor, captopril, or of endothelin does not alter it (Malmström et al. 2002a; Modin et al. 1996; Pernow and Lundberg 1989a). Thus, renal NPY release may be inhibited by prejunctional NPY receptors of the Y_2 subtype and α_2-adrenoceptors and stimulated by prejunctional angiotensin receptors.

3
Renal NPY Receptors and Their Signal Transduction

Five subtypes of NPY receptors have been identified which are designated Y_1, Y_2, Y_4, Y_5 and y_6 (Michel et al. 1998). The gene of the y_6 receptor in humans and primates contains a mutation causing a stop codon, and therefore the y_6 receptor is not functional in these species (Matsumoto et al. 1996). Moreover, the y_6 receptor gene is absent from the rat genome (Burkhoff et al. 1998). Hence this subtype will not be further considered in the following discussion.

The presence of NPY receptor subtypes within the kidney has been studied directly at the mRNA and protein level. Messenger RNA for the Y_1 receptor has been demonstrated in the kidney of mice by Northern blotting (Nakamura et al. 1995), of rats by in situ hybridization (Modin et al. 1999), of dogs by reverse transcriptase (RT)–polymerase chain reaction (PCR) (Holtbäck et al. 1999), in pigs by RT-PCR and in situ hybridization (Malmström et al. 1998), and in human kidney by RT-PCR (Nakamura et al. 1995) or in situ hybridization (Wharton et al. 1993). The latter study has mapped the presence of Y_1 NPY receptor expression in human kidney to collecting ducts, loop of Henle, and the juxtaglomerular apparatus. In contrast, Y_1 receptor mRNA in pig spleen seems to exist primarily in intrarenal arteries (Malmström et al. 1998). Y_2 receptor mRNA was detected by RT-PCR in canine and porcine kidney (Malmström et al. 1998). Northern blotting failed to detect Y_4 receptor mRNA in murine (Gregor et al. 1996b), rat (Lundell et al. 1996) or human kidney (Bard et al. 1995; Lundell et al. 1995; Yan et al. 1996) or Y_5 receptor mRNA in rat (Gerald et al. 1996; Hu et al. 1996) or mouse kidney (Nakamura et al. 1997). Messenger RNA for the y_6 receptor was also not detected by RT-PCR or Northern blotting in rabbit or human kidney (Gregor et al. 1996a; Matsumoto et al. 1996).

At the protein level renal NPY receptor detection has relied on radioligand binding and autoradiography. Numerous studies have characterized NPY binding sites in rabbit kidney and identified them as a predominant population of Y_2 receptors (Blaze et al. 1997; Michel et al. 1992; Price et al. 1991; Schachter et al. 1987; Sheikh et al. 1989; Wieland et al. 1995), but Y_1 and Y_5 receptors can also be demonstrated using suitable radioligands (Parker et al. 1998); moreover, NPY receptors in rabbit kidney appear to have an unusually high affinity for PP (Parker et al. 2001). In contrast, in human or rat kidney only very few if any NPY receptors were detected in radioligand binding studies (Bischoff et al. 1995a; Leys et al. 1987; Price et al. 1991; Schachter et al. 1987). While it was proposed that the lack of NPY receptor detection in rat and human kidney by binding studies may be due to rapid radioligand degradation by phosphoramidon-insensitive endopeptidase-2 (Price et al. 1991), later studies have failed to detect NPY receptor binding in these species even under conditions where radioligand degradation was prevented (Bischoff et al. 1995a). On the other hand, one study using receptor autoradiography rather than radioligand binding to tissue homogenates has detected Y_1 NPY receptors in the papilla of rat kidney (Blaze et al. 1997). Thus, considerable species heterogeneity appears to exist with regard to quantitative renal NPY receptor expression at the protein level with high concentrations in rabbits and much lower ones if any in rat and human kidney. The high density of Y_2 receptor in rabbit kidney has made this tissue a primary source to study the molecular properties of this subtype in detail (Parker et al. 1998).

Functional evidence for NPY receptors on renal cells has been obtained using a number of biochemical approaches. Thus, in rat isolated proximal tubules NPY receptor stimulation enhances Na^+/K^+-ATPase activity (Ohtomo et al. 1994, 1996a, 1996b). In addition to a direct stimulation, NPY can also potentiate the enhancements caused by the preproinsulin-derived C-peptide (Ohtomo et al. 1996a) and by α_1-adrenoceptor stimulation (Holtbäck et al. 1998) and block the deactivation by β-adrenoceptor stimulation (Holtbäck et al. 1998). The direct stimulatory effects of NPY occur via a Y_2 receptor and a pertussis toxin-sensitive G-protein; they can be inhibited by the immunosuppressant drug, tacrolimus (also known as FK 506), indicating a possible involvement of the Ca^{2+}-calmodulin-dependent protein phosphatase (Ohtomo et al. 1996a). The NPY-induced stimulation of Na^+/K^+-ATPase may be age dependent since it was observed in adult but not in juvenile rats (Ohtomo et al. 1996b). In isolated perfused rat cortical collecting tubules NPY attenuates vasopressin-stimulated hydraulic conductivity in a pertussis toxin-sensitive manner (Dillingham and Anderson 1989). Interestingly the α_2-adrenoceptor antagonist, yohimbine, and the α_2-adrenoceptor partial agonist, clonidine, but not the α_1-adrenoceptor antagonist, prazosin, prevented the NPY effect, indicating a possible intermediary involvement of α_2-adrenoceptors (Dillingham and Anderson 1989). Some of these functional effects may occur secondary to inhibition of adenylyl cyclase. Thus, NPY can inhibit adenylyl cyclase in various segments of rat nephrons (Dillingham and Anderson 1989; Edwards 1990; Holtbäck et al. 1998), in rabbit kidney (Gimpl et al. 1991) and in the murine tubular cell line PKSV-PCT (Voisin et al. 1996).

Some of the effects of NPY on renal function may occur indirectly. For example studies with the kidney-derived LLCPK cell line as well as with homogenates from rat renal outer cortical tissue have found that NPY can recruit α_{1A}-adrenoceptors to the cell surface, where they become accessible to agonist and hence functional (Holtbäck et al. 1999). This could explain the enhancement of α_1-adrenoceptor-stimulated Na^+/K^+-ATPase activity (Holtbäck et al. 1998). Moreover, NPY receptor stimulation can generate prostaglandins in the kidney. Thus, in isolated rat kidney NPY has been reported to stimulate formation of prostaglandins, namely prostaglandins E_2 and I_2 (El-Din and Malik 1988). This stimulation of prostaglandin formation was blocked by the Ca^{2+} entry blockers, diltiazem and nifedipine, and by omission of Ca^{2+} from the perfusate; in subthreshold concentrations NPY enhanced the prostaglandin E_2 and I_2 formation stimulated by noradrenaline but not that by vasopressin or angiotensin II. Taken together these molecular, radioligand binding and biochemical data clearly demonstrate the presence of NPY receptors on renal tubular cells. The relationship of these receptors to physiological effects of NPY and related peptides on the kidney, however, is not fully clear (see below).

4
Regulation of Renovascular Function by NPY

Contraction of the renal vasculature is probably the best documented renal effect of NPY and related peptides. It has been shown in vitro with isolated blood vessels and the isolated perfused kidney, and in vivo upon intrarenal and systemic administration. Thus, contraction of isolated segments of the renal vasculature by NPY has directly been demonstrated in rabbits (Owen 1993) and rats (Chen et al. 1997; Torffvit et al. 1999). In the rabbit, direct vasoconstriction was mainly found in small vessels (Owen 1993). In the rat, isolated proximal segments of interlobar arteries had greater responses to NPY than distal segments, that is, those closer to the arcuate arteries (Chen et al. 1997). In isolated perfused rat kidney NPY dose-dependently increased perfusion pressure and/or reduced flow (Allen et al. 1985; El-Din and Malik 1988; Hackenthal et al. 1987; Oberhauser et al. 1999; Oellerich and Malik 1993). In vivo direct intrarenal infusion of NPY reduces renal blood flow in rats (Bischoff et al. 1996), dogs (Persson et al. 1991) and in the primate *Macaca fascicularis* (Echtenkamp and Dandrige 1989). Systemic infusion of NPY and/or PYY causes intrarenal vasoconstriction in rats (Bischoff et al. 1996, 1997a, 1997c, 1998b, 1999; Blaze et al. 1997; Malmström et al. 2001a; Mezzano et al. 1998; Modin et al. 1999; Shin et al. 2000), rabbits (Allen et al. 1986b; Minson et al. 1989, 1990), pigs (Malmström et al. 1997, 1998, 2000, 2001; Malmström and Lundberg 1997; Modin et al. 1991; Pernow and Lundberg 1989a, 1989b), dogs (Malmström et al. 1998), the monkey *Macaca fascicularis* (Echtenkamp and Dandrige 1989) and man (Playford et al. 1995). The potency of NPY to reduce renal blood flow in rats is greater upon intrarenal than upon systemic administration (Bischoff et al. 1996). Taken together these data clearly show that NPY is a potent renal vasoconstrictor under a variety of experimental conditions in all species investigated. Upon systemic administration the renal vasculature seems to be more sensitive to NPY than other vascular beds, for example, the mesenteric and hindlimb vascular bed (Minson et al. 1989).

The contracting effects of NPY on the renal vasculature may be heterogeneous. Thus, systemic infusion of nonpressor doses of NPY in split hydronephrotic kidney rats and mice produces nonuniform alterations of renal vascular reactivity consisting of constriction of proximal and distal arcuate arteries with no change or even dilation of interlobular arteries or the large part of the afferent arteriole—a pattern distinct from that seen with other renal vasoconstrictors such as noradrenaline or angiotensin II (Dietrich et al. 1991; Nobiling et al. 1991).

Whether NPY infusion alters glomerular filtration rate is controversial. A reduced glomerular filtration rate upon intrarenal infusion of NPY has been observed in the primate *Macaca fascicularis* (Echtenkamp and Dandrige 1989) and in dogs (Persson et al. 1991) and upon PYY infusion in man (Playford et al. 1995). A minor reduction in glomerular filtration rate has been observed in the isolated rat kidney perfused with Tyrode's solution by some investigators (Allen et al. 1985) but not by others (Oellerich and Malik 1993). In anesthetized rats intrarenal (Bischoff et al. 1996; Smyth et al. 1988) or systemic infusion of NPY (Bischoff et al. 1996, 1997a, 1997c; Blaze et al. 1997) failed to alter glomerular filtration rate. While different techniques have been used for the assessment of glomerular filtration rate in the various studies, it appears that NPY causes only little alterations of this parameter if any in rats with a possibly greater effect in primates and humans. In face of the uniform reductions of renal blood flow upon NPY administration, the relatively minor if any reductions of glomerular filtration rate indicate that NPY may have greater vasoconstricting effects in the vas efferens than the vas afferens. This conclusion is also supported by the data from the split hydronephrotic kidney studies (Dietrich et al. 1991; Nobiling et al. 1991).

Renovascular NPY effects occur via specific NPY receptors. Thus, the renal vasoconstriction by NPY is not altered by the α-adrenoceptor antagonist, phentolamine (Minson et al. 1989), the angiotensin II receptor antagonist, losartan, the converting enzyme inhibitor, ramiprilat, or the bradykinin receptor antagonist, icatibant (Bischoff et al. 1998b) but it is inhibited by the noncompetitive NPY antagonist, PP56 (Bischoff et al. 1997c).

The receptor subtype mediating renovascular NPY effects has been characterized in vitro and in vivo using Y_1-selective agonists such as [Leu^{31},Pro^{34}]NPY (Bischoff et al. 1997a; Malmström et al. 1998; Modin et al. 1999; Oellerich and Malik 1993), Y_2-selective agonists such as NPY(13–36) (Bischoff et al. 1997a; Modin et al. 1999; Oellerich and Malik 1993), N-acetyl-[Leu^{28},Leu^{31}]NPY(24–36) (Mahns et al. 1999; Malmström et al. 1998) or PYY(3–36) (Bischoff et al. 1997a; Malmström et al. 1998), Y_1-selective antagonists such as BIBP 3226 (Bischoff et al. 1997a, 1997b; Doods et al. 1995; Lundberg and Modin 1995; Malmström et al. 1997, 1998; Malmström and Lundberg 1997; Mezzano et al. 1998; Modin et al. 1999), BIBO 3304 (Shin et al. 2000), GR 231118, formerly known as 1229U91 or GW 1229 (Hegde et al. 1995), SR 120107A (Malmström et al. 1996,1998), H394/84 (Malmström et al. 2001a) or H 409/22(Malmström et al. 2000), and also Y_2-selective antagonists such as BIIE 0246 (Malmström 2001); in some cases inactive analogs of the above such as BIBP 3435 (Lundberg and Modin 1995) or H510/45 (Malmström et al. 2000) were used. Such studies were performed in the isolated perfused kidney (Hegde et al. 1995; Oellerich and Malik 1993) and in vivo in pithed (Doods et al. 1995; Hegde et al. 1995), anesthetized (Bischoff et al. 1997a, 1997b; Mahns et al. 1999; Malmström et al. 1996, 1997, 1998, 2000, 2001; Malmström et al. 2001a; Malmström and Lundberg 1997; Mezzano et al. 1998; Shin et al. 2000) or conscious animals (Shin et al. 2000). Moreover such studies were done in rats (Bischoff et al. 1997a, 1997b; Doods et al. 1995; Hegde et al. 1995; Malmström et al. 2001a; Mezzano et al. 1998; Modin et al. 1991, 1999; Oellerich and Malik 1993; Shin et al. 2000), pigs (Durrett and Ziegler 1978; Lundberg and Modin 1995; Malmström et al. 1997, 1998, 2000; Malmström

2001; Malmström and Lundberg 1997) and dogs (Mahns et al. 1999; Malmström et al. 1998). All of these studies have reached the same conclusion, that is that NPY-induced renal vasoconstriction occurs predominantly if not exclusively via Y_1 receptors. In contrast Y_2 receptor stimulation may, if anything, cause minor relaxation of renal blood vessels, presumably by prejunctionally inhibiting the release of vasoconstricting transmitters (Bischoff et al. 1997a; Mahns et al. 1999).

Several pieces of data demonstrate that endogenously released NPY also reduces renal blood flow via a Y_1 receptor. Thus, the Y_1-selective antagonist H409/22 can attenuate nerve stimulation-induced reductions of renal blood flow in rats (DiBona and Sawin 2001) and in pigs (Malmström et al. 2000); in the latter study the inhibition was not seen with H510/45, an inactive enantiomer of H409/22. Similar inhibition of renal blood flow reduction by endogenously released NPY was seen in pigs with another Y_1-selective antagonist, BIBP 3226 (Malmström et al. 1997). The Y_1-selective antagonist H394/84 inhibited nerve stimulation-induced renal vasoconstriction in both rats and pigs (Malmström et al. 2001a).

Some studies have investigated which signal transduction pathways may be involved in renal vasoconstricting effects of NPY. Thus, pertussis toxin treatment abolishes the NPY-induced renal vasoconstriction in the isolated rat kidney (Hackenthal et al. 1987). NPY-induced vasoconstriction in many extrarenal vascular beds depends on the influx of extracellular Ca^{2+}, particularly via dihydropyridine-sensitive channels (McDermott et al. 1993; Michel and Rascher 1995). Similarly, it has been shown that omission of Ca^{2+} from the perfusate or the organic Ca^{2+} entry blockers, diltiazem and nifedipine, inhibited NPY-stimulated vasoconstriction in the isolated perfused rat kidney (El-Din and Malik 1988). Activation of voltage-operated Ca^{2+} channels by NPY is also likely because NPY can depolarize intrarenal vascular smooth muscle cells in mice; however, this was observed only in arterioles which were more than 200 µm away from the glomeruli but not in those being closer than 50 µm (Nobiling et al. 1991). Moreover, the Ca^{2+} entry blocker nifedipine did not affect peak reductions of renal blood flow induced by systemic NPY infusion in rats (Bischoff et al. 1995b). Thus, the overall role of voltage-operated Ca^{2+} channels in acute renovascular NPY effects remains somewhat unclear. Although NPY-induced vasoconstriction can be blocked by cyclooxygenase inhibitors in some vascular beds, particularly in coronary arteries (Michel and Rascher 1995), NPY-induced reductions of renal blood flow in rats were not affected by indomethacin (Bischoff et al. 1998a). Endogenous and exogenous NO were reported to buffer NPY-induced reduction of renal blood flow in pigs (Malmström et al. 2001b), but it remains unclear whether this is directly related to NPY receptor activation (Table 1).

When NPY is applied by continuous infusion, a biphasic renal blood flow response is observed over time in the anesthetized rat. This consists of a rapid reduction, followed by a sustained phase of a smaller magnitude; this partial recovery despite continued infusion indicates tachyphylaxis of the response

Table 1 Neuropeptide Y receptor subtypes in the kidney

	Y_1	Y_2	Y_4	Y_5	y_6[a]
Receptor mRNA	Human	?	Not detected in rat, mouse or human	Not detected in rat	Not detected in rabbit or human
Receptor protein	?	Rabbit	?	?	?
Receptor function	Vasoconstriction	Stimulation of Na^+/K^+-ATPase	?	Diuresis[b]	?
	Potentiation of vasoconstriction	Inhibition of noradrenaline release		Natriuresis[b]	
	Inhibition of renin release[c]			Calciuresis[b]	
				Inhibition of renin release[c]	

[a] The gene for the y_6 receptor is nonfunctional in primates and humans.
[b] These effects may occur indirectly via an extrarenal receptor.
[c] The available data do not allow us to discriminate whether a Y_1 or Y_5 receptor mediates inhibition of renin release.
?, Not investigated.

(Bischoff et al. 1996, 1997c). This partial tachyphylaxis is more pronounced upon direct intrarenal than upon systemic infusion (Bischoff et al. 1996). It is also accompanied by a partial tachyphylaxis of the renovascular resistance response (A. Bischoff and M.C. Michel, unpublished results). Acute treatment with the Ca^{2+} entry blocker, nifedipine, did not affect the initial peak reduction of renal blood flow but attenuated the sustained phase (Bischoff et al. 1995b). Thus, similar to other vasoconstricting stimuli the phasic intrarenal vasoconstriction may rely mainly on mobilization of Ca^{2+} from intracellular stores while the tonic vasoconstriction may be dependent on influx of extracellular Ca^{2+} through dihydropyridine-sensitive channels. Interestingly, the tachyphylaxis of NPY effects on renal blood flow was fastened by the bradykinin receptor antagonist, icatibant, and prevented by the angiotensin II receptor antagonist, losartan, and the converting enzyme inhibitor, ramiprilat (Bischoff et al. 1998b). Thus, tachyphylaxis of renal blood flow alterations may at least partly result from compensatory changes in the activity of the renin–angiotensin and the kallikrein–kinin system rather than from rapid desensitization of vascular NPY receptors.

In a variety of vascular beds NPY can potentiate the vasoconstricting effects of other agents (McDermott et al. 1993; Michel and Rascher 1995). Potentiation of vasoconstriction by NPY has also been observed for the renal vasculature, for example, for isolated intrarenal arteries of rats (Chen et al. 1997) and rabbits (Owen 1993). In the isolated perfused rat kidney NPY enhances the renal vasoconstriction by noradrenaline, vasopressin and angiotensin II; this potentiation response was also abolished by perfusion with Ca^{2+}-free medium or by the Y_1

antagonist BIBP 3226 (El-Din and Malik 1988; Oberhauser et al. 1999). Similarly, NPY has been shown to potentiate the renal vasoconstricting effects of the α_1-adrenoceptor agonist, methoxamine, in anesthetized rats via a Y_1 receptor (Bischoff et al. 1997b). In one study in control and diabetic rats, however, NPY did not potentiate but rather inhibited noradrenaline-induced contraction of renal blood vessels in vitro (Torffvit et al. 1999). Thus, NPY has not only direct vasoconstricting effects in the renal vasculature but also can potentiate the effects occurring via other receptor systems. The limited available data suggest that direct and potentiating renovascular NPY effects may use similar receptor subtypes and signaling mechanisms.

5
Regulation of Water and Electrolyte Excretion by NPY

Based on the marked reductions of renal blood flow (see above), it could be expected that NPY inhibits water and electrolyte excretion. Indeed NPY infusion has been reported to reduce diuresis, natriuresis and kaliuresis in dogs (Persson et al. 1991). However, studies with isolated perfused rat kidneys (Allen et al. 1985) and with systemic NPY administration in anesthetized and conscious rats have reported an enhanced urine formation upon NPY or PYY administration (Bischoff et al. 1996, 1997a, 1997c, 1998a, 1998b; Smyth et al. 1988). Diuresis has also been observed in humans upon administration of PYY doses corresponding to physiological postprandial plasma levels (Playford et al. 1995). While one study in rats did not confirm PYY-induced diuresis (Blaze et al. 1997), it used a tenfold lower PYY dose than the studies yielding positive results (47 pmol/kg/min compared to 2 µg/kg/min).

In anesthetized rats systemic infusion of NPY enhances urine formation and excretion of sodium and calcium by up to 110%, 110% and 45%, respectively, while effects on potassium excretion were reported to be absent (Bischoff et al. 1996; Smyth et al. 1988) or at least considerably weaker than those on sodium excretion (Bischoff and Michel 2000). These diuretic effects were not accompanied by alterations of creatinine clearance (Bischoff et al. 1996, 1998a, 1998b; Blaze et al. 1997). The NPY-induced diuresis is slow in onset and requires at least 30–45 min of continued agonist infusion to develop fully. Upon termination of agonist infusion it is reversible, but that also requires 30–45 min for a complete recovery. NPY-induced diuresis is observed at agonist doses which are at least 10 times higher than the threshold doses for reductions of renal blood flow in the same animals. The marked enhancements of sodium and calcium excretion with much lower enhancement of potassium excretion (Bischoff et al. 1996; Bischoff and Michel 2000; Smyth et al. 1988) indicate that enhanced urine formation may occur at a site distal to the Na^+/K^+ exchange mechanism in the distal tubules. However, this remains to be confirmed by direct micropuncture studies. Studies with subtype-selective agonists and the Y_1-selective antagonist, BIBP 3226, have proposed that NPY may cause diuresis and natriuresis via a Y_5-like receptor (Bischoff et al. 1997a) but this requires confirmation by Y_5-selective antagonists. Pertussis toxin treatment abolished the NPY-induced enhancements of diuresis and natriuresis in anesthetized rats (Smyth et al. 1988) indicating mediation via a G-protein of the G_i/G_o family. Based on the above data as well as the role of NPY in the regulation of food intake, it has been proposed that NPY may play a crucial role in fasting-induced natriuresis (Sulyok and Tulassay 2001).

The NPY-induced diuresis is a mechanistically complex phenomenon and may represent the integration of a multitude of direct and indirect effects including alterations of tubular function, blood pressure, prejunctional sympatholysis, and of the functional states of the renin-angiotensin, atrial natriuretic peptide and kinin-kallikrein systems. The vast majority of studies in this respect have been done in rats. Therefore, the following primarily refers to rats unless specifically indicated otherwise.

At the tubular level NPY stimulates Na^+/K^+-ATPase activity in proximal tubules (Holtbäck et al. 1998; Ohtomo et al. 1994, 1996a, 1996b). In cortical collecting tubules attenuation of vasopressin-stimulated hydraulic conductivity (Dillingham and Anderson 1989), chloride secretion and cAMP accumulation (Breen et al. 1998) have been observed. The latter two effects were shown to involve pertussis toxin-sensitive G-proteins (Breen et al. 1998). The exact relationship of these phenomena and urine excretion is not clear, but the stimulation of Na^+/K^+-ATPase (Ohtomo et al. 1994) and the inhibition of vasopressin-stimulated chloride secretion (Breen et al. 1998) occur via Y_2 receptors, while Y_2-selective NPY analogs, for example NPY(13-36), do not cause diuresis (Bischoff et al. 1997a).

Sympathetic stimulation may inhibit urine formation, and this may occur predominantly by the action of noradrenaline on tubular α-adrenoceptors (Akpogomeh and Johns 1990). Stimulation of prejunctional NPY receptors in a variety of tissues and species potently inhibits noradrenaline release from sympathetic nerve terminals (Edvinsson et al. 1987). In the isolated perfused rat kidney NPY can also inhibit the nerve stimulation-induced noradrenaline release, but this was observed with high (10-16 Hz) but not low frequency (4 Hz) stimulation (Oellerich and Malik 1993). A more detailed investigation of presynaptic NPY receptors has been presented for rabbit and human kidney (Rump et al. 1997). These studies have shown that inhibition of noradrenaline release occurs via Y_2 receptors and is sensitive to pertussis toxin. In vivo studies have demonstrated that Y_2-selective agonists – NPY(13-36) or PYY(3-36) – slightly enhance renal blood flow in anesthetized rats while NPY, PYY and their Y_1-selective analogs reduce renal blood flow (Bischoff et al. 1997a; Mahns et al. 1999), which is consistent with a Y_2 receptor-mediated inhibition of noradrenaline release. The relationship of this phenomenon to enhanced urine formation is unclear since Y_2-selective NPY analogs did not cause diuresis (Bischoff et al. 1997a).

One study has compared the dose-response curves for renal NPY effects upon systemic and intrarenal NPY administration (Bischoff et al. 1996). As expected it has reported that intrarenal NPY administration causes more potent reductions of renal blood flow than systemic infusion. Surprisingly, NPY-induced diuresis and natriuresis were markedly attenuated or even converted into anti-natriuresis upon intrarenal administration. As systemic NPY infusion causes diuresis via a Y_5-like receptor (Bischoff et al. 1997a), a subtype which is not detectable in rat kidney at the mRNA level (Gerald et al. 1996; Hu et al. 1996), these data indicate that NPY might cause diuresis and natriuresis by acting on an extrarenal receptor. On the other hand, Smyth et al. (1988) have reported NPY-induced diuresis upon intrarenal agonist infusion. To reconcile these contradictory results, two technical differences between the studies should be considered. Firstly, Bischoff et al. (1996) performed retrograde infusions via a catheter placed in the suprarenal artery, while Smyth et al. (1988) placed a needle in the main renal artery via the aorta. Secondly, Smyth et al. (1988) studied only the effects of local infusion and thus cannot exclude the possibility of agonist spillover into the systemic circulation. Considerable NPY spillover into the systemic circulation has been described upon intrarenal infusion in dogs (Persson et al. 1991). Allen et al. (1985) have reported NPY-induced diuresis in isolated perfused rat kidneys. While this preparation clearly excludes extrarenal factors, it has other limitations, for example the preparation is stable for only 2 h, glomerular filtration rate is around 60% of the rate seen in vivo and perfusion pressure and perfusion flow rate are at unphysiological high levels.

In light of the above findings on direct NPY effects on tubular function and on intrarenal sympatholysis, it can be assumed that NPY may cause some diuresis by an intrarenal mechanism. The NPY-induced alterations of intrarenal hemodynamics which are different from those by other renal vasoconstrictors, for example angiotensin II or endothelin, might contribute to these intrarenal diuretic effects (Dietrich et al. 1991). However, these effects may be insufficient to overcome the anti-diuretic effects of pronounced, Y_1 receptor-mediated reductions of renal blood flow (see above). Thus, the major part of NPY-induced diuresis and natriuresis upon systemic administration may result from stimulation of an extrarenal receptor, possibly of the Y_5 subtype. Therefore, the possible mediator of diuretic NPY effects will be discussed below.

Systemic NPY administration elevates blood pressure (Michel and Rascher 1995) and blood pressure alterations can profoundly affect renal function (Firth et al. 1990). Hence, some of the renal NPY effects may occur secondarily to systemic hemodynamic alterations. NPY-induced blood pressure elevations occur via Y_1 receptors (Kirby et al. 1995; Michel et al. 1992). This might enhance urine formation via the pressure natriuresis mechanisms and/or by baroreflex-mediated withdrawal of sympathetic tone from the kidney. The latter mechanism is not very likely, since acute renal denervation does not inhibit NPY-induced diuresis (Bischoff et al. 1998b). The role of the pressure natriuresis mechanism has been investigated by three experimental techniques to inhibit pressure natriuresis. Firstly, experiments have been done in rats with renal decapsulation; this did not modify the NPY-induced diuresis (Bischoff et al. 1996). Secondly, experiments were performed in rats, where renal perfusion pressure was fixed at 100 mmHg with an adjustable mechanical clamp on the abdominal aorta; this partially inhibited NPY-induced diuresis but did not abolish it (Bischoff et al.

1996). Thirdly, NPY-induced blood pressure alterations were pharmacologically prevented by concomitant infusion of nitroprusside sodium; this also failed to abolish NPY-induced diuresis (J. Smits, A. Bischoff and M.C. Michel, personal communication). Moreover, NPY analogs that do not elevate blood pressure, for example PYY(3–36), caused similar diuresis as NPY (Bischoff et al. 1997a). The Y_1 antagonist, BIBP 3226, prevents NPY-induced blood pressure elevations (Doods et al. 1996) but does not inhibit NPY-induced diuresis (Bischoff et al. 1997a); similar data were observed for the combination of BIBP 3226 and the Y_1/Y_5 agonist, [Leu^{31},Pro^{34}]NPY. The noncompetitive NPY antagonist, PP56, also inhibits NPY-induced blood pressure elevations but not diuresis or natriuresis (Bischoff et al. 1997c). Finally, systemically infused PYY causes diuresis in humans at doses where it does not alter blood pressure (Playford et al. 1995). Thus, activation of the pressure natriuresis mechanism by NPY is possible but it is not a predominant factor in diuresis and natriuresis caused by systemic NPY administration.

Therefore, studies have been performed to identify a possible humoral mediator of NPY-induced diuresis. Vasopressin is not a likely candidate since vasopressin increases free water clearance while NPY causes diuresis and natriuresis to a similar degree (Bischoff et al. 1996) and enhances osmolar clearance (Smyth et al. 1988). Moreover, we have not observed alterations of plasma vasopressin concentrations upon NPY infusion (W Rascher, A Bischoff and MC Michel, unpublished results). Another potential mediator would be atrial natriuretic peptide. Systemic infusion of NPY has been reported to enhance release of atrial natriuretic peptide, but this was not mimicked by infusion of similar doses of PYY (Baranowska et al. 1987). On the other hand PYY causes diuresis at least as potent as NPY (Bischoff et al. 1997a). Thus, atrial natriuretic peptide release appear to contribute little if at all to NPY-induced diuresis.

Another possible mediator of the diuretic effects of NPY is the renin–angiotensin system. The juxtaglomerular region is richly endowed with NPY-containing nerves and expresses mRNA for Y_1 receptors (see above). Thus, inhibition of renin release and/or lowering of plasma renin activity by NPY have been observed under a variety of experimental conditions. These include the isolated perfused rat kidney (Hackenthal et al. 1987), systemic infusion in conscious (Aubert et al. 1992) and anesthetized rats (Bischoff et al. 1997a), conscious rabbits (Allen et al. 1986b), anesthetized cats (Corder et al. 1989) and conscious humans (Ahlborg and Lundberg 1994; Playford et al. 1995). On the other hand, studies in conscious dogs (Persson et al. 1991) and in the primate *Macaca fascicularis* (Echtenkamp and Dandrige 1989) did not report alterations of plasma renin activity or renin secretion upon intrarenal or systemic NPY administration. One study in conscious rats has also failed to detect lowering of basal plasma renin activity by NPY infusion, but reported inhibition of captopril-induced elevations (Aubert et al. 1988). The enhancements of renin release which are observed upon β-adrenoceptor stimulation can also be inhibited by NPY (Aubert et al. 1992), although one study on rat isolated renal cortical slices failed to detect NPY-induced inhibition of isoprenaline-induced renin release (Shin et al.

2000). NPY administration can also normalize pathophysiologically elevated plasma renin activities, which can occur following adrenalectomy (Pfister et al. 1986), stenosis of the renal artery (Waeber et al. 1990), or postmyocardial infarction (Zelis et al. 1994). Taken together these data demonstrate that NPY can lower plasma renin activity, presumably by inhibition of renin release, under most experimental conditions. This occurs via a Y_1-like receptor (Aubert et al. 1992; Bischoff et al. 1997a; Evequoz et al. 1996). The subtype-selective agonists which have been used do not allow a clear discrimination between Y_1 and Y_5 receptors, but a participation of the Y_5 receptor is unlikely to be due to the lack of its expression in the kidney (Gerald et al. 1996; Hu et al. 1996). The inhibition of renin release occurs via a pertussis toxin-sensitive G-protein but is not inhibited by the Ca^{2+} entry blocker, methoxyverapamil, or the calmodulin antagonist, calmidazolium (Hackenthal et al. 1987).

While the above data clearly establish the renin–angiotensin system as a potential mediator of diuretic NPY effects, they do not prove its physiological role. Therefore, studies were performed with the angiotensin II receptor antagonist, losartan, and the converting enzyme inhibitor, ramiprilat (Bischoff et al. 1998b). While the applied dose of losartan inhibited the blood pressure-elevating effects of exogenous angiotensin II, it did not affect the NPY-induced diuresis. The applied dose of ramiprilat inhibited the blood pressure-elevating effects of angiotensin I, but also failed to inhibit NPY-induced diuresis. These pharmacological inhibitor data clearly demonstrate that the renin-angiotensin system is not the main mediator of the diuretic NPY effects.

Ramiprilat not only failed to prevent NPY-induced diuresis but actually enhanced it considerably (Bischoff et al. 1998b). In the same study the B2 bradykinin receptor antagonist, icatibant, inhibited NPY-induced diuresis and natriuresis and abolished calciuresis. Since ramiprilat not only prevents the formation of angiotensin II but also prevents the degradation of bradykinin, these data strongly suggest that bradykinin is an important mediator of systemic NPY effects on tubular function. Several other pieces of evidence also support this hypothesis. Thus, bradykinin causes diuresis and natriuresis upon intrarenal infusion in conscious and anesthetized dogs (Blasingham and Nasjletti 1979; Granger and Hall 1985; Lortie et al. 1992; Siragy 1993) and upon systemic infusion in conscious rats, which had been rendered hypertensive by deoxycorticosterone treatment (Pham et al. 1996). The diuretic and natriuretic effects of bradykinin mainly occur in distal nephron segments (Chen and Lokhandwala 1995; Schuster et al. 1984). NPY also appears to cause diuresis mainly by altering the function of distal nephron segments (see above). Finally, bradykinin stimulates renal prostaglandin formation (Grenier et al. 1981; Malik and Nasjletti 1979), and accordingly its diuretic and natriuretic effects can be inhibited by cyclooxygenase inhibitors (Blasingham and Nasjletti 1979; Carretero and Scicli 1980). Similarly, the cyclooxygenase inhibitor, indomethacin, also inhibits the diuretic and natriuretic NPY effects (Bischoff et al. 1998a). These data suggested that bradykinin could be an important mediator of tubular NPY effects. This could occur if stimulation of the extrarenal Y_5 NPY receptors causes bradykinin release, which then reaches the kidney via the bloodstream to act intrarenally on prostaglandin formation. The very short plasma half-life of bradykinin (Regoli and Barabe 1980), however, casts doubt on this mode of action. Alternatively it is possible that the extrarenal Y_5 receptor activates yet another mediator which causes intrarenal bradykinin formation and/or release. To address these possibilities directly two types of studies were performed, measurement of renal bradykinin excretion upon systemic infusion of a diuretic dose of PYY and intrarenal infusion of bradykinin (Bischoff et al. 1999). However, PYY infusion did not enhance but rather significantly reduced urinary bradykinin excretion, and bradykinin infusion did not cause diuresis at doses where it had clear renal hemodynamic effects. While these data do not support a role of bradykinin as mediator of diuretic NPY effects, they cannot exclude a possible intrarenal bradykinin generation following NPY receptor stimulation, which may be sufficient to alter tubular function but is not reflected in urinary bradykinin excretion. In light of the above inhibitor data, particularly those with ramiprilat and icatibant, we still consider this to be the most likely mechanism underlying diuretic and natriuretic NPY effects.

6
A Functional Role for Endogenous NPY-Related Peptides in the Kidney?

The above studies clearly demonstrate that exogenous NPY can affect renal function at the vascular and tubular level. Initial evidence for a role of endogenous NPY in the regulation of renovascular function came from studies with reserpinized pigs (Modin et al. 1994). In those experiments the vasoconstriction response to renal nerve stimulation decreased upon reserpine treatment with only a minor fraction remaining. This fraction gradually decreased further upon repeated stimulation, which occurred in parallel with gradual reductions of NPY release (Modin et al. 1994). While these data have indirectly indicated that NPY may play a physiological role in the regulation of renal blood flow, more direct evidence comes from antagonist studies. Thus, the NPY receptor antagonist SR 120107A inhibits the nerve stimulation-induced vasoconstriction in the kidney but not in several other tissues in pigs in vivo (Malmström and Lundberg 1996). In conscious rats, the Y_1 antagonist H409/22 significantly attenuated reductions of renal blood flow induced by renal nerve stimulation, whereas it did not affect renal hemodynamics in anesthetized rats in the same study (DiBona and Sawin 2001). In another study with anesthetized rats the NPY antagonists, BIBP 3226 (Bischoff et al. 1997a) and PP56 (Bischoff et al. 1997c), both increased renal blood flow over time upon continuous infusion, although this did become apparent only after extended observation periods. These effects occurred at doses which antagonized exogenous NPY and but did not concomitantly alter blood pressure. Taken together, these data indicate that endogenous NPY can also cause renal vasoconstriction and that this mechanism may be tonically active under some conditions such as certain forms of anesthesia. Thus, endogenous NPY may be more important in the regulation of vasoconstriction in the kidney than in other vascular beds.

In the above studies PP56, BIBP 3226 or H409/22 did not alter urine formation or electrolyte excretion (Bischoff et al. 1997a, 1997c; DiBona and Sawin 2001; Zhao et al. 1999). This could be explained by a selective innervation of renal blood vessels relative to the tubules with NPY-containing nerve fibers. In light of the above data on a possible extrarenal location of the NPY receptor which mediates diuresis and natriuresis, it is more likely that plasma levels of NPY are not sufficiently high to tonically activate this extrarenal receptor. This is plausible since NPY is a neurotransmitter that acts mainly in the synaptic cleft, while detectable plasma NPY is considered to reflect spillover (Zukowska-Grojec and Wahlestedt 1993). Thus, endogenous NPY is not very likely to cause diuresis and natriuresis tonically with the possible exception of extreme sympathoadrenal stimulation, for example during open chest surgery (Lundberg et al. 1985) or in patients with phaeochromocytoma (Michel and Rascher 1995). On the other hand, the related peptide PYY is primarily a systemically circulating hormone. Plasma PYY levels in particular increase following a meal (Pironi et al. 1993). This postprandial elevation of plasma PYY is due mainly to an increase in occurrence of PYY(3–36) (Grandt et al. 1994), a peptide which does not increase blood pressure but acts on Y_5 receptors to cause diuresis (Bischoff et al. 1997a). Unfortunately no studies on renal function following a meal have been done with NPY antagonists, but infusion of exogenous PYY at doses corresponding to postprandial plasma levels causes diuresis and natriuresis in human volunteers (Playford et al. 1995). Therefore, we speculate that endogenous NPY may be relevant for the regulation of renovascular function, while endogenous PYY, particularly in the postprandial phase, may participate in the regulation of tubular function.

7
Regulation of Renal NPY Content and Responsiveness

Since our last review on the subject (Bischoff and Michel 1998), a considerable number on studies has been published on the physiological and pathophysiological regulation of renal NPY content and/or NPY responsiveness. With regard to the pathophysiological regulation of renal NPY content most studies have focused on various rat models of arterial hypertension. In the genetic model of spontaneously hypertensive as compared to normotensive Wistar Kyoto rats little if any alteration of renal NPY content was found in study using young adult (3–4 months) and old (17–18 months) rats (Pavia and Morris 1994). On the other hand, Corder (2000) reported a reduced NPY content in kidneys as well as well as atria, spleen and adrenals from spontaneously hypertensive rats. In a model in which blood pressure is chronically elevated by aortic coarctation, the number of renal nerve fibers with NPY-like immunoreactivity as well as the renal NPY content as determined by radioimmunoassay decreased (Ballesta et al. 1987). In models of renal hypertension such as the one-clip-one-kidney and one-clip-two-kidney Goldblatt hypertension, renal NPY content was also reported to be decreased (Allen et al. 1986a; Haefliger et al. 1999); this reduction was detected both at the protein and the mRNA level (Haefliger et al. 1999) and interestingly not only in the kidney directly affected by renal artery stenosis but also, although to a lesser extent, in the non-ischemic kidney (Neri et al. 1991). Since hypertension due to renal artery stenosis belongs to the group of high renin hypertension, it is interesting to note that renal NPY expression was not altered in deoxycorticosterone/salt-induced blood pressure elevation, a model of low renin hypertension (Haefliger et al. 1999). The latter authors interpreted these data to indicate that angiotensin II should play a role in the regulation of renal NPY expression, a notion which is supported by the finding that, at least in atria and skeletal muscle, chronic treatment with the converting enzyme inhibitor lisinopril increased NPY content (Corder 2000). Another hypothesis to explain reduced renal NPY content in some forms of hypertension is based on comparison with alterations of NPY content in other tissues and in plasma in hypertension; this indicates that a reduced renal NPY content may reflect enhanced release which is not fully compensated by de novo synthesis (Michel and Rascher 1995). Finally, an increased NPY content was reported in inflamed kidneys of a mouse model of systemic lupus erythematosus (Bracci-Laudiero et al. 1998), whereas renal NPY content was found to remain unaltered in a rat model of congestive heart failure (Zhang et al. 1999).

With regard to physiological differences in NPY responsiveness, one study has compared the effects of infusion of incremental NPY doses in male and female rats (Bischoff et al. 2000). In these experiments NPY dose-dependently increased renovascular resistance in both sexes; since the responses desensitized faster in males than in females, the cumulative elevations were greater in females. NPY-induced diuresis and natriuresis were similar in male and female rats.

Studies on the regulation of renal NPY responsiveness in disease states have mostly focused on arterial hypertension and congestive heart failure. In spontaneously hypertensive as compared with normotensive Wistar-Kyoto rats infusion of incremental NPY doses caused similar blood pressure elevations in both strains but in the hypertensive rats basal renovascular resistance was greater, and NPY caused considerably greater additional elevations (Bischoff et al. 2000). On the other hand, NPY-induced diuresis and natriuresis appeared smaller in hypertensive than in normotensive rats, but this may be related to a greater basal urine and electrolyte excretion in the hypertensive rats in that study. An enhanced NPY responsiveness of the renal vasculature was also reported in Goldblatt hypertensive rats (Mezzano et al. 1998). Whether this reflects a specific regulation of NPY responsiveness remains unclear since renovascular noradrenaline responses were similarly enhanced in that study. An opposite conclusion was reached in a study with rabbits where renal hypertension was induced by cellophane wrapping of the kidneys (Minson et al. 1990). In this model plasma NPY levels increase, and this was accompanied by a partial desensitization of the renal vasoconstricting effects of NPY. These conflicting results are difficult to interpret due to the very different profile of renal NPY receptor expression in rats and rabbits (see above).

To study the role of NPY in the development of renovascular hypertension, chronic treatment with either NPY or the Y_1 antagonist BIBO 3304 were performed. Despite the ability of acute NPY administration to elevate blood pressure, Waeber et al. (1990) reported that a chronic NPY infusion can prevent the development of renal hypertension in a two-kidney one-clip rat model, possibly secondary to lowering of plasma renin activity (see above). In marked contrast, Shin et al. (2000) used the same rat model of renal hypertension but observed a prevention of blood pressure elevation by the antagonist BIBO 3304. Further studies are clearly necessary to determine whether the blood pressure elevation following renal artery stenosis is due to too much or too little NPY.

Chronic heart failure is a disease state that is accompanied by altered renal function, that is characterized by elevated NPY plasma levels (Feng et al. 1994; Maisel et al. 1989). These elevated NPY plasma levels may indirectly modify renal function since exogenous NPY decreases plasma renin activity in a postmyocardial infarction model of heart failure in rats (Zelis et al. 1994). In rat models of congestive heart failure based upon ligation of a coronary artery, the nonspecific NPY antagonist PP 56 and the Y_1 antagonists BIBP 3326 and H409/22 were reported to lower blood pressure (DiBona and Sawin 2001; Sun et al. 1995; Zhao et al. 1999). This was accompanied by a PP 56-induced diuresis and natriuresis in heart failure rats (Sun et al. 1995), whereas BIBP 3326 or H409/22 did not affect basal renal function (DiBona and Sawin 2001; Zhao et al. 1999). While NPY directly contracted the renal artery and potentiated the contraction induced by noradrenaline in control rats, both responses were largely absent in heart failure rats (Zhao et al. 1999). Since both responses were sensitive to the Y_1 antagonist BIBP 3326, the authors concluded that heart failure was associated with a desensitization of renovascular Y_1 NPY receptors. Nevertheless, NPY may still be an important regulator of renovascular function in heart failure since nerve stimulation-induced renal vasoconstriction was more sensitive to the NPY antagonist H409/22 in heart failure than in control rats (DiBona and Sawin 2001).

Finally, it was reported that reperfusion after 2 h of renal ischemia causes local overflow of noradrenaline but not of NPY in pigs (Malmström and Lundberg 1997). In that study in vivo renal vasoconstriction responses to PYY (but also to the α_1-adrenoceptor agonist phenylephrine and to angiotensin II) were strongly attenuated after the ischemic period, whereas that to $\alpha\beta$-methylene-ATP was maintained to a greater extent. After a 15-min period of ischemia, however, PYY-induced renal vasoconstriction was enhanced (Malmström and Lundberg 1997).

8
NPY System in Renal Disease

End-stage renal failure is a disease state accompanied by numerous abnormalities of autonomic regulation, including elevated plasma catecholamine levels

(Dhein et al. 2000; Ziegler et al. 1990). Since NPY is a co-transmitter of the autonomic nervous system, it seemed logical to assess possible alterations of plasma NPY levels in end-stage renal failure. Thus, elevated plasma NPY concentrations have been reported in various groups of patients with end-stage renal failure including adults and children as well as patients undergoing hemodialysis or peritoneal dialysis (Bald et al. 1997; Crum et al. 1991; Hökfelt 1991; Miller et al. 1990; Takahashi et al. 1987). One study also detected elevated plasma NPY in children with impaired renal function not yet requiring dialysis treatment (Bald et al. 1997). In contrast, Kokot et al. (1999) did not detect elevated plasma NPY in hemodialysis patients as compared to healthy controls, but rather reported enhanced plasma NPY in kidney transplant patients. Another study reported a tendency for increased plasma NPY in hemodialysis patients which failed to reach statistical significance (Klin et al. 1998). Taken together, these data suggest that renal failure is associated with elevated plasma NPY concentrations, but the extent of these elevations is such that it does not consistently reach statistical significance in all studies. The elevated plasma NPY concentrations are not affected to a relevant extent by hemodialysis in children or adults with end-stage renal failure (Bald et al. 1997; Crum et al. 1991; Hegbrant et al. 1995; Klin et al. 1998).

9
Conclusions and Directions for Future Research

In summary NPY is present in and released from the kidneys of various species. It can act on pre- and postjunctional intrarenal receptors to modify renovascular and tubular function. Enhancements of urine and electrolyte excretion following systemic administration of exogenous NPY may mostly occur indirectly. They appear to involve an extrarenal NPY receptor and bradykinin and prostaglandin formation as intermediary steps; PYY may be the physiological agonist at this extrarenal receptor. The exact relationship between the direct and multiple indirect effects of NPY on renal function remains to be established. This will require definition of the anatomical location of the NPY receptor which mediates diuresis and natriuresis. Based upon several antagonist studies it appears that endogenous NPY plays a greater role in the regulation of vascular tone than in that of tubular function. Moreover, its role seems to be minor in conscious healthy animals but may gain importance under pathophysiological conditions. Finally, a number of pathophysiological conditions, most notably congestive heart failure and certain forms of arterial hypertension, can regulate renal NPY expression and/or responsiveness.

Acknowledgements. Work in the author's lab was supported by grants of the Deutsche Forschungsgemeinschaft. The literature search for this article was completed in January 2003.

References

Ahlborg G, Lundberg JM (1994) Inhibitory effects of neuropeptide Y on splanchnic glycogenolysis and renin release in humans. Clin Physiol 14:187–196

Akpogomeh BA, Johns EJ (1990) The α-adrenoceptor mediating the tubular actions of the renal nerves in spontaneously hypertensive and stroke-prone spontaneously hypertensive rats. J Auton Pharmacol 10:201–212

Allen JM, Godfrey NP, Yeats JC, Bing RF, Bloom SR (1986a) Neuropeptide Y in renovascular models of hypertension in the rat. Clin Sci 70:485–488

Allen JM, Hanson C, Lee Y, Mattin R, Unwin RJ (1986b) Renal effects of the homologous neuropeptides pancreatic polypeptide (PP) and neuropeptide Y (NPY) in conscious rabbits. J Physiol (London) 376:24P

Allen JM, Raine AEG, Ledingham JGG, Bloom SR (1985) Neuropeptide Y: a novel renal peptide with vasoconstrictor and natriuretic activity. Clin Sci 68:373–377

Aubert J-F, Walker P, Grouzmann E, Nussberger J, Brunner HR, Waeber B (1992) Inhibitory effect of neuropeptide Y on stimulated renin secretion in awake rats. Clin Exp Pharmacol Physiol 19:223–228

Aubert JF, Burnier M, Waeber B, Nussberger J, Dipette DJ, Burris JF, Brunner HR (1988) Effects of a nonpressor dose of neuropeptide Y on cardiac output, regional blood flow distribution and plasma renin, vasopressin and catecholamine levels. J Pharmacol Exp Ther 244:1109–1115

Bald M, Gerigk M, Rascher W (1997) Elevated plasma concentrations of neuropeptide Y in children and adults with chronic and terminal renal failure. Am J Kidney Dis 30:23–27

Ballesta J, Lawson JA, Pals DT, Ludens JH, Lee YC, Bloom SR, Polak JM (1987) Significant depletion of NPY in the innervation of the rat mesenteric, renal arteries and kidneys in experimentally (aorta coarctation) induced hypertension. Histochemistry 87:273–278

Ballesta J, Polak JM, Allen JM, Bloom SR (1984) The nerves of the juxtaglomerula apparatus of man and other mammals contain the potent peptide NPY. Histochemistry 80:483–485

Baranowska B, Gutkowska J, Lemire A, Cantin M, Genest J (1987) Opposite effects of neuropeptide Y (NPY) and polypeptide YY (PYY) on plasma immunoreactive atrial natriuretic factor (IR-ANF) in rats. Biochem Biophys Res Commun 145:680–685

Bard JA, Walker MW, Branchek TA, Weinshank RL (1995) Cloning and functional expression of a human Y_4 subtype receptor for pancreatic polypeptide, neuropeptide Y, and peptide YY. J Biol Chem 270:26762–26765

Bischoff A, Avramidis P, Erdbrügger W, Münter K, Michel MC (1997a) Receptor subtypes Y_1 and Y_5 are involved in the renal effects of neuropeptide Y. Br J Pharmacol 120:1335–1343

Bischoff A, Erdbrügger W, Smits J, Michel MC (1996) Neuropeptide Y-enhanced diuresis and natriuresis in anaesthetized rats is independent from renal blood flow reduction. J Physiol (London) 495:525–534

Bischoff A, Freund A, Michel MC (1997b) The Y_1 antagonist BIBP 3226 inhibits potentiation of methoxamine-induced vasoconstriction by neuropeptide Y. Naunyn-Schmiedeberg's Arch Pharmacol 356:635–640

Bischoff A, Gerbracht A, Michel MC (2000) Gender and hypertension interact to regulate neuropeptide Y responsiveness. Naunyn-Schmiedeberg's Arch Pharmacol 361:173–180

Bischoff A, Kötting A, Erdbrügger W, Schimiczek M, Grandt D, Michel MC (1995a) Radioligand binding detects NPY receptors in rabbit and guinea pig but not in rat or human kidney. Naunyn-Schmiedeberg's Arch Pharmacol 352 (Suppl):R25

Bischoff A, Limmroth V, Michel MC (1998a) Indomethacin inhibits the natriuretic effects of neuropeptide Y in anesthetized rats. J Pharmacol Exp Ther 286:704–708

Bischoff A, Michel MC (1998) Renal effects of neuropeptide Y. Pflügers Arch 435:443–453

Bischoff A, Michel MC (2000) Neuropeptide Y enhances potassium excretion by mechanisms distinct from those contolling sodium excretion. Can J Physiol Pharmacol 78:93–99

Bischoff A, Neumann A, Dendorfer A, Michel MC (1999) Is bradykinin a mediator of renal neuropeptide Y effects? Pflügers Arch 438:797–803

Bischoff A, Rascher W, Michel MC (1998b) Bradykinin may be involved in neuropeptide Y-induced diuresis, natriuresis, and calciuresis. Am J Physiol 275: F502–F509

Bischoff A, Stickan-Verfürth M, Michel MC (1995b) Interaction between neuropeptide Y and nifedipine in the regulation of rat renal function. Br J Pharmacol 116 (Suppl):153P

Bischoff A, Stickan-Verfürth M, Michel MC (1997c) Renovascular and tubular effects of neuropeptide Y are discriminated by PP56 (D-myo-inositol 1,2,6-triphosphate) in anaesthetized rats. Pflügers Arch 434:57–62

Blasingham MC, Nasjletti A (1979) Contribution of renal prostaglandins to the natriuretic action of bradykinin in the dog. Am J Physiol 237:F182–F187

Blaze CA, Mannon PJ, Vigna SR, Kherani AR, Benjamin BA (1997) Peptide YY receptor distribution and subtype in the kidney: effect on renal hemodynamics and function in rats. Am J Physiol 273:F545–F553

Bracci-Laudiero L, Aloe L, Stenfors C, Theodorsson E, Lundeberg T (1998) Development of systemic lupus erythematosus in mice is associated with alteration of neuropeptide concentrations in inflamed kidneys and immunoregulatory organs. Neurosci Lett 248:97–100

Breen CM, Mannon PJ, Benjamin BA (1998) Peptide YY inhibits vasopressin-stimulated chloride secretion in inner medullary collecting duct cells. Am J Physiol 275:F452–F4457

Burkhoff AM, Linemeyer DL, Salon JA (1998) Distribution of a novel hypothalamic neuropeptide Y receptor gene and its absence in rat. Mol Brain Res 53:311–316

Carretero OA, Scicli AG (1980) The renal kallikrein-kinin system. Am J Physiol 238:F247–F255

Chen C, Lokhandwala MF (1995) Potentiation by enalaprilat of fenoldopam-evoked natriuresis is due to blockade of intrarenal production of angiotensin-II in rats. Naunyn-Schmiedeberg's Arch Pharmacol 352:194–200

Chen H, Bischoff A, Schäfers RF, Wambach G, Philipp T, Michel MC (1997) Vasoconstriction of rat renal interlobar arteries by noradrenaline and neuropeptide Y. J Auton Pharmacol 17:137–146

Chen X, Knuepfer MM, Westfall TC (1990) Hemodynamic and sympathetic effects of spinal administration of neuropeptide Y in rats. Am J Physiol 259:H1674–H1680

Chevendra V, Weaver LC (1992) Distributions of neuropeptide Y, vasoactive intestinal peptide and somatostatin in populations of postganglionic neurons innervating the rat kidney, spleen and intestine. Neuroscience 50:727–743

Corder R (2000) Effect of lisinopril on tissue levels of neuropeptide Y in normotensive and spontaneously hypertensive rats. J Human Hypert 16:381–384

Corder R, Vallotton MB, Lowry PJ, Ramage AG (1989) Neuropeptide Y lowers plasma renin activity in the anaesthetised cat. Neuropeptides 14:111–114

Crum RL, Fairchild R, Bronsther O, Dominic W, Ward D, Fernandez R, Brown MR (1991) Neuroendocrinology of chronic renal failure and renal transplantation. Transplantation 52:818–823

Dhein S, Röhnert P, Markau S, Kotchi-Kotchi E, Becker K, Poller U, Osten B, Brodde O-E (2000) Cardiac beta-adrenoceptors in chronic uremia: studies in humans and rats. J Am Coll Cardiol 36:608–617

DiBona GF, Sawin LL (2001) Role of neuropeptide Y in renal sympathetic vasoconstriction: studies in normal and congestive heart failurerats. J Lab Clin Med 138:119–129

Dietrich MS, Fretschner M, Nobiling R, Persson PB, Steinhausen M (1991) Renovascular effects of neuropeptide-Y in the split hydronephrotic rat kidney: non-uniform pattern of vascular reactivity. J Physiol (London) 444:303–315

Dillingham MA, Anderson RJ (1989) Mechanism of neuropeptide Y inhibition of vasopressin action in rat cortical collecting tubule. Am J Physiol 256:F408–F413

Doods HN, Wieland HA, Engel W, Eberlein W, Willim K-D, Entzeroth M, Wienen W, Rudolf K (1996) BIBP 3226, the first selective NPY Y_1 receptor antagonist: a review of its pharmacological properties. Regul Pept 65:71–77

Doods HN, Wienen W, Entzeroth M, Rudolf K, Eberlein W, Engel W, Wieland HA (1995) Pharmacological characterization of the selective nonpeptide neuropeptide Y Y_1 receptor antagonist BIBP 3226. J Pharmacol Exp Ther 275:136–142

Durrett LR, Ziegler MG (1978) A sensitive radioenzymatic assay for catechol drugs. J Neurosci Res 5:191–194

Echtenkamp SF, Dandrige PF (1989) Renal actions of neuropeptide Y in the primate. Am J Physiol 256:F524–F531

Edvinsson L, Hakanson R, Wahlestedt C, Uddman R (1987) Effects of neuropeptide Y on the cardiovascular system. Trends Pharmacol Sci 8:231–235

Edwards RM (1990) Neuropeptide Y inhibits cAMP accumulation in renal proximal convoluted tubules. Regul Pept 30:201–206

El-Din MMM, Malik KU (1988) Neuropeptide Y stimulates renal prostaglandin synthesis in the isolated rat kidney: Contribution of Ca^{++} and calmodulin. J Pharmacol Exp Ther 246:479–484

Evequoz D, Aubert J-F, Nussberger J, Biollaz J, Diezi J, Brunner HR, Waeber B (1996) Effects of neuropeptide Y on intrarenal hemodynamics, plasma renin activity and urinary sodium excretion in rats. Nephron 73:467–472

Feng QP, Hedner T, Andersson B, Lundberg JM, Waagstein F (1994) Cardiac neuropeptide Y and noradrenaline balance in patients with congestive heart failure. Br Heart J 71:261–267

Firth JD, Raine AEG, Ledingham JGG (1990) The mechanism of pressure natriuresis. J Hypertension 8:97–103

Gerald C, Walker MW, Criscione L, Gustafson EL, Batzl-Hartmann C, Smith KE, Vaysse P, Durkin MM, Laz TM, Linemeyer DL, Schaffhauser AO, Whitebread S, Hofbauer KG, Taber RI, Branchek TA, Weinshank RL (1996) A receptor subtype involved in neuropeptide-Y-induced food intake. Nature 382:168–171

Gimpl G, Wahl J, Lang RE (1991) Identification of a receptor protein for neuropeptide Y in rabbit kidney. G-protein association and inhibition of adenylate cyclase. FEBS Lett 279:219–222

Grandt D, Schimiczek M, Beglinger C, Layer P, Goebell H, Eysselein VE, Reeve JRjr (1994) Two molecular forms of peptide YY (PYY) are abundant in human blood: characterization of a radioimmunoassay recognizing PYY 1–36 and PYY 3–36. Regul Pept 51:151–159

Granger JP, Hall JE (1985) Acute and chronic actions of bradykinin on renal function and arterial pressure. Am J Physiol 248:F87–F92

Gregor P, Feng Y, DeCarr LB, Cornfield LJ, McCaleb ML (1996a) Molecular characterization of a second mouse pancreatic polypeptide receptor and its inactivated human homologue. J Biol Chem 271:27776–27781

Gregor P, Millham ML, Feng Y, DeCarr LB, McCaleb ML, Cornfield LJ (1996b) Cloning and characterization of a novel receptor to pancreatic polypeptide, a member of the neuropeptide Y receptor family. FEBS Lett 381:58–62

Grenier FC, Rollins TE, Smith WL (1981) Kinin-induced prostaglandin synthesis by renal papillary collecting tubule cells in culture. Am J Physiol 241:F94–F104

Grouzmann E, Alvarez-Bolado G, Meyer C, Osterheld MC, Burnier M, Brunner HR, Waeber B (1994) Localization of neuropeptide Y and its C-terminal flanking peptide in human renal tissue. Peptides 15:1377–1382

Hackenthal E, Aktories K, Jakobs KH, Lang RE (1987) Neuropeptide Y inhibits renin release by a pertussis toxin-sensitive mechanism. Am J Physiol 252:F543–F550

Haefliger J-A, Waeber B, Grouzman E, Braissant O, Nussberger J, Nicod LP, Waeber G (1999) Cellular localization, expression and regulation of neuropeptide Y in kidney of hypertensive rats. Regul Pept 82:35–43

Hegbrant J, Thysell H, Ekman R (1995) Circulating neuropeptide Y in plasma from uremic patients consists of multiple peptide fragments. Peptides 16:395–397

Hegde SS, Bonhaus DW, Stanley W, Eglen RM, Moy TM, Loeb M, Shetty SG, Desouza A, Krstenansky J (1995) Pharmacological evaluation of 1229U91, a novel high-affinity and selective neuropeptide Y-Y_1 receptor antagonist. J Pharmacol Exp Ther 275:1261–1266

Holtbäck U, Brismar H, DiBona GF, Fu M, Greengard P, Aperia A (1999) Receptor recruitment: a mechanism for interactions between G protein-coupled receptors. Proc Natl Acad Sci USA 96:7271–7275

Holtbäck U, Ohtomo Y, Förberg P, Sahlgren B, Aperia A (1998) Neuropeptide Y shifts equilibrium between α- and β-adrenergic tonus in proximal tubule cells. Am J Physiol 275:F1–F7

Hökfelt T (1991) Neuropeptides in perspective: the last ten years. Neuron 7:867–879

Hu Y, Bloomquist BT, Cornfield LJ, DeCarr LB, Flores-Riveros JR, Friedman L, Jiang P, Lewis-Higgins L, Sadlowski Y, Schaefer J, Velazquez N, McCaleb ML (1996) Identification of a novel hypothalamic neuropeptide Y receptor associated with feeding behavior. J Biol Chem 271:26315–26319

Kirby DA, Koerber SC, May JM, Hagaman C, Cullen MJ, Pelleymounter MA, Rivier JE (1995) Y_1 and Y_2 receptor selective neuropeptide Y analogues: evidence for a Y_1 receptor subclass. J Med Chem 38:4579–4586

Klin M, Waluga M, Rudka R, Madej A, Janiszewska M, Grzebieniak E, Wesolowky A (1998) Plasma catecholamines, neuropeptide Y and leucin-enkephalin in uremic patients before and after dialysis during rest and handgrip. Boll Chim Farm 137:306–313

Knight DS, Fabre RD, Beal JA (1989) Identification of noradrenergic nerve terminals immunoreactive for neuropeptide Y and vasoactive intestinal peptide in rat kidney. Am J Anatomy 184:190–204

Kokot F, Adamczak M, Wiecedilcek A, Spiechowicz U, Mesjasz J (1999) Plasma immunoreactive leptin and neuropeptide Y levels in kidney transplant patients. Am J Nephrol 19:28–33

Leys K, Schachter M, Sever P (1987) Autoradiographic localisation of NPY receptors in rabbit kidney: comparison with rat, guinea-pig and human. Eur J Pharmacol 134:233–237

Lortie M, Regoli D, Rhaleb N-E, Plante GE (1992) The role of B_1- and B_2-kinin receptors in the renal tubular and hemodynamic response to bradykinin. Am J Physiol 262:R72–R76

Lundberg JM, Modin A (1995) Inhibition of sympathetic vasoconstriction in pigs in vivo by the neuropeptide Y-Y_1 receptor antagonist BIBP 3226. Br J Pharmacol 116:2971–2982

Lundberg JM, Torssell L, Sollevi A, Pernow J, Theodorsson-Norheim E, Anggard A, Hamberger B (1985) Neuropeptide Y and sympathetic vascular control in man. Regul Pept 13:41–52

Lundell I, Blomquist AG, Berglund MM, Schober DA, Johnson D, Statnick MA, Gadski RA, Gehlert DR, Larhammar D (1995) Cloning of a human receptor of the NPY receptor family with high affinity for pancreatic polypeptide and peptide YY. J Biol Chem 270:29123–29128

Lundell I, Statnick MA, Johnson D, Schober DA, Starbäck P, Gehlert DR, Larhammar D (1996) The cloned rat pancreatic polypeptide receptor exhibits profound differences to the orthologous human receptor. Proc Natl Acad Sci USA 93:5111–5115

Mahns DA, Kelly C, McCloskey DI, Potter EK (1999) NPY Y_2 receptor agonist, N-acetyl[Leu28,Leu31]NPY24–36, reduces renal vasoconstrictor activity in anaesthetized dogs. J Auton Nerv System 78:10–17

Maisel AS, Scott NA, Motulsky HJ, Michel MC, Boublik JH, Rivier JE, Ziegler M, Allen RS, Brown MR (1989) Elevation of plasma neuropeptide Y levels in congestive heart failure. Am J Med 86:43–48

Malik KU, Nasjletti A (1979) Attenuation by bradykinin of adrenergically-induced vasoconstriction in the isolated perfused kidney of the rabbit: relationship to prostaglandin synthesis. Br J Pharmacol 67:269–274

Malmström RE (2001) Vascular pharmacology of BIIE0246, the first selective non-peptide neuropeptide Y Y_2 receptor antagonist, in vivo. Br J Pharmacol 133:1073–1080

Malmström RE, Alexandersson A, Balmer KC, Weilitz J (2000) In vivo characterization of the novel neuropeptide Y Y_1 receptor antagonist H409/22. J Cardiovasc Pharmacol 36:516–525

Malmström RE, Balmer KC, Lundberg JM (1997) The neuropeptide Y (NPY) Y_1 receptor antagonist BIBP 3226: equal effects on vascular responses to exogenous and endogenous NPY in the pig in vivo. Br J Pharmacol 121:595–603

Malmström RE, Balmer KC, Weilitz J, Nordlander M, Sjölander M (2001a) Pharmacology of H394/84, a dihydropyridine neuropeptide Y Y_1 receptor antagonist, in vivo. Eur J Pharmacol 418:95–104

Malmström RE, Björne H, Alving K, Weitzberg E, Lundberg JON (2001b) Nitric oxide inhibition of renal vasoconstricor responses to sympathetic cotransmitters in the pig in vivo. Nitric Oxide Biol Chem 5:98–104

Malmström RE, Hökfelt T, Bjorkman J-A, Nihlen C, Bystrom M, Ekstra AJ, Lundberg JM (1998) Characterization and molecular cloning of vascular neuropeptide Y receptor subtypes in pig and dog. Regul Pept 75/76:55–70

Malmström RE, Lundberg JM (1996) Effects of the neuropeptide Y Y_1 receptor antagonist SR 120107A on sympathetic vascular control in pigs in vivo. Naunyn-Schmiedeberg's Arch Pharmacol 354:633–642

Malmström RE, Lundberg JM (1997) Time-dependent effects of ischaemia on neuropeptide Y mechanisms in pig renal vascular control in vivo. Acta Physiol Scand 161:327–338

Malmström RE, Lundberg JM, Weitzberg E (2002a) Effects of the neuropeptide Y Y_2 antagonist BIIE0246 on sympathetic transmitter releas in the pig in vivo. Naunyn-Schmiedeberg's Arch Pharmacol 365:106–111

Malmström RE, Lundberg JON, Weitzberg E (2002b) Autoinhibitory function of the sympathetic prejunctional neuropeptie Y Y_2 receptor evidenced by BIIE0246. Eur J Pharmacol 439:113–119

Malmström RE, Modin A, Lundberg JM (1996) SR 120107A antagonizes neuropeptide Y Y_1 receptor mediated sympathetic vasoconstriction in pigs in vivo. Eur J Pharmacol 305:145–154

Martin JR, Knuepfer MM, Beinfeld MC, Westfall TC (1989) Mechanism of pressor response to posterior hypothalamic injection of neuropeptide Y. Am J Physiol 257:H791–H798

Matsumoto M, Nomura T, Momose K, Ikeda Y, Kondou Y, Akiho H, Togami J, Kimura Y, Okada M, Yamaguchi T (1996) Inactivation of a novel neuropeptide Y/peptide YY receptor gene in primate species. J Biol Chem 271:27217–27220

Matsumura K, Tsuchihashi T, Abe I (2000) Central cardiovascular action of neuropetie Y in conscious rabbits. Hypertension 36:1040–1044

McDermott BJ, Millar BC, Piper HM (1993) Cardiovascular effects of neuropeptide Y: receptor interactions and celular mechanisms. Cardiovasc Res 27:893–905

Mezzano V, Donoso V, Capurro D, Huidobro-Toro JP (1998) Increased neuropeptide Y pressor activity in Goldblatt hypertensive rats: in vivo studies with BIBP 3226. Peptides 19:1227–1232

Michel MC, Beck-Sickinger AG, Cox H, Doods HN, Herzog H, Larhammar D, Quirion R, Schwartz TW, Westfall TC (1998) XVI. International Union of Pharmacology recommendations for the nomenclature of neuropeptide Y, peptide YY and pancreatic polypeptide receptors. Pharmacol Rev 50:143–150

Michel MC, Gaida W, Beck-Sickinger AG, Wieland HA, Doods H, Dürr H, Jung G, Schnorrenberg G (1992) Further characterization of neuropeptide Y receptor subtypes using centrally truncated analogs of neuropeptide Y: Evidence for subtype-differentiating effects on affinity and intrinsic efficacy. Mol Pharmacol 42:642–648

Michel MC, Rascher W (1995) Neuropeptide Y—a possible role in hypertension? J Hypertension 13:385–395

Miller MA, Sagnella GA, Markandu ND, MacGregor GA (1990) Radioimmunoassay for plasma neuropeptide-Y in physiological and physiopathological states and responses to sympathetic stimulation. Clin Chim Acta 192:47–54

Minson R, McRitchie R, Chalmers J (1989) Effects of neuropeptide Y on the renal, mesenteric and hindlimb vascular beds of the conscious rabbit. J Auton Nerv System 27:139–146

Minson RB, McRitchie RJ, Morris MJ, Chalmers JP (1990) Effects of neuropeptide Y on cardiac performance and renal blood flow in conscious normotensive and renal hypertensive rabbits. Clin Exptl Hypert A12:267–284

Modin A, Malmström RE, Meister B (1999) Vascular neuropeptide Y Y_1-receptors in the rat kidney: vasoconstrictor effects and expression of Y_1-receptor mRNA. Neuropeptides 33:253–259

Modin A, Pernow J, Lundberg JM (1991) Evidence for two neuropeptide Y receptors mediating vasoconstriction. Eur J Pharmacol 203:165–171

Modin A, Pernow J, Lundberg JM (1994) Repeated renal and splenic sympathetic nerve stimulation in anaesthetized pigs: maintained overflow of neuropeptide Y in controls but not after reserpine. J Auton Nerv System 49:123–134

Modin A, Pernow J, Lundberg JM (1996) Prejunctional regulation of reserpine-resistant sympathetic vasoconstriction and release of neuropeptide Y in the pig. J Auton Nerv System 57:13–21

Nakamura M, Sakanaka C, Aoki Y, Ogasawara H, Tsuji T, Kodama H, Matsumoto T, Shimizu T, Noma M (1995) Identification of two isoforms of mouse neuropeptide Y-Y_1 receptor generated by alternative splicing. Isolation, genomic structure, and functional expression of the receptors. J Biol Chem 270:30102–30110

Nakamura M, Yokoyama M, Watanabe H, Matsumoto T (1997) Molecular cloning, organization and localization of the gene for the mouse neuropeptide Y-Y_5 receptor. Biochim Biophys Acta 1328:83–89

Neri G, Andreis PG, Malendowicz LK, Nussdorfer GG (1991) Acute action of polypeptide YY (PYY) on rat adrenoceortical cells: in vivo versus in vitro effects. Neuropeptides 19:73–76

Nobiling R, Gabel M, Persson PB, Dietrich MS, Bührle CP (1991) Differential effect of neuropeptide-Y on membrane potential of cells in renal arterioles of the hydronephrotic mouse. J Physiol (London) 444:317–327

Norvell JE, MacBride RG (1989) Neuropeptide Y (NPY)-like immunoactive nerve fibers in the human and monkey (Macaca fascicularis) kidney. Neurosci Lett 105:63–67

Oberhauser V, Vonend O, Rump LC (1999) Neuropeptide Y and ATP interact to control renovascular resistance in the rat. J Am Soc Nephrol 10:1179–1185

Oellerich WF, Malik KU (1993) Neuropeptide Y modulates the vascular response to periarterial nerve stimulation primarily by a postjunctional action in the isolated perfused rat kidney. J Pharmacol Exp Ther 266:1321–1329

Ohtomo Y, Aperia A, Sahlgren B, Johansson B-L, Wahren J (1996a) C-peptide stimulates rat renal tubular Na^+,K^+-ATPase activity in synergism with neuropeptide Y. Diabetologia 39:199–205

Ohtomo Y, Meister B, Hökfelt T, Aperia A (1994) Coexisting NPY and NE synergistically regulate renal tubular Na^+,K^+-ATPase activity. Kidney Int 45:1606–1613

Ohtomo Y, Ono S, Sahlgren B, Aperia A (1996b) Maturation of rat renal tubular response to α-adrenergic agonists and neuropeptide Y: a study on the regulation of Na$^+$,K$^+$-ATPase. Pediatric Research 39:534–538

Owen MP (1993) Similarities and differences in the postjunctional role for neuropeptide Y in sympathetic vasomotor control of large vs. small arteries of rabbit renal and ear vasculature. J Pharmacol Exp Ther 265:887–895

Parker MS, Berlund MM, Lundell I, Parker SL (2001) Blockade of pancreatic polypeptide-sensitive neuropeptide Y (NPY) receptors by agonist peptides is prevented by modulators of sodium transport. Implications for receptor signaling and regulation. Peptides 22:887–898

Parker SL, Parker MS, Crowley WR (1998) Characterization of Y_1, Y_2 and Y_5 subtypes of the neuropeptide Y (NPY) receptor in rabbit kidney. Sensitivity of ligand binding to guanine nucleotides and phospholipase C inhibitors. Regul Pept 75/76:127–143

Pavia JM, Morris MJ (1994) Age-related changes in neuropeptide Y content in brain and peripheral tissues of spontaneously hypertensive rats. Clin Exp Pharmacol Physiol 21:335–338

Pernow J, Lundberg JM (1989a) Modulation of noradrenaline and neuropeptide Y (NPY) release in the pig kidney in vivo: involvement of alpha$_2$, NPY and angiotensin II receptors. Naunyn-Schmiedeberg's Arch Pharmacol 340:379–385

Pernow J, Lundberg JM (1989b) Release and vasoconstrictor effects of neuropeptide Y in relation to non-adrenergic sympathetic control of renal blood flow in the pig. Acta Physiol Scand 136:507–517

Persson PB, Ehmke H, Nafz B, Lang R, Hackenthal E, Nobiling R, Dietrich MS, Kirchheim HR (1991) Effects of neuropeptide-Y on renal function and its interaction with sympathetic stimulation in conscious dogs. J Physiol (London) 444:289–302

Pfister A, Waeber B, Nussberger J, Brunner HR (1986) Neuropeptide Y normalizes renin secretion in adrenalectomized rats without changing blood pressure. Life Sci 39:2161–2167

Pham I, Gonzalez W, Doucet J, Fournier-Zaluski M-C, Roques BP, Michel J-B (1996) Effects of angiotensin-converting enzyme and neutral endopeptidase inhibitors: influence of bradykinin. Eur J Pharmacol 296:267–276

Pironi L, Stanghellini V, Miglioli M, Corinaldesi R, De Giorgio R, Ruggeri E, Tosetti C, Poggioli G, Labate AMM, Monetti N, Gozzetti G, Barbara L, Go VLW (1993) Fat-induced ileal brake in humans: a dose-dependent phenomenon correlated to the plasma levels of peptide YY. Gastroenterology 105:733–739

Playford JR, Mehta S, Upton P, Rentch R, Moss S, Calam J, Bloom S, Payne N, Ghatei M, Edwards R, Unwin R (1995) Effect of peptide YY on human renal function. Am J Physiol 268:F754–F759

Price JS, Kenny AJ, Huskisson NS, Brown MJ (1991) Neuropeptide Y (NPY) metabolism by endopeptidase-2 hinders characterization of NPY receptors in rat kidney. Br J Pharmacol 104:321–326

Regoli D, Barabe J (1980) Pharmacology of bradykinin and related kinins. Pharmacol Rev 32:1–42

Reinecke M, Forssmann WG (1988) Neuropeptide (neuropeptide Y, neurotensin, vasoactive intestinal polypeptide, substance P, calcitonin gene-related peptide, somatostatin) immunohistochemistry and ultrastructure of renal nerves. Histochem 89:1–9

Roberts TJ, Caston-Balderrama A, Nijland MJ, Ross MG (2000) Central neuropeptide Y stimulates ingestive behavior and increases urine output in the ovine fetus. Am J Physiol 279:E494–E500

Rump LC, Riess M, Schwertfeger E, Michel MC, Bohmann C, Schollmeyer P (1997) Prejunctional neuropeptide Y receptors in human kidney and atrium. J Cardiovasc Pharmacol 29:656–661

Schachter M, Miles CMM, Leys K, Sever PS (1987) Characterization of neuropeptide Y receptors in rabbit kidney: preliminary comparisons with rat and human kidney. J Cardiovasc Pharmacol 10 (Suppl. 12):S157–S162

Schuster VL, Kokko JP, Jacobson HR (1984) Interactions of lysyl-bradykinin and antidiuretic hormone in the rabbit cortical collecting tubule. J Clin Invest 73:1659–1667

Sheikh SP, Sheikh MI, Schwartz TW (1989) Y_2-type receptors for peptide YY on renal proximal tubular cells in the rabbit. Am J Physiol 257:F978–F984

Shin LH, Dovgan PS, Nypaver TJ, Carretero OA, Beierwaltes WH (2000) Role of neuropeptide Y in the development of two-kidney, one-clip renovascular hypertension in the rat. J Vasc Surg 32:1015–1021

Siragy HM (1993) Evidence that intrarenal bradykinin plays a role in regulation of renal function. Am J Physiol 265:E648–E654

Smyth DD, Blandford DE, Thom SL (1988) Disparate effects of neuropeptide Y and clonidine on the excretion of sodium and water in the rat. Eur J Pharmacol 152:157–162

Sulyok E, Tulassay T (2001) Natriuresis of fasting: the possible role of leptin-neuropeptide Y system. Med Hypotheses 56:629–633

Sun XY, Feng QP, Zhao X, Edvinsson L, Hedner T (1995) Cardiovascular and renal effects of alpha-trinositol in ischemic heart failure rats. Life Sci 57:1197–1211

Takahashi K, Toraichi M, Keiichi I, Masahiko S, Ohneda M, Murakami O, Nozuki M, Tachibana Y, Yoshinaga K (1987) Increased plasma neuropeptide Y. Concentrations in phaeochromocytoma and chronic renal failure. J Hypertension 5:749–753

Torffvit O, Adamsson M, Edvinsson L (1999) Renal arterial reactivity to potassium, noradrenaline, and neuropeptide Y and association with urinary albumin excretion in the diabetic rat. J Diabetes Complications 11:279–286

Turman MA, Apple CA (1998) Human proximal tubular epithelial cells express somatostatin: regulation by growth factors and cAMP. Am J Physiol 274:F1095–F1101

Turman MA, O'Dorisio MS, O'Dorisio T, Apple CA, Albers AR (1997) Somatostatin expression in human renal cortex and mesangial cells. Regul Pept 68:15–21

Voisin T, Lorinet A-M, Maoret J-J, Couvineau A, Laburthe M (1996) $G\alpha_i$ RNA antisense expression demonstrates the exclusive coupling of peptide YY receptors to G_{i2} proteins in renal proximal tubule cells. J Biol Chem 271:574–580

Waeber B, Evequoz D, Aubert J-F, Flückiger J-P, Juillerat L, Nussberger J, Brunner HR (1990) Prevention of renal hypertension in the rat by neuropeptide Y. J Hypertension 8:21–25

Wharton J, Gordon L, Byrne J, Herzog H, Selbie LA, Moore K, Sullivan MHF, Elder MG, Moscoso G, Taylor KM, Shine J, Polak JM (1993) Expression of the human neuropeptide tyrosine Y_1 receptor. Proc Natl Acad Sci USA 90:687–691

Wieland HA, Willim K, Doods HN (1995) Receptor binding profiles of NPY analogues and fragments in different tissues and cell lines. Peptides 16:1389–1394

Yan H, Yang J, Marasco J, Yamaguchi K, Brenner S, Collins F, Karbon W (1996) Cloning and functional expression of cDNAs encoding human and rat pancreatic polypeptide receptors. Proc Natl Acad Sci USA 93:4661–4665

Zelis R, Nussberger J, Clemson B, Waeber B, Grouzmann E, Brunner HR (1994) Neuropeptide Y infusion decreases plasma renin activity in postmyocardial infarction rats. J Cardiovasc Pharmacol 24:896–899

Zhang W, Lundberg JM, Thoren P (1999) The effect of a neuropeptide Y antagonist, BIBP 3226, on short-termn arterial pressure control in conscious unrestrained rats with congestive heart failure. Life Sci 65:1839–1844

Zhao XH, Sun XY, Bergdahl A, Edvinsson L, Hedner T (1999) Renal and cardiovascular role of the neuropeptide Y Y_1 receptor in ischaemic heart failure rats. J Pharm Pharmacol 51:1257–1265

Ziegler MG, Kennedy B, Morrissey E, O'Connor DT (1990) Norepinephrine clearance, chromogranin A and dopamine β hydroxylase in renal failure. Kidney Int 37:1357–1362

Zukowska-Grojec Z, Wahlestedt C (1993) Origin and actions of neuropeptide Y in the cardiovascular system. 315–388

NPY-Like Peptides, Y Receptors and Gastrointestinal Function

N. P. Hyland · H. M. Cox

Centre for Neuroscience Research, King's College London, Guy's Campus, London, SE1 1UL, UK
e-mail: helen.m.cox@kcl.ac.uk

1	Location of NPY, PYY and PP.	390
2	Co-Localization of NPY with Other Prominent Enteric Nervous System Peptides.	392
3	Factors Influencing Alterations in NPY, PYY and PP Expression and Release.	392
4	Differential Localization of Y Receptor Types in the Gastrointestinal Tract.	394
4.1	Location of Y_1 Receptors.	394
4.2	Location of Y_2 Receptors.	394
4.3	Location of Y_4, Y_5 and y_6 Receptors.	395
5	Insights into Y Receptor Function in the Small and Large Intestine.	396
5.1	NPY, PYY and PP Effects upon Gastrointestinal Ion Secretion.	396
5.2	NPY, PYY and PP Effects upon Gastrointestinal Motility.	397
6	Gastrointestinal Function in Mutant Mice Lacking Y_1, Y_2 and Y_4 Receptors and NPY.	398
6.1	Overlapping Pharmacology.	399
6.2	Enteric Nerve Depolarization by Veratridine in $Y_1^{-/-}$, $Y_2^{-/-}$ and $NPY^{-/-}$ Colon Mucosae.	400
6.3	Y_1 Receptor-Mediated Inhibitory Tone.	401
6.4	Y Receptor-Mediated Smooth Muscle Contraction in $Y_2^{-/-}$ and $Y_4^{-/-}$ Colon.	401
7	NPY and PYY in Intestinal Disorders and Gastrointestinal Pathology.	402
8	PYY: A Trophic Factor in the Intestine?	403
9	Conclusions.	403
References.		404

Abstract This chapter aims to discuss the role of neuropeptide Y (NPY)-like peptides, peptide YY (PYY), PYY(3–36) and pancreatic polypeptide (PP) with respect to gastrointestinal function, specifically in terms of electrolyte transport and motility. First we summarize the localization of each peptide and the Y receptors as indicated by recent immunocytochemical studies. The availability of antibodies for Y_1 and more recently, Y_4 receptors have provided precise information on the cellular localization of these receptors in human and rodent intestine, and most importantly the colocalization of the Y_1 receptor with NPY and other prominent enteric neuropeptides. We will also discuss the significant progress made to date in ascribing precise functions for specific Y receptors made possible by the availability of selective Y_1 (BIBO 3304) and Y_2 (BIIE 0246) receptor antagonists, and most recently by the availability of NPY and Y receptor knockout ($^{-/-}$) mice. Selective antagonists have been crucial for the accurate characterization of the different Y receptors and their roles in peripheral and central targets. The Y_4 receptor however, still remains a challenge and the timely generation of mutant mice lacking this single receptor type has enabled assessment of this receptor's functional significance. We will refer to the association of PYY and NPY with gastrointestinal disease and the potential role these peptides, their analogues and Y receptor antagonists may have as novel therapeutics. Readers are referred to earlier chapters in collected volumes or review journals which provide significant background information (Cox 1993; Cox 1998; Furness 2000; Playford and Cox 1996).

Keywords Neuropeptide Y · Peptide YY · Pancreatic polypeptide · Intestinal function

1
Location of NPY, PYY and PP

Neuropeptide Y (NPY) is located within both intrinsic enteric ganglia, namely the myenteric and submucosal plexi. The two intrinsic interconnecting ganglia form the majority of the enteric nervous system (ENS) providing a continuous neuronal network along the length of the intestine. In the mammalian intestine NPY provides an extensive innervation of multiple targets, being expressed in several different intrinsic enteric neurone types each with different functions. In the guinea pig small intestine for example, NPY is located in: (a) inhibitory myenteric neurones innervating the circular smooth muscle; (b) in descending myenteric interneurones; and (c) cholinergic secretomotor (non-vasodilator) neurones that innervate the mucosa (for details see Furness 2000). In the mouse small intestine, approximately 26% of small intestinal myenteric cell bodies contain NPY while NPY-positive fibres only are present in the myenteric plexus of murine large intestine (Sang and Young 1996). Sandgren et al. (2002) found fewer NPY-positive neurones (3%–10%) in myenteric ganglia compared with 35% of submucosal ganglia containing NPY-positive neurons in mouse ileum. The neuropeptide is therefore localized in cell bodies and fibres within submucosal plexi in both small and large intestine and dense fibre networks are present in the mucosa of both regions. In the proximal colon, 2%–16% of myenteric and 4%–14% of submucosal neurones are NPY positive, compared with 1%–10% of myenteric neurones and 2%–9% of submucosal neurones in the distal colon (Sandgren et al. 2002). NPY-containing, putative inhibitory motor neurones project anally, while NPY-positive interneurones project locally (i.e., circumferentially with no anal or oral bias (Sang et al. 1997). A similarly extensive NPY innervation (together with vasoactive intestinal polypeptide; VIP) has been described in rat small and large intestine but in the latter NPY is co-localized in only a proportion of VIP-positive enteric neurones (Ekblad et al. 1987, 1988).

Peptide YY (PYY) on the other hand is present in endocrine cells in the epithelial lining of all mammals studied to date, occurring most frequently in the large bowel (Böttcher et al. 1986). Endocrine cells (at least in human and feline colon) have basal and apical processes, the latter extending in to the lumen (Böttcher et al. 1986), implicating a bidirectional mechanism of action for released PYY. PYY has been detected by radio-immunoassay of extracts of colonic tissue from mouse, rat, guinea pig, cat and pig (Böttcher et al. 1993b). The concentration of this peptide is higher in mammalian colon and ileum compared to the duodenum and jejunum (for a review see Ekblad and Sundler 2002). PYY is also present in pancreatic islet cells from several species, and in mouse pancreatic islets it is co-localized with pancreatic polypeptide (PP, Böttcher et al. 1993b). Surprisingly, PYY has also been localized in discrete populations of enteric neurones, mainly in the fundus, antrum and small intestine with only very occasional fibres in the large intestine, of the cat, ferret and pig (for a review see Ekblad and Sundler 2002). In the rat, however PYY-positive nerve fibres are rare, exclusive to the antrum, and absent from other intestinal regions studied (Böttcher et al. 1993a).

Circulating PYY is hydrolysed in plasma by aminopeptidase-P and dipeptidylpeptidase-IV, two peptidases which are active in the plasma membrane of the intestinal brush border, resulting in the production of local and circulating PYY(3–36) (Medeiros and Turner 1994). In the canine colon approximately 40% of PYY immunoreactivity can be attributed to PYY(3–36) (discriminated from full length PYY by high performance liquid chromatography (HPLC) and using an antibody with little or no cross-reactivity to human PP or NPY (Grandt et al. 1992). In plasma samples from fasted and postprandial human volunteers, PYY(3–36) accounted for 37% and 63% of total PYY respectively (Grandt et al. 1994), indicating a more prominent functional role for the fragment (and thus the Y_2 receptor) following ingestion of a meal. Recent studies by Batterham et al. (2002) have shown that the fragment is a major satiety factor in man and rodents; infusion of PYY(3–36) at concentrations comparable to postprandial circulating peptide levels, significantly reduced food intake in healthy human volunteers.

The third full-length member of the NPY peptide family, PP is a pancreatic hormone also released in response to food and under vagal control (for a review see Schwartz 1983). PP-containing endocrine cells are more numerous in the duodenal type of pancreatic islets than in those located in the splenic portion of the pancreas (Sundler et al. 1993). However, PP-positive endocrine cells have also been identified in human colon and rectum and in the duodenum of dog. Endocrine PP is expressed postnatally in the rat, detected first at 2 days in the colon, and temporarily between 7 and 21 days in the pylorus and rectum (for a review see Ekblad and Sundler 2002). The different patterns of peptide localization in the mammalian gastrointestinal tract implicate different functional roles elaborated by co-release of a range of co-localized unrelated peptides, neurotransmitters and/or neuromodulators.

2
Co-Localization of NPY with Other Prominent Enteric Nervous System Peptides

In the myenteric ganglia of the mouse small intestine NPY is co-localized with VIP and nitric oxide synthase (NOS), and is not localized with either the excitatory neurotransmitter, substance P or the calcium binding protein, calretinin (Sang and Young 1996). In the rat small and large intestine, myenteric VIP/NPY containing neurones are common, innervating smooth muscle and myenteric ganglia, while submucous VIP/NPY neurones also provide a dense innervation of the mucosa, the submucosa and submucosal ganglia (Ekblad et al. 1987). These two unrelated neuropeptides are co-localized in the same neurosecretory vesicles (in rat jejunum; Cox et al. 1994) and are presumably co-released into the lamina propria. The co-release of a combination of potent secretagogue (VIP) with a similarly potent but anti-secretory NPY (for review see, Cox 1998) remains the most prominent pairing of functional antagonists in the rodent ENS.

A large proportion of myenteric and submucosal neurones in the small intestine and myenteric neurones in the large intestine (approximately 43%–63%) are cholinergic, however of the total neuronal specific enolase immunoreactive neurones in colonic submucosal plexi only approximately 20% are cholinergic (Sang and Young 1998). This finding highlights the importance of noncholinergic neurotransmission, particularly in submucosal neurones of the large intestine of the mouse.

3
Factors Influencing Alterations in NPY, PYY and PP Expression and Release

The ENS receives extrinsic innervation and thus manipulation of extrinsic nerves to detect changes in gastric motility or nuclear protein, c-Fos (produced by activation of the early proto-oncogene c-*fos*) expression in intrinsic enteric neurones can be used to investigate the role of the central nervous system (CNS) in regulation of the ENS. For example, stimulation of the vagal nerve results in c-Fos expression in gastric myenteric plexus neurones (Zheng and Berthoud 2000). Denervation of sympathetic fibres in the lumbar colonic and hypogastric nerves can also result in an increase in c-Fos immunoreactive neurones in guinea pig colon (Yuyama et al. 2002). Similarly, loss of extrinsic sympathetic or sensory nerves also increases c-Fos expression in guinea pig ileum (Yunker et al. 1999). All of these observations suggest that the respective enteric neurones receive inhibitory inputs from extrinsic nerves. It has been suggested that extrinsic innervation is not necessarily a prerequisite for the appearance of colonic contractile activity, but it is required for the propagation and modulation of motility (Tanabe et al. 2002).

With respect to the vagal regulation of NPY or PYY release, neither right nor left cervical vagotomy alters the concentration of NPY (2–8 weeks after vagotomy) in mouse antrum, duodenum and colon, nor PYY levels in mouse ileum (El Salhy et al. 2000). This demonstrates the lack of tonic vagal regulation of NPY/PYY release in these tissues. Nevertheless, PYY release is induced by electrical stimulation of the vagus nerve in pig and dog, and this release is sensitive to atropine and hexamethonium in both species (for a review see Onaga et al. 2002) indicating significant cholinergic control. PYY release in rat duodenum also appears to be neurally regulated (Anini et al. 1999). In response to intraduodenal oleic acid, peptide release is sensitive to hexamethonium and L-NAME demonstrating a role, not only for acetylcholine (and ganglionic nicotinic receptors) but also for NO in the regulation of PYY release (Anini et al. 1999).

Following perfusion of human ileum with fat, PYY is released into the circulation and delays gastric emptying in a dose-dependent manner, contributing to the fat-induced phenomenon, 'ileal brake' (Pironi et al. 1993). Inhibition of gastric emptying appears to be vagally mediated, as subdiaphragmatic vagotomy reverses the inhibition of gastric emptying induced by PYY in the rat (Chen et al. 1996). Oleic acid inhibits intestinal motility in canine ileum (with or without intact extrinsic innervation) and this response is accompanied by an increase in serum PYY. However, the inhibition of small intestinal motility elicited by triglycerides is dependent upon intact extrinsic innervation and has no effect in the antrum whether extrinsic nerves are intact or not. This mechanism is not accompanied by an increase in circulating PYY (Ohtani et al 2001), indicative of PYY-independent, neuronal mechanisms contributing to the triglyceride-induced 'ileal brake'. This suggests that PYY [and potentially, PYY(3–36)] inhibit gastric motility in response to specific fats. The mechanisms are mediated by extrinsic nerves, although this is not the sole mechanism responsible for the fat-induced 'ileal brake'. Interestingly, canine plasma NPY levels are also increased significantly following ingestion of a standard meal (Balasubramaniam et al. 1989).

Furthermore, PP release is also regulated by vagal mechanisms, electrical stimulation of the vagal nerve elicits significantly increased levels of PP release with a time-dependency different from the PYY response. An increase in PP is observed within 10 min of ingesting food, and lasts for up to 4 h in humans (Hornnes et al. 1980). 'Sham feeding' and hypoglycaemia also induce PP release, which is blocked by vagotomy and atropine treatment (for a review see Schwartz 1983). Recent studies utilizing Y_2 receptor knockout mice, suggest that hypothalamic Y_2 receptors regulate plasma levels of PP by stimulating secretion of adrenal glucocorticoids in response to hypoglycaemia (Sainsbury et al. 2002).

4
Differential Localization of Y Receptor Types in the Gastrointestinal Tract

4.1
Location of Y_1 Receptors

Y_1 receptor immunocytochemistry has identified Y_1 receptor positive staining in rat intestine in endothelial cells, endocrine cells (negative for PYY, and most numerous in the small intestine), and on myenteric and submucosal neurones (Jackerott and Larsson 1997). Y_1 labelling is co-localized with VIP in submucosal neurones most numerous in the rat small intestine, while a fraction of Y_1 positive nerves in the myenteric plexus also co-store NOS (and the latter is true also, but to a lesser extent in submucosal neurons; Jackerott and Larsson 1997). In human colon, the majority of Y_1 receptor positive cell bodies and nerve fibres in the myenteric plexus are also positive for NOS and vice versa, however the Y_1 positive neurones in the submucosal plexus are NOS negative (Peaire et al. 1997). In Henle's plexus of human colon Y_1 labelling is co-localized with NPY positive cell soma, indicating a potential pre-synaptic/pre-junctional role for the Y_1 receptor in regulating NPY release, similar to that reported for the Y_2 receptor centrally (King et al. 2000). In the myenteric plexus however, there is no co-localization between NPY and Y_1 rather, Y_1 positive cell bodies are surrounded by NPY positive nerve fibres (Peaire et al. 1997).

Furthermore in human colon, Y_1 receptors are present on neuronal varicosities which extend directly to the mucosa, close to the basolateral domain of crypt epithelia, as well as on the basolateral domain of epithelial cells themselves (Mannon et al. 1999). Consistent with previous studies described above, both submucosal and myenteric neurones are Y_1 receptor-positive in human colon (Mannon et al. 1999). These studies implicate a potential neurogenic role for NPY and PYY activation of Y_1 receptors as well as a direct activation of epithelial Y_1 receptors both influencing ongoing mucosal fluid secretion/absorption in the human and rat intestine. However, it is clear that activation of different populations of Y_1-positive neurones will result in very different responses, dependent upon the excitatory or inhibitory character of the intrinsic neurone and the subcellular localization of the receptor, that is, cell soma versus varicosity/terminal location.

4.2
Location of Y_2 Receptors

The absence of commercially available Y_2 receptor antibodies has meant that studies on this receptor type in gastrointestinal tissues lag somewhat behind those for the Y_1 receptor. Conclusions on the cellular location of probable Y_2 receptors are therefore drawn from autoradiographic, reverse transcription (RT)–polymerase chain reaction (PCR) (both types of investigation are discussed here) and functional studies (discussed below, see Sect. 5). In human colon, ^{125}I-PYY differential displacement by PYY(3–36) (Y$_2$ preferred agonist) and [Leu^{31}, Pro^{34}]NPY (Y$_1$ preferred agonist) identified both fragment-preferring (assumed to be Y$_2$ binding) and Y$_1$ binding sites (from [Leu^{31}, Pro^{34}]NPY displacement) in myenteric and submucosal plexi, and in circular and longitudinal smooth muscle with only 'Y$_1$' binding in blood vessels (Rettenbacher and Reubi 2001). Human intestinal segments subjected to in situ RT–PCR have shown Y$_2$ receptor mRNA in the muscle of the ileum and left colon as well as in the mucosal layers of the ileum and right colon (Ferrier et al. 2002). No Y$_2$ receptor mRNA was detected in muscle preparations from rat intestine (Ferrier et al. 2002) or in nerve–muscle preparations from rat proximal colon (Ferrier et al. 2000). In a separate study Y$_2$ mRNA was found to be highly expressed in rat proximal and distal colon (while Y$_1$ mRNA levels were intermediate; Feletou et al. 1998), specifically in colonic epithelium and 'non-epithelial' tissue presumed to be muscle layers from rat jejunum (Goumain et al. 1998). Taking into account species differences and tissue-specific variations in Y$_2$ receptor expression, these studies indicate that Y$_2$ receptors are predominantly but not exclusively, neuronal. Their localization on myenteric and submucosal neurones (the latter extending to the mucosa) indicates a neuromodulatory role in the ENS as well as direct effects on epithelial and smooth muscle targets. However, little if any information on the subcellular distribution of this receptor type exists and the availability of specific Y$_2$ receptor antibodies is long overdue.

4.3
Location of Y$_4$, Y$_5$ and y$_6$ Receptors

Here too immunohistochemical studies are few due to lack of selective antibodies and thus the patterns of localization are limited. RT–PCR and northern analyses have however consistently identified mRNA for the Y$_4$ receptor in rat proximal colon (Feletou et al. 1998; Ferrier et al. 2000, 2002), distal colon (Ferrier et al. 2002) and 'non- epithelial' tissue from rat jejunum (Goumain et al. 1998); but not apparently in human ileum, right or left colon or rectum (Ferrier et al. 2002). Y$_4$ receptor mRNA is present in both crypt and villus epithelia (of the rat small intestine), with highest levels of expression in colonic epithelium (Goumain et al. 1998). Recently, aided by the first selective antibody, Y$_4$ receptor immunoreactivity has been detected in goblet cells and along the basal lamina of rat intestinal villi (Campbell et al. 2003). The Y$_4$ receptor therefore is predicted to have significant roles in mucosal function, for example, epithelial ion transport and mucus secretion, as well as modulating smooth muscle activity.

The peripheral expression of Y$_5$ receptors in comparison with Y$_1$, Y$_2$ and Y$_4$ receptor expression, is limited and Y$_5$ mRNA has not been detected in rat fundus, antrum, duodenum, jejunum, ileum, caecum or distal colon, or in human ileum, colon, and rectal tissues (Ferrier et al. 2002). Nevertheless Y$_5$ mRNA has reportedly been detected (albeit at low levels) in rat proximal colon (Feletou et al. 1998) and in colonic 'non-epithelial' (presumed to be muscle) layers and jejunal crypts (Goumain et al. 1998). Thus the possibility of as yet unidentified Y$_5$

receptor stimulated effects in these intestinal areas remains but is less prominent than Y_1, Y_2 or Y_4 receptor-mediated functions.

The y_6 receptor is likely to be functionally inactive in humans and primates (Matsumoto et al. 1996) yet it is expressed full-length and is therefore potentially functional in mouse developing embryo (Gregor et al. 1996). Messenger RNA for y_6 receptor has also been identified in adult human and rabbit small intestine and colon (Gregor et al. 1996; Matsumoto et al. 1996) but was apparently absent from adult mouse tissues (Gregor et al. 1996).

5
Insights into Y Receptor Function in the Small and Large Intestine

Studies using specific Y receptor preferring agonists (for a review see Michel et al. 1998) not only give us an insight into the potential location but also identify the functional roles of specific Y receptors in the intestine.

5.1
NPY, PYY and PP Effects upon Gastrointestinal Ion Secretion

By virtue of their primary G_i-protein coupling in epithelia, NPY and PYY are both anti-secretory peptides, causing an inhibition of ongoing epithelial cAMP (Servin et al. 1989) which underpins the anti-secretory effects of both peptides. NPY and PYY cause reductions in electrogenic ion transport, (measured as short-circuit current; I_{sc}) in rat colon and jejunum predominantly as a result of attenuated apical chloride secretion (Cox et al. 1988). Electrogenic responses in the rat jejunum were insensitive to blockade of intrinsic neurogenic activity by tetrodotoxin (TTX), suggesting that PYY/NPY receptors are exclusively located post-junctionally on epithelial cells in this tissue (Cox et al. 1988). The Y_2 preferring fragment, NPY(13–36) also inhibited basal I_{sc} [as did PYY(3–36)] and also significantly reduced electrically-evoked electrogenic responses in rat jejunum (both in the presence and absence of atropine and hexamethonium) indicating the possibility of pre-junctional Y_2 receptors regulating cholinergic and non-cholinergic neurotransmission (Cox and Cuthbert 1990).

The selective Y_2 antagonist, BIIE 0246 has confirmed that NPY and PYY-stimulated inhibitory effects in the rat jejunum and colon (Cox and Tough 2000; Goumain et al. 2001) are solely and partially Y_2 receptor-mediated, respectively. Also, similar long-lasting anti-secretory responses to PYY(3–36) are observed in the wild-type mouse descending colon and these are virtually abolished by BIIE 0246 (Cox et al. 2001a) and are absent from $Y_2^{-/-}$ mouse colon mucosae (up to 100 nM; Hyland et al. 2003). Thus the Y_2 receptor provides major anti-secretory mechanisms, both epithelial (post-junctional) and neurogenic (pre-junctional) dependent upon both the species and intestinal site.

Use of BIBP 3226 first confirmed that Y_1 receptors are involved in NPY and PYY anti-secretory responses in rat colon and the TTX sensitivity of [Leu^{31}, Pro^{34}] NPY responses, indicates significant pre-junctional Y_1 receptor-mediated effects in this tissue (Tough and Cox 1996). In contrast, Y_2 receptor-stimulated effects are less TTX-sensitive and therefore more likely to be post-junctional in origin in the rat intestine (Cox et al. 1988; Tough and Cox 1996). In mouse descending colon [Leu^{31}, Pro^{34}]PYY responses were abolished by the selective Y_1 antagonist BIBO 3304, while PYY responses were only partially attenuated, indicating that the native peptide co-activates both Y_1 and Y_2 receptors in this tissue (Holliday et al. 2000). No functional activity for any species PP has been observed in rat intestinal mucosae.

However PP, together with NPY, PYY and their [Leu^{31} Pro^{34}] analogues (Y_1 receptor preferring), are all anti-secretory in mouse colon (Holliday et al. 2000) and further characterization using Y_1 and Y_2 antagonists demonstrates that collectively these responses are mediated by Y_1, Y_2 and Y_4 receptors, with the Y_1 receptor effects dominating. A Y_5 receptor preferring agonist, Ala^{31},Aib^{32}hNPY, was only active at micro-molar concentrations suggesting that Y_5 receptors do not have a role in mucosal responses in mouse colon (Cox et al. 2001a). Recent data suggest that Y_2 receptors in mouse (and human) colon are predominately located pre-junctionally, as responses to both the Y_2 preferring agonist PYY(3–36), and blockade of endogenous inhibitory Y_2 tone by BIIE 0246 were significantly reduced or abolished by TTX (Hyland et al. 2003, Cox et al. 2003).

In human colon mucosa, Y_1, Y_2 and Y_4 are also the receptor types that mediate NPY, PYY and PP effects (Cox and Tough 2002). Functional studies with isolated human colonic mucosa indicate that Y_1 receptors are present pre- and post-junctionally, Y_2 receptors are pre-junctional on enteric neurones and Y_4 receptors are exclusive to basolateral epithelial surfaces (Cox and Tough 2002). Interestingly, these conclusions are borne out by work with human colonic adenocarcinoma cell lines, which despite being heterogeneous, rarely or never constitutively express Y_1 or Y_2 receptors (respectively) but frequently express Y_4 receptors (H.M. Cox, I.R. Tough and N.D. Holliday, unpublished results; Cox et al. 2001b). In the small and large intestine of the rat, different species PP (and thus presumably Y_4 receptors) do not alter mucosal ion transport (Cox et al. 1988). Recent studies using a polyclonal Y_4 receptor antibody have shown selective staining of goblet cells in the rat duodenum (Campbell et al. 2003), thus alternative Y_4 receptor-mediated effects, such as the regulation of mucus secretion may yet be described.

5.2
NPY, PYY and PP Effects upon Gastrointestinal Motility

As already discussed Y_1, Y_2 and Y_4 receptors mediate the anti-secretory effects of NPY, PYY and PP in human and murine colon, by differentially acting at both pre-junctional neuronal Y receptors or directly acting upon epithelial Y receptors. Excitatory, contractile effects have also been observed in the gastrointestinal tract in response to these polypeptides. In rat proximal colon both PYY and NPY induce muscle contraction (Feletou et al. 1998; Ferrier et al. 2000; Pheng et al. 1999), although there are conflicting reports concerning the combination of Y receptors that mediate these effects. Feletou et al. (1998) suggest that three Y receptors are involved in the rat, namely Y_2, Y_4 and Y_5 receptors. In terms of TTX sensitivity, neither NPY nor PYY contractile responses were sensitive to pre-treatment with the toxin, while Y_2-, Y_4-, and Y_5-preferring agonist responses were abolished by TTX (Feletou et al. 1998). In contrast, Pheng et al. (1999) found that TTX abolished contractile responses to PYY and human PP (hPP). PYY contractile responses were also partially sensitive to pre-treatment with atropine, suggesting neural regulation of the PYY contractile response via a cholinergic mechanism (Pheng et al. 1999). Ferrier et al. (2000) suggest that the contractile effects of NPY and PYY in rat proximal colon are predominantly mediated by Y_1 receptors on muscle cells and neuronal Y_4 receptors. However, the Y_1 receptor antagonist BIBO 3304 does not alter contractile responses in rat (Pheng et al. 1999) or wild-type mouse colon (Tough et al. 2002). The Y_2 antagonist used by Pheng et al. (1999) T_4-[NPY-(33–36)]$_4$, reduced, and BIIE 0246 abolished PYY(3–36) responses (Dumont et al. 2000) leaving [Leu^{31},Pro^{34}]NPY and hPP effects unchanged (Dumont et al. 2000) indicating a definite role for Y_2 receptors in mediating this contractile response, and a potential role for both Y_1 and Y_4 receptors. While PP effects are predominantly mediated by Y_4 receptors, residual effects of this peptide upon Y_1 receptors (as seen in $Y_4^{-/-}$ tissue, and discussed in Sect. 6; Tough et al. 2002) probably account for the conflicting information on Y receptor modulation of contractile activity.

Centrally administered (intracerebroventricular; icv) NPY and PYY also induce changes in gastric motility (Fujimiya et al. 2000). When administered to fed rats NPY returned duodenal activity from a fed state back to a fasted state, via potential central vagal mechanisms involving Y_2 and possibly Y_4 receptors (Fujimiya et al. 2000). It was also speculated by these authors that such a mechanism may lead to the onset of feeding behaviour. However, it has since been shown that the Y_2 preferring agonist PYY(3–36) can directly influence hypothalamic circuits to inhibit food intake (Batterham et al. 2002). In complete contrast, and in a different species (the guinea pig colon), NPY and PYY inhibit motility via Y_1 and Y_2 receptor activation and both effects are mediated via an inhibition of acetylcholine release (Sawa et al. 1995).

6
Gastrointestinal Function in Mutant Mice Lacking Y_1, Y_2 and Y_4 Receptors and NPY

Mice lacking either NPY or a single Y receptor population, that is Y_1, Y_2 or Y_4 receptors have confirmed the functional roles for this neuropeptide and its receptors in the murine gastrointestinal tract. It is noteworthy that we and others have not found differences in Y agonist effects, potency or pharmacology, either between pure mouse strains or the mixed background strains (C57BL6/129 Sv) used to generate conditional and unconditional knockouts (Cox et al. 2001a; Hyland et al. 2003; H.M. Cox, N.P. Hyland, I.R. Tough, E.L. Pollock and H. Herzog, unpublished results). Y receptor mutant mice have also provided useful models for the complete characterization of Y agonists, which have been used to ascribe physiological function to Y receptors in the past. As previously discussed in this chapter both NPY and PYY together with PP are anti-secretory peptides, and all three can significantly alter intestinal motility. These two functional end points have been studied in detail in isolated tissues from wild-type and unconditional knockouts, namely $Y_1^{-/-}$, $Y_2^{-/-}$, $NPY^{-/-}$ and $Y_4^{-/-}$ mice, and our findings are summarized below.

6.1
Overlapping Pharmacology

Anti-secretory mucosal responses mediated by the Y_1 preferring agonist [Leu^{31}, Pro^{34}]PYY and the Y_2 preferring PYY(3–36) are abolished in $Y_1^{-/-}$ (H.M. Cox and N.P. Hyland, unpublished results) and $Y_2^{-/-}$ (Hyland et al. 2002; Fig. 1) tissues

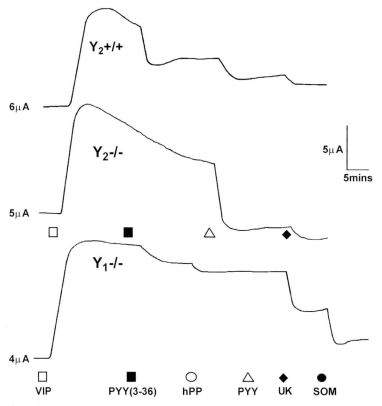

Fig. 1 Representative responses to basolateral VIP (30 nM), PYY(3–36) (100 nM), PYY (10 nM) and UK (1 μM) in descending colon mucosal sheets from $Y_2^{+/+}$, $Y_2^{-/-}$ and $Y_1^{-/-}$ mice. In addition to this sequence of agonists, hPP (100 nM) and somatostatin (*SOM*, 100 nM) were also added to $Y_1^{-/-}$ tissue (lowest trace) only. Values (in μA) to the left of each trace record the basal I_{sc} for each 0.2 cm^2 preparation. Note the loss of PYY(3–36) responses in $Y_2^{-/-}$ tissue compared with those observed in $Y_2^{+/+}$ colon; the absence of PYY responses in $Y_1^{-/-}$ colon (lowest trace only), and the decrease in I_{sc} observed following hPP in $Y_1^{-/-}$ colon

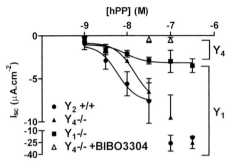

Fig. 2 Concentration–response (*CR*) curves for basolateral hPP added to descending colon mucosae from $Y_2^{+/+}$, $Y_4^{-/-}$ and $Y_1^{-/-}$ tissue. Each point is the mean±1 SE from pooled data and is quoted as "$\mu A.cm^{-2}$". Note the rightward shift in the CR curves from $Y_4^{-/-}$ (and the loss of hPP responses in this mutant plus BIBO 3304, 300 nM) and $Y_1^{-/-}$ colon, and the monophasic character of the hPP curve in $Y_1^{-/-}$ colon (n=2–19)

respectively, although at higher concentrations (300 nM, data not shown) PYY(3–36) decreased I_{sc} in $Y_2^{-/-}$ tissue, an effect that was blocked in the presence of the Y_1 antagonist BIBO 3304, indicating that this fragment co-stimulates Y_1 receptors at higher concentrations (Hyland et al. 2002). The human sequence of PP has been thought to be a Y_4 preferring agonist but we have clear evidence that this species PP can co-stimulate Y_1 and Y_4 receptors at low concentrations (1–30 nM; Fig. 2) and more so at 100 nM and 300 nM. Residual mucosal hPP responses are also present in $Y_4^{-/-}$ tissue and these are abolished by BIBO 3304 (Tough et al. 2002; Fig. 2). In the mouse colon, Y_1-stimulated I_{sc} responses are significantly greater than either Y_2 or Y_4-mediated anti-secretory effects. Thus hPP can stimulate the very large Y_1-mediated decreases in I_{sc} (at 100 nM and 300 nM) in control and knockout colon mucosae ($NPY^{-/-}$; $Y_2^{-/-}$, Hyland et al. 2002; $Y_4^{-/-}$, Tough et al. 2002) but it does not do so in $Y_1^{-/-}$ colon mucosa (Fig. 2). These studies demonstrate the ability of hPP [and to a lesser extent PYY(3–36)] to co-stimulate murine Y_1 receptors.

6.2
Enteric Nerve Depolarization by Veratridine in $Y_1^{-/-}$, $Y_2^{-/-}$ and $NPY^{-/-}$ Colon Mucosae

The depolarizing agent veratridine causes an increase in I_{sc} (Hyland and Cox 2002) in mouse colon as a result of elevated Cl⁻ secretion initiated by cholinergic and non-cholinergic neurotransmitter release (Sheldon et al. 1990) and this neurogenic response is reversed by TTX (Fig. 3A). The loss of the enteric neuropeptide, NPY, caused an unexpected decrease in the resultant secretory response (in the absence of the anti-secretory neurotransmitter), and indicates an inhibitory role for NPY in normal colonic submucosal neurotransmission. Veratridine-induced depolarization of enteric neurones in tissue pre-treated with the Y_2 antagonist, BIIE 0246 resulted in significantly increased mucosal I_{sc} responses

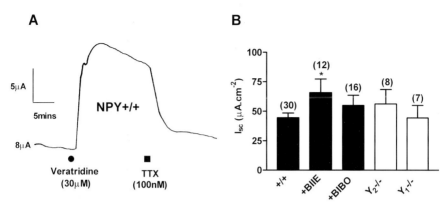

Fig. 3. A A representative I_{sc} response to basolateral veratridine (30 μM) in wild-type (NPY$^{+/+}$) mouse descending colon, followed by a subsequent inhibition of elevated I_{sc} by TTX (100 nM). The value to the left of the trace is the basal I_{sc} prior to stimulation. **B** Histogram showing pooled data of peak height veratridine responses (at 5 min) following, in order; wild-type tissue pre-treated with either BIIE 0246 (1 μM for 15 min) or BIBO 3304 (300 nM for 15 min), then responses in $Y_2^{-/-}$ tissue and $Y_1^{-/-}$ colonic tissues. Note the significant increase (P<0.05, Students' unpaired t test) observed in $^{+/+}$ tissues plus BIIE 0246 compared with controls ($^{+/+}$). The numbers of observations are given in parentheses

(Fig. 3B) in mouse colon (and to a lesser degree in $Y_2^{-/-}$ tissue), while responses were unchanged in $Y_1^{-/-}$, compared with $Y_1^{+/+}$ (with or without BIBO 3304) or $Y_2^{+/+}$ tissues (Hyland and Cox 2003; Fig. 3B). These data indicate that the Y_2 receptor (and to a lesser extent Y_1 receptors) is located pre-junctionally on colonic submucosal neurones and that their stimulation modulates non-cholinergic neurotransmission in the mouse colon (Hyland and Cox 2003).

6.3
Y_1 Receptor-Mediated Inhibitory Tone

The mucosal Y_1 receptor has recently been reported to mediate inhibitory tone in mouse (Hyland and Cox 2002; Hyland et al. 2003) and in isolated human colon mucosae (Cox and Tough 2002). BIBO 3304 alone can stimulate an increase in basal mucosal ion transport in the colon from both species, suggesting loss of endogenous inhibitory (anti-secretory) tone which could be mediated by either NPY or PYY. However, a similar increase in I_{sc} in response to BIBO 3304 is also observed in NPY$^{-/-}$ mouse colon, implicating either circulating PYY and/or PYY(3–36) as the important mediators of this inhibitory tone in the mouse (Hyland and Cox 2002).

6.4
Y Receptor-Mediated Smooth Muscle Contraction in $Y_2^{-/-}$ and $Y_4^{-/-}$ Colon

Longitudinal smooth muscle contractile effects of NPY appear primarily to be mediated by Y_2 receptors in wild-type mouse colon, nevertheless a small increase in tone is observed in $Y_2^{-/-}$ tissue, which is sensitive to pre-treatment with BIBO 3304 (Hyland et al. 2003) and is therefore Y_1 receptor-mediated. Both hPP and porcine (pPP) increase basal smooth muscle tone (and frequency and amplitude of spontaneous contraction) in control and $Y_4^{-/-}$ mouse ascending colon (pre-treated with TTX) and these responses are only abolished by the Y_1 antagonist (BIBO 3304) showing that both species PP can stimulate murine Y_1 receptors.

7
NPY and PYY in Intestinal Disorders and Gastrointestinal Pathology

PYY at postprandial circulating levels, profoundly inhibits VIP-stimulated electrolyte secretion in the small intestine of healthy human volunteers (Playford et al. 1990). Similarly intravenous infusion of NPY into healthy human volunteers has been shown to inhibit prostaglandin-E_2 stimulated fluid and electrolyte secretion (Holzer-Petsche et al. 1991). Such studies strongly suggest that both PYY and NPY have an anti-diarrhoeal role in the human intestine. Conversely, elevated colonic PYY has been associated with chronic idiopathic slow transit constipation (for review see, El Salhy et al. 2002) but there are also reports that serum PYY and PP levels remain unchanged, following ingestion of a meal, in patients suffering from this disease (Van et al. 1998).

Colonic biopsies taken from patients suffering from inflammatory bowel disease (IBS) show changes in both NPY and PYY content when compared to control samples (Simren et al. 2003). Those patients who suffered from diarrhoea predominant IBS (IBS-D) showed a significant decrease in NPY levels. In a similar study the amount of PYY present in biopsies taken from constipation predominant IBS (IBS-C) patients and IBS-D patients displayed tissue specific changes in PYY levels with a significant decrease in the descending colon and no associated changes in the ascending colon (Simren et al. 2003). Thus, while changes in NPY and PYY expression in the intestinal wall may contribute to the pathogenesis of colonic disorders, it may be envisaged that in cases of IBS-D for example NPY and PYY analogues may be beneficial clinically, while Y receptor antagonists may be of use in the treatment of IBS-C predominant disease.

In contrast, biopsies taken from patients suffering from coeliac disease, show an elevation in intestinal NPY, however the functional significance of this increase is as yet unclear (Sjölund and Ekman 1989). Both basal and postprandial levels of PYY are elevated in coeliac disease, and it has been suggested this may be a compensatory response to prolong gastrointestinal transit time and to alleviate diarrhoea (for a review see El Salhy et al. 2002). Circulating levels of PYY also appear to increase following resection of the small and large intestine and this is likely to be a similar compensatory response in order to slow gastrointestinal transit through the resected gut (for a review see El Salhy et al. 2002). Furthermore, PYY has been reported to have a gastric protective effect by inhibiting ethanol-induced gastric injury suggesting a physiological role for PYY in protecting the gastric mucosa from injury following ingestion of a meal

(Kawakubo et al. 2002). It would seem therefore that changes in NPY or PYY may be associated with disease i.e. IBS-C and peptic ulcer formation. PYY (and NPY) have a positive effect in slowing gastrointestinal transit and potentially acting as an anti-diarrhoeal in those patients with functionally or anatomically compromised colons (i.e. as a result of IBS or surgery).

8
PYY: A Trophic Factor in the Intestine?

^{125}I-labelled PYY binding studies suggest that these binding sites (probably Y_2 receptors) are preferentially expressed in crypt cells rather than in villus cells in rat small intestine (Voisin et al. 1990). Interestingly in situ hybridization studies also reveal PYY mRNA positive cells within the epithelial lining of the crypts as well as in the lamina propria of rat colon (Gomez et al. 1995). Thus the close localization of endocrine cell PYY and Y_2 receptors in epithelial cells at the base of crypts where stem cells undergo maturation and differentiation, might indicate a role for PYY in regulating these mechanisms. The Y_2 receptor has been implicated in increasing alkaline phosphatase activity and decreasing dipeptidase enzymes in Caco-2 cells. These cells exhibit morphological and biochemical characteristics of normal small and large intestinal epithelium, demonstrating a role for PYY and/or PYY(3-36) in promoting colonic cell differentiation (Sgambati et al. 1997). Exogenous PYY, when administered (either intraperitoneally or subcutaneously) to rats and mice, results in a significant increase in intestinal weight, DNA content and protein content in small intestine and colon. Species differences and differences in PYY levels within different intestinal regions were cited as reasons for the variation between rat and mouse data, and that between the small intestine and the colon (Gomez et al. 1995). These effects may well be mediated by the Y_1 receptor, which is shown to couple mitogen-activated protein kinase phosphorylation and result in cell growth in gut epithelial cells (Mannon and Mele 2000).

9
Conclusions

This chapter describes the differential localization of NPY, PYY, PYY(3-36) and PP in the gastrointestinal tract of human, rat and mouse intestine (with some reference to guinea pig), as well as the immunocytochemical localization of Y_1 and Y_4 receptors in rat and human tissue, and the detection of Y_2, Y_5 and y_6 receptors using RT-PCR or northern blot analysis respectively. As antibodies for all the Y receptor types are developed and become commercially available, more information about their discrete cellular and subcellular localization will be obtained. Specifically we have focused on two physiological mechanisms here, gastrointestinal ion secretion and motility and the regulation of these responses by the pancreatic polypeptides with particular reference to those studies where the selective antagonists for the Y_1 and Y_2 receptors, and most recently Y receptor knockout mice have been used. These mutant mice have proved to be particularly useful in confirming various physiological functions and also in characterizing the pharmacology of different species peptide agonists. In terms of clinical applications, while NPY and PYY may be associated with some gastrointestinal disorders (where Y receptor antagonists may be useful), there is clear evidence to suggest that development of NPY, PYY or PP agonists would be beneficial as anti-diarrhoeal agents. Indeed Y_4 preferring agonists or antagonists might prove to be the most useful in the development of novel anti-diarrhoeals or anti-constipatory drugs as this receptor type is not widely expressed in peripheral tissues and is functionally significant in human colon (Cox and Tough 2002).

References

Anini Y, Fu-Cheng X, Cuber JC, Kervran A, Chariot J, Roz C (1999) Comparison of the postprandial release of peptide YY and proglucagon-derived peptides in the rat. Pflugers Arch 438:299-306

Balasubramaniam A, McFadden DW, Rudnicki M, Nussbaum MS, Dayal R, Srivastava LS, Fischer JE (1989) Radioimmunoassay to determine postprandial changes in plasma neuropeptide Y levels in awake dogs. Neuropeptides 14:209-212

Batterham RL, Cowley MA, Small CJ, Herzog H, Cohen MA, Dakin CL, Wren AM, Brynes AE, Low MJ, Ghatei MA, Cone RD, Bloom SR (2002) Gut hormone PYY(3-36) physiologically inhibits food intake. Nature 418:650-654

Böttcher G, Alumets J, Håkanson R, Sundler F (1986) Co-existence of glicentin and peptide YY in colorectal L-cells in cat and man. An electron microscopic study. Regul Pept 13:283-291

Böttcher G, Ekblad E, Ekman R, Håkanson R, Sundler F (1993a) Peptide YY: a neuropeptide in the gut. Immunocytochemical and immunochemical evidence. Neuroscience 55:281-290

Böttcher G, Sjöberg J, Ekman R, Håkanson R, Sundler F (1993b) Peptide YY in the mammalian pancreas: immunocytochemical localisation and immunochemical characterization. Regul Pept . 43:115-130

Campbell RE, Smith MS, Allen SE, Grayson BE, Ffrench-Mullen JM, Grove KL (2003) Orexin neurons express a functional pancreatic polypeptide Y_4 receptor. J.Neurosci 23:1487-1497

Chen CH, Rogers RC, Stephens RL, Jr. (1996) Intracisternal injection of peptide YY inhibits gastric emptying in rats. Regul Pept 61:95-98

Cox HM (1993) The role of NPY and Related peptides in the control of gastrointestinal function. In: Colmers WF, Wahlestedt C (eds) The biology of neuropeptide Y and related peptides. Humana Press Inc., Totowa, NJ, pp 273-306

Cox HM (1998) Peptidergic regulation of intestinal ion transport. A major role for neuropeptide Y and the pancreatic polypeptides. Digestion 59:395-399

Cox HM, Cuthbert AW (1990) The effects of neuropeptide Y and its fragments upon basal and electrically stimulated ion secretion in rat jejunum mucosa. Br J Pharmacol 101:247-252

Cox HM, Cuthbert AW, Håkanson R, Wahlestedt C (1988) The effect of neuropeptide Y and peptide YY on electrogenic ion transport in rat intestinal epithelia. J.Physiol 398:65-80

Cox HM, Pollock EL, Tough IR, Herzog H (2001a) Multiple Y receptors mediate pancreatic polypeptide responses in mouse colon mucosa. Peptides 22:445-452

Cox HM, Rudolph A, Gschmeissner S (1994) Ultrastructural co-localisation of neuropeptide Y and vasoactive intestinal polypeptide in neurosecretory vesicles of submucous neurons in the rat jejunum. Neuroscience 59:469–476

Cox HM, Tough IR (2000) Functional studies with a novel neuropeptide Y_2 receptor antagonist, BIIE0246, in isolated mucosal preparations from the rat gastrointestinal tract. Br J Pharmacol 129:85P

Cox HM, Tough IR (2002) Neuropeptide Y, Y_1, Y_2 and Y_4 receptors mediate Y agonist responses in isolated human colon mucosa. Br J Pharmacol 135:1505–1512

Cox HM, Tough IR, Hyland NP (2003) Neuropeptide Y Y_1 and Y_2 receptors mediate inhibitory tone in murine and human colon mucosae. Neurogastroenterol Motil 15:195–237

Cox HM, Tough IR, Zandvliet DWJ, Holliday ND (2001b) Constitutive neuropeptide Y Y4 receptor expression in human colonic adenocarcinoma cell lines. Br J Pharmacol 132:345–353

Dumont Y, Cadieux A, Doods H, Pheng LH, Abounader R, Hamel E, Jacques D, Regoli D, Quirion R (2000) BIIE0246, a potent and highly selective non-peptide neuropeptide Y Y_2 receptor antagonist. Br J Pharmacol 129:1075–1088

Ekblad E, Ekman R, Håkanson R, Sundler F (1988) Projections of peptide-containing neurons in rat colon. Neuroscience 27:655–674

Ekblad E, Sundler F (2002) Distribution of pancreatic polypeptide and peptide YY. Peptides 23:251–261

Ekblad E, Winther C, Ekman R, Håkanson R, Sundler F (1987) Projections of peptide-containing neurons in rat small intestine. Neuroscience 20:169–188

El Salhy M, Danielsson A, Axelsson H, Qian BF (2000) Neuroendocrine peptide levels in the gastrointestinal tract of mice after unilateral cervical vagotomy. Regul Pept 88:15–20

El Salhy M, Suhr O, Danielsson A (2002) Peptide YY in gastrointestinal disorders. Peptides 23:397–402

Feletou M, Rodriguez M, Beauverger P, Germain M, Imbert J, Dromaint S, Macia C, Bourrienne A, Henlin JM, Nicolas JP, Boutin JA, Galizzi JP, Fauchere JL, Canet E, Duhault J (1998) NPY receptor subtypes involved in the contraction of the proximal colon of the rat. Regul Pept 75/76:221–229

Ferrier L, Segain J, Bonnet C, Cherbut C, Lehur P, Jarry A, Galmiche J, Blottiere H (2002) Functional mapping of NPY/PYY receptors in rat and human gastro-intestinal tract. Peptides 23:1765–1771

Ferrier L, Segain JP, Pacaud P, Cherbut C, Loirand G, Galmiche JP, Blottiere HM (2000) Pathways and receptors involved in peptide YY induced contraction of rat proximal colonic muscle in vitro. Gut 46:370–375

Fujimiya M, Itoh E, Kihara N, Yamamoto I, Fujimura M, Inui A (2000) Neuropeptide Y induces fasted pattern of duodenal motility via Y_2 receptors in conscious fed rats. Am.J.Physiol Gastrointest.Liver Physiol 278: G32–G38

Furness JB (2000) Types of neurons in the enteric nervous system. J Auton Nerv Syst 81:87–96

Gomez G, Zhang T, Rajaraman S, Thakore KN, Yanaihara N, Townsend CM, Jr., Thompson JC, Greeley GH (1995) Intestinal peptide YY: ontogeny of gene expression in rat bowel and trophic actions on rat and mouse bowel. Am J Physiol 268: G71–G81

Goumain M, Voisin T, Lorinet AM, Ducroc R, Tsocas A, Roze C, Rouet-Benzineb P, Herzog H, Balasubramaniam A, Laburthe M (2001) The peptide YY-preferring receptor mediating inhibition of small intestinal secretion is a peripheral Y_2 receptor: pharmacological evidence and molecular cloning. Mol Pharmacol 60:124–134

Goumain M, Voisin T, Lorinet AM, Laburthe M (1998) Identification and distribution of mRNA encoding the Y_1, Y_2, Y_4, and Y_5 receptors for peptides of the PP-fold family in the rat intestine and colon. Biochem Biophys Res Commun 247:52–56

Grandt D, Schimiczek M, Beglinger C, Layer P, Goebell H, Eysselein VE, Reeve JR, Jr. (1994) Two molecular forms of peptide YY (PYY) are abundant in human blood: characterization of a radioimmunoassay recognizing PYY 1–36 and PYY 3–36. Regul Pept 51:151–159

Grandt D, Teyssen S, Schimiczek M, Reeve JR, Jr., Feth F, Rascher W, Hirche H, Singer MV, Layer P, Goebell H (1992) Novel generation of hormone receptor specificity by amino terminal processing of peptide YY. Biochem Biophys Res Commun 186:1299–1306

Gregor P, Feng Y, DeCarr LB, Cornfield LJ, McCaleb ML (1996) Molecular characterization of a second mouse pancreatic polypeptide receptor and its inactivated human homologue. J Biol Chem. 271:27776–27781

Holliday ND, Pollock EL, Tough IR, Cox HM (2000) PYY preference is a common characteristic of neuropeptide Y receptors expressed in human, rat, and mouse gastrointestinal epithelia. Can J Physiol Pharmacol 78:126–133

Holzer-Petsche U, Petritsch W, Hinterleitner T, Eherer A, Sperk G, Krejs GJ (1991) Effect of neuropeptide Y on jejunal water and ion transport in humans. Gastroenterology 101:325–330

Hornnes PJ, Kuhl C, Holst JJ, Lauritsen KB, Rehfeld JF, Schwartz TW (1980) Simultaneous recording of the gastro-entero-pancreatic hormonal peptide response to food in man. Metabolism 29:777–779

Hyland NP, Cox HM (2003) Differential role of Y_1 and Y_2 receptors mediating neuropeptide Y's contribution to veratridine-induced ion transport across mouse colon. Br J Pharmacol 138:90P

Hyland NP, Sjöberg F, Tough IR, Herzog H, Cox HM (2003) Functional consequences of neuropeptide Y Y_2 receptor knockout and Y_2 antagonism in mouse and human colonic tissues. Br J Pharmacol 139:863–871

Hyland NP, Cox HM (2002) Effects of veratridine upon electrogenic ion transport across NPY knockout and wildtype mouse descending colon mucosa. Br J Pharmacol 135:231P

Hyland NP, Herzog H, Cox HM (2002) Decreased sensitivity to pancreatic polypeptide in colonic mucosa from Y_2 receptor knockout mice. Br J Pharmacol 135:43P

Jackerott M, Larsson LI (1997) Immunocytochemical localisation of the NPY/PYY Y_1 receptor in enteric neurons, endothelial cells, and endocrine-like cells of the rat intestinal tract. J Histochem Cytochem 45:1643–1650

Kawakubo K, Yang H, Tache Y (2002) Gastric protective effect of peripheral PYY through PYY preferring receptors in anesthetized rats. Am J Physiol Gastrointest Liver Physiol 283: G1035–G1041

King PJ, Williams G, Doods H, Widdowson PS (2000) Effect of a selective neuropeptide Y Y_2 receptor antagonist, BIIE0246 on neuropeptide Y release. Eur J Pharmacol 396: R1–R3

Mannon PJ, Kanungo A, Mannon RB, Ludwig KA (1999) Peptide YY/neuropeptide Y Y_1 receptor expression in the epithelium and mucosal nerves of the human colon. Regul Pept 83:11–19

Mannon PJ, Mele JM (2000) Peptide YY Y_1 receptor activates mitogen-activated protein kinase and proliferation in gut epithelial cells via the epidermal growth factor receptor. Biochem J 350:655–661

Matsumoto M, Nomura T, Momose K, Ikeda Y, Kondou Y, Akiho H, Togami J, Kimura Y, Okada M, Yamaguchi T (1996) Inactivation of a novel neuropeptide Y/peptide YY receptor gene in primate species. J Biol Chem 271:27217–27220

Medeiros MD, Turner AJ (1994) Processing and metabolism of peptide-YY: pivotal roles of dipeptidylpeptidase-IV, aminopeptidase-P, and endopeptidase-24.11. Endocrinology 134:2088–2094

Michel MC, Beck-Sickinger A, Cox H, Doods HN, Herzog H, Larhammar D, Quirion R, Schwartz T, Westfall T (1998) XVI International Union of Pharmacology recommendations for the nomenclature of neuropeptide Y, peptide YY, and pancreatic polypeptide receptors. Pharmacol Rev 50:143–150

Ohtani N, Sasaki I, Naito H, Shibata C, Matsuno S (2001) Mediators for fat-induced ileal brake are different between stomach and proximal small intestine in conscious dogs. J Gastrointest Surg 5:377–382

Onaga T, Zabielski R, Kato S (2002) Multiple regulation of peptide YY secretion in the digestive tract. Peptides 23:279–290

Peaire AE, Krantis A, Staines WA (1997) Distribution of the NPY receptor subtype Y_1 within human colon: evidence for NPY targeting a subpopulation of nitrergic neurons. J Auton Nerv Syst 67:168–175

Pheng LH, Perron A, Quirion R, Cadieux A, Fauchere JL, Dumont Y, Regoli D (1999) Neuropeptide Y-induced contraction is mediated by neuropeptide Y Y_2 and Y_4 receptors in the rat colon. Eur J Pharmacol 374:85–91

Pironi L, Stanghellini V, Miglioli M, Corinaldesi R, De Giorgio R, Ruggeri E, Tosetti C, Poggioli G, Morselli Labate AM, Monetti N. (1993) Fat-induced ileal brake in humans: a dose-dependent phenomenon correlated to the plasma levels of peptide YY. Gastroenterology 105:733–739

Playford RJ, Cox HM (1996) Peptide YY and neuropeptide Y: two peptides intimately involved in electrolyte homeostasis. Trends Pharmacol Sci 17:436–438

Playford RJ, Domin J, Beacham J, Parmar KB, Tatemoto K, Bloom SR, Calam J (1990) Preliminary report: role of peptide YY in defence against diarrhoea. Lancet 335:1555–1557

Rettenbacher M, Reubi JC (2001) Localisation and characterization of neuropeptide receptors in human colon. Naunyn Schmiedebergs Arch Pharmacol 364:291–304

Sainsbury A, Schwarzer C, Couzens M, Fetissov S, Furtinger S, Jenkins A, Cox HM, Sperk G, Hökfelt T, Herzog H (2002) Important role of hypothalamic Y_2 receptors in body weight regulation revealed in conditional knockout mice. Proc Natl Acad Sci USA 99:8938–8943

Sang Q, Williamson S, Young HM (1997) Projections of chemically identified myenteric neurons of the small and large intestine of the mouse. J Anat 190: 209–222

Sang Q, Young HM (1996) Chemical coding of neurons in the myenteric plexus and external muscle of the small and large intestine of the mouse. Cell Tissue Res 284:39–53

Sang Q, Young HM (1998) The identification and chemical coding of cholinergic neurons in the small and large intestine of the mouse. Anat Rec 251:185–199

Sawa T, Mameya S, Yoshimura M, Itsuno M, Makiyama K, Niwa M, Taniyama K (1995) Differential mechanism of peptide YY and neuropeptide Y in inhibiting motility of guinea-pig colon. Eur J Pharmacol 276:223–230

Schwartz TW (1983) Pancreatic polypeptide: a hormone under vagal control. Gastroenterology 85:1411–1425

Servin AL, Rouyer-Fessard C, Balasubramaniam A, Saint PS, Laburthe M (1989) Peptide-YY and neuropeptide-Y inhibit vasoactive intestinal peptide- stimulated adenosine 3',5'-monophosphate production in rat small intestine: structural requirements of peptides for interacting with peptide-YY-preferring receptors. Endocrinology 124:692–700

Sgambati SA, Turowski GA, Basson MD (1997) Peptide YY Selectively Stimulates Expression of the Colonocytic Phenotype. J Gastrointest Surg 1:561–568

Sheldon RJ, Malarchik ME, Burks TF, Porreca F (1990) Effects of nerve stimulation on ion transport in mouse jejunum: responses to Veratrum alkaloids. J Pharmacol Exp Ther 252:636–642

Simren M, Stotzer PO, Sjovall H, Abrahamsson H, Bjornsson ES (2003) Abnormal levels of neuropeptide Y and peptide YY in the colon in irritable bowel syndrome. Eur J Gastroenterol Hepatol 15:55–62

Sjölund K, Ekman R (1989) Increased tissue concentration of neuropeptide Y in the duodenal mucosa in coeliac disease. Scand J Gastroenterol 24:607–612

Sundler F, Böttcher G, Ekblad E, Håkanson R (1993) PP, PYY and NPY occurance and distribution in the periphery. In: Colmers WF, Wahlestedt C (eds) The biology of neuropeptide Y and related peptides. Humana Press Inc., Totowa, NJ, pp 157–196

Tanabe Y, Hotokezaka M, Mibu R, Tanaka M (2002) Effects of extrinsic autonomic denervation and myenteric plexus transection on colonic motility in conscious dogs. Surgery 132:471–479

Tough IR, Cox HM (1996) Selective inhibition of neuropeptide Y Y_1 receptors by BIBP3226 in rat and human epithelial preparations. Eur J Pharmacol. 310:55–60

Tough IR, De Souza RJ, Herzog H, Cox HM (2002) Pancreatic polypeptide responses in colonic mucosal and smooth muscle preparations from wild type and Y_4 receptor knockout mice. Br J Pharmacol 135:44P

Van DS Jr, Kamm MA, Nightingale JM, Akkermans LM, Ghatei MA, Bloom SR, Jansen JB, Lennard-Jones JE (1998) Circulating gastrointestinal hormone abnormalities in patients with severe idiopathic constipation. Am J Gastroenterol 93:1351–1356

Voisin T, Rouyer-Fessard C, Laburthe M (1990) Distribution of common peptide YY-neuropeptide Y receptor along rat intestinal villus-crypt axis. Am J Physiol 258: G753–G759

Yunker AM, Paupore EJ, Galligan JJ (1999) C-Fos in enteric nerves after extrinsic denervation of guinea pig ileum. J Surg Res 82:324–330

Yuyama N, Mizuno J, Tsuzuki H, Wada-Takahashi S, Takahashi O, Tamura K (2002) Effects of extrinsic autonomic inputs on expression of c-Fos immunoreactivity in myenteric neurons of the guinea pig distal colon. Brain Res. 948:8–16

Zheng H, Berthoud HR (2000) Functional vagal input to gastric myenteric plexus as assessed by vagal stimulation-induced Fos expression. Am J Physiol Gastrointest Liver Physiol 279: G73–G81

NPY and Immune Functions: Implications for Health and Disease

S. Bedoui · R. Pabst · S. von Hörsten

Department of Functional and Applied Anatomy, OE 4120,
Carl-Neuberg Str.1, 30625 Hannover, Germany
e-mail: Hoersten.Stephan.von@MH-Hannover.de

1	Introduction	410
2	Bidirectional Communication Between NPY and the Immune System	411
2.1	NPY-Positive Fibers in Spleen, Thymus, Lymph Nodes and Bone Marrow	411
2.2	Coexistence and Corelease of NPY and CA in Lymphatic Organs	412
2.3	NPY Receptors in the Immune System	413
2.4	Effects of the Immune System on Expression and Release of NPY	413
3	Actions of NPY on Innate Immune Functions	414
3.1	Phagocyte Function	414
3.2	Natural Killer Cell Activity	416
3.3	Humoral Immune Functions and Mediator Release	417
3.3.1	Respiratory Burst	417
3.3.2	Cytokine Release	418
3.3.3	Histamine Release by Mast Cells	419
3.3.4	Antimicrobial Activities of NPY	420
4	Lymphocyte Biology	421
4.1	T Lymphocytes	421
4.2	B Lymphocytes	422
5	Migration and Trafficking of Leukocytes	424
6	Comparison of Immunological Studies on NPY in Different Species	427
7	NPY Acts as a Transmitter in Neuroimmune Crosstalk	427
8	Clinical Implications	429
8.1	Autoimmunity	429
8.1.1	T Lymphocytes in Autoimmunity	429
8.1.2	Autoantibodies and CD5 B Lymphocytes in Autoimmunity	430
8.1.3	Involvement of NPY in Autoimmune Disorders	431
8.1.4	Possible Mechanisms for NPY in Autoimmunity	435
8.2	A Role for NPY in the Control of Infectious Diseases	435
9	Conclusion	436
	References	437

Abstract It is now well established that the nervous system and the immune system are connected in a reciprocal manner. A major pathway in the neuroimmune crosstalk is provided by the sympathetic nervous system (SNS). Though this interaction is thought to be predominantly mediated by catecholamines (CA), compelling evidence is now available indicating that neuropeptide Y (NPY), released from sympathetic nerves within immune organs, also exerts profound influence on both cellular and humoral immune functions. NPY interferes with critical immune functions, such as differentiation of T helper cells, monocyte mediator release, natural killer (NK) cell activation, and immune cell redistribution. Additionally, NPY also modulates the outcome of CA induced immune alterations in vitro and in vivo, thereby acting as a neuroimmune cotransmitter. In this chapter, we summarize key findings from the interaction between NPY and immune functions and discuss implications for the clinical situation. In addition, fields of future research on this interesting topic are outlined.

Keywords Immune functions · Macrophages/monocytes · NK cells · Lymphocytes · Autoimmunity

1
Introduction

The sympathetic nervous system (SNS) is critical for the maintenance of homeostasis. In addition to its involvement in the regulation of crucial networks of the organism, such as the cardiovascular or gastrointestinal system, it is now well established that a functional interaction between the SNS and immune system is also present (reviewed by Straub et al. 1998; Elenkov et al. 2000; Downing et al. 2000; Madden 2001). Sympathetic nerves directly terminate in primary and secondary lymphoid organs. SNS transmitters, such as catecholamines (CA), are shown to bind to specific receptors expressed on various types of immune cells and finally numerous immunological functions are modulated by the SNS (reviewed by Elenkov et al. 2000). On the other hand, it is well documented that mediators from the immune system, such as cytokines, can directly interact with critical central nervous system (CNS) structures and functions (Turnbull et al. 1999; Parnet et al. 2002). A more detailed understanding of this bidirectional communication will provide new insights into both physiology and pathophysiology.

The vast majority of these studies have focused on the role of CA (reviewed by Kohm and Sanders 2001). However, as indicated by early studies (Lundberg et al. 1989), CA are not the only sympathetic transmitters being released in the immune system. NPY is costored and coreleased from sympathetic nerves terminating in lymphoid organs (reviewed by Lundberg et al. 1996) and leukocytes express functional NPY receptors (Petitto et al. 1998; Bedoui et al. 2002). Strikingly, a growing number of investigations have proven that NPY also modulates critical immunological functions. Thus, NPY can also be considered a player in the interaction between the SNS and the immune system (Bedoui et al. 2003a).

The aim of this chapter is to summarize key findings from the interaction between NPY and the immune system. The description of the anatomical and functional basis for this interaction will be followed by a detailed overview on the effects of NPY on critical immune functions in vitro and in vivo, such as phagocyte function, cytokine release, natural killer (NK) activity, lymphocyte biology and trafficking of immune cells. Finally, we present evidence for a role of NPY in different pathophysiological conditions and discuss possible mechanisms underlying these observations.

2
Bidirectional Communication Between NPY and the Immune System

Neuroimmune interactions are bidirectional: the nervous system modulates immune mechanisms, whereas the immune system affects functions of the nervous system. Therefore, a physiological role for NPY in the neuroimmune crosstalk depends on certain anatomical and functional prerequisites. First of all, the immune system needs to be supplied with NPY-containing nerve fibers. Secondly, NPY must be released within the local immunological microenvironment. Finally, cells of the immune system should posses NPY receptors in order to enable a direct communication between locally present NPY and immune cells. On the other hand, the expression and the release of NPY should be subject to modulation by the immune system.

2.1
NPY-Positive Fibers in Spleen, Thymus, Lymph Nodes and Bone Marrow

Sympathetic nerve fibers, originating from brain regions that are crucial for homeostasis, such as the paraventricular nucleus in the hypothalamus or the caudal raphe nuclei (reviewed by Sawchenko and Swanson 1982) project to preganglionic neurons in the thoracic and lumbal spinal cord. Myelinated fibers from these neurons branch off the spinal cord through the ventral root to form synapses with cell bodies located in the visceral ganglia, such as the superior mesenteric or the celiac ganglion. From these ganglia, a second projection follows the vasculature to terminate in target organs.

An overwhelming number of studies demonstrated the presence of sympathetic fibers in lymphoid organs (for details see Elenkov et al. 2000). Even though the majority of these studies were designed to detect the expression of CA or CA-specific enzymes, such as tyrosine hydroxylase and dopamine beta hydroxylase, there is also compelling evidence for the presence of NPY-containing fibers in various lymphoid organs. Using immunohistochemistry to detect NPY expression, Lundberg et al. (1985) found several NPY-positive fibers innervating the cat spleen that originated in the celiac ganglion. Furthermore, an analysis using electron microscopy revealed that NPY-positive nerve terminals are found in close apposition to lymphocytes and macrophages in the marginal zone of the rat spleen, indicating a direct interaction between NPY and immune cells (Romano et al. 1991). Fink et al. (1988) investigated the presence of NPY-positive sympathetic nerves supplying lymph nodes in a variety of mammalian species (guinea pig, rat, cat, pig, mouse, and human). Both visceral and somatic lymph nodes are richly innervated by NPY-positive fibers and some fibers were also found to branch off into the lymphoid parenchyma. NPY-positive nerves also innervate other important lymphoid organs, such as the thymus and the bone marrow. Though the majority of NPY-positive fibers were found to surround blood vessels, several fibers also ramified among immune cells within these organs (Cavalotti et al. 1999; Tabarowski et al. 1996). The functional significance of these findings is further substantiated by studies demonstrating that electrical stimulation of superfused pig and murine spleens induces the release of NPY from sympathetic nerves into the local immunological microenvironment (Lundberg et al. 1989; Straub et al. 2000a).

Finally, the presence of NPY-positive sympathetic fibers has been associated with the morphological and functional development of lymphoid organs. Westermann et al. (1998) described that the re-innervation of transplanted splenic tissues with NPY/TH-positive fibers is a critical element for the functional development of the transplants. In addition, thymocyte repopulation following bone marrow transfer into *scid* mice was also demonstrated to dependent on the innervation with NPY (Mitchell et al. 1997).

Thus, the presence of NPY-containing fibers in lymphoid organs and the release of NPY within the local microenvironment allows for a direct interaction between NPY and immune cells. However, a detailed study on a potentially differential innervation of the discrete compartments and on nerve endings next to lymphocyte subsets is still lacking.

2.2
Coexistence and Corelease of NPY and CA in Lymphatic Organs

Initial studies have revealed that NPY and CA can be found within the same nerve terminals. Subsequent ultrastructural investigations further demonstrated a particular storage pattern for NPY and CA in sympathetic nerve terminals. Small vesicles were found to contain NPY only, whereas in large vesicles NPY is present in combination with CAs (Fried et al. 1985; Ekblad et al. 1984). Lundberg et al. (1989) investigated the functional relevance of this costorage in the spleen. It was found that upon electrical stimulation splenic sympathetic nerves can release NPY either alone or in combination with CAs (Lundberg et al. 1989). The process of corelease is regulated by several mechanisms. According to neurophysiological studies NPY is preferentially released under conditions of elevated neuronal activity (5–40 Hz), whereas the release of CAs dominates under moderate stimulation (1–10 Hz) (Hökfeldt et al. 1991). However, the duration of the nerve activation also modifies the composition the sympathetic transmitter output. Prolonged sympathetic stimulation over 1 h decreases the content of NPY in sympathetic fibers innervating the spleen by 58% (Lundberg et al. 1989). Due to the rather slow velocity of the axonal transport of

NPY of approximately 5 mm/h (D'Hooge et al. 1990), a complete resupply of the nerve terminal with NPY can take up to 11 days (Lundberg et al. 1989).

Thus, NPY and CA can be released together in the immune system. This provides a rationale for a possible interaction between both of these transmitters in affecting immunological functions. Acute, repeated or prolonged nerve stimulation might show different effects due to the surprisingly slow axonal transport of NPY.

2.3
NPY Receptors in the Immune System

Studies characterizing the distribution of the Y_1 receptor have revealed a high density in the spleen of different species (Gehlert et al. 1996; Malström et al. 1998). Unfortunately, since most of these studies used spleen homogenates, a differentiation between vascular components on the one hand and immunological contents, such as the white pulp, on the other was impossible. One study, however, compared binding sites between blood vessels, the red and the white pulp. Even though they found much stronger signals in the red pulp and blood vessels, moderate signals were also obtained in lymphocyte-rich areas of the white pulp (Lundberg et al. 1989). This first indication for the expression of the Y_1 receptor on immune cells was further substantiated by de la Fuente et al. (1993), who demonstrated binding sites for radiolabeled NPY on peritoneal murine macrophages. A more detailed investigation on this matter was conducted by Petitto et al. (1994). They successfully cloned a Y_1 receptor from splenic lymphocytes, which displayed a 100% sequence homology upon comparison to the Y_1 receptor cloned from the frontal cortex. Using reverse transcription (RT)–polymerase chain reaction (PCR) and radioligand binding studies, they demonstrated the functionality of the Y_1 receptor expressed on splenic lymphocytes. In a recent study we extended this finding by the observation that Y_1 receptor transcripts are also detectable in rat peripheral blood mononuclear cells (PBMC) (Bedoui et al. 2002).

In conclusion, with the detection of NPY-containing nerves in lymphatic organs, evidence for the release of NPY within the immunological microenvironment and the demonstration of functional NPY receptors on immune cells, all requirements for an interaction between NPY and the immune system are fulfilled. However, a potential lymphocyte subset-specific expression difference, e.g., CD4, CD8 regulatory cells, memory and naive lymphocytes of the Y_1 receptor has not been studied so far.

2.4
Effects of the Immune System on Expression and Release of NPY

The NPY pathway between the SNS and the immune system is not unidirectional. Immune cell derived messengers, such as interleukin (IL)-1β, are shown to interfere with the production of NPY by human astrocytes (Barnea et al. 2001).

It is also documented that the levels of NPY in specific brain regions are significantly increased in mice transgenic for tumor necrosis factor (TNF)-α (Fiore et al. 2000). Furthermore, leukemia inhibitor factor, another important cytokine for neuroimmune interactions (Arzt et al. 1999), is known to modulate the expression of NPY in sympathetic neurons (Nawa et al. 1991; Zigmond et al. 1997). Finally, a study using coculture of sympathetic ganglion cells (SGC) and splenocytes demonstrated that NPY expression in SGC directly depends upon the activation status of the cocultured splenocytes (Barbany et al. 1991). Thus, several immune derived mediators are capable of interfering with the expression and production of NPY in the nervous system.

There is evidence demonstrating that the immune system is also able to produce NPY. NPY mRNA is found in both murine PBMC (Ericsson et al. 1987) and in human monocytes and lymphocytes (Schwarz et al. 1994). Several treatments, such as dexamethasone and nerve growth factor, upregulate NPY mRNA and NPY in rat PBMC (Ericcson et al. 1991) and human lymphocytes (Bracci-Laudiero et al. 1996), indicating that NPY expression in the immune system is subject to modulation.

In conclusion, there is compelling evidence for a bidirectional communication between NPY and the immune system, thus providing the anatomical and functional basis for NPY to function as transmitter in the neuroimmune crosstalk.

3
Actions of NPY on Innate Immune Functions

NPY exerts critical effects on various immunological parameters, including important actions in both innate and acquired immunity (reviewed by Bedoui et al. 2003a). The aim of this section is to outline the involvement of NPY in critical immune functions, such as phagocytosis, mediator release, NK activity, T helper cell differentiation, B cell homeostasis and leukocyte trafficking.

3.1
Phagocyte Function

Monocytes and neutrophils are key players in the first-line defense of the body. These cells have the ability to phagocytose pathogens and nonbiological particles. The process of phagocytosis, involving engulfment of the pathogen and the subsequent enzymatic degradation within the phagocyte, is mediated through surface molecules that are directed against highly conserved microbial components and is triggered by antibodies or complement fragments bound to the surface of a pathogen.

Several studies have focused on the ability of NPY to interfere with phagocyte function (Fig. 1). An initial study using murine peritoneal monocytes (de la Fuente et al. 1993) compared phagocytotic responses to *Candida albicans* and inert latex beats. The results demonstrated that NPY increases phagocytosis. In-

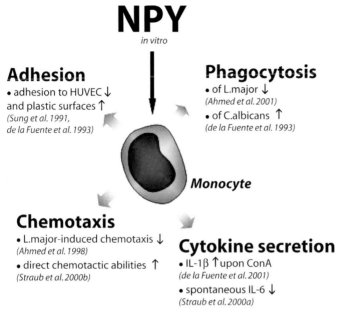

Fig. 1 Summary of the in vitro effects of NPY on monocytes/macrophages. *HUVEC*, human umbilical cord endothelial cells

terestingly, these results are in contrast to a recent study investigating the effect of NPY on the ingestion and degradation of *Leishmania major*, where it was demonstrated that NPY inhibits phagocytosis of these intracellular microbes (Ahmed et al. 2001). The different results can probably be attributed to the various pathophysiological mechanisms by which the microbes and the particles interact with monocytes. Interestingly, phagocytosis of *Leishmania* is a crucial step for the survival of these microbes, since they replicate and proliferate within phagocytes. Thus the prevention of phagocytosis of *Leishmania* might reflect a means by which NPY exerts a protective role during Leishmaniasis.

Another important function of monocytes is the presentation of ingested antigens on the cell surface, thereby fulfilling their role as antigen presenting cells (APCs). APCs are critical to direct immune responses in health and disease. However, until today the possible effects of NPY on this important function have not been investigated. Clearly, there is a need to investigate this matter, in particular as it has been recognized that dendritic cells are several fold more potent APC. The role of NPY on this interesting cell family has not been studied so far.

3.2
Natural Killer Cell Activity

NK cells play an important role in the host response to viruses and tumors. Among several other mechanisms physical and psychological stress, such as physical exercise or mood disorders (Irwin et al. 1987), modulate NK cell activity. However the underlying mechanisms of this interaction are not fully understood. Most likely these alterations are mediated via the SNS. This suggestion is supported by several findings. Chemical sympathectomy is shown to increase NK cell activity (Reder et al. 1989), whereas SNS activation results in an inhibition of NK cell activity (Cunnick et al. 1990). Furthermore, bilateral lesions in the hypothalamic preoptic nucleus result in suppressed NK cell activity (Katafuchi et al. 1993) and these effects can be abrogated by resection of sympathetic splenic nerves (Hori et al. 1995). However, a growing number of studies indicates that sympathetic control on NK cell activity also involves NPY (Fig. 2).

The first detailed study on the association of NPY levels and NK activity was performed in patients undergoing bereavement or other severely threatening life events. It was found that the known decrease in NK activity resulting from such events (Kiecolt-Glaser et al. 1984; Irwin et al. 1987) is inversely correlated with NPY plasma levels. Whereas increased NPY levels result in decreased NK cell activity, decreased NPY levels result in an increased NK cell activity (Irwin et al. 1991). These findings were further emphasized by in vitro experiments demonstrating a direct and concentration-dependent inhibition of NK cell activity by

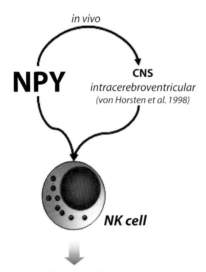

Fig. 2 NPY interacts with NK cells both directly (in vitro) and indirectly (intracerebroventricular). *LN*, lymph node

NPY (Fig. 2). The study clearly demonstrated the specificity of the findings, since anti-NPY sera abrogated the effects and the impaired NK activity was not due to effects of NPY on the target cells themselves (Nair et al. 1993). Interestingly, these suppressive effects on NK activity can be induced not only by direct interaction of NPY with these cells, but also by intracerebroventricular (icv) application (Fig. 2). However, these indirect effects seem to have a rebound effect, since NK activity increased after 24 h (von Hörsten et al. 1998a,b). The peripheral distribution of NK cells is also modulated by NPY. This is of particular importance since immune reactions and inflammations involve the migration and trafficking of leukocytes to their site of action. Recently, we have shown that intravenous application of NPY alters NK cell distribution in the blood in a dose-dependent manner (Bedoui et al. 2001). The effects were bimodal: a low dose NPY injection resulted in decreased numbers of NK cells, whereas high dose NPY was able to mobilize these cells. Interestingly, the distribution of blood NK cells can be also altered by icv administration of NPY. Shortly after injection of NPY a significant increase in NK cell numbers was detected (von Hörsten et al. 1998a,b). The role of a differential expression of adhesion molecules induced or suppressed by NPY or effects on chemokines and their receptors are still unknown.

3.3
Humoral Immune Functions and Mediator Release

3.3.1
Respiratory Burst

Besides the direct ingestion of microbes, phagocytes can also damage pathogens through the generation of toxic metabolites. In addition to numerous different intracellular enzymes and antimicrobial peptides, monocytes and neutrophils produce highly reactive oxygen species, a process termed as respiratory burst. The respiratory burst depends on the enzyme NADPH oxidase and is characterized by the production of hydrogen peroxide, superoxide anion and nitric oxide. The release of these oxidative agents is essential for killing of a number of microorganisms (Hampton et al. 1998).

During phagocytosis, the respiratory burst is strongly activated. Whereas NPY has no influence on the respiratory burst of resting murine peritoneal monocytes, the stimulation of the respiratory burst resulting from activation of phagocytosis is directly increased by NPY (de la Fuente et al. 1993; Fig. 3). Interestingly, NPY also interferes with other agents that induce the release of oxidative reagents in neutrophils. Formyl-methionyl-leucyl-phenylalanine (fMLP), when administered alone, potently activates this pathway, as indicated by increased oxygen radical formation. Hafström et al. (1993) investigated the influence of NPY on fMLP-induced oxygen radical formation in human neutrophils. Their results displayed a bidirectional pattern. If a very low dose of NPY is coadministered with fMLP the action of fMLP is significantly enhanced. However, us-

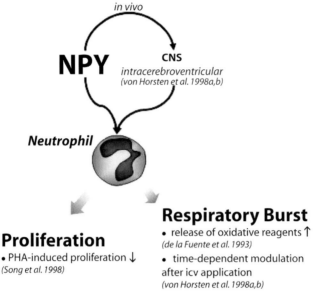

Fig. 3 The effect of NPY on functional aspects of neutrophils. NPY modulates neutrophils function through both direct interactions in vitro and after administration in the brain ventricles (intracerebroventricular). *PHA*, phytohemagglutinin

ing a rather high dose of NPY, fMLP-induced oxygen radical formation is inhibited by NPY. These NPY mediated effects on neutrophil function are not only induced by direct or combined interaction of NPY with these cells, but icv application also induced a similar alteration of blood neutrophil function (von Hörsten et al. 1998a; Fig. 2). Blood neutrophils immediately isolated from icv NPY treated rats exhibited a decrease in zymosan and phorbol-12-myristate 13-acetate-induced oxygen radical formation (von Hörsten et al. 1998a,b; Song et al. 1998). When the same assay was performed 24 h after icv administration of NPY an increase in oxygen radical formation was observed (von Hörsten et al. 1998). Further studies revealed that these effects are mediated via the Y_1 receptor, since the results were mimicked using an Y_1 receptor agonist and were blocked by an Y_1 receptor antagonist (von Hörsten et al. 1998c).

3.3.2
Cytokine Release

Macrophages are a major source of several cytokines released at inflammatory sites. Using electrical stimulation of ex vivo spleen slices (Straub et al 1996), the interaction of CA and NPY in the sympathetic nerve–macrophage interplay was recently examined (Straub et al. 2000a). Administration of exogenous NPY inhibits IL-6 release from splenic macrophages. This process is mediated via the Y_1 receptor, because administration of the Y_1 receptor antagonist BIBP 3226 abrogates this action. Interestingly, coadministration of NPY and norepinephrine (NE) resulted in an even stronger inhibition of the IL-6 release, indicating a potentiation of NE-mediated effect. Because NE mainly acts through α-adrenoceptors (ARs), but at higher concentrations also involves β-adrenergic signaling, the effect of the β-AR agonist isoproterenol was also investigated. Isoproterenol increases the release of IL-6 in a dose-dependent fashion and thus operates in an opposite manner to a α_2-AR agonist. Surprisingly, the combined action of NPY and isoproterenol intensifies the increased IL-6 release. One possible explanation for the potentiating effects of NPY can be a modulation of intracellular cyclic adenosine monophosphate (cAMP) formation. Since NPY is shown to induce both an increase and an inhibition of cAMP formation (Michel et al. 1998), we speculate that the NPY induced effects depend on the catecholaminergic pathway that is activated. In case of β-adrenergic stimulation NPY augments the increase of cAMP, whereas upon α_2-AR activation NPY further inhibits cAMP formation. In conclusion, NPY potentiates the outcome of adrenergic stimuli in the nerve–macrophage interplay, depending on which AR subtype is activated (Straub et al. 2000a). This provides evidence for cotransmission between NPY and CA during neuroimmune interactions.

NPY treatment of a fraction of gut-derived human macrophages and lymphocytes results in a significant proliferative effect on these lymphocytes (Elitsur et al. 1994). This effect seems to be mediated by IL-1β released from macrophages, since depletion of adherent cells abolishes this effect and it can be restored by the addition of exogenous IL-1β. This conclusion is supported by the in vitro capacity of NPY to increase IL-1β production from whole blood (Hernanz et al. 1996) and murine monocytes (de la Fuente et al. 2001).

The effects of NPY on the release of cytokines, such as IL-1β and IL-6 (Fig. 1), and the interaction of NPY with the release of oxidative reagents may be of major importance for host defense. Furthermore, after being secreted from monocytes these cytokines diffuse throughout the body. Since systemically released cytokines can induce a wide spectrum of actions, it is of major importance to investigate to what extent locally NPY evoked cytokine release contributes to systemic levels of cytokines.

3.3.3
Histamine Release by Mast Cells

Mast cells also belong to the innate branch of the immune system and their mediators are crucial components in the first line of immune responses against microbes. Murine connective tissue mast cells respond to NPY by releasing histamine (Arzubiaga et al 1991), whereas NPY has no effect on human mucosa mast cells (T. Gebhardt and S. Bedoui, unpublished results). The effect on connective tissue mast cells appears to be different from other NPY evoked effects on the immune system. Investigations focusing on the involvement of NPY receptors reveal that degranulation in rat peritoneal mast cells can be evoked by C-terminal fragments of NPY, which, however, do not have proper affinities for NPY receptors (Shen et al. 1991). This receptor independent effect is most likely due to positive charges at the C-terminal end of centrally truncated NPY analogues (Cross et al. 1996), and may represent a receptor independent effect of NPY.

The findings on the interaction between NPY and mast cells suggest unspecific effects on the release of histamine that are associated with structural components of NPY receptor-specific analogs. Future development of drugs based on NPY analogs should carefully consider this receptor-independent mechanism.

3.3.4
Antimicrobial Activities of NPY

The skin and the mucosa form the interface between the organism and the outside world. Numerous microbes colonize these sites and therefore the skin and the mucosa must function as barriers. However, with the exchange of nutrients, the export of certain substances and the excretion of waste products, there is also an imperative need for the skin and the mucosa to be permeable. With the dermal and mucosal immune system, the organism has achieved a balance of keeping up the barrier function, yet, ensuring the necessary permeability. In addition to patrolling phagocytes and lymphocytes, these barriers are also covered by a large number of microbicidal proteins and peptides. Small antimicrobial peptides, such as defensins, have the capacity to kill or inactivate several different microbial organisms through disruption of their membranes (reviewed by de Yang et al. 2002).

Isolating biologically active peptides from the skin of a South American frog, Mor et al. (1994) detected a peptide exhibiting 94% similarity with the NPY-related peptide YY (PYY). Microbiological testing of the compound referred to as skin-PYY, revealed remarkable microbicidal activities. A broad variety of bacteria and fungi are irreversibly destroyed upon incubation with skin-PYY for 1 h (Vouldoukis et al. 1996). Interestingly, the study also demonstrated that NPY exhibits similar antibiotic activities. The mechanism appears to involve electrostatic interactions between positive charges of the peptide and negative charges of the microbial membrane, finally resulting in osmolysis of the microbe (Shimizu et al. 1998). The relevance of antimicrobial activities exerted by NPY is further strengthened by a study demonstrating that *Aspergillus niger* is able to secrete a factor that inhibits binding of NPY to NPY receptor expressing cells (Kodukula et al. 1995).

In addition to the antimicrobial activity, NPY and defensins share further functional similarities. NPY and defensins are released from immune cells (de Yang et al. 2002; Schwarz et al. 1994), they both exert strong chemotactic activities for monocytes (Yang et al. 2000; Straub et al. 2000b) and their actions are mediated through $G_{\alpha 1}$-protein-coupled receptors (de Yang et al. 2002; Michel et al. 1998). In light of these apparent similarities we conclude that NPY functions as an antimicrobial peptide. However, several points need to be clarified. What are the concentrations of NPY on the very surface of these compartments? What is the source of NPY? Is it of neuronal origin or is it secreted from non-neuronal cells, possibly immune cells?

4
Lymphocyte Biology

4.1
T Lymphocytes

The main characteristic of lymphocytes is their ability to specifically react against a certain antigen. This unique feature is mediated by the interaction of an antigen with T cell receptor (TCR) or B cell receptor (BCR), respectively. T lymphocytes form two major subpopulations: cytotoxic T lymphocytes and T helper lymphocytes (THL). Whereas cytotoxic T lymphocytes can directly destroy tumor bearing or virus infected cells, THL regulate immune responses and the action of other immune cells, such as the production of antibodies by B lymphocytes. A very important characteristic of THL in health and disease is the functional differentiation of naive THL upon antigen recognition (Th1/Th2 balance). The T_H1 phenotype is characterized by the production of cytokines, such as IL-2 and interferon (IFN)-γ, whereas the production of IL-4 is an important feature of T cells biased towards a T_H2 subtype (Abbas et al. 1997).

Crucial lymphocyte functions are also profoundly influenced by NPY (Fig. 4). Using lymphocytes isolated from BALB/c mouse spleens, it was demonstrated that NPY elevates IL-4 production and decreases IFN-γ production upon

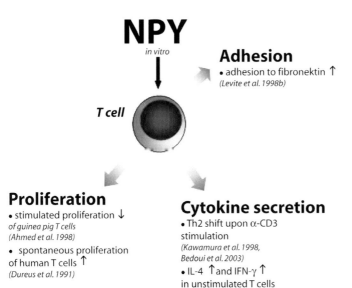

Fig. 4 Summary of the in vitro effects of NPY on T cells

stimulation with a plate-bound anti-CD3 antibody (Kawamura et al. 1998). To confirm these findings, several different T_H1 and T_H2 clones were stimulated with their specific antigens in the presence or absence of NPY. These results showed that NPY suppresses differentiated T_H1 cells in their production of IFN-γ and stimulates the production of IL-4 by T_H2 cells. Thus, a typical T_H2 shift is induced by NPY in vitro. Currently we are addressing the question of whether NPY also induces a T_H2 shift in vivo. In agreement with the in vitro data, intraperitoneal application of NPY significantly inhibits the ex vivo production of IFN-γ in antigen-specific murine lymphocytes. The induction of a T_H2 shift in vivo is further substantiated by the observation that NPY also selectively inhibits the production of antigen-specific IgG2a (Bedoui et al. 2003b).

Direct in vitro stimulation of T_H1 and T_H2 cells by NPY without any additional stimulus was recently shown to produce unexpected results (Fig. 4). In addition to the expected cytokine pattern, NPY was also shown to induce the secretion of IFN-γ by Th2 cells and IL-4 by T_H1 cells (Levite 1998a). However, when these cells were stimulated in the presence of the specific antigen and syngenic APCs, NPY elevated IL-4 in T_H2 cells and inhibited IFN-γ in T_H1 cells, indicating a typical T_H2 shift. Thus, this study demonstrated that under physiological circumstances of antigen stimulation NPY induces a T_H2 shift in vitro.

In addition to these effects on the functional characteristics of lymphocytes, NPY has also been shown to interfere with lymphocyte proliferation (Fig. 4). For instance, in vitro treatment of murine lymphocytes with NPY increases their ability for spontaneous proliferation (Medina et al. 2000). However, in the same study another interesting finding was reported. When NPY is coadministered with concanavalin A (ConA), a strong activation signal, NPY inhibits the ConA-induced proliferatory effect. Thus, an interesting picture is found in NPY/mitogen interaction: NPY administered alone induces completely different results from the effects resulting from interaction with a mitogen. However, the number of lymphocytes in a lymphoid organ depends not only on lymphocyte proliferation but also on programmed cell death, apoptosis. A potential effect of NPY on receptors for cell death has not been studied so far.

4.2
B Lymphocytes

The main function of B lymphocytes is the production of antibodies. Antibodies also have a unique specificity towards an antigen and the antibody produced by individual B lymphocytes can be considered a soluble form of the BCR expressed on the surface this cell. As a part critical component of the humoral immune response, antibodies mediate several immunological effector mechanisms, such as opsonization and complement-mediated lysis.

Though NPY-positive sympathetic fibers form close contacts with B cells in lymphoid organs (Felten et al. 1985), providing an anatomical basis for a direct influence of NPY on B cells, little is known about such an interaction. Indirect evidence for NPY-induced mobilizing effects on B cells comes from a study

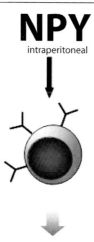

Antibody production
- KLH-specific IgM and IgG ↓
  (Friedmann et al. 1993)
- selective inhibition of
  MOG$_{35-55}$-specific IgG2a ↓
  (Bedoui et al. 2003)

Fig. 5 Scheme to illustrate the influence of NPY on antibody production in vivo. *KLH*, keyhole limpet hemocyanin

using L-deprenyl, a drug capable of reversing chemical sympathectomy (ThyagaRajan et al 1998). These authors demonstrated that the L-deprenyl-induced re-loading of sympathetic nerve fibers with CAs and NPY causes an increase in circulating IgM B cells (ThyagaRajan et al. 1998). However, this study only demonstrates that re-loading of sympathetic nerves induced an increase in circulating B cells, and therefore provides no direct evidence for involvement of NPY. More direct data come from a study investigating the distribution of B cells after intravenous application of NPY. Intravenous infusion of a low dose of NPY results in a significant decrease in the number of circulating IgM B cells (Bedoui et al. 2001). This work also revealed the dose-dependent mobilization of a previously undetected B cell subpopulation in the rat, referred to as B-1-like B cells (Bedoui et al. 2001). B-1 B cells form a distinct B cell subpopulation and are well known in other species (Herzenberg et al. 2000). Even though their function is mostly unclear, these cells are believed to represent an early developmental stage of the immune system (Paul 1998). Finally, intraperitoneal application of NPY in rats results in a dose-dependent inhibition of IgM and IgG antibody responses against keyhole limpet hemocyanin (Friedman et al. 1995, Fig. 5). We have recently found that intraperitoneally administered NPY induces a selective inhibition of anti-myelin oligodendrocyte glycoprotein (MOG)$_{35-55}$ IgG2a (Bedoui et al. 2003b; Fig. 5). It is clear that the influence of NPY and B cells needs to be investigated more intensively.

In conclusion, recent findings provide clear evidence that critical elements in lymphocyte biology are modulated by NPY. However, important questions remain unanswered. What are the mechanisms underlying the effects on the T_H1/T_H2 balance? Which NPY receptor subtypes are involved? What is the physiological role of this interaction? Furthermore, we need to extend our understanding on the effects of NPY on B lymphocytes. Does NPY interfere with the differentiation of naive B lymphocyte into antibody secreting cells or does the inhibitory action results from direct effects on preformed cells? Finally, do the effects of NPY on THL change their abilities to regulate the production of antibodies by B lymphocytes?

5
Migration and Trafficking of Leukocytes

Trafficking and migration of white blood cells are crucial steps in the distribution of leukocytes throughout the body (Westermann and Pabst 1991). This process is regulated by several mechanisms, including the interaction of leukocytes with adhesion molecules expressed on the endothelium and attraction of leukocytes by chemotaxis. The elucidation of mechanisms regulating leukocyte trafficking and migration will have great impact on our understanding of critical immunological phenomena, including inflammation and antigen distribution within the immune system.

Determining the number and the phenotype of peripheral blood leukocytes after intravenous injections of NPY allows investigation of the role of NPY in regulating leukocyte trafficking. Recent experiments have shown that NPY dose-dependently regulates trafficking of B cells, monocytes and NK cells in vivo (Bedoui et al. 2001). In addition, to inducing these quantitative effects, NPY also changes the composition of blood leukocytes. The major fraction of monocytes mobilized by NPY expresses the NK cell typical marker CD161 (Bedoui et al. 2001). CD161-positive monocytes represent a subpopulation that was first described after challenging rats intravenously with IFN-γ (Scriba et al. 1997) and are regarded as activated monocytes (Grau et al. 2000). Thus, among other effects, NPY induces a preferential mobilization of activated monocytes (Bedoui et al. 2001). Interestingly, NPY also exerts mobilizing effects on leukocytes after icv application. Shortly after the administration of NPY into the CNS a moderate leukocytosis is observed, which is mainly composed of neutrophils and NK cells (von Hörsten et al. 1998b).

In addition to these solitary effects of NPY on leukocyte distribution, we have recently conducted a study to investigate the effects of a combined action of NPY and CA on leukocyte numbers in the blood. If NPY is coadministered with epinephrine (EPI) to mimic the naturally occurring corelease of both SNS neurotransmitters, the effects are very different compared to the effects of either EPI or NPY alone. Single administration of low-dose EPI alone does not mobilize blood leukocytes but when administered in combination with NPY, a significant increase in leukocyte numbers is observed. In contrast, the combination of NPY with a high dose of EPI, which by itself induces a strong leukocytosis, results in completely opposite effects. In that case NPY inhibits EPI-induced leukocytosis (Bedoui et al. 2002). Furthermore, this interaction between EPI and NPY is receptor specific, since the inhibitory action is mediated via Y_1 receptor signaling, whereas the potentiation results from Y_5 receptor activation (Bedoui et al. 2002). Thus, a bimodal effect of NPY on EPI-mediated leukocyte mobilization is effective: low dose EPI effects are facilitated, whereas high dose effects are inhibited. These observations are not only due to purely additional effects of NPY upon the EPI effects, since a similar dosage of NPY alone results in different effects.

In conclusion, these findings provide compelling evidence for a role of NPY in regulating the number and composition of blood leukocytes. This action can be induced by both intravenous and icv application of NPY. NPY also modulates the effects of CA on blood leukocyte numbers and composition.

These findings raise some important questions. What is the source of the leukocytes mobilized by NPY? Adrenaline-induced leukocytosis, serving as the prototype of SNS-induced leukocyte mobilization, has been intensively investigated. However, a definite source of mobilized cells could not yet be identified, but most likely cells are derived from lungs, spleen, bone marrow and the marginal pool of blood vessels (Benschop et al. 1996). However, taking into consideration that the observed effects are present within a few minutes and that recruiting cells from lymph nodes or the bone marrow requires more time, we suggest that the interaction between NPY and leukocytes takes place at the endothelium.

What possible molecular mechanisms could be involved in the effects of NPY on leukocyte distribution? With respect to the kinetics described, it is unlikely that changes in leukocyte cellular expression markers influence their migratory capacity. Thus, NPY most probably interacts with preformed leukocytes; this suggests that NPY affects the capabilities of leukocytes to adhere to the endothelium. Indeed, Levite et al. (1998b) and Sung et al. (1991) both demonstrated that NPY enhances leukocyte adhesion to extracellular molecules in vitro, and proposed that the NPY-induced enhancement of cell adhesion is due to increased overall avidity of adhesion molecule receptors on lymphocytes. Since Y_1 receptor antagonism by BIBO 3304 increases leukocyte numbers and a injection of a Y_1 receptor decreases the blood leukocyte count (Bedoui et al. 2002), we propose that NPY increases the adhesion of leukocytes to the endothelium via activation of the Y_1 receptor (Fig. 6). For a better understanding of the underlying mechanisms of this phenomenon, future investigation should examine carefully the involvement of adhesion molecules in NPY-induced leukocyte mobilization.

Another possible mechanism underlying these effects could be an interference of NPY with the chemotactic properties of leukocytes. Chemotaxis is a process that involves guiding neutrophil and monocytes from the blood into the tissues, where they principally encounter their antigens. The action of NPY on human monocytes results in increased chemotaxis, as demonstrated by a chemotactic in vitro assay (Straub et al 2000b). fMLP, a strong chemotactic stimulus, exerts a two- to threefold stronger effect than NPY (Straub et al. 2000b). However, the combined action of fMLP and NPY did not result in additional ef-

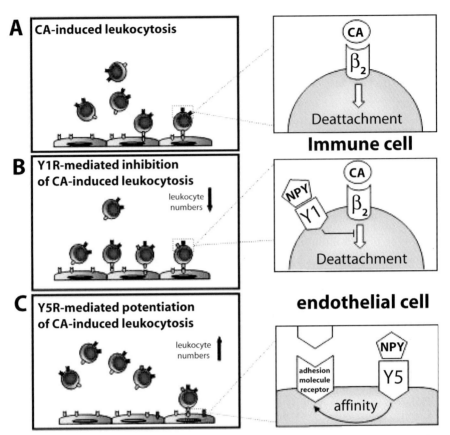

Fig. 6A–C Proposed model for the regulation of CA-induced leukocyte mobilization by NPY. **A** Stimulation of the β_2-AR expressed on leukocytes increases the number of circulating leukocytes in the blood by decreasing their abilities to adhere to the endothelium (deattachment). **B** Activation of the Y_1 receptor inhibits CA-induced leukocyte mobilization. It is proposed that the stimulation of the Y_1 receptor on leukocytes directly inhibits or ameliorates signaling through the β_2-AR, thereby reducing the mobilizing effect of CA. **C** In the presence of Y_5 receptor stimulation the mobilizing effect of CA is potentiated. This effect is likely to take place at the endothelium, where it may reduce the affinity of adhesion molecule receptors

fects: NPY inhibited fMLP-induced chemotaxis of human monocytes (Dureus et al. 1991). Thus, chemotaxis, a critical step in leukocyte migration into and within an organ, is not only profoundly altered by NPY directly, but NPY also modulates the outcome of other chemotactic stimuli.

The findings presented above demonstrate the ability of NPY to regulate leukocyte trafficking and migration in vitro and in vivo. This process most likely involves a receptor-specific modulation of the capabilities of leukocytes to adhere to the endothelium (Fig. 6) and an interaction of NPY with the chemotactic properties of leukocytes. In addition, it has been demonstrated that NPY exerts its immunological action through at least three pathways: direct action on immune cells, effects on leukocytes after injection into the CNS and a receptor-specific interaction with CA in mobilizing leukocytes.

6
Comparison of Immunological Studies on NPY in Different Species

The studies mentioned so far provide a large variety of data on the role of NPY in the immune system. However, since these studies were mostly conducted in different species it is important to differentiate at least between the murine and the human system. The importance of such differentiation is indicated by several findings (see Figs. 1–5).

In humans and mice, CD5 B cells form a constant subpopulation of the B cell pool (reviewed by Herzenberg 2000). In rats, however, the situation is different: several studies have failed to detect a constitutively present subpopulation in this species (de Boer et al. 1992; Vermeer et al. 1994). It is therefore rather surprising that NPY dose-dependently mobilizes a B cell subpopulation in the rat that is characterized by the expression of CD5 and CD11b. Differences between species also become obvious when the effects of NPY on T cell proliferation are examined: NPY increases spontaneous proliferation in human T cells (Elitsur et al. 1994), whereas NPY decreases proliferation in the murine model (Soder et al. 1987). Similar differences are also present with respect to the influence of NPY on monocyte chemotaxis. Human monocytes respond to NPY with increased chemotaxis (Straub et al. 2000b; de la Fuente et al. 1993), while NPY inhibits murine monocyte chemotaxis (Ahmed et al. 1998). Nevertheless, several functions are regulated similarly in the two systems. The NPY induced production of cytokines, such as IL-1β is upregulated in both human and mouse immune cells (de la Fuente et al. 2001; Hernanz et al. 1996). Furthermore, NPY increases in vitro adhesion of monocytes in both the human and the mouse system (Sung et al. 1991; Medina et al. 1998).

Thus, some effects of NPY are specific to the species being investigated, but importantly, other functions also occur in different species. Furthermore, strain differences may add another facet to this picture. However, several immune modulating functions of NPY have been investigated in only one species and/or strain. It is therefore necessary to conduct experiments comparing these effects between different species and strains. This approach will help in understanding the significance of these effects.

7
NPY Acts as a Transmitter in Neuroimmune Crosstalk

The findings presented here have consistently shown that NPY interacts with critical immune functions. Its influence on functions, such as phagocytosis and mediator release, suggests that NPY is involved in the rapid first-line defense against microorganisms. However, NPY is involved not only in innate immune functions, but also interferes with important mechanisms of the acquired limb of the immune system. By altering the T_H1/T_H2-balance and antibody production by B cells, NPY can have a profound influence on directing and maintaining immune responses. Finally, the impact of NPY on trafficking and migration of leukocytes further emphasizes the role of NPY in immune functions.

But does that also mean that NPY is an endogenous player in the interaction between the SNS and the immune system? In order to act as an endogenous neuroimmune transmitter, NPY must meet the criteria for chemically mediated transmission (Eruklar et al. 1994). (a) NPY fulfills the first requirement for a role as a neuroimmune transmitter, since it is synthesized and stored in postganglionic nerves terminating in lymphatic organs (Madden 2001). (b) With its release into lymphatic organs and its presence in the vicinity of these nerves (Lundberg et al. 1989; Straub et al. 2000a) NPY also meets the second criterion. (c) PBMC express functional NPY receptor (NPY-R) (Petitto et al. 1994; Bedoui et al. 2002) and a large number of studies has revealed direct in vitro effects of NPY on different immune cells. (d) The final prerequisite for a neuroimmune transmitter is that the endogenous action of the transmitter can be blocked by administration of competitive antagonists. Administration of an Y_1 receptor antagonist elevates blood leukocyte numbers (Bedoui et al. 2002) and inhibits the release of IL-6 from splenic macrophages evoked by electric stimulation (Straub et al. 2000a). Thus, NPY meets the criteria for chemically mediated transmission between the CNS and the immune system and can, as such, be regarded as a neuroimmune transmitter.

Interestingly, the action of NPY in the immune system is not only restricted to its function as a direct neuroimmune transmitter. Moreover, NPY also modulates the outcome of adrenergic immune functions, such as the release of IL-6 and the mobilization of blood leukocytes. These functions are receptor specific and probably involve G-protein dependent pathways. Upon activation of their specific receptors both CA and NPY modulate the level of intracellular cAMP levels. Therefore, potentiating or inhibiting effects of NPY on CA induced functions might be due to additional effects on cAMP formation or inhibition. These findings are in line with several studies conducted in the cardiovascular system indicating that NPY serves as a cotransmitter (Lundberg et al. 1996). For example, NPY modulates the outcome of EPI-mediated vasoconstriction in a bimodal manner; since both a potentiation (Wahlestedt et al. 1985) and an inhibition were demonstrated (Ohia et al. 1990). In addition, NPY also potentiates EPI-induced growth of cardiomyocytes (Pellieux et al. 2000). In the same manner, we hypothesize that NPY acts a cotransmitter during neuroimmune interactions.

In the in vivo situation, particularly within the local microenvironment, it is unlikely that any single mediator interferes with only one effector system. Many different mediators from several regulatory super-systems, such as the endocrine system and the nervous system, influence the immune system simultaneously. Resulting effects are due to balances between permissive and inhibitory effects exerted by different mediators and their interactions. It is obvious that cotransmitters, which regulate effects of other neurotransmitters, are important players in the local microenvironment. The observations mentioned so far provide a role for NPY, a sympathetic transmitter, to be involved in fine-tuning of immunological functions.

8
Clinical Implications

8.1
Autoimmunity

In the following section a brief introduction into the role of T lymphocytes and antibody producing B lymphocytes in autoimmunity is followed by a detailed discussion of recent findings regarding the involvement of NPY in certain autoimmune disorders. Finally, we will discuss the relevance to autoimmunity of the effects of NPY on lymphocyte function.

8.1.1
T Lymphocytes in Autoimmunity

Each T cell has a unique receptor, the TCR, which enables these cells to specifically recognize antigens. During the T cell development in the thymus, several mechanisms exist to ensure that T cells expressing TCRs that are reactive towards self-antigens are deleted. In some cases these autoreactive T cells escape thymic deletion and are present in the periphery. However, the pure presence of autoreactive T cells in the body does not necessarily lead to the induction of autoimmunity, since autoreactive T cells can also be detected in healthy individuals (Goebels et al. 2000). Obviously, there are mechanisms controlling the activation of these cells in the periphery, hence preventing the onset of autoimmunity.

A critical step in the activation of autoreactive T cells is the differentiation into different T cell subtypes. The T_H1 phenotype is characterized by the production of cytokines, such as IL-2 and IFN-γ, whereas the production of IL-4 is an important feature of T cells biased towards a T_H2 subtype (Abbas et al. 1997). In autoimmune disorders, such as rheumatoid arthritis (RA) and multiple sclerosis (MS), autoreactive T cells are predominantly biased towards the T_H1 phenotype. With the production of pro-inflammatory cytokines, such as IL-2 and IFN-γ, these cells initiate an immune response directed against joint tissues and myelin sheaths, respectively (van Roon et al. 2000; Neumann et al. 2001). On the other hand, other autoimmune disorders, such as systemic lupus erythematodes (SLE), are characterized by a predominance of autoreactive T_H2 cells (Furukawa et al. 1997). These T_H2 cells activate humoral immune responses and the subsequent production of autoantibodies is a critical pathophysiological element in SLE. The significance of the T_H1/T_H2 balance in autoimmunity is further demonstrated by therapeutic strategies seeking to modify this balance. Shifting T_H1

cells towards the T_H2 phenotype is shown to inhibit RA and MS, both experimentally and clinically (van Roon et al. 2000; Neumann et al. 2001). On the other hand, inducing a T_H1 shift cells in SLE has also proven a beneficial therapeutic intervention. Thus, the T_H1/T_H2 balance is a major element in the pathophysiology of autoimmune disorders (Davidson et al. 2001).

8.1.2
Autoantibodies and CD5 B Lymphocytes in Autoimmunity

Another important pathophysiological factor in autoimmunity is the production of antibodies that are directed against self antigens. Autoantibodies are involved in the pathogenesis of RA, MS, SLE and other autoimmune disorders, such as myasthenia gravis or Guillain-Barré Syndrome. In 95% of patients with clinically definite MS, autoantibodies directed against myelin structures, such as MOG, can be detected in the cerebrospinal fluid (CSF) (Correale et al. 2002). The significance of these antibodies for the disease was recently demonstrated by a study reporting that MOG-specific antibodies bind directly to disintegrating myelin around axons in lesions of acute MS (Genain et al. 1999). In RA it is well established that binding of autoantibodies to joint structures initiates macrophage-mediated tissue destruction (reviewed by Burmester et al. 1997). Hence, the production of autoreactive antibodies is another important pathophysiological feature of autoimmune disorders.

CD5 B lymphocytes comprise a special subset of B lymphocytes, which differ in several aspects from normal B lymphocytes. Numerous studies have investigated the involvement of this subset in certain autoimmune disorders, since CD5 B cells were initially shown to produce autoantibodies. However, the situation appears to be not as easy as to ascribe a definite role for CD5 B lymphocytes in autoimmunity. On the one hand, there are investigations demonstrating an increase of CD5 B lymphocytes in the CSF of MS patients (Mix et al. 1991; Correale et al. 1991) and increased numbers of CD5 B lymphocytes are associated with elevated autoantibodies in RA (Arinbjarnarson et al. 1997). On the other hand, several studies reported on protective effects of CD5 B cells in autoimmune disorders. Sellebjerg et al. (2002) have recently shown that the number of CD5 B lymphocytes in the CSF of MS patients is negatively correlated to CSF cells secreting myelin-specific antibodies. Furthermore, after treatment with linodamide, a drug that reduces lesions in MS brains, the percentage of CD5 B lymphocytes in peripheral blood of MS patients is increased (Lehmann et al. 1997). Finally, one study suggested that the dramatically decreased percentage of CD5 B lymphocytes in patients with RA and IgA reflects the abilities of CD5 B lymphocytes to inhibit autoantibody production (Jones et al. 1993). Hence, despite these conflicting observations, an involvement of CD5 B lymphocytes can be concluded. However, to understand fully the role of the cells more thorough investigation of the regulation of these cells in autoimmunity is required.

The mechanisms underlying the endogenous regulation of the T_H1/T_H2 balance and the production of autoantibodies are only poorly understood. Interestingly, in addition to the autochthonous modulation by immune mechanisms, such as the mode of antigen presentation, cytokine milieus and B cell activation, there is consistent evidence indicating that the SNS is also involved the regulation of these functions in health and disease.

8.1.3
Involvement of NPY in Autoimmune Disorders

This section will discuss findings on the involvement of NPY in RA, MS and SLE exemplarily. For a detailed overview on the association of NPY and other autoimmune diseases see Bedoui et al. (2003a).

RA is an autoimmune disorder that is dominated by joint inflammation. The persistent inflammation leads to the destruction of joint tissues, in particular bones and cartilage, finally resulting in severely impaired joint function. A well documented phenomenon in joints of RA patients is the loss of sympathetic nerve fibers (Miller et al. 2000). An initial hypothesis on the involvement of the SNS in RA (Levine et al. 1985) has been substantiated by several investigators. Using antisera against CA and NPY the distribution of sympathetic fibers in RA joints was compared to that in healthy controls. Despite the presence of sympathetic fibers in all layers of healthy joints, it was found that sympathetic nerves were absent in the synovium of RA joints. Since most of these studies used antisera against CA and NPY, it was argued that depleted transmitter stores could account for the failure to detect sympathetic fibers. However, this possibility was ruled out by the use of the neuronal marker protein gene product 9.5 in several of these studies (Mapp et al. 1990; Pereira da Silva et al. 1991). Thus, it is evident that a loss of sympathetic fibers is present in joint tissues of RA patients.

The question 'What role does the apparent sympathetic depletion play in the pathophysiology of RA?' thus arises To address this question several investigations have focused on the role of CA in the pathophysiology of RA. Chemical depletion of sympathetic transmitters in lymph nodes draining the hind limb of rats was recently shown to result in aggravated joint inflammation and increased paw diameter in a model of adjuvans-induced arthritis (Lorton et al. 1999). On the other hand, continued treatment using salbutamol, a β-adrenergic agonist, inhibits collagen-induced arthritis in mice (Malfait et al. 2000). These findings strongly suggest a protective role for the SNS in RA that is probably mediated by CA.

A more detailed look at the involvement of CA in RA and experimental models makes some interesting observations. Studies comparing the density of adrenoceptors on synovial lymphocytes from RA patients and healthy controls revealed a significantly reduced expression of adrenergic receptors on the surface of lymphocytes from RA patients (Baerwald et al. 1997). Furthermore, the suppressive effect of CA on the proliferation of these cells is reduced in synovial lymphocytes of RA patients. Finally, there is a direct correlation between disease severity and the number of adrenergic binding sites: patients with high systemic disease activity have lower numbers of adrenoceptors than patients with low disease activity (Baerwald et al. 1992). These findings indicate that in addition to the loss of sympathetic innervation, the responsiveness towards CA is also decreased in RA joint tissues.

In conclusion, there is consistent evidence that the SNS exerts a protective effect in RA. However, since the above mentioned studies also demonstrate a reduced efficacy of CA and a decreased expression of adrenoceptors in RA patients, it is apparent that the protective effect of the SNS on RA can not be attributed entirely to the action of CA. Instead the integrity of the SNS itself and not just the presence of CA appears to be the critical element.

A difference between intact sympathetic innervation and the solitary presence of CA in joints of RA patients is that the activation of sympathetic nerves results not only in the secretion of CA, but also in the release of NPY.

Several lines of evidence suggest a role for NPY in the regulation of RA. As indicated above a specific loss of NPY-containing fibers is found in the synovium from RA patients (Mapp et al. 1990; Pereira da Silva et al. 1990). Investigations to determine the concentration of NPY in inflamed arthritic joints have revealed some unexpected observations. Despite of the lack of NPY-containing nerves, different studies have reported elevated NPY levels in RA joints, such as the knee, the ankle or the temporomandibular joint (Larson et al. 1991; Appelgren et al. 1993; Ahmed et al. 1995). One study also found a correlation between NPY levels in the temporomandibular joint and plasma NPY levels (Holmlund et al. 1990).

What is the significance of elevated NPY levels in inflamed joints? The important question to address is whether elevated NPY levels are a secondary, reactive phenomenon or whether the increase in NPY is involved in causing joint inflammation. Appelgren et al. (1993) conducted a study to correlate symptom severity, intra-articular NPY levels and intra-articular temperature in the temporomandibular joint of RA patients. It was found that intra-articular NPY levels and the intra-articular temperature are negatively correlated, indicating that high levels of NPY are associated with decreased inflammation. Furthermore, it was found that high levels of NPY correspond to a shorter duration of pain. A similar study produced conflicting data, since it found that increased pain and impaired mandibular mobility are correlated with elevated levels of NPY and calcitonin gene-related peptide (CGRP) (Appelgren et al. 1995). However, it is not clear to what extent a possible interaction between NPY and CGRP might interfere with the solitary action of NPY and also these findings were not correlated to general indicators of inflammation, such as the intra-articular temperature. We therefore conclude that elevated NPY levels can be considered a secondary, reactive mechanism.

In addition, the question of how NPY levels can be elevated when a dramatic reduction of sympathetic innervation is present should be addressed. Ahmed et al. (1995) investigated possible mechanisms underlying elevated NPY levels in the ankle joint of arthritic Lewis rats. If increased activity of NPY-positive fibers would account for this phenomenon, altered NPY levels in the dorsal root ganglia should be expected. However, NPY levels in the dorsal root ganglia revealed no changes. To investigate extraneuronal sources of NPY, they determined the number of megakaryocytes in the bone marrow of the tibial bone and found a dramatic increase of NPY-positive megakaryocytes. These findings suggest the interesting possibility that the increased concentration of NPY could be derived from megakaryocytes, a cell type that contains high amounts of NPY. This suggestion is further emphasized by a study seeking to determine the source of NE in RA synovial tissues. Miller et al. (2000) reported that despite the loss of sympathetic fibers the concentration of NE is increased in synovial tissues from RA patients. The study revealed that the additional NE amount is produced by CD163-positive macrophages residing in the synovium. Hence, in analogy with the possibility that megakaryocytes release NPY in inflamed RA tissues, extraneuronal NE is released by local macrophages in the synovium of RA patients.

Taken together, it appears that upon the loss of sympathetic nerves in RA and the resulting lack of neuronal-derived sympathetic transmitters in synovial tissues, other cells types, such as macrophages and megakaryocytes, attempt to re-supply the synovium with extraneural sympathetic transmitters. Thus, it is important to ask why this re-supply is not sufficient enough to suppress RA. We believe the critical differences between the presence of neuronal and extraneuronal transmitters is the way in which they are released. The release of transmitters from intact nerves is a well regulated interactive process, involving supraspinal control, interaction with other nerves and transmitters, and time and concentration-specific modulation of the final transmitter output. Several mechanisms, such as the frequency dependent release pattern, ensure that the sympathetic output is well orchestrated. The secretion of NE and NPY from macrophages and megakaryocytes, however, does not involve such highly fine-tuned regulatory mechanisms. This is probably the reason why replacing neuronal-derived by extraneuronal NE and NPY secreted from macrophages or megakaryocytes is inefficient in maintaining the suppressive effect of the SNS in RA.

Impaired sympathetic function appears to be a common feature in other autoimmune disorders also. In MS, a disease characterized by the destruction of myelin sheaths, several lines of evidence suggest impaired sympathetic functions. Using experimental autoimmune encephalitis (EAE), an animal model for MS, initial studies demonstrated that depletion of peripheral SNS transmitters by 6-hydroxydopamine augments the severity of EAE (Chelmicka-Schorr et al. 1988), whereas treatment using β-adrenergic agonists suppresses EAE (Chelmicka-Schorr et al. 1989). Clinical observations further strengthened the concept of sympathetic dysfunction in MS. A variety of sympathetic dysfunctions are reported in MS patients, such as pupillary disturbances (de Seze et al. 2001), impaired sympathetic skin response (Yokota et al. 1991) and cardiovascular abnormalities (Nordenbo et al. 1989). Furthermore, it was shown that the normal decrease in blood pressure following the administration of clonidine, an α_2-agonist, is absent in MS patients (Zoukos et al. 1992). Interestingly, the density of adrenergic receptor on lymphocytes from MS patients is also altered. The upregulation of β-adrenoceptors on T cells from MS patients (Karazewski et al.

1991) is believed to reflect the loss of sympathetic influence, since chemical sympathectomy results in a similar receptor upregulation (Miles et al. 1981). Hence, in MS we are left with a picture that is comparable to RA. A sympathetic dysfunction is accompanied by altered responsiveness of lymphocytes. This, again, makes a solitary action of CA as a mediator for the sympathetic influence unlikely. Indeed changes in the concentration of NPY in the CSF and the plasma have been documented. Maeda et al. (1994) found decreased NPY concentrations in the CSF of MS patients in comparison to healthy subjects. These findings were confirmed by others and in addition it was found that a more severe form of MS is associated with even lower CSF NPY levels than observed in milder forms (Gallai et al. 1994). We have recently conducted a study to investigate the effects of NPY on EAE (Bedoui et al. 2003b). Female B6 mice, immunized with MOG_{35-55}, were subjected to a continued treatment with exogenous NPY. The results demonstrate a dose-dependent suppression of actively induced EAE. Pharmacological assessment of the receptor involved revealed that the suppression is mediated via the Y_1 receptor. Comparing T cell responses and anti-MOG_{35-55} antibody titers from treated vs. control animals we found that the suppression is due to a Y_1 receptor-mediated induction of a T_H2 shift of autoreactive lymphocytes in vivo. These findings indicate for the first time that NPY has a direct suppressive effect in a model of autoimmunity.

In contrast to MS and RA, SLE is driven by autoreactive T cells that are biased towards the T_H2 subtype. These cells induce the production of autoantibodies, which in turn mediate the typical tissue destruction. Whereas decreased sympathetic functions are a feature of T_H1 driven autoimmune disorders, such as MS and RA, observations in SLE patients revealed several indications for increased sympathetic activity. Prolonged pupillary reaction and increased maximal pupillary areas, as well as increases in blood pressure and heart rate are found in SLE patients (Gluck et al. 2000; Nakajima et al. 1998). Comparing stress-induced alterations in the composition of leukocytes in the peripheral blood of SLE patients and healthy controls, Jacobs et al. (2001) observed an interesting difference between SLE patients and healthy subjects. Whereas in the control group the number of both IFN-γ (T_H1) and IL-4 (T_H2) producing cells increased upon stress, in SLE patients only IL-4 producing cells increased, suggesting a selective inhibition of stress-induced T_H1 responses in SLE patients.

So the question of whether there is also an involvement of NPY in SLE arises. Indeed, observations demonstrated elevated NPY levels in animal models of SLE. Ericsson et al. (1987) compared the expression of NPY mRNA in the spleen, the bone marrow and the peripheral blood in different mouse strains and found largely elevated NPY levels in mice that develop SLE. Furthermore, it was demonstrated that the onset of glomerulonephritis in lupus-prone mice is accompanied by increased quantities of NPY in the kidney (Bracci-Laudiero et al. 1998). Taken together these findings indicate that the T_H2 shift in SLE is associated with SNS hyperactivity and increased NPY levels in immune organs.

8.1.4
Possible Mechanisms for NPY in Autoimmunity

The observations presented above demonstrate an interesting association between NPY and autoimmunity. Autoimmune disorders that are characterized by the predominance of T_H1 cells are associated with decreased NPY levels, whereas a T_H2 driven autoimmune disease is accompanied by elevated NPY levels. The lack of NPY in RA and MS might contribute to the predominance of autoreactive T_H1 cells and elevated NPY concentrations in SLE may be at least partly responsible for the T_H2 shift.

The proliferative capabilities of lymphocytes are also crucial in autoimmunity, since a reduced number of autoreactive T lymphocytes will ameliorate tissue damage. The ability of NPY to inhibit T lymphocyte proliferation induced by activation signals provides another basis for a role of NPY in autoimmunity.

As indicated above, the production of autoantibodies by B lymphocytes is a crucial factor in the pathophysiology of various autoimmune disorders. Since we have shown that NPY induces a selective inhibition of anti-MOG_{35-55} IgG2a (Bedoui et al. 2003b), it is obvious that the production of autoantibodies in RA and MS is affected by NPY. This would also help to explain the association between decreased NPY levels and increased autoantibody production in RA and MS. This notion, however, is in contrast to the observation of increased autoantibody production and elevated NPY levels in SLE. Remarkably, there is consistent evidence demonstrating that B lymphocyte function in SLE is abnormal. For example, SLE B lymphocytes differ in their functional phenotype (Higuchi et al. 2002), exhibit altered intracellular signaling (Huck et al. 2001) and it is well documented that SLE is associated with B cell malignancy (Xu et al. 2001). Thus, the apparent B lymphocyte abnormalities in SLE might also involve impaired responsiveness to NPY, which in turn would explain this obvious discrepancy between elevated NPY levels and increased autoantibody production.

The dose-dependent mobilization of CD5/CD11b B lymphocytes could provide another means by which NPY is involved in autoimmunity. Bearing in mind the observation that increased numbers of CD5 B lymphocytes in the CSF of MS patients are correlated with decreased secretion of myelin-specific antibodies, it can be speculated that the lack of NPY observed in MS patients contributes to decreased numbers of CD5 B lymphocytes. Clearly, further studies on the effect of NPY on the function of this unique cell type are needed and will undoubtedly shed light into the role of NPY in autoimmunity.

8.2
A Role for NPY in the Control of Infectious Diseases

The immune system is prepared to mount immediate responses against certain highly conserved bacterial fragments. Lipopolysacharides, major components of the membrane of Gram negative bacteria, are readily recognized by specific receptors, such as Toll-like receptors or CD14, that are expressed on the surface of phagocytes. The subsequent release of pro-inflammatory cytokines, including IL-1 and TNF-α, induces a strong activation of cellular and humoral immune functions, which in turn initiate the complex pathophysiological state of sepsis. Sepsis is associated with high mortality and involves alterations of all major homeostatic functions. Cardiovascular and gastrointestinal malfunctions are accompanied by severe reactions within the endocrine system and the CNS. In particular the SNS is affected by sepsis. A common feature during sepsis is a hypo-responsiveness to the SNS, indicated by impaired cardiovascular functions and resistance to exogenous CA treatment (Jarek et al. 1993; Jindal et al 2000).

Using lipopolysaccharide-induced septic shock in rats, Hauser et al. (1993) demonstrated that NPY infusion improves survival and also restores cardiovascular responsiveness to CA. They concluded that the beneficial effects of NPY result from both its vasoconstrictor abilities, as well as a potentiation of catecholaminergic vasopressor effects, since the observed effects were partially blocked using α-AR antagonists. Since NPY improves critical immune functions that are important in protecting the host from overwhelming infections it can be speculated that the beneficial effects from NPY in experimental sepsis can also be attributed to the influence of NPY on the immune system.

Dramatic pathological changes can also result from excessive SNS input on the immune system. Major injuries or other severe pathological states, such as trauma or surgery, can lead to a condition often referred to as 'sympathetic storm'. Under such circumstances the exaggerated SNS activation inhibits pro-inflammatory cytokines such as TNF-α, IFN-γ and IL-12 (Severn et al. 1992; Andrade-Mena 1997; Panina-Bordignon et al. 1997, Straub et al. 2000c) and increases the amount of systemically released IL-10 (van der Poll et al. 1996). Characterized by severe immunosuppression, this response is mediated mainly by the action of CAs, which thereby contribute to detrimental complications, such as sepsis and death (Woiciechowsky et al. 1998, Straub et al. 2000c). Bearing in mind that NPY is also released upon SNS activation and exerts profound effects on the immune system, an involvement of NPY in this critical condition is likely.

9
Conclusion

The above mentioned findings clearly indicate the crucial role of NPY in immune homeostasis. Furthermore, since several immune disorders are associated with impaired NPY levels or functions, the immunological action of NPY might prove an interesting therapeutic target. However, fundamental questions remain unanswered (Table 1). What are the precise mechanisms underlying the effect of NPY on the immune system? Which NPY receptor subtypes mediate the effects? What second messenger systems are activated? What is the cellular and molecular basis for neuroimmune cotransmission? And finally, what are the precise effects of NPY on diseases characterized by either overwhelming or suppressed SNS activity, such as sepsis or autoimmunity?

Table 1 Areas of future research. Summary of several important issues in the field of NPY and immune functions that should be addressed

Topic	Open questions
Innervation of immune organs	Are NPY-positive fibers present in all species? Do synapse-like contacts occur between nerve terminals and immune cells in all immune organs and their different compartments?
NPY receptor expression on immune cells	Are receptors other than the Y1 receptor present on immune cells? Which leukocyte subsets express NPY receptors?
Significance of NPY produced by immune cells	What is the role of NPY secreted locally from immune cells? Does immune cell-derived NPY affect other immune cells?
Modulation of cytokine secretion by NPY	What is the effect on the production of cytokines, such as IL-10 or TGF-β? Does NPY affect the abilities of nonimmune cells to produce cytokines? What are the molecular mechanisms underlying the modulation of IL-4 and IFN-γ secretion?
Role of NPY in antigen presentation	Does NPY modulate antigen presentation? Are dendritic cells affected by NPY? Does NPY administration/treatment alter the number of MHC molecules on the surface of antigen presenting cells?
Interaction with other transmitters	Which immunological functions of catecholamines are modulated by NPY? What is the molecular basis for this interaction?
NPY in autoimmunity	What is the role of NPY in animal models of RA? What is the significance of lowered NPY in the CSF/plasma of MS patients? Does NPY alter cytokine secretion of immune cells from MS or RA patients?
NPY in infectious diseases	Are viral diseases influenced by NPY? Does NPY alter immune responses to viruses? What is the reason for contradictory findings regarding NPY-induced phagocytosis of microbes?

MHC, major histocompatibility complex.

Acknowledgements. The authors are grateful for the helpful comments and language correction of Kylie Bruce. The Volkswagen Foundation supported the experiments of the authors: I/75169.

References

Abbas AK, Murphy KM, Sher A (1996) Functional diversity of helper T lymphocytes. Nature 383:787–793

Ahmed AA, Wahbi A, Nordlind K, Kharazmi A, Sundqvist KG, Mutt V, Liden S (1998) In vitro Leishmania major promastigote-induced macrophage migration is modulated by sensory and autonomic neuropeptides. Scand J Immunol 48:79–85

Ahmed AA, Wahbi AH, Nordlin K (2001) Neuropeptides modulate a murine monocyte/macrophage cell line capacity for phagocytosis and killing of Leishmania major parasites. Immunopharmacol Immunotoxicol 23:397–409

Ahmed M, Bjurholm A, Theodorsson E, Schultzberg M, Kreicbergs A (1995) Neuropeptide Y- and vasoactive intestinal polypeptide-like immunoreactivity in adjuvant arthritis: effects of capsaicin treatment. Neuropeptides 29:33–43

Andrade-Mena CE (1997) Inhibition of gamma interferon synthesis by catecholamines. J Neuroimmunol 76:10–14

Appelgren A, Appelgren B, Kopp S, Lundeberg T, Theodorsson E (1993) Relation between the intra-articular temperature of the temporomandibular joint and the presence of neuropeptide Y-like immunoreactivity in the joint fluid. A clinical study. Acta Odontol Scand 51:1–8

Appelgren A, Appelgren B, Kopp S, Lundeberg T, Theodorsson E (1995) Neuropeptides in the arthritic TMJ and symptoms and signs from the stomatognathic system with special consideration to rheumatoid arthritis. J Orofac Pain 9:215–225

Arinbjarnarson S, Jonsson T, Steinsson K, Sigfusson A, Jonsson H, Geirsson A, Thorsteinsson J, Valdimarsson H (1997) IgA rheumatoid factor correlates with changes in B and T lymphocyte subsets and disease manifestations in rheumatoid arthritis. J Rheumatol 24:269–274

Arzt E, Pereda MP, Castro CP, Pagotto U, Renner U, Stalla GK (1999) Pathophysiological role of the cytokine network in the anterior pituitary gland. Front Neuroendocrinol 20:71–95

Arzubiaga C, Morrow J, Roberts LJ, Biaggioni I (1991) Neuropeptide Y, a putative cotransmitter in noradrenergic neurons, induces mast cell degranulation but not prostaglandin D2 release. J Allergy Clin Immunol 87:88–93

Baerwald C, Graefe C, von Wichert P, Krause A (1992) Decreased density of beta-adrenergic receptors on peripheral blood mononuclear cells in patients with rheumatoid arthritis. J Rheumatol 19:204–210

Baerwald CG, Laufenberg M, Specht T, von Wichert P, Burmester GR, Krause A (1997) Impaired sympathetic influence on the immune response in patients with rheumatoid arthritis due to lymphocyte subset-specific modulation of beta 2-adrenergic receptors. Br J Rheumatol 36:1262–1269

Barbany G, Friedman WJ, Persson H (1991) Lymphocyte-mediated regulation of neurotransmitter gene expression in rat sympathetic ganglia. J Neuroimmunol 32:97–104

Barnea A, Roberts J, Keller P, Word RA (2001) Interleukin-1 beta induces expression of neuropeptide Y in primary astrocyte cultures in a cytokine-specific manner: induction in human but not rat astrocytes. Brain Res 896:137–145

Bedoui S, Kuhlmann S, Nave H, Drube J, Pabst R, von Hörsten S (2001) Differential effects of neuropeptide Y (NPY) on leukocyte subsets in the blood: mobilization of B-1-like B-lymphocytes and activated monocytes. J Neuroimmunol 117:125–132

Bedoui S, Lechner S, Gebhardt T, Nave H, Beck-Sickinger AG, Straub RH, Pabst R, von Hörsten S (2002) NPY modulates epinephrine-induced leukocytosis via Y-1 and Y-5 receptor activation in vivo: sympathetic co-transmission during leukocyte mobilization. J Neuroimmunol 132:25–33

Bedoui S, Kawamura N, Straub RH, Pabst R, Yamamura T, von Hörsten S (2003a) Contributions of NPY to the neuroimmune crosstalk. J Neuroimmunol 134:1–11

Bedoui S, Miyake S, Lin Y, Miyamoto K, Oki S, Kawamura N, Beck-Sickinger A, Hörsten Sv, Yamamura T (2003) Neuropeptide Y (NPY) suppresses experimental autoimmune encephalomyelitis: Neuropeptide Y1 receptor-specific inhibition of autoreactive Th1 responses in vivo. J Immunol 171:3451–3458

Benschop RJ, Rodriguez-Feuerhahn M, Schedlowski M (1996) Catecholamine-induced leukocytosis: early observations, current research, and future directions. Brain Behav Immun 10:77–91

Bracci-Laudiero L, Aloe L, Stenfors C, Theodorsson E, Lundeberg T (1998) Development of systemic lupus erythematosus in mice is associated with alteration of neuropeptide concentrations in inflamed kidneys and immunoregulatory organs. Neurosci Lett 248:97–100

Burmester GR, Stuhlmuller B, Keyszer G, Kinne RW (1997) Mononuclear phagocytes and rheumatoid synovitis. Mastermind or workhorse in arthritis? Arthritis Rheum 40:5-18

Cavallotti C, Artico M, Cavallotti D (1999) Occurrence of adrenergic nerve fibers and of noradrenaline in thymus gland of juvenile and aged rats. Immunol Lett 70:53-62

Chelmicka-Schorr E, Checinski M, Arnason BG (1988) Chemical sympathectomy augments the severity of experimental allergic encephalomyelitis. J Neuroimmunol 17:347-350

Chelmicka-Schorr E, Kwasniewski MN, Thomas BE, Arnason BG (1989) The beta-adrenergic agonist isoproterenol suppresses experimental allergic encephalomyelitis in Lewis rats. J Neuroimmunol 25:203-207

Correale J, Mix E, Olsson T, Kostulas V, Fredrikson S, Hojeberg B, Link H (1991) CD5+ B cells and CD4-8-T cells in neuroimmunological diseases. J Neuroimmunol 32:123-132

Correale J, Mix E, Olsson T, Kostulas V, Fredrikson S, Hojeberg B, Link H (1991) CD5+ B cells and CD4-8-T cells in neuroimmunological diseases. J Neuroimmunol 32:123-132

Correale J de Los Milagros Bassani Molinas (2002) Oligoclonal bands and antibody responses in Multiple Sclerosis. J Neurol 249:375-389

Cross LJ, Beck-Sickinger AG, Bienert M, Gaida W, Jung G, Krause E, Ennis M (1996) Structure activity studies of mast cell activation and hypotension induced by neuropeptide Y (NPY), centrally truncated and C-terminal NPY analogues. Br J Pharmacol 117:325-332

Cunnick JE, Lysle DT, Armfield A, Rabin BS (1988) Shock-induced modulation of lymphocyte responsiveness and natural killer activity: differential mechanisms of induction. Brain Behav Immun 2:102-113

D'Hooge R, De Deyn PP, Verzwijvelen A, De Block J, De Potter WP (1990) Storage and fast transport of noradrenaline, dopamine beta-hydroxylase and neuropeptide Y in dog sciatic nerve axons. Life Sci 47:1851-1859

Davidson A, Diamond B (2001) Autoimmune diseases. N Engl J Med 345:340-350

de Boer NK, Ammerlaan WA, Meedendorp B, Kroese FG (1992) CD5 B cells in the rat? Ann NY Acad Sci 651:157-159

de la Fuente M, Bernaez I, Del Rio M, Hernanz A (1993) Stimulation of murine peritoneal macrophage functions by neuropeptide Y and peptide YY. Involvement of protein kinase C. Immunology 80:259-265

de la Fuente M, Del Rio M, Medina S (2001) Changes with aging in the modulation by neuropeptide Y of murine peritoneal macrophage functions. J Neuroimmunol 116:156-167

de Seze J, Arndt C, Stojkovic T, Ayachi M, Gauvrit JY, Bughin M, Saint MT, Pruvo JP, Hache JC, Vermersch P (2001) Pupillary disturbances in multiple sclerosis: correlation with MRI findings. J Neurol Sci 188:37-41

Downing JE Miyan JA (2000) Neural immunoregulation: emerging roles for nerves in immune homeostasis and disease. Immunol Today 21:281-289

Dureus P, Louis D, Grant AV, Bilfinger TV, Stefano GB (1993) Neuropeptide Y inhibits human and invertebrate immunocyte chemotaxis, chemokinesis, and spontaneous activation. Cell Mol Neurobiol 13:541-546

Ekblad E, Edvinsson L, Wahlestedt C, Uddman R, Hakanson R, Sundler F (1984) Neuropeptide Y co-exists and co-operates with noradrenaline in perivascular nerve fibers. Regul Pept 8:225-235

Elenkov IJ, Wilder RL, Chrousos GP, Vizi ES (2000) The sympathetic nerve-an integrative interface between two supersystems: the brain and the immune system. Pharmacol Rev 52:595-638

Elitsur Y, Luk GD, Colberg M, Gesell MS, Dosescu J, Moshier JA (1994) Neuropeptide Y (NPY) enhances proliferation of human colonic lamina propria lymphocytes. Neuropeptides 26:289–295

Ericsson A, Schalling M, McIntyre KR, Lundberg JM, Larhammar D, Seroogy K, Hökfelt T, Persson H (1987) Detection of neuropeptide Y and its mRNA in megakaryocytes: enhanced levels in certain autoimmune mice. Proc Natl Acad Sci USA 84:5585–5589

Ericsson A, Hemsen A, Lundberg JM, Persson H (1991) Detection of neuropeptide Y-like immunoreactivity and messenger RNA in rat platelets: the effects of vinblastine, reserpine, and dexamethasone on NPY expression in blood cells. Exp Cell Res 192:604–611

Eruklar SD (1994). Molecular, cellular and medical aspects. In: Siegel G, Albers J, Agranoff BW, Albers R, Molinoff PB (eds) Basic Neurochemistry. Raven Press, New York, pp 181–208

Felten DL, Felten SY, Carlson SL, Olschowka JA, Livnat S (1985) Noradrenergic and peptidergic innervation of lymphoid tissue. J Immunol 135:S755–S765.

Fink T, Weihe E (1988) Multiple neuropeptides in nerves supplying mammalian lymph nodes: messenger candidates for sensory and autonomic neuroimmunomodulation? Neurosci Lett 90:39–44

Fiore M, Angelucci F, Alleva E, Branchi I, Probert L, Aloe L (2000) Learning performances, brain NGF distribution and NPY levels in transgenic mice expressing TNF-alpha. Behav Brain Res 112:165–175

Fried G, Terenius L, Hökfelt T, Goldstein M (1985) Evidence for differential localization of noradrenaline and neuropeptide Y in neuronal storage vesicles isolated from rat vas deferens. J Neurosci 5:450–458

Friedman EM, Irwin MR, Nonogaki K (1995) Neuropeptide Y inhibits in vivo specific antibody production in rats. Brain Behav Immun 9:182–189

Furukawa F (1997) Animal models of cutaneous lupus erythematosus and lupus erythematosus photosensitivity. Lupus 6:193–202

Gallai V, Sarchielli P, Firenze C, Trequattrini A, Paciaroni M, Usai F, Franceschini M, Palumbo R (1994) Neuropeptide Y plasma levels and serum dopamine-beta-hydroxylase activity in MS patients with and without abnormal cardiovascular reflexes. Acta Neurol Belg 94:44–52

Gehlert DR, Gackenheimer SL (1996) Unexpected high density of neuropeptide Y Y1 receptors in the guinea pig spleen. Peptides 17:1345–1348

Genain CP, Cannella B, Hauser SL, Raine CS (1999) Identification of autoantibodies associated with myelin damage in multiple sclerosis. Nature Med 5:170–175

Gluck T, Oertel M, Reber T, Zietz B, Schölmerich J, Straub RH (2000) Altered function of the hypothalamic stress axes in patients with moderately active systemic lupus erythematosus. I. The hypothalamus- autonomic nervous system axis. J Rheumatol 27:903–910

Goebels N, Hofstetter H, Schmidt S, Brunner C, Wekerle H, Hohlfeld R (2000) Repertoire dynamics of autoreactive T cells in multiple sclerosis patients and healthy subjects: epitope spreading versus clonal persistence. Brain 123(3):508–518

Grau V, Scriba A, Stehling O, Steiniger B (2000) Monocytes in the rat. Immunobiology 202:94–103

Hafström I, Ringertz B, Lundeberg T, Palmblad J (1993) The effect of endothelin, neuropeptide Y, calcitonin gene-related peptide and substance P on neutrophil functions. Acta Physiol Scand 148:341–346

Hampton MB, Kettle AJ, Winterbourn CC (1998) Inside the neutrophil phagosome: oxidants, myeloperoxidase, and bacterial killing. Blood 92:3007–3017

Hauser GJ, Myers AK, Dayao EK, Zukowska-Grojec Z (1993) Neuropeptide Y infusion improves hemodynamics and survival in rat endotoxic shock. Am J Physiol 265:H1416–H1423

Hernanz A, Tato E, De la F M, de Miguel E, Arnalich F (1996) Differential effects of gastrin-releasing peptide, neuropeptide Y, somatostatin and vasoactive intestinal peptide on interleukin-1 beta, interleukin-6 and tumor necrosis factor-alpha production by whole blood cells from healthy young and old subjects. J Neuroimmunol 71:25–30

Herzenberg LA (2000) B-1 cells: the lineage question revisited. Immunol Rev 175:9–22

Higuchi T, Aiba Y, Nomura T, Matsuda J, Mochida K, Suzuki M, Kikutani H, Honjo T, Nishioka K, Tsubata T (2002) Cutting Edge: Ectopic expression of CD40 ligand on B cells induces lupus-like autoimmune disease. J Immunol 168:9–12

Holmlund A, Ekblom A, Hansson P, Lind J, Lundeberg T, Theodorsson E (1991) Concentrations of neuropeptides substance P, neurokinin A, calcitonin gene-related peptide, neuropeptide Y and vasoactive intestinal polypeptide in synovial fluid of the human temporomandibular joint. A correlation with symptoms, signs and arthroscopic findings. Int J Oral Maxillofac Surg 20:228–231

Hori T, Katafuchi T, Take S, Shimizu N, Niijima A (1995) The autonomic nervous system as a communication channel between the brain and the immune system. Neuroimmunomodulation 2:203–215

Hökfelt T (1991) Neuropeptides in perspective: the last ten years. Neuron 7:867–879

Huck S, Le Corre R, Youinou P, Zouali M (2001) Expression of B cell receptor-associated signaling molecules in human lupus. Autoimmunity 33:213–224

Irwin M, Daniels M, Bloom E T, Smith T L, Weiner H (1987) Life events, depressive symptoms, and immune function. Am J Psychiatry 144:437–441

Irwin M, Brown M, Patterson T, Hauger R, Mascovich A, Grant I (1991) Neuropeptide Y and natural killer cell activity: findings in depression and Alzheimer caregiver stress. FASEB J 5:3100–3107

Jacobs R, Pawlak CR, Mikeska E, Meyer-Olson D, Martin M, Heijnen CJ, Schedlowski M, Schmidt RE (2001) Systemic lupus erythematosus and rheumatoid arthritis patients differ from healthy controls in their cytokine pattern after stress exposure. Rheumatology (Oxford) 40:868–875

Jarek MJ, Legare EJ, McDermott MT, Merenich JA, Kollef MH (1993) Endocrine profiles for outcome prediction from the intensive care unit. Crit Care Med 21:543–550

Jindal N, Hollenberg SM, Dellinger RP (2000) Pharmacologic issues in the management of septic shock. Crit Care Clin 16:233–249

Jones BM, Cheng IK, Wong RW, Kung AW (1993) CD5-positive and CD5-negative rheumatoid factor-secreting B cells in IgA nephropathy, rheumatoid arthritis and Graves' disease. Scand J Immunol 38:575–580

Karaszewski JW, Reder AT, Anlar B, Arnason GW (1993) Increased high affinity beta-adrenergic receptor densities and cyclic AMP responses of CD8 cells in multiple sclerosis. J Neuroimmunol 43:1–7

Katafuchi T, Ichijo T, Take S, Hori T (1993) Hypothalamic modulation of splenic natural killer cell activity in rats. J Physiol 471:209–221

Kawamura N, Tamura H, Obana S, Wenner M, Ishikawa T, Nakata A, Yamamoto H (1998) Differential effects of neuropeptides on cytokine production by mouse helper T cell subsets. Neuroimmunomodulation 5:9–15

Kiecolt-Glaser JK, Ricker D, George J, Messick G, Speicher CE, Garner W, Glaser R (1984) Urinary cortisol levels, cellular immunocompetency, and loneliness in psychiatric inpatients. Psychosom Med 46:15–23

Kodukula K, Arcuri M, Cutrone JQ, Hugill RM, Lowe SE, Pirnik DM, Shu YZ, Fernandes PB, Seethala R (1995) BMS-192548, a tetracyclic binding inhibitor of neuropeptide Y receptors, from Aspergillus niger WB2346. I. Taxonomy, fermentation, isolation and biological activity. J Antibiot (Tokyo) 48:1055–1059

Kohm AP, Sanders VM (2001) Norepinephrine and Beta2-adrenergic receptor stimulation regulate CD4+ T and B lymphocyte function in vitro and in vivo. Pharmacol Rev 53:487–525

Konttinen YT, Hukkanen M, Kemppinen P, Segerberg M, Sorsa T, Malmstrom M, Rose S, Itescu S, Polak JM (1992) Peptide-containing nerves in labial salivary glands in Sjogren's syndrome. Arthritis Rheum 35:815–820

Larsson J, Ekblom A, Henriksson K, Lundeberg T, Theodorsson E (1991) Concentration of substance P, neurokinin A, calcitonin gene-related peptide, neuropeptide Y and vasoactive intestinal polypeptide in synovial fluid from knee joints in patients suffering from rheumatoid arthritis. Scand J Rheumatol 20:326–335

Lehmann D, Karussis D, Mizrachi-Koll R, Linde A S, Abramsky O (1997) Inhibition of the progression of multiple sclerosis by linomide is associated with upregulation of CD4+/CD45RA+ cells and downregulation of CD4+/CD45RO+ cells. Clin Immunol Immunopathol 85:202–209

Levine JD, Collier DH, Basbaum AI, Moskowitz MA, Helms CA (1985) Hypothesis: the nervous system may contribute to the pathophysiology of rheumatoid arthritis. J Rheumatol 12:406–411

Levite M (1998a) Neuropeptides, by direct interaction with T cells, induce cytokine secretion and break the commitment to a distinct T helper phenotype. Proc Natl Acad Sci USA 95:12544–12549

Levite M, Cahalon L, Hershkoviz R, Steinman L, Lider O (1998b) Neuropeptides, via specific receptors, regulate T cell adhesion to fibronectin. J Immunol 160:993–1000

Lorton D, Lubahn C, Klein N, Schaller J, Bellinger DL (1999) Dual role for noradrenergic innervation of lymphoid tissue and arthritic joints in adjuvant-induced arthritis. Brain Behav Immun 13:315–334

Lundberg JM, Saria A, Franco-Cereceda A, Theodorsson-Norheim E (1985) Mechanisms underlying changes in the contents of neuropeptide Y in cardiovascular nerves and adrenal gland induced by sympatholytic drugs. Acta Physiol Scand 124:603–611

Lundberg JM, Rudehill A, Sollevi A, Fried G, Wallin G (1989) Co-release of neuropeptide Y and noradrenaline from pig spleen in vivo: importance of subcellular storage, nerve impulse frequency and pattern, feedback regulation and resupply by axonal transport. Neuroscience 28:475–486

Lundberg JM (1996) Pharmacology of cotransmission in the autonomic nervous system: integrative aspects on amines, neuropeptides, adenosine triphosphate, amino acids and nitric oxide. Pharmacol Rev 48:113–178

Madden KS (2001) Catecholamines, sympathetic nerves, and immunity. In: Ader R, Felten DL, Cohen N (eds) Psychoneuroimmunology. Academic Press, San Diego, pp 197–230

Maeda K, Yasuda M, Kaneda H, Maeda S, Yamadori A (1994) Cerebrospinal fluid (CSF) neuropeptide Y- and somatostatin-like immunoreactivities in man. Neuropeptides 27:323–332

Malfait AM, Malik AS, Marinova-Mutafchieva L, Butler DM, Maini RN, Feldmann M (1999) The beta2-adrenergic agonist salbutamol is a potent suppressor of established collagen-induced arthritis: mechanisms of action. J Immunol 162:6278–6283

Malmström RE, Hökfelt T, Bjorkman JA, Nihlen C, Bystrom M, Ekstrand AJ, Lundberg JM (1998) Characterization and molecular cloning of vascular neuropeptide Y receptor subtypes in pig and dog. Regul Pept 75/76:55–70

Mapp PI, Kidd BL, Gibson SJ, Terry JM, Revell PA, Ibrahim NB, Blake DR, Polak JM (1990) Substance P-, calcitonin gene-related peptide- and c-flanking peptide of neuropeptide Y ir-fibers are present in normal synovium but depleted in patients with rheumatoid arthritis. Neuroscience 37:143–153

Medina S, Del Rio M, Hernanz A, de la Fuente M (2000) Age-related changes in the neuropeptide Y effects on murine lymphoproliferation and interleukin-2 production. Peptides 21:1403–1409

Michel MC, Beck-Sickinger A, Cox H, Doods HN, Herzog H, Larhammar D, Quirion R, Schwartz T, Westfall T (1998) XVI. International Union of Pharmacology recommendations for the nomenclature of neuropeptide Y, peptide YY, and pancreatic polypeptide receptors. Pharmacol Rev 50:143–150

Miles K, Quintans J, Chelmicka-Schorr E, Arnason BG (1981) The sympathetic nervous system modulates antibody response to thymus- independent antigens. J Neuroimmunol 1:101–105

Miller LE, Justen HP, Schölmerich J, Straub RH (2000) The loss of sympathetic nerve fibers in the synovial tissue of patients with rheumatoid arthritis is accompanied by increased norepinephrine release from synovial macrophages. FASEB J 14:2097–2107

Mitchell B, Kendall M, Adam E, Schumacher U (1997) Innervation of the thymus in normal and bone marrow reconstituted severe combined immunodeficient (SCID) mice. J Neuroimmunol 75:19–27

Mor A, Chartrel N, Vaudry H, Nicolas P (1994) Skin peptide tyrosine-tyrosine, a member of the pancreatic polypeptide family: isolation, structure, synthesis, and endocrine activity. Proc Natl Acad Sci USA 91:10295–10299

Nair MP, Schwartz SA, Wu K, Kronfol Z (1993) Effect of neuropeptide Y on natural killer activity of normal human lymphocytes. Brain Behav Immun 7:70–78

Nakajima A, Sendo W, Tsutsumino M, Koseki Y, Ichikawa N, Akama H, Taniguchi A, Terai C, Hara M, Kashiwazaki S (1998) Acute sympathetic hyperfunction in overlapping syndromes of systemic lupus erythematosus and polymyositis. J Rheumatol 25:1638–1641

Nawa H, Nakanishi S, Patterson PH (1991) Recombinant cholinergic differentiation factor (leukemia inhibitory factor) regulates sympathetic neuron phenotype by alterations in the size and amounts of neuropeptide mRNAs. J Neurochem 56:2147–2150

Neumann H, Medana I M, Bauer J, Lassmann H (2002) Cytotoxic T lymphocytes in autoimmune and degenerative CNS diseases. Trends Neurosci 25:313–319

Nordenbo AM, Boesen F, Andersen EB (1989) Cardiovascular autonomic function in multiple sclerosis. J Auton Nerv Syst 26:77–84

Ohia SE, Jumblatt JE (1990) Inhibitory effects of neuropeptide Y on sympathetic neurotransmission in the rabbit iris-ciliary body. Neurochem Res 15:251–256

Panina-Bordignon P, Mazzeo D, Lucia PD, D'Ambrosio D, Lang R, Fabbri L, Self C, Sinigaglia F (1997) Beta2-agonists prevent Th1 development by selective inhibition of interleukin 12. J Clin Invest 100:1513–1519

Parnet P, Kelley KW, Bluthe RM, Dantzer R (2002) Expression and regulation of interleukin-1 receptors in the brain. Role in cytokines-induced sickness behavior. J Neuroimmunol 125:5–14

Paul WE (1998) Fundamental Immunology, Lippincott-Raven, New York, p 245

Pellieux C, Sauthier T, Domenighetti A, Marsh DJ, Palmiter RD, Brunner HR, Pedrazzini T (2000) Neuropeptide Y (NPY) potentiates phenylephrine-induced mitogen-activated protein kinase activation in primary cardiomyocytes via NPY Y5 receptors. Proc Natl Acad Sci USA 97:1595–1600

Pereira da Silva JA, Carmo-Fonseca M (1990) Peptide containing nerves in human synovium: immunohistochemical evidence for decreased innervation in rheumatoid arthritis. J Rheumatol 17:1592–1599

Petitto JM, Huang Z, McCarthy DB (1994) Molecular cloning of NPY-Y1 receptor cDNA from rat splenic lymphocytes: evidence of low levels of mRNA expression and [125I]NPY binding sites. J Neuroimmunol 54:81–86

Reder A, Checinski M, Chelmicka-Schorr E (1989) The effect of chemical sympathectomy on natural killer cells in mice. Brain Behav Immun 3:110–118

Romano TA, Felten SY, Felten DL, Olschowka JA (1991) Neuropeptide-Y innervation of the rat spleen: another potential immunomodulatory neuropeptide. Brain Behav Immun 5:116–131

Sawchenko PE, Swanson LW (1982) The organization of noradrenergic pathways from the brainstem to the paraventricular and supraoptic nuclei in the rat. Brain Res 257:275–325

Schwarz H, Villiger PM, von Kempis J, Lotz M (1994) Neuropeptide Y is an inducible gene in the human immune system. J Neuroimmunol 51:53–61

Scriba A, Schneider M, Grau V, van der Meide P H, Steiniger B (1997) Rat monocytes upregulate NKR-P1A and down-modulate CD4 and CD43 during activation in vivo: monocyte subpopulations in normal and IFN-gamma- treated rats. J Leukoc Biol 62:741–752

Sellebjerg F, Jensen J, Jensen C V, Wiik A (2002) Expansion of CD5 - B cells in multiple sclerosis correlates with CD80 (B7–1) expression. Scand J Immunol 56:101–107

Severn A, Rapson NT, Hunter CA, Liew FY (1992) Regulation of tumor necrosis factor production by adrenaline and beta- adrenergic agonists. J Immunol 148:3441–3445

Shen GH, Grundemar L, Zukowska-Grojec Z, Hakanson R, Wahlestedt C (1991) C-terminal neuropeptide Y fragments are mast cell-dependent vasodepressor agents. Eur J Pharmacol 204:249–256

Shimizu M, Shigeri Y, Tatsu Y, Yoshikawa S, Yumoto N (1998) Enhancement of antimicrobial activity of neuropeptide Y by N-terminal truncation. Antimicrob Agents Chemother 42:2745–2746

Soder O Hellström PM (1987) Neuropeptide regulation of human thymocyte, guinea pig T lymphocyte and rat B lymphocyte mitogenesis. Int Arch Allergy Appl Immunol 84:205–211

Song C Leonard B E (1998) Comparison between the effects of sigma receptor ligand JO 1784 and neuropeptide Y on immune functions. Eur J Pharmacol 345:79–87

Straub RH, Westermann J, Schölmerich J, Falk W (1998) Dialogue between the CNS and the immune system in lymphoid organs. Immunol Today 19:409–413

Straub RH, Schaller T, Miller LE, von Hörsten S, Jessop DS, Falk W, Schölmerich J (2000a) Neuropeptide Y cotransmission with norepinephrine in the sympathetic nerve-macrophage interplay. J Neurochem 75:2464–2471

Straub RH, Mayer M, Kreutz M, Leeb S, Schölmerich J, Falk W (2000b) Neurotransmitters of the sympathetic nerve terminal are powerful chemoattractants for monocytes. J Leukoc Biol 67:553–558

Straub RH, Herfarth H, Falk W, Andus T, Schölmerich J (2002) Uncoupling of the sympathetic nervous system and the hypothalamic–pituitary–adrenal axis in inflammatory bowel disease? J Neuroimmunol 126:116–125

Sung CP, Arleth AJ, Feuerstein GZ (1991) Neuropeptide Y upregulates the adhesiveness of human endothelial cells for leukocytes. Circ Res 68:314–318

Tabarowski Z, Gibson-Berry K, Felten SY (1996) Noradrenergic and peptidergic innervation of the mouse femur bone marrow. Acta Histochem 98:453–457

ThyagaRajan S, Felten SY, Felten DL (1998) Restoration of sympathetic noradrenergic nerve fibers in the spleen by low doses of L-deprenyl treatment in young sympathectomized and old Fischer 344 rats. J Neuroimmunol 81:144–157

Turnbull AV Rivier CL (1999) Regulation of the hypothalamic-pituitary-adrenal axis by cytokines: actions and mechanisms of action. Physiol Rev 79:1–71

van der Poll T, Coyle SM, Barbosa K, Braxton CC, Lowry SF (1996) Epinephrine inhibits tumor necrosis factor-alpha and potentiates interleukin 10 production during human endotoxemia. J Clin Invest 97:713–719

van Roon JA, Lafeber FP, Bijlsma JW (2001) Synergistic activity of interleukin-4 and interleukin-10 in suppression of inflammation and joint destruction in rheumatoid arthritis. Arthritis Rheum 44:3–12

Vermeer LA, de Boer NK, Bucci C, Bos NA, Kroese FG, Alberti S (1994) MRC OX19 recognizes the rat CD5 surface glycoprotein, but does not provide evidence for a population of CD5bright B cells. Eur J Immunol 24:585–592

von Hörsten S, Nave H, Ballof J, Helfritz F, Meyer D, Schmidt RE, Stalp M, Exton NG, Exton MS, Straub RH, Radulovic J, Pabst R (1998a) Centrally applied NPY mimics immunoactivation induced by non-analgesic doses of met-enkephalin. Neuroreport 9:3881–3885 von Hörsten S, Ballof J, Helfritz F, Nave H, Meyer D, Schmidt RE, Stalp M, Klemm A, Tschernig T, Pabst R (1998b) Modulation of innate immune functions by intracerebroventricularly applied neuropeptide Y: dose and time dependent effects. Life Sci 63:909–922

von Hörsten S, Exton NG, Exton MS, Helfritz F, Nave H, Ballof J, Stalp M, Pabst R (1998c) Brain NPY Y1 receptors rapidly mediate the behavioral response to novelty and a compartment-specific modulation of granulocyte function in blood and spleen. Brain Res 806:282–286

Vouldoukis I, Shai Y, Nicolas P, Mor A (1996) Broad spectrum antibiotic activity of the skin-PYY. FEBS Lett 380:237–240

Wahlestedt C, Edvinsson L, Ekblad E, Hakanson R (1985) Neuropeptide Y potentiates noradrenaline-evoked vasoconstriction: mode of action. J Pharmacol Exp Ther 234:735–741

Westermann J, Pabst R (1990) Lymphocyte subsets in the blood: a diagnostic window on the lymphoid system? Immunol Today 11:406–410

Westermann J, Michel S, Lopez-Kostka S, Bode U, Rothkötter HJ, Bette M, Weihe E, Straub RH, Pabst R (1998) Regeneration of implanted splenic tissue in the rat: re-innervation is host age-dependent and necessary for tissue development. J Neuroimmunol 88:67–76

Woiciechowsky C, Asadullah K, Nestler D, Eberhardt B, Platzer C, Schoning B, Glockner F, Lanksch WR, Volk HD, Docke WD (1998) Sympathetic activation triggers systemic interleukin-10 release in immunodepression induced by brain injury. Nature Med 4:808–813

Xu Y Wiernik PH (2001) Systemic lupus erythematosus and B-cell hematologic neoplasm. Lupus 10:841–850

Yang D, Biragyn A, Kwak LW, Oppenheim JJ (2002) Mammalian defensins in immunity: more than just microbicidal. Trends Immunol 23:291–296

Yokota T, Matsunaga T, Okiyama R, Hirose K, Tanabe H, Furukawa T, Tsukagoshi H (1991) Sympathetic skin response in patients with multiple sclerosis compared with patients with spinal cord transection and normal controls. Brain 114(3):1381–1394

Zoukos Y, Thomaides T, Pavitt DV, Leonard JP, Cuzner ML, Mathias CJ (1992) Up-regulation of beta-adrenoceptors on circulating mononuclear cells after reduction of central sympathetic outflow by clonidine in normal subjects. Clin Auton Res 2:165–170

Transgenic and Knockout Models in NPY Research

H. Herzog

Garvan Institute of Medical Research, 384 Victoria St.,
2010 Darlinghurst, Sydney, NSW, Australia
e-mail: h.herzog@garvan.org.au

1	**Generation of Transgenic Models**	449
1.1	Transgenic Mice	449
1.1.1	NPY-Overexpressing Mice	449
1.1.2	Y_1-Receptor-Transgenic Mice	449
1.1.3	Pancreatic Polypeptide-Overexpressing Mice	449
1.2	NPY-Overexpressing Rats	450
2	**Generation of Knockout Models**	450
2.1	NPY Knockout Mice	450
2.2	Single Y Receptor Knockout Mice	451
2.2.1	Y_1 Receptor Knockout	451
2.2.2	Y_2 Receptor Knockout	452
2.2.3	Y_4 Receptor Knockout	453
2.2.4	Y_5 Receptor Knockout	453
2.3	Conditional Y Receptor Knockout Mice	454
2.4	Combinatorial NPY and Y Receptor Knockout Mice	455
3	**Phenotypes of Mutant Mice**	455
3.1	Feeding and Energy Homeostasis	456
3.1.1	NPY Transgenic Mice	458
3.1.2	PP Transgenic Mice	458
3.1.3	NPY Knockout Mice	459
3.1.4	Y_1 Knockout Mice	459
3.1.5	Y_2 Knockout Mice	460
3.1.6	Y_4 Receptor Knockout Mice	461
3.1.7	Y_5 Receptor Knockout	461
3.1.8	Double Knockout Models	462
3.1.9	Summary on Y Receptor Involvement in Energy Homeostasis	463
3.2	Reproduction	463
3.2.1	NPY Knockout	464
3.2.2	Y_1 Receptor Knockout	464
3.2.3	Y_2 Receptor Knockout	464
3.2.4	Y_4 Receptor Knockout	465
3.2.5	Y_5 Receptor Knockout	465
3.3	Cardiovascular Phenotypes	465
3.3.1	Transgenic Rat	466
3.3.2	Y_1 Receptor Knockout	466
3.3.3	Y_2 Receptor Knockout	466

3.3.4 Y_4 Receptor Knockout... 467
3.4   Seizure Susceptibility 467
3.4.1 Transgenic Rat 468
3.4.2 NPY Knockout Mice... 468
3.4.3 Y_5 Receptor Knockout Mice.. 468
3.5   Anxiety-Related Behaviors 469
3.5.1 NPY Knockout Mice... 469
3.5.2 Y_2 Knockout Mice 470
3.6   Other Behavioral Phenotypes..................................... 470
3.6.1 Voluntary Alcohol Intake 470
3.6.2 Memory.. 471
3.6.3 Nociception.. 471
3.6.4 Regulation of Bone Homeostasis 472

4   **Concluding Remarks** 473

References 474

Abstract The neuropeptide Y system consists of three peptide precursor genes encoding neuropeptide Y (NPY), peptide YY (PYY) and pancreatic polypeptide (PP) and at least five receptor genes encoding the Y_1, Y_2, Y_4, Y_5 and y_6 receptor making it complex to work with. In the past, delineating the function of the NPY system relied mainly on pharmacological approaches using modified peptide ligands and a few synthetic molecules. However, interpretation of the results was difficult due to the poor knowledge of the *in vivo* selectivity and the distinct metabolic or pharmacokinetic properties of the compounds used. One can never be sure that ligands which are highly selective in *in vitro* systems, demonstrated by their binding properties on brain tissue or recombinant receptor preparations, do not crossreact with other known or yet unidentified receptors *in vivo*. Furthermore, their action can vary strongly depending on the dose, the site and the mode (acute versus chronic) of administration of the compound. This leaves unanswered many questions regarding the functional contributions of the different Y receptors to a variety of important physiological processes. Modern molecular biology technologies such as the generation of transgenic or gene-targeted rodent models offer valid alternatives to these pharmacological approaches. Over recent years studies describing the overexpression of NPY in transgenic mice and rat models as well as the inactivation of the NPY, Y_1, Y_2, Y_4 and Y_5 gene by homologous recombination have been published. The analysis of the phenotypes of all these animal models has revealed significant and distinct roles of each gene in modulating feeding behavior, fertility, seizure susceptibility, pain perception, cardiovascular function and emotional behaviors.

Keywords Conditional · Knockout · Mice · Neuropeptide Y · Y Receptor

1
Generation of Transgenic Models

1.1
Transgenic Mice

1.1.1
NPY-Overexpressing Mice

The Thy-1 promoter was used to drive the expression of the mouse neuropeptide Y (NPY) transgene, which restricts expression strictly to neurons in the central nervous system (CNS) (Inui et al. 1998). However, only a modest 115% increase in NPY immunoreactivity could be detected by semiquantitative immunohistochemical analysis and by radioligand binding assay in these transgenic mice. Transgene expression was also detected by *in situ* hybridization analysis that demonstrated transgene-derived NPY expression in neurons in the hippocampus, cerebral cortex, and the arcuate nucleus of the hypothalamus. The mice expressing the NPY transgene are on a BDF1 background.

Thiele et al. also generated a NPY transgenic mouse model (Thiele et al. 1998). These investigators used a construct containing 10 kb of the endogenous mouse NPY promoter plus 3 kb of 3'-flanking sequence. Insertion of an oligonucleotide in the 3'-untranslated sequence of the NPY gene allowed for screening of transgene expression and estimation of copy numbers. Approximately five times higher NPY mRNA levels could be detected in a line with germline transmission of the transgene. Immunocytochemistry identified more abundant NPY protein expression in the cortex, amygdala and hippocampus but interestingly not in the arcuate nucleus of the hypothalamus. This NPY transgenic mouse line is on FVB background.

1.1.2
Y$_1$-Receptor-Transgenic Mice

A 1.3-kb fragment upstream of the transcription initiation sites of the murine Y$_1$ receptor gene has been shown to direct specific expression of reporter genes in neuronal cell cultures. This putative promoter region linked to the LacZ reporter gene was also shown to express, in a temporal and spatial manner, the transgene in the CNS of mice (Oberto et al. 2000). However, no Y$_1$ receptor overexpressing transgenic mouse model has been reported so far.

1.1.3
Pancreatic Polypeptide-Overexpressing Mice

The cytomegalovirus immediate early enhancer chicken-β-actin hybrid promoter was fused to the mouse PP cDNA followed by the 3'-flanking sequence of the rabbit β-globin gene. Transgenic animals were generated by injecting this construct into fertilized eggs from the BDF1 strain and founder mice crossed onto the C57BL/6 background (Ueno et al. 1999) from which one line containing 14 copies of the mouse PP gene was derived. The expression of the transgene was specific and limited to the pancreas with strong increases in immunreactivity in islets particularly in insulin positive cells. Serum levels of PP reached 20 times higher levels in the transgenic animals compared to controls and half of the offspring of the transgenic animals died within 2 weeks of birth. No gross abnormalities were observed in these casualties.

1.2
NPY-Overexpressing Rats

Experimental procedures to investigate physiological parameters are well established in rat models. Therefore, the generation of transgenic rats, despite being more costly, has certain advantages. In order to develop NPY overexpressing rats, Michalkiewicz and Michalkiewicz (2000) performed experiments using a mouse metallothionin and a human cytomegalovirus driven NPY cDNA construct. However, no founders could be produced by this strategy suggesting that overexpression of the NPY gene in early development might be lethal. Subsequently a strategy was adopted to use a 14.5-kb fragment that contained the entire rat NPY gene including 5 kb of 5′ and 1.5 kb of 3′ flanking sequences. It was presumed that this should ensure tissue specific expression of the transgene. The successful insertion of the transgene and the copy number achieved in the various lines was evaluated by an *Eco*RI restriction fragment length polymorphism artificially generated in the transgene. The amount of overexpression of NPY in different tissues was assessed by quantitative radioimmunoassay, immunocytochemistry and *in situ* hybridization in homozygous animals on a Sprague-Dawley background.

2
Generation of Knockout Models

2.1
NPY Knockout Mice

The first report of a knockout of a member of the NPY gene family was published in 1996 (Erickson et al. 1996a). The targeting vector was designed to replace the coding sequence of the pre-pro-NPY peptide with the lacZ gene and a neomycin-resistant cassette for selection (Fig. 1). The successful deletion of the NPY gene has been verified by the absence of immunostaining for NPY and the expression and functional activity of the LacZ gene in place of it. AB1 embryonic stem cells were used to target the NPY gene locus but the background of the knockout animals has not been described in this initial publication. In a subsequent report the background of the mice is stated as being 129 Sv (Erickson et al. 1996b).

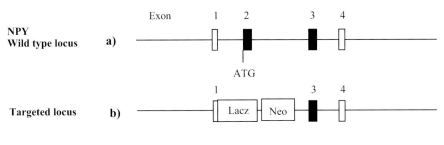

Erickson et al 1996

Fig. 1 Schematic representation of (**a**) the mouse NPY locus and (**b**) the targeting strategy employed. *Open boxes* represent noncoding exons or sequences. *Black boxes* represent coding sequence of the NPY gene

2.2
Single Y Receptor Knockout Mice

With the exception of the Y_1 receptor gene, which contains a small, approximately 100-bp-long intron following the fifth putative transmembrane-spanning domain of the coding sequence, all other Y receptors contain their entire coding region on one exon (Herzog et al. 1992). An alternative splice variant of the Y_1 receptor gene leading to a truncated receptor molecule has also been described (Li et al. 1992). However, all of the Y receptor genes have been shown to have variable numbers of exons encoding the 5'-untranslated region potentially allowing tissue or cell specific regulation of transcription (Ball et al. 1995; Herzog et al. 1997). In Figs. 2, 3, 4, and 5 the mutated Y receptor gene loci produced by homologous recombination are shown.

2.2.1
Y_1 Receptor Knockout

Four laboratories have reported the generation of Y_1 receptor deficient mice (Howell et al. 2003; Kushi et al. 1998; Naveilhan et al. 2001c; Pedrazzini et al. 1998) (Fig. 2b–e). Two of the constructs are designed to interrupt the coding sequence with either a 'IRES Tau–LacZ–Neo' (Naveilhan et al. 2001c) or a single neo cassette (Pedrazzini et al. 1998), with the third one actually deleting a part of the gene – including the start codon– –and replacing it with a neo cassette (Kushi et al. 1998). A fourth report describes the removal of the entire coding sequence of the Y_1 receptor gene by using cre/loxP technology. (For a detailed description of this conditional knockout approach see Sect. 3.3). The Y_1 knockout strains generated by Naveilhan is on a mixed 129SV/Balb/c background. The Y_1 knockout mice generated by Pedrazzini originally on a 129 Sv/C57BL/6 background were backcrossed onto the C57BL/6 strain for at least four generations. The other two Y_1 knockout mice strains are on a mixed 129SvJ/C57BL/6 background. Absence of functional Y_1 receptors were demonstrated by lack of binding to the Y_1 antagonist BIBP3226 (Pedrazzini et al. 1998), lack of Y_1 recep-

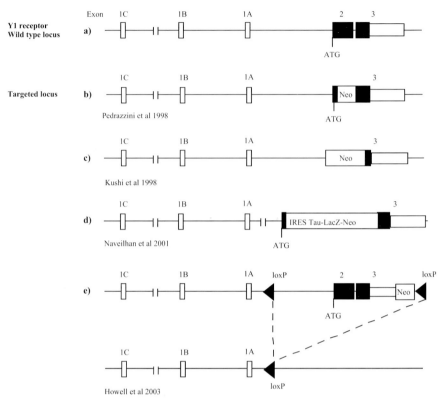

Fig. 2 Schematic representation of (**a**) the mouse Y1 receptor locus and (**b–e**) different targeting strategies used. *Open boxes* represent noncoding exons or sequences. *Black boxes* represent coding sequence of the receptor gene. *Black triangles* indicate the locations of loxP sites

tor mRNA (Kushi et al. 1998; Naveilhan et al. 2001c) and replacement of it by β-galactosidase expression. In the conditional model, *in situ* hybridization, radioligand binding and genomic southern blotting was used to confirm the removal of the Y_1 receptor gene (Howell et al. 2003).

2.2.2
Y_2 Receptor Knockout

Again two different strategies were used to generate Y_2 knockout mice. The first one operates by disrupting the coding sequence of the Y_2 receptor by the insertion of a cassette containing the IRES Tau–LacZ–Neo genes and leaving the translation initiation codon intact (Naveilhan et al. 1999) (Fig. 3b). Homozygous mutant mice were generated on a mixed 129SV×Balb/c background. Verification of the knockout was by Northern analysis and radioligand binding on brain membranes. The second Y_2 knockout line was generated by using the cre/loxP technology, allowing the entire coding sequence of the Y_2 receptor including the

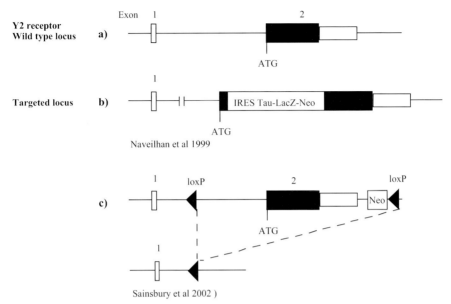

Fig. 3 Schematic representation of (**a**) the mouse Y2 receptor locus and (**b, c**) different targeting strategies used. *Open boxes* represent noncoding exons or sequences. *Black boxes* represent coding sequence of the receptor gene. *Black triangles* indicate the location of loxP sites

initiation start site eventually to be removed (Sainsbury et al. 2002a) (Fig. 3c). This strain was propagated on a mixed 129SvJ/C57BL/6 background. The deletion of the Y_2 receptor gene was confirmed by *in situ* hybridization, radioligand binding on brain sections as well as Southern analysis (Sainsbury et al. 2002a).

2.2.3
Y_4 Receptor Knockout

Only one report so far describes the generation and characterization of a Y_4 deficient mouse strain on a mixed 129SvJ/C57BL/6 background (Sainsbury et al. 2002c). Similar to the Y_1 and Y_2 described above this knockout was also created by using cre/loxP technology to establish germline and conditional Y_4 receptor knockout in one step (Fig. 4b). Genomic Southern analysis was used to prove the successful deletion.

2.2.4
Y_5 Receptor Knockout

Marsh et al. are the only group to report the generation of Y_5 receptor deficient mice (Marsh et al. 1998). The targeting construct was designed to replace the coding sequence of the Y_5 receptor with a Tau–LacZ cassette for visualization of gene expression (Fig. 5b). However, due to incorrect sequence information at

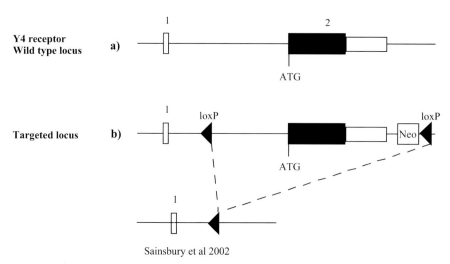

Fig. 4 Schematic representation of (**a**) the mouse Y4 receptor locus and (**b**) the conditional targeting strategy used. *Open boxes* represent noncoding exons or sequences. *Black boxes* represent coding sequence of the receptor gene. *Black triangles* indicate the location of loxP sites

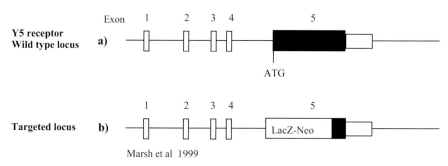

Fig. 5 Schematic representation of (**a**) the mouse Y5 receptor locus and (**b**) the targeting strategy used. *Open boxes* represent noncoding exons or sequences. *Black boxes* represent coding sequence of the receptor gene

the time (Gerald et al. 1996), the cassette was placed into the first intron rather then the first coding exon of the Y_5 receptor gene and expression of the reporter gene was minimal. The mice were bred on a mixed 129 Sv/C57BL/6 background and gene knockout was confirmed by Southern analysis.

2.3
Conditional Y Receptor Knockout Mice

Three conditional Y receptor knockout mice lines have been described so far. All three, $Y_1^{lox/lox}$, $Y_2^{lox/lox}$ and $Y_4^{lox/lox}$, were made using the same strategy which includes the introduction of a 34-base pair long loxP sequence upstream of the initiation codon, and the insertion of a neomycin cassette followed by another loxP sequence 3' to the coding exon. No genomic DNA was removed or rearranged in this process ensuring minimal disturbance to the targeted gene. Functionality of the loxP system was demonstrated by the absence of the respective gene after crossing the chimeric mice with transgenic mice which express Cre-recombinase under the control of an oocyte specific promoter leading to the deletion of the gene in the first cell stage (Schenk et al. 1995). In order to remove the targeted gene in the adult animal an alternative strategy was the injection of Cre-recombinase expressing adenovirus into the target tissue (Sainsbury et al. 2002a). Successful deletion of the Y_2 receptor in the target tissue was shown by *in situ* PCR and PCR on genomic DNA (Sainsbury et al. 2002a).

2.4
Combinatorial NPY and Y Receptor Knockout Mice

A series of double knockout models have been generated by interbreeding single homozygous mutant mice deficient of the NPY, Y-receptor(s), agouti-related peptide (AgRP) or leptin genes. So far double knockouts for $NPY^{-/-}, ob/ob$ (Erickson et al. 1996b), $NPY^{-/-}, AgRP^{-/-}$ (Qian et al. 2002), $NPY^{-/-}, A^y$ (Hollopeter et al. 1998), $Y_1^{-/-}, ob/ob$ (Pralong et al. 2002), $Y_2^{-/-}, ob/ob$ (Naveilhan et al. 2002; Sainsbury et al. 2002b), $Y_4^{-/-}, ob/ob$ (Sainsbury et al. 2002c), $Y_5^{-/-}, ob/ob$ (Marsh et al. 1998), $Y_1^{-/-}, Y_2^{-/-}$ (Naveilhan et al. 2001b), and $Y_2^{-/-}, Y_4^{-/-}$ (Sainsbury et al. 2003) have been reported. The background of the double knockout mice is difficult to verify but all mutant mice with the leptin deletion must contain variable percentages of the C57BL/6 genome. All of these double knockout mutants are viable and show different degrees of improvement over the obese, diabetic and infertile phenotype of the leptin deficient mice demonstrating the importance of NPY as a downstream mediator of leptin function.

3
Phenotypes of Mutant Mice

It is not surprising that mice lacking NPY and Y-receptors are reported not to have any gross abnormalities. It is plausible that removing one component of the NPY family system can lead to adaptive changes during development in order to maintain homeostasis within the NPY system. Such modification may be subtle and needs to be investigated further in these mutant mice. Y-receptors in particular show overlapping expression and do also have similar affinity for NPY. It is therefore possible that this redundancy in the system could lead to compensatory responses when one of the components is missing. The investigation of the local expression levels of the remaining Y receptors at the mRNA as well as the protein level in defined areas in such mice is an important prerequisite to defining the role of the particular Y receptor that is missing. So far only a few studies on that topic have been reported.

In the NPY knockout mice, the expression and binding of NPY receptors were investigated by *in situ* hybridization and receptor autoradiography using $[^{125}I]$-(Leu^{31},Pro^{34})PYY and $[^{125}I]$-PYY(3–36) as radioligands (Trivedi et al. 2001). No significant change could be detected for Y_1, Y_4, Y_5 and y_6 receptors. However, a sixfold increase in Y_2 receptor mRNA was observed in the CA1 region of the hippocampus. This is confirmed by a 60%–400% increase of Y_2 receptor binding in a variety of brain areas. Y_1 receptor binding was also increased but only in the hypothalamus confirming that the lack of the ligand gene can lead to alterations in Y receptor levels (Trivedi et al. 2001). Interestingly, radioligand binding experiments on brain sections of Y_2 knockout animals, using the Y_2/Y_5 preferring ligand $[^{125}I]$-PYY(3–36), revealed an almost complete lack of signal indicating a very low level of Y_5 receptor in these mice (Sainsbury et al. 2002b).

In a similar way to the NPY system, other neurotransmitter systems could have been altered to help balance the functional loss of one component of a defined pathway. The NPY system and several other neuropeptides such as AgRP, pro-opiomelanocortin (POMC) and cocaine and amphetamine-regulated transcript (CART) are known to interact closely. The loss of control of regulatory functions on some of these other pathways after deletion of a particular Y receptor can trigger other pathways to take over these regulatory functions by increasing or decreasing the expression of a particular neurotransmitter. Studies on the expression of NPY, AgRP, POMC, CART, thyrotropin-releasing-hormone and corticotropin-releasing-hormone in Y_2 and Y_4 receptor knockout animals have revealed such changes and can help explain the observed or missing phenotypical changes (Sainsbury et al. 2002a, 2002b, 2002c).

Gene array technology may also be helpful in investigating changes of expression pattern in a variety of neuronal pathways. However, it is important to remember that whole brain analysis may not be conclusive as downregulation of a particular transcript in one brain nucleus could be canceled out by the simultaneous increase of this transcript in another nucleus. Dissecting out defined subareas in the brain for this type of analysis is therefore necessary and verification of the observed changes by *in situ* hybridization analysis is essential.

The advances made to date by using transgenic technologies for investigating a variety of aspects of physiology from the NPY point of view are summarized below. An overview of the phenotypes of male animals under basal conditions for all the transgenic models known at this time is shown in Table 1.

3.1
Feeding and Energy Homeostasis

A complex system has evolved to regulate food intake and to maintain energy homeostasis. Meal size is controlled by a series of short-term hormonal and neural signals that derive from the gastrointestinal tract, such as cholecystokinin, PP and the recently discovered peptide PYY3–36 (Batterham et al. 2002).

Table 1 Summary of phenotypes of NPY and Y receptor transgenic models

	Transgenic models			Single knockouts						Double knockouts						
	NPY (rat)	NPY mouse	PP mouse	NPY	Y1	Y2	Y2 cond	Y4	Y5	Y1/Y2	NPY/AgRP	NPY/ob	Y1/ob	Y2/ob	Y4/ob	Y5/ob
Food intake	◄	⧖	►	⧖	◄	◄►	◄►	►	⧖		⧖	►	⧖	⧖	◄	⧖
Body weight	◄	⧖	►	⧖	⧖	⧖	►	►	⧖		⧖	►	►	⧖	◄	⧖
Heart rate	◄				⧖			►								
Blood pressure	◄							►								
Anxiety				◄	◄	►										
Nociception				◄		◄	◄			◄						
Bone Mass					◄	◄			◄							
Memory	►			⧖	⧖	⧖		⧖	⧖							⧖
Fertility		►		◄												
Alcohol consumption				◄								◄	◄		◄	
Seizure susceptibility	⧖								⧖							
References	Michal-kiewicz 2001; Vezzani 2002	Thiele 1998	Ueno 1996	Bannon 2000; Erickson 1996; Thiele 1998	Kushi 1998	Baldock 2002; Naveilhan 1999; Redrobe 2003; Sainsbury 1996; Smith-White 2002	Sainsbury 2002	Sainsbury 2002; Smith-White 2002	Guo 2002; Marsh 1999; Marsh 1998	Naveilhan 2001	Qian 2002	Erickson 1996	Pralong 2002	Naveilhan 2002; Sainsbury 2002	Sainsbury 2002	Marsh 1998

Alterations of parameters or changes in phenotypes under basal conditions in male animals of various NPY transgenic and knockout models. Symbols: ⧖, normal or unaffected; ▲, increased or ▼, decreased phenotype or parameter compared to wild-type animals in case of the transgenic and single knockout models or compared to *ob/ob* animals in the double knockout models.

Others initiate meals, such as ghrelin. Insulin and leptin, together with circulating nutrients, indicate long-term energy stores. All these signals act on CNS sites which converge on the hypothalamus, an area that contains a large number of peptide and other neurotransmitters that influence food intake. Energy deficit compromises survival, and it is not surprising that the most powerful of these pathways are those that increase food intake and decrease energy expenditure when reserves are low. In such a situation circulating leptin concentrations fall, leading to increased production of hypothalamic neurotransmitters that strongly increase food intake—such as NPY and AgRP—and decreased levels of α-melanocyte-stimulating hormone (α-MSH) and CART, peptides that reduce food intake and increase energy expenditure. However, due to the complexity and limited availability of selective agonists and antagonists many important questions are still unanswered. Transgenic overexpressing and knockout animal models can help to shed more light on this area.

3.1.1
NPY Transgenic Mice

Intracerebroventricular (icv) injection of NPY results in a significant increase in food intake and when administered chronically it leads to the development of obesity, both in freely feeding as well as in pair-fed animals (Sainsbury et al. 1996b). This can also be seen in the natural mutant *ob/ob* and *db/db* mice which both have strongly elevated NPY levels and are grossly obese. It is therefore not to surprising that NPY transgenic mice with an only 18% increase in arcuate NPY levels do not show such a severe change in bodyweight (Kaga et al. 2001). However, when such homozygous mice are fed a sucrose loaded diet an obese phenotype can be induced. These NPY-overexpressing mice exhibit significantly increased body weight gain with transiently increased food intake and later (approximately 1 year) develop hyperglycemia and hyperinsulinemia although without altered glucose excursion (Kaga et al. 2001). This confirms that elevated hypothalamic NPY expression levels are important for the development of obesity but also shows a strong influence of the type of food needed to induce such a phenotype.

No reports concerning the effects of feeding and energy homeostasis have been made for the second NPY overexpressing model generated by Thiele et al. (1998).

3.1.2
PP Transgenic Mice

Opposing effects on food intake have been reported for PP when it is administered centrally rather than by peripheral injections (Katsuura et al. 2002). However, these data are difficult to compare as differences are apparent when PPs from different species are used (e.g., human PP exhibits specificities on rat Y-receptors that are different from those of rat PP). Overexpressing PP in mice led to the development of a lean phenotype with a reduction of fat mass (Ueno et al. 1999). This phenotype was much more pronounced in male than in female transgenic PP mice. These mice also exhibited a significant decrease in food intake, during both dark and light phases. The rate of gastric emptying was also reduced in these mice and this, and the reduction in food intake, was reversible by intraperitoneal injection of PP antiserum. Overexpression of PP does not have any effect on either water intake or oxygen consumption. Together these data confirm the role of PP as a postprandial satiety signal.

3.1.3
NPY Knockout Mice

The germline deletion of the NPY gene in mice did not lead to a reduction in food intake and body weight under normal conditions (Erickson et al. 1996a). However, these mice showed hyperphagic behavior after fasting. Treatment with leptin reduced food intake and bodyweight to a greater extent in the NPY$^{-/-}$ mice compared to the controls. Another study by Bannon et al. also showed that the NPY knockout animals do have a reduced food intake in response to fasting compared to controls (Bannon et al. 2000). Again, this is evidence of an involvement of NPY in the regulation of energy balance.

More importantly, crossing of the NPY knockout mice onto the leptin deficient *ob/ob* background demonstrated a clear function of the NPY system in the regulation of food intake and energy homeostasis as these double knockout mice showed a significant reduction of the severe obese phenotype of *ob/ob* mice (Erickson et al. 1996b). This was accompanied by reduced food intake, increased energy expenditure and improved serum parameters influencing the development of diabetes. This also shows that central NPY is acting downstream of leptin.

3.1.4
Y$_1$ Knockout Mice

An icv injection of the Y$_1$/Y$_5$ preferring ligand Leu^{31}/Pro^{34}NPY strongly stimulates feeding behavior. In contrast, Y$_1$ receptor knockout models, similar to NPY knockout models, do not show any major abnormalities in food intake or energy homeostasis. However, subtle changes can be seen in all Y$_1$ receptor knockouts analyzed. The report by Pedrazinni et al. describes a phenotype of slightly diminished food intake in Y$_1$ receptor knockouts, both in freely feeding and NPY induced feeding mice. Fasting induced re-feeding on the other hand is strongly decreased (Pedrazzini et al. 1998). Interestingly, when adults these mice show an increased body fat content with no change in protein content. It is suggested that this is at least in part due to the reduced locomotor activity and decreased metabolic rate of these mice.

The Y$_1$ knockout mice generated by Kushi et al. (1998) show a similar phenotype with no change in basal food intake. Again, these mice develop mild hyperinsulinemia and obesity with a significant increase in fat mass particular in females. Furthermore, the increased levels of uncoupling protein (UCP) 1 in brown adipose tissue and the reduced levels of UCP2 in white adipose tissue indicate a decrease in energy expenditure. This further highlights that a global knockout of a particular Y receptor might not reveal its true physiological function in a specific pathway as a secondary mechanism may become activated which can mask the primary effect.

3.1.5
Y_2 Knockout Mice

In contrast to all other Y receptors, the Y_2 receptor is expressed predominately presynaptically and has therefore been proposed to regulate the synthesis and release of NPY and other important neurotransmitters involved in energy homeostasis (King et al. 2000). However, its high level expression in an area of the arcuate nucleus with a permeable blood–brain barrier, accessible to circulating factors, makes it an ideal candidate to mediate peripheral signals for the regulation of energy homeostasis. Furthermore, injections of Y_2 preferring ligands into hypothalamic nuclei have been shown to reduce food intake and meal sizes (Leibowitz and Alexander 1991). Two different phenotypes have been reported on germline Y_2 knockout mice. Naveilhan et al. describes a phenotype of increased body weight, food intake and fat deposition accompanied by an attenuated response to leptin in female mice (Naveilhan et al. 1999). NPY induced food intake and the re-feeding responses after 24-h starvation were normal. In contrast, the germline Y_2 knockout model described by Sainsbury et al. shows a reduced bodyweight gain and adiposity in male mice (Sainsbury et al. 2002a). Food intake was unaltered in male and increased in female knockout mice with re-feeding after starvation strongly elevated in both sexes. The different background of the two models, as well as the different way the targeting of the Y_2 gene has been performed, may explain some of the differences.

Most importantly, compared to wild-type mice the Y_2 receptor preferring endogenously produced and postprandially released hormone PYY(3–36) fails to inhibit food intake in germline Y_2 deficient mice when administrated intraperitoneally (ip) (Batterham et al. 2002). This strongly suggests a gut–hypothalamic pathway that regulates food intake via arcuate Y_2 receptors.

Hypothalamus-specific deletion of Y_2 receptors in conditional Y_2 knockout mice also show a significant decrease in body weight and a significant increase in food intake that is associated with increased mRNA levels for the orexigenic NPY and AgRP, as well as the anorexic POMC and CART in the arcuate nucleus (Sainsbury et al. 2002a). These hypothalamic changes persist over a period of 4 weeks after Y_2 deletion, yet the effect on body weight and food intake subsided within this time. The transience of the observed effects on food intake and body weight in the hypothalamus-specific Y_2 knockout mice compared to germline Y_2 knockout mice, underlines the importance of conditional models of gene deletion, as developmental, secondary, or extra-hypothalamic mechanisms may mask such effects in germline knockouts.

3.1.6
Y$_4$ Receptor Knockout Mice

The functions of the Y$_4$ receptor are among the less well understood of the Y receptor family. Although expressed predominantly in peripheral tissues there are significant amounts present in areas known to be important in the regulation of food intake and energy homeostasis such as the paraventricular nucleus and some brain stem nuclei, and central injections of PP have been shown to stimulate food intake (Katsuura et al. 2002). However, as mentioned above, this could also be attributed to cross reactivities on different Y receptors due to non-species conforming PP ligands. Germline deletion of this Y receptor generates mice with a slightly but significant reduced bodyweight gain accompanied by reduced food intake (Sainsbury et al. 2002c). White adipose tissue mass was also reduced in male knockout mice and plasma levels for PP, the proposed endogenous high affinity ligand for this Y receptor, was strongly elevated. All other serum parameters important in the regulation of energy homeostasis were not significantly altered. Peptide mRNA levels for NPY, AgRP, CART and POMC in the arcuate nucleus were also unaffected by the ablation of the Y$_4$ receptor. Changes in other Y receptor systems were not analyzed. The increases in plasma PP levels in these animals, which reached similar values as seen in the lean PP transgenic animals (Ueno et al. 1999), are therefore the most likely candidate to explain the reduced food intake and bodyweight.

3.1.7
Y$_5$ Receptor Knockout

Pharmacological data have also implicated the Y$_5$ receptor in the regulation of food intake. Germline deletion of this Y receptor, however, does not provide any such evidence (Marsh et al. 1998). These knockout mice feed and grow normally when young but develop late onset (>30 weeks) obesity accompanied by increases in food intake and bodyweight. In contrast to the Y$_1$ knockout mice these Y$_5$ knockout mice do not show any changes in the fasting induced re-feeding pattern. However, responses to icv NPY or NPY analogs are either reduced or missing. This is particularly obvious when a higher dose of NPY is used. The combined administration of NPY and the Y$_1$ antagonist 1229U91 completely abolishes any effect of NPY induced feeding suggesting that an integrated action of both Y$_1$ and Y$_5$ receptors is necessary to elicit the strong stimulatory effect of NPY on food intake.

3.1.8
Double Knockout Models

Hypothalamic NPY levels are strongly regulated by leptin. Lack of this predominantly adipocyte produced hormone, or its receptor, causes severe obesity syndromes showing extreme high expression levels of hypothalamic NPY. In order to identify potential pathways, double knockout mice models have been generated by crossing the leptin deficient *ob/ob* background onto the NPY knockout mice. Reduced food intake and increased energy expenditure are the main reasons for the clear reduction in obesity (40%) and all other obesity related parameters in the NPY$^{-/-}$/*ob/ob* double knockout animals (Erickson et al. 1996b) confirming NPY as an important downstream mediator of leptin action.

To identify the particular Y-receptor(s) mediating this response, Y-receptor deficient mice additionally carrying a nonfunctional leptin gene have been generated. Pralong et al. describes a $Y_1^{-/-}$,*ob/ob* double knockout mouse model that shows a significantly reduced bodyweight in both males and female double knockout mice compared to $Y_1^{-/-}$ mice (Pralong et al. 2002). However, only the bodyweight gain over the first 7 weeks of life has been reported. As shown by the same group as well as others, this is the period before a strong increase in bodyweight gain, particular in females, takes place in the $Y_1^{-/-}$ animals (Kushi et al. 1998; Pedrazzini et al. 1998). It will be important to monitor the development of these double knockout mice over an extended period to identify the role of the Y_1 receptor under these conditions. The main reason for the reduced bodyweight seen in these mice was a reduction in hyperphagia. Interestingly, despite the reduced bodyweight the total fat mass (as a percentage of body weight) in the double knockout mice was still elevated compared to *ob/ob* mice. This suggests that although food intake is regulated by Y_1 receptor activity, energy homeostasis and peripheral fat deposition might be mediated by another mechanism(s).

Crossing the $Y_2^{-/-}$ mice onto the *ob/ob* background attenuates the increased adiposity, hyperinsulinemia, hyperglycemia, and increased hypothalamo–pituitary–adrenal (HPA) axis activity of *ob/ob* mice without affecting food intake and bodyweight gain (Sainsbury et al. 2002b). Y_2 deletion in *ob/ob* mice also significantly increases the hypothalamic POMC mRNA expression with no effect on NPY, AgRP, or CART expression. This suggests that Y_2 receptors mediate the obese, type 2 diabetes phenotype of *ob/ob* mice, possibly via alterations in melanocortin tonus in the arcuate nucleus, and/or effects on the HPA axis. Interestingly, the reduction in white adipose tissue mass in $Y_2^{-/-}$,*ob/ob* double knockout mice is compensated for by increased lean mass.

Y_4 deficiency has no beneficial effects in terms of a reduction in body weight or excessive adiposity of *ob/ob* mice neither does it reduce food intake in these mice (Sainsbury et al. 2002c). Interestingly, male $Y_4^{-/-}$,*ob/ob* double knockout mice actually seem to have a tendency to even higher bodyweight compared to *ob/ob* mice. As the WAT mass in the $Y_4^{-/-}$,*ob/ob* double knockout mice is not different from that of the *ob/ob* mice the additional weight might be contributed by lean mass, which is consistent with the elevated testosterone levels in these mice.

Mice lacking a functional Y_5 receptor have also been crossed onto the *ob/ob* background. Marsh et al. reports that although the single Y_5 knockout shows reduced responses to central NPY induced feeding stimulation this effect has no beneficial consequences in the $Y_5^{-/-}$,*ob/ob* double knockout animals (Marsh et al. 1998). In both sexes food intake, bodyweight and adiposity was not different from those of *ob/ob* mice suggesting that the Y_5 receptor is not required for NPY's ability to induce obesity in leptin deficient *ob/ob* mice.

3.1.9
Summary on Y Receptor Involvement in Energy Homeostasis

Taken together these results indicate that the Y_1 receptor is most likely to mediate the strong stimulatory activity of NPY on food intake, whereas the Y_2 and Y_4 receptors have either an inhibitory or no effect on this parameter. Y_5 receptors do seem to contribute slightly to the stimulation of food intake but do not have any significant role under conditions in which leptin signaling is absent. It still remains to be seen what role the y_6 receptor, which is a functional entity in the mouse system, plays in this and other processes of the regulation of energy homeostasis. Investigations of additional double knockout models including Y_1/Y_5 double mutant mice may also help to shed more light onto this question.

3.2
Reproduction

NPY is a major central regulator of sexual behavior and reproductive functions. Intracerebroventricular administration of NPY to sex steroid-primed ovariectomized (OVX) rats increases secretion of luteinizing hormone (LH) and gonadotropin releasing hormone (GnRH; Sabatino et al. 1990; Urban et al. 1996). In contrast, in sex steroid-deficient OVX rats, or intact male and female rats, NPY markedly inhibits reproductive function (Pierroz et al. 1996; Xu et al. 1993), with testicular and seminal vesicle or ovarian weights being reduced, and sexual maturation and estrous cyclicity in female rats being disrupted. These inhibitory effects of NPY on reproductive function particularly obvious under conditions of negative energy balance, such as food restriction, heavy exercise, lactation, and insulin-dependent diabetes mellitus, all of which are associated with elevated hypothalamic NPY expression (Aubert et al. 1998; Krysiak et al. 1999). In this way NPY coordinates energy availability with reproduction, inhibiting procreation during unfavorable metabolic conditions.

Obesity is also associated with reproductive defects and reduced fertility (Caprio et al. 2001). Genetically obese *ob/ob* mice are also infertile due to insufficient hypothalamo-pituitary–gonadal drive, underdevelopment of reproductive organs, and impaired spermatogenesis (Caprio et al. 2001). In *ob/ob* mice, the lack of leptin-mediated inhibition of NPY expression and secretion in the hypothalamus (Schwartz et al. 1996; Widdowson and Wilding 2000) leads to chronically elevated hypothalamic NPY-ergic activity. Treatment of *ob/ob* mice with leptin reduces NPY mRNA expression and peptide levels in the hypothalamus, and restores fertility of male and female mice by improved function of the hypothalamo-pituitary–gonadal axis (Chehab et al. 1996). Insights into the regulatory role of the different Y receptors in reproduction as gained from investigation of the different Y receptor knockout models are summarized below.

3.2.1
NPY Knockout

NPY deletion per se does not have any adverse effects on reproductive functions in the mouse. However, as with most of the side effects of leptin deficient *ob/ob* mice fertility was improved when the NPY null locus was combined with the *ob/ob* locus (Erickson et al. 1996b). This includes increased fertility rates for males (33%) and females (20%) as well as increases in the size of the reproductive organs in both sexes of the $NPY^{-/-},ob/ob$ double knockout mice compared to the *ob/ob* mice. These improvements are most likely due to an increased hypothalamic pituitary activity with consequent elevation of sex steroids.

3.2.2
Y_1 Receptor Knockout

Y_1 receptor deficient animals produced by Pedrazzini et al. showed increased pituitary levels of LH and increased seminal vesicle size after 48 h of starvation, parameters that are otherwise not altered in $Y_1^{-/-}$ mice under basal conditions (Pedrazzini et al. 1998). The increased levels of NPY caused by the starvation and the lack of Y_1 receptor signaling are likely to be responsible for this effect. The inhibitory effect of leptin on NPY expression has been reported to be one of the major modulators for the onset of puberty. Daily injections of leptin into juvenile $Y_1^{-/-}$ female mice triggered a marked advancement in puberty compared to wild-type control mice. Consistent with increased activity of the gonadotropin axis the uterine weight of the leptin treated Y_1 knockout mice was significantly higher than that of age-matched controls. Further support for an involvement of the Y_1 receptor in reproductive function comes from the observation of improved function of the gonadotropic axis in $Y_1^{-/-},ob/ob$ double knockout mutant mice (Pralong et al. 2002). Both pituitary LH concentration and seminal vesicle weight was increased twofold in the double knockout animals compared to *ob/ob* mice.

3.2.3
Y_2 Receptor Knockout

No effects on the reproductive functions of mice missing the Y_2 receptor have been reported. This is also true for $Y_2^{-/-},ob/ob$ double knockout mice where the deletion of both genes leads to an improvement of the side effects of an obesity phenotype but does not have any positive effects on fertility—suggesting that different leptin-dependent pathways for the regulation of feeding behavior and reproduction exist (Sainsbury et al. 2002b).

3.2.4
Y$_4$ Receptor Knockout

Recently it has been demonstrated that icv administration of the Y$_1$ antagonist and Y$_4$ agonist 1229U91 (Schober et al, 1998) to estrogen-primed OVX or intact male rats rapidly increased follicle stimulating hormone and/or LH secretion (Jain et al. 1999; Raposinho et al. 2000). These effects were attributed to Y$_4$ activation since rat PP, which is specific for the rat Y$_4$ receptor, also induced a similar profile of LH secretion in estrogen-primed OVX rats. Conformation of a major role in reproduction for this Y receptor comes from the analysis of Y$_4^{-/-}$ mice which have been shown to have elevated GnRH expression in forebrain neurons accompanied by increased testosterone levels in male Y$_4^{-/-}$ mice (Sainsbury et al. 2002c). Furthermore, female Y$_4^{-/-}$ mice show a strong advancement in mammary gland development with significant increases in ductile branching both in virgin and in pregnant animals (Sainsbury et al. 2002c). Moreover, ablation of the Y$_4$ receptor in *ob/ob* mice restores fertility to 100% in male mice and improves fertility in female double knockout mice by 50%. Reproductive organ sizes are also normalized and testosterone levels are similar to wild-type control levels. As the high affinity ligand for the Y$_4$ receptor, PP, cannot cross the blood–brain barrier to reach the central Y$_4$ receptors it has been proposed that the lower affinity ligand NPY is the prime candidate to mediate the effects on the reproductive axis when levels reach high enough values—as in the case of starvation or leptin deficiency. In contrast, other aspects of the obese phenotype of the *ob/ob* mice are unaffected by Y$_4$ deletion again suggesting that different leptin dependent pathways exist for the control of reproduction and energy homeostasis.

3.2.5
Y$_5$ Receptor Knockout

No effects on the reproductive axis have been reported in either Y$_5^{-/-}$ or Y$_5^{-/-}$/ *ob/ob* double knockout animals. Consistent with the low abundance of Y$_5$ receptor binding sites in the brain it can be concluded that this Y receptor does not play a major role in fertility mediated effects of NPY.

3.3
Cardiovascular Phenotypes

NPY is known to mediate potent effects, both centrally as well as peripherally, on the cardiovascular system thereby regulating heart rate and blood pressure (Lundberg et al. 1983). Peripherally, NPY is a putative sympathetic cotransmitter, colocalized with noradrenaline in most postganglionic sympathetic nerve terminals. It is released together with noradrenaline during sympathetic nerve stimulation to act at both prejunctional and postjunctional receptors. The prejunctional effects of NPY are thought to be mediated by the Y_2 receptor whereas the postjunctional effects are most likely due to activation of Y_1 receptors expressed on smooth muscle cells. NPY released during sympathetic stimulation is also known to attenuate the cardiac responses to vagal activity, probably by inhibiting the release of acetycholine (Potter 1987). Effects on overexpressing NPY and consequences of deleting Y receptors from the mouse genome are summarized below.

3.3.1
Transgenic Rat

NPY overexpressing rats generated by Michalkiewicz et al. show significant elevated NPY levels in several internal organs including heart and blood vessels (Michalkiewicz et al. 2001). This increase in NPY concentration in the transgenic animals does not affect basal arterial blood pressure or heart rate; however, total vascular resistance was significantly increased. In addition, the potentiation activity of NPY on noradrenaline induced blood pressure increases was also significantly elevated in the transgenic animals suggesting that NPY is not a major regulator of basal blood pressure but has important functions under more stressful situations.

3.3.2
Y_1 Receptor Knockout

The deletion of the Y_1 receptor generated mice that were no longer responsive to the vasoconstrictor action of NPY (Pedrazzini et al. 1998). Basal blood pressure and heart rate were unaffected in these mice. NPY infusion in $Y_1^{-/-}$ mice no longer triggered an effect but the effects on blood pressure and heart rate of other colocalized neurotransmitters such as noradreanlin were still present in $Y_1^{-/-}$ mice. Also the normally seen strong potentiation of noradrenalin action on blood pressure by NPY was lost. The findings from this study suggest that NPY is not essential for the basal regulation of blood pressure, but demonstrates that the majority of blood pressure effects of elevated NPY are mediated by the Y_1 receptor.

3.3.3
Y_2 Receptor Knockout

Similar to Y_1 knockout the removal of the Y_2 receptor did not affect basal blood pressure, however, heart rate was increased in these mice (Naveilhan et al. 1999; Smith-White et al. 2002a). Infusion of NPY still produced a similar response on blood pressure in the $Y_2^{-/-}$ mice as in control mice confirming that Y_2 receptors are not responsible for the pressure activity of NPY. However, the ability of NPY to attenuate the decrease in heart rate evoked by cardiac vagal stimulation was completely lost in the $Y_2^{-/-}$ mice. Vagotomy in control mice causes an increase in heart rate but this parameter remains unchanged in $Y_2^{-/-}$ mice. The results of this investigation demonstrate that the presynaptically located Y_2 receptors are responsible for the NPY mediated attenuation of parasympathetic activity to the heart.

3.3.4
Y_4 Receptor Knockout

Interestingly, in contrast to the deletion of the Y_1 or the Y_2 receptor, mice missing the Y_4 receptor display lower basal blood pressure compared to control mice (Smith-White et al. 2002b). In addition, $Y_4^{-/-}$ mice also show a significantly reduced heart rate (Smith-White et al. 2002b). Interestingly, the endogenous high affinity ligand for the Y_4 receptor, PP, had no effect on blood pressure or cardiac vagal activity in $Y_4^{-/-}$ or control mice suggesting that other NPY family members may activate this Y receptor to mediate the effects on blood pressure. Consistent with this hypothesis is that injection of NPY evokes an increase in blood pressure in control mice but this response was significantly reduced in Y_4 receptor knockout mice. Similarly, vagotomy increased heart rate in both $Y_4^{-/-}$ and control mice, however, this increase was significantly lower in the $Y_4^{-/-}$ mice.

The reduced vasoconstrictor and vagal inhibitory activity evoked by NPY in $Y_4^{-/-}$ mice may be due to a reduction in sympathetic activity, possibly resulting from altered NPY activity in other central brain areas such as the nucleus of the tractus solitarii (NTS) thereby affecting adrenergic transmission. Deletion of Y_4 receptor located in the NTS may therefore disrupt the autonomic balance within the cardiovascular system.

3.4
Seizure Susceptibility

It is known that NPY influences the appearance of epileptic seizures and the manifestation of epilepsy (Hokfelt et al. 1998; Vezzani et al. 1999). NPY and its receptors are also undergoing adaptive changes after acute seizures and in chronic epilepsy suggesting a pivotal role of this neuropeptide system in the modulation of seizure activity. In the hippocampus, NPY is constitutively expressed in gamma-aminobutyric acid (GABA) interneurons. Y_1 receptors are located on granule cell dendrites and Y_2 receptors are preferentially located presynaptically on mossy fibers and on Schaffer collaterals. Y_2 receptors mediate a potent presynaptic inhibition of glutamate release with Y_1 receptor stimulation causing a block of N-type calcium channels. Receptor adaptations such as the increase of Y_2 and Y_5 mRNA and the decrease in expression of Y_1 mRNA together with the upregulation of NPY levels in mossy fibers may therefore represent an important protective mechanism for epilepsy (Klapstein and Colmers 1997).

3.4.1
Transgenic Rat

Seizure susceptibility in transgenic rats overexpressing the rat neuropeptide Y gene under the control of its natural promoter is normal under basal conditions (Vezzani et al. 2002). However, induction of seizures in adult transgenic male rats by icv injection of kainic acid or by electrical kindling showed a significant reduction in the number and duration of electroencephalographic seizures compared to wild-type rats. Transgenic rats were also less susceptible to epileptogenesis than wild-type littermates suggesting that endogenous NPY overexpression in the rat hippocampus is protective against seizures and epileptogenesis.

3.4.2
NPY Knockout Mice

Erickson et al. describe that young adult mice deficient of the NPY gene show mild seizures when exposed to an unfamiliar environment such as on the top of the cage (Erickson et al. 1996a). Such short convulsion episodes occurred in about 30% of the NPY knockout animals and was only observed in the age range 6–8 weeks but not in older mice. However, hyperexcitability of $NPY^{-/-}$ mice was observed when they were challenged with a convulsant agent such as penetetraxole, a GABA antagonist. Eighty percent of $NPY^{-/-}$ mice developed motor convulsions compared to only 30% of control mice. In addition, the time until the start of the seizure was reduced and the severity of the seizure was increased in the $NPY^{-/-}$ mice consistent with the lack of inhibitory action of NPY on excitatory transmission. The most likely Y receptor candidate(s) mediating these inhibitory signals is the pre-synaptically located Y_2 receptor and potentially the Y_5 receptor. Results from Y_5 knockout animals confirm this suspicion and so do reports on $Y_2^{-/-}$ mice that have been presented at several meetings but have not yet been published.

3.4.3
Y$_5$ Receptor Knockout Mice

A report by Marsh et al. showed that $Y_5^{-/-}$ mice do not exhibit spontaneous seizure-like activity; however, they are more sensitive to kainic acid-induced seizures (Marsh et al. 1999). The analysis of hippocampal slices from these mutant mice by electrophysiological techniques revealed normal function, but the anti-epileptic effects of NPY were absent. In a different study on the same $Y_5^{-/-}$ mice it was shown that NPY-ergic inhibition of excitatory CA3 synaptic transmission is absent (Guo et al. 2002). The effect on seizure susceptibility in $Y_5^{-/-}$ seems to be strongly dependent on the genetic background of the mice as mice on a 129/

Sv background show much greater responses to kainic acid-induced seizures than $Y_5^{-/-}$ mice on a mixed 129/Sv x C57BL/6 J background.

3.5
Anxiety-Related Behaviors

One of the major biological actions of NPY is believed to be its regulatory role in anxiolytic-like responses (Hokfelt et al. 1998). NPY has been shown to reduce experimental anxiety in several animal models including the elevated plus-maze, the social interaction test and the fear-potentiated startle paradigm (Broqua et al. 1995). The major center where NPY exerts its anti-anxiety related behavior is the amygdala where strong alterations in NPY mRNA and peptide expression can be observed after acute or repeated restrain stress (Redrobe et al. 2002). The majority of evidence suggests that the anxiolytic actions of NPY are mediated by the Y_1 receptor (Redrobe et al. 2002). However, recent studies have also shown a major role for the Y_2 receptor as intra-amygdalar administrations of Y_2 preferring ligands induce anxiogenic-like responses (Nakajima et al. 1998).

3.5.1
NPY Knockout Mice

An extensive analysis of behavioral phenotypes on $NPY^{-/-}$ mice was performed by Bannon and colleagues (Bannon et al. 2000). Several lines of evidence suggest an anxiogenic-like phenotype of these mice. In particular the experiments in the open-field test detected significantly less activity in the center area of the apparatus for the $NPY^{-/-}$ mice compared to the control animals. This was not due to a general decrease in activity as the total distance traveled in $NPY^{-/-}$ and wild-type mice was not different. Further evidence for an anxiogenic-like behavior in $NPY^{-/-}$ mice comes from experiments investigating the arcustic startle response where the knockout animals showed a significant increase in amplitude.

In the situation of elevated NPY expression as in the transgenic NPY rat model anxiety behavior was also influenced (Thorsell et al. 2000). Although no differences between control and transgenic animals can be observed in the elevated plus-maze under basal conditions, preceding restrain stress causes a significant reduction in exploratory behavior in the wild-type rats but had no effect on the NPY overexpressing rats. Furthermore, NPY transgenic rats also showed an increase in the number of drinking episodes in the punished drinking test compared to control animals. Interestingly, NPY transgenic mice do not seem to differ from control mice when exposed to the elevated plus-maze or another anxiety paradigm, fear conditioning (Thiele et al. 1998).

3.5.2
Y$_2$ Knockout Mice

So far reports on anxiety-related behavior of Y receptor knockout mice are only available for the Y$_2^{-/-}$ mice (Redrobe et al. 2003; Tschenett et al. 2003). Both studies suggest that the Y$_2$ receptor has an inhibitory role in the anxiolytic-like effects of NPY. This is demonstrated in two models of anxiety, the elevated plus-maze and the open-field test. In both tests Y$_2^{-/-}$ mice had more entries and spent more time in the open arms of the elevated plus-maze and also had more crossings and spent more time in the central area of the open-field test compared to control animals. The suggested mechanism by which Y$_2$ receptors mediate their effects on anxiety might be via negative inhibitory action on NPY release. A lack of feedback inhibition of NPY release in the Y$_2^{-/-}$ mice might lead to increased NPY mediated activation of other receptors, particularly Y$_1$ receptors and consequently induce—through this pathway—anxiolytic-like effects.

3.6
Other Behavioral Phenotypes

3.6.1
Voluntary Alcohol Intake

Voluntary alcohol intake can be viewed as an anxiety related behavior; genetic linkage analysis in P rats, which have an increased alcohol preference, identified a locus that includes the NPY gene (Carr et al. 1998). The expression of NPY and Y receptors in the CNS of these rats is also altered suggesting an important role for NPY in alcohol consumption.

Experiments in NPY$^{-/-}$ and NPY overexpressing mice have shown that voluntary ethanol consumption is inversely related to NPY levels: NPY$^{-/-}$ mice exhibit significantly higher ethanol consumption and lower sensitivity to ethanol than control mice, and NPY overexpressing mice show the opposite effects (Thiele et al. 1998). However, there seems to be genetic background effect causing differences in locomotor activity in NPY$^{-/-}$ mice. For example, NPY$^{-/-}$ mice on a mixed C57BL/6 J x 129/SvEv background showed increased sensitivity to locomotor activation caused by intraperitoneal injection of ethanol, and were resistant to sedation caused by ethanol. In contrast, NPY$^{-/-}$ mice on an inbred 129/SvEv background exhibited normal locomotor activation following injection of ethanol, and displayed normal sedation in response to ethanol.

The Y receptor that is most likely to be responsible for mediating voluntary alcohol consumption is the Y$_1$ receptor. Experiments on mice that lack the Y$_1$ receptor demonstrated that male and female Y$_1^{-/-}$ mice showed increased consumption of ethanol when compared with wild-type mice (Thiele et al. 2002). Male Y$_1^{-/-}$ mice were found to be less sensitive to the sedative effects of alcohol with a more rapid recovery from ethanol-induced sleep compared to wild-type mice.

Y_5 receptor knockout $Y_5^{-/-}$ mice on an inbred 129/SvEv background showed normal ethanol-induced locomotor activity and normal voluntary ethanol consumption, but displayed increased sleep time caused by injection of ethanol (Thiele et al. 1998). No results are yet available for the other Y receptor knockout mice.

3.6.2
Memory

One of the biological actions of NPY, which has received little attention so far, is its regulatory role in learning and memory processing (Redrobe et al. 1999). Injection of NPY into the rostral hippocampus and the septal area was shown to enhance memory retention, whereas NPY injection into the amygdaloid body and the caudal hippocampus induced amnesia (Flood et al. 1989). The development of an NPY-transgenic rat with significantly elevated NPY levels in the hippocampus has helped to study this aspect of NPY action in greater detail (Thorsell et al. 2000). Surprisingly, these NPY transgenic rats were shown to display deficits in both the acquisition and retention of a spatial memory task when tested in the Morris Water Maze (Thorsell et al. 2000).

The identity of the specific Y receptor subtype(s) involved in NPY-induced effects on learning and memory processing remain unclear, although some evidence suggests that the ability of NPY to enhance learning and memory may be linked to activation of Y_2 receptors (Heilig 1993). *In situ* hybridization studies have shown that Y_2 receptor mRNA is discretely localized in the rat brain, including comparatively high levels in the hippocampus (Dumont et al. 1997, 2000). In addition, radioligand binding experiments have demonstrated that Y_2 receptors seem to be the predominant NPY receptor subtype in the hippocampal formation (Dumont et al. 1997, 2000), where they have generally been assumed to be presynaptic receptors which negatively modulate glutamatergic neurons (Colmers and Bleakman 1994) and possibly NPY release (King et al. 2000). Further studies are needed to clarify the role of the different Y receptors in this physiological process.

3.6.3
Nociception

NPY is involved in pain processing at the level of the spinal cord as structural and functional studies have shown (Abdulla and Smith 1999; Zhang et al. 2000). A particularly high density of NPY-containing fibers can be found in laminae I and II of the dorsal horn suggesting that NPY may control transmitter release from the central terminals of primary afferents. NPY has been shown to induce biphasic effects on the nociceptive flexor reflex in rats when injected intrathecally (White 1997). In addition, NPY is able to block substance P release from dorsal root ganglion neurons *in vitro* as well as *in vivo* and has anti-nociceptive effects after intrathecal administration (White 1997).

The lack of NPY in NPY$^{-/-}$ mice causes an exaggerated autotomy, a self-mutilation behavior possibly related to pain sensation, in agreement with analgesic effect of NPY (Shi et al. 1998). Alterations in the levels of Y_1 and especially Y_2 receptor mRNAs were observed in the spinal cord of NPY-deficient mice suggesting these two Y receptors are the major mediators of the analgesic effect of NPY. Consistent with these results, reduced anti-nociception and plasma extravasation has been shown in mice deficient in the Y_1 receptor (Naveilhan et al. 2001c). These $Y_1^{-/-}$ animals develop hyperalgesia to acute thermal, cutaneous and visceral chemical pain and also show mechanical hypersensitivity. Experiments with these animals also confirm the requirement of Y_1 receptors for the release of substance P. Furthermore, the Y_1 receptor has been implicated in mediating the effect of NPY on sedation (Naveilhan et al. 2001a).

3.6.4
Regulation of Bone Homeostasis

The most recent function added to the already extensive list of the processes mediated by NPY is its role in the regulation of bone homeostasis. Intriguingly, 4 weeks of central NPY infusion in wild-type mice had an inhibitory effect on bone formation (Ducy et al. 2000). However, it is unclear from this study whether the inhibitory effects of central NPY infusion on bone are a direct consequence of hypothalamic NPY action, or a secondary effect of the resultant increase in expression (Sainsbury et al. 1996a) and circulating concentrations (Sainsbury and Herzog 2001) of leptin which also has an negative effect on bone formation.

As NPY is a downstream modulator of leptin action at the level of the arcuate nucleus where NPY neurons are known to express both leptin receptors and Y_2 receptors it is likely that this is the pathway through which NPY mediates is function on bone.

Analysis of bones from Y_2 receptor deficient mice has demonstrated a twofold increase in trabecular bone volume as well as greater trabecular number and thickness compared with control mice (Baldock et al. 2002). Moreover, this study also demonstrated that central Y_2 receptors are crucial for this process, since selective deletion of hypothalamic Y_2 receptors in mature conditional Y_2 knockout mice results in an identical increase in trabecular bone volume within 5 weeks of deletion of the Y_2 receptor. The lack of any significant changes in plasma content of hypothalamo-pituitary–corticotropic, –thyrotropic, –somatotropic, or –gonadotropic related parameters suggests that Y_2 receptors do not modulate bone formation by humoral mechanisms, but it is more likely that alteration of autonomic function through hypothalamic Y_2 receptors play a key role in a major central regulatory circuit of bone formation. Additional experiments will be necessary to elucidate the full scale of NPY's involvement and the function of possible other Y receptors in this process.

4
Concluding Remarks

Over the last couple of years studies on transgenic models of NPY and its receptors have provided a wealth of information about molecular and behavioral phenotypes with finally some more defined roles for the individual Y receptors emerging. Particularly in the coordinated regulation of food intake and energy homeostasis important new insights have been revealed with the help of these knockout and transgenic models. However, specific roles for the individual receptors have also been found in other areas including reproduction, anxiety, epilepsy and cardiovascular regulation. Furthermore, with the help of these models completely new and unexpected functions of the NPY system have been revealed such as the central regulation of bone formation. It also became clear that Y_4 receptors are not only responsive to PP, and in particular centrally located Y_4 receptors might be activated by other NPY family members.

However, some of the findings in the transgenic models did not show the expected 'severe' phenotypes, which would have been predicted from previously conducted pharmacological intervention studies. In particular the relative 'normal' feeding phenotype of the NPY$^{-/-}$ mouse was a surprise considering the massive increase in food intake and body weight gain observed after acute and chronic NPY administration. Some of these might be due to the limitations (lack of high enough specificity for one particular Y receptor) of the pharmacological tools used in the *in vitro* and *in vivo* studies. One other explanation for the discrepancies between pharmacological and transgenic research outcomes could be the redundancy in the system, which could lead to compensatory changes in knockout animals during development. In addition, germline deletion of a gene causes the loss of function of that gene in all tissues where it is normally expressed leading to a phenotype which is compiled by the sum of all lost functions. Germline deletion of a gene can also produce secondary effects not directly linked to the actual function of the deleted gene.

Future strategies will therefore need some refinement of the *in vivo* targeting mutagenesis techniques used. In particular the use of knock-in strategies, which allow modification of, rather than complete inactivation of, receptor or ligand functions, will be required. More importantly, the generation of a full set of conditional knockout models for all Y receptors and ligands is needed to overcome the inherent problems of developmental influences on phenotypes, as well as allowing selective deletion of the genes in only a defined tissue or nucleus in an adult animal (thereby avoiding complication due to whole body ablation of the gene).

There are still some transgenic models for members of the NPY gene and Y receptor family missing. These include knockout models for PYY and PP as well as y_6. Although the later is considered to be a pseudogene in primates it is still a functional entity in the mouse system. As such it also needs to be analyzed with the same rigor as all other members of this family to answer completely all the questions about the functional contributions of all the different Y receptors to physiological processes.

Nevertheless, the data accumulated from the models generated so far are of great value for future, more detailed studies on the complex nature of NPY physiology. These knockout models are also an invaluable resource for the testing of the specificity of agonists and antagonists and for determining the specificity of antibodies.

Finally, with the completion of the human and mouse genome sequencing programs and the rapid identification of additional genes paralleled by the development of gene array expression technologies, these NPY and Y receptor knockout and transgenic mice models will be extremely useful in the identification of molecular networks that contribute to certain complex physiological processes. Carefully conducted studies using this technology may identify a great number of new candidate genes involved in many different NPY mediated functions not yet known or associated with these systems.

References

Abdulla FA, PA Smith (1999) Nerve injury increases an excitatory action of neuropeptide Y and Y_2-agonists on dorsal root ganglion neurons, Neuroscience 89:43–60

Aubert ML, Pierroz DD, Gruaz NM, d'Alleves V, Vuagnat BA, Pralong FP, Blum WF, Sizonenko PC (1998) Metabolic control of sexual function and growth: role of neuropeptide Y and leptin. Mol Cell Endocrinol 140:107–13

Baldock PA, Sainsbury A, Couzens M, Enriquez RF, Ghomas GP, Gardiner EM, Herzog H (2002) Hypothalamic Y_2 receptors regulate bone formation. J Clin Invest 109: 915–921

Ball H, Shine J, Herzog H (1995) Multiple promoters regulate tissue-specific expression of the human NPY-Y_1 receptor gene. J Biol Chem 270: 27272–27276

Bannon AW, Seda J, Carmouche M, Francis JM, Norman MH, Karbon B, McCaleb ML (2000) Behavioral characterization of neuropeptide Y knockout mice. Brain Res 868:79–87

Batterham RL, Cowley MA, Small CJ, Herzog H, Cohen MA, Dakin CL, Wren AM, Brynes AE, Low MJ, Ghatei MA, Cone RD, Bloom SR (2002) Gut hormone PYY3-36 physiologically inhibits food intake. Nature 418:650–654

Blomqvist AG, Herzog H (1997) Y-receptor subtypes-how many more? Trends Neurosci 20: 294–298

Broqua P, Wettstein JP, Rocher MN, Gauthier-Martin B, Junien JL (1995) Behavioral effects of neuropeptide Y receptor agonists in the elevated plus-maze and fear-potentiated startel procedure. Behav Pharmacol 6:215–222

Caprio M, Fabbrini E, Isidori AM, Aversa A, Fabbri A (2001) Leptin in reproduction. Trends Endocrinol Metab 12:65–72

Carr LG, Foroud T, Bice P, Gobbett T, Ivashina J, Edenberg H, Lumeng L, Li TK (1998) A quantitative trait locus for alcohol consumption in selectively bred rat lines. Alcoholism: Clin Exp Res 22: 884–887

Chehab FF, Lim ME, Lu R (1996) Correction of the sterility defect in homozygous obese female mice by treatment with the human recombinant leptin. Nature Genet 12: 318–320

Colmers WF, Bleakman D (1994) Effects of neuropeptide Y on the electrical properties of neurons. Trends Neurosci 17: 373–379

Ducy P, Amling M, Takeda S, Priemel M, Schilling AF, Beil FT, Shen J, Vinson C, Rueger JM, Karsenty G (2000) Leptin inhibits bone formation through a hypothalamic relay: a central control of bone mass. Cell 100:197–207

Dumont Y, Jacques D, St-Pierre JA, Quirion R (1997) Neuropeptide Y receptor types in the mammalian brain: species differences and status in the human central nervous system. In: Grundemar L, Bloom SR (eds) Neuropeptide Y. Academic Press, London, pp 57–86

Dumont Y, Jacques D, St-Pierre JA, Tong Y, Parker R, Herzog H, Quirion R (2000) Neuropeptide Y, peptide YY and pancreatic polypeptide receptor proteins and mRNAs in mammalian brain. In: Quirion R, Bjorklund A, Hokfelt T (eds) Handbook of chemical neuroanatomy. Elsevier, London, pp 375–475

Erickson JC, Clegg KE, Palmiter RD (1996a) Sensitivity to leptin and susceptibility to seizures of mice lacking neuropeptide Y. Nature 381:415–418

Erickson JC, Hollopeter G, Palmiter RD (1996b) Attenuation of the obesity syndrome of ob/ob mice by the loss of neuropeptide Y. Science 274:1704–1707

Flood JF, Baker ML, Hernandez EN, Morley JE (1989) Modulation of memory processing by neuropeptide Y varies with brain injection site. Brain Res 503:73–82

Gerald C, Walker MW, Criscione L, Gustafson EL, Batzl-Hartmann C, Smith KE, Vaysse P, Durkin MM, Laz TM, Linemeyer DL, Schaffhauser AO, Whitebread S, Hofbauer KG, Taber RI, Branchek TA, Weinshank RL (1996) A receptor subtype involved in neuropeptide-Y-induced food intake. Nature 382:168–171

Guo H, Castro PA, Palmiter RD, Baraban SC (2002) Y_5 receptors mediate neuropeptide Y actions at excitatory synapses in area CA3 of the mouse hippocampus. J Neurophysiol 87:558–566

Heilig M (1993) Neuropeptide Y in relation to behavior and psychiatric disorders. In: Colmers WF, Wahlestedt C (eds) The biology of neuropeptide Y and related peptides. Humana Press, Totowa, pp 511–554

Herzog H, Baumgartner M, Vivero C, Selbie LA, Auer B, Shine J (1992) Genomic organisation, localisation and allelic differences in the gene for the human neuropeptide Y Y_1 receptor. J Biol Chem 268:6703–6707

Herzog H, Darby K, Ball H, Hort Y, Beck-Sickinger A, Shine J (1997) Overlapping gene structure of the human neuropeptide Y receptor subtypes Y_1 and Y_5 suggests coordinate transcriptional regulation. Genomics 41:315–319

Hokfelt T, Broberger C, Zhang X, Diez M, Kopp J, Xu Z, Landry M, Bao L, Schalling M, Koistinaho J, DeArmond SJ, Prusiner S, Gong J, Walsh JH (1998) Neuropeptide Y: some viewpoints on a multifaceted peptide in the normal and diseased nervous system. Brain Res Brain Res Rev 26:154–166

Hollopeter G, Erickson JC, Palmiter RD (1998) Role of neuropeptide Y in diet-, chemical- and genetic-induced obesity of mice. Int J Obesity 22:506–512

Howell OW, Scharfman HE, Beck-Sickinger AG, Herzog H, Gray WP (2003) Neuropeptide Y is neuroproliferative in primary hippocampal cultures. Journal of Neurochemistry 86:646–659

Inui A, Okita M, Nakajima M, Momose K, Ueno N, Teranishi A, Miura M, Hirosue Y, Sano K, Sato M, Watanabe M, Sakai T, Watanabe T, Ishida K, Silver J, Baba S, Kasuga M (1998) Anxiety-like behavior in transgenic mice with brain expression of neuropeptide Y. Proc Assn Am Physicianc 110:171–182

Jain MR, Pu S, Kalra PS, Kalra SP (1999) Evidence that stimulation of two modalities of pituitary luteinizing hormone release in ovarian steroid-primed ovariectomized rats may involve neuropeptide Y Y_1 and Y_4 receptors. Endocrinology 140:5171–5177

Kaga T, Inui A, Okita M, Asakawa A, Ueno N, Kasuga M, Fujimiya M, Nishimura N, Dobashi R, Morimoto Y, Liu I, Cheng J (2001) Modest overexpression of neuropeptide Y in the brain leads to obesity after high-sucrose feeding. Diabetes 50:1206–1210

Katsuura G, Asakawa A, Inui A (2002) Roles of pancreatic polypeptide in regulation of food intake, Peptides 23:323–329

King PJ, Williams G, Doods H, Widdowson PS (2000) Effect of a selective neuropeptide Y Y(2) receptor antagonist, BIIE0246 on neuropeptide Y release. Eur J Pharmacol 396:R1–R3

Klapstein GJ, Colmers WF (1997) Neuropeptide Y suppresses epileptiform activity in rat hippocampus in vitro. J Neurophysiol 78: 1651–1661

Krysiak R, Obuchowicz E, Herman ZS (1999) Interactions between the neuropeptide Y system and the hypothalamic- pituitary-adrenal axis. Eur J Endocrinol 140: 130–136

Kushi A, Sasai H, Koizumi H, Takeda N, Yokoyama M, Nakamura M (1998) Obesity and mild hyperinsulinemia found in neuropeptide Y- Y_1receptor-deficient mice. Proc Natl Acad Sci USA 95:15659–15664

Leibowitz SF, Alexander JT (1991) Analysis of neuropeptide Y-induced feeding: dissociation of Y_1 and Y_2 receptor effects on natural meal patterns. Peptides 12: 1251–1260

Li XJ, Wu YN, North RA, Forte M, Source J (1992) Cloning, functional expression, and developmental regulation of a neuropeptide Y receptor from Drosophila melanogaster. J Biol Chem 267:9–12

Lundberg JM, Terenius LU, Hokfelt T, Goldstein M (1983) High levels of neuropeptide Y in peripheral noradrenergic neurons in various mamals including man. Neurosci Lett 42:167–172

Marsh DJ, Baraban SC, Hollopeter G, Palmiter RD (1999) Role of the Y_5 neuropeptide Y receptor in limbic seizures. Proc Natl Acad Sci USA 96:13518–13523

Marsh DJ, Hollopeter G, Kafer KE, Palmiter RD (1998) Role of the Y_5 neuropeptide Y receptor in feeding and obesity. Nature Med 4:718–721

Michalkiewicz M, Michalkiewicz T (2000) Developing transgenic neuropeptide Y rats. Methods Mol Biol 153:73–89

Michalkiewicz M, Michalkiewicz T, Kreulen DL, McDougall SJ (2001) Increased blood pressure responses in neuropeptide Y transgenic rats. Am J Physiol - Regul Integrative Comp Physiol 281:R417–R426.

Nakajima M, Inui A, Asakawa A, Momose K, Ueno N, Teranishi A, Baba S, Kasuga M (1998) Neuropeptide Y alters sedation through a hypothalamic Y_1-mediated mechanism. Peptides 19:359–363

Naveilhan P, Canals J, Valjakka A, Vartiainen J, Arenas E, Ernfors P (2001a) Neuropeptide Y alters sedation through a hypothalamic Y_1mediated mechanism. Eur J Neurosci 13:2241–2246

Naveilhan P, Canals JM, Arenas E, Ernfors P (2001b) Distinct roles of the Y_1 and Y_2 receptors on neuropeptide Y-induced sensitization to sedation. J Neurochem 78:1201–1207

Naveilhan P, Hassani H, Canals JM, Ekstrand AJ, Larefalk A, Chhajlani V, Arenas E, Gedda K, Svensson L, Thoren P, Ernfors P (1999) Normal feeding behavior, body weight and leptin response require the neuropeptide Y Y_2 receptor. Nature Med 5:1188–1193

Naveilhan P, Hassani H, Lucas G, Blakeman KH, Hao JX, Xu XJ, Wiesenfeld-Hallin Z, Thoren P, Ernfors P (2001c) Reduced antinociception and plasma extravasation in mice lacking a neuropeptide Y receptor. Nature 409:513–517

Naveilhan P, Svensson L, Nystrom S, Ekstrand AJ, Ernfors P (2002) Attenuation of hypercholesterolemia and hyperglycemia in ob/ob mice by NPY Y_2 receptor ablation. Peptides 23:1087–1091

Oberto A, Panzica G, Altruda F, Eva C (2000) Chronic modulation of the GABA(A) receptor complex regulates Y_1 receptor gene expression in the medial amygdala of transgenic mice. Neuropharmacology 39:227–234

Pedrazzini T, Seydoux J, Kunstner P, Aubert JF, Grouzmann E, Beermann F, Brunner HR (1998) Cardiovascular response, feeding behavior and locomotor activity in mice lacking the NPY Y_1 receptor. Nature Med 4:722–726

Pierroz DD, Catzeflis C, Aebi AC, Rivier JE, Aubert ML (1996) Chronic administration of neuropeptide Y into the lateral ventricle inhibits both the pituitary-testicular axis and growth hormone and insulin-like growth factor I secretion in intact adult male rats. Endocrinology 137:3–12

Potter EK (1987) Presynaptic inhibition of cardiac vagal postganglionic nerves by neuropeptide Y. Neurosci Lett 83:101–106

Pralong FP, Gonzales C, Voirol MJ, Palmiter RD, Brunner HR, Gaillard RC, Seydoux J, Pedrazzini T (2002) The neuropeptide Y Y_1 receptor regulates leptin-mediated control of energy homeostasis and reproductive functions. FASEB J 16:712–714

Qian S, Chen H, Weingarth D, Trumbauer ME, Novi DE, Guan X, Yu H, Shen Z, Feng Y, Frazier E, Chen A, Camacho RE, Shearman LP, Gopal-Truter S, MacNeil DJ, Van der Ploeg LH, Marsh DJ (2002) Neither agouti-related protein nor neuropeptide Y is critically required for the regulation of energy homeostasis in mice. Mol Cell Biol 22:5027–5035

Raposinho PD, Broqua P, Hayward A, Akinsanya K, Galyean R, Schteingart C, Junien J, Aubert ML (2000) Stimulation of the gonadotropic axis by the neuropeptide Y receptor Y_1 antagonist/ Y_4 agonist 1229U91 in the male rat. Neuroendocrinology 71:2–7

Redrobe JP, Dumont Y, Fournier A, Quirion R (2002) The neuropeptide Y (NPY) Y_1 receptor subtype mediates NPY-induced antidepressant-like activity in the mouse forced swimming test. Neuropsychopharmacology 26:615–624

Redrobe JP, Dumont Y, Herzog H, Quirion R (2003) Neuropeptide Y (NPY) Y_2 receptors mediate behaviour in two animal models of anxiety: Evidence from Y_2 receptor knockout mice. Behav Brain Res 141: 251–255

Redrobe JP, Dumont Y, St-Pierre JA, Quirion R (1999) Multiple receptors for neuropeptide Y in the hippocampus: putative roles in seizures and cognition. Brain Res 848:153–166

Sabatino FD, Collins P, McDonald JK (1990) Investigation of the effects of progesterone on neuropeptide Y-stimulated luteinizing hormone-releasing hormone secretion from the median eminence of ovariectomized and estrogen-treated rats. Neuroendocrinology 52:600–607

Sainsbury A, Baldock PA, Schwarzer C, Ueno N, Enriquez RF, Couzens M, Inui A, Herzog H, Gardiner EM (2003) Synergistic reduction in adiposity and increase in bone mass in Y_2 / Y_4 receptor double knockout mice. Mol Cell Biol 23: 5225–5233

Sainsbury A, Cusin I, Doyle P, Rohner-Jeanrenaud F, Jeanrenaud B (1996a) Intracerebroventricular administration of neuropeptide Y to normal rats increases *obese* gene expression in white adipose tissue. Diabetologia 39:353–356

Sainsbury A, Herzog H (2001) Inhibitory effects of central neuropeptide Y on the somatotropic and gonadotropic axes in male rats are independent of adrenal hormones. Peptides 22:467–471

Sainsbury A, Rohner-Jeanrenaud F, Grouzmann E, Jeanrenaud B (1996b) Acute intracerebroventricular administration of neuropeptide Y stimulates corticosterone output and feeding but not insulin output in normal rats. Neuroendocrinology 63:318–326

Sainsbury A, Schwarzer C, Couzens M, Fitissov S, Furtinger S, Jenkins A, Cox HM, Sperk G, Hokfelt T, Herzog H (2002a) Important role of hypothalamic Y_2 receptors in body-weight regulation revealed in conditional knockout mice. Proc Natl Acad Sci USA 99:8938–8943

Sainsbury A, Schwarzer C, Couzens M, Herzog H (2002b) Y_2 receptor deletion attenuates the type 2 diabetic syndrome of ob/ob mice. Diabetes 51:3420–3427

Sainsbury A, Schwarzer C, Couzens M, Jenkins S, Oakes SR, Ormandy CJ, Herzog H (2002c) Y_4 receptor knockout rescues fertility in ob/ob mice. Genes Dev 16:1077–1088

Schenk F, Baron U, Rajewsky K (1995) A cre-transgenic mouse strain for the ubiquitous deletion of loxP-flanked gene segments including deletion in germ cells. Nucleic Acids Res 23:5080–5081

Schober DA, Van Abbema AM, Smiley DL, Bruns RF, Gehlert DR (1998) The neuropeptide Y Y_1 antagonist, 1229U91, a potent agonist for the human pancreatic polypeptide-preferring (NPY Y_4) receptor. Peptides 19:537–542

Schwartz MW, Baskin DG, Bukowski TR, Kuijper JL, Foster D, Lasser G, Prunkard DE, Porte DJ, Woods SC, Seeley RJ, Weigle DS (1996) Specificity of leptin action on elevated blood glucose levels and hypothalamic neuropeptide Y gene expression in *ob/ob* mice. Diabetes 45:531–535

Shi T, Zhang X, Berge O, Erickson J, Palmiter R, Hokfelt T (1998) Effect of peripheral axotomy on dorsal root ganglion neuron phenotype and autonomy behaviour in neuropeptide Y-deficient mice. Regul Peptides 75/76:161–173

Smith-White M, Herzog H, Potter E (2002a) Role of neuropeptide Y Y(2) receptors in modulation of cardiac parasympathetic neurotransmission. Regul Peptides 103:105–111

Smith-White MA, Herzog H, Potter EK (2002b) Cardiac function in neuropeptide Y Y_4 receptor-knockout mice. Regul Peptides 110:47–54

Thiele T, Koh M, Pedrazzini T (2002) Voluntary alcohol consumption is controlled via the neuropeptide Y Y_1 receptor. J Neurosci 22:208–212

Thiele T, Marsh D, Ste Marie L, Bernstein I, Palmiter R (1998) Ethanol consumption and resistance are inversely related to neuropeptide Y levels. Nature 396:366–369

Thorsell A, Michalkiewicz M, Dumont Y, Quirion R, Caberlotto L, Rimondini R, Mathe A, Heilig M (2000) Behavioral insensitivity to restraint stress, absent fear suppression of behavior and impaired spatial learning in transgenic rats with hippocampal neuropeptide Y overexpression. Proc Natl Acad Sci USA 97:12852–12857

Trivedi P, Yu H, Trumbauer M, Chen H, Van der Ploeg L, Guan X (2001) Differential regulation of neuropeptide Y receptors in the brains of NPY knock-out mice. Peptides 22:395–403

Tschenett A, Singewald N, Carli M, Balducci C, Salchner P, Vezzani A, Herzog H, Sperk G (2003) Reduced anxiety and improved stress coping ability in mice lacking neuropeptide Y-Y_2 receptors. Eur J Neurosci 18:143–148

Ueno N, Inui A, Iwamoto M, Kaga T, Asakawa A, Okita M, Fujimiya M, Nakajima Y, Ohmoto Y, Ohnaka M, Nakaya Y, Miyazaki JI, Kasuga M (1999) Decreased food intake and body weight in pancreatic polypeptide-overexpressing mice. Gastroenterology 117:1427–1432

Urban JH, Das I, Levine JE (1996) Steroid modulation of neuropeptide Y-induced luteinizing hormone releasing hormone release from median eminence fragments from male rats. Neuroendocrinology 63:112–119

Vezzani A, Michalkiewicz M, Michalkiewicz T, Moneta D, Ravizza T, Richichi C, Aliprandi M, Mule F, Pirona L, Gobbi M, Schwarzer C, Sperk G (2002) Seizure susceptibility and epileptogenesis are decreased in transgenic rats overexpressing neuropeptide Y. Neuroscience 110:237–243

Vezzani A, Sperk G, Colmers WF (1999) Neuropeptide Y: emerging evidence for a functional role in seizure modulation. Trends Neurosci 22:25–30

White DM (1997) Intrathecal neuropeptide Y exacerbates nerve injury-induced mechanical hyperalgesia. Brain Res 750:141–146

Widdowson PS, Wilding JP (2000) Hypothalamic neuropeptide Y and its neuroendocrine regulation by leptin. Front Hormone Res 26:71–86

Xu B, Sahu A, Crowley WR, Leranth C, Horvath T, Kalra SP (1993) Role of neuropeptide-Y in episodic luteinizing hormone release in ovariectomized rats: an excitatory component and opioid involvement. Endocrinology 133:747–754

Zhang Y, Lundeberg T, Yu L (2000) Involvement of neuropeptide Y and Y_1 receptor in antinociception in nucleus raphe magnus of rats. Regul Peptides 95:109–113

Structure–Activity Relationship of Peptide-Derived Ligands at NPY Receptors

K. Mörl · A. G. Beck-Sickinger

Institute of Biochemistry, University of Leipzig, Brüderstrasse 34, 04103 Leipzig, Germany
e-mail: moerl@rz.uni-leipzig.de

1	Introduction	480
2	NPY and Peptide Derived Ligands at the Y_1 Receptor	483
3	NPY and Peptide Derived Ligands at the Y_2 Receptor	490
4	NPY and Peptide Derived Ligands at the Y_4 Receptor	493
5	NPY and Peptide Derived Ligands at the Y_5 Receptor	495
6	NPY and Peptide Derived Ligands at the y_6 Receptor	497
7	Concluding Remarks	498
	References	498

Abstract NPY receptors belong to the large superfamily of heptahelical G-protein coupled receptors. They are part of a network consisting of receptor subtypes (Y_1, Y_2, Y_3, Y_4, Y_5, y_6) and their ligands, the hormones neuropeptide Y (NPY), peptide YY (PYY) and pancreatic polypeptide (PP). Structure–affinity and structure–activity relationship studies of peptide analogs are useful tools not only for the identification of the bioactive conformation of the ligands at the various receptor subtypes, but also for the development of analogs selective for the individual receptors. Such compounds will help to correlate biological actions to the interaction of the three hormones to the distinct receptor subtypes. In this article we summarize characteristic features corresponding to the specific ligand–receptor subtype interactions elaborated so far by structure-activity and structure-affinity studies.

Keywords Neuropeptide Y · Agonist · Antagonist · Receptor selectivity · Y Receptors

1
Introduction

Neuropeptide Y receptors belong to the large superfamily of heptahelical G-protein coupled receptors. Several NPY receptor subtypes have been characterized pharmacologically and presently five distinct types of NPY family receptors are cloned in mammals (Y_1, Y_2, Y_4, Y_5 and y_6). All of them are correlated to a distinct pharmacological profile. An additional putative receptor (Y_3) has so far only been described pharmacologically and no specific agonists or antagonists are known. The y_6 receptor has been found in mice and rabbits, but not in primates although its mRNA is present in various tissues. Its pharmacological profile is still controversial and no physiologically relevant actions could be attributed to the y_6 receptor (Michel et al. 1998).

These receptors are activated by a family of neuroendocrine hormones called the NPY hormone family, comprising neuropeptide Y (NPY), peptide YY (PYY) and pancreatic polypeptide (PP) (Cerda-Reverter and Larhammar 2000). All three peptides consist of 36 residues and are C-terminally amidated. X-ray crystallography of avian PP resulted in crystals of symmetric dimers, with the monomer consisting of an extended type II polyproline helix (residues 1–8) that is followed by a turn (residues 9–13) and an amphipathic α-helix (residues 14–31). Based on the high sequence homology, this three-dimensional hairpin-like structure, the so-called PP-fold, was assumed to be the structural feature common to the whole NPY family. Accordingly, although PYY and NPY are thought to be entirely monomeric at physiological concentrations, NMR studies revealed a PP-like fold for monomeric PYY (Keire et al. 2002, 2000a). A similar model was suggested for human NPY, a polyproline stretch (residues 1–10) followed by a tight hairpin structure (residues 11-14) and two short α-helices (residues15–26 and 28–35). N- and C-terminal ends are thought to be kept together in close proximity by hydrophobic interaction. Other NMR studies however highlight a dimeric NPY structure, which consists of an antiparallel, hydrophobic packing of two helical units (residues 11–36 or 15–36), with the structure of the N-terminal residues poorly defined (Cowley et al. 1992; Darbon et al. 1992; Glover et al. 1983; Nordmann et al. 1999). Recent studies exhibit the existence of dimers only at high, millimolar concentrations. Whereas the existence of an N-terminal polyproline helix was excluded, the C-terminal part of the molecule adopts an α-helical conformation which is stabilized in the dimer by intermolecular hydrophobic interactions. The α-helical segments are concomitantly present in parallel and antiparallel orientation in the dimer. Decrease of the concentration causes a shift of the dimerization equilibrium towards the monomeric form of NPY. Fluorescence resonance energy transfer studies used to investigate the conformation of the monomeric peptide could not support the existence of a PP-fold motif, characterized by a folding back of the N-terminal tail onto the C-terminal α-helix. Correspondingly, the solution structure of NPY in the presence of lipid mimetic dodecylphosphocholine (DPC) micelles revealed a conformation of micelle associated NPY with a flexible N-terminal segment (residue 1–13) and an α-helical C terminus (residues 14–36) oriented

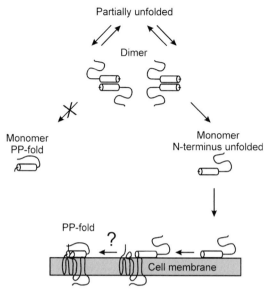

Fig. 1 Suggested model for receptor activation by NPY. Two different types of dimers exist in equilibrium with partially unfolded monomer in solution. The peptide binds to the cell membrane as a monomer and is brought to bind the receptor by two-dimensional diffusion. Changes in the secondary and tertiary structure upon interaction with the receptor are so far unclear

parallel to the micelle surface. The C-terminal residues 32–36 which are flexible in solution become α-helical in the membrane bound molecule (Bader et al. 2001; Bettio et al. 2002; Bettio and Beck-Sickinger 2002). Further changes in the secondary and tertiary structure upon interaction of the ligand with the corresponding receptor subtype could so far not be elucidated (Fig. 1).

Whereas NPY and PYY bind with high affinity to the Y_1, Y_2 and Y_5 receptor subtypes, the Y_4 receptor is preferentially bound by PP. NPY is widely distributed within the central and peripheral nervous system and is one of the most abundant neuropeptides in the brain. It is involved in the regulation of many physiological activities, such as food intake, energy expenditure, learning and memory, nociception, blood pressure, and circadian rhythms. PYY and PP are mainly synthesized and released by pancreatic and intestinal cells and act as hormones in an endo- and exocrine fashion, for example by regulating pancreatic and gastric secretion (Cerda-Reverter and Larhammar 2000; Michel et al. 1998). The wide range of actions influenced by the NPY family of peptides, together with the existence of multiple receptors with overlapping binding patterns for the three peptide ligands, makes it difficult to correlate the extent to which different receptor subtypes contribute to a distinct physiological effect. Different receptor subtypes may redundantly mediate identical actions in a given system. For example Y_1 and Y_5 receptors are both involved in the regulation of food intake and interplay of these receptors is suggested in this system (Inui 1999; Iyengar et al. 1999; Soll et al. 2001). The clarification and dissection of the complicated network of NPY receptor subtypes and their ligands demands the characterization of each receptor subtype relative to its interaction with the ligand, as well as to its biological function. Structure–affinity and structure–activity relationship studies of peptide analogs, combined with studies based on site-directed mutagenesis and anti-receptor antibodies are useful tools to approach this question.

The situation is even more complicated, since the Y_4 receptor is one of the most rapidly evolving G-protein receptors known and shows a very high species diversity compared to the Y_1 and Y_2 receptor subtypes. The variability is not only restricted to its sequence, but also concerns its pharmacology and tissue distribution. For example NPY is able to act as an antagonist on the guinea pig Y_4 receptor, but not on the rat Y_4 receptor, where it has no inhibitory effect on cAMP synthesis (Eriksson et al. 1998; Larhammar 1996; Lundell et al. 1996). Species-specific differences are also observed for the other receptors with regard to their distribution and density in the brain. This may point to different functional roles for the NPY receptors within each species (Dumont et al. 1993, 1998; Gackenheimer et al. 2001; Gehlert and Gackenheimer 1997; Gehlert et al. 1997; Widdowson 1993). To address this problem, not only specific antibodies directed against different receptor subtypes but also the development of selective ligands can help to localize the distinct receptor subtypes in a given tissue.

Taken together, structure–activity relationship studies are the basis for the identification of the bioactive conformation of a ligand and the understanding of its interaction at a distinct receptor subtype. For this purpose a variety of peptides has been designed, that are modifications of the natural ligands in primary, and consequently, secondary and tertiary structure. The analysis of their affinity and activity in connection with their conformation gives insight into receptor binding properties and is informative for the pharmacological characterization of the receptor subtypes. It will help to characterize signal transduction pathways connected to the distinct receptor subtypes as well as their physiological role. In addition selective ligands for each receptor subtype would be useful tools to localize the distribution of each receptor in vivo.

Exchange of single or multiple amino acids, introduction of conformationally constraining amino acids, or C-terminal, N-terminal or central truncations of the native peptide segment are used to create receptor-selective peptide analogs. N-terminal, C-terminal or internal truncation identifies domains of a natural ligand, which are important for the receptor binding. Single exchanges of each residue with L-alanine, so called L-Ala scans, are a first step towards the valuation of the importance of each position with respect to binding affinity. They allow clarification of the significance of chemical properties of the corresponding side chains at these positions. Additional D-isomer scans give information on the role of the orientation of a certain side chain. The introduction of special amino acid units, spacer templates, cyclization or backbone modification can limit the conformational space available to a peptide chain. If the induced or stabilized conformation is similar to that adopted from the native ligand when it interacts with a single receptor subtype this will lead to selectivity. If not, there will be loss of affinity at all receptor subtypes.

2
NPY and Peptide Derived Ligands at the Y_1 Receptor

The Y_1 NPY receptor subtype was the first member of the PP-fold peptide receptor family to be cloned. This subtype displays high affinity for both PYY and NPY with little affinity for PP (Herzog et al. 1992; Larhammar et al. 1992). The contribution of each amino acid side chain of NPY to the receptor binding was investigated by exchanging systematically each single residue of NPY by L-Ala. This L-Ala scan revealed that the most sensitive positions in this receptor are Pro^2, Pro^5, Arg^{19}, Tyr^{20} and the C-terminal decapeptide, namely positions 27–36. In particular the two arginine residues at positions 33 and 35 turned out to be the most important residues for Y_1 binding, since their replacement by alanine provoked the most efficient loss of affinity (Beck-Sickinger et al. 1994b; Eckard and Beck-Sickinger 2001) (Table 1).

The importance of the tyrosine amide at position 36 of NPY for binding to the Y_1 receptor was confirmed by the investigation of the affinity of NPY analogs containing different chemical modifications at the C-terminus: Even the change of the peptide amide to the corresponding peptide acid resulted in a complete loss of affinity at the Y_1 receptor (Hoffmann et al. 1996).

Table 1 Results of the l-Ala scan: effects of single amino acid replacement on the binding affinity of NPY at the human Y-receptors (in parentheses the affinity of the peptide divided by the affinity of pNPY is given)

Peptide	hY_1		hY_2		hY_4		hY_5	
	IC_{50} [nM]	IC_{50} (Pep) IC_{50} (NPY)	IC_{50} [nM]	IC_{50} (Pep) IC_{50} (NPY)	IC_{50} [nM]	IC_{50} (Pep) IC_{50} (NPY)	IC_{50} [nM]	IC_{50} (Pep) IC_{50} (NPY)
pNPY	0.23	(1)	0.04	(1)	5.5	(1)	0.8	(1)
$[A^1]$-pNPY	21	(91)	0.2	(5)	5.8	(1.1)	2.2	(2.8)
$[A^2]$-pNPY	114	(496)	0.3	(8)	7.8	(1.4)	5.5	(7)
$[A^5]$-pNPY	228	(991)	24	(600)	25	(4.5)	32	(40)
$[A^8]$-pNPY	32	(139)	0.7	(18)	60	(11)	55	(69)
$[A^{11}]$-pNPY	8	(35)	0.2	(5)	3.1	(0.6)	0.5	(0.6)
$[A^{13}]$-pNPY	7.5	(33)	0.1	(3)	37	(6.7)	17	(21)
$[A^{19}]$-pNPY	282	(1226)	1.6	(40)	4.1	(0.7)	1.4	(1.8)
$[A^{20}]$-pNPY	71	(309)	1.2	(30)	161	(29)	19	(24)
$[A^{21}]$-pNPY	5.5	(24)	0.2	(5)	66	(12)	32	(40)
$[A^{25}]$-pNPY	11	(48)	0.7	(18)	201	(37)	80	(100)
$[A^{27}]$-pNPY	250	(1087)	1.4	(35)	340	(62)	370	(463)
$[A^{28}]$-pNPY	64	(278)	0.1	(2.5)				
$[A^{29}]$-pNPY	58	(252)	1.4	(35)				
$[A^{30}]$-pNPY	26	(113)	0.1	(2.5)				
$[A^{31}]$-pNPY	365	(1587)	0.3	(7.5)				
$[A^{32}]$-pNPY	723	(3143)	45	(1125)	380	(69)	7.7	(9.5)
$[A^{33}]$-pNPY	>1000	(>4348)	54	(1350)	>1000	(>182)	94	(118)
$[A^{34}]$-pNPY	94	(409)	6	(150)	7.4	(1.3)	1.3	(1.6)
$[A^{35}]$-pNPY	>1000	(>4348)	>1000	(>25000)	>1000	(>182)	>1000	(>1250)
$[A^{36}]$-pNPY	970	(4217)	48	(1200)	141	(26)	68	(85)

A D-amino acid scan of NPY showed an affinity profile very similar to that of the L-Ala scan (Kirby et al. 1993a). This suggests that the most important positions are sensitive not only to the exchange of the side chain but also to the orientation of the side chain itself.

To gain further insight into structure–affinity relationship, residues identified as sensitive in the L-Ala and D-amino acid scans are further exchanged with other amino acids: Substitution of Pro^2 and Pro^5 by L-Ala led to a more than 500- or 1,000-fold lower affinity at the Y_1 receptor, respectively, and the introduction of the more hydrophobic side chain of leucine or phenylalanine at these positions improved the binding only partially. Thus at both positions not only hydrophobicity but also the presence of a turn inducing residue seems to be required. The mutation $Tyr^{20}Ala$ and $Tyr^{27}Ala$ also led to a 450- or more than 1,000- fold decrease, respectively, in affinity while here the introduction of the hydrophobic residues Bpa (p-benzoyl-L-phenylalanine) or phenylalanine still resulted in good affinity. This suggests that the hydrophobic character of the side chain at this position is important for the ligand to adopt the bioactive conformation. Together $Pro^{2,5}$ and $Tyr^{20,27}$ may be important in stabilizing the hairpin-like structure of NPY by hydrophobic interaction. In agreement with this observation, perturbation of the PP-fold by substitution of Tyr^{20} with the helix breaking amino acid proline led to a loss of binding to Y_1 receptors (Fuhlendorff et al. 1990b). Hydrophobicity also is required at position 28 since here the substitution of isoleucine by alanine resulted in a more than 300-fold decrease in affinity. The side chain of Asn^{29} might play a role in interacting with the receptor or in stabilizing the bioactive conformation since introduction of alanine resulted in 290-fold lower affinity, whereas replacement of asparagine with its homologue glutamine led to an affinity that was only 55-fold lower than that of NPY. At position Leu^{30} the chemistry, size and orientation of the side chain are all of major importance for the binding, since here replacement by alanine, phenylalanine or D-tryptophan all reduced the binding affinity to the receptor. Replacement of Gln^{34} by alanine led to a more than 300-fold decrease in affinity. However, the introduction of Pro^{34} gave a ligand which was as potent as NPY at the Y_1 receptor, in contrast to the Y_2 receptor. This suggests that the turn inducing residue at this position favors the bioactive conformation of the peptide at the Y_1 receptor. Furthermore, the orientation of the C-terminal turn turned out to be important, as the analog containing D-Pro^{34} even bound with lower affinity than the $Gln^{34}Ala$ mutant (Beck-Sickinger et al. 1994b; Cabrele and Beck-Sickinger 2000). In agreement with this, substitution of Glu^{34} by proline in PYY does not alter the binding at the Y_1 receptor, but abolishes binding to the Y_2 receptor (Keire et al. 2002). A hydrophobic side chain seems to be favored at the C terminus of NPY. The introduction of Ala^{36} in place of Tyr^{36} led to a severe loss of affinity, which could be limited by the introduction of Phe^{36} at this position. However, incorporation of the large, highly hydrophobic residue Bpa or the imidazole ring of histidine were poorly tolerated, showing that the size of the hydrophobic side chain is determinant as well. In contrast, at the N-terminus, Bpa and alanine substitution of tyrosine at position 1 resulted in analogs that had the same reduced affinity. This suggests that here the loss of affinity is due to the lack of the phenolic group of Tyr^1, which might be involved in the formation of an intramolecular hydrogen bond. However, the comparably good affinity of Bpa^1-NPY is also indicative of the flexibility and spatial availability of the N-terminus of NPY (Beck-Sickinger et al. 1994b; Cabrele and Beck-Sickinger 2000).

To develop Y_1 receptor specific agonists, the results emerging from the scans mentioned above are also important in distinguishing crucial residues from those which might be modified without impairing Y_1 receptor binding. The later might be replaced to gain agonists with improved receptor selectivity, without loss of Y_1 receptor affinity. Accordingly almost all Y_1 analogs are synthesized with proline in position 34. While the presence of proline is tolerated at the Y_1 and Y_5 receptor subtypes, and even enhances the affinity for the Y_4 receptor, it leads to complete loss of affinity at the Y_2 receptor. It is suggested that Pro^{34} disrupts the C-terminal helix that is required for Y_2 binding, but does not disturb the juxtaposition of the N- and C-terminus and therefore allows potent Y_1 binding (Cabrele and Beck-Sickinger 2000; Fuhlendorff et al. 1990a; Keire et al. 2000b, 2002). Similarly modifications at positions 6 and 7 were interesting because of their moderate influence on the binding of NPY to the Y_1 receptor, as shown by the L-Ala scan. Variations at these positions yielded the agonists [Phe^7,Pro^{34}] pNPY and [Arg^6,Pro^{34}] pNPY with significant Y_1 receptor preference (Beck-Sickinger et al. 1994b; Soll et al. 2001). Another example is the substitution of the corresponding D-amino acid at either position 25 or 26 of NPY, which results in loss of Y_2 affinity but maintains Y_1 affinity. The resulting agonists are also selective for the Y_1 receptor relative to the Y_4 and Y_5 receptors. The loss of Y_2, Y_4 and Y_5 receptor binding upon changing the chirality of a single amino acid in the C-terminal α-helical segment of the molecule suggests that these receptors have more stringent requirements for the native conformation of this region of NPY than the Y_1 receptor. Such agonists are promising tools for the evaluation of the physiological role of the Y_1 receptor in vivo. For example it was shown that these two compounds stimulate food intake dose responsively in Long-Evans rats for at least 4 h after intracerebroventricular administration (Boublik et al. 1989; Kirby et al. 1993a; Mullins et al. 2001).

Also the results of the L-Ala scan showed that single amino acid exchanges at central positions of NPY are not essential for binding to the Y_1 receptor, the replacement of larger segments in the central region led, in almost all cases, to a significant reduction or complete loss of affinity at the Y_1 receptor (Rist et al. 1995). Analogs containing the N- and C-terminal NPY segments connected by a spacer, i.e., 6-amino hexanoic acid (Ahx), showed only moderate affinity. Only the putative β-turn (amino acids 10–17) constituting the PP-fold can be deleted without loss of affinity, provided that the N- and C-terminal helices are held in close proximity by a covalent linkage. Furthermore the location of the disulfide bridge and the chirality of the cysteine residues influence the affinity. For maintaining Y_1 receptor affinity a disulfide linkage between Cys^7 and Cys^{21} was favored above other interchain bridges examined (Kirby et al. 1993b, 1995, 1997). These observations were combined with an exchange of Pro^{34} to decrease the affinity at the Y_2 receptor and D-His^{26} to decrease affinity at the Y_4 receptor. The resulting cyclic peptide Des-AA$^{11-18}$[Cys^{7,21},D-Lys^9(Ac), D-His^{26}, Pro^{34}]NPY showed excellent selectivity and adequate affinity for the Y_1 receptor. It represents an agonist of NPY at this receptor with orexigenic activity in vivo (Mullins et al. 2001).

Other investigations are based on the comparison of structural features of the members of the NPY hormone family. Albeit the common hairpin like structure, NPY and PP reveal a completely different affinity at the Y_1 receptor, which is 500-fold lower for PP compared to NPY. This might be due to a different folding of the N- and C-terminal segments in the two peptides, leading to a more stable helical conformation in PP than in NPY. Chimeras based on the combination of NPY and PP segments are used as a tool to answer the question whether the primary structure is crucial for the peptide either to fold in a specific way or to be recognized from the various receptor subtypes. Substitution of the pig NPY sequence 19-23 (RYYSA) by the corresponding human or rat PP (QYAAD/QYETQ) led to a decrease in affinity and was structurally characterized by helix destabilization and a higher content of unordered structure, as suggested by circular dichroism (CD) studies. The poorest Y_1 receptor analog in these studies [rPP$^{19-23}$,H^{34}] pNPY was characterized by a helical content of only 9%, compared to 18% helicity in pNPY. Some of the lost affinity was recovered by the introduction of proline at position 34 in place of glutamine, indicating that the presence of a turn inducing element at the C-terminus may favor the binding to the Y_1 receptor. This is in agreement with the results of the single amino acid replacements. These results suggest that the central region of NPY plays an important role in inducing and/or stabilizing the helical peptide conformation and consequently confers high affinity. Inverse exchanges, meaning human PP (hPP) analogs containing the central NPY sequence, increased the hPP affinity at the receptors Y_1, Y_2 and Y_5. The hPP analogs are all as highly helical as the unmodified hPP; however, they probably adopt a different tertiary structure, which may be more similar to that of NPY than that of hPP. Thus the central sequence might also play a role in the formation and orientation of the C-terminal helix with respect to the N-terminus (Cabrele et al. 2001).

Introduction of the hPP segment 1-7 or 1-17 in pNPY resulted in potent analogs. [hPP$^{1-7}$] pNPY showed a CD spectrum that was very similar to that of NPY, and [hPP$^{1-17}$] pNPY was characterized by a highly stable helix (64% versus 18% found for NPY). These results support the hypothesis that the N-terminal part of the molecule is important for the stabilization of the C-terminal helix, especially by interdigitation of the Pro^2 and Pro^5 with the tyrosine side chains at positions 20 and 27, since these positions are conserved between pNPY and hPP (Cabrele et al. 2001).

The simultaneous presence of the pNPY sequences 1-7 and 19-23 led to an analog of hPP that is as potent as NPY at the Y_1 receptor, comparable to hPP at the Y_4 receptor and even better than NPY at the Y_5 receptor. This suggests that the NPY segments 1-7 and 19-23 drive the formation of the bioactive conformation of the ligand (Cabrele et al. 2001).

The exchange of the C-terminal hexapeptide, which is very conserved among PP-fold peptides, led to the development of [Leu^{31},Pro^{34}]NPY and [Leu^{31}, Pro^{34}] PYY. This analog lost affinity for the Y$_2$ receptor and had originally been described as a Y$_1$ selective agonist. However, after the identification of the Y$_4$ and Y$_5$ receptor subtypes, it was found that [Leu^{31},Pro^{34}]NPY and [Leu^{31}, Pro^{34}]-PYY also have significant affinity and potency at these receptors (Dumont et al. 1994; Fuhlendorff et al. 1990a; Gerald et al. 1996; Hu et al. 1996; Michel et al. 1998).

The most striking feature of the Y$_1$ receptor as compared to the other receptor subtypes is its low affinity for analogs of NPY and PYY lacking the N-terminal sequence. Deletion of the N-terminal tyrosine in NPY already resulted in a 75-fold decrease in affinity. Increasing the extent of the deletion further decreased the affinity of the analogs to the micromolar range (Beck-Sickinger and Jung 1995). Despite this observation a C-terminal cyclic peptide of NPY, YM-42454 (Ac-Cys-Leu-Ile-Thr-Arg-Cys-Arg-Tyr-NH$_2$) showed high affinity and selectivity for Y$_1$ versus Y$_2$ receptors in SK-N-MC cells. Amino acid replacement revealed similar crucial positions for YM-42454 as it had been shown for the full-length NPY molecule: The hydrophobic side chains of Leu^{30} and Ile^{31}, the guanidinium groups of Arg^{33} and Arg^{35} and the C-terminal amide are critical for the binding affinity of YM-42454 to the Y$_1$ receptor. Accordingly the binding mode of YM-42454 to the Y$_1$ receptor might be similar to that of the C terminus of NPY. ^{1}H-NMR studies for YM-42454 and its derivatives have suggested that the critical residues are involved in the direct interaction with the Y$_1$ receptor rather than in maintaining the bioactive conformation (Takebayashi et al. 2000). On the other hand the conformation of peptide analogs that mimic the C-terminal region of NPY might play an important role in modifying the affinities at NPY receptors: Peptides corresponding to the C-terminal nonapeptide of NPY, with modifications at position 30, 32 and 34 were found to antagonize the Y$_1$ receptor. The high activity of these analogs compared to the corresponding unmodified NPY segment (amino acid 28–36) was correlated to their ability to adopt a stable helix, which is initiated by the turn-inducing sequence Asn^{29}–Pro^{30}. Surprisingly some of these nonapeptides, which were covalently linked by lactam or disulfide bridges, bound to the receptor with subnanomolar affinity. One example is the dimer of Ile-Glu-Pro-Dpr-Tyr-Arg-Leu-Arg-Tyr-NH$_2$ that contains two interchain lactam bridges between glutamine and Dpr (2,3-diaminopropionic acid). This antagonist is known as GW1229, GR231118 or 1229U91, and binds to the Y$_1$ receptor in the picomolar range (Daniels et al. 1995). It has therefore found widespread application in defining the role and distribution of Y$_1$ receptors (Dumont and Quirion 2000; Ishihara et al. 1998; Kanatani et al. 1996). However, GW1229 also turned out to be a potent agonist at the Y$_4$ receptor, which limits its use as a pharmacological tool (Kanatani et al. 1998; Parker et al. 1998; Schober et al. 1998, 2000). Replacement of the C-terminal amides of GW1229 with methyl ester or replacement of the peptide bond between Arg^{35} and Tyr^{36} with Ψ(CH$_2$-NH) deduced compounds that retained subnanomolar affinity for Y$_1$ receptors and exhibited high selectivity for Y$_1$ receptors relative to Y$_2$, Y$_4$ and Y$_5$ receptors. These compounds were confirmed to be potent and selective Y$_1$ receptor compet-

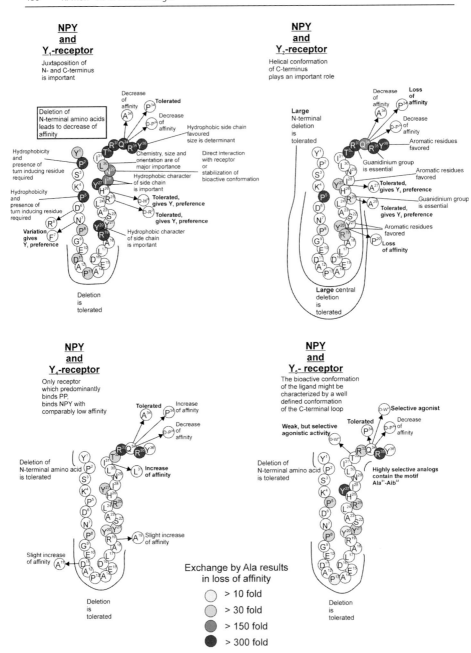

Fig. 2 Summary of amino acid modifications in NPY and their effect on binding to the human receptors Y_1, Y_2, Y_4 or Y_5

itive antagonists with respect to forskolin-stimulated cAMP synthesis in cell culture. It is suggested that, due to the loss of hydrogen bonding ability, the C-terminal modifications induce conformational changes and consequently provoke receptor selectivity (Balasubramaniam et al. 2001).

The only linear, conformationally constrained and Y_1 receptor selective analog contains β-amino cyclopropyl residues at position 34. In this case even the C-terminal octapeptide shows significant affinity (Koglin et al. 2003).

In conclusion, the most striking feature of the Y_1 receptor compared to the other subtypes is its low affinity for analogs of NPY and PYY lacking the N-terminal amino acid. Binding to the Y_1 receptor supposes intramolecular interactions between the juxtaposed N- and C-termini to stabilize the tertiary structure of the ligand. The central region of the ligand is comparably insensitive to single amino acid replacement, yet it influences the binding affinity of the ligand. This part of the peptide confers high affinity by inducing and/or stabilizing the C-terminal α-helical segment. It is also of importance to orientate the C terminus with respect to the N terminus. Whereas the C-terminal NPY residues Arg^{33} and Arg^{35} are essential for high-affinity binding to the receptor, the analogs $[Pro^{34}]NPY$ and $[Leu^{31},Pro^{34}]NPY$ retain full activity on the Y_1 receptor, despite the disruption of the C-terminal helix. Surprisingly short cyclic peptides or nonapeptides linked by lactam or disulfide bridges, which are imitating the C-terminal region of NPY, or linear peptides containing constraint amino acids, bind with high affinity to the receptor and even led to the development of highly selective Y_1 receptor antagonists.

The effects of modifications in the peptide sequence of NPY on the binding to the receptor subtypes are illustrated in Fig. 2 and examples for peptide derived ligands with selective binding affinity are given in Table 2.

Table 2 Peptide-derived ligands with selective binding affinity for the NPY receptors Y_1, Y_2 and Y_5

Y_1	$[Arg^6,Pro^{34}]NPY$	Agonist	Söll et al. 2001
	$[Phe^7,Pro^{34}]NPY$		
	$[d\text{-}Arg^{25}]NPY$		Mullins et al. 2001
	$[d\text{-}His^{26}]NPY$		
	$Des\text{-}AA^{11\text{-}18}[Cys^9(Ac),d\text{-}His^{26},Pro^{34}]NPY$		
	GW1229 with C-terminal methyl ester	Antagonist	Balasubramaniam et al. 2001
	GW1229 with peptide bond between $Arg^{35}\text{-}Tyr^{36}$ replaced with $\psi(CH_2\text{-}NH)$		
Y_2	$[Ala^{25}]NPY$	Agonist	Eckard and Beck-Sickinger 2001
	$[Ala^{27}]NPY$		
	NPY(13–36)		Wahlestedt et al. 1986
	$[Ahx^{5\text{-}24}]NPY$		Beck et al. 1989
	cyclo-(2/30)$[Ahx^{5\text{-}24},Glu^2,Lys^{30}]NPY$		Beck-Sickinger et al. 1992
Y_5	$[d\text{-}Trp^{34}]NPY$	Agonist	Parker et al. 2000
	$[Ala^{31},Pro^{32}]NPY$		Cabrele et al. 2002
	$[Ala^{31},Aib^{32}]NPY$		Cabrele et al. 2000
	$[cPP^{1\text{-}7},NPY^{19\text{-}23},Ala^{31},Aib^{32},Gln^{34}]PP$		Cabrele et al. 2000

3
NPY and Peptide Derived Ligands at the Y_2 Receptor

Like the Y_1 receptor, the Y_2 receptor displays high affinity for NPY and PYY. However, differences in binding affinities for ligand analogs can be observed.

The L-Ala scan of NPY showed that the exchange of single residues by L-alanine affects binding to the Y_2 receptor less than binding to the Y_1 receptor. With only a few exceptions, the exchange of single residues by L-alanine only provoked a 2–20-fold loss of affinity to the Y_2 receptor in human neuroblastoma cell lines: The substitution Pro^5Ala was the only one in the N-terminal part that led to a more severe change, by reducing the binding affinity 600-fold. Exchange of positions Arg^{19}, Tyr^{20}, Tyr^{27} and Asn^{29} in the central region of NPY resulted in a 30–40-fold reduction of affinity. Highest sensitivity was located to the C-terminal positions 32–36, where the mutation $Tyr^{34}Ala$ resulted in a 150-fold loss of affinity and substitution of Thr^{32}, Arg^{33}, Arg^{35} and Tyr^{36} provoked a loss of affinity more than 1,000-fold (Beck-Sickinger et al. 1994b; Eckard and Beck-Sickinger 2001). The activity of several analogs was tested in mucosal preparations of rat jejunum, where NPY exerts an antisecretory function. All analogs showed agonistic potential, with the only exception of $[Ala^{34}]NPY$ that was inactive despite the observation that binding affinity to the human receptors was reduced only 150-fold (Cox et al. 1998). In comparison the $[Ala^5]NPY$ mutant that provoked a 600-fold loss of affinity still showed 40% activity. The replacement of Pro^8, Pro^{13}, Tyr^{27}, Ile^{28} and Asn^{29} resulted in analogs with high Y_2 receptor affinity, but only partial activity. These differences might be attributed to species-specific differences in receptor binding. Despite this observation the relative activities of the Ala-NPY analogs closely resemble their order of affinity to the Y_2-expressing human neuroblastoma cell line (Beck-Sickinger et al. 1994b; Cabrele and Beck-Sickinger 2000) (Table 1).

The L-Ala scan, as well as the D-amino acid scan showed that the most important part of NPY as a ligand at the Y_2 receptor may be located in the C-terminal fragment. The side chains in this region apparently play a role in either stabilizing the bioactive conformation of the ligand or in directly interacting with the receptor. Interestingly, the orientation of the aromatic side chain at position 36 is not very important, as demonstrated by the still high affinity of the analog $[D-Tyr^{36}]NPY$ (Kirby et al. 1993a).

Furthermore the presence of an aromatic residue is favored at positions Tyr^{20}, Tyr^{21}, Tyr^{27} and is most evident at position Tyr^{36}, since here the replacement with phenylalanine or Bpa resulted in minor changes compared to replacement by alanine. As already mentioned, replacement of Gln^{34} by proline is tolerated at the Y_1 receptor, but leads to loss of affinity at the Y_2 receptor, by disrupting the C-terminal α-helix (Keire et al. 2002, 2000b; Nordmann et al. 1999). Similarly substituting Tyr^{20} by proline reduced the α-helical content of NPY and diminished affinity to the Y_2 receptor (Fuhlendorff et al. 1990b). Substitution of Arg^{25} and Arg^{33} and particularly Arg^{35} with lysine shows the importance of the guanidinium group at these positions for functional activity (Fournier et al. 1994).

Comparable to the Y_1 receptor, a centrally truncated analog composed of residues 1–4 linked by a spacer molecule to residues 31–36, did not show binding to Y_2 receptors or biological activity (Fournier et al. 1994). Nevertheless, in contrast to the Y_1 receptor, the Y_2 receptor is able to bind the centrally truncated analog [Ahx$^{5-24}$]NPY with high affinity (Beck et al. 1989; Eckard and Beck-Sickinger 2001). The binding affinity of this analog could be slightly improved by introducing the hydrophobic residue Tic (tetrahydroisoquinoline-3-carboxylic acid) at position1 or by using Pac (1-phenyl-2-aminomethyl-cyclopropanoic acid) as a spacer molecule (Beck-Sickinger et al. 1994a). Cyclization of the linear peptide in cyclo-(2/30)[Ahx$^{5-24}$,Glu^2,Lys^{30}]-NPY slightly reduced the binding affinity to the Y_2 receptor, but resulted in higher activity with respect to the inhibition of cAMP accumulation in cell culture experiments. Unlike the linear peptide, the cyclopeptide was unable to increase blood pressure via the Y_1 receptor in rats. This suggests that cyclization improves Y_2 selectivity by conferring the right orientation of the C-terminus, which is required to activate the Y_2 but not the Y_1 receptor subtype (Beck-Sickinger et al. 1992; Cabrele and Beck-Sickinger 2000).

The influence of the central region also becomes evident by the introduction of the pNPY segment 19–23 in hPP, which resulted in Y_2 receptor affinity in the nanomolar range. As expected from the substitution of single amino acids in NPY, the replacement of Pro^{34} by glutamine in the N-terminal PP segment of this analog further increased its binding potency (Cabrele et al. 2001). Exchange of the PP-fold of NPY with the corresponding segment of PP in the chimeric molecule PP(1–30)-NPY(31–36) resulted in binding affinity to Y_2 receptors which was comparable to that of NPY. This indicates that the PP-fold is an important but purely structural element in the interaction of NPY with the Y_2 receptor (Fuhlendorff et al. 1990b).

Unlike the Y_1 receptor the Y_2 receptor is relatively resistant to N-terminal deletion: Deletion of Tyr^1 and Tyr^1-Pro^2 does not affect the binding to the Y_2 receptor and even shorter fragments [e.g., NPY(13–36), NPY(18–36) and NPY(22–36)] still bind the receptor with subnanomolar affinity (Beck-Sickinger and Jung 1995; Fuhlendorff et al. 1990b; Wahlestedt et al. 1986). In contrast the acetylated dodecapeptid NPY(25–36) showed only minor affinity. However, the binding affinity of this analog could be improved by the introduction of amino acids with hydrophobic and conformationally restricted side chains. For example compounds with cyclohexylalanine (Cha) or Tic at position Leu^{30} and/or Cha, Tic or β-naftyl-alanine (Nal) at position Ile^{31} are able to strongly activate the Y_2 receptor in a rat vas deferens assay. Together with the finding that [Tyr^{32},Leu^{34}] NPY(25–36) binds to the Y_2 receptor with nanomolar affinity, this supports the hypothesis that large and hydrophobic side chains in the region 28–32 favor Y_2 receptor binding. Further shortening of these analogs showed that the nonapeptide is the minimal sequence required for receptor binding (Cabrele and Beck-Sickinger 2000).

Another possibility to obtain C-terminal analogs with full agonistic properties is the stabilization of the dodecapeptide by the introduction of lactam bridges. The binding affinity of the resulting analogs is dependent on the position, the orientation and the length of the lactam bridges, with the most potent ligands corresponding to lactamizations Glu^{27}–Lys^{31} and Lys^{28}–Glu^{32}. Two-dimensional NMR and molecular dynamics studies showed that high affinity in lactam bridged analogs is correlated with the presence of a short helix (residues 29–34) ending with a turn at the C-terminus and facing to the N-terminal fragment as in a hairpin-like structure (Rist et al. 1997). The importance of a helical structure was also shown for the agonist N-acetyl [Leu^{28,31}]NPY(24–36), where substitutions were introduced to stabilize the α-helical structure of the molecule. The presence of the amphipathic α-helix extending from residue 24 to residue 32 in this molecule is crucial for potent cardiac vagal inhibition, representing binding to Y$_2$ receptors. In addition, some degree of flexibility must be retained within the C-terminal extension (Barden et al. 1994). Furthermore it was shown that, according to the L-Ala scan of the full-length peptide, Arg^{25}, Arg^{33}, Arg^{35}, Asn^{29} and Tyr^{36} are critical residues for maintaining the functional activity of the C-terminal peptide. Whereas a maximum response of cardiac vagal inhibition could be reached for some analogs, the duration of the response was decreased for all active analogs. This might be explained by either enhanced sensitivity of the NPY analogs to enzyme degradation or by the absence of additional conformational parameters in the truncated peptides. As expected these analogs are active at the Y$_2$ receptor but do not bind to the Y$_1$ receptor, whereas binding to the receptors Y$_4$, Y$_5$ and y$_6$ remains to be tested (Smith-White and Potter 1999). Furthermore, the occurrence of self-association of this peptide has been described in aqueous trifluoroethanol but its meaning for the formation of the biologically active conformation remains to be established (Barnham et al. 1999; Potter et al. 1994).

To summarize, in comparison to the Y$_1$ receptor, the Y$_2$ receptor is less affected by single amino acid replacements in the L-Ala scan. However, high-affinity binding to this receptor has more stringent requirements on the C-terminal region of the ligand. The side chains in this region either stabilize the bioactive conformation of the ligand or are involved in direct interactions with the receptor. The helical conformation of the C-terminus plays an important role for binding to this receptor. For example, the insertion of proline at position 34 of NPY disrupts the C-terminal helix and impedes ligand binding to the Y$_2$ receptor in contrast to the other receptors. Unlike the Y$_1$ receptor, the Y$_2$ receptor still binds N-terminally truncated fragments like NPY(13–36) or NPY(22–36) and centrally truncated analogs like [Ahx$^{5-24}$]NPY with high affinity. The N-terminal and central regions are of importance in stabilizing and orientating the functional part of the molecule represented by the C-terminus (Fig. 2 and Table 2).

4
NPY and Peptide Derived Ligands at the Y_4 Receptor

The Y_4 receptor is preferentially bound by the PP compared to NPY and PYY. However, at least for the human and guinea pig Y_4 receptors, binding of NPY and PYY may still occur in the physiological range (Berglund et al. 2001; Cerda-Reverter and Larhammar 2000). As already mentioned in the beginning of this chapter the Y_4 receptor is one of the most rapidly evolving G-protein coupled receptors and comparably high sequence diversity between species can be observed. Despite this, PP from human (h), cow (b), rat (r) and guinea pig (gp) bind equally well to human and rat Y_4 receptors. For bPP a slightly higher affinity (two- to threefold) for the gpY_4 receptor could be observed with respect to hPP, rPP and gpPP (Berglund et al. 2001).

An L-Ala scan of NPY binding at human, rat and guinea pig Y_4 receptors revealed that, comparable to the Y_2 receptor, the C-terminus is the most sensitive part of the molecule (Table 1). As it had already been shown for the other receptors Arg^{33} and Arg^{35} provoke the most pronounced loss of binding affinity at the Y_4 receptor. The effect of the Arg^{35}Ala mutation on the guinea pig receptor was slightly less dramatic than on human and rat receptors. Furthermore, Gln^{34} is of interest as it is the only amino acid which differs between NPY and PP segments 33–36. Substitution of this residue by proline, which is the corresponding amino acid in PP, increased the affinity 200-fold for the rat Y_4 receptor (Lundell et al. 1996). However, binding to human and guinea pig receptors was unaffected by this modification (Berglund et al. 2001). Replacement of Gln^{34} by leucine increased the affinity 18-fold, but replacement by D-proline induced a 49-fold loss of affinity at the human Y_4 receptor (Cabrele and Beck-Sickinger 2000). The introduction of alanine instead of aspartic acid at position 11 and arginine at position 19 slightly increased the affinity, which might be attributed to the removal of charged side chains. In PP the polar but neutral side chains of glutamine and asparagine are present at these positions and are thought to be involved in direct interactions with the receptor. In contrast the charged side chains of NPY at these positions might lead to electrostatic repulsion (Cabrele and Beck-Sickinger 2000).

Although C-terminal fragments of NPY, PYY and PP have reduced binding potency compared with the native peptides, bPP(13–36) displays only 13-fold lower affinity to the human Y_4 receptor (Balasubramaniam et al. 1990). Interestingly it was shown that the Y_1 receptor antagonist GW1229, a dimer of modified nonapeptides corresponding to the C terminus of NPY, linked by two interchain lactam bridges, represents a potent agonist at the Y_4 receptor (Kanatani et al. 1996; Parker et al. 1998; Schober et al. 1998).

In contrast to N-terminal deletions, truncation of the central sequence in analogs like $[Ahx^{5-24}]$-pNPY reduces the affinity to human, rat and guinea pig receptors more than 100-fold. Interestingly deletion of the central residues 8–20 only affected binding to human and guinea pig Y_4 receptors, but not to the rat receptor. Confining the deletion to $[Ahx^{9-17}]$pNPY, which still leads to a 65-fold decrease at the Y_1 receptor, only provokes a 8.2-fold loss of binding affinity at the human Y_4 receptor. Correspondingly deletion of central sequences 5–24 or 5–20 of hPP reduces the affinity even more (3600- and 5400-fold) and could be improved only partially by substitution of Ahx by tyrosine (Berglund et al. 2001; Cabrele and Beck-Sickinger 2000; Eriksson et al. 1998; Gehlert et al. 1996). Thus the presence of the central part induces and stabilizes the bioactive conformation of NPY as well as PP. However, species-specific differences could be observed concerning the degree of the central deletion tolerated.

The role of the central region was further investigated by studying binding affinities of PP/NPY chimeras: introduction of the central hPP segment 19–23 into pNPY resulted in a 181-fold loss of affinity versus the full length NPY peptide and could only be recovered partially by the replacement Gln^{34}Pro. The exchange of the central peptide between the two peptides might induce conformational changes, thereby causing an unfavorable orientation of the C-terminus. This orientation might be improved by the introduction of proline at position 34. Similarly substitution of the central segment of hPP by the corresponding pNPY sequence led to a 12-fold loss of affinity and could be recovered completely by the introduction of glutamine at position 34 (Cabrele et al. 2001).

In contrast the presence of the N-terminal segments hPP(1–7) or hPP(1–17) in pNPY increased the binding affinities at the human Y_4 receptor, which could be further improved by the introduction of histidine at position 34. This substitution was correlated with an increase of helix content, and suggests that the N-terminal fragment stabilizes the C-terminal helix by hydrophobic, intramolecular interactions. This is in accordance with the X-ray structure of avian PP. On the other hand introduction of pNPY(1–7) in hPP resulted in a more than 100-fold reduced affinity, but did not change the helix content in comparison with hPP. However, this modification might induce a different folding of the peptide backbone. Furthermore combining N-terminal and central replacement in [pNPY$^{1-7,19-23}$]hPP yields a molecule with similar affinity to the Y_4 receptor as hPP (Cabrele et al. 2001).

This suggests that N-terminal and central regions of hPP are important structural elements for the PP-fold. While their substitution by the corresponding pNPY regions induces a conformational change that can lead to significant loss of affinity, it can be recovered by using the right combination of replacements along the sequence.

As expected, since the C-terminus was shown to play a crucial role for the binding, the exchange of the C-terminal hexapeptide in [Leu^{31},Pro^{34}]NPY or [Leu^{31},Pro^{34}]PYY enhanced the affinity of these analogs at Y_4 receptors (Gehlert et al. 1997; Gerald et al. 1996; Hu et al. 1996).

Interestingly, binding of several ligands to human, rat and guinea pig Y_4 receptors occurs according to a two-site binding model, as has been shown for NPY and PYY in competition experiments with the radioligand [^{125}I]hPP at Y_4 receptors expressed in Chinese hamster ovary cells. Similarly porcine [Ala^{34}]NPY and [Ahx$^{8-20}$]NPY showed binding to rat Y_4 receptors characterized by a two-site binding model. However, these mutations completely abolish high-affinity binding to human and guinea pig Y_4 receptors. These results suggest that these agonists can distinguish between different active conformations of the human and guinea pig, but not the rat receptor. It is speculated that a cell can host only a limited number of Y_4 receptors in a high-affinity state, possibly due to G-protein depletion or loss of interaction with other cytoplasmic or membrane-bound proteins. A very efficient expression system could thereby mask the high-affinity binding. This could explain different values of binding affinity depending on the assay used to study the ligand–receptor interaction (Berglund et al. 2001).

Taken together, the Y_4 receptor is the only receptor which predominantly binds PP. Also at this receptor the C-terminal part of the ligand and a high helix content are important for high-affinity binding and receptor activation. Similar to the Y_2 receptor N-terminal deletions of the ligand, as for example in bPP(13–36), are tolerated at the Y_4 receptor. However, the centrally truncated peptide [Ahx$^{5-24}$]pNPY shows more than 100-fold reduced affinity to the Y_4 receptor, whereas it still binds the Y_2 receptor with high affinity. Interestingly studies on PP/NPY chimeras showed that at this receptor interplay of the central region with the corresponding amino acid at position 34 is important for stabilizing the bioactive conformation of the ligand. But also the combination of N-terminal and central NPY segments in hPP still results in high affinity. Thus stabilization of the PP-fold by the right combination of structural elements confers high affinity at this receptor (Fig. 2).

5
NPY and Peptide Derived Ligands at the Y_5 Receptor

Like the Y_1 and Y_2 receptors, the Y_5 receptor binds NPY and PYY with higher affinity than PP. Besides the Y_1 receptor, the Y_5 receptor is thought to play a role in mediating the stimulation of feeding behavior by NPY (Gehlert 1999; Inui 1999; Soll and Beck-Sickinger 2001). Therefore its structural and biological properties relative to the other receptors, especially to the Y_1 receptor, are of great interest.

The L-Ala scan of NPY showed that N-terminal positions like Tyr^1,Pro^2 are relatively insensitive to replacement by L-alanine (Cabrele and Beck-Sickinger 2000). However, amino acid substitutions at position Asp^6 and Asn^7 of NPY are better tolerated at the Y_1 receptor compared to the Y_2 and Y_5 receptor. For example the introduction of the basic residue arginine in [Arg^6,Pro^{34}]NPY increased the affinity at the Y_1 receptor twofold, whereas a sevenfold decrease in affinity to the Y_5 receptor was provoked by this mutation. Moreover the analog [Phe^7,Pro^{34}]NPY displayed even higher significant Y_1 receptor preference (Soll et al. 2001). The replacement of Arg^{25} by L-alanine, which affected binding to the Y_1 and Y_2 receptors only moderately, decreased binding to the Y_5 receptor more than 100-fold. The mutation Tyr^{27}Ala, which had a comparably small effect on Y_2 binding, reduced the affinity to the Y_5 receptor 600-fold. In contrast to the Y_1 and Y_2 receptors, replacement of Gln^{34} by alanine led to only a twofold decrease of the affinity to the Y_5 receptor. However, similar to the Y_1 receptor, the introduction of D-Pro at position 34 was poorly tolerated at the Y_5 receptor. As it had been shown for the other receptors, Arg^{33} and Arg^{35} are essential residues also for the binding to the Y_5 receptor (Cabrele and Beck-Sickinger 2000) (Table 1).

[D-Trp^{32}]NPY displayed weak, but selective agonistic activity at the Y_5 receptor with respect to inhibition of forskolin stimulated cAMP increase in cell culture. In vivo orexigenic properties have been attributed to this NPY mutant; however, antagonism against NPY induced increase in food intake has been observed as well (Balasubramaniam et al. 1994; Gerald et al. 1996; Matos et al. 1996; Small et al. 1997; Wyss et al. 1998). In comparison [D-Trp^{34}]NPY showed slightly higher affinity for the Y_5 receptor in radioligand binding assays and was also identified as a moderate Y_5 selective agonist in cell culture studies. Furthermore [D-Trp^{34}]NPY was shown to have significant orexigenic activity after central administration in rats, which could be blocked by administration of the Y_5 selective nonpeptide antagonist CGP 71683A (Parker et al. 2000).

As already suggested by the small effect of L-alanine replacement at position 1, unlike the Y_1 receptor, the Y_5 receptor tolerates deletion of the N-terminal amino acid in NPY(2–36) (Borowsky et al. 1998; Lundell et al. 2001).

Deletion of the central region generally resulted in low affinity, for example [Ahx$^{5-24}$]NPY had more than 1,000-fold reduced binding affinity. The most potent analog in this series [Ahx$^{9-17}$]NPY still displayed 14-fold reduction of affinity to the Y_5 receptor. Comparably a more than 1,000-fold decrease of Y_5 receptor binding was observed for centrally truncated hPP analogs [Ahx$^{5-24}$]hPP, [Ahx$^{5-20}$]hPP and [Tyr$^{5-20}$]hPP (Cabrele and Beck-Sickinger 2000; Eckard and Beck-Sickinger 2001).

In agreement, substitution of the central region of pNPY(19–23) by the corresponding segments of human or rat PP resulted in 300-fold decrease of affinity. As it had already been observed for the Y_1 and Y_4 receptors, unlike the Y_2 receptor, the binding was partially recovered by the introduction of Pro^{34}. Thus the central segment is important for the orientation of the C-terminal helix also at the Y_5 receptor. However, the structural requirements differ between the receptors, since these analogs display different binding potencies at all receptors (Cabrele et al. 2001).

N-terminal replacement in [hPP$^{1-17}$]NPY, which displays a highly helical content, slightly increased the affinity to the Y_5 receptor, in contrast to a decrease of Y_1 receptor binding. However, [hPP$^{1-7}$]NPY, which is characterized by changes in the folding of the peptide backbone without affecting the hPP helix, reduced the binding affinities for both the Y_5 and Y_1 receptors (Cabrele et al. 2001).

Replacement of single or multiple positions in the central region of hPP resulted in ligands with good affinity to the Y_5 receptor. Interestingly the simultaneous introduction of the segment pNPY(1–7) or cPP(1–7) and pNPY(19–23) in human PP led to analogs, which displayed even higher affinity to the Y_5 receptor than human NPY. These data suggest that the N-terminal, central and C-terminal region contribute to induce and stabilize the tertiary structure that is adopted from the ligand at the Y_5 receptor (Cabrele et al. 2001).

NPY analogs and PP/NPY chimeras containing the motif Ala^{31},Pro^{32} or even better Ala^{31},Aib^{32}-motif (Aib: alpha-amino isobutyric acid) represent very potent and highly selective Y_5 agonists. Remarkably, the peptide $[cPP^{1-7},NPY^{19-23}, Ala^{31},Aib^{32},Gln^{34}]hPP$ is more potent than NPY in stimulating food intake in rats. Even centrally truncated analogs and C-terminal fragments containing this motif, like $[Ahx^{5-24},Ala^{31},Aib^{32}]NPY$ and $[Ala^{31}, Aib^{32}]NPY(18-36)$ show comparably good affinity to the Y_5 receptor. Therefore the motif Ala^{31},Aib^{32} seems to be sufficient to induce and stabilize the required bioactive conformation of the C-terminal part of the ligand. Solution structure in water revealed the presence of a C-terminal α-helix ending with a 3_{10}-helical turn of the residues 28–31, followed by an apparently not well-defined structure of the last five residues. Surprisingly the inverse motif Aib^{31},Ala^{32}, which also improved binding of NPY to the Y_5 receptor significantly, led to loss of selectivity relative to the Y_2 receptor. This suggests that a specific type of turn structure, representing the bioactive conformation of the C-terminus at the Y_5 receptor, is induced and stabilized more correctly by the motif Ala^{31},Aib^{32} than Aib^{31},Ala^{32}, whereas we find the inverse situation at the Y_2 receptor (Bader et al. 2002; Cabrele and Beck-Sickinger 2000; Cabrele et al. 2000, 2002).

In conclusion, unlike the Y_1 receptor, the Y_5 receptor tolerates deletion of the first N-terminal amino acid. In contrast to the Y_2 receptor, the centrally truncated analog $[Ahx^{5-24}]NPY$ only displays low affinity binding to the Y_5 receptor. A stable α-helix is an important but not sufficient structural requirement for high-affinity binding to the Y_5 receptor. Comparably to the other receptors the central and N-terminal segments of NPY are important for inducing and stabilizing the tertiary structure of the ligand at this receptor. Highly Y_5 receptor selective analogs containing the motif Ala^{31},Aib^{32} are characterized by a C-terminal α-helix ending with a 3_{10}-helical turn. Therefore the bioactive conformation of the ligand at this receptor might be characterized by a well-defined conformation of the C-terminal loop (Fig. 2 and Table 2).

6
NPY and Peptide Derived Ligands at the y_6 Receptor

The y_6 receptor has been identified in mice and rabbits, but is a pseudogene in primates and is absent in rats (Burkhoff et al. 1998; Gregor et al. 1996; Matsumoto et al. 1996; Rose et al. 1997). The pharmacological properties of the mouse y_6 receptor are controversial; however, two reports suggest a pharmacological profile that most closely resembles those of the Y_1 receptor (Gregor et al. 1996; Mullins et al. 2000; Weinberg et al. 1996).

Like the Y_1 receptor, the y_6 receptor binds the dimeric nonapeptide 1229U91 with high affinity. However, as has been shown for the Y_4 receptor, this analog exhibits agonistic properties for the y_6 receptor. In contrast 1229U91 represents an antagonist for the Y_1 receptor. Although N-terminal truncation reduces affinity at the y_6 receptor, the requirement of N-terminal amino acids is less stringent than for the Y_1 receptor. Unlike the Y_2 receptor, the y_6 receptor binds NPY(13–36) with low and [Leu^{31},Pro^{34}]NPY with high affinity. Low affinity has also been shown for the Y_5 receptor selective agonist [D-Trp^{32}]NPY. In contrast to the Y_4 receptor the y_6 receptor only weakly binds hPP and rPP (Mullins et al. 2000).

7
Concluding Remarks

The structure–affinity and structure–activity studies elaborated so far elucidate that the bioactive conformation of the ligands differs at the various receptor subtypes. Some characteristic features corresponding to the specific ligand–receptor subtype interactions could be identified. However, further investigations are necessary for a full understanding of the structures adopted by the bioactive ligand–receptor complexes. Continuing mutational analysis on both receptor and ligand sites will be of particular interest to gain further insight into these structural characteristics and their relationship to physiological reactions.

References

Bader R, Bettio A, Beck-Sickinger AG, Zerbe O (2001) Structure and dynamics of micelle-bound neuropeptide Y: comparison with unligated NPY and implications for receptor selection. J Mol Biol 305:307–329

Bader R, Rytz G, Lerch M, Beck-Sickinger AG, Zerbe O (2002) Key motif to gain selectivity at the neuropeptide Y$_5$-receptor: structure and dynamics of micelle-bound [Ala^{31}, Pro^{32}]-NPY. Biochemistry 41:8031–8042

Balasubramaniam A, Dhawan VC, Mullins DE, Chance WT, Sheriff S, Guzzi M, Prabhakaran M, Parker EM (2001) Highly selective and potent neuropeptide Y (NPY) Y$_1$ receptor antagonists based on [Pro^{30}, Tyr^{32}, Leu^{34}]NPY(28–36)-NH2 (BW1911U90). J Med Chem 44:1479–1482

Balasubramaniam A, Sheriff S, Johnson ME, Prabhakaran M, Huang Y, Fischer JE, Chance WT (1994) [D-TRP^{32}]neuropeptide Y: a competitive antagonist of NPY in rat hypothalamus. J Med Chem 37:811–815

Balasubramaniam A, Sheriff S, Rigel DF, Fischer JE (1990) Characterization of neuropeptide Y binding sites in rat cardiac ventricular membranes. Peptides 11:545–550

Barden JA, Cuthbertson RM, Potter EK, Selbie LA, Tseng A (1994) Stabilized structure of the presynaptic (Y$_2$) receptor-specific neuropeptide Y analog N-acetyl[Leu^{28}, Leu^{31}]NPY(24–36). Biochim Biophys Acta 1206:191–196

Barnham KJ, Catalfamo F, Pallaghy PK, Howlett GJ, Norton RS (1999) Helical structure and self-association in a 13 residue neuropeptide Y Y$_2$ receptor agonist: relationship to biological activity. Biochim Biophys Acta 1435:127–137

Beck A, Jung G, Gaida W, Koppen H, Lang R, Schnorrenberg G (1989) Highly potent and small neuropeptide Y agonist obtained by linking NPY 1–4 via spacer to alpha-helical NPY(25–36). FEBS Lett 244:119–122

Beck-Sickinger AG, Grouzmann E, Hoffmann E, Gaida W, van Meir EG, Waeber B, Jung G (1992) A novel cyclic analog of neuropeptide Y specific for the Y$_2$ receptor. Eur J Biochem 206:957–964

Beck-Sickinger AG, Hoffmann E, Paulini K, Reissig HU, Willim KD, Wieland HA, Jung G (1994a) High-affinity analogues of neuropeptide Y containing conformationally restricted non-proteinogenic amino acids. Biochem Soc Trans 22:145–149

Beck-Sickinger AG, Jung G (1995) Structure-activity relationships of neuropeptide Y analogues with respect to Y_1 and Y_2 receptors. Biopolymers 37:123–142

Beck-Sickinger AG, Wieland HA, Wittneben H, Willim KD, Rudolf K, Jung G (1994b) Complete L-alanine scan of neuropeptide Y reveals ligands binding to Y_1 and Y_2 receptors with distinguished conformations. Eur J Biochem 225:947–958

Berglund MM, Lundell I, Eriksson H, Soll R, Beck-Sickinger AG, Larhammar D (2001) Studies of the human, rat, and guinea pig Y_4 receptors using neuropeptide Y analogues and two distinct radioligands. Peptides 22:351–356

Bettio A, Beck-Sickinger AG (2001) Biophysical methods to study ligand-receptor interactions of neuropeptide Y. Biopolymers 60:420–37

Bettio A, Dinger MC, Beck-Sickinger AG (2002) The neuropeptide Y monomer in solution is not folded in the pancreatic-polypeptide-fold. Protein Sci 11:1834–44

Borowsky B, Walker MW, Bard J, Weinshank RL, Laz TM, Vaysse P, Branchek TA, Gerald C (1998) Molecular biology and pharmacology of multiple NPY Y_5 receptor species homologs. Regul Pept 75/76:45–53

Boublik JH, Scott NA, Brown MR, Rivier JE (1989) Synthesis and hypertensive activity of neuropeptide Y fragments and analogues with modified N- or C-termini or D-substitutions. J Med Chem 32:597–601

Burkhoff A, Linemeyer DL, Salon JA (1998) Distribution of a novel hypothalamic neuropeptide Y receptor gene and it's absence in rat. Brain Res Mol Brain Res 53:311–316

Cabrele C, Beck-Sickinger AG (2000) Molecular characterization of the ligand-receptor interaction of the neuropeptide Y family. J Pept Sci 6:97–122

Cabrele C, Langer M, Bader R, Wieland HA, Doods HN, Zerbe O, Beck-Sickinger AG (2000) The first selective agonist for the neuropeptide Y Y_5 receptor increases food intake in rats. J Biol Chem 275:36043–36048

Cabrele C, Wieland HA, Koglin N, Stidsen C, Beck-Sickinger AG (2002) Ala[31]-Aib[32]: identification of the key motif for high affinity and selectivity of neuropeptide Y at the Y_5-receptor. Biochemistry 41:8043–8049

Cabrele C, Wieland HA, Langer M, Stidsen CE, Beck-Sickinger AG (2001) Y-receptor affinity modulation by the design of pancreatic polypeptide/neuropeptide Y chimera led to Y_5-receptor ligands with picomolar affinity. Peptides 22:365–378

Cerda-Reverter JM, Larhammar D (2000) Neuropeptide Y family of peptides: structure, anatomical expression, function, and molecular evolution. Biochem Cell Biol 78:371–392

Cowley DJ, Hoflack JM, Pelton JT, Saudek V (1992) Structure of neuropeptide Y dimer in solution. Eur J Biochem 205:1099–1106

Cox HM, Tough IR, Ingenhoven N, Beck-Sickinger AG (1998) Structure-activity relationships with neuropeptide Y analogues: a comparison of human Y_1-, Y_2- and rat Y_2-like systems. Regul Pept 75/76:3–8

Daniels AJ, Matthews JE, Slepetis RJ, Jansen M, Viveros OH, Tadepalli A, Harrington W, Heyer D, Landavazo A, Leban JJ, et al. (1995) High-affinity neuropeptide Y receptor antagonists. Proc Natl Acad Sci USA 92:9067–9071

Darbon H, Bernassau JM, Deleuze C, Chenu J, Roussel A, Cambillau C (1992) Solution conformation of human neuropeptide Y by ^{1}H nuclear magnetic resonance and restrained molecular dynamics. Eur J Biochem 209:765–771

Dumont Y, Cadieux A, Pheng LH, Fournier A, St-Pierre S, Quirion R (1994) Peptide YY derivatives as selective neuropeptide Y/peptide YY Y_1 and Y_2 agonists devoided of activity for the Y_3 receptor sub-type. Brain Res Mol Brain Res 26:320–324

Dumont Y, Fournier A, St-Pierre S, Quirion R (1993) Comparative characterization and autoradiographic distribution of neuropeptide Y receptor subtypes in the rat brain. J Neurosci 13:73–86

Dumont Y, Jacques D, Bouchard P, Quirion R (1998) Species differences in the expression and distribution of the neuropeptide Y Y_1, Y_2, Y_4, and Y_5 receptors in rodents, guinea pig, and primates brains. J Comp Neurol 402:372-384

Dumont Y, Quirion R (2000) [(125)I]-GR231118: a high affinity radioligand to investigate neuropeptide Y Y_1 and Y_4 receptors. Br J Pharmacol 129:37-46

Eckard CP, Beck-Sickinger AG (2001) Characterisation of Neuropeptide Y receptor subtypes by synthetic NPY analogues and by anti-receptor antibodies. Molecules 6:448-467

Eriksson H, Berglund MM, Holmberg SK, Kahl U, Gehlert DR, Larhammar D (1998) The cloned guinea pig pancreatic polypeptide receptor Y_4 resembles more the human Y_4 than does the rat Y_4. Regul Pept 75/76:29-37

Fournier A, Gagnon D, Quirion R, Cadieux A, Dumont Y, Pheng LH, St-Pierre S (1994) Conformational and biological studies of neuropeptide Y analogs containing structural alterations. Mol Pharmacol 45:93-101

Fuhlendorff J, Gether U, Aakerlund L, Langeland-Johansen N, Thogersen H, Melberg SG, Olsen UB, Thastrup O, Schwartz TW (1990a) [Leu^{31}, Pro^{34}]neuropeptide Y: a specific Y_1 receptor agonist. Proc Natl Acad Sci USA 87:182-186

Fuhlendorff J, Johansen NL, Melberg SG, Thogersen H, Schwartz TW (1990b) The antiparallel pancreatic polypeptide fold in the binding of neuropeptide Y to Y_1 and Y_2 receptors. J Biol Chem 265:11706-11712

Gackenheimer SL, Schober DA, Gehlert DR (2001) Characterization of neuropeptide Y Y_1-like and Y_2-like receptor subtypes in the mouse brain. Peptides 22:335-341

Gehlert DR (1999) Role of hypothalamic neuropeptide Y in feeding and obesity. Neuropeptides 33:329-338

Gehlert DR, Gackenheimer SL (1997) Differential distribution of neuropeptide Y Y_1 and Y_2 receptors in rat and guinea-pig brains. Neuroscience 76:215-224

Gehlert DR, Schober DA, Beavers L, Gadski R, Hoffman JA, Smiley DL, Chance RE, Lundell I, Larhammar D (1996) Characterization of the peptide binding requirements for the cloned human pancreatic polypeptide-preferring receptor. Mol Pharmacol 50:112-118

Gehlert DR, Schober DA, Gackenheimer SL, Beavers L, Gadski R, Lundell I, Larhammar D (1997) [^{125}I]Leu^{31}, Pro^{34}-PYY is a high affinity radioligand for rat PP1/Y_4 and Y_1 receptors: evidence for heterogeneity in pancreatic polypeptide receptors. Peptides 18:397-401

Gerald C, Walker MW, Criscione L, Gustafson EL, Batzl-Hartmann C, Smith KE, Vaysse P, Durkin MM, Laz TM, Linemeyer DL, Schaffhauser AO, Whitebread S, Hofbauer KG, Taber RI, Branchek TA, Weinshank RL (1996) A receptor subtype involved in neuropeptide-Y-induced food intake. Nature 382:168-171

Glover I, Haneef I, Pitts J, Wood S, Moss D, Tickle I, Blundell T (1983) Conformational flexibility in a small globular hormone: x-ray analysis of avian pancreatic polypeptide at 0.98-A resolution. Biopolymers 22:293-304

Gregor P, Feng Y, DeCarr LB, Cornfield LJ, McCaleb ML (1996) Molecular characterization of a second mouse pancreatic polypeptide receptor and its inactivated human homologue. J Biol Chem 271:27776-27781

Herzog H, Hort YJ, Ball HJ, Hayes G, Shine J, Selbie LA (1992) Cloned human neuropeptide Y receptor couples to two different second messenger systems. Proc Natl Acad Sci USA 89:5794-5798

Hoffmann S, Rist B, Videnov G, Jung G, Beck-Sickinger AG (1996) Structure-affinity studies of C-terminally modified analogs of neuropeptide Y led to a novel class of peptidic Y_1 receptor antagonist. Regul Pept 65:61-70

Hu Y, Bloomquist BT, Cornfield LJ, DeCarr LB, Flores-Riveros JR, Friedman L, Jiang P, Lewis-Higgins L, Sadlowski Y, Schaefer J, Velazquez N, McCaleb ML (1996) Identification of a novel hypothalamic neuropeptide Y receptor associated with feeding behavior. J Biol Chem 271:26315-26319

Inui A (1999) Neuropeptide Y feeding receptors: are multiple subtypes involved? Trends Pharmacol Sci 20:43–46

Ishihara A, Tanaka T, Kanatani A, Fukami T, Ihara M, Fukuroda T (1998) A potent neuropeptide Y antagonist, 1229U91, suppressed spontaneous food intake in Zucker fatty rats. Am J Physiol 274: R1500–R1504

Iyengar S, Li DL, Simmons RM (1999) Characterization of neuropeptide Y-induced feeding in mice: do Y_1-y_6 receptor subtypes mediate feeding? J Pharmacol Exp Ther 289:1031–1040

Kanatani A, Ishihara A, Asahi S, Tanaka T, Ozaki S, Ihara M (1996) Potent neuropeptide Y Y_1 receptor antagonist, 1229U91: blockade of neuropeptide Y-induced and physiological food intake. Endocrinology 137:3177–3182

Kanatani A, Ito J, Ishihara A, Iwaasa H, Fukuroda T, Fukami T, MacNeil DJ, Van der Ploeg LH, Ihara M (1998) NPY-induced feeding involves the action of a Y_1-like receptor in rodents. Regul Pept 75/76:409–415

Keire DA, Bowers CW, Solomon TE, Reeve JR (2002) Structure and receptor binding of PYY analogs. Peptides 23:305–321

Keire DA, Kobayashi M, Solomon TE, Reeve JR, Jr. (2000a) Solution structure of monomeric peptide YY supports the functional significance of the PP-fold. Biochemistry 39:9935–9942

Keire DA, Mannon P, Kobayashi M, Walsh JH, Solomon TE, Reeve JR, Jr. (2000b) Primary structures of PYY, [Pro^{34}]PYY, and PYY-(3–36) confer different conformations and receptor selectivity. Am J Physiol Gastrointest Liver Physiol 279: G126–G131

Kirby DA, Boublik JH, Rivier JE (1993a) Neuropeptide Y: Y_1 and Y_2 affinities of the complete series of analogues with single D-residue substitutions. J Med Chem 36:3802–3808

Kirby DA, Britton KT, Aubert ML, Rivier JE (1997) Identification of high-potency neuropeptide Y analogues through systematic lactamization. J Med Chem 40:210–215

Kirby DA, Koerber SC, Craig AG, Feinstein RD, Delmas L, Brown MR, Rivier JE (1993b) Defining structural requirements for neuropeptide Y receptors using truncated and conformationally restricted analogues. J Med Chem 36:385–393

Kirby DA, Koerber SC, May JM, Hagaman C, Cullen MJ, Pelleymounter MA, Rivier JE (1995) Y_1 and Y_2 receptor selective neuropeptide Y analogues: evidence for a Y_1 receptor subclass. J Med Chem 38:4579–4586

Koglin N, Zorn C, Beumer R, Cabrele C, Bubert C, Sewald N, Reiser O, Beck-Sickinger AG (2003) Analogues of neuropeptide Y containing beta-aminocyclopropane carboxylic acids are the shortest linear peptides that are selective for the Y_1 receptor. Angew Chem Int Ed Engl 42:202–205

Larhammar D (1996) Structural diversity of receptors for neuropeptide Y, peptide YY and pancreatic polypeptide. Regul Pept 65:165–174

Larhammar D, Blomqvist AG, Yee F, Jazin E, Yoo H, Wahlested C (1992) Cloning and functional expression of a human neuropeptide Y/peptide YY receptor of the Y_1 type. J Biol Chem 267:10935–10938

Lundell I, Eriksson H, Marklund U, Larhammar D (2001) Cloning and characterization of the guinea pig neuropeptide Y receptor Y_5. Peptides 22:357–363

Lundell I, Statnick MA, Johnson D, Schober DA, Starback P, Gehlert DR, Larhammar D (1996) The cloned rat pancreatic polypeptide receptor exhibits profound differences to the orthologous receptor. Proc Natl Acad Sci USA 93:5111–5115

Matos FF, Guss V, Korpinen C (1996) Effects of neuropeptide Y (NPY) and [D-Trp^{32}]NPY on monoamine and metabolite levels in dialysates from rat hypothalamus during feeding behavior. Neuropeptides 30:391–398

Matsumoto M, Nomura T, Momose K, Ikeda Y, Kondou Y, Akiho H, Togami J, Kimura Y, Okada M, Yamaguchi T (1996) Inactivation of a novel neuropeptide Y/peptide YY receptor gene in primate species. J Biol Chem 271:27217–27220

Michel MC, Beck-Sickinger A, Cox H, Doods HN, Herzog H, Larhammar D, Quirion R, Schwartz T, Westfall T (1998) XVI. International Union of Pharmacology recommendations for the nomenclature of neuropeptide Y, peptide YY, and pancreatic polypeptide receptors. Pharmacol Rev 50:143–150

Mullins D, Kirby D, Hwa J, Guzzi M, Rivier J, Parker E (2001) Identification of potent and selective neuropeptide Y Y_1 receptor agonists with orexigenic activity in vivo. Mol Pharmacol 60:534–540

Mullins DE, Guzzi M, Xia L, Parker EM (2000) Pharmacological characterization of the cloned neuropeptide Y y_6 receptor. Eur J Pharmacol 395:87–93

Nordmann A, Blommers MJ, Fretz H, Arvinte T, Drake AF (1999) Aspects of the molecular structure and dynamics of neuropeptide Y. Eur J Biochem 261:216–226.

Parker EM, Babij CK, Balasubramaniam A, Burrier RE, Guzzi M, Hamud F, Mukhopadhyay G, Rudinski MS, Tao Z, Tice M, Xia L, Mullins DE, Salisbury BG (1998) GR231118 (1229U91) and other analogues of the C-terminus of neuropeptide Y are potent neuropeptide Y Y_1 receptor antagonists and neuropeptide Y Y_4 receptor agonists. Eur J Pharmacol 349:97–105

Parker EM, Balasubramaniam A, Guzzi M, Mullins DE, Salisbury BG, Sheriff S, Witten MB, Hwa JJ (2000) [D-Trp^{34}] neuropeptide Y is a potent and selective neuropeptide Y Y_5 receptor agonist with dramatic effects on food intake. Peptides 21:393–399

Potter EK, Barden JA, McCloskey MJ, Selbie LA, Tseng A, Herzog H, Shine J (1994) A novel neuropeptide Y analog, N-acetyl [Leu^{28},Leu^{31}]neuropeptide Y-(24–36), with functional specificity for the presynaptic (Y_2) receptor. Eur J Pharmacol 267:253–262

Rist B, Ingenhoven N, Scapozza L, Schnorrenberg G, Gaida W, Wieland HA, Beck-Sickinger AG (1997) The bioactive conformation of neuropeptide Y analogues at the human Y_2- receptor. Eur J Biochem 247:1019–1028

Rist B, Wieland HA, Willim KD, Beck-Sickinger AG (1995) A rational approach for the development of reduced-size analogues of neuropeptide Y with high affinity to the Y_1 receptor. J Pept Sci 1:341–348

Rose PM, Lynch JS, Frazier ST, Fisher SM, Chung W, Battaglino P, Fathi Z, Leibel R, Fernandes P (1997) Molecular genetic analysis of a human neuropeptide Y receptor. The human homolog of the murine 'Y_5' receptor may be a pseudogene. J Biol Chem 272:3622–3627

Schober DA, Gackenheimer SL, Heiman ML, Gehlert DR (2000) Pharmacological characterization of ^{125}I-1229U91 binding to Y_1 and Y_4 neuropeptide Y/Peptide YY receptors. J Pharmacol Exp Ther 293:275–280

Schober DA, Van Abbema AM, Smiley DL, Bruns RF, Gehlert DR (1998) The neuropeptide Y Y_1 antagonist, 1229U91, a potent agonist for the human pancreatic polypeptide-preferring (NPY Y_4) receptor. Peptides 19:537–542

Small CJ, Morgan DG, Meeran K, Heath MM, Gunn I, Edwards CM, Gardiner J, Taylor GM, Hurley JD, Rossi M, Goldstone AP, O'Shea D, Smith DM, Ghatei MA, Bloom SR (1997) Peptide analogue studies of the hypothalamic neuropeptide Y receptor mediating pituitary adrenocorticotrophic hormone release. Proc Natl Acad Sci USA 94:11686–11691

Smith-White MA, Potter EK (1999) Structure-activity analysis of N-acetyl [Leu^{28,31}] NPY (24–36): a potent neuropeptide Y Y_2 receptor agonist. Neuropeptides 33:526–533

Soll RM, Beck-Sickinger AG (2001) On the role of neuropeptides in the hypothalamic regulation of food intake. Curr Med Chem—Imm Endoc and Metab Agents 1:151–169

Soll RM, Dinger MC, Lundell I, Larhammer D, Beck-Sickinger AG (2001) Novel analogues of neuropeptide Y with a preference for the Y_1-receptor. Eur J Biochem 268:2828–2837

Takebayashi Y, Koga H, Togami J, Kurihara H, Furuya T, Tanaka A, Murase K (2000) Structure-affinity relationships of C-terminal cyclic analogue of neuropeptide Y for the Y_1-receptor. Chem Pharm Bull (Tokyo) 48:1925–1929

Wahlestedt C, Yanaihara N, Hakanson R (1986) Evidence for different pre- and post-junctional receptors for neuropeptide Y and related peptides. Regul Pept 13:307–318

Weinberg DH, Sirinathsinghji DJ, Tan CP, Shiao LL, Morin N, Rigby MR, Heavens RH, Rapoport DR, Bayne ML, Cascieri MA, Strader CD, Linemeyer DL, MacNeil DJ (1996) Cloning and expression of a novel neuropeptide Y receptor. J Biol Chem 271:16435–16438

Widdowson PS (1993) Quantitative receptor autoradiography demonstrates a differential distribution of neuropeptide-Y Y_1 and Y_2 receptor subtypes in human and rat brain. Brain Res 631:27–38

Wyss P, Stricker-Krongrad A, Brunner L, Miller J, Crossthwaite A, Whitebread S, Criscione L (1998) The pharmacology of neuropeptide Y (NPY) receptor-mediated feeding in rats characterizes better Y_5 than Y_1, but not Y_2 or Y_4 subtypes. Regul Pept 75/76:363–371

Structure–Activity Relationships of Nonpeptide Neuropeptide Y Receptor Antagonists

A. Brennauer · S. Dove · A. Buschauer

Institute of Pharmacy, University of Regensburg, 93040 Regensburg, Germany
e-mail: armin.buschauer@chemie.uni-regensburg.de

1	Introduction	506
2	First Nonspecific NPY Receptor Antagonists and Structurally Derived Compounds	508
2.1	α-Trinositol	508
2.2	Benextramine and Related Compounds	509
2.3	Y_1 Antagonists Related to Arpromidine	509
3	Potent and Selective Nonpeptide NPY Y_1 Receptor Antagonists	511
3.1	BIBP 3226 and Other (R)-Argininamides	511
3.1.1	Design and Pharmacology of BIBP 3226	511
3.1.2	Structure–Activity Relationships of BIBP 3226 Derivatives	512
3.1.3	The Y_1 Receptor Binding Site for BIBP 3226: In Vitro Mutagenesis Results and Computer Models	514
3.1.4	N^G-Substituted (R)-Argininamides with Reduced Basicity	517
3.2	Benzamidine-Type Y_1 Antagonists SR 120819A and SR 120107A	517
3.3	Indoles, Benzimidazoles, and Benzothiophenes	518
3.4	Y_1 Antagonists Based on Other Common Structures	521
3.4.1	6-Benzylsulfonyl-5-nitroquinolines	521
3.4.2	Phenylpiperazines	521
3.4.3	Bis[diamino(phenyl)triazines]	522
3.4.4	Benzazepines and Benzodiazepines	522
3.4.5	Morpholinopyridines J-104870 and J-115814	523
3.4.6	Dihydropyridines and Dihydropyrazines	524
4	Selective Nonpeptide Y_2 Receptor Antagonists	524
5	NPY Y_5 Receptor Antagonists	526
5.1	Arylsulfonamide-Type and Related NPY Y_5 Receptor Antagonists	527
5.1.1	Sulfonamides with Tetraline or Homologous Cyclic Hydrocarbon Moieties	529
5.1.2	Heterocyclic Analogs of CGP 71683	531
5.2	Various Heterocyclic NPY Y_5 Receptor Antagonists	532
5.2.1	Azoles, Pyridines and Diazines	532
5.2.2	Carbazoles, Fluorenones and Phenylureas	534
5.2.3	Tetrahydroxanthene-1-ones	536
6	Conclusion	537
	References	537

Abstract Reports in 1990 on some weakly to moderately active nonpeptides which were not originally designed for neuropeptide Y (NPY) receptors, followed by the discovery of the first highly potent and selective Y_1 receptor antagonists—the (R)-argininamide BIBP 3226 and the benzamidine derivative SR 120819A—as well as raising hope for novel drug treatment of hypertension, obesity and metabolic diseases stimulated the search for NPY-blocking compounds. Most of the currently known nonpeptidic NPY antagonists are ligands of Y_1 or Y_5 receptors, whereas only one class of Y_2 selective antagonists around the (S)-arginine derivative BIIE 0246 has been disclosed. Nonpeptidic ligands of the Y_4 receptor are not known. In some cases the design of Y_1 antagonists followed rational strategies considering amino acids which are essential for binding to Y_1 and/or Y_2 receptors according to results of a complete alanine scan of NPY. Typical Y_1 antagonists (e.g., compounds of the argininamide, benzamidine, benzimidazole, indole and aminopyridine series) have one or two basic groups which—according to the working hypothesis—could mimic Arg^{33} and/or Arg^{35} in NPY. Binding models derived for some compounds (e.g., BIBP 3226 and J-104870) based on investigations with Y_1 receptor mutants suggest key interactions between the basic group(s) and acidic residues of the Y_1 receptor protein, especially Asp^{287}. Compared to Y_1 antagonists the known Y_5 antagonists are often based on hits from screening of libraries and show a considerably higher degree of structural diversity. Nevertheless many highly active Y_5 antagonists represent a common structural pattern suggesting at least overlapping binding sites.

Keywords NPY · Y_1 receptor · Y_2 receptor · Y_5 receptor · Nonpeptidic antagonist

1
Introduction

Since Rudolf et al. (1994a) published the D-argininamide BIBP 3226 as the first highly potent and selective nonpeptide Y_1 receptor antagonist, speculations about the therapeutic potential of neuropeptide Y (NPY) blocking agents (Grundemar and Bloom 1997; Silva et al. 2002), for example as anti-hypertensive or anti-obesity drugs, and the discovery of additional receptor subtypes (Michel et al. 1998) stimulated tremendously the search for new drug candidates in the NPY field. Some of the early successful approaches in the design of NPY antagonists were more or less rational starting from the structure of the natural ligand NPY. Regardless of the fact that the three-dimensional structure of NPY and its active conformation(s) at the different NPY receptors is still a matter of debate, the putative pancreatic polypeptide (PP)-fold structure of NPY (Allen et al. 1987) was used by many groups as a model to develop working hypotheses, in particular in the field of Y_1 receptor antagonists. The principle of drug design resulting in BIBP 3226 as mimic of the C terminus of NPY appeared to be generally a promising approach, in particular, as a complete L-alanine scan has provided valuable information which residues of NPY are important for binding at Y_1 and Y_2 receptors (Beck-Sickinger et al. 1994). Meanwhile, the initial optimism concerning the impact of nonpeptide NPY receptor antagonists as new therapeutics in the near future was dampened in some respect. This is certainly, at least in part, due to the complexity (and even redundancy) of NPY mediated physiological and pathophysiological effects resulting from the interaction of NPY with differently localized receptor subtypes and its interplay with a multitude of other neurotransmitters and hormones, for example in the regulation of food intake, metabolic processes, blood pressure, hormone release, modulation of emotional processing, and sexual and cognitive function (Silva et al. 2002). Although highly potent and selective substances, mainly Y_1 and Y_5 receptor antagonists, have been developed, no NPY receptor ligand was launched onto the market up to now. However, selective nonpeptide NPY antagonists proved to be indispensable pharmacological tools for the investigation of the physiological and pathophysiolgial role of NPY and the contribution of receptor subtypes, especially Y_1, Y_2, and Y_5, to complex biological responses, for example in feeding-related metabolic processes.

Within the last decade the number of known nonpeptide NPY ligands has largely increased (for recent reviews see Carpino 2000; Zimanyi and Poindexter 2000; Hammond 2001). However, comparing the described compounds a strong imbalance is obvious. With respect to the possible market for anti-obesity drugs the pharmaceutical companies were focusing their research on Y_1 and Y_5 antagonists, whereas only one class of potent nonpeptidic Y_2 (BIIE 0246) and no selective Y_4 receptor antagonist has been described so far. The structures of known Y_1 antagonists are less diverse than those of Y_5 antagonists, but the design strategies were rational and ligand based in some cases, leading to well explored structure–activity relationships in different series. The overall diversity of nonpeptide NPY antagonists is not surprising if one considers the large spread of the putative NPY binding sites covering extracellular and transmembrane regions of the receptor protein, as indicated, for instance, by in vitro mutagenesis of the Y_1 receptor and modeling approaches (Walker et al. 1994; Sautel et al. 1996; Du et al. 1997a; Robin-Jagerschmidt et al. 1998; Sylte et al. 1998, 1999). Therefore, different partially overlapping or even nonoverlapping antagonistic binding sites are possible which may or may not reproduce key interactions of NPY, for example those between crucial Arg residues and acidic amino acids of the Y_1 receptor. This is reflected by the diversity especially of the Y_5 antagonistic leads and has offered the chance of finding structurally distinct leads by high throughput screening of large and diverse compound libraries.

Compared to antagonists the structural requirements which must be fulfilled by an agonistic pharmacophore to induce (or stabilize) the active conformation of the receptor are much more stringent. Nonpeptidic agonists, which could be extremely useful pharmacological tools and potential drugs as well, are not reported in the literature. In the following sections nonpeptide NPY antagonists are summarized according to their Y_1, Y_2, and Y_5 selectivity and with focus on their structure–activity relationships.

2
First Nonspecific NPY Receptor Antagonists and Structurally Derived Compounds

Prior to the discovery of the first highly active nonpeptide NPY receptor ligands and the application of rational approaches to design such compounds, some substances originally described to act on other targets were found to be weak or moderate NPY antagonists: D-*myo*-inositol-1,2,6-trisphosphate (α-trinositol or pp56, **1**), an isomer of the second messenger inositol-1,4,5-trisphosphate, benextramine (**2a**), an irreversible α_1-adrenergic antagonist, and BU-E-76 (He 90481, **3a**), a highly potent histamine H_2 receptor agonist.

2.1
α-Trinositol

It has been reported, that α-trinositol (**1**, Fig. 1)—apart from its anti-inflammatory and analgesic effects—noncompetitively inhibits NPY induced vasoconstriction and pressor responses in several in vitro and in vivo assays (Edvinsson et al. 1990). Since the compound is not able to displace radiolabeled NPY from Y_1 or Y_2 binding sites it is suggested that α-trinositol acts at a point in the NPY-activated signaling pathways downstream from the receptors (Heilig et al. 1991; Feth et al. 1993). Thus α-trinositol is not an NPY receptor ligand, but it was the first nonpeptide agent described to inhibit some NPY mediated effects—

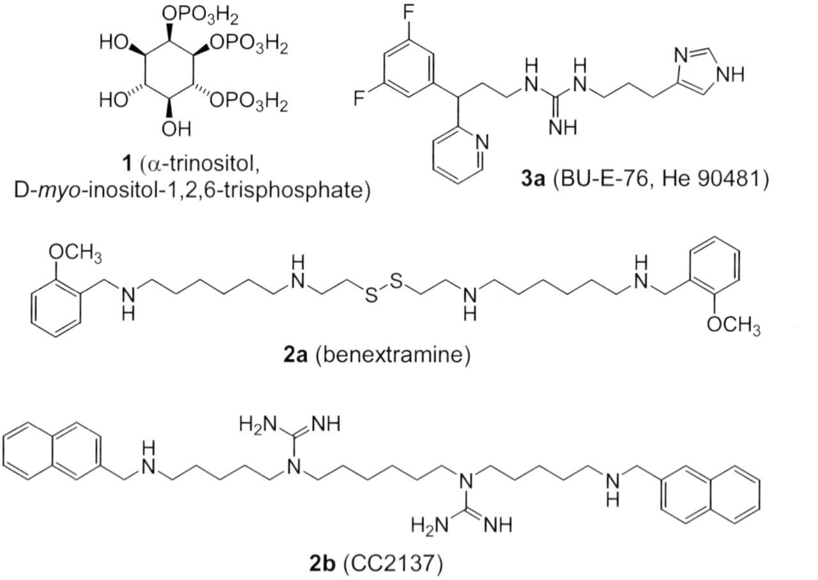

Fig. 1 First nonspecific NPY antagonists—compounds with different main pharmacological effects

amongst them the NPY induced stimulation of food intake (for a review see Bell and McDermott 1998).

2.2
Benextramine and Related Compounds

Benextramine (**2a**, Fig. 1), an irreversible α_1-adrenoceptor antagonist (Benfey 1982), produces a long-lasting antagonism of NPY induced pressor effects (Doughty et al. 1990) and was presented as the first nonpeptide inhibiting specific binding of [^{3}H]NPY to a NPY receptor population in rat brain membranes (Doughty et al. 1992). Since the inhibition of NPY binding is irreversible, the authors suggested a covalent linkage of benextramine to a cysteine residue of the receptor protein via a thiol-disulfide exchange. Whereas functional assays indicated Y_1 selectivity (Palea et al. 1995), binding studies with cloned human NPY receptors resulted in K_i values of 2 µM at the Y_1 and the Y_4, 7.5 µM at the Y_2 (Wright 1997) and 5 µM at the Y_5 subtype (Islam et al. 2002). A lead optimization approach was based on the hypothesis that the terminal benzylic moieties of benextramine possibly mimic Tyr^1 and/or Tyr^{36} of NPY. Analogs lacking a benzylic portion did not displace [^{3}H]NPY from rat brain membranes. 3-Hydroxy or 3-methoxy substituted benzyl as well as naphthyl groups are favorable (Doughty et al. 1993). Reversible antagonists were obtained when the central disulfide moiety was replaced with an ethylene bridge. Functional experiments with CC2137 (**2b**) indicated a shift towards Y_2 (vs. Y_1) receptor selectivity (Chaurasia et al. 1994).

2.3
Y_1 Antagonists Related to Arpromidine

The potent histamine H_2 agonist BU-E-76 (**3a**, also named HE 90481; Fig. 1), an analog of arpromidine (Buschauer 1989), is a weak competitive NPY Y_1-antagonist (Michel and Motulsky 1990; Michel et al. 1991). As **3a** and related substances displayed some Y_1 receptor selectivity these imidazolylpropylguanidines were considered as model compounds and investigated for inhibition of the NPY-induced Ca^{2+} mobilization in HEL cells to elaborate structure–activity relationships and to derive a pharmacophore model. Compared to BU-E-76 the Y_1 antagonistic activity could be increased by a factor of about 100 or more by increasing lipophilicity, for example by introduction of two chlorine substituents (Michel et al. 1993), and/or by vicinal instead of geminal arrangement of the aromatic rings. For instance, the halogenated benzyl(2-pyridyl)aminoalkylguanidines **3b,c** achieve Y_1 antagonistic activities in the submicromolar range (**3b**: K_b 0.47 µM, **3c**: K_b 0.36 µM; calcium assay in HEL cells; Dove et al. 2000) (Fig. 2).

Based on the assumption that two basic groups, mimicking Arg^{33} and Arg^{35} in NPY, are beneficial for Y_1-receptor affinity, the imidazole ring of the arpromidine analogs, which is essential for histamine H_2 receptor agonism, was replaced by a different basic heterocycle or a second guanidino group. Active bisguani-

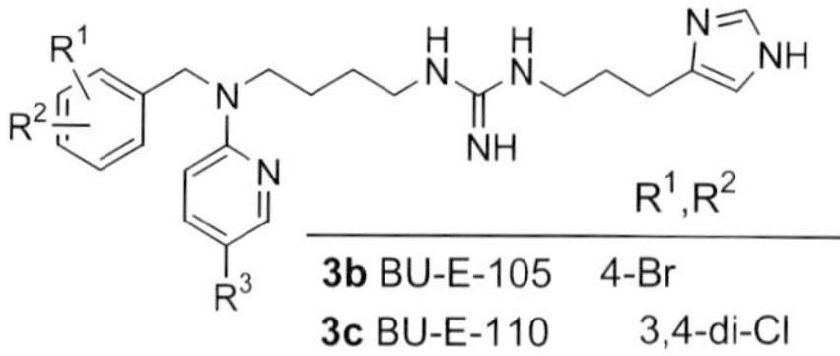

Fig. 2 Guanidine-type NPY antagonists derived from arpromidine

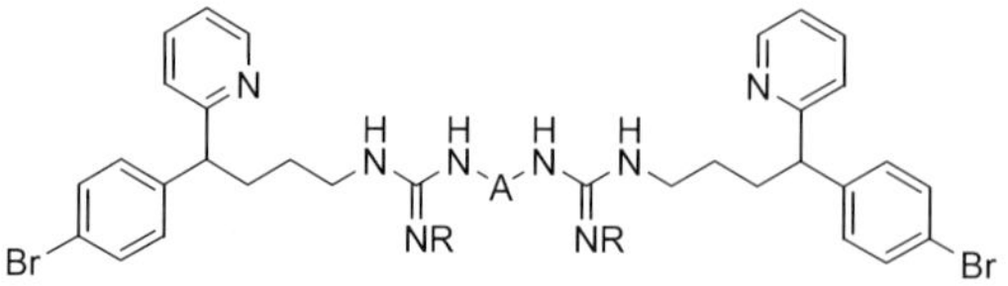

dines (e.g., **3d**; Knieps et al. 1996; Fig. 2) with *trans*-cyclohexane-1,4-diyl spacers point to an optimal distance between the guanidine groups of about 8 Å in agreement with the Arg^{33}–Arg^{35} side-chain distance postulated for the Y_1 active conformation of NPY (Beck-Sickinger et al. 1994; Beck-Sickinger 1997). The derivative **3e** (SK 48; Fig. 2) displayed also Y_2 receptor binding (K_i 1 µM) and, surprisingly, a Y_2 agonist-like profile in the isolated electrically stimulated rat vas deferens (EC_{50} 2.7 µM, inhibition of the twitch response) which could, however, not be unequivocally attributed to Y_2 receptor stimulation (Meister 1999).

In another arpromidine-based series the basic center was moved to the terminus of a flexible side chain in order to better mimic Arg^{35} of NPY (Müller et al. 1997). Surprisingly, *N*-imidazolylethyl-*N*-diphenyl-alkanoic acid amides with a terminal amino group (e.g., **3f** with submicromolar activity; Fig. 3) are considerably more potent than the corresponding guanidines in the functional Y_1 assay (Ca^{2+} assay, HEL cells). By contrast, when the imidazole moiety is replaced with phenol to imitate the C-terminal tyrosine of NPY, highest activity is found in combination with a guanidine (e.g., **3g**; Fig. 3). These inverse structure–activ-

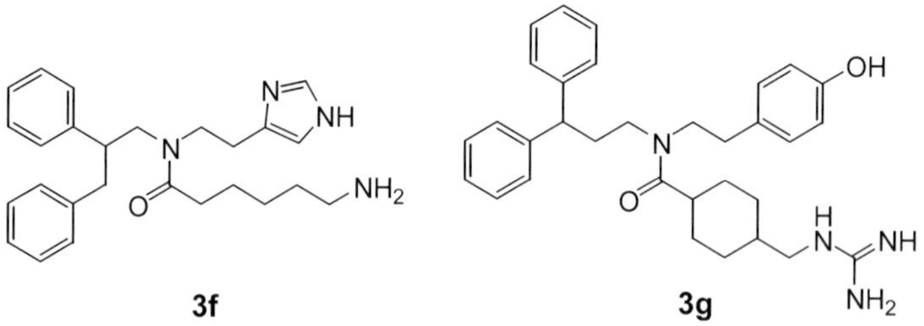

Fig. 3 *N*-imidazolylalkyl- and *N*-(hydroxyphenyl)ethyl-*N*-diphenylalkyl-alkanoic acid amides with terminal basic functions

ity relationships suggest different binding modes for NPY Y_1 antagonists with one and with two basic sites.

3
Potent and Selective Nonpeptide NPY Y_1 Receptor Antagonists

3.1
BIBP 3226 and Other (R)-Argininamides

3.1.1
Design and Pharmacology of BIBP 3226

A rational mimetic strategy based on the structure of NPY led to the synthesis of the first highly active and Y_1 selective nonpeptidic antagonist, BIBP 3226 (**4a**, Fig. 4) at Boehringer Ingelheim Pharma (Rudolf et al. 1994a; Doods et al. 1995; Wieland et al. 1995). The complete alanine scan of NPY (Beck-Sickinger et al. 1994) revealed that the C-terminal tetrapeptide, in particular Arg^{35} and Tyr^{36}, is most important for Y_1 receptor binding. Deletion of the carboxamide terminus and, surprisingly, replacing of L-arginine by its D-enantiomer proved to reproduce this pharmacophoric pattern. Lead optimization with hundreds of analogs resulted in BIBP 3226, (R)-N^2-(diphenylacetyl)-N-[(4-hydroxyphenyl)methyl]argininamide, a highly potent and selective Y_1 receptor antagonist (K_i 5.1 and 6.8 nM at human and rat Y_1 receptors, respectively; Rudolf et al. 1994a).

BIBP 3226 was found to be active in numerous functional in vitro tests, for example on rabbit vas deferens, rat renal tissue (Doods et al. 1995), guinea-pig vena cava (Doods et al. 1996) and HEL cells (Aiglstorfer et al. 1998). Except on human cerebral arteries (pK_b 8.5, Abounader et al. 1995), in vitro activity (pK_b 7–7.6) was lower than binding affinity. The receptor selectivity was also confirmed in functional tests for NPY antagonism. For example, using rat vas deferens for Y_2 and Y_4 (Doods et al. 1995, 1996) and rat colon for Y_3 receptors (Jacques et al. 1995) the compound was found to be inactive at concentrations ≤10 μM. Interestingly, BIBP 3226 also binds in a 50–100 nM range to human neuropeptide FF (NPFF) receptors and antagonizes the anti-opioid effect of NPFF (Mollereau et al. 2001, 2002), probably since the ligand fits with the C terminus of the octapeptide NPFF, Pro^5-Gln^6-Arg^7-Phe^8-amide, as with the analogous NPY terminus.

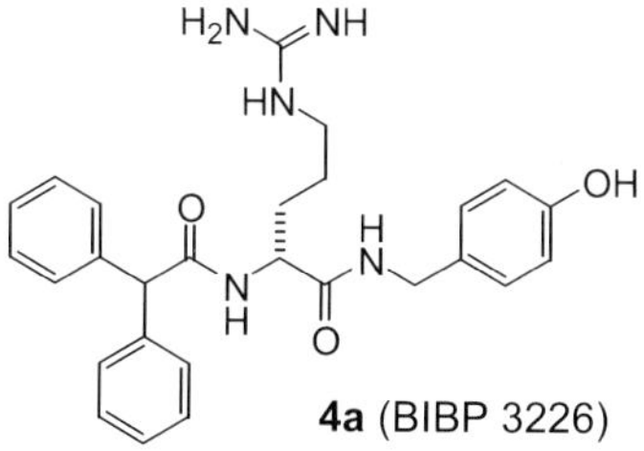

4a (BIBP 3226)

Fig. 4 Structure of the Y_1 receptor antagonist BIBP 3226 (**4a**)

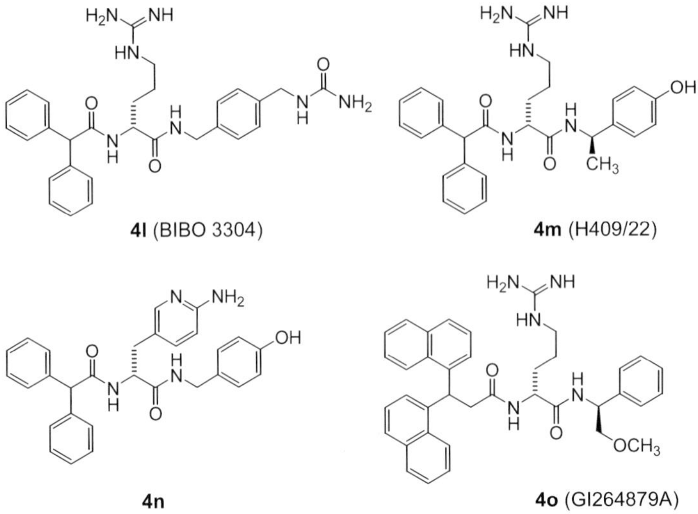

Fig. 5 NPY Y_1 antagonists from different sources based on BIBP 3226 as lead

In vivo, BIBP 3226 does not influence the basal blood pressure, but inhibits the hypertensive effect induced by administration of NPY, stimulation of the sympathetic nervous system or stress (Doods et al. 1996; Malmström et al. 1997). Though the compound is not an appropriate drug candidate due to, for example lack of oral bioavailability and inability to cross the blood–brain barrier, BIBP 3226 was used as a pharmacological tool in more than 100 studies to investigate Y_1 receptor mediated peripheral and central effects of NPY. Investigations of the effect of BIBP 3226 on the central regulation of feeding revealed contradictory results (Kask et al. 1998; Morgan et al. 1998). Morgan et al. (1998) and Iyengar et al. (1999) reported for both, BIBP 3226 and its inactive (S)-enantiomer BIBP 3435, the ability to block NPY induced food intake after injection into the paraventricular nucleus (PVN) of the hypothalamus or intracerebroventricular (icv) injection, so that a Y_1 specific mechanism is questionable. However, the closely related and more potent Y_1 antagonist BIBO 3304 (**4l**, Fig. 5) does exhibit central anorexigenic effects after icv or PVN administration (Wieland et al. 1998; Dumont et al. 2000b; Kask and Harro 2000; Polidori et al. 2000).

3.1.2
Structure–Activity Relationships of BIBP 3226 Derivatives

Some pharmacological data reflecting the structure–activity relationships of BIBP 3226 analogs are summarized in Table 1. First studies (Rudolf et al. 1997; Doods et al. 1996) indicated that the fit of BIBP 3226 to the Y_1 receptor binding site is highly stereospecific and nearly optimal, hardly leaving degrees of freedom for structural variation (but see below for N^G-substituted analogs). The (S)-enantiomer **4b** (BIBP 3435) is almost inactive. Moderate affinity remains if

Table 1 NPY Y_1 receptor binding of BIBP 3226 derivatives (Rudolf et al. 1997)

No.	R^1	R^2	X	n	*[a]	IC_{50} (nM)[b]
4a[c]	$CH(C_6H_5)_2$	$CH_2C_6H_4$-4-OH	$NHC(=NH)NH_2$	3	(R)-	5
4b[d]	$CH(C_6H_5)_2$	$CH_2C_6H_4$-4-OH	$NHC(=NH)NH_2$	3	(S)-	>10,000
4c	$CH_2C_6H_5$	$CH_2C_6H_4$-4-OH	$NHC(=NH)NH_2$	3	(R)-	370
4d	CH_3	$CH_2C_6H_4$-4-OH	$NHC(=NH)NH_2$	3	(R)-	>10,000
4e	9H-Fluoren-9-yl	$CH_2C_6H_4$-4-OH	$NHC(=NH)NH_2$	3	(R)-	72
4f	$CH(C_6H_5)_2$	$CH_2C_6H_4$-4-OH	$NHC(=NH)NH_2$	4	(R)-	220
4g	$CH(C_6H_5)_2$	$CH_2C_6H_4$-4-OH	NH_2	3	(R)-	>10,000
4h	$CH(C_6H_5)_2$	$CH_2C_6H_4$-4-OH	NH_2	4	(R)-	>10,000
4i	$CH(C_6H_5)_2$	$CH_2C_6H_5$	$NHC(=NH)NH_2$	3	(R)-	70
4j	$CH(C_6H_5)_2$	$(CH_2)_2C_6H_4$-4-OH	$NHC(=NH)NH_2$	3	(R)-	290
4k	$CH(C_6H_5)_2$	$CH_2C_6H_{10}$-4-OH	$NHC(=NH)NH_2$	3	(R)-	9,000

[a] Configuration of Arg.
[b] Receptor affinity determined by radioligand binding studies on SK-N-MC cells.
[c] BIBP 3226.
[d] BIBP 3435.

the (R)-arginine side chain is extended by one CH_2 group (**4f**), but, independent of the chain length, an exchange of the guanidine against an amine function results in complete loss of affinity. With respect to better pharmacokinetic properties various basic groups such as benzamidines or aminopyridines (cf. **4n**) were incorporated as mimics of the arginine side chain, usually resulting in compounds with reduced Y_1 receptor affinity compared to that of the reference compound **4a** (Rudolf et al. 1994b; Aiglstorfer et al. 1998; Antonsson et al. 2001). The backbone is open to modification only at the argininamide nitrogen; N-methylation reduces affinity by a factor of not more than five. As indicated by the weak binding of the monophenyl analog **4c**, the diphenylacetyl moiety is essential and should be sufficiently flexible since rigidization within a fluorene nucleus (**4e**) results in about 15-fold lower affinity. The para-OH substituent of the phenylmethyl moiety directly contributes to the high affinity of BIBP 3226. The nonhydroxylated analog **4i** is 14 times less active. However, the 4-(ureidomethyl) derivative BIBO 3304 (**4l**) has subnanomolar affinity for both the human and the rat Y_1 receptor [50% inhibitory concentration (IC_{50}), 0.38 and 0.72 nM, respectively] and is nearly inactive at Y_2, Y_4 and Y_5 receptors (IC_{50} >1,000 nM, Wieland et al. 1998). The chain length of the amide substituent is optimal with one methylene group as in **4a**, although a 2-(4-hydroxyphenylethyl) residue as in **4j** should be a better mimic of the C-terminal tyrosinamide in NPY.

Additional substituents at the benzylic carbon may be tolerated as demonstrated with H409/22 (**4m**, Fig. 5) and related compounds (Aiglstorfer et al. 2000; Malmström 2000; Malmström et al. 2000). The higher potency of the (*R*)-enantiomers is characteristic of the argininamide series of Y_1 antagonists [cf. BIBP 3226 (**4a**) vs. BIBP 3435 (**4b**); BIBO 3304 (**4l**) vs. its inactive enantiomer BIBO 3457]. In case of the α-methylated compound, highest activity resides in the (*R,R*)-configured stereoisomer **4m**, H409/22, which was tested in man, whereas the (*S,S*)-enantiomer is inactive (Malmström et al. 2000; Malmström 2002). Other examples of BIBP 3226-like Y_1 antagonists are **4n** (Antonsson et al. 2001) and GI264879A (**4o**) (Daniels et al. 2001). **4o** weakly binds in the micromolar range to Y_1, Y_4 and Y_5 receptors, but reduces food intake and body weight gain in obese animals, suggesting that interaction with more than one NPY receptor and/or other mechanisms may contribute to the inhibition of NPY mediated hyperphagia (Daniels et al. 2001).

Further structure–activity relationships of Y_1 antagonists related to BIBP 3226 were explored by functional investigations on HEL cells (inhibition of intracellular calcium mobilization induced by 10 nM NPY; Aiglstorfer et al. 1998; Aiglstorfer et al. 2000). Introduction of a *p*-Cl substituent at the diphenylacetyl group is tolerated and may be even favorable. The 3,3-diphenylpropionyl homolog of BIBP 3226 (IC_{50} 510 nM compared to 17 nM for BIBP 3226) is much more active than the 2,3-diphenylpropionyl analog. Relatively open to the introduction of substituents is again the (4-hydroxyphenyl)methylamide moiety which may be incorporated into a tetrahydro-1*H*-benzo[c]azepine nucleus (IC_{50} 280 nM; Aiglstorfer et al. 2000). A methylation at the hydroxybenzyl α-carbon leads to compounds with activities comparable to that of **4a**, indicating that a certain bulk is tolerated in this position. The backbone conformations of the NPY C-terminus and of BIBP 3226 should therefore be different so that the corresponding guanidino and *para*-hydroxyphenyl groups may similarly interact with the Y_1 receptor.

3.1.3
The Y_1 Receptor Binding Site for BIBP 3226:
In Vitro Mutagenesis Results and Computer Models

The obvious suggestion that NPY and BIBP 3226 share an overlapping binding site at the human Y_1 receptor has been extensively investigated by in vitro mutagenesis and computer modeling (Sautel et al. 1996; Du et al. 1997a). Reduced affinity of the antagonist to the respective alanine mutants indicates which residues might contribute to BIBP 3226 binding. Most of these positions, namely W163, F173, Q219, N283, F286, D287 (Sautel et al. 1996) and additionally Q120, F282, H306 (Du et al. 1997a) are important for NPY and BIBP 3226 affinity and thus thought to form an overlapping binding region of both ligands. Positions Y211 (Sautel et al. 1996), Y47, W276, H298 and F302 (Du et al. 1997a) seem to participate only in binding of BIBP 3226, but Y47 and H298 were demonstrated in another in vitro mutagenesis study (Kanno et al. 2001) to interact with peptide YY (PYY). These experimental results have been considered in computer models of

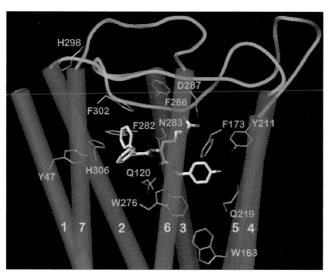

Fig. 6 Computer model of the human neuropeptide Y Y_1 receptor, based on the crystal structure of bovine rhodopsin, in complex with BIBP 3226. TM regions are numbered and shown as *blue cylinders*. Labeled residues (C atoms: *orange*): weak or no binding of BIBP 3226 after mutation). The model was generated by the software package SYBYL 6.8 (Tripos Inc., St. Louis)

the Y_1 receptor complexed with BIBP 3226, but the proposed binding modes are rather different due to the mutants taken into account. Moreover, the homology modeling based on bacteriorhodopsin and the electron microscopy map of rhodopsin, respectively, could not represent the very recent progress resulting from the high resolution crystal structure of bovine rhodopsin (Palczewski et al. 2000).

Recently, a new and more reliable model of BIBP 3226 binding to the Y_1 receptor was generated (S. Dove, unpublished results) on the basis of an unambiguous sequence alignment of the transmembrane (TM) regions with those of bovine rhodopsin, using the crystal structure of the latter as template and taking into consideration all published results with Y_1 receptor mutants. The suggested topology of the BIBP 3226 binding site within the novel, rhodopsin-based alignment of the TM and extracellular regions becomes obvious from the important residues highlighted in Fig. 6. The binding mode derived from the mutants reported by Sautel et al. (1996) could be reproduced with the new model. All key interactions occur within a deep pocket between TM domains 4–7. However, Y47 (TM1) and Q120 (TM3) (Du et al. 1997a) cannot approach BIBP 3226 in this mode. To include the highest possible number of responding mutants, another mode is suggested which, in principle, retains interactions of the D-argininamide and the (4-hydroxyphenyl)methyl moiety as previously proposed, but extends the diphenylacetyl site towards TM domains 1 and 3 (see Fig. 6). With respect to the number and quality of interactions, this mode is superior to that suggested by Du et al. (1997a) where essentially the (4-hydroxyphenyl)methyl and diphenylacetyl sites were exchanged. Interestingly, it is never possible to include W163

Table 2 Pharmacological data of N^G-acylated BIBP 3226 derivatives (Hutzler 2001)

[Structure of 4a, 5a-j]

No	X	Y	R^1	R^2	Y_1 antagonism[a] IC_{50} (nM)	Binding data K_i (nM)[b] Y_1	Y_2	Y_5
4a	H	H	H	H	14	2	8,000	52,300
5a	H	H	H	COMe[c]	45.4	11.9	2,1100	9,350
5b	H	H	H	CO_2Et[c]	2.5	4.5	19,100	1,4500
5c	H	F	H	CO_2Et	0.91	8.5	5,080	12,300
5d	H	H	H	CO_2CH_2Ph[c]	0.98	48.6	4,200	21,400
5e	H	H	H	CONHEt	1.18	0.06	19,500	21,300
5f	H	H	H	$CONHCH_2CO_2Et$	1.65	0.06	2480	17,700
5g	H	F	H	$CONHCH_2CO_2Et$	0.86	0.31	2,340	44000
5h	Cl	Cl	H	$CONHCH_2CO_2Et$	0.6	0.53	650	24,100
5i	H	H	H	$CONH(CH_2)_5CO_2Et$	0.64	0.72	550	7,500
5j	H	H	CO_2Et	CO_2Et	8200	–	–	–

[a] Inhibition of NPY (10 nM) stimulated Ca^{2+} mobilization in HEL cells.
[b] Determined on SK-N-MC cells (Y_1), SMS-KAN cell membrane preparations (Y_2) and hY_5-transfected HEC-1B cells (Moser et al. 2000); radioligand: [^{3}H]propionyl-NPY (1 nM).
[c] Me=CH_3; Et=C_2H_5; Ph=C_6H_5.

(TM4) into binding of BIBP 3226. The inability of the W163A mutant to bind the antagonist and NPY might be due to rearrangement of the transmembrane regions since the indole nitrogen probably forms a hydrogen bond with N81 (TM2) like the identical residues in the rhodopsin crystal structure.

The suggested key interactions are depicted in Fig. 6. The D-argininamide backbone oxygen is hydrogen bonded to the side chain of N283 (TM6). The guanidino group interacts with the carboxylate of D287 (at the top of TM6 in the rhodopsin-based alignment). Also the suggested hydrogen bond between the amide nitrogen of Q219 (TM5) and the (4-hydroxyphenyl)methyl oxygen (Sautel et al. 1996) is retained. Y211 (TM5) might form another hydrogen bond to the 4-OH group. The diphenylacetyl moiety extends, with one phenyl ring, towards Y47 (TM1) and H306 (TM7). The model suggests that a p-Cl substituent should be slightly favorable for interaction with Y47 as indicated by structure–activity relationships (Aiglstorfer et al. 2000; see also **5h**, Table 2). Q120 (TM3) is supposed to form an additional H bond with the diphenylacetyl oxygen. This pattern is completed by aromatic-aromatic and π-cation interactions within a large pocket aligned by the side chains of F173 (TM4), W276 (TM6), F282 (TM6), F286 (TM6), F302 (TM7) and H306 (TM7), comprising all terminal groups of BIBP 3226.

3.1.4
N^G-Substituted (R)-Argininamides with Reduced Basicity

Recently, the Y_1 receptor binding models of BIBP 3226 were used to suggest that appropriate N^G-substituents at the D-arginine side chain will retain or even increase antagonistic activity (Hutzler 2001; Hutzler et al. 2001). With single alkyl or arylalkyl groups, no improvement was achieved. Radioligand binding studies on SK-N-MC cells resulted in K_i values of 2 nM (BIBP 3226), 2.6 nM (N^G-methyl), 27 nM (N^G-propyl) and 48 nM (N^G-phenylpropyl). With the intention to reduce the basicity of the guanidino group and, by this, to increase the hydrophobicity of the ligands for better blood–brain passage, electron-withdrawing substituents were introduced. Selected N^G-acylated derivatives are presented in Table 2 together with results of calcium assays on HEL cells and with binding data on Y_1, Y_2 and Y_5 receptors. Some of the compounds are up to 20 times more active in the functional test and show more than 30 times higher Y_1 receptor affinity than BIBP 3226. Y_1 selectivity is even increased in most cases. The basicity of the guanidino group is reduced to pK_a values of about 8, indicating that considerable amounts of the N^G-acylated argininamides are uncharged under physiological conditions. Probably, the ionic interaction of BIBP 3226 with Asp^{287} can be replaced by a charge-assisted hydrogen bond. Long N^G-substituents may interact with residues in TM domains 5 and 6 and project towards the extracellular loops. The N^G-ester substituted compounds **5a–d** as such are active as Y_1 antagonists (see Table 2), but they are also prodrugs which may be enzymatically cleaved by esterases to form the unsubstituted guanidine **4a** (BIBP 3226) as demonstrated for some alkoxycarbonyl derivatives in vitro. The inactive diester **5j** is stepwise (via **5b**) converted to **4a** (Hutzler 2001; Hutzler et al. 2001).

3.2
Benzamidine-Type Y_1 Antagonists SR 120819A and SR 120107A

The potent NPY Y_1 receptor antagonist SR 120819A (**6a**; Fig. 7), designed at Sanofi (Serradeil-Le Gal et al. 1995; Serradeil-Le Gal 1997), was published shortly after BIBP 3226 as the first orally active Y_1 antagonist. The backbone of this arylsulfonyl substituted peptide mimetic resembles that of the benzamidine-type thrombin inhibitor NAPAP. The (R,R)-cis-configured compound SR 120819A and its less active trans-diastereoisomer SR 120107A (**6b**; Fig. 7) (Serradeil-Le Gal 1997) are based on the C terminus of NPY and provided with two basic centers (benzamidine and tertiary amine) presumably mimicking Arg^{33} and Arg^{35}. Radioligand binding studies revealed high affinity at rat, guin-

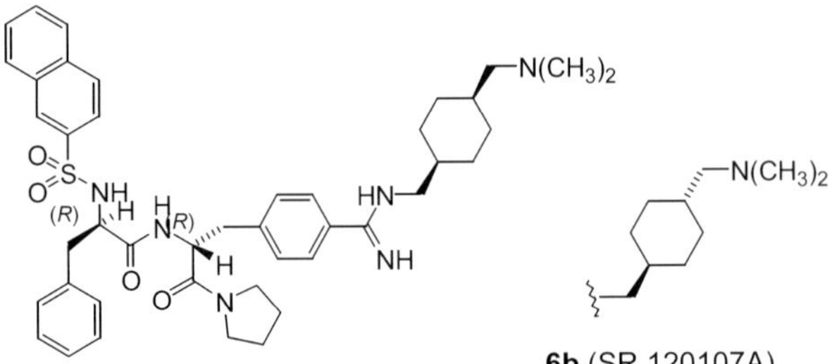

6a (SR 120819A) **6b** (SR 120107A)

Fig. 7 Structure of the benzamidine derivatives SR 120819A (**6a**) and SR 120107A (**6b**)

ea pig and human Y_1 receptors (**6a:** K_i 11–22 nM, **6b:** K_i 11–80 nM; Serradeil-Le Gal 1997; Malmström 2002). At a dosage of 5 and 10 mg/kg **6a** inhibited the rise in diastolic blood pressure induced by [Leu^{31}, Pro^{34}]NPY (5 μg/kg iv) in anesthetized guinea pigs with a long duration of action of more than 4 h (Serradeil-Le Gal, 1995).

3.3
Indoles, Benzimidazoles, and Benzothiophenes

By library screening and similarity searches at Lilly Research Laboratories the trisubstituted indole **7a** (Fig. 8) was discovered as NPY Y_1 antagonistic lead with low affinity at human Y_1 receptors expressed in AV-12 cells (K_i 2.1 μM for displacement of [^{125}I]PYY; Hipskind et al. 1997). This structure was optimized in different positions, leading to some of the most potent Y_1 antagonists known so far. First attempts maintained the 1-methyl-2-(4-chlorophenoxy)methylindole scaffold. Variation of the 3-substituent resulted in markedly improved activity with a 1,4'-bipiperidine group linked by two C atoms to C-3 (**7b:** K_i 93 nM; **7c:** K_i 26 nM). Based on the C terminus of NPY, the introduction of an additional basic moiety at *N*-1 was suggested. Alkylpiperidine side chains with a free NH

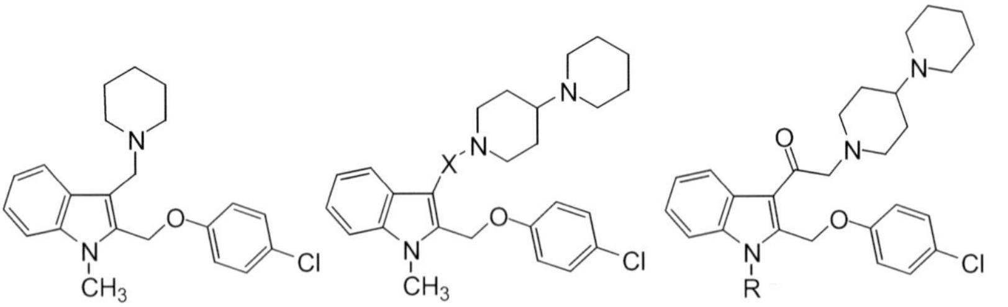

7a

7b: X = CH_2CH_2
7c: X = $C(O)CH_2$

7d: R = 2-(piperidin-4-yl)ethyl
(*R*)-**7e**: } R = 3-(piperidin-
(*S*)-**7e**: } 3-yl)propyl

Fig. 8 Optimization of 1,3-substituted 2-[(4-chlorophenoxy)methyl]indoles

Table 3 In vitro Y_1 receptor binding (human Y_1 in AV-12 cells) and Y_1 antagonistic activity (cAMP assay in SK-N-MC cells) of selected 2-[(4-chlorophenoxy)methyl]benzimidazoles **8** (Zarrinmayeh et al. 1998)

No.	R	Y_1 binding data K_i (nM)	Y_1 antagonism K_i (nM)
4a[a]		4.6	11
8a	H	700	–
8b	CH_3	97	–
8c	[2-(piperidin-1-yl)ethyl]oxy	43	240
8d	[3-(piperidin-1-yl)propyl]oxy	1.7	2.7
8e	[2-(piperidin-2-yl)ethyl]oxy	29	–
8f	[3-(piperidin-2-yl)propyl]oxy	16	53
8g	[2-(piperidin-3-yl)ethyl]oxy	18	91
8h	[3-(piperidin-3-yl)propyl]oxy	30	77
8i	[2-(piperidin-4-yl)ethyl]oxy	7	15
8j	[3-(piperidin-4-yl)propyl]oxy	152	119
(S,S)-8h	[3-(piperidin-3-yl)propyl]oxy	6	87
(R,R)-8h	[3-(piperidin-3-yl)propyl]oxy	41	153
(R,S)-8h	[3-(piperidin-3-yl)propyl]oxy	27	137
(S,R)-8h	[3-(piperidin-3-yl)propyl]oxy	17	65

[a] BIBP 3226.

were optimal in this position. K_i values in the low nanomolar and subnanomolar range were obtained in binding studies with the compounds **7d** (K_i 1.9 nM), (R)-**7e** (K_i 1.4 nM), and (S)-**7e** (LY 357897, K_i 0.75 nM). The activity of (S)-**7e** in different functional assays was in a similar range: K_i 1.8 nM for reversal of NPY-induced inhibition of forskolin-stimulated cAMP, 3.2 nM for inhibition of NPY-induced Ca^{2+} mobilization in SK-N-MC cells. The compounds proved to be highly selective for the Y_1 receptor (Y_2, Y_4, Y_5: K_i values >10 μM). (S)-**7e** blocked the food consumption in mice, elicited by a submaximal (230 pmol) icv administered dose of NPY, with an ED_{50} of 17 nmol.

Attempts to replace the indole moiety by other nuclei resulted in a series of 2-[(4-chlorophenoxy)methyl]benzimidazoles with an optimal 1-[3-(piperidin-3-ylpropyl)] substituent (Zarrinmayeh et al. 1998, 1999). Some of the analogs and their NPY Y_1 antagonistic potencies are summarized in Table 3. The parent compound **8a** was weakly active. Comparison with the most potent indoles suggested that their structure may be best matched with appropriate 4-substituents at the benzimidazole moiety. Introduction of a methyl group in **8b** produced a seven-fold increase in receptor affinity. 3-Piperidinylpropoxy and 2-piperidiny-

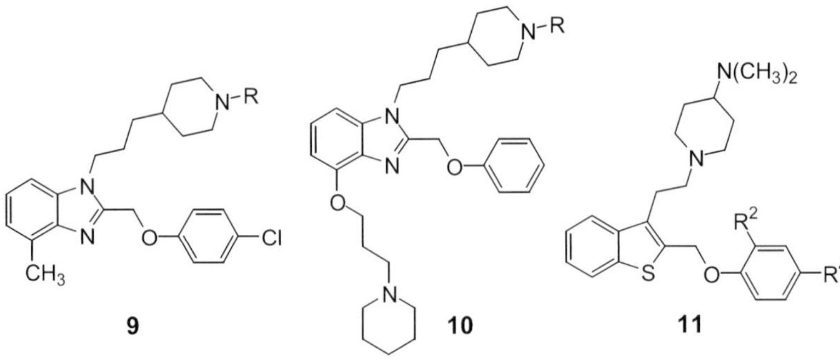

Fig. 9 General structures of highly active NPY Y_1 antagonistic benzimidazole (**9, 10**) and benzothiophene (**11**) derivatives. Example structures: **9a**, R=3-(piperidin-1-yl)propyl; **9b**, R=4-(piperidin-1-yl)butyl; **10a**, R=isobutyl; **10b**, R=cyclohexylmethyl; **10c**, R=2-phenylethyl; **10d**, R=3-phenylprop-2-en-1-yl; **10e**, R=3-(piperidin-1-yl)propyl; **10f**, R=4-oxo-4-phenylbutyl; **10g**, R=2-(4-iodophenyl)ethyl; **11a**, $R^2=CH_2OH$, $R^4=Br$; **11b**, $R^2=CH_2OCH_3$, $R^4=Br$; **11c**, $R^2=CN$, $R^4=Br$

lethoxy substituents are more favorable. The structure–activity relationships of the piperidine isomers are not uniform and point to an optimal position of the basic nitrogen relative to the benzimidazole nucleus: 3-(piperidin-1-yl)- and 3-(piperidin-2-yl)propoxy derivatives are more active, but 3-(piperidin-3-yl)- and 3-(piperidin-4-yl)propoxy substituted compounds are less potent than their ethoxy analogs. Among the diastereomers of **8h**, highest affinity was found with the (S,S)-configuration. The most potent compound in Table 3, **8d**, approached the nanomolar range in NPY Y_1 receptor binding as well as in the functional data for Y_1 antagonism in SK-N-MC cells.

Optimization of substituents at the piperidine nitrogen in 2-[(4-chlorophenoxy)methyl]-4-methyl-1-[3-(piperidin-4-yl)propyl]benzimidazoles **9** (Fig. 9) also resulted in very active Y_1 antagonists (Zarrinmayeh et al. 1998). The highest NPY Y_1 receptor affinity was obtained by introduction of an additional basic nitrogen separated by 3–4 C atoms from the piperidine-N (Fig. 9; **9a**: K_i 5 nM; **9b**: K_i 6 nM). It was a reasonable extension of this work to combine such 1-substituents with the 4-[3-(piperidin-1-yl)propoxy] group present in **8d** (Zarrinmayeh et al. 1998). Indeed, a number of compounds of the common structure **10** displayed Y_1 receptor affinities in the subnanomolar range (Zarrinmayeh et al. 1999). K_i values lower than 0.3 nM were found for derivatives with higher alkyl groups (see Fig. 9; **10a,b**). Phenylalkyl or phenylalkenyl substitution (**10c,d**) or the attachment of a moiety with a polar group as in **10e,f** led to even higher affinities (K_i 0.1–0.2 nM). The derivative with a p-iodophenylethyl substituent R (**10g**) is among the most potent nonpeptide NPY Y_1 receptor ligands known so far (K_i 0.05 nM).

With benzothiophene derivatives a third nucleus was used at Lilly as a scaffold for the design of NPY Y_1 antagonists (Britton et al. 1999). Optimization of both side chains in 2- and 3-position resulted in the common structure **11** (Fig. 9). A 4-chlorophenoxymethyl group, as present in the indole and benzimidazole series, leads only to moderate affinity (K_i 310 nM). However, the potency may be significantly increased by appropriate multiple substitution at the phenyl ring. The most active Y_1 antagonists in this series (K_i values: 11–15 nM) are those with polar groups in *ortho* position (R^2=CH$_2$OH, CH$_2$OMe, CN) and an additional *para*-Br substituent (**11a–c**).

3.4
Y_1 Antagonists Based on Other Common Structures

3.4.1
6-Benzylsulfonyl-5-nitroquinolines

Arylsulfonyl compounds from Parke-Davis/Warner-Lambert with a nitroquinoline nucleus are only weakly basic and do not obviously overlap with NPY (Wright et al. 1996). The 8-amino-5-nitroquinoline and the phenylsulfonyl moieties in the general structure **12** (Fig. 10) are essential for Y_1 antagonistic activity. Whereas the parent compound PD 9262 (**12a**) is not very potent (K_i 282 nM for displacement of [^{125}I]PYY from SK-N-MC membranes), *ortho*-alkyl or halogen substituents at the phenyl ring enhance affinity up to a K_i value of 48 nM for the *ortho*-isopropyl derivative PD 160170 (**12b**).

3.4.2
Phenylpiperazines

Neurogen and Pfizer have patented 1-(1-phenylcyclohexyl)-4-phenyl-piperazines with amide, amine and ether substituents R (**13**, Fig. 10) as novel class of NPY Y_1 specific ligands (Peterson et al. 1996; Blum et al. 2000). No biological

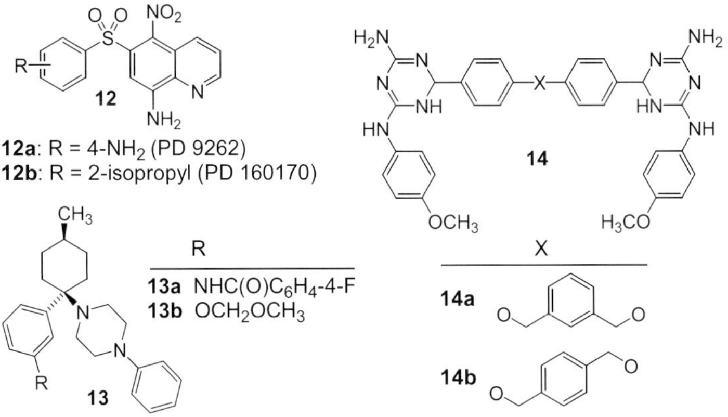

12a: R = 4-NH$_2$ (PD 9262)
12b: R = 2-isopropyl (PD 160170)

R
13a NHC(O)C$_6$H$_4$-4-F
13b OCH$_2$OCH$_3$

Fig. 10 6-Benzylsulfonyl-5-nitroquinolines **12**, phenylpiperazines **13** and bis[diamino(phenyl)triazines] **14**

data were given for the new amides (e.g., **13a**). An ether derivative (**13b**) displaced [^{125}I]PYY with an IC$_{50}$ value of 30 nM.

3.4.3
Bis[diamino(phenyl)triazines]

At Alanex different structures were identified as nonpeptide NPY Y$_1$ receptor antagonists by pharmacophore-based approaches and by screening of combinatorial libraries (Rabinovich et al. 1997). Examples of discovered compounds are symmetric bis[diamino(phenyl)triazines] **14** with a disubstituted central benzene ring as spacer (Fig. 10). The K_i values for Y$_1$ receptor binding are 117 nM (*meta* derivative AXC01829 **14a**) and 150 nM (*para* analog AXC011018 **14b**). A certain similarity of the compounds to benextramine (**2a**) and analogs is obvious.

3.4.4
Benzazepines and Benzodiazepines

Hybrid compounds combining a CCK-B receptor antagonistic benzodiazepine and a histamine H$_2$ receptor blocking roxatidine-like moiety were synthesized at Shionogi and Co. (Shigeri et al. 1998), for example the derivative **15** (Fig. 11), which was about equiactive with BIBP 3226 as NPY Y$_1$ antagonist (K_i 6.4 nM in radioligand binding studies, IC$_{50}$ 95 and 320 nM in functional Ca^{2+} and cAMP assays in SK-N-MC cells). No binding to Y$_2$ and Y$_5$ receptors was observed up to concentrations of 1 μM. The hybrid molecule maintains the CCK-B and histamine H$_2$ antagonistic potency of the components, which were, however, both inactive at NPY Y$_1$ receptors. Other series of Y$_1$ antagonists from Shionogi and Co. are based on a 1,3-disubstituted benzazepine nucleus (Murakami et al. 1999a, 1999b). The common structure **16** was optimized at both positions. Generally, derivatives with urea moieties (R^1=NH-alkyl) are about 10 times more potent than the corresponding carbamates (R^1=O-alkyl). Maximal Y$_1$ receptor binding affinity (K_i 2.9 nM) was observed for the 3-guanidino derivative

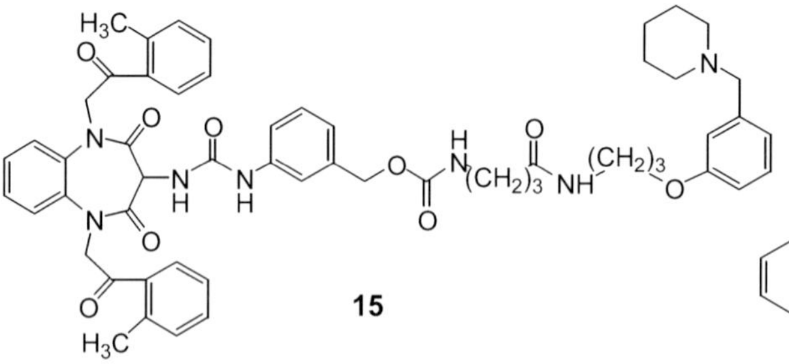

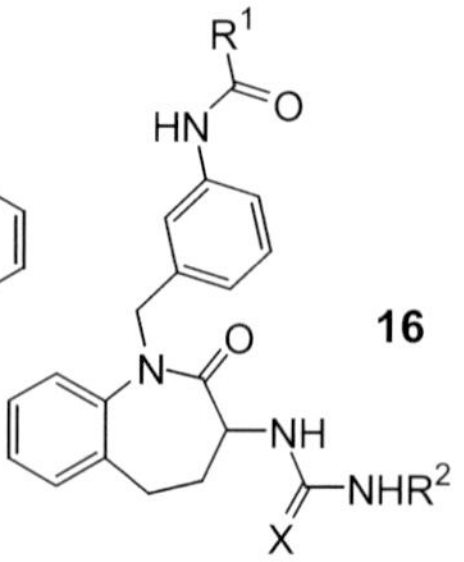

Fig. 11 Benzodiazepine **15** and general structure of benzazepines **16** (X=O; R^1=NH-alkyl, O-alkyl; R^2=aryl, heteroaryl)

(X=NH) with an isopropylamino and a 4-hydroxyphenyl group as R^1 and R^2, respectively, whereas the 3-ureido analog (X=O) was much less potent (K_i 82 nM). Further optimization of R^2 in a 3-ureido series (R^1=NH-isopropyl) also led to compounds with K_i values lower than 10 nM (R^2=6-benzofuryl, 6-benzothienyl, 6-benzothiazolyl, 2-F-phenyl, 2,4-di-F-phenyl) (Murakami et al. 1999b). The 6-benzothiazolyl derivative (K_i 5.1 nM) was functionally characterized as a Y_1 antagonist and did not show any effects on Y_2, Y_4 and Y_5 receptors.

3.4.5
Morpholinopyridines J-104870 and J-115814

Two highly potent NPY Y_1 receptor antagonists, the morpholinopyridines J-104870 (**17a**) and J-115814 (**17b**, Fig. 12) were disclosed by Banyu (Kanatani et al. 1999; Kanatani et al. 2001). J-104870 displaced [^{125}I]PYY binding to cloned human and rat Y_1 receptors with K_i values of 0.29 and 0.54 nM, respectively, and inhibited the NPY-induced intracellular calcium mobilization (IC_{50} 3.2 nM). K_i values determined for the binding at other NPY receptors were greater than 5 μM. Anorexigenic effects on NPY-mediated feeding of rats were demonstrated by both intracerebroventricular and oral administration of the compound (Kanatani et al. 1999). J-115814 (K_i 1.4–1.8 nM) was nearly as potent as J-104870. Feeding induced by icv NPY was unaffected by intraperitoneally injected J-115814 in Y_1(−/−) mice, but suppressed in wild-type and Y_5(−/−) mice (Kanatani 2001). Together these findings suggest the contribution of Y_1 receptors in the regulation of food intake. In vitro mutagenesis studies on the human Y_1 receptor (Kanno et al. 2001) resulted in reduced affinity of J-104870 at alanine mutants of amino acids Trp^{163}, Phe^{173}, Asn^{283}, Asp^{287} and Leu^{303}, indicating

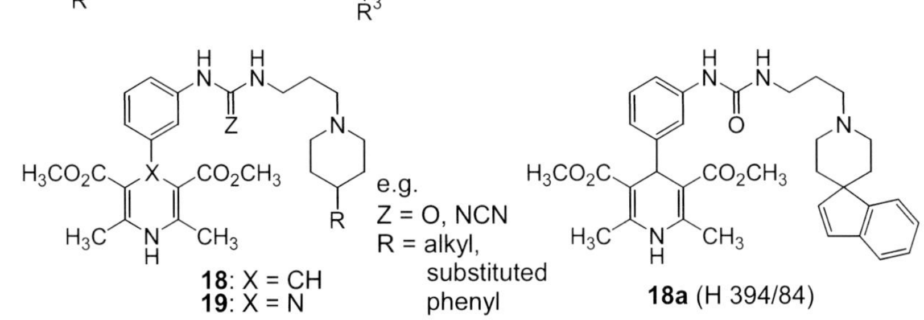

Fig. 12 Morpholinopyridines **17**, dihydropyridines **18** and dihydropyrazines **19**

that the compound recognizes a pocket formed by TM domains 4, 5 and 6 which only partially overlaps with the binding site of other antagonists like BIBP 3226 or the peptide 1229U91.

3.4.6
Dihydropyridines and Dihydropyrazines

Recently, dihydropyridine (**18**) and dihydropyrazine derivatives (**19**, Fig. 12) from Bristol-Myers Squibb (Poindexter et al. 1996, 2002; Sit et al. 2002) were described as NPY Y_1 antagonists. Generally the dihydropyridines **18** were up to about 100 times more potent in displacing [^{125}I]PYY from human Y_1 receptors than the corresponding dihydropyrazine analogs **19**. Highly active compounds **18** (K_i 2–5 nM) are urea derivatives (Z=O) with 2-methoxy-, 3-methoxy- or 3-hydroxy-substituted phenyl rings as R. Replacement of the urea with a cyanoguanidine group (Z=NCN) results in a further increase in activity (e.g., with R=*tert*-butyl: K_i <1 nM). For the derivative BMS-193885 (Z=O, R=2-methoxyphenyl) full functional Y_1 antagonism (K_b 4.5 nM) was observed in a cAMP assay using human Y_1 receptor expressing CHO cells. The compounds are Y_1 selective and specific in spite of the presence of α_1 adrenoceptor and calcium channel blocking pharmacophores. H 394/84 (**18a**) antagonized vascular responses to exogenous and endogenous, neuronally released NPY with similar potency already at plasma levels of 29 nM with a long duration of action in vivo (Malmström et al. 2001; Malmström 2002).

4
Selective Nonpeptide Y_2 Receptor Antagonists

For a long time Y_2 receptor blocking agents were eagerly awaited as pharmacological tools. First approaches to the design of NPY Y_2 receptor antagonists, described by Grouzman et al. (1997), were based on the structure of the C-terminal tetrapeptide in NPY. A template-assembled synthetic protein (TASP), T_4-[NPY(33–36)]$_4$, consisting of four NPY(33–36) residues bound via spacer groups to a cyclic template (T_4) was reported to display Y_2 receptor binding in the submicromolar range (Grouzmann et al. 1997). However, it was not before 1999 that a nonpeptide ligand, the L-arginine derivative BIIE 0246 (**20a**, Fig. 13), was disclosed (Boehringer Ingelheim Pharma; Doods et al. 1999). BIIE 0246 proved to be a highly potent and selective Y_2 receptor antagonist in binding experiments as well as in functional pharmacological studies (Doods et al. 1999; Dumont et al. 2000a).

Radioligand binding studies on Y_2 receptors (SMS-KAN cells) revealed an IC$_{50}$ value of 3.3 nM, whereas no displacement of radiolabeled NPY was observed in Y_1, Y_4 and Y_5 receptor assays (Doods et al. 1999). Competitive Y_2 antagonism of **20a** was demonstrated, for instance, in pharmacological investigations on the isolated electrically stimulated rat vas deferens. The compound was used in numerous studies to investigate the contribution of NPY Y_2 receptors to

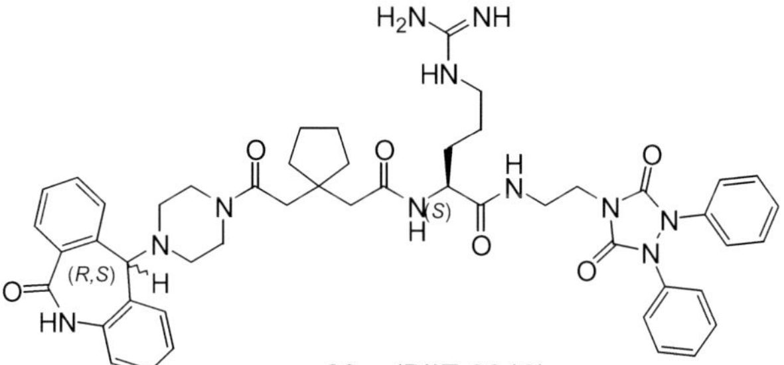

Fig. 13 Structure of the nonpeptide Y$_2$ receptor antagonist BIIE 0246 (**20a**)

Table 4 Structure–activity relationships of BIIE 0246 related argininamides (Dollinger et al. 1999; Esser et al. 1999; Rudolf et al. 1999)

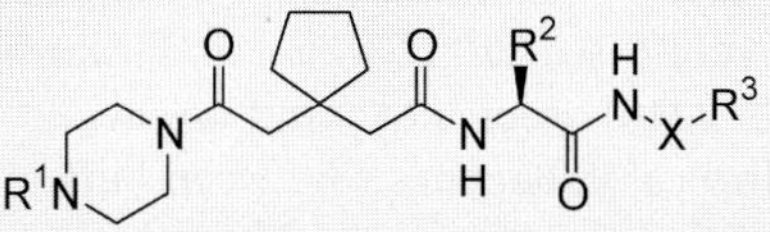

No.	R^1	R^2	X	R^3	IC$_{50}$ (nM)[a]
20a[b]	i	(CH$_2$)$_3$NHC(=NH)NH$_2$	(CH$_2$)$_2$	iii	7.5
20b	i	(CH$_2$)$_3$NHC(=NH)NH$_2$	(S)-CH(CONH$_2$)	iv	36
20c	i	4-C$_6$H$_4$-C(=NH)NH$_2$	(S)-CH(CONH$_2$)	iv	1 000
20d	ii	(CH$_2$)$_3$NHC(=NH)NH$_2$	(S)-CH(CONH$_2$)	iv	40
20e	ii	(CH$_2$)$_3$NHC(=NH)NH$_2$	(CH$_2$)$_2$	iii	32
20f	ii	(CH$_2$)$_3$NHC(=NH)NH$_2$	(CH$_2$)$_2$	v	220
20g	ii	(CH$_2$)$_3$NHC(=NH)NH$_2$	(CH$_2$)$_2$	vi	340
20h	ii	(CH$_2$)$_3$NHC(=NH)NH$_2$	(CH$_2$)$_4$	iii	68

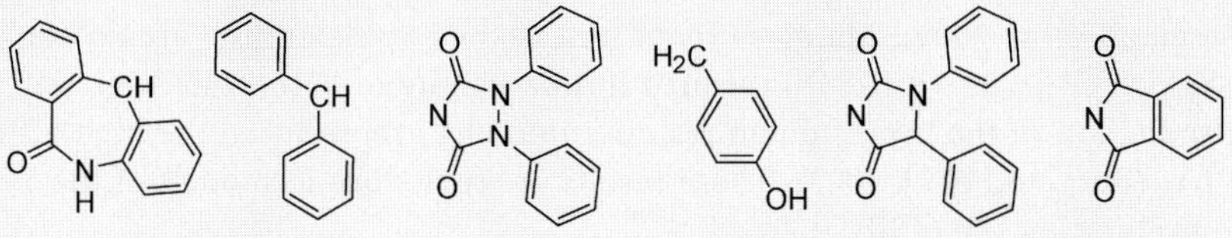

[a] Displacement of ^{125}I labeled NPY from rabbit kidney preparations.
[b] BIIE 0246.

complex physiological effects of NPY, for example in tissues with heterogenous receptor populations (Dumont et al. 2000a; Malmström 2001a, 2001b; Smith-White et al. 2001; Malmström et al. 2002a, 2002b).

Similar to the discovery of the Y$_1$-selective antagonist BIBP 3226 (**4a**, see Sect. 3.1), BIIE 0246 was synthesized as a member of an extensive set of related peptidomimetics which were designed as putative mimics of the C terminus in

NPY. For some of these BIIE 0246 analogs Y_2 receptor binding data are given in the patent literature (for examples see Table 4) (Dollinger et al. 1999; Esser et al. 1999; Rudolf et al. 1999). The 5,11-dihydrodibenzo[b,e]azepin-6-one group, though obviously representing the best suited substructure, can be replaced by an α-diphenylmethyl residue resulting in a relatively small decrease in affinity at the Y_2 receptor by a factor of 4–5 (**20e**) (Dollinger et al. 1999). By contrast, exchange of the L-Arg side chain, which is presumably mimicking Arg^{35} of NPY, by an isosteric p-benzamidino group is not tolerated (**20c**). Another interesting observation is that the replacement of the 4-(2-aminoethyl)-1,2-diphenyl-1,2,4-triazolidine-3,5-dione group by an L-tyrosinamide residue (**20b**) does not strongly alter affinity. This may be interpreted to mean that both substructures are bioisosters mimicking the C terminus in NPY. In contrast to BIIE 0246, other compounds of this series, for example **20d**, were found to produce an elevation of blood pressure in anesthetized rats which was attributed to different qualities of action (agonism/antagonism) at NPY receptors (Esser et al. 1999). Variations of the 4-(2-aminoethyl)-1,2-diphenyl-1,2,4-triazolidine-3,5-dione motif (see **20f–20h**) led to diminished Y_2 affinity.

Interestingly, **20a** (BIIE 0246) does not bind to avian NPY Y_2 receptors (Salaneck et al. 2000). Reciprocal mutagenesis between human (hY_2) and chicken Y_2 receptor (chY_2) revealed that three amino acids in hY_2 are especially important for BIIE 0246 binding: Gln^{135} in TM3, Leu^{227} in TM5, and Leu^{284} in TM6 (Berglund et al. 2002). Mutagenesis of hY_2 to the corresponding amino acids in chY_2 (Q135H, L227Q, L284F) resulted in low affinity of BIIE 0246. Inversely, the introduction of the three human residues into chY_2 reproduced the high affinity to hY_2 (Berglund et al. 2002). These results are first clues to models of the Y_2 receptor binding site of BIIE 0246.

5
NPY Y_5 Receptor Antagonists

The cloning of Y_5 receptors (Gerald et al. 1996) and the discovery of first low molecular weight Y_5-blocking compounds stimulated an intensive and successful search for potent and selective nonpeptide ligands. The interest in such compounds was not surprising as the Y_5 receptor—previously referred to as 'Y_1-like' or the putative 'feeding receptor'—has been considered as a key target for the control of body weight. Meanwhile various Y_1 and Y_5 receptor antagonists have been investigated as potential anti-obesity agents. The results do not allow us to conclude that the orexigenic effect of NPY depends on a single receptor subtype (Duhault et al. 2000), and there is some doubt concerning a significant contribution of the Y_5 receptor (Turnbull et al. 2002). Investigations of the NPY receptor mediated feeding response in Y_1 and Y_5 receptor knockout mice indicated a dominant role of Y_1 receptors (Kanatani et al. 2000b). Nevertheless, the following selection of compounds reflects the enormous efforts of pharmaceutical companies in the Y_5 field: all nonpeptide Y_5 receptor ligands mentioned in this section have been disclosed in patents over the last few years. The novel structures were summarized in detailed reviews by Ling (1999), Hammond (2001), and Dax (2002). Published pharmacological data for these compounds are mostly restricted to IC_{50} values for the displacement of radioligands, and information on structure–activity relationships is limited to a few series.

Although the diversity of the known Y_5 antagonists is very large, many of the highly active structures seem to represent a common pattern which may be roughly characterized as 'barbell-shaped': a nonbulky central group containing heteroatoms connects two larger terminal moieties consisting of polar or hydrophobic, aromatic or alicyclic, hetero- and/or polycyclic rings. Although this very general and simplified pattern does not necessarily suggest common ligand–receptor interactions, overlapping binding sites with a similar orientation of the structures may be assumed. The superposition of electrostatic and hydrophobic potentials would further elucidate this point.

5.1
Arylsulfonamide-Type and Related NPY Y_5 Receptor Antagonists

The first nonpeptide Y_5 receptor antagonists with binding affinities in the micromolar range, for example diarylalkanediamines such as JCF 104 (**21a**) and JCF 105 (**21b**, Fig. 14), were disclosed in patents by Synaptic Pharma and Eli Lilly & Co. (Gerald et al. 1997; Fritz et al. 1998). Moreover, diarylalkanediamines structurally similar to benextramine, which has some affinity to the hY_5 receptor (K_i 5 µM; Islam et al. 2002), were prepared, for example **21c** (Y_5: K_i 1.7 µM,

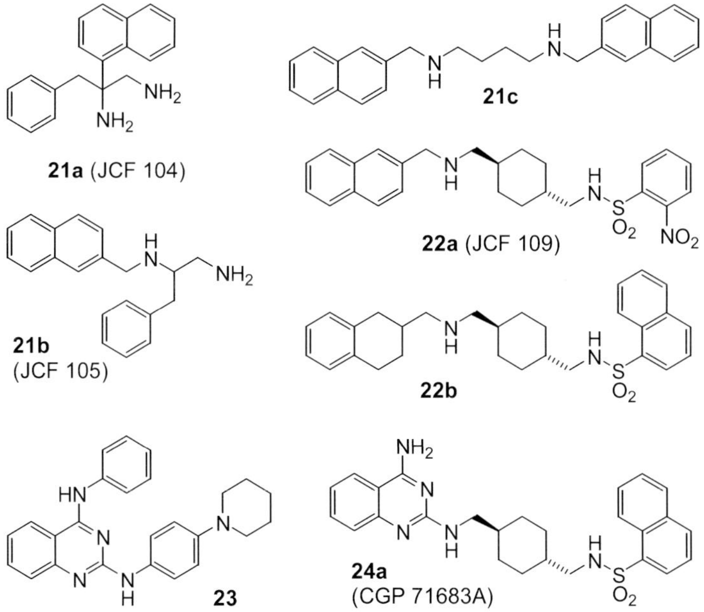

Fig. 14 Examples of diarylalkanediamine-, diaminoquinazoline- and arylsulfonamide-type Y_5 antagonists

Islam et al. 2002), and optimized in subsequent work. Y_5 receptor affinity and selectivity could be considerably improved by replacement of one amine center with a (nonbasic) arylsulfonamide group and optimization of the connecting chain. A *trans*-cyclohexane-1,4-diyldimethyl spacer was found to be the most favorable linker between the two nitrogen centers (Islam et al. 2002). Examples of the resulting arylsulfonamides are **22a** (JCF 109; hY_5: K_i10 nM radioligand: [^{125}I]PYY; Duhault et al. 2000) and **22b** (hY_5: K_i11 nM, radioligand: [^{125}I]PYY; Du et al. 1997b). Based on homology modeling and ligand binding data obtained from studies with a set of receptor mutants, a ligand–receptor interaction model was generated (Islam et al. 2002). According to this model a hydrogen bond is possible between the sulfonamide NH and His^{398} in TM6, which is absent in all other NPY receptor subtypes. Furthermore, the model suggests a salt bridge between the basic amino function of the ligand and Glu^{211} as well as hydrophobic interactions of the terminal aromatic rings with the receptor protein.

The 2,4-diaminoquinazoline motif was the scaffold of some moderate Y_5 ligands with slight Y_1/Y_5 selectivity (Rüeger et al. 2000). Structural modifications led to compounds with nanomolar affinity for Y_5 receptors (e.g., **23**, hY_5: K_i 10 nM, radioligand: [^{125}I][Pro^{34}]hPYY) (Rüeger et al. 1997a). A crucial step towards further increase in Y_5 receptor affinity and selectivity was the combination of the diaminoquinazoline moiety with *N*-cyclohexylmethyl-arylsulfonamide substructures like those present in compounds **22**. This approach resulted in the synthesis of the Y_5 receptor antagonist CGP 71683A (**24a**, Fig. 14) and

Table 5 Selected structures and binding data of substituted 2,4-diaminoquinazolines with Y_5 affinity (Rüeger et al. 2000)

No.	R^1	R^2	X	R^3	IC$_{50}$ (nM)^a
24a^b	H	H	CH$_2$-*t*-chx^c-CH$_2$NH	1-Naphthyl	2.9
24b	H	CH$_3$	CH$_2$-*t*-chx^c-CH$_2$N(CH$_3$)	1-Naphthyl	710
24c	H	H	CH$_2$-*p*-C$_6$H$_4$-CH$_2$NH	1-Naphthyl	290
24d	Phenyl	H	*p*-C$_6$H$_4$-CH$_2$	C$_2$H$_5$	0.6
24e	Phenyl	H	*p*-C$_6$H$_4$-CH$_2$	N(CH$_3$)$_2$	0.9
24f	Phenyl	H	*t*-chx^c-CH$_2$NH	CH$_3$	2
24g	H	H	*t*-chx^c-CH$_2$NH	4-Methylphenyl	4
24h	H	H	(CH$_2$)$_6$	1-Naphthyl	28

^{a} Binding affinities to human NPY Y_5 receptors stably expressed in LM(tk$^-$) cells.
^{b} CGP 71683A.
^{c} *Trans*-cyclohexane-1,4-diyl.

analogs at Novartis Pharma (Criscione et al. 1997; Rüeger et al. 1997a, 1997b, 1997c). Substitution patterns and Y_5 receptor binding data of some diaminoquinazolines are given in Table 5.

JCF 104 (**21a**), JCF 109 (**22a**) and CGP 71683A (**24a**, Fig. 14 and Table 5) were among the first Y_5 antagonistic tools available for the exploration of the occurrence and the pharmacological role of Y_5 receptors—especially their influence on feeding (e.g., Criscione et al. 1998; Feletou et al. 1998, 1999; Duhault et al. 2000; Dumont et al. 2000b; Polidori et al. 2000; Kask et al. 2001; Yokosuka et al. 2001; Lecklin et al. 2002). CGP 71683A binds to Y_5 receptors with >1,000-fold higher affinity than to the Y_1, Y_2 and Y_4 subtypes. It is probably the most intensively studied Y_5 antagonist so far. In fact, CGP 71683A (**24a**) is able to reduce food intake in animals. However, application of the substance in vivo is limited due to unfavorable properties (poor solubility, induction of local inflammatory changes) and significant affinity to other neurotransmitter receptors (e.g., muscarinic acetylcholine receptors) and to the serotonin transporter (5-HT reuptake) which may interfere with the regulation of food consumption (Della Zuana et al. 2001). Reports on the peripheral effects of CGP 71683A and other Y_5 antagonists were contradictory. Inhibition of PP-induced relaxation of rabbit ileum preparations was originally ascribed to Y_5 receptors (Pheng et al. 1997), but further studies suggest that other NPY receptor subtypes (Y_4) are involved in this biological response (Feletou et al. 1999).

The combination of the aforementioned structural motifs proved to be very successful. Neither a sulfonamide moiety nor a cyclic hydrocarbon or a quinazoline ring is essential for Y_5 receptor affinity. Consequently, the substructures of the Y_5 antagonists **22** and **24** were used as scaffolds by many groups to synthesize new potent and selective nonpeptide Y_5 antagonists. In the following, sulfonamides related to **22** and **24** are subdivided into two chemical classes:

a. Analogs of **22**, that is, compounds having a (partially hydrogenated) hydrocarbon system, for example tetraline or a homolog;
b. Analogs of **24a** (CGP 71683), that is, compounds with other heterocyclic rings in place of the quinazoline.

Similar structures with other groups in place of the sulfonamide are included as analogs in Sect. 5.1, whereas the majority of structurally diverse heterocyclic compounds is summarized in Sect. 5.2.

5.1.1
Sulfonamides with Tetraline or Homologous Cyclic Hydrocarbon Moieties

Recently, Itani et al. (2002b) described the synthesis of compounds in which, for instance, the tetrahydronaphthalene moiety of **22b** was expanded to a benzo[a]cycloheptene one. Additional exchange of the cyclohexane-1,4-diyl group against a piperidine containing central spacer results in **25a** (FR 226928, Fig. 15) as the most potent compound (Y_5 IC_{50} 16 nM). Further optimization led to the

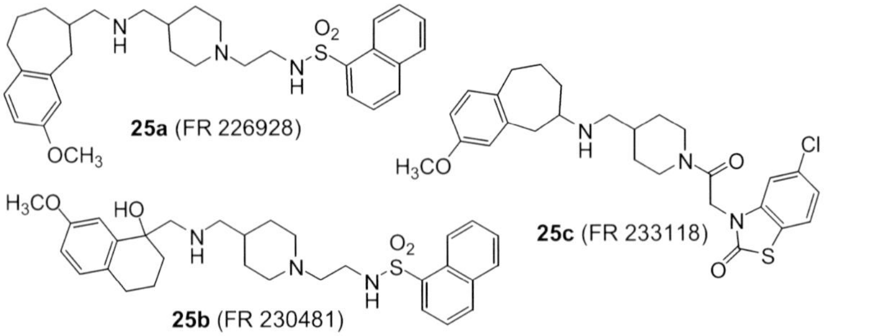

Fig. 15 Sulfonamide-type and related Y$_5$ antagonists with piperidine-containing spacer group

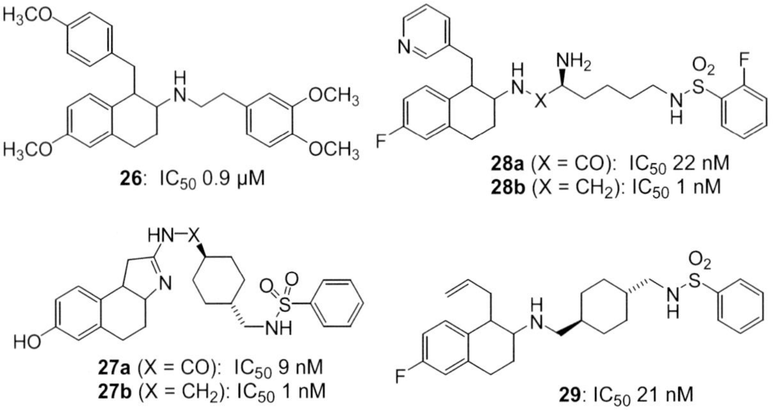

Fig. 16 Structures of *cis*-configured 1-substituted 2-aminotetralines and hY$_5$ receptor binding data (IC$_{50}$ values; displacement of [^{125}I]PYY (80 pM), HEK 293 cells) of some Y$_5$ antagonists with aminotetraline portion. (McNally et al. 2000b; Youngman et al. 2000)

structures **25b** and the nonsulfonamide **25c** with subnanomolar affinities (Itani et al. 2002a).

NPY Y$_5$ antagonists with a common 2-aminotetraline motif were disclosed by Ortho-McNeil/R. W. Johnson in patent applications (Dax et al. 2000, 2001; Dax and McNally 2000). The structure–activity relationships of some α-substituted *N*-(sulfonamido)alkyl-β-aminotetralines have been the subject of subsequent journal papers (McNally et al. 2000a, 2000b; Youngman et al. 2000). Some representative structures (**26–29**) are depicted in Fig. 16.

The substituted *trans*-cyclohexane-1,4-diyl scaffold is present in many of the derivatives, though there are equipotent Y$_5$ antagonists with flexible alkyl chains such as *n*-pentyl instead of cyclohexylmethyl as central spacer. Moreover, analogs of, for example, **28** with a carboxylic acid amide in place of the sulfonamide group were also found to have high Y$_5$ receptor affinity (Dax et al. 2001; Dax 2002). Though not exactly matching the features of compounds **22a,b** or **24a** a large series of sulfonamides and sulfinamides covered in a patent application by

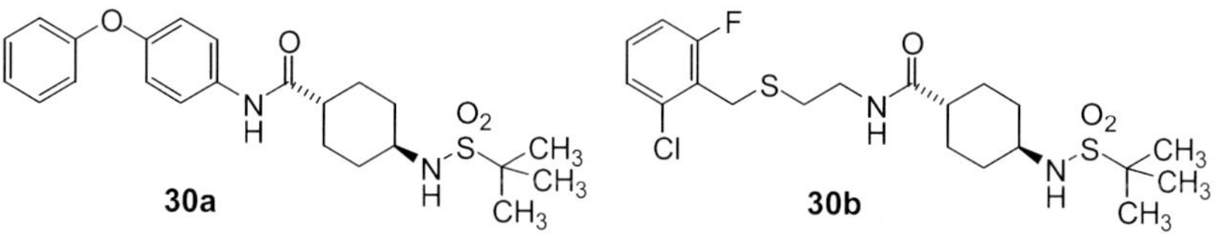

Fig. 17 4-(Sulfonylamino)cyclohexanecarboxylic acid amides

Shionogi (Kawanishi et al. 2001) may be subsumed in this group of Y_5 antagonists due to their structural design. Highly potent representative examples are **30a** and **30b** (Fig. 17) with IC_{50} values of 0.3 and 0.17 nM for binding affinity (Y_5 receptor expressing CHO cells) and 8.4 and 2.6 nM for antagonistic activity in the cAMP assay, respectively (Kawanishi et al. 2001).

5.1.2
Heterocyclic Analogs of CGP 71863

Numerous heterocyclic analogs of **24a** (CGP 71683A), for instance, compounds with an aminothiazole or aminotriazine group or tricyclic ring systems (Marzabadi et al. 2000a, 2000b, 2001; Schmidlin et al. 2001) in place of the aminoquinazoline moiety, were disclosed as Y_5 antagonists with high affinity and selectivity for the human Y_5 receptor. Examples of such compounds (**31–35**) with Y_5 receptor binding data are given in Fig. 18. For some examples, e.g.

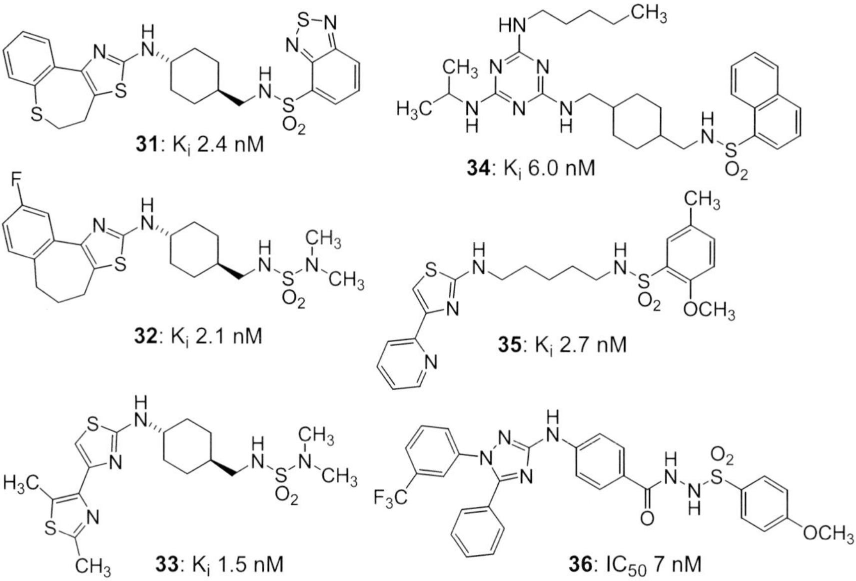

Fig. 18 Heterocyclic analogs of CGP 71683A. hY_5 K_i values for compounds **31–36** determined in radioligand binding studies on cell membrane preparations. (Marzabadi et al. 2000a,b, 2001)

31, 32 and 35, hY$_5$ selectivity (vs. hY$_1$, hY$_2$, and hY$_4$ receptors) was demonstrated and the Y$_5$ antagonistic activity was confirmed in a cAMP assay.

Aminotriazoles such as compound **36** were claimed as Y$_5$ antagonists in a patent application by Adir (Fauchere et al. 2000). Substance **36** was reported to displace radiolabeled [^{125}I]PYY from Y$_5$ receptors with an IC$_{50}$ value of 7 nM (assay not specified) and to lower food consumption and body weight in ob/ob mice (5 mg/kg intraperitoneally twice a day for 3 days) (Fauchere et al. 2000).

5.2
Various Heterocyclic NPY Y$_5$ Receptor Antagonists

5.2.1
Azoles, Pyridines and Diazines

Nitrogen-containing heterocycles are recurring structural elements in many nonpeptide Y$_5$ ligands. For instance, several series of aminopyrazoles (e.g., **37a–d**, Fig. 19) were presented by Banyu Pharm. (Fukami et al. 1998a, 1998b, 1998c, 1998d, 2001a). Compounds **37a** (JCF 114) and **37c** are reported to have Y$_5$ receptor affinities in the low nanomolar range (IC$_{50}$ values, **37a**: 8.3 nM; **37c**: 2.5 nM). Very recently, the (−)-enantiomer of **37d** was described as an orally available and brain-penetrating Y$_5$ antagonist (Sato et al. 2003). The compound displaces [^{125}I]PYY from human recombinant Y$_5$ receptors in LM(tk$^-$) cells with a K_i value of 3.5 nM. Though **37d** can be detected in the brain of SD rats after

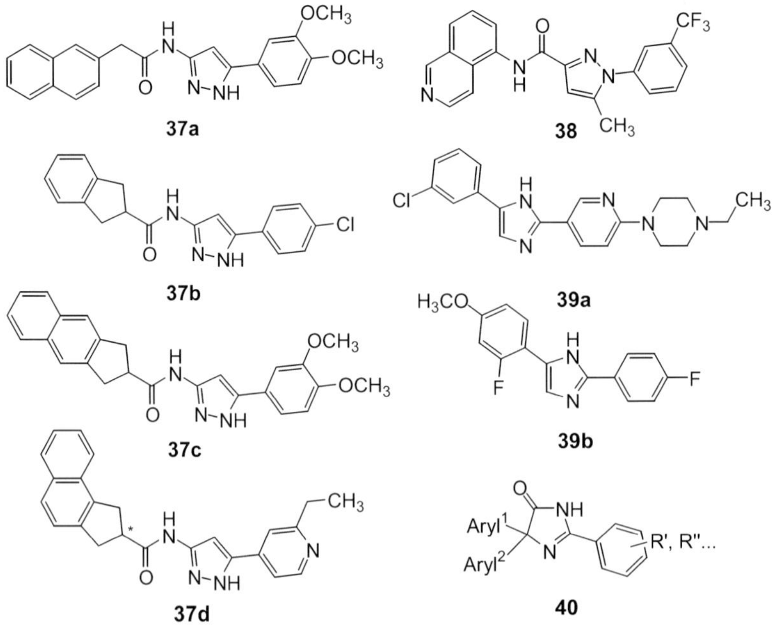

Fig. 19 Pyrazoles and imidazoles described as Y$_5$ antagonists

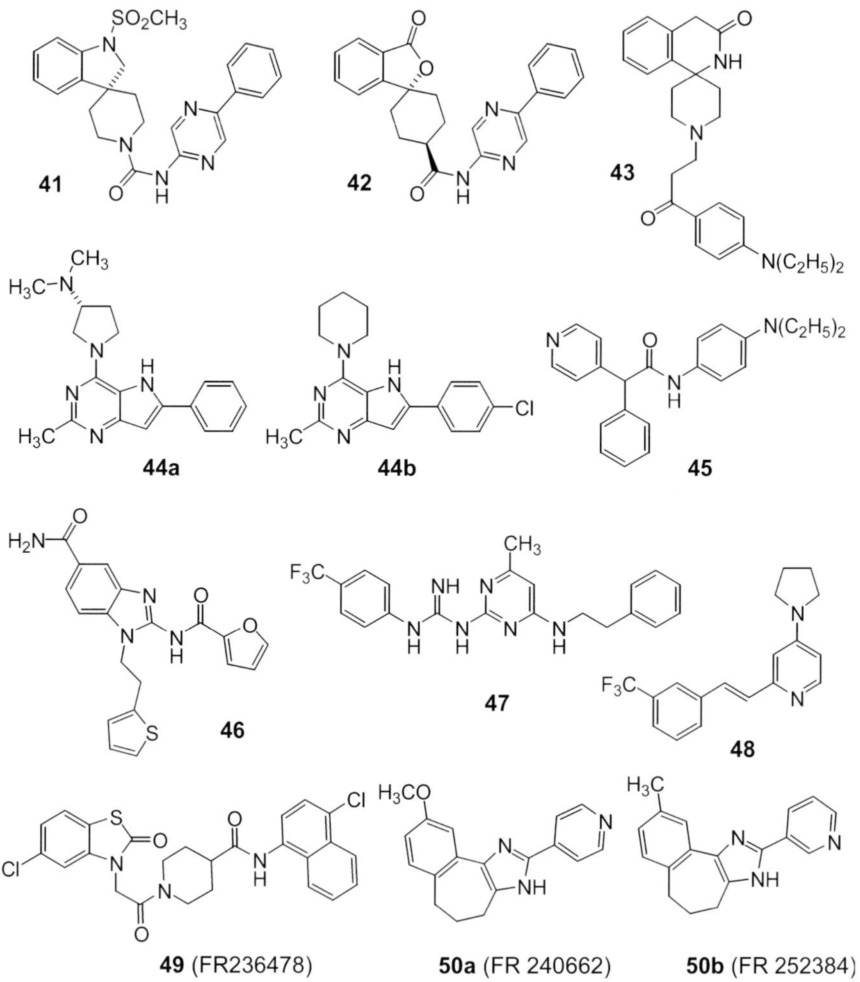

Fig. 20 Spiro compounds, annelated azoles and pyridines described as Y_5 antagonists

oral administration, its ability to suppress bovine pancreatic polypeptide (bPP)-induced food-uptake is only moderate (Sato et al. 2003).

Among some pyrazole-3-carboxamides claimed by Ortho McNeil compound **38** (IC_{50} 80 nM) was reported to produce a 39% reduction in food intake in fasted rats within the first 6 h after resumption of feeding, relative to control rats (Kordik et al. 2000, 2001).

Imidazole derivatives (see examples **39a** and **39b**, Fig. 19) were described as Y_5 antagonists by Neurogen and Pfizer (Thurkauf et al. 1999; Elliott et al. 2000, 2003), and a large collection of 5,5-diaryl-substituted imidazolones of general formula **40** was disclosed by Bristol-Myers Squibb without specified pharmacological data (Poindexter et al. 1999; Poindexter and Gillman 1999a, 1999b).

The spiroindoline **41** (Fig. 20) from Merck and Banyu suppressed bPP-induced food intake in rats at an oral dosage of 3 mg/kg (Gao et al. 2000). Related spirolactones were disclosed by Banyu; **42** inhibited binding of [^{125}I]PP with an IC_{50} value of 0.48 nM (Fukami et al. 2001b). Spiroisoquinolines such as **43** from Bristol-Myers Squibb belong to the same type of scaffold (Poindexter et al. 2001). Pfizer introduced some 4-aminopyrrole[3,2-d]pyrimidines (e.g., **44a**) as Y_5 antagonists (Dow and Hammond 2001). Similar bicyclic pyridine and pyrimidine derivatives (e.g., **44b**, furano- and thienopyrimidines) were reported by Amgen (Norman et al. 1999). Within a series of pyrrolo[3,2-d]pyrimidines, structure–activity relationships of Y_5 receptor binding were analyzed and a pharmacophore model was derived (Norman et al. 2000). Potent Y_5 antagonists of this series have IC_{50} values in the subnanomolar range (e.g., **44b**, Y_5: IC_{50} <0.1 nM).

Phenylacetamide derivatives (e.g., **45**) are the subject of a further patent application by Pfizer (Carpino et al. 2000), and a series of benzimidazole-based Y_5 antagonists has been reported by the GlaxoSmithKline group (Akwabi-Ameyaw et al. 2001; Fang et al. 2001; Heyer et al. 2001a, 2001b; Linn et al. 2001; Daniels et al. 2002). The latter series covers orally available compounds such as **46** which penetrate into the central nervous system (CNS) and reduce food intake in fasted rats (**46**: Y_5 IC_{50} 7.5 nM). Diarylguanidines (e.g., **47**: Y_5 IC_{50} 6 nM) are reported to be active in Zucker rats (Ramanjulu et al. 2001). 4-Aminopyridines such as **48** from Banyu are described as Y_5 antagonists with affinity in the nanomolar range (**48**: IC_{50} 4.1 nM, Fukami et al. 1998e).

Screening of an in-house library at Fujisawa Pharm. led to the discovery of benzothiazolone derivatives (e.g., **49**, FR 236478) with high Y_5 affinity but poor bioavailability (Tabuchi et al. 2002) and of tetrahydrodiazabenzazulenes like **50a** and **50b** which combine nanomolar affinity to the Y_5 receptor with oral absorption and penetration into brain (Satoh et al. 2002).

5.2.2
Carbazoles, Fluorenones and Phenylureas

AstraZeneca disclosed carbazole derivatives (Fig. 21) as NPY Y_5 antagonists (Block et al. 2001; Donald et al. 2002). Recently, the discovery of these compounds and the optimization of their pharmacokinetic and toxicological properties were described (Block et al. 2002). **51a** is reported to bind with high affinity and selectivity to the human Y_5 receptor (IC_{50},Y_5: 2 nM; Y_1, Y_2, and Y_4: >10 µM). But poor phamacokinetic properties (half-life ~15 min, oral bioavailability ~1%) and potential mutagenicity and carcinogenicity disqualify the compound for the use in vivo. Replacement of the pyridylpropionamide side chain by a morpholinocarboxamide (i.e., a urea group) as in **51b** greatly improved the oral bioavailability and increased the half-life to 3 h (Block et al. 2002). Moreover, compound **51b** is able to penetrate into the CNS. However, further modifications were necessary to suppress the mutagenic potential. In this context an isopropyl substituent at the carbazole-nitrogen turned out to be superior to eth-

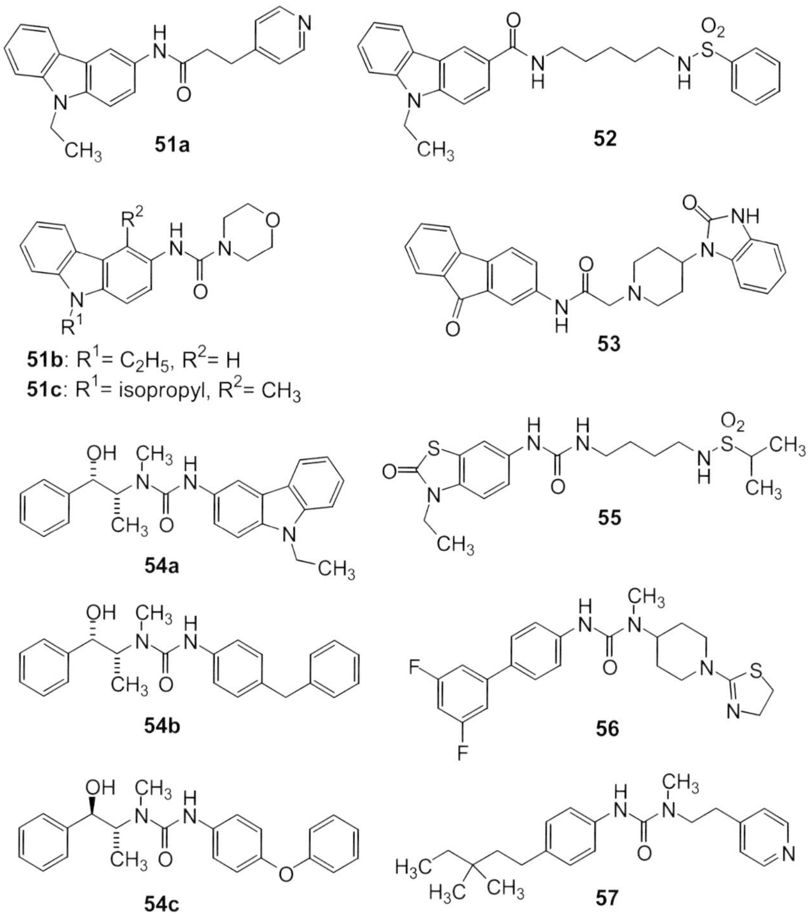

Fig. 21 Carbazole, fluorenone and phenylurea derivatives described as Y_5 antagonists

yl, and further improvement was achieved by an additional methyl group *ortho* to the aniline function. The resulting optimized compound **51c** is reported to combine high Y_5 receptor affinity (IC_{50} 3 nM) with high selectivity, good bioavailability and central activity (Block et al. 2002). **51c** (3 mg/kg, dosed orally) is able to completely block food uptake provoked by injection of Y_5 selective agonists into the third ventricle of the brain of rats. By contrast, the compound has no or only little effect on fasting-induced feeding or on NPY-induced feeding.

Carbazole derivatives with inverted amide function were also described: the carbazole carboxamide **52** (Meiji Seika Kaisha Ltd) was reported to completely displace radiolabeled NPY from membranes prepared from insect cells in which the human Y_5 receptor was expressed at a concentration of 10 µM (no detailed data included) (Nishikawa et al. 2000).

Within a series of Y_5 selective amides some fluorenone derivatives resembling the aforementioned carbazoles were presented by Bayer (Connell et al. 2000). Compound **53** displaced ^{125}I-labeled pPP from human NPY Y_5 receptors with an IC_{50} value of 0.47 nM.

At Amgen, some tri-substituted phenylurea derivatives with subnanomolar affinity were designed (Fotsch et al. 2001). The starting point was structure **54c**, which was identified as a hit by random screening. Lead optimization resulted in highly potent Y_5 antagonists, for example **54b** (IC_{50} <0.1 nM) and **54a**, a carbazolylurea which resembles **51a–c**.

Similar approaches combined a central urea group with many other substituents to obtain Y_5 antagonists such as the ureidobenzothiazolone **55** (Meiji Seika Kaisha Ltd; Aoki et al. 2001) or the ureas **56** and **57** which belong to a large series of NPY Y_5 receptor antagonists recently claimed by Schering-Plough (Stamford et al. 1999; 2001, 2002a, 2002b; McCombie et al. 2002). For instance **57** was found to have a K_i value of 0.4 nM (Y_5), and **56** inhibited [D-Trp^{34}]NPY-stimulated food intake in a dose-dependent manner (ID_{50} 0.5 mg/kg) (McCombie et al. 2002).

5.2.3
Tetrahydroxanthene-1-ones

The tetrahydroxanthene-1-one derivatives **58** (L-152,804) and **59** (Fukuroda et al. 2001) were introduced by Banyu as new orally active and selective neuropeptide Y Y_5 antagonists (Kanatani et al. 2000a) (Fig. 22). Compared to all the other potent NPY antagonists the structures look rather exotic as nitrogen atoms are completely lacking. No information on the binding mode of L-152,804 (**58**) or the molecular ligand–receptor interactions has been published so far. Nevertheless, IC_{50} values of 52 and 14 nM for the inhibition of [^{125}I]PYY binding to a NPY Y_5 receptor membrane preparation are reported for compounds **58** and **59**, respectively (Fukuroda et al. 2001), and a significant reduction of bPP-induced food consumption was found both after intracerebroventricular (30 µg) and oral (10 mg/kg) administration of L-152,804 (**58**) in Sprague-Dawley rats. Interest-

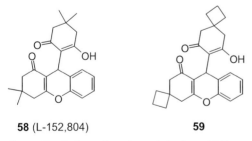

58 (L-152,804)   **59**

Fig. 22 Tetrahydroxanthene-1-one derivatives with Y_5 antagonistic activity

ingly, enhanced feeding, evoked by NPY, could not be influenced by application of L-152,804.

6
Conclusion

Since the discovery of the first nonpeptide NPY antagonists the number of known highly potent and selective small molecular ligands and the structural diversity has considerably increased, mainly due to strategies based on screening of in-house libraries followed by rational lead optimization approaches based on pharmacophore hypotheses. As for other G-protein-coupled receptors, structure-based (receptor-based) design methods are not yet generally applicable. However, site-directed mutagenesis of NPY receptors and binding studies supported by molecular modeling have led to insight into ligand–receptor interactions on the molecular level and to the construction of receptor models complexed with ligands. The potential market for Y_1 and/or Y_5 antagonists as antiobesity drugs has been the major stimulus of industrial NPY research. Considering the effort which has been spent on the discovery of neuropeptide Y Y_5 receptor ligands by various research groups in the last few years, some disillusion arises since no agent has entered human clinical trial, although many compounds with good bioavailability and reducing effects on food consumption in animal experiments have been identified. Unfortunately, there is increasing evidence that blocking a single receptor, in particular the Y_5 receptor (Kanatani et al. 2000b; Turnbull et al. 2002), is not suitable for long term body weight control as there are many redundant mechanisms in the regulation of food uptake and energy expenditure (Kirkpatrick 2002; Parker et al. 2002). The role of NPY in feeding is by far not the only interesting therapeutic implication. Selective antagonists and agonists for all known NPY receptors are required as pharmacological tools to study the contribution of NPY and NPY receptor subtypes to different central and peripheral effects. Obviously, for NPY receptor antagonists (and agonists) as drug candidates the most interesting targets are located in the CNS, that is, optimization of absorption, distribution, metabolism, elimination (ADME) parameters affecting pharmacokinetics, in particular penetration across the blood–brain barrier, is the bottle neck rather than further increasing binding affinity, provided that sufficiently potent and selective ligands for the respective NPY receptor subtypes are already available.

References

Abounader R, Villemure JG, Hamel E (1995) Characterization of neuropeptide Y (NPY) receptors in human cerebral arteries with selective agonists and the new Y_1 antagonist BIBP 3226. Br J Pharmacol 116:2245–2250

Aiglstorfer I, Uffrecht A, Gessele K et al. (1998) NPY Y_1 antagonists: structure-activity relationships of arginine derivatives and hybrid compounds with arpromidine-like partial structures. Regul Pept 75/76:9–21

Aiglstorfer I, Hendrich I, Moser C et al. (2000) Structure-activity relationships of neuropeptide Y Y_1 receptor antagonists related to BIBP 3226. Bioorg Med Chem Lett 10:1597–1600

Akwabi-Ameyaw A, Heyer D, Aquino C et al. (2001) Synthesis and SAR of substituted 5-acylamino benzimidazoles as potent neuropeptide Y Y_5 antagonists. 222nd ACS National Meeting. Chicago, August 2001 (abstract MEDI-033)

Allen J, Novotny J, Martin J et al. (1987) Molecular structure of mammalian neuropeptide Y: analysis by molecular cloning and computer-aided comparison with crystal structure of avian homologue. Proc Natl Acad Sci USA 84:2532–2536

Antonsson T, Bergman N-A, Linschoten M et al., AstraZeneca (2001) Preparation and effect of diphenylacetylaminoalkylamides as neuropeptide Y antagonists. Patent WO 2001002364

Aoki K, Nishikawa N, Ikekawa A et al., Meiji Seika Kaisha, Japan (2001) Preparation of benzothiazolone derivatives having affinity to neuropeptide Y receptor. Patent Jp 2001122865

Beck-Sickinger AG, Wieland HA, Wittneben H et al. (1994) Complete L-alanine scan of neuropeptide Y reveals ligands binding to Y_1 and Y_2 receptors with distinguished conformations. Eur J Biochem 225:947–958

Beck-Sickinger AG (1997) The importance of various parts of the NPY molecule for receptor recognition. In: Grundemar L, Bloom SR (eds) Neuropeptide Y and drug development. Academic Press, London, San Diego, pp 107–126

Bell D, McDermott BJ (1998) D-myo inositol 1,2,6, trisphosphate (α-trinositol, pp56): selective antagonist at neuropeptide Y (NPY) Y-receptors or selective inhibitor of phosphatidylinositol cell signaling? Gen Pharmacol 31:689–696

Benfey BG (1982) Two long acting alpha-adrenoceptor blocking drugs: benextramine and phenoxybenzamine. Trends Pharmacol Sci 8:231

Berglund MM, Fredriksson R, Salaneck E et al. (2002) Reciprocal mutations of neuropeptide Y receptor Y_2 in human and chicken identify amino acids important for antagonist binding. FEBS Lett 518:5–9

Block MH, Donald CS, Foote KM et al., AstraZeneca UK Ltd (2001) Preparation of carbazoles as neuropeptide Y_5 receptor ligand. Patent WO 0107409

Block MH, Boyer S, Brailsford W et al. (2002) Discovery and optimization of a series of carbazole ureas as NPY_5 antagonists for the treatment of obesity. J Med Chem 45:3509–3523

Blum CA, Desimone R, Hutchison A et al., Neurogen Corp., USA (2000) Certain amido- and amino-substituted benzylamine derivatives; a new class of neuropeptide Y_1 specific ligands. Patent US 6133265

Britton TC, Spinazze PG, Hipskind PA et al. (1999) Structure-activity relationships of a series of benzothiophene-derived NPY Y_1 antagonists: optimization of the C-2 side chain. Bioorg Med Chem Lett 9:475–480

Buschauer A (1989) Synthesis and in vitro pharmacology of arpromidine and related phenyl(pyridylalkyl)guanidines, a potential new class of positive inotropic drugs. J Med Chem 32:1963–1970

Carpino PA, Hammond M, Hank RF, Pfizer Products Inc., USA (2000) Amide derivatives useful as Neuropeptide Y (NPY) antagonists. Patent EP 1033366

Carpino PA (2000) Patent focus on new anti-obesity agents: September 1999–February 2000. Expert Opin Ther Pat 10:819–831

Chaurasia C, Misse G, Tessel R et al. (1994) Nonpeptide peptidomimetic antagonists of the neuropeptide Y receptor: benextramine analogs with selectivity for the peripheral Y_2 receptor. J Med Chem 37:2242–2248

Connell RD, Lease TG, Ladouceur GH et al., Bayer Corp., USA (2000) Amide derivatives and methods for using the same as selective neuropeptide Y receptor antagonists. Patent US 6048900

Criscione L, Yamaguchi Y, Mah R et al., Novartis AG, Switzerland (1997) Receptor Antagonists. Patent WO 9720823

Criscione L, Rigollier P, Batzl-Hartmann C et al. (1998) Food intake in free-feeding and energy-deprived lean rats is mediated by the neuropeptide Y_5 receptor. J Clin Invest 102:2136–2145

Daniels AJ, Chance WT, Grizzle MK et al. (2001) Food intake inhibition and reduction in body weight gain in rats treated with GI264879A, a non-selective NPY-Y_1 receptor antagonist. Peptides 22:483–491

Daniels AJ, Grizzle MK, Wiard RP et al. (2002) Food intake inhibition and reduction in body weight gain in lean and obese rodents treated with GW438014A, a potent and selective NPY-Y_5 receptor antagonist. Regul Pept 106:47–54

Dax SL, Lovenberg TW, Baxter EW et al., Ortho-McNeil (2000) N-Aralkylaminotetralins as ligands for the neuropeptide Y Y_5 receptor. Patent WO 0020376

Dax SL, McNally JJ, Ortho-McNeil (2000) 3a,4,5,9b-tetrahydro-1H-benz[e]indol-2-yl amine-derived neuropeptide Y receptor ligands useful in the treatment of obesity and other disorders. Patent WO 0068197

Dax SL, McNally JJ, Youngman MA, Ortho-McNeil (2001) Amine and amide derivatives as ligands for the neuropeptide Y Y_5 receptor useful in the treatment of obesity and other disorders. Patent WO 0109120

Dax SL (2002) Small-molecule neuropeptide Y Y5 antagonists. Drugs Fut 27:273–287

Della Zuana O, Sadlo M, Germain M et al. (2001) Reduced food intake in response to CGP 71683A may be due to mechanisms other than NPY Y_5 receptor blockade. Int J Obes Relat Metab Disord 25:84–94

Dollinger H, Esser F, Mihm G et al., Boehringer Ingelheim Pharma K.-G., Germany (1999) Preparation of novel peptides for use as NPY antagonists. Patent DE 19816929

Donald CS, Foote KM, Schofield P et al., AstraZeneca, UK (2002) Carbazole derivatives and their use as neuropeptide Y_5 receptor ligands. Patent WO 02051806

Doods H, Gaida W, Wieland HA et al. (1999) BIIE0246: A selective and high affinity neuropeptide Y Y_2 receptor antagonist. Eur J Pharmacol 384:R3–R5

Doods HN, Wienen W, Entzeroth M et al. (1995) Pharmacological characterization of the selective nonpeptide neuropeptide Y_1 receptor antagonist BIBP 3226. J Pharmacol Exp Ther 275:136–142

Doods HN, Wieland HA, Engel W et al. (1996) BIBP 3226, the first selective neuropeptide Y_1 receptor antagonist: a review of its pharmacological properties. Regul Pept 65:71–77

Doughty MB, Chu SS, Miller DW et al. (1990) Benextramine: a long-lasting neuropeptide Y receptor antagonist. Eur J Pharmacol 185:113–114

Doughty MB, Li K, Hu L et al. (1992) Benextramine-neuropeptide Y (NPY) binding site interactions: characterization of ^{3}H-NPY binding site heterogeneity in rat brain. Neuropeptides (Edinburgh) 23:169–180

Doughty MB, Chaurasia CS, Li K (1993) Benextramine-neuropeptide Y receptor interactions: contribution of the benzylic moieties to [^{3}H]neuropeptide Y displacement activity. J Med Chem 36:272–279

Dove S, Michel MC, Knieps S et al. (2000) Pharmacology and quantitative structure-activity relationships of imidazolylpropylguanidines with mepyramine-like substructures as non-peptide neuropeptide Y Y_1 receptor antagonists. Can J Physiol Pharmacol 78:108–115

Dow RL, Hammond M, Pfizer Products Inc., USA (2001) 4-aminopyrrolo[3,2-d]pyrimidines as neuropeptide Y receptor antagonists. Patent US 6187778

Du P, Salon JA, Tamm JA et al. (1997a) Modeling the G-protein-coupled neuropeptide Y Y_1 receptor agonist and antagonist binding sites. Protein Eng 10:109–117

Du P, Finn JM, Jeon YT et al., Synaptic Pharmaceutical Corp. (1997b) Aryl sulfonamide and sulfamide derivatives and uses thereof. Patent WO 9719682

Duhault J, Boulanger M, Chamorro S et al. (2000) Food intake regulation in rodents: Y_5 or Y_1 NPY receptors or both? Can J Physiol Pharmacol 78:173–185

Dumont Y, Cadieux A, Doods H et al. (2000a) BIIE0246, a potent and highly selective non-peptide neuropeptide Y Y_2 receptor antagonist. Br J Pharmacol 129:1075–1088

Dumont Y, Cadieux A, Doods H et al. (2000b) Potent and selective tools to investigate neuropeptide Y receptors in the central and peripheral nervous systems: BIBO3304 (Y_1) and CGP71683A (Y_5). Can J Physiol Pharmacol 78:116–125

Edvinsson L, Adamsson M, Jansen I (1990) Neuropeptide Y antagonistic properties of D-myo-inositol-1.2.6-trisphosphate in guinea pig basilar arteries. Neuropeptides 17:99–105

Elliott RL, Hammond M, Hank RF, Pfizer Products Inc. (2000) 2-Pyridyl-4-heteroarylimidazoles for the treatment of obesity. Patent WO 0066578

Elliott RL, Oliver RM, Hammond M et al. (2003) In vitro and in vivo characterization of 3-[2-[6-(2-*tert*-butoxyethoxy)pyridin-3-yl]-1H-imidazol-4-yl]benzonitrile hydrochloride salt, a potent and selective NPY_5 receptor antagonist. J Med Chem 46:670–673

Esser F, Schnorrenberg G, Dollinger H et al., Boehringer Ingelheim Pharma K.-G., Germany (1999) Preparation of novel peptides for use as NPY antagonists. Patent DE 19816932

Fang J, Aquino C, Heyer D et al. (2001) 2-Aminobenzimidazoles as neuropeptide Y Y_5 antagonists: Solution phase synthesis and structure-activity relationships. 222nd ACS National Meeting. Chicago, August 2001 (abstractMEDI-026)

Fauchere J-L, Ortuno J-C, Duhault J et al., Adir et Compagnie, France (2000) Aminotriazole compounds useful as neuropeptide Y receptor ligands, process for their preparation, and pharmaceutical compositions containing them. Patent EP 1044970

Feletou M, Rodriguez M, Beauverger P et al. (1998) NPY receptor subtypes involved in the contraction of the proximal colon of the rat. Regul Pept 75/76:221–229

Feletou M, Nicolas JP, Rodriguez M et al. (1999) NPY receptor subtype in the rabbit isolated ileum. Br J Pharmacol 127:795–801

Feth F, Erdbrugger W, Rascher W et al. (1993) Is PP56 (D-myo-inositol-1,2,6-trisphosphate) an antagonist at neuropeptide Y receptors? Life Sci 52:1835–1844

Fotsch C, Sonnenberg JD, Chen N et al. (2001) Synthesis and structure-activity relationships of trisubstituted phenyl urea derivatives as neuropeptide Y_5 receptor antagonists. J Med Chem 44:2344–2356

Fritz JE, Kaldor SW, Scott WL et al., Eli Lilly & Co. (1998) 3-Arylpropylamino Neuropeptide Y Receptor Antagonists. Patent WO 9852890

Fukami T, Fukuroda T, Kanatani A et al., Banyu Pharm. (1998a) Preparation of urea moiety-containing pyrazole derivatives as neuropeptide Y antagonists. Patent WO 9824768

Fukami T, Fukuroda T, Kanatani A et al., Banyu Pharm. (1998b) Preparation of pyrazole derivatives for treatment of bulimia, obesity, and diabetes. Patent WO 9825907

Fukami T, Fukuroda T, Kanatani A et al., Banyu Pharm. (1998c) Preparation of aminopyrazole derivatives for the treatment of bulimia, obesity, and diabetes. Patent WO 9827063

Fukami T, Fukuroda T, Kanatani A et al., Banyu Pharm. (1998d) Preparation of novel aminopyrazole derivatives as neuropeptide Y antagonists. Patent WO 9825908

Fukami T, Okamoto O, Fukuroda T et al., Banyu Pharm. (1998e) Preparation and formulation of aminopyridine derivatives as neuropeptide Y receptor antagonists. Patent WO 9840356

Fukami T, Kanatani A, Ishihara A et al. (2001a) Synthesis and pharmacology of potent and selective aminopyridine NPY Y_1 receptor antagonists. 222nd ACS National Meeting. Chicago, August 2001 (abstract MEDI-282)

Fukami T, Kanatani A, Ishihara A et al., Banyu Pharm. (2001b) Preparation of spiroisoindolinepiperidines, spiroisoquinolinepiperidines, spiroisobenzofuranpiperidines and related compounds. Patent WO 0114376

Fukuroda T, Kanatani A, Fukami T et al., Banyu Pharmaceutical Co. Ltd. (2001) Novel neuropeptide Y receptor antagonists. Patent WO 9915516

Gao Y, McNeil D, Yang L et al., Merck; Banyu Pharm. (2000) Preparation of spiroindolines as Y_5 receptor antagonists. Patent WO 0027845

Gerald C, Walker MW, Criscione L et al. (1996) A receptor subtype involved in neuropeptide-Y-induced food intake. Nature 382:168–171

Gerald C, Weinschank RL, Walker MW et al., Synaptic Pharma Corp. (USA) (1997) DNA encoding a hypothalamic atypical neuropeptide Y/peptide YY receptor (Y_5) and uses thereof. Patent WO 9616542

Grouzmann E, Buclin T, Martire M et al. (1997) Characterization of a selective antagonist of neuropeptide Y at the Y_2 receptor. Synthesis and pharmacological evaluation of a Y_2 antagonist. J Biol Chem 272:7699–7706

Grundemar L, Bloom SR (eds) (1997) Neuropeptide Y and drug development. London, San Diego, Academic Press

Hammond M (2001) Neuropeptide Y receptor antagonists. IDrugs 4:920–927

Heilig M, Edvinsson L, Wahlestedt C (1991) Effects of intracerebroventricular D-myo-inositol-1,2,6-trisphosphate (PP56), a proposed neuropeptide Y (NPY) antagonist, on locomotor activity, food intake, central effects of NPY and NPY-receptor binding. Eur J Pharmacol 209:27–32

Heyer D, Marron B, Miyasaki Y et al. (2001a) Discovery of a novel series of benzimidazole-based neuropeptide Y Y_5 antagonists from a 7-TM targeted chemical library. 222nd ACS National Meeting, Chicago, August 2001 (abstract MEDI-030)

Heyer D, Akwabi-Ameyaw A, Aquino C (2001b) Aminobenzimidazoles: A class of orally bioavailable and brain permeable neuropeptide Y Y5 antagonists. 222nd ACS National Meeting, Chicago, August 2001: (abstract MEDI-284)

Hipskind PA, Lobb KL, Nixon JA et al. (1997) Potent and selective 1,2,3-trisubstituted indole NPY Y-1 antagonists. J Med Chem 40:3712–3714

Hutzler C (2001) Synthesis and pharmacological activity of new neuropeptide Y receptor ligands: from N,N-disubstituted alkanamides to highly potent argininamide-type Y_1 antagonists. Doctoral thesis, University of Regensburg, Regensburg, Germany

Hutzler C, Kracht J, Mayer M et al. (2001) N^G-Acylated argininamides: highly potent selective NPY Y_1 receptor antagonists with special properties, Arch. Pharm. Pharm. Med. Chem. 334, Suppl. 2, 17 (2001)

Islam I, Dhanoa D, Finn J et al. (2002) Discovery of potent and selective small molecule NPY Y_5 receptor antagonists. Bioorg Med Chem Lett 12:1767–1769

Itani H, Ito H, Sakata Y et al. (2002a) Novel potent antagonists of human neuropeptide Y Y_5 receptors. Part 3:7-methoxy-1-hydroxy-1-substituted tetraline derivatives. Bioorg Med Chem Lett 12:799–802

Itani H, Ito H, Sakata Y et al. (2002b) Novel potent antagonists of human neuropeptide Y Y_5 receptors. Part 2: substituted benzo[a]cycloheptene derivatives. Bioorg Med Chem Lett 12:757–761

Iyengar S, Li DL, Simmons RM (1999) Characterization of neuropeptide Y-induced feeding in mice: do Y_1-Y_6 receptor subtypes mediate feeding? J Pharmacol Exp Ther 289:1031–1040

Jacques D, Cadieux A, Dumont Y et al. (1995) Apparent affinity and potency of BIBP3226, a non-peptide neuropeptide Y receptor antagonist, on purported neuropeptide Y Y_1, Y_2 and Y_3 receptors. Eur J Pharmacol 278:R3–5

Kanatani A, Kanno T, Ishihara A et al. (1999) The novel neuropeptide Y Y_1 receptor antagonist J-104870: a potent feeding suppressant with oral bioavailability. Biochem Biophys Res Commun 266:88–91

Kanatani A, Ishihara A, Iwaasa H et al. (2000a) L-152,804: orally active and selective neuropeptide Y Y_5 receptor antagonist. Biochem Biophys Res Commun 272:169–173

Kanatani A, Mashiko S, Murai N et al. (2000b) Role of the Y_1 receptor in the regulation of neuropeptide Y-mediated feeding: comparison of wild-type, Y_1 receptor-deficient, and Y_5 receptor-deficient mice. Endocrinology 141:1011–1016

Kanatani A, Hata M, Mashiko S et al. (2001) A typical Y_1 receptor regulates feeding behaviors: effects of a potent and selective Y_1 antagonist, J-115814. Mol Pharmacol 59:501–505

Kanno T, Kanatani A, Keen SLC et al. (2001) Different binding sites for the neuropeptide Y Y_1 antagonists 1229U91 and J-104870 on human Y_1 receptors. Peptides 22:405–413

Kask A, Rago L, Harro J (1998) Evidence for involvement of neuropeptide Y receptors in the regulation of food intake: studies with Y_1-selective antagonist BIBP3226. Br J Pharmacol 124:1507–1515

Kask A, Harro J (2000) Inhibition of amphetamine- and apomorphine-induced behavioural effects by neuropeptide Y Y(1) receptor antagonist BIBO 3304. Neuropharmacology 39:1292–1302

Kask A, Vasar E, Heidmets LT et al. (2001) Neuropeptide Y Y(5) receptor antagonist CGP71683A: the effects on food intake and anxiety-related behavior in the rat. Eur J Pharmacol 414:215–224

Kawanishi Y, Takenaka H, Hanasaki K et al., Shionogi + Co., Ltd, Japan (2001) Preparation of sulfonamides and sulfinamides as NPY Y_5 antagonists. Patent WO 0137826

Kirkpatrick P (2002) Dead end for NPY Y_5-receptor antagonists? Nature Rev Drug Discov 1:932–933

Knieps S, Dove S, Michel MC et al. (1996) ω-Phenyl-ω-(2-pyridyl)alkyl-substituted bisguanidines are moderate neuropeptide Y antagonists. Pharm Pharmacol Lett 6:27–30

Kordik CP, Lovenberg TW, Reitz A, Ortho- McNeil Pharm. (2000) Preparation of pyrazole carboxamides for the treatment of obesity and other disorders. Patent WO 0069849

Kordik CP, Luo C, Zanoni BC et al. (2001) Aminopyrazoles with high affinity for the human neuropeptide Y_5 receptor. Bioorg Med Chem Lett 11:2283–2286

Lecklin A, Lundell I, Paananen L et al. (2002) Receptor subtypes Y_1 and Y_5 mediate neuropeptide Y induced feeding in the guinea-pig. Br J Pharmacol 135:2029–2037

Ling AL (1999) Neuropeptide Y receptor antagonists. Expert Opin Ther Pat 9:375–384

Linn J, Aquino C, Dezube M (2001) Benzimidazole neuropeptide Y Y_5 antagonists: rapid SAR development using a solid phase approach. 222nd ACS National Meeting. Chicago, August 2001: (abstract MEDI-027)

Malmström RE, Balmer KC, Lundberg JM (1997) The neuropeptide Y (NPY) Y_1 receptor antagonist BIBP 3226: equal effects on vascular responses to exogenous and endogenous NPY in the pig in vivo. Br J Pharmacol 121:595–603

Malmström RE, Alexandersson A, Balmer KC et al. (2000) In vivo characterization of the novel neuropeptide Y Y_1 receptor antagonist H 409/22. J Cardiovasc Pharmacol 36:516–525

Malmström RE (2000) Neuropeptide Y Y_1 receptor mediated mesenteric vasoconstriction in the pig in vivo. Regul Pept 95:59–63

Malmström RE (2001a) Vascular pharmacology of BIIE0246, the first selective non-peptide neuropeptide Y Y_2 receptor antagonist, in vivo. Br J Pharmacol 133:1073–1080

Malmström RE, Balmer KC, Weilitz J et al. (2001) Pharmacology of H 394/84, a dihydropyridine neuropeptide Y Y(1) receptor antagonist, in vivo. Eur J Pharmacol 418:95–104.

Malmström RE (2001b) Existence of both neuropeptide Y, Y_1 and Y_2 receptors in pig spleen: evidence using subtype-selective antagonists in vivo. Life Sci 69:1999–2005

Malmström RE, Lundberg JN, Weitzberg E (2002a) Effects of the neuropeptide Y Y_2 receptor antagonist BIIE0246 on sympathetic transmitter release in the pig in vivo. Naunyn Schmiedebergs Arch Pharmacol 365:106–111

Malmström RE, Lundberg JO, Weitzberg E (2002b) Autoinhibitory function of the sympathetic prejunctional neuropeptide Y Y(2) receptor evidenced by BIIE0246. Eur J Pharmacol 439:113–119

Malmström RE (2002) Pharmacology of neuropeptide Y receptor antagonists. Focus on cardiovascular functions. Eur J Pharmacol 447:11–30

Marzabadi MR, Wong WC, Noble SA, Synaptic Pharm. Corp., USA (2000a) Preparation of benzo[2,3]thiepino[4,5-d]thiazol-2-ylaminoalkylsulfonamides and related compounds as selective neuropeptide Y (Y_5) antagonists. Patent US 6124331

Marzabadi MR, Wong WC, Noble SA et al., Synaptic Pharm. Corp., USA. (2000b) Preparation of triazineamines, thiazolamines, and benzo[2,3]thiepino[4,5-d][1,3]thiazol-2-ylamines as selective NPY (Y_5) antagonists. Patent WO 0064880

Marzabadi MR, Wong WC, Noble SA et al., Synaptic Pharm. Corp., USA; Novartis A.-G. (2001) Preparation of selective NPY (Y5) antagonists and pharmaceutical compositions thereof for treating an abnormality modulated by human Y5 receptor activity. Patent WO 0102379

McCombie S, Stamford AW, Huang Y et al., Schering-Plough Corp., USA (2002) Substituted urea neuropeptide Y Y_5 receptor antagonists. Patent US 2002165223

McNally JJ, Youngman MA, Lovenberg TW et al. (2000a) N-acylated alpha-(3-pyridylmethyl)-beta-aminotetralin antagonists of the human neuropeptide Y Y_5 receptor. Bioorg Med Chem Lett 10:1641–1643

McNally JJ, Youngman MA, Lovenberg TW et al. (2000b) N-(sulfonamido)alkyl[tetrahydro-1H-benzo[e]indol-2-yl]amines: potent antagonists of human neuropeptide Y Y_5 receptor. Bioorg Med Chem Lett 10:213–216

Meister A (1999) Pharmacological characterization of new histamine and neuropeptide Y receptor ligands: set-up and validation of functional test models on isolated organs. Doctoral thesis, University of Regensburg, Regensburg, Germany

Michel MC, Motulsky HJ (1990) He 90481: A competitive nonpeptidergic antagonist at neuropeptide Y receptors. Ann N Y Acad Sci 611:392–394

Michel MC, Moersdorf JP, Engler H et al., Heumann Pharma G.m.b.H. und Co., Germany (1991) Use of guanidine derivatives for the manufacture of a medicament with neuropeptide Y-antagonistic activity. Patent EP448765

Michel MC, Dove S, Buschauer A (1993) Arpromidine and related histamine H_2 agonists as chemical leads for the development of non-peptide neuropeptide Y antagonists. Arch Pharm (Weinheim) 326:664

Michel MC, Beck-Sickinger AG, Cox H et al. (1998) XVI. International Union of Pharmacology Recommandations for the Nomenclature of Neuropeptide Y, Peptide YY, and Pancreatic Polypeptide Receptors. Pharmacological Reviews 50:143–150

Mollereau C, Gouarderes C, Dumont Y et al. (2001) Agonist and antagonist activities on human NPFF(2) receptors of the NPY ligands GR231118 and BIBP3226. Br J Pharmacol 133:1–4

Mollereau C, Mazarguil H, Marcus D et al. (2002) Pharmacological characterization of human NPFF(1) and NPFF(2) receptors expressed in CHO cells by using NPY Y(1) receptor antagonists. Eur J Pharmacol 451:245–256

Morgan DGA, Small CJ, Abusnana S et al. (1998) The NPY Y_1 receptor antagonist BIBP 3226 blocks NPY induced feeding via a non-specific mechanism. Regul Pept 75/76:377–382

Moser C, Bernhardt G, Michel J et al. (2000) Cloning and functional expression of the hNPY Y_5 receptor in human endometrial cancer (HEC-1B) cells. Can J Physiol Pharmacol 78:134–142

Müller M, Knieps S, Gessele K et al. (1997) Synthesis and neuropeptide Y Y_1 receptor antagonistic activity of N,N-disubstituted ω-guanidino- and ω-aminoalkanoic acid amides. Arch Pharm (Weinheim) 330:333–342

Murakami Y, Hagishita S, Okada T et al. (1999a) 1,3-disubstituted benzazepines as neuropeptide Y Y_1 receptor antagonists. Bioorg Med Chem 7:1703–1714

Murakami Y, Hara H, Okada T et al. (1999b) 1,3-Disubstituted benzazepines as novel, potent, selective neuropeptide Y Y_1 receptor antagonists. J Med Chem 42:2621–2632

Nishikawa N, Aoki K, Suzuki M et al., Meiji Seika Kaisha Ltd., Japan (2000) Tricyclic compounds. Patent WO 0063171

Norman MH, Chen N, Han N et al., Amgen Inc., USA (1999) Preparation of bicyclic pyridine and pyrimidine derivatives as neuropeptide Y receptor antagonists. Patent WO 9940091

Norman MH, Chen N, Chen Z et al. (2000) Structure-Activity Relationships of a Series of Pyrrolo[3,2-d]pyrimidine derivatives and related compounds as neuropeptide Y_5 receptor antagonists. J Med Chem 43:4288–4312

Palczewski K, Kumasaka T, Hori T et al. (2000) Crystal structure of rhodopsin: A G protein-coupled receptor. Science 289:739–745

Palea S, Corsi M, Rimland JM et al. (1995) Discrimination by benextramine between the NPY-Y_1 receptor subtypes present in rabbit isolated vas deferens and saphenous vein. Br J Pharmacol 115:3–10

Parker E, Van Heek M, Stamford A (2002) Neuropeptide Y receptors as targets for anti-obesity drug development: perspective and current status. Eur J Pharmacol 440:173–187

Peterson JM, Blum CA, Cai G et al., Pfizer Inc., USA (1996) Certain substituted benzylamine derivatives: a new class of neuropeptide Y_1 specific ligands. Patent WO 9640660

Pheng LH, Quirion R, Iyengar S et al. (1997) The rabbit ileum: a sensitive and selective preparation for the neuropeptide Y Y_5 receptor. Eur J Pharmacol 333:R3–5

Poindexter GS, Bruce M, Johnson G et al., Bristol-Myers Squibb Company, USA (1996) Dihydropyridine NPY antagonists: piperazine derivatives. Patent EP 747356

Poindexter GS, Gillman K, Bristol-Myers Squibb Company, USA (1999a) Imidazolone anorectic agents: III. heteroaryl derivatives. Patent WO 9948873

Poindexter GS, Gillman K, Bristol Myers Squibb Co., USA (1999b) Imidazolone anorectic agents: I. acyclic derivatives. Patent WO 9948888

Poindexter GS, Antal I, Guipponi LM et al., Bristol-Myers Squibb Co. (1999) Imidazolone anorectic agents: II. phenyl derivatives. Patent WO 9948888

Poindexter GS, Antal I, Giupponi LM et al., Bristol-Myers Squibb Company, USA (2001) NPY antagonists: spiroisoquinolinone derivatives. Patent WO 0113917

Poindexter GS, Bruce MA, LeBoulluec KL et al. (2002) Dihydropyridine neuropeptide Y Y(1) receptor antagonists. Bioorg Med Chem Lett 12:379–382

Polidori C, Ciccocioppo R, Regoli D et al. (2000) Neuropeptide Y receptor(s) mediating feeding in the rat: characterization with antagonists. Peptides 21:29–35

Rabinovich A, Mazurov A, Krokhina G et al. (1997) Discovery of Neuropeptide Y Receptor Antagonists. In: Grundemar L, Bloom S (eds) Neuropeptide Y and drug development. Academic Press, London, San Diego, pp 191–201

Ramanjulu J, Aquino C, Palazzo F et al. (2001) Synthesis and structure activity studies of bisheteroaryl guanidines as NPY-Y_5 antagonists. 221st ACS National Meeting. San Diego, April 2001 (abstract MEDI-210)

Robin-Jagerschmidt C, Sylte I, Bihoreau C et al. (1998) The ligand binding site of NPY at the rat Y_1 receptor investigated by site-directed mutagenesis and molecular modeling. Mol Cell Endocrinol 139:187–198

Rudolf K, Eberlein W, Wieland HA et al. (1994a) The first highly potent and selective non-peptide NPY Y_1 receptor antagonist: BIBP 3226. Eur J Pharm 271:R11–13

Rudolf K, Eberlein W, Engel W et al., (Dr. Karl Thomae GmbH, Germany). (1994b) Preparation of amino acid derivatives as neuropeptide Y antagonists. Patent WO 19940118

Rudolf K, Eberlein W, Engel W et al., Boehringer Ingelheim Pharma K.-G., Germany (1999) Preparation of piperazine-containing peptidomimetics for use as NPY antagonists. Patent DE 19816889

Rudolf K, Eberlein W, Engel W et al. (1997) BIBP3226, a potent and selective neuropeptide Y Y_1-receptor antagonist. Structure-activity studies and localization of the human Y_1 receptor binding site. In: Grundemar L, Bloom S (eds) Neuropeptide Y and drug development. Academic Press, London, San Diego, pp 175–190

Rüeger H, Schmidlin T, Rigollier P et al., Novartis AG (1997a) Quinazoline-2,4-diazirines as NPY receptor antagonists. Patent WO 9720822

Rüeger H, Schmidlin T, Rigollier P et al., Novartis AG (1997b) Quinazoline derivatives useful as antagonists of NPY receptor subtype Y_5. Patent WO 9720820

Rüeger H, Schmidlin T, Rigollier P et al., Novartis AG (1997c) Heteroaryl Derivatives. Patent WO 9720821

Rüeger H, Rigollier P, Yamaguchi Y et al. (2000) Design, synthesis and SAR of a series of 2-substituted 4-amino-quinazoline neuropeptide Y Y_5 receptor antagonists. Bioorg Med Chem Lett 10:1175–1179

Salaneck E, Holmberg SK, Berglund MM et al. (2000) Chicken neuropeptide Y receptor Y2: structural and pharmacological differences to mammalian Y2(1). FEBS Lett 484:229–234

Sato N, Takahashi T, Shibata T et al. (2003) Design and Synthesis of the Potent, Orally Available, Brain-Penetrable Arylpyrazole Class of Neuropeptide Y_5 Receptor Antagonists. J Med Chem 46:666–669

Satoh Y, Hatori C, Ito H (2002) Novel potent antagonists of human neuropeptide Y-Y5 receptor. Part 4: tetrahydrodiazabenzazulene derivatives. Bioorg Med Chem Lett 12:1009–1011

Sautel M, Rudolf K, Wittneben H et al. (1996) Neuropeptide Y and the nonpeptide antagonist BIBP 3226 share an overlapping binding site at the human Y_1 receptor. Mol Pharmacol 50:285–292

Schmidlin T, Rüeger H, Gerspacher M, Novartis A.-G., Switzerland (2001) Preparation of condensed thiazolamines as neuropeptide Y_5 antagonists. Patent WO 0164675

Serradeil-Le Gal C, Valette G, Rouby P-E et al. (1995) SR120819A, an orally-active and selective neuropeptide Y Y_1 receptor antagonist. FEBS Lett 362:192–196

Serradeil-Le Gal C (1997) SR 120819A or the first generation of orally active Y_1 receptor antagonists. In: Grundemar L, Bloom SR (eds) Neuropeptide Y and drug development. Academic Press, London, San Diego, pp 157–174

Shigeri Y, Ishikawa M, Ishihara Y et al. (1998) A potent nonpeptide neuropeptide Y Y_1 receptor antagonist, a benzodiazepine derivative. Life Sci 63:PL 151–160

Silva AP, Cavadas C, Grouzmann E (2002) Neuropeptide Y and its receptors as potential therapeutic drug targets. Clin Chim Acta 326:3–25

Sit SY, Huang Y, Antal-Zimanyi I et al. (2002) Novel dihydropyrazine analogues as NPY antagonists. Bioorg Med Chem Lett 12:337–340

Smith-White MA, Hardy TA, Brock JA et al. (2001) Effects of a selective neuropeptide Y Y_2 receptor antagonist, BIIE0246, on Y_2 receptors at peripheral neuroeffector junctions. Br J Pharmacol 132:861–868

Stamford AW, Dugar S, Wu Y et al., Schering Corp., USA (1999) Neuropeptide Y_5 receptor antagonists. Patent WO 9964394

Stamford AW, Dugar S, Wu Y et al., Schering Corp., USA (2001) Neuropeptide Y_5 receptor antagonists. Patent US 6329395

Stamford AW, Huang Y, Kelly JM et al., Schering Corp., USA (2002a) Substituted urea neuropeptide Y Y_5 receptor antagonists. Patent WO 0222592

Stamford AW, Dong Y, McCombie S, Schering Corp., USA (2002b) Heteroaryl urea neuropeptide Y Y_5 receptor antagonists. Patent WO 0249648

Sylte I, Robin-Jagerschmidt C, Bihoreau C et al. (1998) Molecular modeling of the NPY binding site on the Y_1 receptor. J Mol Model 4:221–233

Sylte I, Andrianjara CR, Calvet A et al. (1999) Molecular dynamics of NPY Y_1 receptor activation. Bioorg Med Chem 7:2737–2748

Tabuchi S, Itani H, Sakata Y et al. (2002) Novel potent antagonists of human neuropeptide Y Y_5 receptor. Part 1:2-oxobenzothiazolin-3-acetic acid derivatives. Bioorg Med Chem Lett 12:1171–1175

Thurkauf A, Yuan J, Hutchison A et al., Neurogen Corp.; Pfizer Inc. (1999) Preparation of diarylimidazoles as neuropeptide Y receptor ligands. Patent WO 9901128

Turnbull AV, Ellershaw L, Masters DJ et al. (2002) Selective antagonism of the NPY Y_5 receptor does not have a major effect on feeding in rats. Diabetes 51:2441–2449

Walker P, Munoz M, Martinez R et al. (1994) Acidic residues in extracellular loops of the human Y_1 neuropeptide Y receptor are essential for ligand binding. J Biol Chem 269:2863–2869

Wieland HA, Willim KD, Entzeroth M et al. (1995) Subtype selectivity and antagonistic profile of the nonpeptide Y_1 receptor antagonist BIBP 3226. J Pharmacol Exp Ther 275:143–149

Wieland HA, Engel W, Eberlein W et al. (1998) Subtype selectivity of the novel nonpeptide neuropeptide Y Y_1 receptor antagonist BIBO 3304 and its effect on feeding in rodents. Br J Pharmacol 125:549–555

Wright J, Bolton G, Creswell M et al. (1996) 8-Amino-6-(arylsulfonyl)-5-nitroquinolines: novel nonpeptide neuropeptide Y_1 receptor antagonists. Bioorg Med Chem Lett 6:1809–1814

Wright J (1997) Design and discovery of neuropeptide Y_1 receptor antagonists. Drug Discovery Today 2:19–24

Yokosuka M, Dube MG, Kalra PS et al. (2001) The mPVN mediates blockade of NPY-induced feeding by a Y_5 receptor antagonist: a c-FOS analysis. Peptides 22:507–514

Youngman MA, McNally JJ, Lovenberg TW et al. (2000) alpha-Substituted N-(sulfonamido)alkyl-beta-aminotetralins: potent and selective neuropeptide Y Y_5 receptor antagonists. J Med Chem 43:346–350

Zarrinmayeh H, Nunes AM, Ornstein PL et al. (1998) Synthesis and evaluation of a series of novel 2-[(4-chlorophenoxy)methyl]benzimidazoles as selective neuropeptide Y Y_1 receptor antagonists. J Med Chem 41:2709–2719

Zarrinmayeh H, Zimmerman DM, Cantrell BE et al. (1999) Structure-activity relationship of a series of diaminoalkyl substituted benzimidazole as neuropeptide Y Y_1 receptor antagonists. Bioorg Med Chem Lett 9:647–652

Zimanyi IA, Poindexter GS (2000) NPY-ergic agents for the treatment of obesity. Drug Dev Res 51:94–111

Subject Index

Acetylcholine (Ach) 35, 159
Acquired immunity 414
Action potential 144
Active zone 33
Adenosine 156
Adenosine diphosphate 31
Adenosine-5′-triphosphate (ATP) 138
Adenylyl cyclase 365
Adhesion molecule 424
Adrenal medulla 30, 140
Adrenalectomy 296
Adrenals 295
Adrenergic C1 group 189
Adrenergic C2 group 189
Adrenergic C3 group 189
α-adrenergic antagonist 508, 509
α-adrenoceptor 142
α_2-adrenoceptor 145
β-adrenoceptor 142
β_2-adrenoceptor 156
Adrenocorticotropic hormone 188
Aedes aegypti 78
Ageing 342
Agonists 308
Agouti-related peptide 189
AgRP 456, 460-462
Alarm systems 259
Alcohol 470
– consumption 2
Alcoholism 15, 252
Amidation 27
α-amidation 29
γ-aminobutyric acid 189

Aminopeptidase P 24, 36
Aminopyrazole 532
4-aminopyridines 534
4-aminopyrrole[3,2-d]pyrimidine 534
2-aminotetraline 530
Aminothiazole 531, 532
Aminotriazine 531
Amniotes 86
Amphibian 77
Amygdala 469
Angiogenesis 338
Angiotensin 156
Anopheles gambiae 86
Antagonists 305, 314
Anterior pituitary 187
Anti- MOG$_{35-55}$ IgG2a 435
Antibiotic activity 420
Antibody 422
Anticonvulsant action 166
Antidepressant-like drug 267
Antigen presenting cell (APC) 415
Antimicrobial activity 420
Anti-myelin oligodendrocyte
   glycoprotein (MOG)$_{35-55}$ 423
Anti-NPY antibodies 299
Antisense ODNs 299, 313
Anxiety 2, 7, 13, 38, 108, 111-117,
   122, 252, 469
APC (antigen presenting cell) 415
Apomorphine 164
Appetitive phase 290
Arcuate nucleus (ARC) 188, 286, 287,
   339

Arcuate nucleus-paraventricular nucleus 222
Argininamides 511
Arpromidine 509
Arteries 332
Artiodactyls 79
Arylsulfonamide 527
Astrocyte 413
Atherosclerosis 338
ATP (adenosine-5′-triphosphate) 138
Atrial natriuretic peptide 373
Autoantibody 430
Autoimmune disorder 429, 433
Autoimmunity 429
Autonomic 150
Autoradiography 364
Autoreceptor 144
AXC011018 522
AXC01829 522

B cell 424
B cell receptor (BCR) 421
B lymphocyte 422
B-1-like B cell 423
Baboon 79
Bacterial fragment 435
Baroreceptor 339
Baroreflex 337
BCR (B cell receptor) 421
Behavior 252
Benextramine 509
Benzazepine 522
Benzimidazoles 518
Benzodiazepine 270, 522
Benzothiazolone derivatives 534
Benzothiopenes 518
6-Benzylsulfonyl-5-nitroquolines 521
Bereavement 416
BIBO 3304 155, 257, 297, 306, 512-514
BIBP 3226 256, 306, 367, 370, 373, 375, 376, 378, 511-513, 517
– binding site 514
Bidirectional 411
BIIE 0246 93, 149, 310, 367, 524, 526
– binding site 514
Binge eating 292
Bis[diamino](phenyl)triazines] 522
Blood-brain barrier 460
Blood pressure 2, 335, 372, 466

Blood vessel 146
Bone 472
Bone marrow 412
Bradykinin 374
Brain 362
Brain NPY levels 292-294
Brain region 164
Brainstem 341
BU-E-76 509

C terminus 29
CA1 region 456
Ca^{2+} channel 151, 368
CA3 468
Caenorhabditis elegans 86
Calcitonin gene related peptide 160
Carbazole derivatives 534
Cardiac failure 344
Cardiomyocyte 338
Cardiovascular effects 328
Cardiovascular response 5, 6, 12
Cardiovascular system 465
CART 456, 458, 460-462
Cat spleen 411
Catecholamine 141, 410
– synthesis 155
Cattle 79
CC2137 509
CCK-B receptor antagonists benzodiazepine 522
CDE5/CD11b 435
cDNA 271
Cell membrane 33
Cellular localization 4
Central nervous system (CNS) 2, 5, 449
c-Fos 207
CGP 71683A 155, 314, 528
Chemical depletion 431
Chemotactic property 425
Chemotaxis 425
Chicken 77
Cholinergic 150
Cholinergic neuron 160
Chromaffin 30
Ciona intestinalis 78
Circadian rhythms 2, 6, 12
Citrated blood 31
Clonidine 157
Cloning 9

Cluster 33
CNS (central nervous system) 2, 5, 449
Coelacanth 91
Collagen 31
Colon 33
Computer models 514
Conditional knockout 309
ψ-conotoxin GVIA 154
Consumption phase 291
Coronary sinus 344
Coronary vessel 336
Corticosterone 199
Corticotropin-releasing hormone (CRH) 189, 199
Cotransmitter 139, 188, 191, 339, 428
Cre/loxP 452
Cre-recombinase 455
CRH (corticotropin-releasing hormone) 189, 199
C-terminal fragment of NPY 419
CyclicAMP 8
Cyclohexaladenosine 156
Cyclooxygenase 374
Cytokine 418
Cytosolic Ca^{2+} 194

DA neuron 161
DA neurotransmission 163
Danio rerio 79
Degranulation 419
Dentate gyrus 165
Depeptidyl peptidas IV see DP IV
Depression 2, 7, 13, 108-112, 122, 252
Desensitization 55-58, 60, 63, 67
2,4-diaminoquinazoline 528
Diarylguanidines 534
Diet
– high-calory 296
– palatable high-calorie 296
Dihydropyrazines 524
Dihydropyridine 524
Dihydroxyphenylalanine (DOPA) 153
Dimerization 50, 51, 67
Dipeptidyl-peptidase IV-like enzyme 24
Distal ileum 33
Distribution (of the Y_1 receptor) 413
Diuresis 370
DMN (dorsomedial nucleus) 286
DOPA (dihydroxyphenylalanine) 153

Dopamine 189
Dorsal root ganglia 433
Dorsomedial nucleus (DMN) 188, 286
DP IV (depeptidyl peptidase IV) 24, 36
Drosophila melanogaster 78
D_2 receptor antagonist 164
Drugs 15

Eating disorder 7
Electrical stimulation 412
Electron microscopic 165
Electrophysiological 162
Electrophysiological activity 272
Elevated plus-maze 469
Emotional behavior 252
Endocrine 167
Endogenous neuroimmune transmitter 428
Endogenous peptide 292
Endopeptidase 24
Endopeptidase-24.18 35
Endoplasmatic reticulum 24
β-endorphin 193
Endothelin 3 157
End-stage renal failure 379
Energy balance 187, 292, 293
Energy expenditure 6, 12
Epilepsy 108, 120, 121, 123
Epilepsy 15
Epileptic seizures 467
Estrogen-primed OVX 465
Ethanol 270
– consumption 14
Exercise 344
Exocytotic release 24
Exogenous NPY 292
Extracellular concentrations of K^+ 35
Extra-hypothalamic sites 289

Fatty acid acylation 27
Feeding disorder 2
Fine-tuning 429
Flinder's Sensitive Line 263
Fluorenone derivatives 536
Food intake 2, 6, 12, 37, 285, 288, 293, 295, 304, 308
– spontaneous 305
Food, rewarding properties 291

FR 226928   529
Frontal cortex   161
Fugu rubripes   79

γ_2 antagonist   149
γ_1 receptor   166
GABA   152, 467
GABAergic transmission   164
Garter snake   92
Gastric mucosa   32
Gastrointestinal function   398
Geller-Seifter   255
Gene array   474
Gene loss   83
Germline knockout   309
GHRH (growth hormone-releasing hormone)   189, 205
GI264879A   514
Glomerula filtration rate   366
Glucagons-like peptide I   33
Glucagons-like peptide II   33
Glucocorticoids   295
Glucose   288
– tolerance   38
Glutamate   152
Glutamatergic   165
Glycolysylation   27
Gly-Lys-Arg cleavage   28
Gnathostomes   75, 77
GnRH (gonadotropin relasing hormone)   463, 465
Goldblatt   377
– hypertension   377
Goldfish   92
Golgi apparatus   24
Gonadotropin   187
Gonadotropin relasing hormone (GnRH)   463, 465
GR 231118   367
Growth hormone-releasing hormone (GHRH)   189, 205
Guanethidine   159
Guinea pig   85
Gut   309
GW438014A   314

H 394/84   524
H_2 receptor blocking   522

H409/22   375, 376
Half-life   25
HE 90481   509
Heart   30, 336
– failure   378
Hemodialysis   379
Hemorrhage   340
Hermorrhagic hypotension   141
Heteroreceptor   144
Highly specific peptidase   35
Hindbrain   290
Hippocampus   150, 290
Histamine   420
Homeobox   80
Homeostatic buffer   273
Homology modeling   528
Hormone secretion   2, 6, 7, 13
Hormones   287
HOXB   79
HPA (hypothalamo-pituitary-adrenal)   462
hPP   167
HPT (hypothalamic-pituitary-thyroid)   203
Hydroxycopamine   433
6-hydroxydopamine   163, 363
Hyperinsulinemia   295, 304
Hyperphagia   196, 304
Hypertension   2, 335, 377
Hypertensive rats, spontaneously   376
Hypertrophy   338
Hypothalamic   150
Hypothalamic circuitries   222
Hypothalamic NPY levels   293
Hypothalamic-pituitary-adrenal (HPA) axis   199, 264, 462
Hypothalamic-pituitary-gonadal axis   186, 464
Hypothalamic-pituitary-thyroid (HPT)   203
Hypothalamus   161, 222, 223, 235, 237, 285, 286, 288

Idazoxan   157
IgG antibody response   423
IL-6   418
Iloprost   158
IgM   423

Subject Index

Imidazole derivatives 533
Imidazolone 533
Immune function 410
Immune system 410, 411
Immunohistochemistry 145, 187, 362
In situ hybridization 5, 187
In vitro 146
In vitro mutagenesis 514, 523
In vitro studies 5
In vivo release 37
In vivo situation 428
Indoles 518
Inflammation 424
Influx of Ca^{++} 34
Inhibit DP IV activity 36
Innate acquired immunity 414
Inositol phosphate/Ca^{+2} messenger system 194
Insulin 197, 287
– resistance 295
Interaction (between NPY and the immune system) 411
Interleukin (IL)-1β 413
Internalization 58, 60, 61, 63, 67
Intestinal polypeptide see PYY
Intestinal PYY 36
Intracellular calcium 8
Intracellular signaling 8
Intracellular sorting mechanism 26
Intracerebroventricular 162
Ion channel 34
Ischemia 337
Isolation 2
Isoproterenol 156

J-104870 523
J-115914 (Banyu) 306, 523
JCF 104 527
JCF 105 527
JCF 109 528
JCF 114 532
Joint inflammation 431
Juxtaglomerular region 363

K^+ channel 151
Kaliuresis 370
Knockout mouse 298

L channel 151
L-152,804 297, 315, 536
Lactation 192
LacZ reporter gene 449
Lampetra fluviatilis 81
Lamprey 81
Large dense core vesicle 24, 26
Large precursor molecule 24
Latimeria chalumnae 91
Learning and memory 107, 108, 117-119, 123, 471
Leptin 187, 287
[Leu^{31} Pro^{34}]NPY 153, 337
Leukocyte 410
– distribution 425
LH (luteinizing hormone) 463-465
– release 7
Ligand binding 145
Lipopolysacharides 435
Local action 24
Local microenvironment 428
Locomotor activity 459
Locus coeruleus 162
Lupus erythematosus 377
Luteinizing hormone (LH) 463-465
LY 357897 519
Lymnaea stagnalis 78
Lymphocyte 421
Lymphoid organ 411

Macrophage 418
Major depression 262
MAPK (mitogen-activated protein kinase) 64, 65, 66
Mast cell 419
MC-4 receptor 196
Medial preoptic area-median eminence 222
Median eminence 188
Medulla oblongata 161
Megakaryocate 31, 433
Memory 471
Meprin 24
Mesenteric artery 140
Metabolic rate, decreased 304
Metabolism 37
Microtubuli 26
Migration 417, 424

Mitogen-activated protein kinase see MAPK
MOG_{35-55} 434
Monocyte 414, 424
Morpholinopyridines 523
Morris Water Maze 471
Mosquito 78
Mossy fiber/C3 synapses 165
Motivation to eat 290
MS (multiple sclerosis) 429
– patient 430
α-MSH 196, 458
Mucosa 420
Multiple sclerosis (MS) 429
Myenteric nerve 33
Myocardium 337

N channel 151
N^6-substituted R-Argininamides 517
NA (noradrenaline) 138
Na^+/K^+-ATPase activity 365, 371
N-acetyl [$Leu^{28}Leu^{31}$]NPY(24-36) 149
N-acetyleted 22-36 fragment 36
Nascent peptide 27
Neprilysin-like enzyme 24
Nerve activity, sympathetic 345
Nerve stimulation 331
Neural circuitry 222
Neuroendocrine 167
Neurohypophysis 188
Neuroimmune crosstalk 410
Neuron 144
Neuropeptide 3, 222, 223
Neuropeptide Y see NPY
Neurotransmitter 146, 287
Neutrophil 414, 417
NGF 163
Nifedipine 154
NK cell 416, 424
Noradrenalin (NA) 140, 138, 329, 465
– release 371
A2 noradrenergic cell 189
A6 noradrenergic cell 189
NPFF receptors 511
NPR-1 86
NPY (Neuropeptide Y) 2, 102-117-123, 138, 223-237, 239, 252, 285, 287, 328, 393-404, 411, 462

– degradation 24
– exogenous 288, 292
– femtomolar dose 269
– infusion 295
– isolation 2
– mRNA 4
– overexpressing 297
– primary structures 2
– release 24, 363
– sedative effectiveness 253
– storage 24
– synthesis 24
NPY
– germline receptor knockout mouse 297
– receptor 300
– receptor agonists 8, 11
– receptor antagonists 8, 11
– receptor number 294
– receptor subtypes 2, 12, 332
NPY transgenic rat model 261
NPY Y_1 300
– agonists 301
– antagonists 305
– receptors 302
– subtype antagonists 305
NPY Y_2 300
– knockout mice 309
– knockouts 309
– receptor knockouts 312
– subtype agonists 308
NPY Y_2 receptor
– agonists 309
– antagonists 310
NPY Y_4 300
NPY Y_5 300, 312
NPY Y_5 313
– antagonists 314
NPY Y_6 300
NPY(13-36) 147, 337
Y_1 NPY receptor subtype 191
Y_5 NPY receptor subtype 195
NPY/AgRP 287
NPY5RA-972 315
NPY-like immunoreactivity 4
NTS (nucleus of the tractus solitarii) 339, 467
N-type channel 140
Nucleus of the tractus solitarii (NTS) 339, 467

Ob/ob mice   298, 304
Obesity   15, 285, 295
Orexigenic peptide   196
OT (oxytocin)   188
Overexpression   450
Overflow   146, 344
Oxygenic activity   307
Oxytocin (OT)   188

P/Q channel   151
Pain   2, 7, 13
– disorder   2, 15
Pancreas   32, 33
Pancreatic polypeptide (PP)   2, 3
Paralogon   81
Paraventricular nucleus   188, 339
PBMC (peripheral blood mononuclear cell)   413
PC12 cell 151
PD 160170   521
PD 9262   521
Peccary   92
Peptidase   24, 35
Peptide, endogenous   292
Peptide hormones   3
Peptide YY (PPY)   2, 3
Periaqueductal gray matter   258
Peripheral blood mononuclear cell (PBMC)   413
Peripheral distribution of NK cell   417
Peripheral hormones   288
Peripheral metabolism   288
Peripheral nervous system   2, 4
Periventricular nucleus   205
PGE_1 159
Phenylacetamide derivatives   534
Phenylpiperazine   521
Phenylurea derivatives   536
Pheochromocytoma   343
Phosphorylation   27
Pig   85
Pithed rat   146
PKC   154
PMY   81
Polymorphic variants   301
Polymorphism   266
POMC (proopiomelanocortin)   193, 287, 456, 460-462

Post-arginine hydrolysing endoprotease   24
Postganglionic sympathetic nerve fiber   29
Postjunction   139
Postnatal period   32
Postprandial satiety signal   459
Posttranslational step   24
Potentiation   369
PP (Pancreatic polypeptide)   2, 397-400, 402-404
– containing cell   32
– metabolism   37
PPY (Peptide YY)   2, 3
Precursor   24, 28
Prejunction   139
Pre-pro NPY   28, 187, 266
Pre-pro-PP   28
PreproTRH mRNA   204
Pressor   334
Pressure natriuresis mechanism   372
Presynaptic   144
Prohormone convertases   29
Prolactin   188
Proopiomelanocortin (POMC)   193, 287, 456, 460-462
Pro-ortholog   85
Prostaglandins   365
Prostanoid   158
Proteolytic cleavage   27
Pseudogene   77
Pufferfish   79
Purinergic   143
Pyrazole-3-carboxamide   533
Python   77
PYY (intestinal polypeptide)   147, 393-404
– expressing cell   32
PYY(13-36)   147
PYY(3-36)   37, 309

QTL (quantitative trait loci)   267

RA (rheumatoid arthritis)   429
Rabbit   85
Rat models   450
Rat platelet   31
Rat vas deferens   146

Receptor agonists 2
Receptor antagonists 2
Receptor selectivity 25
Receptor subtypes 2
D_2 receptor antagonist 164
γ_1 receptor 166
MC-4 receptor 196
Y_1 receptor 10, 33, 334, 364, 368, 372, 413
Y_2 receptor 10, 36, 364, 365
Y_4 receptor 10, 364
Y_5 receptor 11, 364
Y_6 receptor 11
Y_1 receptor transcript 412
Y_2(-/-)receptor knockout mouse 160
Re-innervation 412
Renal blood flow 366, 371
Renal hypertension 377
Renin 373
Renin-angiotensin system 373, 374
Reproduction 2, 7, 13, 207
Reserpine 143
Respiratory burst 417
Resupply 413
Rewarding properties of food 291
Rheumatoid arthritis (RA) 429
rPP 167

Salmon 92
Sarcopterygian 91
SCN (supra chiasmatic nucleus) 164
Secretory granule 32
Secretory pathway 26
Seizures 2, 8, 14, 166
Seminalplasmin 79
Septic shock 6, 436
^{3}H-serotonin 166
Sexual behavior 7
Sexual development 208
Shaffer collateral/C1 synapses 165
Shark 75, 77
Sheep 79
Skin 420
SLE (systemic lupus erythematodes) 429
Smooth muscle 33
SNAP-25 153
SNARE (soluble N-ethyl-maleimide-sensitive fusion protein) 34, 153

Soluble N-ethyl-maleimide-sensitive fusion protein (SNARE) 34
Somatostatin 189
Sparrow 92
Species 427
- differences 363
Specific assay 37
Spiny dogfish 85
Spiroindoline 534
Spiroisoquinolines 534
Spirolactone 534
Spleen 30
Splice variant, alternative 451
Spontaneously hypertensive rat 162, 376
Squalus acanthias 85
SR 120107A 517
SR 120819A 517
Strain difference 427
Stress 2, 7, 13, 252
Striatum 161
Subfunctionalization 78
Substance P 160
Suicide victims 265
Sulfinamide 530
Sulfonamide 530
Sulprostone 158
Supersensitivity to NPY 143
Supra chiasmatic nucleus (SCN) 164
Supraoptic nucleus 190
Sympathectomy 336
Sympathetic nerve 328, 466
– fiber 411
Sympathetic nervous system 410
Sympathetic neuroeffector junction 138, 139, 141
Sympathetic neuron 138
Synaptic vesicle 24, 27
Synaptobrevin 153
Synaptostagmin 34
Synovial lymphocyte 431
Syntaxin 1 153
Systemic lupus erythematodes (SLE) 429

T cell receptor (TCR) 421
$T_4[NPY(33-36)]_4$ 167, 524
Tachyphylaxis 368
TCR (T cell receptor) 421
Testosterone 463
Tetrahydrodiazabenzazulenes 534

Tetrahydroxanthene-1-ones 536
Tetraploidization 79
Th1 cell 422
Th1/Th2 balance 421
Th2 cell 422
Th2 shift 434
Thymus 412
Thyroid hormone 204
Thyroid stimulating hormone 188
Thyrotropin-releasing hormone 189
Tissue distribution 2
TN (trans Golgi network) 24
Toll-like receptor or CD14 435
Tractus solitarii, nucleus 339
Trafficking 424
– of leukocytes 417
Trans Golgi network (ZGN) 24
α-Trinositol 508
True hunger state 291
Tuberoinfundibular dopamine 189
Tubular structure 27
Tumor necrosis factor (TNF)-α 414

UCP (uncoupling protein) 460
UCP2 460
Ureidobenzothiazolone 536

Vagal effect 159
Vagal inhibition 337
Vagotomy 467
Vascular response 524
Vascular smooth muscle 338
Vasoconstriction 5, 142, 329, 366-368, 375
Vasoconstrictor 330
Vasodilator 160, 330
Vasopressin (VP) 188, 373

Vasospasm 337
Veins 332
Velocity 26
Ventricle 290
Vesicle 157
Vesicle release complex 151
Vesicular transport 27
Vogel punishing drinking conflict 254

Withdrawal responses 271

Xenopus laevis 79

Y receptor 46-58, 61, 62, 64-67
Y receptor knockout mice, conditional 454
Y_1 mediation 255
Y_1 receptor 10, 333, 334, 364, 368, 372, 413
– antagonists 511
– knockout mice 269
– suppression 260
– transcript 412
Y_2 receptor 10, 36, 364, 365
– antagonists 524
Y_2(-/-)receptor knockout mouse 160
Y_4 receptor 10, 364
Y_5 antagonists
– benzimidazole-based 534
Y_5 receptor 11, 364
– antagonists 526
Y_6 receptor 11
Yohimbine 157

Zebrafish 79